RECENT PROGRESS IN
HORMONE RESEARCH

*Proceedings of the
1976 Laurentian Hormone Conference*

Edited by
ROY O. GREEP

VOLUME 33

PROGRAM COMMITTEE

E. Alpert
G. D. Aurbach
J. C. Beck
I. S. Edelman
L. L. Engel
I. Geschwind
R. O. Greep
M. M. Grumbach

D.T. Krieger
E. E. McGarry
H. Papkoff
J. E. Rall
K. J. Ryan
K. Savard
N. B. Schwartz
A. White

1977

ACADEMIC PRESS   New York   San Francisco   London
A Subsidiary of Harcourt Brace Jovanovich, Publishers

Copyright © 1977, by Academic Press, Inc.
ALL RIGHTS RESERVED.
NO PART OF THIS PUBLICATION MAY BE REPRODUCED OR
TRANSMITTED IN ANY FORM OR BY ANY MEANS, ELECTRONIC
OR MECHANICAL, INCLUDING PHOTOCOPY, RECORDING, OR ANY
INFORMATION STORAGE AND RETRIEVAL SYSTEM, WITHOUT
PERMISSION IN WRITING FROM THE PUBLISHER.

ACADEMIC PRESS, INC.
111 Fifth Avenue, New York, New York 10003

United Kingdom Edition published by
ACADEMIC PRESS, INC. (LONDON) LTD.
24/28 Oval Road, London NW1

LIBRARY OF CONGRESS CATALOG CARD NUMBER: Med. 47-38

ISBN 0-12-571133-6

PRINTED IN THE UNITED STATES OF AMERICA

Recent Progress in

HORMONE RESEARCH

The Proceedings of the Laurentian Hormone Conference

VOLUME 33

This volume is respectfully dedicated
to the memory of
Edwin Bennett Astwood, M.D., C. M., Ph.D.
(1909–1976)

CONTENTS

List of Contributors and Discussants ix

Preface xi

Obituary—Edwin B. Astwood xiii

1. The Expanding Significance of Hypothalamic Peptides, or, Is Endocrinology a Branch of Neuroendocrinology?
   Roger Guillemin 1
   Discussion by *Bogdanove, Chretien, Edgren, Greep, Guillemin, Hedge, Korenman, Leeman, Munson, Parsons, Savard, Sterling, Sulman, Thompson, and W. F. White* 20

2. LATS in Graves' Disease
   J. M. McKenzie and M. Zakarija 29
   Discussion by *Bullock, Dobyns, Geller, Greer, Korenman, Kowal, McKenzie, M. M. Martin, Rall, Schwartz, Sulman, Volpe, Weisz, A. White, and Yalow* .. 53

3. Ontogenesis of Hypothalamic–Pituitary–Thyroid Function and Metabolism in Man, Sheep, and Rat
   Delbert A. Fisher, Jean H. Dussault, Joseph Sack, and Inder J. Chopra 59
   Discussion by *Bernstein, Dratman, Fisher, Guillemin, Guyda, Jacobs, McKenzie, Murphy, Nathanielsz, Sterling, Tulchinsky, and Volpe* 107

4. The Antimüllerian Hormone
   Nathalie Josso, Jean-Yves Picard, and Dien Tran 117
   Discussion by *Bardin, Cohen, Donahoe, Falvo, Greep, Josso, Leeman, Licht, McKenzie, M. M. Martin, Nathanielsz, Orr, Pasqualini, Roberts, Rodbard, Thompson, and Vaitukaitis* 162

5. Evolution of Gonadotropin Structure and Function
   Paul Licht, Harold Papkoff, Susan W. Farmer, Charles H. Muller, Hing Wo Tsui, and David Crews 169
   Discussion by *Callard, Condliffe, Dobyns, Farmer, Frisch, Furth, Korenman, Licht, Nureddin, Rice, Sairan, Schwartz, Segaloff, and Vaitukaitis* 243

6. Biosynthesis of Parathyroid Hormone
   Joel F. Habener, Byron W. Kemper, Alexander Rich, and John T. Potts, Jr. 249
   Discussion by *Brown, Habener, Kemper, Korenman, Lippman, McKenzie, Papkoff, Potts, Raacke, Rall, Rubenstein, Sterling, Thompson, and Yalow* 299

7. Carbon-13 Nuclear Magnetic Resonance Investigations of Hormone Structure and Function
   Ian C. P. Smith and Roxanne Deslauriers 309
   Discussion by *Andreoli, Engel, Korenman, Monder, Orr, Papkoff, Savard, Scoggins, Smith, and Vallotton* 329

8. The Regulation of Vasopressin Function in Health and Disease
   Gary L. Robertson 333
   Discussion by *Andreoli, Bartter, Bernstein, Fisher, Flouret, Frisch, Gill, Little, Martin, Nathanielsz, Robertson, Sawyer, Sterling, and Vallotton* 374

CONTENTS

9. Some Considerations of the Role of Antidiuretic Hormone in Water Homeostasis
   Thomas E. Andreoli and James A. Schafer 387
   Discussion by *Andreoli, Greep, Korenman, Orr, Roberts, Rodbard, Rogol, Smith, and Sterling* 431

10. Clinical Significance of Circulating Proinsulin and C-Peptide
    Arthur H. Rubenstein, Donald F. Steiner, David L. Horwitz, Mary E. Mako, Marshall B. Block, Jerome I. Starr, Hideshi Kuzuya, and Franco Melani 435
    Discussion by *Bodganove, Bullock, Cahill, Carter, Dratman, Fisher, Guillemin, Habener, M. M. Martin, Parsons, Rubenstein, Steiner, Sterling, Unger, and Yalow* 468

11. Glucagon and the A Cells
    R. H. Unger, P. Raskin, C. B. Srikant, and L. Orci 477
    Discussion by *Bernstein, Bogdanove, Cahill, Condliffe, Dupré, Geller, Orci, Patel, Rall, Rubenstein, and Unger* 511

12. A Newly Recognized Pancreatic Polypeptide; Plasma Levels in Health and Disease
    John C. Floyd, Jr., Stefan S. Fajans, Sumer Pek, and Ronald E. Chance .. 519
    Discussion by *Cahill, Chance, Donahoe, Floyd, Frisch, Hazelwood, Jacobs, Kimmel, Korenman, Licht, Lippman, Macdonald, Marx, Orci, Papkoff, Roberts, and Spiegel* 556

13. Steroid Hormone Actions in Tissue Culture Cells and Cell Hybrids—Their Relation to Human Malignancies
    E. Brad Thompson, Michael R. Norman, and Marc E. Lippman 571
    Discussion by *Bardin, Bradlow, Furth, Hollander, Lippman, A. Martin, Marx, Nekola, Pasqualini, Segaloff, Thompson, and A. White* 608

14. The Role of Hormones on Digestive and Urinary Tract Carcinogenesis
    Richard S. Yamamoto and Elizabeth K. Weisburger 617
    Discussion by *Bullock, Furth, Schwartz, and Yamamoto* 651

Subject Index 655

Cumulative Index 662

LIST OF CONTRIBUTORS AND DISCUSSANTS

T. E. Andreoli
C. W. Bardin
F. C. Bartter
R. S. Bernstein
M. B. Block
E. M. Bogdanove
H. L. Bradlow
E. M. Brown
L. Bullock
G. S. Cahill
I. Callard
J. Carter
R. E. Chance
I. J. Chopra
M. Chretien
S. Cohen
P. G. Condliffe
D. Crews
R. Deslauriers
B. M. Dobyns
P. K. Donahoe
M. Dratman
J. Dupré
J. H. Dussault
R. A. Edgren
L. L. Engel
S. S. Fajans
R. Falvo
S. W. Farmer
D. A. Fisher
G. Flouret
J. C. Floyd
R. E. Frisch
J. Furth
J. Geller
J. R. Gill
R. O. Greep
M. A. Greer
R. Guillemin
H. Guyda
J. F. Habener
R. L. Hazelwood
G. A. Hedge
V. P. Hollander
D. L. Horwitz
L. S. Jacobs
N. Josso

B. W. Kemper
J. R. Kimmel
S. Korenman
J. Kowal
H. Kuzuya
S. Leeman
P. Licht
M. Lippman
B. Little
J. M. McKenzie
G. J. Macdonald
M. E. Mako
A. Martin
M. M. Martin
S. J. Marx
F. Melani
C. Monder
C. H. Muller
P. Munson
B. E. P. Murphy
P. W. Nathanielsz
M. V. Nekola
M. R. Norman
A. Nureddin
L. Orci
J. C. Orr
H. Papkoff
J. A. Parsons
J. R. Pasqualini
Y. C. Patel
S. Pek
J.-Y. Picard
J. T. Potts, Jr.
I. D. Raacke
J. E. Rall
P. Raskin
B. F. Rice
A. Rich
J. S. Roberts
G. L. Robertson
D. Rodbard
A. D. Rogol
A. Rubenstein
J. Sack
M. Sairam
K. Savard
W. H. Sawyer

LIST OF CONTRIBUTORS AND DISCUSSANTS

J. A. Schafer
N. B. Schwartz
B. A. Scoggins
A. Segaloff
I. C. P. Smith
A. M. Spiegel
C. B. Srikant
J. I. Starr
D. F. Steiner
K. Sterling
F. G. Sulman
E. B. Thompson
D. Tran

H. W. Tsui
D. Tulchinsky
R. H. Unger
J. Vaitukaitis
M. B. Vallotton
R. Volpe
E. K. Weisburger
J. Weisz
A. White
W. F. White
R. S. Yalow
R. S. Yamamoto
M. Zakarija

PREFACE

With the death of Dr. Edwin Bennett Astwood on February 17, 1976, the world community of endocrinologists and The Laurentian Hormone Conference lost one of their ablest and most respected members. Dr. Astwood had long been prominently identified with the Conference and served as its administrative leader for five years following the death of Dr. Gregory Pincus in 1967. The 1976 Conference opened with a brief tribute to Dr. Astwood and an expression of sympathy to his family. No tribute could have been more in keeping with the life and memory of Ted Astwood than the high standard of excellence set by the ensuing program.

The Gregory Pincus Memorial Lecture, presented by Dr. Roger Guillemin, detailed the remarkable recent progress made in isolating central nervous system peptides, some of which have startling biological properties. This set the tone for a week of reports on other fields having such far-reaching implications that one might justly question whether wonders will ever cease. The discussion of each paper was spirited and contributed valuable insights, as any reader of this volume will quickly discover. A program of such uniform excellence could not have been put together without the help of the many members of the Conference who regularly make suggestions as to speakers and topics appropriate to this proud gathering of workers from the most advanced forefronts of hormone research.

It is a pleasure to acknowledge our great debt to Drs. Robert Volpe, Susan Leeman, Rosalyn S. Yalow, James C. Orr, Wilbur H. Sawyer, George F. Cahill, Janet W. McArthur, and Vincent Hollander for chairing the several sessions with skill and finesse. I am deeply grateful to the members of the Program Committee and the Board of Directors who give of themselves so freely to make this Conference what it is. To the staff of the Mont Tremblant Lodge who make this meeting a pleasant affair, and to our Executive Secretary, Martha Wright, whose cheerful attention to every logistic detail makes things go smoothly, I extend the gratitude of all. The amazing efficiency of Lucy Felicissimo and Linda Passalalpi in transcribing and deciphering the taped discussions is acknowledged with sincere appreciation. Working with the staff of the Academic Press in bringing the proceedings of this Conference to publication has, as always, been carried out with a commonness of purpose and cordiality that eases the burden of all concerned.

EDWIN BENNETT ASTWOOD

When death comes to a man with the brilliance of mind, eminence in medicine, and achievements in science of Dr. Edwin B. Astwood, the loss to all mankind is incalculable. To family, friends, and admirers the death of a man with the personal charm and endearing human qualities of Dr. Astwood tears at the heartstrings and leaves a wound that does not easily heal. As one of his close friends wrote: "It may sound maudlin, but when Ted died a part of me went with him." Many will share those sentiments. Ted Astwood was that kind of person. The admiration, love, and respect accorded him was indeed out of the ordinary. In the hearts and minds of all who knew him personally, he stood alone as someone special.

Edwin B. Astwood, known to many as Ted, and to his family and intimate friends as Teddy, was born in Hamilton, Bermuda. There he spent a normal, happy youth excelling in school and athletics, roaming the byways and beaches of that fair isle, and exploring the wonders of nature. The ingeniousness and the ever reaching out into the unknown that were to mark his entire professional career came to bloom early in life. As a boy, he built his own telescope and came to know the stars and their movement; studying the fathomless universe filled him with wonder and ponderment. This intense and insatiable curiosity, this yearning to know and determination to understand never faltered. He was one of those rare individuals whose knowledge was simply encyclopedic. The range and accuracy of his factual information concerning the natural world, science, medicine, mechanics, and electronics seemed almost uncannily limitless. No matter what the problem or the question, the response was always, "Ask Ted."

Ted's contributions to basic and clinical endocrinology, cited below, were monumental, and it is for this reason alone that he must be numbered among the great men of his era. Despite the fame and kudos that came to him, he remained—to his enduring credit—an unassuming and kindly gentleman. He was always generous in his praise and admiration of the talents and accomplishments of others. Being an ardent advocate of scientific truths and an exceptionally able experimentalist, he was quick to detect any flaw in his own evidence or that of others. Though intolerant of slipshod work and ill-founded claims, his criticisms were always constructive. Many will recall that in open scientific meetings, as at the Laurentian Hormone Conferences, his probings were delivered in such a deceptively kindly tone of voice, and with such complete detachment, that the victim might not immediately detect the turn of the knife.

In deference to the deep religious convictions of his mother, Ted attended Washington Missionary College and entered the College of Medical Evangelists at Loma Linda. Midway through medical school, Ted could not face the folly and futility of following a course that he could no longer adhere to in good conscience. The need for liberty of mind, and the free spirit that was his by nature, led him to take his future into his own hands. To make the break clean Ted transferred to McGill University's Faculty of Medicine and received the M.D. degree in 1934. In the following year, as a medical house officer at the Royal Victoria Hospital in Montreal, he came under the guidance of J.S.L. Browne, cementing a lifelong friendship. The succeeding two years were spent as a Fellow in the Surgical-Pathology Laboratory at Johns Hopkins Hospital. There he met and married Sarah Ruth Merritt, a nurse. Still thirsting for experience in basic research, and with the aid of a Rockefeller Foundation Fellowship, Ted spent the next three years as a graduate student working in the laboratory of Professor Frederick L. Hisaw at Harvard University. These were fruitful years that earned him a Ph.D. degree and launched him into a distinguished research career in biology and medicine. Parenthetically, it was there that our paths met and joined with the formation of an interacting closeness in all things that only death did part.

Armed with degrees in medicine and in science, and with a substantial record of success in research, Ted returned to Hopkins as Associate in Obstetrics and worked in collaboration with Drs. Georgeanna Jones, Eleanor Delfs, and Charles G. Geschickter. In 1941, Soma Weiss enticed him back to Boston with a joint appointment as Associate in Medicine at the Peter Bent Brigham Hospital and Assistant Professor of Pharmacotherapy in Otto Krayer's Department of Pharmacology, Harvard Medical School. Four years later, and ready for a show of his own, Ted accepted a position as Research Professor of Medicine at Tufts University School of Medicine and Physician and Endocrinologist to the New England Center Hospital and the Boston Dispensary. The hospital provided him with a generous allocation of research facilities immediately adjacent and open to the clinic. This environment was to serve his needs in a highly rewarding manner over the next 27 years. Promotions came in rapid succession, boosting Ted to Senior Physician at the hospital and Professor of Medicine at Tufts University where he held one of the much-coveted Research Career Awards of the National Institutes of Health.

Dr. Astwood got his first taste of research working as a student helper in J. B. Collip's laboratory at Montreal. In his early days at Hopkins he made noteworthy observations on pigment changes in amphibia and on endocrine factors influencing mammary growth and tumor formation in rats. The observations that were to mark him as a man of destiny were made as a graduate student at Harvard. Having noted the rapid action of estrogen on uterine imbibition of water, he quickly developed a six-hour quantitative bioassay for estrogenic substances that served to speed research in an area of rampant clinical concern, ovarian estrogen output. Next came the disclosure of a hitherto unrecognized luteal sustaining principle in extracts of rat placenta and a similarly acting hypophysial gonadotropin, distinct from FSH or LH, that prove to be identical with the lactogenic hormone, prolactin. For this he coined the term luteotrophin. These were seminal observations that have long continued to stimulate research.

Ted's expertise in the realms of physiology, biochemistry, pharmacology, and medicine enabled him to work effectively in diverse areas of endocrinology. His research greatly enriched our understanding of the secretions and regulatory mechanisms of the ovary, pituitary, adrenal, thyroid, and parathyroid glands. Mention can be made here of only a few of his further major contributions. Working with the group at Hopkins, he succeeded in developing a simple and rapid chemical procedure for the measurement of pregnanediol in human urine that remains a standard technique. The work that was to bring him international recognition and many distinguished honors and awards was the beautiful series of studies on the chemical manipulation of thyroid function by antithyroid drugs. This work was initiated in 1942 as the result of interest aroused by earlier observations that goiter could be induced in experimental animals by sulfaguanidine and phenylthiourea. He recognized the potential of this model for the study of mechanisms involved in the regulation of thyroid function and as an approach to the treatment of thyroid diseases that were persistently confronting him in medical practice.

What Astwood did was to demonstrate in rats that antithyroid drugs interfered with the synthesis of thyroid hormones and that the resulting enlargement of the thyroid gland was due to compensatory hypersecretion of TSH. This being so he reasoned that if patients with hyperthyroidism were treated with thiourea, thereby blocking the action of TSH, they might be benefited. They were, but some toxic side effects appeared and the search was on for less toxic antithyroid compounds. Propylthiourea proved much less toxic and suitable for long-term therapy, but the search continued. Since plants of the genus *Brassica* were believed to be goitrogenic, cabbages and turnips loomed as a likely source of antithyroid agents. Indeed Ted and associates proceeded to isolate from the yellow turnip (rutabaga), a progoitrin, L-5-vinyl-2-thiooxazolidone that proved to be highly effective and safe. Since the goitrogenic action of all these antithyroid agents could be neutralized or reversed by giving thyroid hormone, Ted deduced that patients with simple or nodular goiter might benefit from the administration of thyroid hormone in the form of thyroid powder and again his hunch was verified. Some patients with hyperthyroidism did not respond to antithyroid drugs, but he found that they could be treated successfully with radioactive iodine. Thus some form of chemical therapy was made available for the management of the entire spectrum of thyroid dyscrasias.

As is well known, Astwood's work on the use of antithyroid drugs was to revolutionize our understanding of the biochemistry of the thyroid gland and the therapeutic management of goiter and hyperthyroidism, commonest of all endocrine diseases. This new evidence ran counter to the established surgical approach to treatment of thyroid disorders. Ted was soon to contend with an entrenched interest that did not welcome this introduction of chemical therapy by some upstart from Boston. The opposition was doomed from the beginning, however, as Ted's evidence of successful treatment was overwhelming and bound to prevail, as it did on a worldwide scale. As understanding of thyroid mechanisms mounted from his laboratory studies, further triumphs in therapy were forthcoming such as the paradoxical use of thyroid hormone in the treatment of simple and nodular goiter and the use of radioactive iodine in the treatment of certain forms of purported carcinomas of the thyroid. Some of his famous dictums relative to these matters are worth repeating. Once when asked for his views on the role of surgery in the treatment of thyroid disease, his cryptic retort was, "Historical." There is also Ted's challenging comment that "an encounter between a physician and a thyroid nodule is not cause for panic."

The indelible stamp of Astwood's ingenuity is also to be found in improved methods for the extraction and purification of pituitary hormones. In the heyday of research on the pituitary adrenocorticotrophic hormone (ACTH), sparked by the 1949 bombshell suggesting that this might have a role in the treatment of arthritis, Ted and Maurice Raben, an able associate and mainstay, jumped into the fray and soon came up with a radically new and simplified method of extracting ACTH in greatly improved yield and purity. This soon became the standard procedure used in the commercial preparation of ACTH for clinical use and for later isolation studies. A few years later, research on pituitary growth hormone was given a shot in the arm through the finding that primates respond only to growth hormone prepared from primate pituitary glands. Again Raben and Astwood came through with a simplified extraction procedure that yielded growth hormone from human pituitaries in suitable purity and potency for the successful treatment of hypopituitary dwarfism. The growth hormone that yielded the first positive response in a dwarfed child was prepared in Astwood's laboratory. Likewise, the isolation of parathyroid hormone was achieved there as was the development of a standard method for the assay of LATS, and a method for the extraction and purification of placental lactogen. In the final years of his experimental work, Dr. Astwood engaged the major challenge of obesity. He felt quite certain that pituitary hormones were involved in lipid metabolism and indeed he proceeded to isolate two lipolytic peptides from porcine and human pituitary glands. Designated as peptides I and II they had the ability not only to promptly mobilize free fatty acids in experimental animals, but to also produce in turn gross lipemia. The problems in this area were not resolved but Ted did lead the way in opening the field to exploratory study. Much as he had hoped to find something that would "burn up fat" that was not to be. As Carl Cassidy, one of Ted's long-time associates remarked in a commemorative tribute "He'd done enough."

Although the Astwood laboratory was equipped with highly sophisticated instruments, his relationship to them was as master, not slave. Except for his fascination with their operating principles and electronic gadgetry, he had rather a disdain for fancy technology when some homemade substitute would do the job as well. The same conservative attitude pervaded his relationship with patients. Having a knowledgeable respect for nature's defense mechanisms and holding that most patient complaints had no organic basis, he would as often let aspirin, bedrest, and some reassuring words work their magic; but he was also quick to utilize any of today's specific remedies when the need was evident. Once when called to examine a spreading rash on the abdomen of one of my young daughters, he told her that if she would stop scratching it the rash would go away. She did and the next morning it was gone.

Honors and awards were showered upon Dr. Astwood. First came the Ciba Award in 1944 from the Association for the Study of Internal Secretions (U.S. Endocrine Society); then in 1948 the Cameron Prize from the University of Edinburgh; followed a year later by the John Phillips Memorial Award from the American College of Physicians, thence the Borden Award in 1952 from the Association of American Medical Colleges, the Claude Bernard Medal from the University of Montreal in 1953, and the Lasker Award in 1954 from the American Public Health Association. After a brief letup, the shower resumed with the award of the Gordon Wilson Medal in 1966 from the American Clinical and Climatological Association and a year later came both the Koch Award from the Endocrine Society and an honorary Doctor of Science degree from the University of Chicago. In the final year of his life, Dr. Astwood was named Distinguished Thyroid Scientist by the VII International Thyroid Conference and given the Distinguished Leadership Award by the Endocrine Society.

Obviously, any man of such distinction could have develoted most of his time and effort to activities outside his office and laboratory, but not Ted. The demand was there but he seldom yielded. He did serve as President of The Endocrine Society and was editor of its journal, *Endocrinology,* for a brief period. His service otherwise included goodly stints as a member of the Advisory Council of the National Institute of Arthritis and Metabolic Disease and as Chairman of the Committee on Arrangements of the Laurentian Hormone Conference. Dr. Astwood was happier working in his own private laboratory than in following the lecture circuit, but again the demand was overwhelming. The worlds of science and medicine wanted very much to hear him. Over the years, he delivered many named lectures and made distinguished appearances in many parts of the world. Fame has its price.

Writing was different; it could be done at home in the early morning hours, it was more creative and Ted did willingly devote a large amount of time and effort to this demanding pursuit. His protestations that writing did not come easily is hardly in keeping with the elegant style of his published works. He was ever an advocate of brevity, but not at the expense of thoroughness. A few of his major contributions to the scientific literature, in addition to articles on original research, include the section on endocrines in Goodman and Gilman's *Textbook of Pharmacology;* a comprehensive review on Growth Hormone and Corticotrophin in Volume III of *The Hormones;* A paper on Chemistry of Corticotrophins in *Recent Progress in Hormone Research,* Volume 7; and a chapter on Clinical Use of Antithyroid Drugs in Soskin's *Progress in Clinical Endocrinology.* Ted also served as editor or coeditor for two multiauthored volumes, *Clinical Endocrinology,* I and II, and Volumes IV and V of *The Hormones,* and lastly the multivolume Section on Endocrinology of the *Handbook of Physiology.*

One of the great legacies of the Astwood laboratory is the large number of distinguished endocrinologists who took their postdoctoral training with him. His laboratory was considered a mecca of opportunity. The master was available for counsel and encouragement but not for advice on what to do. He also refrained from utilizing their talents to his own ends and did not as a policy lend his name to joint authorship of a paper unless he had made a substantial contribution to the work with his own hands. Training came by experience and helping one another. These young aspirants had but one goal in mind and that was to measure up in a setting where the standards were high. It was a sink or swim proposition, and swim they did. At last count, this group of 86 former followers included 33 full professors, of whom 8 were also departmental chairmen and 23 were Chiefs of Services or Chiefs of Medicine. Ted's modesty did not permit him to acknowledge the remarkable success of his training program, but none could doubt that the results were both pleasing and a matter of pride.

Ted's home life and advocational activities brought much enjoyment as did his children Philip Merritt and Nancy Bennett. He looked after their budding interests and entertainment needs and enlarged upon their educational experience at every step of the way. No task about the house or garden was too menial or laborious for him to undertake. He had boundless energy and always had a project going, whether it be uprooting trees, rewiring the house, painting, plumbing, woodworking or building hi-fi sets of superb quality in his basement workshop.

Ted had a keen wit and a lively sense of humor. He loved to entertain at home and was a most hospitable host. The Astwood home was a gathering place for distinguished visitors from far and near. Somehow it was all taken in stride. The pillar of the home was not Ted but his gracious and charming wife, Sally, to whom he was devoted and on whom he was heavily dependent. It was Sally's calm judgment, her wise management of the affairs of a home and family, her understanding of Ted's needs and her steadying influence that enabled him to accomplish so much despite some frustrations along the way.

In 1971, Ted and Sally raised anchor and moved back to Ted's homeland, Bermuda, where he engaged in the practice of internal medicine until the day of his death from cancer on February 17, 1976. The courage and fortitude with which he faced these final months of illness cannot be overstated. He ignored the discomfort, carried on with what strength he could muster and never uttered a hint of concern as to his fate. Again, he was just that kind of man. The likes of Ted Astwood pass this way with a rarity that is matched only by the preciousness of their being.

Roy O. Greep

The Expanding Significance of Hypothalamic Peptides, or, Is Endocrinology a Branch of Neuroendocrinology?*

ROGER GUILLEMIN

Laboratories for Neuroendocrinology, The Salk Institute, La Jolla, California

I am honored to have been asked to deliver this year's Gregory Pincus Memorial Lecture.

The time elapsed from the date of publication of the first paper describing in a definitive manner the existence in crude hypothalamic extracts of a substance with all the activities to be expected from a thyrotropin-releasing factor (TRF) (Guillemin *et al.*, 1962) to the date of the elucidation of its structure was seven years (Burgus *et al.*, 1969), 1962 to 1969. Another seven years has passed since then.

I would like to review some of the major achievements of these seven years since the isolation of TRF. Rather than discussing technical details or procedures, I will present what I consider to be these major achievements, in a somewhat dogmatic manner perhaps, but essentially to draw from these many new facts the new concepts and working hypotheses that, as a physiologist, I see in them. This text will thus be that of a short review-lecture, very different from the traditional exhaustive text of a publication in *Recent Progress in Hormone Research*.

Since the isolation of TRF (Z H P), the following biologically active peptides have been isolated from hypothalamic extracts and characterized: LRF, (Z H W S Y G L R P G); somatostatin (A G C K N F F W K T F T S C); α-endorphin (Y G G F M T S E K S Q T P L V T), γ-endorphin (Y G G F M T S E K S Q T P L V T L), β-endorphin (Y G G F M T S E K S Q T P L V T L F K N A I V K N A H K K G Q), substance P (R P K P Q Q F F G L M), neurotensin (Z L Y E N K P R R P Y I L); also vasopressin (C Y F Q N C P K G) and α-MSH (S Y S M E H F R W G K P V) have been isolated and sequenced from ovine hypothalamic extracts.

I. Thyrotropin Releasing Factor (TRF) and Luteinizing Hormone Releasing Factor (LRF)

Somewhat to the surprise of researchers in the field, of the many analogs of TRF which have been synthesized and studied biologically, the only one which

---

*The Gregory Pincus Memorial Lecture.

has an increased specific activity over that of the native compound, is that described by our group and synthesized by Rivier (Vale *et al.*, 1971) a few years ago. It is the analog [3-*N*-Methyl-His]-TRF. Its specific activity is approximately 10 times that of the native molecule, on the secretion of TSH as well as of prolactin. Also surprisingly, of the several hundreds of TRF analogs synthesized, none has been found so far to be even a partial antagonist. They are all agonists with full intrinsic activity but variable specific activity; no true antagonist of TRF has been reported.

In contradistinction to TRF, antagonist as well as superagonist analogs of LRF have been prepared by a number of laboratories. We now have available preparations from various groups of a series of what we may accurately call super LRFs, analogs having as much as 150 times the specific activity of the native compound. In fact, in certain assays, such as ovulation, they may have 1000 times the specific activity of the native peptide. All the agonist analogs of LRF possess structural variations around two major modifications of the amino acid sequence of native LRF: They all have a modification of the C-terminal glycine, as originally reported for a series of analogs by Fujino *et al.* (1974). The Fujino modification consists of deletion of Gly^{10}-NH_2 and replacement by primary or secondary amide on the (now C-terminal) Pro^9. In addition to the Fujino modification, they have an additional modification at the Gly^6 position by substitution of one of several D-amino acids (Monahan *et al.*, 1973).

The most potent of the LRF analogs agonist prepared are [D-Trp^6]-LRF; des-Gly^{10}-[D-Trp^6-Pro^9-*N*-Et]-LRF (Vale *et al.*, 1976), [D-Leu^6, Pro^9-*N*-Et]-LRF (Vilchez-Martinez *et al.*, 1975).

In an *in vitro* assay in which the peptides stimulate release of LH and FSH by surviving adenohypophysial cells in monolayer cultures, or in surviving pituitary fragments, these analogs of LRF have a specific activity 50 to 100 times greater than that of the synthetic replicate of native LRF. There is no evidence of dissociation of the specific activity for the release of LH from that of FSH. All agonist analogs release LH and FSH in the same ratio (in that particular assay system) as native LRF. Probably because of their much greater specific activity, when given in doses identical in weight to the reference doses of LRF, the super-LRFs are remarkably long acting. While the elevated secretion of LH (or FSH) induced by LRF is returned to normal in 60 minutes, identical amounts in weight, of [D-Trp^6-des-Gly^{10}]-*N*-Et-LRF leads to statistically elevated levels of LH up to 24 hours. These analogs are ideal agents to stimulate ovulation (Vilchez-Martinez *et al.*, 1975). Marks and Stern (1975) have reported that these analogs are considerably more resistant than the native structures to degradation by tissue enzymes.

Catherine Rivier and Wylie Vale in our laboratories have observed that the injection of 1–10 µg of the analogs [D-Trp^6-des-Gly^{10}]-*N*-ethylamide-LRF in pregnant rats, either once or on consecutive days, over the first 7 days of gestation causes resorption of the fetuses and prevents normal pregnancy. Johnson et al. (1976) have recently reported similar observations. The mechanism involved in these observations is not clear at the moment.

All the antagonist LRF analogs as originally found by our group (Vale et al., 1972) and later as reported by others are of a series which have deletion or a D-amino acid substitution of His^2. For reasons not clearly understood, addition of the Fujino modification on the C-terminal does not increase the specific activity (as antagonists) of the antagonist. Their potency has been described in ratios ranging from 5:1 to 15:1. The recently developed antagonist analogs of LRF with low weight ratio for antagonist activity are inhibitory of LRF not only *in vitro*, but also in various tests *in vivo*. They inhibit the release of LH and FSH induced by a dose of LRF administered acutely; they also inhibit endogenous release of LH−FSH and thus prevent ovulation in laboratory animals. The clinical testing of some of these LRF antagonists prepared in our laboratory has recently started in collaboration with Samuel Yen at UCSD.

II. Somatostatin

It is now well recognized that somatostatin has many biological effects other than the one on the basis of which we isolated it in extracts of the hypothalamus, i.e., as an inhibitor of the secretion of growth hormone (Brazeau et al., 1973). Somatostatin inhibits the secretion of thyrotropin, but not of prolactin, normally stimulated by TRF; it also inhibits the secretion of glucagon, insulin, gastrin, and secretin by acting directly on the secretory elements of these peptides. It inhibits the secretion of pepsin, HCl, the exocrine secretion of the pancreas, also by direct action on the secretory elements of these peptides. I have recently shown that somatostatin also inhibits the secretion of acetylcholine from the (electrically stimulated) myenteric plexus of the guinea pig ileum, probably at a presynaptic locus—thus explaining, at least in part, the reportedly inhibitory effects of somatostatin on gut contraction, both *in vivo* and *in vitro*.

It is also now well recognized that somatostatin is to be found in many locations other than the hypothalamus, from which we originally isolated it. Somatostatin has been found in neuronal elements and axonal fibers in multiple locations in the central nervous system, including the spinal cord. It has been found also in discrete secretory cells of classical epithelial appearance in all the parts of the stomach, gut, and pancreas in which it had been first recognized to have an inhibitory effect. [For references, see the two recent reviews by Vale *et al.* (1975) and Guillemin and Gerich (1976).]

Somatostatin does not inhibit indiscriminately the secretion of everything or anything. For instance, as I said above, somatostatin does not inhibit the secretion of prolactin concomitant to that of thyrotropin stimulated by a dose of TRF; this is true *in vivo* when dealing with normal animals or *in vitro* when dealing with normal pituitary tissue. Somatostatin does not inhibit the secretion of either gonadotropin LH and FSH, the secretion of calcitonin, the secretion of ACTH in normal animals or from normal pituitary tissues *in vitro*; it does not inhibit the secretion of steroids from adrenal cortex or gonads under any circumstances. Regarding the secretion of polypeptides or proteins from nonnormal tissues, such as pituitary adenomas, gastrinomas, insulinomas, somatostatin has been shown to be inhibitory according to its normal pattern of activity or being now nondiscriminative. The latter must reflect one of the differences between a normal and a neoplastic tissue. I have long thought of acromegaly as a disease of the (plasma membrane) receptors of the somatotrophs, in which any and all stimuli attaching somehow to these (diseased) receptors will lead to the secretion of growth hormone. This is in keeping with the observation that TRF, or LRF, can stimulate release of growth hormone from acromegalic pituitaries, though that never happens with normal tissues. I am aware that there are other possible explanations for the observation.

With no exceptions, studies in clinical medicine have confirmed, in man, all observations obtained in the laboratory. The powerful inhibitory effects of somatostatin on the secretion not only of growth hormone, but also of insulin and glucagon, have led to extensive studies over the last three years of a possible role of somatostatin in the management or treatment of juvenile diabetes. First of all, the ability of somatostatin to inhibit insulin and glucagon secretion has provided a useful tool for studying the physiological and pathological effects of these hormones on human metabolism. Infusion of somatostatin lowers plasma glucose levels in normal man despite concomitant lowering of both plasma insulin and glucagon levels (Gerich *et al.*, 1975; Mortimer *et al.*, 1974; Alford *et al.*, 1974). These observations provided the first clear-cut evidence that glucagon has an important physiological role in human carbohydrate homeostasis. Somatostatin itself has no direct effect on either hepatic glucose production or peripheral glucose utilization, since the fall in plasma glucose levels could be prevented by exogenous glucagon (Gerich *et al.*, 1974).

In juvenile-type diabetics, somatostatin diminishes fasting hyperglycemia by as much as 50% in the complete absence of circulating insulin (Gerich *et al.*, 1974, 1975). Although somatostatin impairs carbohydrate tolerance after oral or intravenous glucose challenges in normal man by inhibiting insulin secretion, carbohydrate tolerance after ingestion of balanced meals is improved in patients with insulin-dependent diabetes mellitus through the suppression of excessive glucagon responses (Gerich *et al.*, 1974, 1975). The combination of somatostatin and a suboptimal amount of exogenous insulin (which by itself had prevented neither excessive hyperglycemia nor hyperglucagonemia in response to meals) completely prevents plasma glucose levels from rising after meal ingestion in insulin-dependent diabetics (Gerich *et al.*, 1975). Through its suppression of glucagon and growth hormone secretion, somatostatin has also been shown to moderate or prevent completely the development of diabetic ketoacidosis after the acute withdrawal of insulin from patients with insulin-dependent diabetes mellitus (Gerich et al., 1975).

Pancreatic tumors secreting either glucagon or insulin (Lorenzi et al., 1975) have been reported to be responsive to somatostatin. Curiously, in patients with insulinomas, basal and glucagon-stimulated insulin secretion is inhibited by somatostatin, but that due to tolbutamide characteristically is not (Lorenzi et al., 1975). Infusion of somatostatin in such patients must be performed cautiously since, despite lowering of basal insulin levels, plasma glucose levels fall precipitously.

In a report lacking adequate controls, adverse hematological effects have been reported in baboons (Koerker et al., 1975) chronically implanted with multiple catheters and receiving large amounts of somatostatin. In contradistinction to this report, no serious toxicity has been encountered in man during a two-year clinical experience in a single metabolic ward in over 300 subjects. The most frequent side effects encountered are transient nausea (certainly much less acute and not as certain to occur as that accompanying an intravenous pyelography), occasional diarrhea, and abdominal cramps. Transient falls in blood pressure (10 mm Hg for 10 minutes) and rises in pulse (10 beats per minute lasting 10 minutes) are sometimes observed. No significant hypoglycemia has been found, except in diabetics given exogenous insulin. No changes in white blood cell count, platelet count, bleeding time, prothrombin time, partial thromboplastin time, serum electrolyte levels, or renal or hepatic function have been encountered either in acute studies or after prolonged (12 hour) infusions of somatostatin. All these values have been found to be normal after intermittent administration of somatostatin as frequently as once weekly for over a year (Mielke et al., 1975). Although a recent report (Besser et al., 1975) suggested that somatostatin might diminish platelet aggregation in man, similar studies using higher doses of somatostatin found no effect (Mielke et al., 1975); these discrepant results might have been due to the use of different preparations of somatostatin. Since bleeding times were unaffected in both studies, the clinical significance of the reported platelet abnormalities is unclear. At the moment, clinical studies with somatostatin of high purity as provided by our group at the Salk Institute are proceeding in several clinical centers in the United States with the concurrence of the FDA having granted INDs to several groups of investigators.

From the foregoing description of the ability of somatostatin to inhibit the secretion of various hormones, it would appear that this agent may be of therapeutic use in certain clinical conditions such as acromegaly, pancreatic islet cell tumors, and diabetes mellitus. With regard to endocrine tumors, it must be emphasized that while somatostatin will inhibit hormone secretion by these tissues, it would not be expected to diminish tumor growth. Thus, in these conditions it is unlikely that somatostatin will find use other than as a symptomatic or temporizing measure.

In diabetes mellitus, however, somatostatin might be of considerable clinical value. First, it has already been demonstrated that it can acutely improve fasting as well as postprandial hyperglycemia in insulin-requiring diabetics by inhibiting glucagon secretion. Second, since growth hormone has been implicated in the development of diabetic retinopathy, the inhibition of growth hormone secretion by somatostatin may lessen this complication of diabetes. Finally, through suppression of both growth hormone and glucagon secretion, somatostatin may prevent or diminish the severity of diabetic ketoacidosis and find application in "brittle diabetes." These optimistic expectations must be considered with the facts that the multiple effects of somatostatin on hormone secretions and its short duration of action make its clinical use impractical at the present time and that its long-term effectiveness and safety have not been established as yet. For complete references to all the statements regarding clinical use of somatostatin, please see the recent review by Guillemin and Gerich (1976).

With the considerable interest in somatostatin as a part of the treatment of diabetics, "improved" analogs of somatostatin have been in the mind of clinicians and investigators. Analogs of somatostatin have been prepared in attempts to obtain substances of longer duration of activity than the native form of somatostatin; this has not been very successful so far. Other analogs have been sought that would have dissociated biological activity on one or more of the multiple recognized target of somatostatin. Remarkable results have recently been obtained. The first such analog so recognized by the group of the Wyeth Research Laboratories was [des-Asn^5]-somatostatin, an analog with approximately 4%, 10%, and 1% the activity of somatostatin to inhibit, respectively, secretion of growth hormone, insulin, and glucagon (Sarantakis *et al.*, 1976). Although such an analog is not of clinical interest, it showed that dissociation of the biological activities of the native somatostatin on three of its receptors could be achieved. The most interesting analogs with dissociated activities reported so far, all prepared and studied by J. Rivier, M. Brown, and W. Vale in our laboratories are [D-Ser^{13}]-somatostatin, [D-Cys^{14}]-somatostatin, and [D-Trp^8, D-Cys^{14}]-somatostatin. When compared to somatostatin, this latest compound has ratios of activity such as 300%, 10%, 100% to inhibit the secretions, respectively, of growth hormone, insulin, and glucagon (Brown *et al.*, 1976). These and other analogs are obviously of much clinical interest and are being so investigated at the moment in several laboratories.

III. The Endorphins

The concept and the demonstration a couple of years ago of the existence in the brain of mammalians of (synaptosomal) opiate receptors (see for a review, Marx, 1975) has led to the search of what has been termed the endogenous ligand(s) of these opiate receptors. The generic name *endorphins* (from endogenous and morphine) was proposed for these (then hypothetical substances) by Eric Simon. Sometime last year, I became interested in these early observations because like morphine, endorphins might stimulate the secretion of growth hormone; indeed the nature of the growth hormone-releasing factor remains unknown.

I first confirmed that dilute acetic acid–methanol extracts of whole brain (ox, pig, rat) contain substances presumably peptidic in nature, with naloxone-reversible morphinelike activity in a bioassay (mouse vas deferens or myenteric-plexus longitudinal muscle of guinea pig ileum) or in rat-brain synaptosomal opiate-receptor assays. Evidence of such biological activity in our laboratory was in agreement with earlier results of Hughes *et al.* (1975b), of Terenius and Wahlstrom (1975), and from the laboratories of Goldstein (Teschemacher *et al.*, 1975) and of Snyder (Pasternak *et al.*, 1975). I found, however, that similar extracts of (porcine) hypothalamus-neurohypophysis contain much greater concentrations of this morphinelike activity, than extracts of whole brain. From such an extract of approximately 1/4 million fragments of (pig) hypothalamus–neurohypophysis, I rapidly isolated several oligopeptides named *endorphins*, with opioid activity (Guillemin *et al.*, 1976; Lazarus *et al.*, 1976). Met^5-enkephalin and Leu^5-enkephalin recently isolated by Hughes *et al.* (1975b) have not been observed in these extracts. The primary structure of α-*endorphin* was established by Nicholas Ling by mass spectrometry of the enzymically cleaved peptide and by classical Edman degradation by Roger Burgus and is H-Tyr-Gly-Gly-Phe-Met-Thr-Ser-Gly-Lys-Ser-Gln-Thr-Pro-Leu-Val-Thr-OH. The primary structure of γ-*endorphin* was similarly established; it has the same primary structure as α-endorphin with one additional Leu as the COOH-terminal residue in position 17. Thus Met^5-enkephalin is the N-terminal pentapeptide of the endorphins, which have the same amino acid sequence as β-lipotropin (β-LPH)[61–76 and 61–77]. It is to the credit of Hughes *et al.* (1975a) to have recognized that the primary sequence of Met^5-enkephalin is identical to that of Tyr^{61} to Met^{65} of the β-lipotropins. β-LPH[61–91], a fragment of β-LPH isolated earlier on the basis of its chemical characteristics (Bradbury *et al.*, 1975; Li and Chung, 1976) was shown to have opiatelike activity (Cox *et al.*, 1976; Bradbury *et al.*, 1976; Lazarus *et al.*, 1976) and was named β-endorphin by Li and Chung (1976). On the evidence of amino acid composition, β-endorphin has also been isolated by us from the extract of hypothalamus–neurohypophysis from which we isolated α- and γ-endorphin; it has been isolated and similarly characterized also from an extract of whole (sheep) pituitary glands. In all cases, the various substances were purified and isolated on the basis of the response of the guinea pig myenteric plexus bioassay for morphinelike substances.

β-LPH has no opioid activity in any of the tests as above. Incubation of β-LPH at 37°C with the 10^5 *g* supernatant of a neutral sucrose extract of rat brain generates opioid activity. Thus β-LPH may be a prohormone for the opiatelike peptides (Lazarus et al., 1976). This would imply that the biogenesis of endorphins may be similar to that of angiotensin with cleaving enzymes available in the central nervous system. β-LPH[61–63] has no opioid activity at 10^{-4} M; β-LPH[61–64], β-LPH[61–65]-NH$_2$, (Met(O)^{65})-β-LPH[61–65], β-LPH[61–69], β-LPH[61–76], and β-LPH[61–91] all have opioid activity. β-LPH[61–65]-NH$_2$, β-LPH[61–65]-NEt, and all peptides larger than β-LPH[61–65] have longer duration of biological activity than met-enkephalin in the myenteric plexus bioassay. All these peptides were prepared by solid-phase synthesis by Nicholas Ling in our laboratories (1977). β-Endorphin is by far the longest-acting peptide when compared at equimolar ratios with all other fragments of the 61–91 COOH-fragment of β-LPH. In quantitative assays using the myenteric plexus, β-endorphin is about 5 times more potent than Met^5-enkephalin; the two analogs of Met^5-enkephalin amidated on the C-terminal residue have also 2–3 times greater specific activity than the free acid form of the peptide, with 95% fiducial limits of the assays overlapping those of β-endorphin.

When tested in the myenteric plexus–ileum bioassay, β-LPH[61–64], Phe^1-Met^5-enkephalin, (O-Me)Tyr^1-enkephalin, though of low specific activity when compared to Met^5-enkephalin, all have full intrinsic activity. [Arg-Tyr]^1-Met^5-enkephalin, i.e., β-LPH[60–65] is equipotent to Met^5-enkephalin. Thus an intact Tyr NH^2-terminal is not a requisite for full intrinsic biological activity. Acetyl-Tyr^1-Met^5-enkephalin has very low specific activity, which cannot be further quantitated as the log-dose response function is totally divergent from that of α-endorphin or Met^5-enkephalin as reference standard. The peptide [Arg-Tyr^1]-Met^5-enkephalin, i.e., β-LPH[60–65], reported by Ungar (Ungar and Ungar, 1976) to have antagonistic properties to opiates, was found to be agonist in the *in vitro* bioassay with a specific activity statistically identical to that of Met^5-enkephalin and to be devoid of antagonistic activity against normorphine or β-endorphin *in vivo*, when administered in large doses (1 mg) intravenously or directly into the cisterna (250 μg).

With the use of a highly specific antiserum against α-endorphin, we have recently observed by immunofluorescence techniques that the peptide is present in all cells of the pars intermedia of the rat hypophysis and also in discrete cells of the pars distalis (adenohypophysis); it was not observed in the pars nervosa (neural lobe) (Bloom et al., 1977). Similar results were observed for β-endorphin, though with an antiserum of somewhat less specificity. By the same methodology α- and β-endorphins have been found in discrete axons in the hypothalamus and other parts of the brain of the rat. A complete mapping is in progress. Of interest was the observation that the immunoreactive peptides were found in the brain as late as 5 weeks after total hypophysectomy (rat material).

Administration of synthetic α-, β-, γ-endorphins (doses ranging from 0.5 to 2000 μg) in the cisterna under rapid ether anesthesia or through a permanently implanted cannula in the lateral ventricle in nonanesthetized rats leads to dramatic behavioral changes (Bloom et al., 1976). The analgesia produced by

Met[5]-enkephalin or α-endorphin is always minor, best shown as corneal anesthesia and very short-lived (5 minutes or less). β-Endorphin produces profound analgesia as part of a catatonic state lasting several hours. γ-Endorphin produces no analgesia. The effects of α-, β-, and γ-endorphin are immediately reversed (in seconds) by intravenous administration of naloxone (0.66 mg/100 gm body weight).

These profound behavioral effects are produced by peptides extracted and isolated from hypothalamus–neurohypophysis. α, β, and γ-Endorphins are all subunits of β-lipotropin characterized as a normal constituent of the pituitary gland of several mammalian species [see Yamashiro and Li (1974) for a review of the earlier chemical characterization of β-LPH and fragments]. Although β-LPH has no opiatelike activity per se, it has been shown that its incubation with an aqueous extract of rat brain rapidly generates fragments with opiatelike activity in bioassays or opiate-binding receptor assays (Lazarus et al., 1976). It is thus tempting to consider the results reported here as the basis for a hypothesis that would relate them to a causative mechanism of the mental illnesses of man. Practically nothing is known at the moment of the physiological regulation of the synthesis and release of β-lipotropins; the same can be said of the secretion or metabolism of the endorphins. The hypothesis proposed here has the advantage that it can be tested, confirmed, or disproved or be modified promptly on the basis of results obtained with available drugs. Moreover, and of immediate possible testing, the hypothesis proposed here would imply that the various mental states hypothetically related to abnormally elevated levels of α-, β-, and γ-endorphin should promptly be alleviated by administration of the powerful opiate antagonists, such as (available) naloxone.

IV. Looking to the Future: Two Hypotheses

After this rapid review of the most salient observation out of the expanding field of neuroendocrinology, I would like to conclude this lecture with a look to the future. This will take the form of two rather novel concepts. One of these, I will call the *low voltage processing of information by brain cells*. The other, I will call the *neural origin of endocrine glands* or *is endocrinology a branch of neuroendocrinology*. Both concepts have remarkable implications which, in my own mind, represent what the areas of research in that field will be for the next few years. What we will see also is that these two concepts lead to recognizing a remarkable unity of the mechanisms involved in physiological phenomena as widely separate as the stimulation of the secretion of ACTH or growth hormone by pituitary cells and the inhibition by β-endorphin of the firing pattern of a neuron in the cerebral cortex.

Until recently the neuron has been seen primarily as a one-way communication system with a central processor for proximally received inputs and a one-way cable for output, the axon. The axon is characterized by its self-regenerating ability to conduct waves of high voltage depolarization for rapid transmission of an essentially binary type of information, expressed at the axon terminal. The axon is usually of considerable length, many times the average diameter of the cell body. While the dendritic surface has long been recognized morphologically, and its vastness well observed, it was not granted much of an active role in the performance of the neuron, principally because experimental evidence of such activity was simply lacking. Any integrative ability or capability of the system was located at the axon hillock. One of the major characteristics of this view of the *projection neuron*, well studied for over fifty years, is the high-voltage action potential, from a few millivolts to as much as 100 millivolts. Classically, such a neuron will deliver its ultimate message at essentially a single address in the form of the packets of a discrete neurotransmitter. For most projection neurons we still do not know whether norepinephrine, acetylcholine, dopamine, and in a few instances serotonin, are involved.

No consensus exists as to the ultimate significance of substances like certain amino acids, γ-aminobutyric acid, substance P, and, lately, the small peptides like somatostatin, the hypothalamic releasing factors, and the endorphins, traced by immunoassays or immunocytochemistry to increasing numbers of neuronal fibers and neuronal bodies far removed from the ventral hypothalamus and the pituitary.

While this simplified picture of the projection neuron is still correct, a recent view of other types of neurons which seem to be in the majority appears to welcome this multiplicity of effectors and may well in the next few years permit satisfactory integration of these multiple effectors, with specific functional processes of individual neurons as members of a neuronal network. This new view of the neuron is based on new morphology as seen by the electron microscope. Much of the discussion that follows is based on a recent review by Schmitt *et al.* (1976) and on the text of the proceedings of a Meeting of the Neurosciences Research Program devoted to local circuits (Rakic, 1975). The dendrite is seen no more as a "passive receptor surface," but rather as a locus for transmitting as well as receiving information in traffic with dendrites of other neurons or with extracellular compartments including capillaries. The means to such communications are seen in the release or uptake of diverse small or large molecules. The electric phenomena involved are those of pinpoint depolarizations and are measured in a few microvolts. Such electrotonic currents spread only over distances measured in microns, not millimeters or centimeters. This type of extremely low voltage communication constitute the so-called local circuits. The local circuit neuron can also modify the ultimate behavior of one or more projection neurons, which can send responses to a remote contact by a high voltage, long axon-pathway. Such systems have been well studied in the retina, the olfactory bulb, the lateral geniculate body. There is increasing evidence that such local circuitry is actually present in all parts of the central nervous system. It actually represents the structure of the greatest mass or volume of the central nervous system (CNS), with the projection neurons in their classical anatomical arrangements, probably a minority in number as well as in occupied space.

These multiple dendritic connections have been observed in the electron microscope. Figure 1 is a diagram of possible connections (redrawn from the recent review by Schmitt *et al.*, 1976, mentioned above), between one axon terminal and two dendrites, one of which also is in contact with three other dendrites at about ten other locations. The diagram conveys the observation that dendrites may be both presynaptic and postsynaptic to each other as in reciprocal synapses. There is also electron microscope evidence of *gap junctions* between dendrites. Such electrotonic coupling has been demonstrated in the central nervous system. Several neurons so coupled will respond synchronically with extremely low voltages required. Oscillatory behavior has been observed in populations of neurons in some invertebrates (Gettings and Willows, 1974) when electrically coupled. Such electronic junctions are frequently observed in imme-

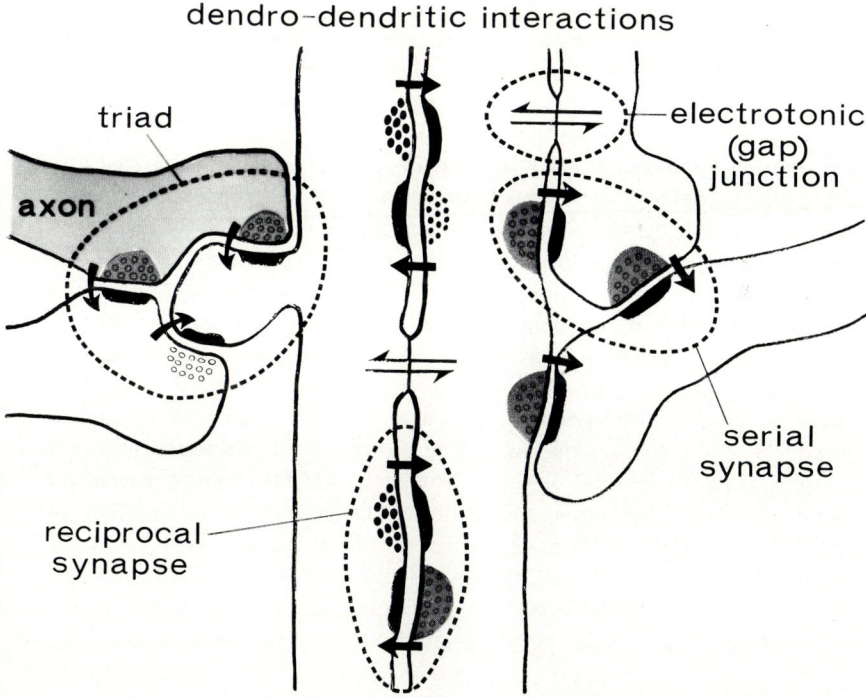

FIG. 1. Diagram of possible connections between one axon terminal and two dendrites. Redrawn from Schmitt *et al.* (1976).

diate proximity to chemical synapses (Schmitt *et al.*, 1976). None of these phenomena require, and none of these structures produce, high voltage spikes. Information transfer by such mechanisms is relatively slow, in seconds or longer, not milliseconds.

What are the relationships between this new view of neurons and the peptides of neuroendocrinology? They become apparent when one asks about the nature of the chemicals involved in these dendro-dendritic contacts? The fragmentary emerging picture is most interesting. It has long been considered that dendrites are involved in uptake of necessary metabolites, such as sugars, free amino acids, adenosine. Such uptake can proceed from extracellular fluid and also from capillary vessels, through endothelial cells. Molecules of much larger size appear similarly to be taken up by dendrites. For instance, an enzyme like acetylcholinesterase, after being released in extracellular compartments, has been reported (Kreutzberg *et al.*, 1975) to be bound to the outer surface of the dendritic membrane and later to be taken up by dendrites.

Thus an interesting *hypothesis* would be for the release at dendritic points of (still to be characterized) enzymes of neuronal origin that would specifically cleave a biologically active peptide, such as one of the endorphins, from a biologically inactive precursor, such as β-lipotropin available in extracellular compartments or perhaps to be found in the axoplasm.

Classical transmitters appear to be released and taken up by the dendrites. Similar release and reciprocal uptake of the small peptides such as substance P, neurotensin, somatostatin, TRF, endorphins has not been demonstrated as yet. That this is possible and actually happens is a working hypothesis worth investigating. It would go a long way to explain the multiplicity of effects of the polypeptides on the CNS. See, for instance (Table I), the multiplicity of effects of the tripeptide TRF on biological events which have nothing to do with the release of pituitary thyrotropin, the well known hypophysiotropic activity of TRF, for which it was originally named and recognized. The possibility of an enormous number of such transfer sites might also explain the psychotropic effects of some of these peptides. Cajal as early as 1899 had made the comment that local-circuit neurons may well play an important role as the substrate of complex behavior because of their "prodigious abundance and unaccustomed wealth of forms." Considering such enormous dendritic trees, with each dendrite ending compounded to the description as in Fig. 1, the number of contacts and control points for a single neuron defies the imagination. Such cellular anatomy when considered with the hypotheses mentioned above for chemical inputs and outputs shows the considerable possible functional significance of an expanded dendritic connection network.

The multiple terminal network hypothesis with the releasing and utilizing of biologically active peptides, may also be involved in another way in explaining some current data. It has been recognized that the total amounts as well as the

TABLE I
Central Nervous System Mediated Actions of Thyrotropin Releasing Factor

---

Increases spontaneous motor activity
Alters sleep patterns
Produces anorexia
Inhibition of condition avoidance behavior
Head-to-tail rotation
Opposes actions of barbiturates on sleeping time, hypothermia, lethality
Opposes actions of ethanol, chloral hydrate, chlorpromazine, and diazepam on sleeping time and hypothermia
Enhances convulsion time and lethality of strychnine
Increases motor activity in morphine-treated animals
Potentiates DOPA-pargyline effects
Amelioration of human behavioral disorders?
Central inhibition of morphine-mediated secretion of growth hormone and prolactin
Alteration of brain cell membrane electrical activity
Increases norepinephrine turnover
Releases norepinephrine and dopamine from synaptosomal preparations
Enhances disappearance of norepinephrine from nerve terminals
Potentiates excitory actions of acetylcholine on cerebral cortical neurons

---

concentrations of TRF, or LRF or somatostatin in the extrahypothalamic CNS as measured by bioassays or radioimmunoassays are considerably higher than can be accounted for by the number of cell bodies shown to contain such peptides by immunocytochemistry. A hypothesis to consider is that there would be relatively few neurons manufacturing, say, TRF, LRF (primarily), located in the hypophysiotropic area of the hypothalamus with perhaps a few more cells in the amygdala and that these neurons have long axons with *multiple axon collaterals*, all with peptide containing and secreting boutons terminals.

There is already evidence that the dendritic traffic of chemicals works both ways with release and uptake. Thus, in a reciprocally functioning system, if the endorphins and enkephalins are enzymatically cleaved extracellularly from β-lipotropin as the circulating precursor in a manner reminiscent of the biogenesis of angiotensin, they could then be picked up by the multiple dendritic endings and carried by retrograde axoplasmic flow whatever distance is necessary for their physiological function.

In summary, *in this hypothesis* we see the small peptides as substances released locally and perhaps produced locally at innumerable possible source points. The functions they would subserve would thus be dependent on the effector cell they would be modulating. Not exclusively secreted at classical synapses, they would be truly *modulators* of neuronal functions rather than true neurotransmitters. They could also be either transmitters or modulators, depending on their locus of release. If the effect of the small peptides is then the activation of the adenylcyclase–cAMP system of their effector neurons, as they may well do in other target tissues, such as the adenohypophysis, their effects in neuronal networks would be amplified, long lasting as well as possibly expanding from their exact source point.

Such a system involving neuronal cAMP has already been demonstrated by Bloom and Siggins for neurons of the locus coeruleus (Siggins *et al.*, 1973).

This concept of a *local* release and *local* immediate effect of the peptides from multiple sources in the CNS is a point to remember for future discussion. It does not belong to substances called hormones in the classical definition of the word. This is true even if each local event may lead to ultimately widespread effects. Needless to say that the technology involved in exploring such secretory functions of the dendrites or of the boutons of axon collaterals will be particularly challenging.

As TRF activity, LRF activity, and somatostatin activity were being found by bioassays, later radioimmunoassays, in the extrahypothalamic central nervous system, reports appeared showing effects of somatostatin in inhibiting secretion of glucagon and insulin by direct action at the level of the endocrine pancreas. Because somatostatin has a short biological half-life upon injection in peripheral blood, it was unlikely that any physiological effect of endogenous somatostatin on the endocrine pancreas would be due to somatostatin of hypothalamic origin. Looked for originally in pancreatic nerve endings, somatostatin was found by immunofluorescence in discrete endocrine *cells* of the pancreas, now well characterized as the D cells (Dubois, 1975; Luft *et al.*, 1974). The same studies showed somatostatin in discrete cells in the jejunum, the colon, the duodenum, and the gastric mucosa. Observations were also made of somatostatin inhibiting the secretion of gastrin, secretin, gastric HCl, and, recently, acetylcholine from the myenteric plexus. TRF, LRF, though found in extrahypothalamic CNS, have not been found to my knowledge in extra-CNS tissues. This is something to look for. As early as 1957 I had observed CRF activity in extracts of gut tissues also containing substance P (Guillemin, 1957). Brodish has also described extrahypothalamic CRF.

Besides somatostatin, other peptides are now known to be present, and most likely synthesized by cellular elements, *both* in the central and peripheral nervous system and also in glandular elements of the gastrointestinal tract. The first peptide so observed was substance P in the remarkable experiments of Ulf von Euler and Gaddum as early as 1931. There is now evidence that neurotensin, gastrin, VIP, GIP, the endorphins and enkephalin(s) are found both in the brain and in the gastrointestinal tract plus the pancreas. This is also true for several of the small peptides, such as bombesin, caerulein, physalaemin, isolated years ago from extracts of the skin of frogs of several species. Furthermore, there are remarkable analogies and homologies between the amino acid sequences of several of these peptides of CNS origin and gastrointestinal origin as well as those isolated from the frog skin. These peptides have been found by immunocytochemistry essentially in two types of cells: (1) cell bodies and nerve fibers, i.e., neural and dendritic processes of *neurons* in brain, spinal cord, in spinal ganglia, in the myenteric plexus; (2) typical *endocrine* cells, for instance, in the islets of Langerhans, in the enterochromaffin cells of the gastrointestinal tract, of the adrenal medulla. Neuroblastomas have been reported to contain high levels of the vasointestinal peptide (VIP) (Said and Rosenberg, 1976). An undifferentiated mediastinoma has been mentioned to contain somatostatin, calcitonin, ACTH, prolactin (private communication from Iain MacIntyre). All these results are based on radioimmunoassays, immunocytochemistry, in some instances also bioassays, with in most cases evidence of parallelism of the responses to the known peptide-reference standard and the crude tissue extracts. In cases involving immunological methods there is a modicum of specificity of the antibodies utilized. More significant, reports are beginning to appear showing identity of the primary structure of the gastrointestinal variety of a peptide when compared to its central nervous system variety as, for instance, neurotensin (Carraway *et al.*, 1976) and substance P. Our laboratory has already reported the complete sequencing of hypothalamic α-MSH which is identical to that of pituitary α-MSH (Burgus *et al.*, 1976). Thus there is every reason to believe that we are dealing with the same peptides regardless of their tissue origin.

What is the message to be read in these observations of startling commonalities between the central nervous system and endocrine tissues, and what does it imply for future research?

There is already an interesting unifying concept: Much credit must go to A. G. E. Pearse for his visionary concept formulated some ten years ago, of the APUD cells: Pearse observed that neurons and some endocrine cells producing polypeptide hormones shared a set of common cytochemical features and ultrastructural characteristics. APUD is an acronym referring to *A*mine content and/or amine *P*recursor *U*ptake and *D*ecarboxylation, as common qualities of these cells (Pearse, 1968). The APUD concept postulated that these endocrine cells were derived from a common neuroectodermal ancestor, the transient *neural crest*. On the basis of these observations, Pearse postulated that a large number of endocrine cells, larger than that known to him at the time, would be found sharing these common properties if one were to explore further, in the adult, endocrine tissues derived from the neural crest. Recent observations with refined techniques, particularly the work of Le Douarin on topical chimeras with chromosomal markers have led Pearse to modify the original APUD concept, but, as we will see, in a remarkable manner. The new evidence of the many multiple sources of the large number of peptides mentioned above showed that tissues were involved that were not of neural crest origin; this is particularly true for the peptide-secreting cells of the gut. All these cells have been shown to arise from specialized neuroectoderm (Pearse and Takor, 1976)—that is, not only the neural crest but also the neural tube, the neural ridges, and the placodes.

The expanded concept now postulates that all peptide hormone-producing cells are derived from the neural ectoderm, as are all neurons. For instance recently Pearse and Takor (1976) have reexamined the early stages of development of both the hypophysis and the hypothalamus. They confirmed and expanded earlier conclusions of Ballard, who as early as 1964 had recognized that Rathke's pouch does not come from the stomodeum (the pharyngeal origin) as is classically written, but originates from the ventral neural ridge (from studies in the chick embryo).

Thus, the hypophysis would share with the hypothalamus the same ventral neural ridge of the neuroectoderm for its origin. Recent work by Ferrand and Hraoui (1973) using the chromosomal markers in topical chimeras have also concluded that there is an exclusively neuroectodermic origin of the adenohypophysis. Thus, Pearse and Takor have recently concluded: "It is therefore necessary to postulate a neuroectodermal derivation for *all* the endocrine cells of the adenohypophysis and to regard the whole hypothalamo-hypophysial complex as a neuroendocrine derivative of the ventral neural ridge."

I mentioned earlier that we have recently observed with a well characterized antiserum to α-endorphin that this peptide can be seen by immunocytochemistry in discrete nerve fibers in the hypothalamus and in all cells of the intermediate lobe plus some cells of the adenohypophysis (Bloom *et al.*, 1977). These same cells in the pars intermedia have long been known from the work of several groups of investigators also to contain ACTH [1–39]; CLIP, which is ACTH [18–39], α-MSH, i.e., ACTH [1–13]; β-LPH [1–91]; γ-LPH [1–48]. Cells of the pars intermedia have also been considered to belong to the APUD series.

The conclusion from all this is that the peptide-secreting cells and tissues appear to be as much part of the nervous system as is the adrenal medulla, or what has been called traditionally the neurohypophysis. The word *neuroendocrinology* is now taking a fuller meaning than ever. Pearse has gone so far as to propose that the nervous system should be recognized as composed of three divisions—somatic, autonomic, and endocrine.

Perhaps the time has come to redefine the word *hormone*. The hypothalamic hypophysiotropic peptides, TRF, LRF, somatostatin, are really not hormones according to the current definition which is still that proposed by Starling in 1905 implying a single source of secretion, and, fundamentally, the direct ingress into blood vessels for distribution to a distant receptor, itself thus triggered to respond by its own discrete secretion. At the level of the dendritic network discussed above, between median eminence and adenohypophysial cells, in the endocrine pancreas or in the gastric mucosa, TRF, LRF, and somatostatin appear to have remarkably localized ranges of extracellular movements ranging from angstroms to microns, at most a few millimeters through the hypothalamo-hypophysial portal capillaries. There is no incontrovertible evidence so far that these peptides circulate in peripheral blood in physiologically significant concentration. They have multiple sources, possibly, as hypothesized above, innumerable. They would thus not be hormones in the classical sense. The concept of *paracrine* secretion first proposed by Feyrter in 1938 seems much more appropriate to describe products from cells which act on immediate neighbors. The distinction with *neurotransmitters* is not obvious. I proposed earlier the name *cybernin* for these substances, the etymology of the word implying "local information" or "local control." I have not pushed that new terminology too boldly as it is still another word, and an entirely new root. As we discussed above, there is good evidence that the small peptides can act as *modulators* of the function of neuronal system. They are not necessarily neurotransmitters in the classical sense and, so far, have not been demonstrated as such. We may want to redefine a *hormone* to be any substance released by a cell and which acts on another cell near or far, regardless of the singularity or ubiquity of the source and regardless of the means of conveyance, blood stream, axoplasmic flow, immediate extracellular space. If we do not do so, because of the old definition, and to be consistent, hormones would be the steroids, the products of the adenohypophysis, of the thyroid, insulin, glucagon, etc.—i.e., those messengers which really circulate wide and far in peripheral blood. The neuroendocrine peptides would have to be something else. I think that it is becoming of heuristic significance to reconsider that terminology. The choice of words or of definitions that will be proposed should take into consideration the remarkable developments I have summarized briefly here.

The remarkable picture that emerges is not only that more and more neurophysiologists and endocrinologists are dealing with similar concepts, but that they are, and have been all along, talking about various forms or embodiments of a single anatomical structure fundamentally devoted to the centrifugal dispatch of information. It comes as the classical neuron or the classical endocrine cell and several overlapping forms. The new information we have discussed here, the new working hypotheses we have proposed are all parts of the concept of neuroendocrinology as proposed thirty years ago by several groups of clinicians and laboratory investigators—Roussy and Mosinger, Muller, Laniel-Lavastine, etc.—and so clearly recognized on the basis of their own data by the late Ernst Scharrer and by Berta Scharrer. While the book as a whole has many shortcomings, the principle of neuroendocrine integration is clearly stated in the Scharrers' monograph "Neuroendocrinology" (Scharrer and Scharrer, 1963). Recent knowledge, as well as recent proposals for future studies, both outlined in this rapid review, are tributes to the far-reaching grasp of the subject that must be credited to these two genial investigators.

From the increasing number of studies using highly specific radioimmunoassays as well as methods of immunocytochemistry (with fluorescein-coupled antibody or with the horseradish peroxidase), more and more neurons are being recognized as containing specific peptides in their cell body or their peripheral elements. Gastrin (immunoreactive gastrin) is found in cells of the cerebral cortex; substance P, angiotensin, somatostatin, vasointestinal peptide (VIP), peptide sequences corresponding to fragments of ACTH or of β-LPH, are found in neurons, their axons and dendrites. When one realizes that all our current classical neurophysiology simply ignored this increasingly overwhelming morphological and biochemical evidence, one wonders what is the significance of these peptides in the physiology of the brain. All these questions, even the simplest one of physiological phenomenology, of neuroanatomy, of cellular biochemistry, should keep us thinking and working for quite a few years. This has to be a new era in our knowledge and perspective of the central nervous system, both normal and diseased.

ACKNOWLEDGEMENTS

Research of the Laboratories for Neuroendocrinology at the Salk Institute is currently supported by research grants from NIH (HD-09690-02 and AM-18811-02), National Foundation (1-411), and the William Randolph Hearst Foundation.

REFERENCES

Alford, F. P., Blood, S. R., Nabarro, J. D., Hall, R., Besser, G., Coy, D., Kastin, J., and Schally, A. (1974). *Lancet* 2, 974–977.
Ballard, W. W. (1964). *In* "Comparative Anatomy and Embryology." Ronald Press, New York.
Besser, G. M., Paxton, A., Johnson, S., Moody, E., Mortimer, C., Hall, R., Gomez-Pan, A., Schally, A., Kastin, A., and Coy, D. (1975). *Lancet* 1, 1166.
Bloom, F. E., Segal, D. S., Ling, N., and Guillemin, R. (1976). *Science* 194, 630.
Bloom, F. E., Battenberg, E., Rossier, J., Ling, N., Leppaluoto, J., Vargo, T. M., and Guillemin, R. (1977). *Life Sci.*
Bradbury, A. F., Smyth, D. G., and Snell, C. R. (1975). *In* "Peptides: Chemistry, Structure and Biology" (R. Walter and G. Meienhofer, eds.), p. 609. Ann Arbor Sci. Publ., Ann Arbor, Michigan.
Bradbury, A. F., Smyth, D. G., Snell, C. R., Birsall, N. T., Hulme, E. C. (1976). *Nature (London)* 260, 793–795.
Brazeau, P., Vale, W., Burgus, R., Ling, N., Butcher, M., Rivier, J., and Guillemin, R. (1973). *Science* 179, 77–79.
Brown, M., Rivier, J., and Vale, W. (1976). *Metab., Clin. Exp.* 25, 1501–1503.
Burgus, R., Dunn, T. F., Desiderio, D., and Guillemin, R. (1969). *C.R. Hebd. Seances Acad. Sci.* 261, 1870.
Burgus, R., Amoss, M., Brazeau, P., Brown, M., Ling, N., Rivier, C., Rivier, J., Vale, W., and Villarreal, J. (1976). *In* "Hypothalamus and Endocrine Functions" (F. Labrie, ed.) p. 355. Plenum, New York.
Carraway, R. E., Kitabgi, P., and Leeman, S. E. (1976). *Proc. Int. Congr. Endocrinol., 5th, 1976* Abstract, p. 178.

Cox, B. M., Goldstein, A., and Li, C. H. (1976). *Proc. Natl. Acad. Sci. U.S.A.* **73**, 1821.
Dubois, M. (1975). *Proc. Natl. Acad. Sci. U.S.A.* **72**, 1340–1343.
Ferrand, R., and Hraoui, S. (1973). *C.R. Seances Soc. Biol. Ses Fil.* **167**, 740.
Feyrter, F. (1938). "Uber Diffuse Endokrine Epitheliale Organe." Barth, Leipzig.
Fujino, M., Yamazaki, I., Kobayashi, S., Fukuda, T., Shinagawa, S., and Nakayama, R. (1974). *Biochem. Biophys. Res. Commun.* **57**, 1248–1256.
Gerich, J. E., Lorenzi, M., Schneider, V., Karam, J. H., Rivier, J., Guillemin, R., and Forsham, P. H. (1974). *N. Engl. J. Med.* **291**, 544.
Gerich, J. E., Lorenzi, M., Gustafson, G., Guillemin, R., and Forsham, P. (1975). *Metabolism* **24**, 175.
Gettings, P. A., and A. O. D. Willows (1974). *J. Neurophysiol.* **37**, 358–361.
Guillemin, R., Hearn, W. R., Check, W. R., and Housholder, D. E. (1957). *Endocrinology* **60**, 488–506.
Guillemin, R., and Gerich, J. E. (1976). *Annu. Rev. Med.* **27**, 379–388.
Guillemin, R., Yamazaki, E., Jutisz, M., and Sakiz, E. (1962). *C.R. Hebd. Seances Acad. Sci.* **255**, 1018–1020.
Guillemin, R., Ling, N., and Burgus, R. (1976). *C.R. Hebd. Seances Acad. Sci.* **282**, 783–785.
Hughes, J., Smith, T. W., Kosterlitz, H. W., Fothergill, L. A., Morgan, B. A., and Morris, H. R. (1975a). *Nature (London)* **258**, 577–579.
Hughes, J., Smith, T., Morgan, B., and Fothergill, L. (1975b). *Life Sci.* **16**, 1753–1758.
Johnson, E. S., Gendrich, R. L., and White, W. F. (1976). *Fertil. Steril.* **27**, 853–860.
Koerker, D. J., Harker, L. A., and Goodner, C. J. (1975). *N. Engl. J. Med.* **293**, 476–479.
Kreutzberg, G. W., Schubert, P., and Lux, H. D. (1975). *In* "Neuroplasmic Transport in Axons and Dendrites." Golgi Centennial Symp., p. 161. Raven, New York.
Lazarus, L., Ling, N., and Guillemin, R. (1976). *Proc. Natl. Acad. Sci. U.S.A.* **73**, 2156–2159.
Li, C. H., and Chung, D. (1976). *Proc. Natl. Acad. Sci. U.S.A.* **73**, 1145–1148.
Ling, N. (1977). *Biochem. Biophys. Res. Commun.* **74**, 248–255.
Lorenzi, M., Gerich, J. E., Karam, J. H., and Forsham, P. H. (1975). *J. Clin. Endocrinol. Metab.* **40**, 1121–1124.
Luft, R., Efendic, S., Hokfelt, T., Johansson, O., and Arimura, A. (1974). *Med. Biol.* **52**, 428–30.
Marks, N., and Stern, F. (1975). *FEBS Lett.* **55**, 220–224.
Marx, Jean L. (1976). *Science* **193**, 1227–1229.
Mielke, C. H., Gerich, J. E., Lorenzi, M., Tsalikian, E., Rodvien, R., and Forsham, P. H. (1975). *N. Engl. J. Med.* **293**, 480.
Monahan, M., Amoss, M., Anderson, H., and Vale, W. (1973). *Biochemistry* **12**, 4616–4620.
Mortimer, C. H., Carr, D., Lind, T., Bloom, S. R., Mallinson, C. N., Schally, A. V., Tunbridge, W. M. G., Yeomans, L., Coy, D. H., Kastin, A., Besser, G. M., and Hall, R. (1974). *Lancet* **1**, 697–701.
Pasternak, G., Goodman, R., and Snyder, S. H. (1975). *Life Sci.* **16**, 1765.
Pearse, A. G. E. (1968). *Proc. R. Soc. London, Ser. B* **170**, 71.
Pearse, A. G. E., and Takor, T. (1976). *Clin. Endocrinol.* **5**, Suppl., 299s-244s.
Rakic, P. (1975). *Neurosci. Res. Program, Bull.* **13**, 291.
Said, S. I., and Rosenberg, R. N. (1976). *Science* **192**, 907.
Sarantakis, D., McKinley, W. A., Jaunakais, I., Clark, D., Grant, N. H. (1976). *Clin. Endocrinol.* **5**, 275s–278s.
Scharrer, E., and Scharrer, B. (1963). *Neuroendocrinology* **1**, 1.
Schmitt, F. O., Dev, P., and Smith, B. H. (1976). *Science* **193**, 114.
Siggins, G. R., Battenberg, E. F., Hoffer, B. J., and Bloom, F. E. (1973). *Science* **179**, 585.

Starling, E., H. (1905). *Lancet* **1**, 340–341.
Terenius, L., and Wahlstrom, A. (1975). *Life Sci.* **16**, 1759.
Teschemacher, H., Opheim, K. E., Cox, B. J., and Goldstein, A. (1975). *Life Sci.* **16**, 1771.
Ungar, G., and Ungar, A. (1976). *Fed. Proc., Fed. Am. Soc. Exp. Biol.* **35**, 309.
Vale, W., Rivier, J., and Burgus, R. (1971). *Endocrinology* **89**, 1485–1488.
Vale, W., Grant, G., Rivier, J., Monahan, M., Amoss, M., Blackwell, R., Burgus, and Guillemin, R. (1972). *Science* **176**, 933–934.
Vale, W., Brazeau, P., Rivier, C., Brown, M., Boss, Rivier, J., Burgus, R., Ling, N., and Guillemin, R. (1975). *Recent Prog. Horm. Res.* **31**, 365–397.
Vale, W., Rivier, C., Brown, M., Leppaluoto, J., Ling, N., Monahan, M., and Rivier, J. (1976). *Clin. Endocrinol.* **5**, Suppl., 261s–273s.
Vilchez-Martinez, J., Coy, D., Coy, E., De la Cruz, A., Nishi, N., and Schally, A. V. (1975). *Endocrinology* **96**, 354A.
von Euler, U. S., and Gaddum, J. H. (1931). *J. Physiol.* **72**, 74.
Yamashiro, D., and Li, C. H. (1974). *Proc. Natl. Acad. Sci. U.S.A.* **71**, 4945–4949.

DISCUSSION

J. A. Parsons: Could you please elaborate on the *in vivo* evidence from morphinelike activity of the peptide you call β-endorphin? I think this is the same substance that Derek Smyth and his colleagues isolated from porcine pituitaries [A. F. Bradbury, D. G. Smyth, and C. R. Snell, in "Peptides: Chemistry, Structure and Biology" (R. Walter and G. Meienhofer, eds.), p. 609. Ann Arbor Sci. Publ., Ann Arbor, Michigan (1975)] and named the C-fragment of lipotropin [A. F. Bradbury, D. G. Smyth, and C. R. Snell, *Polypept. Horm.: Mol. Cell. Aspects, Ciba Found. Symp.* **41**, 61–75 (1976)]. After the appearance of the paper on the encephalins by Hughes *et al.* [J. Hughes, T. Smith, B. Morgan, and L. Fothergill, *Life Sci.* **16**, 1753 (1975)], W. Feldberg and D. G. Smyth [*J. Physiol. (London)* **260**, 30P (1976)] carried out what I suppose pharmacologists would regard as a decisive test for *in vivo* morphinelike properties of the C-fragment. They injected it into the third cerebral ventricle of unanesthetized cats with indwelling cannulas and determined log dose–response regressions for inhibition of the response to graded heavy pressure on the tail. On a molar basis, the C-fragment proved to be 100–200 times more potent an analgesic than morphine sulfate, and it also produced other characteristic effects of morphine, such as shivering, tachypnea, and mydriasis, as well as the catatonia you have just described.

R. Guillemin: We have injected β-endorphin (β-LPH 61–91) in the lateral ventricle of the brain of unanesthetized rats with an implanted cannula or in the cisterna, under ether anesthesia. Doses have ranged from 0.5 μg to 250 μg of peptide. Profound analgesia of the whole body is regularly produced. On a molar basis, β-endorphin is 10–40 times more potent as an analgesic agent than normorphine, which we use routinely. The syndrome of catatonia induced by β-endorphin differs from the immobility of the animal given normorphine by the same route. Animals given morphine usually move about slightly awkwardly; they are not stiff. Animals given β-endorphin never move on their own; they are stiff, the whole body being moved en bloc when one moves one part or another. With 0.5 mg of β-endorphin, the rats will remain in this state for 3–4 hours. If one keeps disturbing the animals by handling, for instance, they will move somewhat, but immediately resume their catatonic state as soon as left undisturbed. As I said in the lecture, if you give them naloxone intravenously (0.6 mg/kg as a single shot), they will wake up and start moving before the needle is out of the vein. They will show a few body shakes, occasionally, and resume immediately, within seconds, normal exploratory behavior. With a large dose of naloxone they will not return to their catatonic state; with a small dose of naloxone, given intravenously, sometihnes they will wake up immediately, remain awake for 5–10 minutes and then resume their catatonic state for what looks like the remainder of the expected syndrome (had they not received naloxone).

J. A. Parsons: I do not know whether it is a question of species or of dose, but I do not believe that catatonia was an important feature of what they were observing with the small doses in the cat.

W. F. White: In the early part of his presentation, Dr. Guillemin was gracious enough to refer to some of our antireproductive work with the high-potency analogs of LRF. I would like to summarize some of this work, much of it reported in an article in the July issue of *Fertility and Sterility* [E. S. Johnson, R. L. Gendrich, and W. F. White, *Fertil. Steril.* **27**, 853 (1976)]. In short, there is a series of these analogs in which a change is made in the 6-position incorporating D-amino acid and which have exceedingly high reproductive activity under normal conditions in nanogram amounts. However, if one uses microgram amounts, two or three orders of magnitude higher than the so-called physiological level, one gets entirely the opposite effect. In the top panel of Fig. A are the ovarian weights of animals started on treatment at 21 days of age and injected twice a day with 1–10 μg of the test compound. In fact, there are two levels shown—the middle being the lower level of hormone, and the bottom the higher level. The top curve shows the normal growth of the ovaries on saline. During the treatment period, ovarian growth is limited and puberty is indefinitely delayed. The animals show no tendency to breed even though they have passed the age when normal breeding activities begin. However, if one later withholds treatment for 6–10 days, the females completely reverse and will accept the male, breed normally, and have normal litters of pups. This general finding is true also of a cycling animal. If one starts treatment in the cycle, within 4–5 days all cycling stops; the animals remain acyclic and they do not breed until treatment is stopped. Again, within 6–10 days, full reproductive competency is recovered.

In the lower panel of Fig. A it is shown that uterine growth too is repressed. This finding led to the hypothesis that perhaps this compound could be used in treating breast tumors, particularly those that are estrogen dependent. In Fig. B are data for the first test made of that theory. Here, an animal in a large group of females was delivered unexpectedly bearing a large tumor in the breast region. Biopsy showed the tumor to be an adenomacarcinoma. Treatment was started on day 1, and within 10 days the tumor had disappeared completely. Then, during the holiday season (about two years ago), the animal was left without treatment for 20 or 30 days. By the time the animal was reexamined, the tumor had actually grown to a larger size than when it was first observed. Again, institution of treatment reduced tumor to the point where it could barely be located. At this point, treatment was again withheld, and after a long time the tumor regrew. At this time, ovariectomy was performed, and this too caused regression of the tumor. These experiments have been repeated in dimethylbenzanthracene (DMBA)-induced tumors; we found that 2.5–10 μg of this LRF analog twice a day would cause the same regression of tumors as did ovariectomy in these animals. These results have led us at Abbott Laboratories to plan to test this compound in human breast cancer.

R. Guillemin: I mentioned that our observations were on entirely different targets; however, with the same synthetic analogs they are very much in keeping with yours, since we were able to show inhibition of pregnancy and resorption of the fetuses. I wish you would tell us what are your ideas about the mechanism of action of this unexpected effect of a super-LRF.

W. F. White: There is not enough time to go over all the evidence that we have, but perhaps I would be safe in saying, in a preliminary way, that I think there is a desensitization of the ovarian receptor sites for FSH, particularly, and possibly also of LH. I say this because, during the course of treatment, large quantities of both LH and FSH are poured out twice a day and the high levels are maintained during the entire course of the treatment. Thus, there is no lack of LH or FSH.

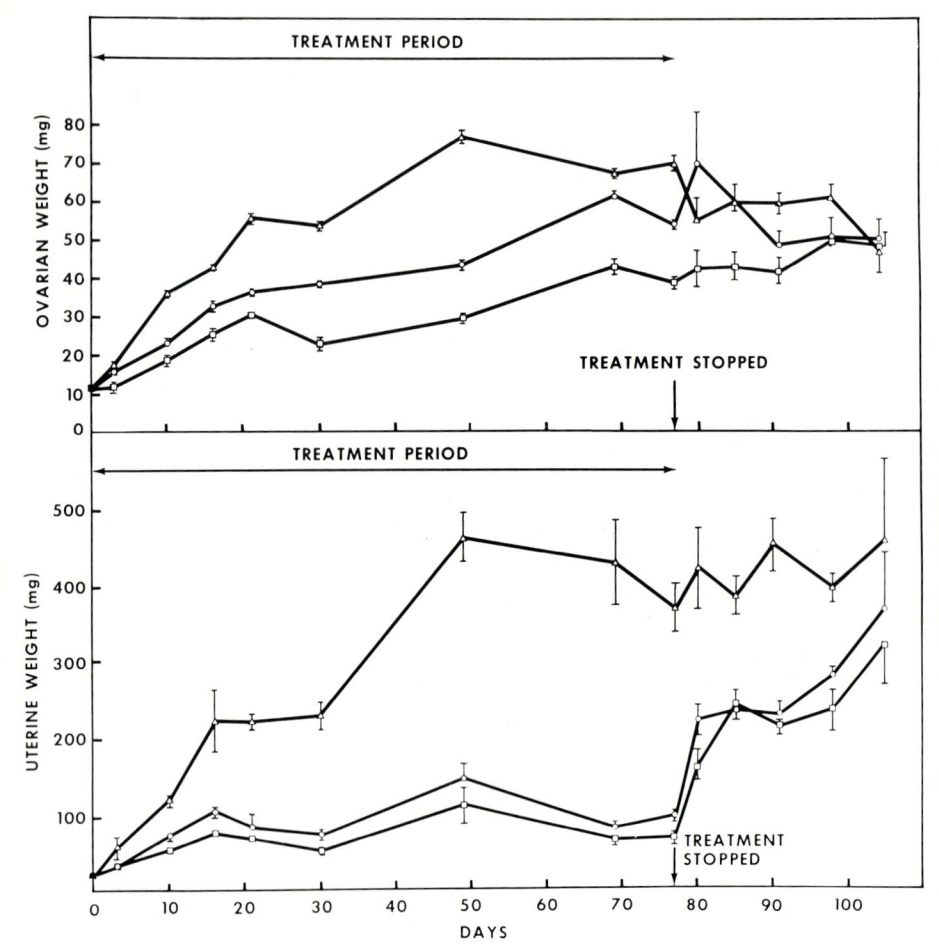

FIG. A. Ovarian and uterine weights during and after treatment with [D-Leu⁶, des-Gly-NH$_2$¹⁰, Pro-ethylamide⁹]-GnRH. △——△, controls; ○——○, 0.5 µg analog b.i.d.; □——□, 3.0 µg analog b.i.d.

G. A. Hedge: I was very interested in the variety of effects of these endorphins, particularly in the effects you mentioned first, those on pituitary hormone secretion. You mentioned growth hormone and prolactin, but what about the other pituitary hormones? Have you tested the endorphins for possible regulatory roles regarding TSH, ACTH, and the gonadotropins?

R. Guillemin: You would expect that they would release ACTH. We have evidence that they do not modify the secretion of LH and FSH; we do not have the results of the experiments with ACTH, but you would expect that in this system there would be a release of ACTH: then the difficulty would be to prove that it is due not to the procedure, but to the particular molecule directly at the pituitary level. You well know how difficult that is.

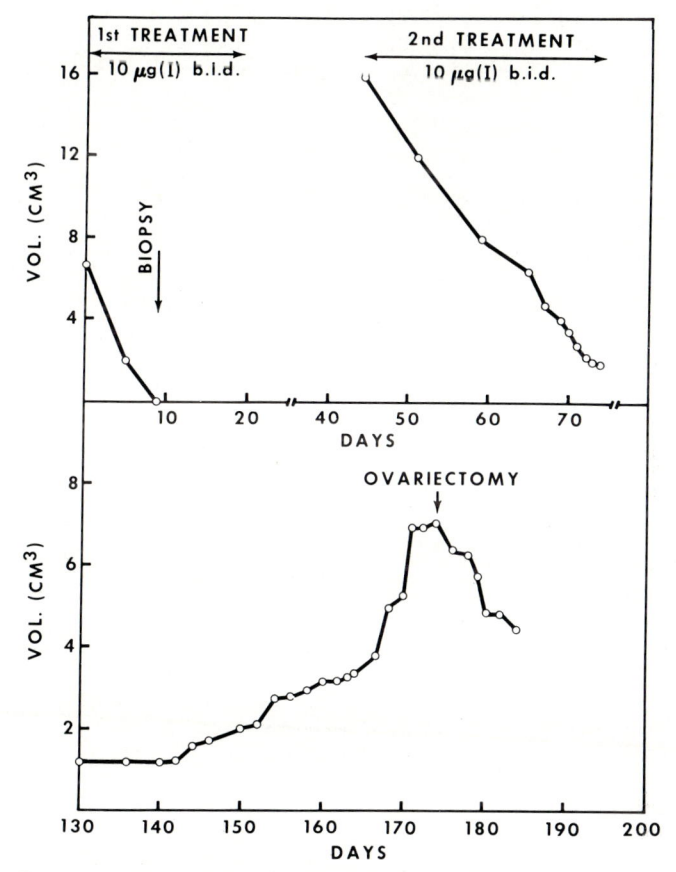

FIG. B. Tumor volumes as measured during two treatment periods with [D-Leu⁶, des-Gly-NH$_2$¹⁰, Pro-ethylamide⁹]-GnRH (upper panel) and before and after ovariectomy (lower panel).

The answer to the question of a direct pituitary effect on the secretion of ACTH will best be approached by an *in vitro* method.

G. A. Hedge: Did you mention TSH?

R. Guillemin: No, we have no data on TSH.

E. B. Thompson: In considering the modifications of somatostatin that led to varying effects on insulin and glucagon, it is your interpretation that this suggests the existence of physically different receptors in each cell type with specific response to somatostatin? Or do you have some other interpretation for those results?

R. Guillemin: I have no good explanation to offer at the moment. All of this has been done in our laboratories over the past few months; we are still developing a series of these analogs, and we are just starting to look at mechanisms of action. It would be somewhat surprising if the receptors were very different in the several target tissues in view of the fact that the original molecule was active on all of them. What is not impossible is that the receptor may have subtle differences revealed only with the structurally modified analogs.

Furthermore, we should not forget that in any of the *in vivo* systems, we are dealing with multicomponent systems, one influencing the other. Thus, the net effects on glucagon secretion, for instance, will reflect effects of the analog not only on the α-cells directly, but also on the β-cells.

M. Chrétien: Although you mentioned that α- and γ-endorphins were found in hypothalamus–neurohypophysis fragments, others have found β-endorphin in the pituitary gland. I would like to report that we have isolated and fully characterized β-endorphin in sheep and in human pituitaries [*Biochem. Biophys. Res. Commun.* 72, 472–478 (1976)].

This last finding is extremely interesting since it is the first report of a morphinomimetic substance found in a human tissue, thus adding much importance to this new system of β-LPH-endorphins.

In your presentation, you mentioned that the hypothalamic fragments are enriched by the neurohypophysis. I wonder whether they are also enriched by the intermediate lobe, which is known to contain almost exclusively β-LPH-containing cells and is closely attached to the posterior lobe. My question is: Is it possible that your peptides come from the intermediate lobe?

R. Guillemin: I call these fragments hypothalamus–neurohypophysis. They are essentially composed of the neurohypophysis, the pituitary stalk attached to it, and the median eminence. Since this is material of porcine origin, most of the intermediate lobe would be sticking with the neural lobe. It is entirely possible that all the endorphin activity is genuinely and exclusively in the intermediate lobe. We have early evidence with immunofluorescence techniques on rat pituitaries that would support such a proposal. On the basis of radioimmunoassays that we are just starting, we find enormous quantities of these opiatelike peptides and, particularly in this case, α-endorphin in the posterior lobe of the rat pituitary, and much less in the anterior lobe. Again the neurohypophysis of the rat would include pars intermedia. I think that the question will be best handled by immunohistofluorescence techniques. Along those lines of thought, I would like to recall the earlier observations of Jacob Furth, and also of Dubois in France as well as the more recent biochemical conclusions of Scott and Lowry in England. They all concluded that most likely the same cells, either in the intermediate lobe and/or in the anterior lobe, that appear to contain β-lipotropin appear also to contain ACTH, α-MSH, and β-MSH. Moreover, Lowry at the Endocrinology Congress in Hamburg reported evidence for the existence of a large molecule that would be a precursor to both lipotropin and ACTH. This would be very much in keeping with the observations based on histology.

K. Savard: You did not have time to elaborate on a subject that I think warrants a little more attention–the subject of the multiplicity of the sources of the somatostatin. I find it fascinating that, at least by the immunofluorescence technique, sources for these are so disseminated throughout the body–the central nervous system, the stomach, the pancreas in particular. If this is a conventional hormone, why is it that it need be elaborated in so many remote tissues? Or, at least the other way around, why is it formed so proximally to the cells which it apparently must affect? Would you care to philosophize a little about this type of new concept in endocrinology?

R. Guillemin: There are two parts to your question. First: there is no definitive evidence at the moment that what I will call peripheral somatostatin is structurally identical to hypothalamic somatostatin as we isolated it from the hypothalamus and as we characterized it. Nobody has isolated as yet enough pancreatic or gut somatostatin to allow its molecular structure to be determined. On the other hand, particularly with all the evidence based on immunochemistry with well characterized antibodies, all the available evidence is in favor of the peripheral somatostatin being identical with hypothalamic somatostatin. Chemical characterization remains to be done and should be done eventually. What is also possible is that, both in the hypothalamus and in the peripheral sources, somatostatin and a prosomatostatin exist. We, and probably another group also, have preliminary evidence for two sizes of somatostatin, both in hypothalamic extract and in pancreatic extracts. It is certainly of interest to characterize prosomatostatin to see how it would relate to other prohormones now well characterized, such as proinsulin or pro-PTH.

For the second part of your question: as many of us in this audience well know, I have never been happy to call these peptides "hormones"; in fact, I have carefully avoided calling these peptides "hormones." I still refer to them as hypophysiotropic factors. I think it is easier and allows for a somewhat cleaner nomenclature. More than that, I think that it was a misleading and superficial proposal to call these substances "hormones." If someone disagrees with this statement I would hope that they come up in this discussion. But to go on: I am not sure that anybody has ever measured either TRF or LRF or somatostatin in peripheral blood in physiological concentrations and in an incontrovertible manner. That statement is made, notwithstanding the several papers in the literature claiming measurements of plasma TRF, LRF, and also of somatostatin. I think that this is a matter still open for further and more careful investigations. There was an interesting workshop at the Endocrinology Congress in Hamburg on that very subject. The consensus, including the opinion of Jeffcoate, was that no incontrovertible evidence exists for claiming measurement of TRF or LRF in peripheral blood.

On what basis do we call these substances "hormones"? Hormones, by the original definition, are substances which, besides having been dear to us for many years, must meet several criteria: they must circulate in the blood in a reasonable concentration, have definite and well recognized targets, be endowed with a biological half-life that is reasonable and compatible with the distance they have to travel. This is not altogether the case for these peptides. It is true that the hypothalamic peptides must travel within the hypothalamohypophysial portal system to reach the adenohypophysis. In fact, I find this requirement difficult to reconcile with knowledge that portal blood inactivates TRF just as rapidly as peripheral blood, as we showed some years ago with John Porter. But, this particular case aside, what do we have? More and more evidence that, in the central nervous system, in the pancreas, in the gastrointestinal tract, the peptides seem to act on targets immediately adjacent to their source of production. This fits remarkably well the concept of *paracrine* secretions (rather than *endocrine* secretions) as described some years ago by the German pathologist Feyrter. Hoping for a way out of this dilemma, I had proposed [see R. Guillemin and J. E. Gerich, *Annu. Rev. Med.* **27**, 379–224 (1976)] to name these peptides *cybernins*. The derivation of the word is as in *cybernetics*: kubernetes means rudder or pilot, i.e., a local or localized means of information or action. That is what governs *a* boat, *a* ship, rather than the fleet. Governing the fleet requires a different type of command. Cybernins are thus the local information, while the hormones are the generalized or farther-reaching information. Cybernin, implying a local information or command is exactly what we are talking about regarding these peptides. We are talking of these peptides as being released and communicating within angstroms either within synaptic spaces or possibly within dendritic spaces; also between epithelial cells either within gap junctions or at any rate within very close extracellular spaces. Unless we redefine the word hormone, I think that we should make the effort of designing a nomenclature for the class of substances we are now discussing. We should have done so several years ago.

P. Munson: What information do you have about development of tolerance to, or of dependence on, compounds of the endorphin-enkephalin series?

R. Guillemin: I have very few data, and some are difficult to interpret. With Floyd Bloom and David Segal, we administered endorphins in the lateral ventricle of the brain in rats, chronically, every day for 7 days. We have no evidence that the catatonic effects of β-endorphin would have been decreasing in intensity. This is, however, somewhat difficult to quantify. Also, the duration of the "chronic" treatment is rather limited. What we have observed, which is puzzling, is the following: when we administer α-endorphin or β-endorphin intracisternally or in the cerebral ventricle to naive animals, i.e., normal animals (rats) which have never been exposed to opiates or opiatelike peptides, within a few minutes after injection, the rats regularly show signs such as wet-dog shakes, teeth grinding, activated grooming, all signs so far always associated with the syndrome of opiate withdrawal. This is most evident with α-endorphin; in the case of β-endorphin the signs are of short duration as the animals enter rapidly the phase of total immobility and catatonia. This observation is difficult to interpret. We never see such signs when we give, in the same conditions, normorphine. There is one report in the literature that administration of β-endorphin to rats inhibits the development of the classical withdrawal syndrome that follows removal of a pellet of morphine. I think that this important question is not properly resolved at the moment. It certainly should be in the near future, with availability of larger amounts of synthetic β-endorphin of proved purity.

May I take this opportunity to say a few words about the problem of purity of the synthetic peptides. Several academic groups and also several commercial laboratories have been engaged for some time in synthesizing peptides like LRF, somatostatin, and now the endorphins and enkephalins, as well as a series of analogs of the native structures. I would like to emphasize the extreme importance of ascertaining the high purity of the peptides so obtained. The peptides are usually prepared by solid-phase technology, a method of great efficiency and remarkable versatility. The end product of the synthesis process must undergo several steps of purification in order to yield a product of high purity. Commercial manufacturers as well as colleagues in the academic world should be ready to demonstrate the high purity of the materials sold or made available for chemical or biological studies. The criteria should be extremely demanding. Probably the best method at the moment [see the report by R. Burgus and J. Rivier, *Proc. Eur. Symp. Peptides, 14th, 1976* pp. 85–94 (1976)] would be to require evidence of homogeneity by high-pressure liquid chromatography (HPLC).

Perhaps some reference standards preparations should be located at some agency of the NIH or the Bureau of Standards and be made available for ultimate comparison. Why do I say that here? We have been faced with reports of toxicity of somatostatin, for instance. In all those cases where we were able to obtain samples of the incriminated material, we could clearly demonstrate (by HPLC) that the materials in question were not pure, but regularly contained contaminants along with the principal product. In the extensive clinical investigations of Gerich using exclusively the preparation of somatostatin made in our laboratory, no evidence of toxicity was ever observed [see also R. Guillemin and J. E. Gerich, *Annu. Rev. Med.* **27**, 379 (1976)]. The same criteria must be employed to ascertain the absence of L-form contaminants when reporting on some analog in which a D-form amino acid has been introduced. I have already heard comments by colleagues who have found it difficult to duplicate one result or another using β-endorphin of commercial origin. While their own technique might be at fault, I think that it is of paramount importance to ascertain that the peptide involved is of the highest purity. Finally there is a modicum of precaution that must be taken in handling and storing peptides that should also be respected by anybody dealing with them.

E. M. Bogdanove: Can you clarify a position that you seemed to take in the discussion a few minutes ago? You pointed out quite correctly that there is, so far, no valid evidence that TRF and LRF and somatostatin are actually present in the peripheral, or even the portal, circulation. Did you go as far as to say, because of this, that these hypothalamic factors may not actually be physiological releasing factors for the anterior lobe, even though they certainly have pharmacologically demonstrable actions on the anterior pituitary? In stressing the impressive interneuronal actions of these compounds, were you also saying that these materials do not actually get to act on the cells of the anterior lobe under physiological conditions?

R. Guillemin: No. We have every reason to believe that the hypothalamic peptides act at the level of the pituitary cells. On the other hand, and as implied in your question, we also have reasons to believe that the same peptides are active also in other systems in the organism. The unified concept that this implies, and that I tried to present in the lecture, is probably one of the most remarkable contributions of those of us working in neuroendocrinology.

K. Sterling: I have one question. As you may recall, I agree that the term TRF may well be preferable to the prevalent TRH. Now the query I want to put to you is whether an animal or human subject addicted or "hooked" on heroin or morphine could substitute one of the endorphines to replace the usual "fix."

R. Guillemin: The substitution of the peptides for morphine has not been done in humans, to my knowledge. In animals I have mentioned before that there are studies going on on that very subject. One report says that β-endorphin will stop withdrawal symptoms in rats upon acute removal of a pellet of morphine.

S. Leeman: I would like to comment on the discussion relating to the presence of hypothalamic peptides in different tissues, particularly gastrointestinal tissue, and their presence in plasma. The two peptides that have been isolated from hypothalamic extracts in may laboratory, substance P and neurotensin, have now both also been isolated, i.e., taken to purity, starting with extracts of small-intestinal tissue. The work on substance P was done earlier by Studer and his colleagues [*Helv. Chim. Acta* **56**, 860–866 (1973)], and the isolation of neurotensin from bovine intestinal tissue has just been accomplished in our laboratory by Kitabgi *et al.* [*J. Biol. Chem.* **251**, 7053–7058 (1976)].

With the help of Dr. Earl Zimmerman we have also demonstrated neurotensin-like immunoreactivity in epithelial cells of the rat small intestine using the immunoperoxidase technique. These cells are morphologically similar to the endocrine cells of the gut. Substance P has been localized in endocrine cells in the intestinal mucosa by A. G. E. Pearse and J. M. Polak [*Histochemistry* **41**, 373 (1973)]

In addition, we have been trying to characterize chemically the immunoreactive substance P and neurotensin that occur in plasma. Although we have not isolated either of these peptides from plasma extracts, we can report that thus far, on Sephadex G-25 and sulfoethyl Sephadex, the immunoassayable material cochromatographs with the synthetic peptides. Thus it may turn out that substance P and neurotensin are hormones in the classical sense, with intestinal tissue as their source, and that they also function within the nervous system, in a different role, participating in neurotransmission. This all agrees very well with what Dr. Guillemin has so excellently described about the ever-broadening field of neuroendocrinology.

R. Guillemin: I am delighted to hear what you said. I was at the meeting in Hamburg, but I did not know that you had isolated neurotensin from the gut. If so, I think it fits perfectly well with what I have been trying to propose here, especially if you have sequenced the gut peptide.

S. Leeman: Dr. Carraway has not yet finished the sequence work on the neurotensin isolated from intestinal tissue. However, since this tridecapeptide has the same amino acid composition and yields the same fragments, of identical composition to those of neurotensin, when submitted to digestion with carboxypeptidase A or papain, it seems highly likely that the sequence of material isolated from gut will be identical to that of material isolated from the hypothalamus.

R. Guillemin: That answers further Dr. Savard's original question. It may thus very well be that the molecules at the periphery are the same molecules that we characterized in the brain. With all the observations showing so much in common between the gut and the brain, I have been more and more convinced that there was some deep sense in that all-American expression "to have guts."

S. Leeman: I remember years ago, when I was a graduate student in Dr. Munson's laboratory, Dr. Gaddum came to visit. At that time I had no idea that I would end up some day working on substance P, and it struck me then as so remarkable that there were peptides from the gut that were the same as in brain. Well, where are your feelings?

R. Guillemin: I think that it is a wonderful outcome. Professor Ulf von Euler visited our laboratories not too long ago. He and Gaddum had shown the existence of substance P more than 30 years ago. We presented him with about 100 mg of pure substance P prepared by total synthesis in our laboratories. The methodology used to synthesize LRF, somatostatin, or their analogs, as in your case, has allowed us to study also those other peptides not originally considered to belong to neuroendocrinology. There is increasing evidence that they do.

R. O. Greep: Owing to my interest in the control of human fertility, I focused on one of your statements reporting an agonist with 140 times the activity of LRF that interrupted pregnancy in its early stages. What is the mechanism?

R. Guillemin: I had hoped that Dr. White would have some proposal for us, for a mechanism of action of the peptide in that series of unexpected observations. At the moment, we do not know how to explain the results.

F. G. Sulman: Working in the laboratory of B. Zondek some 20 years ago, we did exactly what you did with chorionic gonadotropin, and we had immediately the same result. That means we had an abortion in pregnant rats, which started with strong uterine bleeding, then we had the next stage of resorption of fetuses, and when we looked for the mechanism we found a luteolytic effect.

R. Guillemin: That is very interesting.

W. F. White: I think I can give at least a partial answer to the question of the resorption of the fetus as mentioned by Dr. Guillemin. Giving the correct amount of estradiol at that time will totally reverse this effect. I am not so sure concerning the mechanism of all the other aspects of what we are doing. However, that particular action has been completely reversed by giving the proper amount of estradiol.

R. A. Edgren: Steroid effects depend on when the LHRH is administered during pregnancy. In early pregnancy estrogen will reverse the effects of LHRH. During the postnidatory period, progesterone, and only progesterone, is required.

S. Korenman: May I suggest a mechanism by which the actions of morphine, α- and β-endorphins, and competitive inhibitors may be explained. Assume that there are two consequences of receptor occupation, stimulation of neural cell discharge and inhibition of functional receptor due to decreased synthesis, deployment, or structural alteration reducing the number of available binding sites. Assume further that these actions are partially dissociable so that some molecules may be more potent as effectors and some may be more active as receptor inhibitors.

Morphine would stimulate briefly but maintain a low concentration of receptor sites and therefore keep the affected cells in a high threshold condition. By contrast, β-endorphin would be a powerful stimulator producing wet shakes prior to its profound receptor-inhibiting properties. α-Endorphin would be principally a stimulator with weak receptor inhibition properties. Finally, the competitive inhibitors would remove morphine from receptors and permit endogenous activation without inhibiting further receptor binding. In withdrawal, as morphine concentration falls the available receptor sites would increase in number and be bound by native endorphins, causing profound activation until a new equilibrium is reached. Tachyphylaxis to β-adrenergic catecholamines has been related to receptor-site depletion supporting a similar pathogenesis in this case.

LATS in Graves' Disease[1]

J. M. McKenzie[2] and M. Zakarija

*McGill University Clinic, Royal Victoria Hospital,
Montreal, Quebec, Canada*

I. Introduction

The history of the long-acting thyroid stimulator (LATS) dates from 1956, when Adams and Purves described an abnormal response in their bioassay for thyrotropin; serum from a patient with Graves' disease had a greatly prolonged thyroid-stimulating effect when injected into guinea pigs. Shortly thereafter a similar procedure, using mice instead of guinea pigs, was developed for the bioassay of thyrotropin (McKenzie, 1958a), and some Graves' disease sera were found to exert an analogous prolonged effect in that species also (McKenzie, 1958b); Fig. 1 illustrates responses of the mouse thyroid to thyrotropin and to LATS.

Although this fact seems now to be accepted without comment, initially there was reluctance to accept that an IgG could act as a metabolic stimulant (McKenzie, 1967). To document this concept further, we raised antisera to nonthyroid tissues and subsequently examined the effects of such sera on other preparations of the homologous tissue. Figures 2 and 3 show data from two such experiments when antisera were raised to leukocytes of guinea pig peritoneal exudate (Fig. 2) and to rat spleen lymphocytes (Fig. 3). In both sets of data the progressive change of potency of the antiserum is related to the number of injections of antigen that were given at monthly intervals; the rabbits were bled 1 week after each injection. As shown, the antisera to guinea pig leukocytes raised in rabbits Cicero and Socrates and to lymphocytes in rabbit Che progressively stimulated the incorporation of [^{3}H]uridine into the respective cells; the antisera raised in rabbit François (Fig. 3) progressively suppressed the [^{3}H]uridine incorporation into lymphocytes. Thus, metabolism-stimulating antibodies (accepting [^{3}H]uridine incorporation as a metabolic event) can be prepared experimentally against not only the thyroid (see below) but also nonthyroid tissue. The implications of the inhibition of incorporation observed in François (Fig. 3), in the absence of complement, have not been further explored.

To return to the main theme, in the 1960s positive assays were reported in a

---

[1] Financially supported by grants from the Medical Research Council of Canada (MT884 and MA 5190) and the U.S. Public Health Service (AM 04121).
[2] Research Associate, Medical Research Council of Canada.

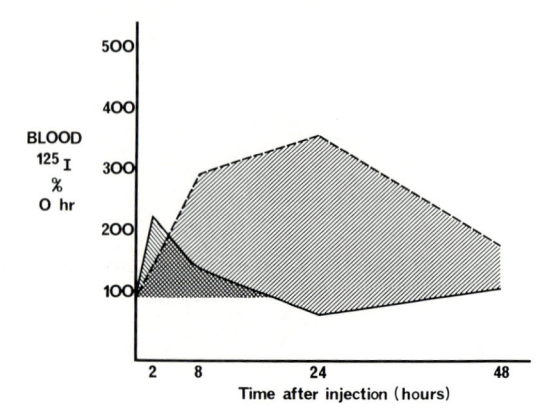

FIG. 1. Time course of increase in blood radioactivity in assay of thyrotropin and long-acting thyroid stimulator (LATS). Preparation of LATS or 0.2 mU of thyrotropin was injected into two groups of six mice at 0 hour. Animals were bled immediately before and 2, 8, 24, and 48 hours thereafter. The mean of each set of observations was plotted, and the area under the curve shaded. Small area at left is TSH; large area at right is LATS. Response at each time (n) was calculated as (blood ^{125}I, n hour)/(blood ^{125}I, 0 hour) × 100%, so that 100% indicates no change. For more details of the assay procedure see McKenzie and Zaharija (1972).

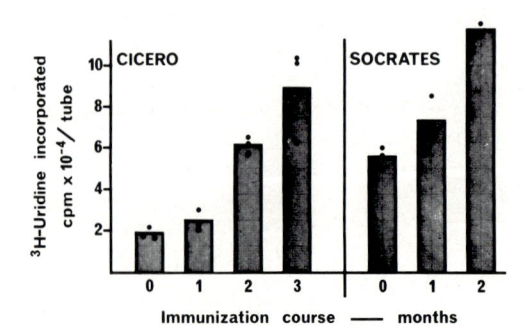

FIG. 2. Influence of antisera to guinea pig leukocytes on their incorporation of [^{3}H]uridine. Cells were obtained as peritoneal exudate after injection of casein and were incubated for 24 hours in Eagle's minimal essential medium that had decomplemented serum included to a concentration of 2%; [^{3}H]uridine was added for the last 60 minutes, and incorporation was measured in material insoluble in trichloroacetic acid–acetone–ether. Individual values are shown as dots, and the bars indicate the means. Immunization was by injection of 10^9 leukocytes, initially subcutaneously with complete Freund's adjuvant and subsequent intraperitoneally without adjuvant; blood was obtained before the immunization course and 1 week after each booster. Cicero and Socrates were the two rabbits out of six to produce stimulatory antisera; the remaining 4 rabbits' antisera had no effect. Data from McKenzie (1968a).

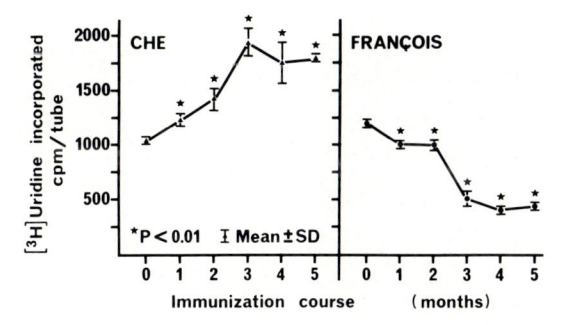

FIG. 3. Influence of antisera to rat spleen lymphocytes on their incorporation of [^{3}H]uridine. The experimental design was as in Fig. 2 except that rat spleen lymphocytes were used for immunization and subsequent incubations. Four additional rabbits, similarly immunized, produced antisera without effect on the incorporation of [^{3}H]uridine. Data from McKenzie (1968a).

variable percentage of Graves' disease patients (McKenzie, 1968b). The higher rates of positive assays were found with up to 10-fold concentration of serum IgG (Carneiro et al., 1966), and LATS was considered by some to be the cause of the hyperthyroidism of Graves' disease; insensitivity of the bioassay was offered as explanation for the failure to obtain 100% positive results in the syndrome. However several arguments were developed against this thesis. A few patients were found to have LATS in the blood but had unequivocally normal thyroid function, including suppression of thyroid uptake of radioiodine on administration of thyroid hormone (Chopra and Solomon, 1970; Wong and Doe, 1972). Suppressible thyroid function despite the presence of LATS in the blood was also described in several patients being treated with antithyroid drugs (Chopra et al., 1970; Silverstein and Burke, 1970), but the definition and interpretation of "suppression" in these circumstances is too debatable for such observations to be particularly telling. In addition, however, some members of families of patients with active Graves' disease were shown to have LATS circulating (Bonnyns et al., 1973; Wall et al., 1969); unfortunately, there was no rigorous documentation of entirely normal thyroid function in these subjects. Nonetheless, these various reports led to LATS being considered to be perhaps of little importance in Graves' disease (Chopra et al., 1970; Hollingsworth and Mabry, 1976).

In more recent years various new procedures have been described for the measurement of thyroid-stimulating IgG, and these have in common either binding to, or stimulation of, preparations of human thyroid gland; Table I lists and briefly describes these assays. Experience with these techniques has led to suggestions that there may be in Graves' disease thyroid-stimulating IgG molecules that are specific for the human thyroid, whereas LATS acts only on the mouse gland (Adams, 1975).

TABLE I
Newer Methods for Assay of Thyroid-Stimulating IgG of Graves' Disease

Reference	Acronym	Name	Method
Adams and Kennedy (1967)	LATS-P	Long-acting thyroid stimulator-protector	Binding to human thyroid, preventing ("protecting") subsequent binding of LATS
Onaya et al. (1973)	HTS	Human thyroid stimulator	Colloid droplet formation or increase in cyclic AMP concentration in human thyroid slices
Smith and Hall (1974)	TSI	Thyroid-stimulating immunoglobulins	Competition with TSH in TSH receptor assay using human thyroid membranes.
Orgiazzi et al. (1976)	H-TACS	Human thyroid adenyl cyclase stimulator	Stimulation of adenyl cyclase in human thyroid membranes

We developed a method for assay of thyroid-stimulating IgG, using an increase in the concentration of cyclic AMP (cAMP) in human thyroid slices as the end point (McKenzie and Zakarija, 1976), and our experience with this procedure has led to the following thesis. In Graves' disease there is a circulating polyclonal IgG that stimulates the human thyroid by interaction with a homologous antigen; variably the IgG may cross-react with a similar molecule in the thyroid of a distant species, e.g., the mouse (in which case a positive LATS assay response may be obtained); hyperthyroidism in Graves' disease is due to the presence of the thyroid-stimulating IgG in blood, and its persistence during antithyroid drug therapy presages relapse on cessation of the medication. This report offers the evidence to support this thesis and touches on a topic that evolves from it: If an IgG causes hyperthyroidism, what controls the production of that IgG?

II. Human Thyroid Slice cAMP Assay of Thyroid-Stimulating IgG

This procedure was recently described in detail (McKenzie and Zakarija, 1976) and has been slightly modified as follows: Slices of normal human thyroid, obtained at operation for benign nontoxic thyroid disease or at parathyroidectomy, were incubated at 37°C in 0.3 ml of Krebs–Ringer bicarbonate buffer (containing one-third the original concentration of Ca^{2+}), pH 7.4, containing 20 mM theophylline, 0.1% glucose, 0.2% human serum albumin, and 0.3 ml solution of test material, with 95% oxygen, 5% CO_2 as the gas phase. After 2 hours the slices were homogenized in 6% trichloracetic acid and cAMP was extracted for subsequent radioimmunoassay using Schwartz-Mann kits. The test material was usually a crude 3-fold concentrate of serum IgG obtained by precipitation with 1.64 M $(NH_4)_2SO_4$; all experiments included the concurrent testing of IgG from normal serum and thyrotropin, 50 mU/ml. Examples of results, indicating the reason for our choosing an incubation period of 2 hours, are shown in Fig. 4. Thyrotropin reached approximately maximum effect within 7 minutes, a potent (LATS-positive) IgG preparation by 30 minutes and a less potent (LATS-negative) IgG showed a statistically significant effect only after 60 minutes, and at 2 hours the concentration of cAMP had not yet "plateaued."

The delay in the effect of thyroid-stimulating IgG reaching a maximum (compared with the time required for thyrotropin to have a maximum effect) might be due to a slow rate of binding of the IgG to a receptor or antigen (assuming this to be an essential step in its action) or to a slow and continued activation of adenyl cyclase despite rapid binding. Therefore we studied, with both thyrotropin and thyroid-stimulating IgG, over 2 hours of incubation, the effect of changing the medium to one that contained no stimulator. The results, depicted in Fig. 5, show that, for thyrotropin, 10-minute exposure of slices to the hormone is sufficient for a maximum effect to be exerted and maintained over the succeeding 110 minutes; on the other hand, 10 minutes is not sufficient exposure of the slices to IgG, although an effect exerted in 60 minutes is maintained for the additional 60 minutes. The data are compatible with thyrotropin's action involving rapid and complete occupancy of available receptor sites, and the action of thyroid-stimulating IgG reflecting a slower rate of binding to the homologous receptor or antigen.

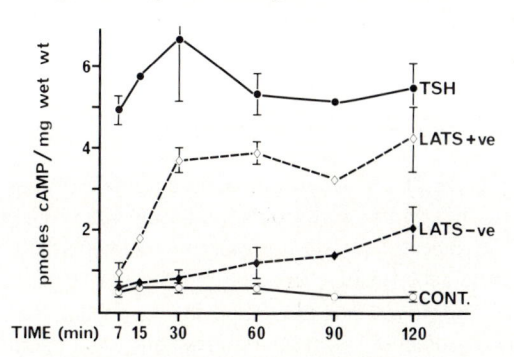

FIG. 4. Cyclic AMP (cAMP) in human thyroid slices: effects of thyroid-stimulating IgG or thyrotropin. Slices were incubated *in vitro* (McKenzie and Zakarija, 1976) for the time periods and with materials shown. Thyrotropin (TSH) was 50 mU per milliliter of medium; thyroid-stimulating IgG was a crude preparation, $(NH_4)_2SO_4$ precipitate, from sera known to be LATS-positive (+ve) or LATS-negative (−ve) and the control (cont) was a similar IgG preparation from normal human serum. Mean and SD of 4 observations per point are shown; the response to LATS−ve IgG became statistically significant ($P < 0.02$) only by 60 minutes of incubation.

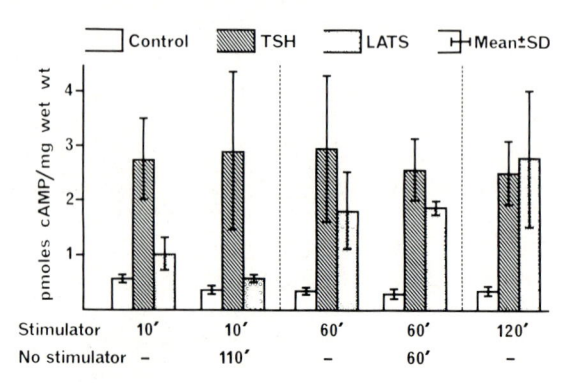

FIG. 5. Cyclic AMP (cAMP) in human thyroid slices: effect of limited exposure to thyroid stimulators. For details of the incubation system, see text. On the abscissa the figures indicate that the stimulator [thyrotropin (TSH) or LATS] or control solution was present for 10, 60, or 120 minutes (') and then half the slices in the case of the first two time periods were transferred to stimulator-free vials for the remainder of the 120-minute period, i.e., 110 minutes and 60 minutes. The group incubated for 10 minutes with LATS and then 110 minutes without LATS was significantly greater than control ($P < 0.05$) but also significantly less than the group incubated with LATS for only 10 minutes. $n=4$.

In Fig. 6 results of assay of IgG from 11 Graves' disease subjects are depicted, together with data for the bioassay of LATS. All IgG preparations increased the thyroid concentrations of cAMP, and three of them also gave positive LATS responses. Our total experience with the use of this assay for thyroid-stimulating IgG is described below (see Table VII).

III. Zoological Specificity of Human Thyroid-Stimulating IgG

With techniques similar to that which uses human thyroid slices, described above, we tested the ability of thyroid-stimulating IgG of Graves' disease sera to increase cAMP in slices of canine thyroid and fragments of guinea pig thyroid. The time–response relationship, using three different preparations of IgG, for these two species of thyroid is illustrated in Fig. 7; as may be seen, there was with both species a delay of 60–120 minutes before the maximum (for the time period studied) effect was realized. Testing of IgG from 22 Graves' disease sera are summarized in Table II, which includes also data for responses of human thyroid slices and LATS bioassay of the same material. Table III further analyzes these results by grouping the responses in the three nonhuman species according to an approximation (cAMP increase due to the thyroid-stimulating IgG expressed as a percentage increase over the concentration of cAMP in concomi-

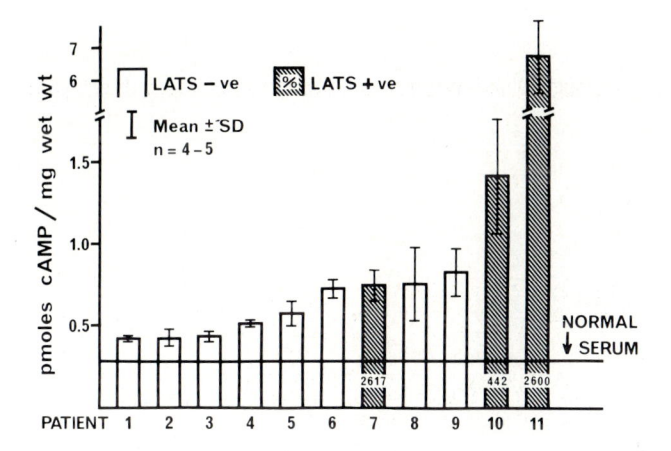

FIG. 6. Cyclic AMP (cAMP) in human thyroid slices: effect of serum IgG from 11 patients with Graves' disease. For details of the techniques of incubation and of preparing the IgG, see text. All values of cAMP are significantly ($P < 0.01$) greater than that of control slices incubated with IgG of normal human serum. The figures within the columns for sera from patients 7, 10, and 11 indicate the long-acting thyroid stimulator (LATS) bioassay response as a percentage at 24 hours after injection.

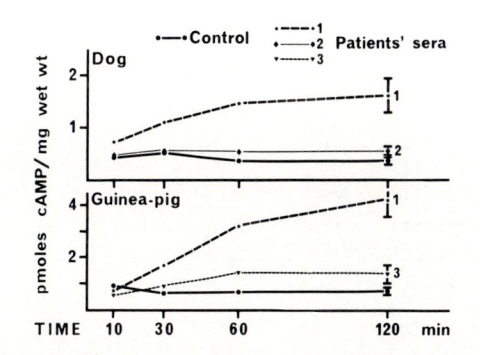

FIG. 7. Cyclic AMP (cAMP) in thyroid slices (dog) or fragments (guinea pig). Incubation conditions were as described in the text for human thyroid slices. Response to serum from patient 1 was statistically significant ($P < 0.01$) at 10 minutes (dog only) and at 30, 60, and 120 minutes for both species; for serum 2 and dog slices, only responses at 60 and 120 minutes were significant ($P < 0.02$), and for serum 3 (guinea pig segments) 30-, 60-, and 120-minute responses were significant ($P < 0.01$).

TABLE II
Summary of Cross-reactions[a]

No. of sera	Man	Mouse	Guinea pig	Dog
2	+	+	+	+
2	+	+	−	+
2	+	−	+	+
3	+	+	−	−
1	+	−	+	−
3	+	−	−	+
9	+	−	−	−
22	22	7	5	9

[a] A plus sign (+) indicates that the IgG of the number of sera listed increased *in vitro* the concentration of cAMP in the thyroid (slice or fragment, see text) of man, guinea pig, or dog, or gave a positive response in the long-acting thyroid stimulator assay (mouse).

TABLE III
Correlation of Human Thyroid-Stimulating IgG with Cross-Reactions in Three Species[a]

Patients' sera	Human thyroid cAMP increase (%)	Responsive thyroid Mouse	Dog	Guinea pig
1, 2	2389, 2882	++++	+++	++
3	1230	+++	++	−
4, 5, 6	1132, 1004, 610	−	+	−
7	749	++	+	−
8, 9	739, 503	−	+	+
10, 11	504, 438	+	−	−
12	268	−	−	+
13	264	+++	−	−
14–22	150–273	−	−	−

[a] The cAMP increase in the human thyroid slice assay is expressed as a percentage of the value obtained with slices, in the same assay run, incubated with normal serum IgG. The responses in the other species are given in a semiquantitative fashion (− to +++) reflecting the fact that maximum increases in these species varied markedly when expressed similarly as a percentage over control.

tantly processed control slices) of the potency of the effect exerted on human slices; although the relationship is not absolute, it appears that the more potent the IgG effect on human thyroid, the more likely that preparation is to react with nonhuman thyroids. A more-detailed quantitative presentation of cross-reaction data is given in Fig. 8; in this experiment, two sera were tested in four species of thyroid (human, guinea pig, canine, and bovine) as well as for LATS in the mouse; it is apparent that cross-reaction with the thyroid of one nonhuman species does not signify that there will inevitably be an effect on another.

IV. Isoelectric Focusing of Thyroid-Stimulating IgG of Graves' Disease Sera

The application of the technique of isoelectric focusing to these studies was carried out as summarized in Table IV; an example of the elution data from a typical experiment is shown in Fig. 9 and, in Fig. 10, the immunoelectrophoresis patterns of the proteins in the pools of eluate. In Figs. 11–13 are depicted the distribution of human thyroid-stimulating activity according to pI for three sera, two LATS-negative and the other LATS-positive; the distribution of LATS, antithyroglobulin, antibody to thyroid microsomal antigen, and antibody to gastric pareital cells, when these immunoglobulins were present in the sera, is also shown. Figure 14 summarizes our experience with the technique of isoelectric focusing and shows that while an effect on the human thyroid slice may occur over a pI range of 4–10, the maximum potency, as illustrated more

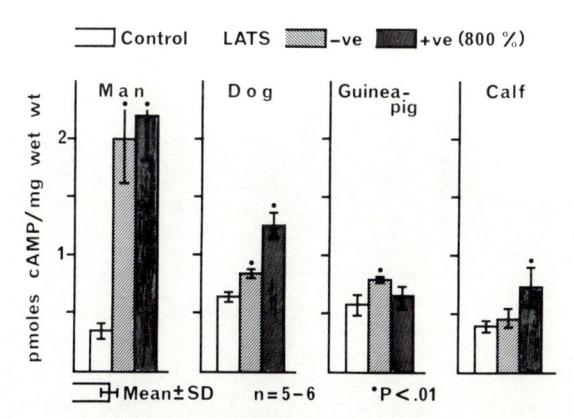

FIG. 8. Assay of human thyroid-stimulating IgG in five species. Long-acting thyroid stimulator (LATS) was by the mouse bioassay (McKenzie and Zakarija, 1976); for man, dog, and calf, an increase in cAMP in thyroid slices was measured; for guinea pig, thyroid segments were used. For details see text.

TABLE IV
Isoelectric Focusing (I.F.) of Long-Acting Thyroid Stimulator

1. 40% saturation $(NH_4)SO_4$ precipitate of serum
2. I.F. pH 3–10 or 6–9.5
3. Fractions = 1 or 0.5 pH unit
4. Testing
   a. Mouse bioassay
   b. Human thyroid slices
   c. Other antibody systems
   d. Immunoelectrophoresis
   e. Polyacrylamide gel electrophoresis

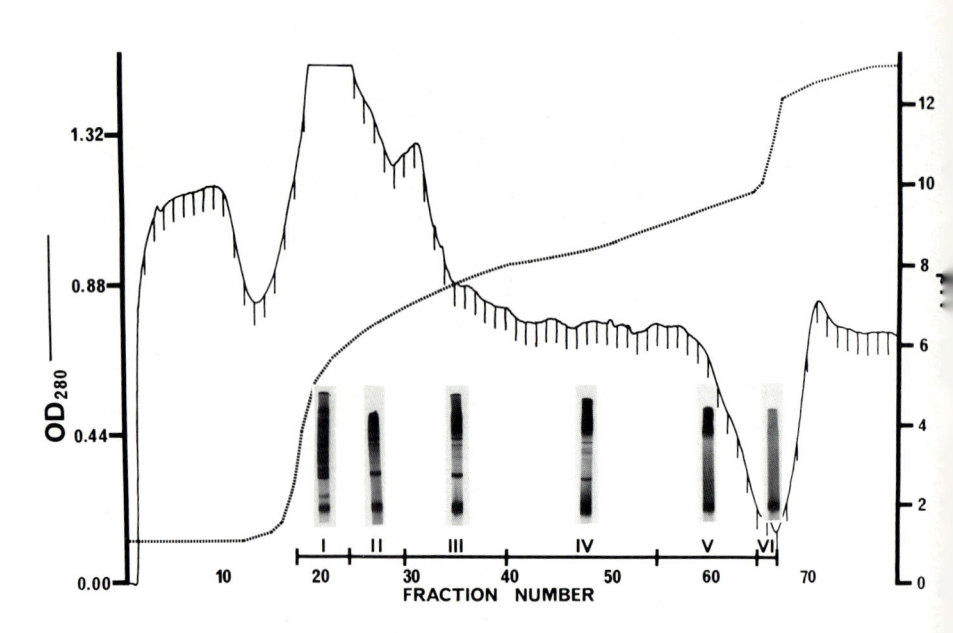

FIG. 9. Elution pattern of isoelectric focusing of $(NH_4)_2SO_4$ precipitate of serum proteins. Focusing was carried out as described in Table IV over the range of pH 3–10. Fractions were pooled as identified by Roman numerals according to pH, i.e., pH 4–6 (I), –7 (II), –8 (III), –9 (IV), –10 (V), and –11.2 (VI). Polyacrylamide gel electrophoresis patterns are shown for each fraction; the dark band at the bottom of each gel is the tracking dye.

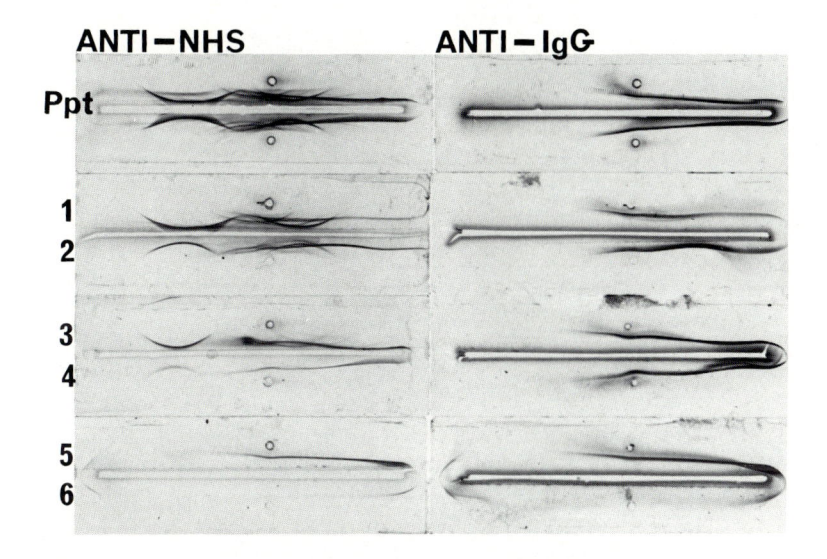

FIG. 10. Immunoelectrophoresis patterns of fractions of serum protein obtained by isoelectric focusing. The fractions were obtained as described in Table IV and Fig. 9. Immunoelectrophoresis was in agar with the anode to the left of the figure; antisera were to normal human serum (anti-NHS) and to normal human IgG (anti-IgG). Ppt = $(NH_4)_2SO_4$ precipitate of proteins loaded on the column; Arabic figures on left indicate fractions of eluate shown in Roman numerals in Fig. 9.

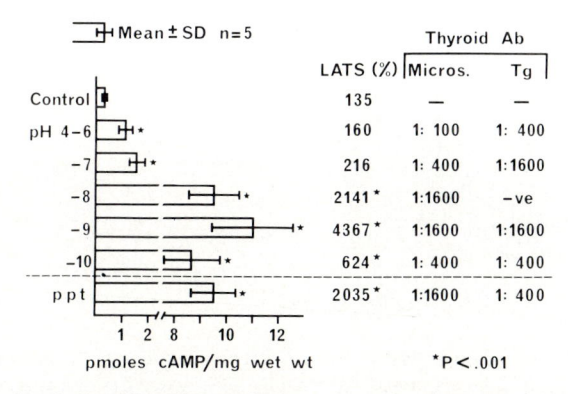

	LATS (%)	Micros.	Tg
Control	135	—	—
pH 4–6	160	1: 100	1: 400
–7	216	1: 400	1:1600
–8	2141*	1:1600	–ve
–9	4367*	1:1600	1:1600
–10	624*	1: 400	1: 400
ppt	2035*	1:1600	1: 400

pmoles cAMP/mg wet wt   *P < .001

FIG. 11. Isoelectric focusing of Graves' disease serum IgG: long-acting thyroid stimulator (LATS)-positive. Fractions of eluate are indicated by the pH figures on left of the figure; ppt = $(NH_4)_2SO_4$ precipitate of proteins loaded on the column. The cyclic AMP (cAMP) data on the left show the responses of human thyroid slices to the fractions; on the right, are responses in the LATS bioassay and the titers of antibody (Ab) to thyroid microsomal antigen (micros) and to thyroglobulin (Tg).

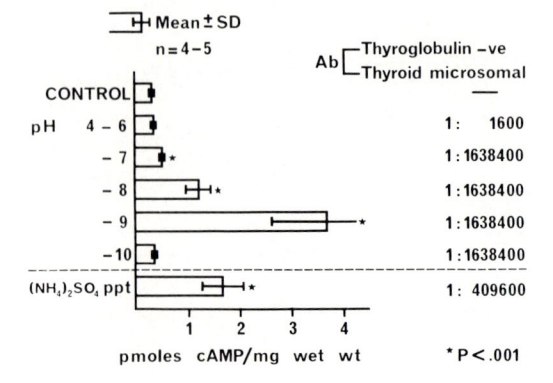

FIG. 12. Isoelectric focusing of Graves' disease serum IgG: long-acting thyroid stimulator-negative. Serum did not have measurable antibody to thyroglobulin, and titers of antibody to thyroid microsomal antigen are shown at the right of the figure. Other details are as in Fig. 11.

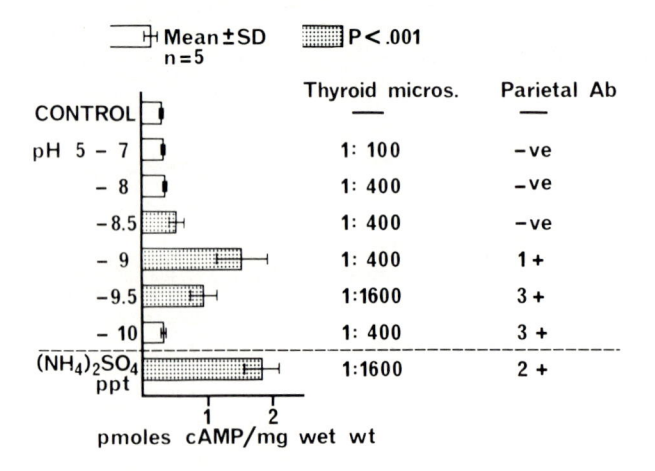

FIG. 13. Isoelectric focusing of serum IgG from a patient with Graves' disease and pernicious anemia. For details of the experimental design, see Fig. 11. Gastric parietal cell antibodies (parietal Ab) were measured for us by Dr. C. K. Osterland using a cytofluorescent technique.

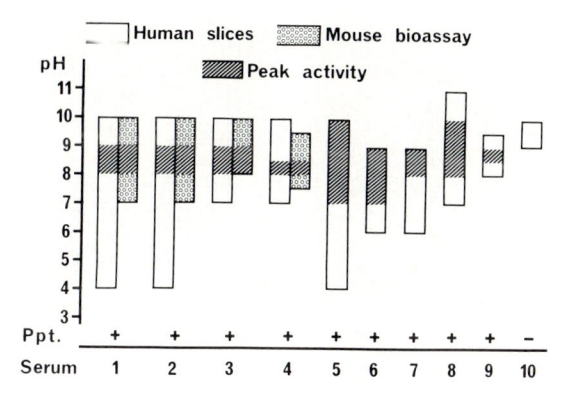

FIG. 14. Summary of data from isoelectric focusing of serum IgG from patients with Graves' disease. "Human slices" indicates the fractions increasing cyclic AMP (c-AMP) in human thyroid slices, and "mouse bioassay" those that are long-acting thyroid stimulator (LATS)-positive. The heavily hatched areas show peak activity in either assay; in several, as in Fig. 11, responses to various fractions were not significantly different. With serum 10 the precipitate (ppt) was negative in both the human slice and LATS assays, but one fraction, pH 9–10, significantly ($p < 0.02$) increased the concentration of cAMP in the human thyroid slice.

quantitatively in Figs. 11–13, occurred within a narrow range of pI around pH 9. As shown also with 4 LATS-positive sera, data for which are incorporated in Fig. 14, this activity followed a closely similar pattern of distribution, although within a more-restricted range of pI. On the other hand, the other immunoglobulins analyzed followed no such uniformity of distribution (Figs. 11–13).

One preparation of IgG, known to affect human, mouse, canine, and guinea pig thyroids, was fractionated by isoelectric focusing, and the fractions were tested for their influence on these four species of thyroid; the data, shown graphically in Fig. 15, indicate that the distribution of thyroid-stimulating activity is constant, no matter which species of thyroid is responding.

V. Cytochemical Bioassay of LATS

At last year's Laurentian Hormone Conference, Chayen, Bitensky, and Daly reviewed the techniques of cytochemical bioassay of hormones, and we heard how the procedure had been applied to the bioassay of LATS as an offshoot of the method for assay of thyrotropin. The technique uses fragments of guinea pig thyroid *in vitro*, and the end point of thyroid stimulation is an enzyme reaction (hydrolysis of leucine naphthylamide monitored by scanning microdensitometry) that reflects thyrotropin-induced increased permeability of lysosomes in

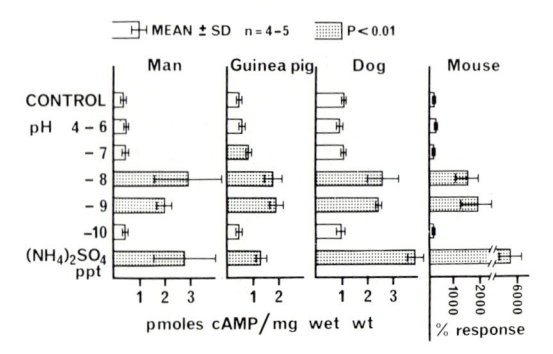

FIG. 15. Isoelectric focusing of IgG from Graves' disease serum: Influence of fractions on cyclic AMP (cAMP) in slices of thyroid of man and dog, in guinea pig thyroid fragments, and on the mouse thyroid (long-acting thyroid stimulator bioassay). For details of the various assays, see text and legend to Fig. 11.

individual follicular cells; for thyrotropin the assay can detect about 10^{-4} μU/ml and for LATS there is similar exquisite sensitivity (Bitensky et al., 1974). To date, there are only two reports describing the application of the cytochemical assay to measurement of the thyroid-stimulating IgG of Graves' disease, and neither refers to study of LATS-negative sera (Bitensky et al., 1974; Petersen et al., 1975). Through collaboration with Bitensky and Chayen, we have made preliminary use of this technique and, in particular, analyzed in detail two sera pertinent to the main thrust of this report. Results of the cytochemical assay of these sera are shown in Fig. 16; the two sera had a difference in potency in this system of at least 10-fold. The point of interest is that the less-potent serum (cytochemical assay) was LATS-positive but was equipotent with the other in the human slice assay (Fig. 8). Furthermore when the two guinea pig thyroid assays were used, similar ratios of potency for the two sera were found; that is, the LATS-negative serum was the more potent in both the cAMP (Fig. 8) and cytochemical assays (Fig. 16). These data, obtained in five different assay systems, are summarized in Table V.

VI. Relationship of Thyroid-Stimulating IgG to Hyperthyroidism

We have not attempted direct analyses of thyroid-stimulating IgG, either measured by the human thyroid-slice technique or by LATS bioassay, to thyroid function, but rather concentrated on the significance of the presence in serum of the activity when hyperthyroidism was not overt. Specifically, for LATS, we are able to relate the occurrence of a positive bioassay, using serum obtained at the end of a course of antithyroid drug therapy, to the incidence of relapse of the

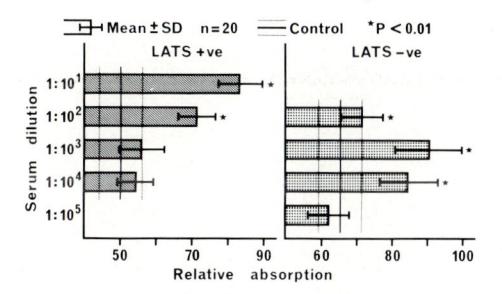

FIG. 16. Cytochemical bioassay of two sera (3-fold IgG concentrate was used) from patients with Graves' disease: The procedure is described in brief in the text and in detail in Bitensky *et al.* (1974). The low response of the LATS-negative (–ve) serum (right) at a dilution of $1:10^2$ is typical of the biphasic response seen with this assay.

hyperthyroidism; these data are summarized in Table VI, showing that although the overall incidence of an LATS-positive assay was only 18.6%, its occurrence at this stage in the course of the syndrome was a clear forecaster of relapse of hyperthyroidism. Our experience with thyroid-stimulating IgG as measured in the human thyroid slice is of necessity different, since we have not had sufficient time to follow an adequate number of patients long enough throughout the course of their disease to make a similar analysis. Rather the occurrence of positive assays in relation to the type of thyroid abnormality existing in 60 patients has been summarized in Table VII. This shows that all sera from freshly diagnosed cases of Graves' hyperthyroidism were positive in the assay, although sera from patients in apparent remission were uniformly negative. The group "euthyroid or hypothyroid 'abnormal'" consists of 5 patients treated for

TABLE V
Multiple Bioassays of Sera from Two Patients with Graves' Disease[a]

	Response	
Assay LATS (mouse)	Serum 1 –ve	Serum 2 +ve
cAMP in human thyroid	++	++
cAMP in guinea pig	+	–
cAMP in dog	+	+
cAMP in calf	–	+
Cytochemical in guinea pig thyroid	++	+

[a] These data summarize results shown in Figs. 8 and 16.

TABLE VI
Long-Acting Thyroid Stimulator (LATS) and Relapse in Medically Treated Graves' Disease[a]

Patients	No.	%	LATS +ve	LATS −ve
In remission	30	35	0	30
Relapsed	56	65	16	40
	86	100	16	70

[a]Incidence of LATS = 18.6%. The patients were treated with antithyroid drugs, usually propylthiouracil, for at least 6 months and were observed for at least 2 years after cessation of therapy; all but one of the relapsed patients were diagnosed within 6 months.

hyperthyroidism by radioiodine or subtotal thyroidectomy but who had persisting, clinically progressive, exophthalmos; 3 other patients presented as "euthyroid exophthalmos" but with abnormal results of thyroid-suppression tests or of analyses with thyrotropin-releasing hormone (McKenzie *et al.*, 1975); the ninth patient came to our attention because she delivered a child with neonatal Graves' disease, having become hypothyroid following treatment with ^{131}I for hyperthyroidism 2 years previously. Of 3 sera from patients with Hashimoto's thyroiditis, 1 was positive; this seems to us to be in accord with the concept that Graves' disease and Hashimoto's disease represent extremes of a single spectrum of pathology within which some subjects might produced the thyroid-stimulating IgG coincident with or subsequent to the development of Hashimoto's thyroiditis—a disease process that might be expected to make the thyroid unresponsive to any form of stimulation (Liddle *et al.*, 1965).

The group of patients listed as "relapsed" or "recurrent" hyperthyroidism all had, with one important exception, serum that gave a positive assay response. Additional experimental data relevant to that exceptional patient are the topic of the next section.

VII. Circulating Immune Complexes in Graves' Disease

Circulating immune complexes (combinations of antigen and antibody) have been measured in various autoimmune diseases (Sobel *et al.*, 1975; Theofilopoulos *et al.*, 1976) including Graves' disease (Calder *et al.*, 1974). For similar studies Cano and co-workers recently reported from this school (1976) a sensitive radiometric assay based on the binding of ^{125}I-labeled Clq (a component of complement) to sheep erythrocytes that had been sensitized with subagglutinating amounts of rabbit antisheep hemolysin and the inhibition of

TABLE VII
Summary of Testing Patients' Sera

Graves' disease	n	Human thyroid slice assay Positive	Negative
Fresh	16	16	0
Relapsed	21	20	1
Eu- or hypothyroid "abnormal"[a]	9	9	0
Euthyroid p̄ propylthiouracil	7	0	7
Euthyroid p̄ ^{131}I or surgery	4	0	4
Hashimoto's	3	1	2
	60	46	14
Normal	5	0	5

[a]The definition of "eu- or hypothyroid 'abnormal' " is given in the text.

this binding by serum containing immune complexes. Eleven of 24 Graves' disease sera were found to contain immune complexes. [The significance of this finding is discussed elsewhere (Cano et al., 1976)]. Most of the 11 sera were also tested in our thyroid slice assay (see above), and all but one enhanced the concentration of cAMP; the exception [the same serum as that listed negative (relapsing disease) in Table VII] contained the greatest concentration of immune complexes in the series studied, and consequently this serum was subjected to further study. Table VIII illustrates that there was no evidence for the presence of an active inhibitor of thyroid-stimulating IgG in this serum, in that its

TABLE VIII
Cyclic AMP (cAMP) in Human Thyroid Slices: Effect of IgG from Patients with Graves' Disease[a]

Test preparation[b]	cAMP (pmoles/mg wet wt)	P
Normal IgG	0.15 ± 0.02 (4)	—
Normal IgG + L.C.IgG	3.30 ± 0.12 (4)	0.001
M.S.IgG + L.C.IgG	3.54 ± 0.63 (5)	0.001
Normal IgG + M.S.IgG	0.15 ± 0.04 (5)	NS[c]

[a]Details of the incubation system are given in the text. The mixtures of normal and patients' IgG were in the ratio of 1:1.
[b]L.C.IgG, standard thyroid-stimulating IgG.
M.S.IgG, IgG from patient under study.
[c]NS, not significant.

46   J. M. McKENZIE AND M. ZAKARIJA

addition to an aliquot of a potent serum failed to diminish the effect of the latter. Therefore in our series of Graves' disease patients there remains one whose serum (IgG) did not stimulate the human thyroid slice *in vitro*; the relationship of this finding, and the overall relevance of immune complexes (which were in high concentration in this particular serum) to the pathogenetic mechanisms in Graves' disease remains to be elucidated.

VIII. The Etiology of Graves' Disease

The first patient described in the English literature having the syndrome we now call Graves' disease was Parry's (1825); Fig. 17 illustrates the events of that case report and accentuates the "trigger" of emotional stress that has been the

FIG. 17. Illustration of the anamnesis described by Parry (1825) in his first patient with "Graves" disease. By Mrs. Judy Smith.

subject of debate ever since. Now that the autoimmune aspects of the syndrome are widely recognized, psychic trauma as an etiological factor may be considered from the point of view of its influencing antibody formation. There are a few pieces of experimental observation implicating emotion in affecting either autoimmune disease or antibody formation (for review, see McKenzie and Zakarija, 1977). We have tackled this problem partly with experiments in animals (Joasoo and McKenzie, 1976; Bonnyns and McKenzie, 1977a) and partly by longitudinal studies in patients with Graves' disease (Bonnyns *et al.*, 1977). Some of the data from these experiences are presented here to illustrate a few points perhaps relevant to the present discussion.

Figure 18 illustrates that the nontraumatic stress of either overcrowding or isolation influenced the spleen cells of immunized rats in their *in vitro* response to antigen. Results of a different experimental model are shown in Fig. 19. For this a microtechnique was developed permitting the use of serial samples of venous blood from rats to measure lymphocyte responsiveness to phytomitogens (Bonnyns and McKenzie, 1977b). By this procedure it was shown (Fig. 19) that *in vitro* lymphocyte responsiveness to mitogens could be affected by nontraumatic stress and, as seen in the previous model (Fig. 18), the result was influenced by the sex of the animal; moreover, the nature of the stress is clearly another important variable (Fig. 19).

Using a similar microtechnique for human blood, Bonnyns (1977b) and his colleagues (1977) followed lymphocyte responsiveness in patients undergoing antithyroid drug therapy for hyperthyroidism of Graves' disease. A number of abnormalities of lymphocyte function were identified, especially when data were compared with values obtained using a control population carefully matched for sex and age (Bonnyns *et al.*, 1977). One especially intriguing finding was the

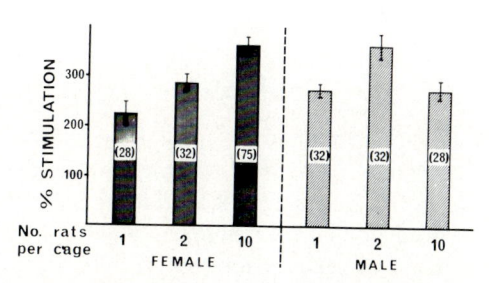

FIG. 18. Effect of immunogen on spleen lymphocytes from immunized rats and influence of isolation or overcrowding. From data of Joasoo and McKenzie (1976). Rats were immunized with thyroglobulin while isolated (1 per cage) or crowded (10 per cage). Spleen lymphocytes were incubated *in vitro* with the immunogen, and the incorporation of [^{3}H] thymidine into DNA was measured. For both sexes the responses of cells from isolated and crowded rats were significantly different ($p < 0.05$) from the response of cell from control animals (2 per cage).

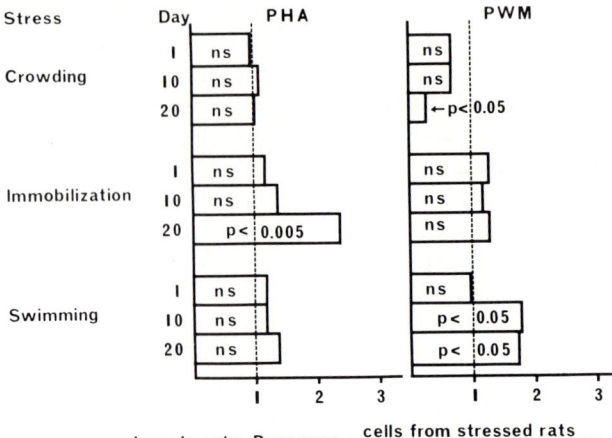

FIG. 19. Peripheral blood lymphocyte responses to phytomitogens in male rats submitted to nontraumatic stress. The three types of stress were maintained for 10 days, blood for testing being taken immediately before onset (day 1), immediately after (day 10), and 10 days later (day 20). Responses to phytohemagglutinin (PHA) and pokeweed mitogen (PWM) are expressed as a ratio to values obtained with blood from control (nonstressed) rats. Students' t test was applied to individual pairs of data before calculation of these ratios. ns, Not significant.

relationship of lymphocyte responsiveness to phytohemagglutinin to the concentration of serum thyroxine (Fig. 20). Whether these data reflect a dependence of lymphocyte function on thyroid hormone, or signify that, as the disease comes under control there are changes in the immune system that might lead to remission, is still to be evaluated.

IX. Discussion

We have presented data relevant to the topic: Why is the thyroid overactive in Graves' disease? It seems to us that the evidence for the universal presence, at least when there is hyperthyroidism, of a thyroid-stimulating IgG is impressive. To a degree it is unfortunate that the mouse bioassay for LATS came so early on the scene, in that issues would probably be much clearer today if the initial bioassays had used human tissue to measure the stimulator. However, appropriate techniques were not available in 1958 (McKenzie, 1958b), so that many data accumulated that were based on the mouse bioassay. Now that assays based on stimulation of, or binding to, the human thyroid have become prevalent, a certain amount of confusion has arisen regarding the interrelationship of data by

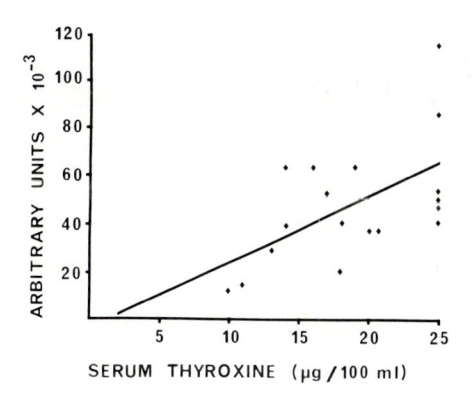

FIG. 20. Correlation of lymphocyte response to phytohemagglutinin (PHA) with serum concentration of thyroxine in Graves' disease. "Arbitrary units" on the ordinate denotes the responses to a range of doses of PHA, after 1 day of culture of peripheral blood lymphocytes; the incorporation of [^{3}H]thymidine was measured. The cluster of values at 25 µg/100 ml represents those with thyroxine measured as > 20 µg per 100 ml of serum. $r = 0.57; p = 0.01$.

the different methods. We have attempted to show that the immunoglobulin stimulating the human thyroid is the same as that which variably stimulates the nonhuman thyroid; when the mouse thyroid responds to the IgG, then a positive LATS assay is obtained. Although the putative antigen has not yet been isolated, thyroid-stimulating IgG preparations have been achieved by various groups (McKenzie, 1968c; Solomon et al., 1970; Ong et al., 1976), and it seems highly likely that the spontaneous thyroid-stimulating IgG, circulating in Graves' disease, is truly an antibody, presumably to an antigen residing in the plasma membrane of the follicular cell. [There is accumulating circumstantial evidence that the antigen is either the receptor for thyrotropin or a closely related molecule (Smith and Hall, 1974; Tate et al., 1976)]. In that case, it seems to us reasonable to expect that such an antibody will cross-react with molecules similar to its homologous antigen, even though they are present in a distant species. Similar cross-reactions are well recognized in experiments with synthetic haptens (Eisen, 1974) and, even more pertinent in immunization with a zoologically specific protein, such as thyroglobulin (Rose and Witebsky, 1955).

The data we have acquired, showing the constancy of pI of the thyroid-stimulating activity, whatever the species of thyroid being studied, and the random nature of the cross-reaction, are in accord with this concept. Furthermore, as would be expected (Eisen, 1974), the more potent the IgG in the homologous system, the more frequently cross-reaction occurred with heterologous thyroids (although this relationship was of necessity only approximate, see Table III).

Thus, although proof must await isolation of the putative antigen, there is in Graves' disease circumstantial evidence supporting a concept of formation of an antibody to a single thyroid antigen and this thyroid-stimulating IgG having similar immunochemical properties whenever it occurs.

The evidence that the thyroid-stimulating IgG is indeed the cause of hyperthyroidism in Graves' disease is also, at present, circumstantial. Munro et al. (1977) and Shishiba (1977) recently reviewed their wide experience with the use, in particular, of the LATS-P assay and the HTS assay. It appears that the sensitivity of these two assays is such that the blood of the majority of patients studied gives a positive result in one or other test. It is an assumption that negative results reflect only insensitivity of the method being applied to the problem; in our experience (Table VII) only one patient with active hyperthyroidism of Graves' disease has had a negative result in the assay system we use. That patient, considering the high level of circulating immune complexes measured, may reflect an aspect of the pathogenetic mechanism that has still to be uncovered.

Rare examples have been described of thyroid-stimulating IgG occurring in blood of patients with normal thyroid function. We have noted some such instances (Table VII), but the explanation appears obvious in these cases; the thyroid probably is no longer capable of responding to the action of the stimulator, just as was hypothesized, and tested, for a few patients studied with the LATS assay (Liddle et al., 1965). More difficult to understand are patients, such as one described by Chopra and Solomon (1970) and one described by Wong and Doe (1972), who had LATS in the blood but apparently completely normal thyroid function; a few similar patients were recently reported by Solomon et al. (1977), who used the LATS-P assay. We have not had the opportunity to study such patients and feel that they must be quite rare. It is possible that they represent, in the case of the LATS assays, an unusual degree of cross-reaction before the affinity for the homologous antigen is sufficiently high to produce effects in the patient; this could readily be tested by application of appropriate assays. In the case of data depending upon the LATS-P assay, since that involves binding to, rather than stimulation of, human thyroid, there is obvious need for caution in interpreting the data in relation to the patient's thyroid function.

If a thyroid-stimulating IgG is accepted as the cause of hyperthyroidism in this syndrome, one must ask what controls its production. For instance, approximately 50% of patients treated with antithyroid drugs enter a permanent remission, and this appears to be related to the thyroid-stimulating IgG no longer circulating (although it must be admitted that proof of this statement requires more prolonged longitudinal studies than are currently available). The explanation probably lies in the realm of the function of the lymphocyte, the cell synthesizing humoral IgG. As shown initially in Europe (Berthaux et al., 1970;

Mahieu and Winand, 1972), but in most detail by Volpé and his colleagues (1974), there are disorders of cell-mediated immunity in Graves' disease. We described 10 years ago at the Laurentian Hormone Conference that the addition of phytohemagglutinin to patients' lymphocytes *in vitro* led to the production of LATS (McKenzie, 1967). The fact that only cells from subjects with LATS-positive serum gave such results can now be explained in terms of the cross-reaction thesis we have expounded above; presumably lymphocytes programmed *in vivo* to synthesize an IgG that does not cross-react with the mouse thyroid cannot produce anything different *in vitro*. More recently it has been shown (Knox *et al.*, 1976) that lymphocytes from hyperthyroid Graves' disease subjects will synthesize thyroid-stimulating IgG, as measured by a human thyroid slice technique, in response to the addition of an extract (antigen?) of human thyroid. Consequently, the chain of pathogenesis can be followed backward by one link to the production of thyroid-stimulating IgG *in vivo* by the patient's lymphocytes.

The control of lymphocyte function is uncertain despite many recent advances regarding the specifics of antibody formation and the knowledge of the types of interacting cells (Eisen, 1974). There is some evidence that infections and autoimmune disease (Amkraut *et al.*, 1971; Allison, 1973; Solomon *et al.*, 1974), both involving lymphocyte function, can be modified by nontraumatic (emotional?) stress and, as described above, we and our colleagues have attempted to add to the relevant body of data. There is very little reported experience, however, bearing on the topic as it applies to man, and even less with regard to Graves' disease (for review, see McKenzie and Zakarija, 1977). Data accumulated by Bonnyns and McKenzie (1977b) are therefore very exploratory; they found in Graves' disease changes in peripheral blood responsiveness to phytohemagglutinin that regressed during therapy of hyperthyroidism with antithyroid drugs. As shown in Fig. 20, the response to phytohemagglutinin in a group of such patients could be correlated statistically with the serum concentration of thyroxine, the response being greater the higher the level of thyroxine. Clearly at least two interpretations of these data are possible. First, both changes (decreased lymphocyte responsiveness and concentration of thyroxine) might reflect a trend toward remission, whatever the underlying mechanism; second, it may be that thyroxine per se (or triiodothyronine that was not measured in these studies but might be presumed to change in parallel with the thyroxine concentration) has an influence on lymphocyte function, as has been reported in the experimental animal (Fabris, 1973). An extrapolation of the last conjecture might be that to some degree hyperthyroidism is a self-perpetuating disease that, once started, continues, owing to a maintained effect of the hyperthyroidism on immune mechanisms. This is blatantly too simplistic a concept to fit all that is known about the clinical course of such patients, but it is one possible aspect, bearing on etiological concepts, that lends itself to experimental testing.

REFERENCES

Adams, D. D. (1975). *N. Z. Med. J.* **81**, 22.
Adams, D. D., and Kennedy, T. H. (1967). *J. Clin. Endocrinol. Metab.* **27**, 173.
Adams, D. D., and Purves, H. D. (1956). *Proc. Univ. Otago Med. Sch.* **34**, 11.
Allison, A. C. (1973). *Ann. Rheum. Dis.* **32**, 283.
Amkraut, A., Solomon, G. F., and Kraemer, H. C. (1971). *Psychosom. Med.* **33**, 203.
Berthaux, P., Moulias, R., and Goust, J. M. (1970). *Rapp. Reun. Endocrinol. Langue Fr., 11th, 1970* p. 271.
Bitensky, L., Alaghband-Zadeh, J., and Chayen, J. (1974). *Clin. Endocrinol. (Oxford)* **3**, 363.
Bonnyns, M., and McKenzie, J. M. (1977a). Submitted for publication.
Bonnyns, M., and McKenzie, J. M. (1976b). Submitted for publication.
Bonnyns, M., Vanhaelst, L., Golstein, J., Cauchie, C., Ermans, A. M., and Bastenie, P. A. (1973). *Clin. Endocrinol. (Oxford)* **2**, 227.
Bonnyns, M., Cano, P., Osterland, R., and McKenzie, J. M. (1977). Submitted for publication.
Calder, E. D., Penhale, W. J., Barnes, E. W., and Irvine, W. J. (1974). *Br. Med. J.* **2**, 30.
Cano, P., Chertman, M. M., Jerry, L. M., and McKenzie, J. M. (1976). *Endocrinol. Res. Commun.* **3**, 307.
Carneiro, L., Dorrington, K. J., and Munro, D. S. (1966). *Clin. Sci.* **31**, 215.
Chopra, I. J., and Solomon, D. H. (1970). *Ann. Intern. Med.* **73**, 985.
Chopra, I. J., Solomon, D. H., Johnson, D. E., and Chopra, U. (1970). *Metab., Clin. Exp.* **19**, 760.
Eisen, H. N. (1974). "Immunology, an Introduction to Molecular and Cellular Principles of the Immune Responses." Harper, New York.
Fabris, N. (1973). *Clin. Exp. Immunol.* **15**, 601.
Hollingsworth, D. R., and Mabry, C. C. (1976). *Am. J. Dis. Child.* **130**, 148.
Joasoo, A., and McKenzie, J. M. (1976). *Int. Arch. Allergy Appl. Immunol.* **50**, 659.
Knox, A. J. S., von Westarp, C., Row, V. V., and Volpe, R. (1976). *J. Clin. Endocrinol. Metab.* **43**, 330.
Kriss, J. P., Pleshakov, V., and Chien, J. R. (1964). *J. Clin. Endocrinol. Metab.* **24**, 1005.
Liddle, G. W., Heyssel, R. M., and McKenzie, J. M. (1965). *Am. J. Med.* **39**, 845.
McKenzie, J. M. (1958a). *Endocrinology* **63**, 372.
McKenzie, J. M. (1958b). *Endocrinology* **62**, 865.
McKenzie, J. M. (1967). *Recent Prog. Horm. Res.* **23**, 1.
McKenzie, J. M. (1968a). *Schweiz. Akad. Med. Wiss.* **24**, 408.
McKenzie, J. M. (1968b). *Physiol. Rev.* **48**, 252.
McKenzie, J. M. (1968c). *J. Clin. Endocrinol. Metab.* **28**, 596.
McKenzie, J. M., and Zakarija, M. (1972). *In* "Methods in Investigative and Diagnostic Endocrinology" (S. A. Berson, and R. S. Yalow, eds.), Vol. 1, p. 275. North-Holland Publ., Amsterdam.
McKenzie, J. M., and Zakarija, M. (1976). *J. Clin. Endocrinol. Metab.* **4**, 778.
McKenzie, J. M., and Zakarija, M. (1977). *In* "Metabolic Basis of Endocrinology" (L. J. De Groot *et al.*, eds). Grune & Stratton, New York, in press.
McKenzie, J. M., Zakarija, M., D'Amour, P., and Joasoo, A. (1975). *N. Z. Med. J.* **81**, 18.
Mahieu, P., and Winand, P. (1972). *J. Clin. Endocrinol. Metab.* **34**, 1090.
Munro, D. S., Dirmikis, S. M., and Kendall-Taylor, P. (1977). *Proc. Int. Congr. Endocrinol., 5th, 1976* (in press).
Onaya, T., Kotani, M., Yamada, T., and Ochi, Y. (1973). *J. Clin. Endocrinol. Metab.* **36**, 859.
Ong, M., Malkin, D., Tay, S. K., and Malkin, A. (1976). *Endocrinology* **98**, 880.

Orgiazzi, J., Williams, D. E., Chopra, I. J., and Solomon, D. H. (1976). *J. Clin. Endocrinol. Metab.* **42**, 341.
Parry, C. H. (1825). "Collections from the Unpublished Medical Papers of the Late Caleb Hillier Parry, M.D., F.R.S., etc., etc. etc.," Vol. 2, p. 111. Underwoods, London.
Petersen, V., Smith, B. R., and Hall, R. (1975). *J. Clin. Endocrinol. Metab.* **41**, 199.
Rose, N. R., and Witebsky, E. (1955). *J. Immunol.* **75**, 282.
Shishiba, Y. (1977). *Proc. Int. Congr. Endocrinol., 5th, 1976* (in press).
Silverstein, G. E., and Burke, G. (1970). *Arch. Intern. Med.* **126**, 615.
Smith, B. R., and Hall, R. (1974). *Lancet* **2**, 427.
Sobel, A. T., Bokisch, V. A., and Müller-Eberhard, H. J. (1975). *J. Exp. Med.* **142**, 139.
Solomon, D. H., Beall, G. N., and Chopra, I. J. (1970). *J. Clin. Endocrinol. Metab.* **5**, 603.
Solomon, D. H., Chopra, I. J., Chopra, U., and Smith, F. J. (1977). *N. Engl. J. Med.* **296**, 186.
Solomon, G. F., Amkraut, A. A., and Kasper, P. (1974). *Ann. Clin. Res.* **6**, 313.
Tate, R. L., Winand, R. J., and Kohn, L. D. (1976). *Thyroid Res., Proc. Int. Thyroid Conf., 7th, 1975* p. 57.
Theofilopoulos, A. N., Wilson, C. B., and Dixon, F. J. (1976). *Clin. Invest.* **57**, 169.
Volpé, R., Farid, N. R., von Westarp, C., and Row, V. V. (1974). *Clin. Endocrinol. (Oxford)* **3**, 239.
Wall, J. R., Good, B. F., and Hetzel, B. S. (1969). *Lancet* **2**, 1024.
Wong, E. T., and Doe, R. P. (1972). *Ann. Intern. Med.* **76**, 77.

DISCUSSION

J. Weisz: I wonder if you would care to tell us how you visualize the sequence of events that lead to the generation of antibodies that can act as a hormone. You carefully avoided applying the term hormone to these immunoglobulins, but rather referred to them as "metabolic stimulators." Yet, from what you have shown us about the behavior of these antibodies, in particular in your *in vitro* systems, they seem to act as the genuine hormone; TSH; they bind, or at least compete for, the receptors designed for TSH and then produce many of the same effects.

J. M. McKenzie: I cannot give you a complete answer, but I can certainly speculate with the background that is being furnished outside the thyroid, particularly by Jesse Roth and his colleagues, who are looking at antibodies to insulin receptors; they have shown that some antibodies will mimic, and some will antagonize, insulin. If we go back to Che and François (text Fig. 3), you have perhaps the same sort of system, although we immunized; with whole cells, not receptors. We can now recognize that antibodies to receptors can be different, one from another, even though the receptor is a single antigen; what is different is the way the animal sees the receptor and makes an antibody relatively unique for that animal. In Graves' disease we tend to get a stimulating antibody, but how this is explained in molecular terms, how the antibody combines with the receptor in one case to stimulate, in the other case to inhibit, I do not think I can even speculate at this stage.

J. E. Rall: Do you have any data, now that you have some sort of purification of the antibody, as to how it reacts with purified thyroid cell membrane and what will inhibit it in terms of the gangliosides in which Kohn has been interested? A comment concerns the interesting finding that none of the patients in remission after propylthiouracil had the persistence of the antibody. This makes me wonder whether active Graves' disease is not in a sense autocatalytic, in that the thyroid-stimulating antibody in the process of reacting with the thyroid plasma membrane pulls off some of the receptor so that this receptor can continue to stimulate the immune system. Hence when you stop the hyperthyroidism you break this autocatalytic phenomenon. Do you have any data on what happens when antibody is incubated with slices of plasma membrane? Does the antibody eventually debind from the membrane, carrying with it part of the receptor?

J. M. McKenzie: We have no experience as yet of the influence of gangliocides on the thyroid response to LATS. Regarding your suggestion of an autocatalytic effect, or self-perpetuating as we have thought of it, it could be as you are suggesting that there is a tendency for the patients to boost themselves by the very action of the antibody, and make more antibody; or it might be that thyroid hormone is stimulating the immune system. This has been shown experimentally; that is, the state of thyroid function is important to the immune responsiveness of an experimental animal, and that may be in part responsible for the remission of hyperthyroidism; i.e., as the patients become euthyroid, they stop stimulating their immune system. So we tend to view a clinical remission as really as immunological remission, but we have insufficient data to be dogmatic on this concept.

R. Volpé: This may be some part of another alternative possibility of why this disease is self-perpetuating: hyperthyroidism may itself be a stress. There is increase in adrenal cortical size, increase to adrenal cortical metabolic activity, and that would have to be a fourth alternative possibility.

N. B. Schwartz: Could tell us something about the endocrinology that accounts for more women than men having Graves' disease?

J. M. McKenzie: Not really. It is interesting that the female rat, among many obvious differences from the male, has on average fewer lymphocytes and there is less of a response of peripheral blood lymphocytes to phytohemagglutinin (M. Bonnyns and J. M. KcKenzie, unpublished). So whether endocrinologically based or not, the sex difference may well be a difference of the immune systems.

M. A. Greer: You said that, when people were in remission, essentially all had negative responses whereas a positive response occurred only in those with active disease. How does this tie in with data that other people have obtained, and I think you too, in euthyroid Graves' disease with only eye involvement where the thyroid is capable of responding to TSH, yet there is no hyperthyroidism. In this condition, there may or may not be positive antibodies that one can detect.

J. M. McKenzie: The data that Dr. Greer is referring to, for those less familiar, are that there are a few patients, and I should emphasize that they are really very few, who have apparently normal thyroid function, with responsiveness to TSH, and yet have a thyroid stimulator in the blood. Most of these reports have been with use of the LATS-P and TSI assays, i.e., two assays using binding to a thyroid membrane as an index of stimulation, although that is not necessarily so. The insulin story that I referred to earlier bears that out, so that although these patients are difficult to accommodate in a simple theory— and obviously the theory that we have offered is just a little too simplistic—I think until we get evidence that there is a potent thyroid stimulator in these patients who do not have hyperthyroidism, it is still a defensible thesis.

A. White: First, I'd like to point out that the female is not the weaker sex immunologically speaking. It has been known for many years that the female is more responsive than the male to the antigenic challenge, i.e., with antibody production. This apparently can be related to the proliferative effect of estrogens on lymphoid tissue and the effects of estrogen in increasing the phagocytic activity of the reticuloendothelial system. Whether or not this is related to the higher incidence of Graves' disease in the female may possibly be relevant in relation to the question of whether the disease has an autoimmune component. In this regard, have you had the opportunity in the course of Graves' disease and in remission to look at peripheral blood lymphocytes and perhaps obtain an index of the numbers of suppressor T cells that might be present in these individuals? In the various assays that you showed, did you have evidence that as you enriched fractions for your specific IgG component, e.g., by isoelectric focusing, the specific activity—i.e., the activity per milligram of purified protein—increased in the bioassays? Perhaps I might also comment in relation to something Dr. McKenzie has already mentioned, namely, the data of Fabris, and also data of Pierpaoli and Sorkin; these strongly indicate that both thyroid hormone and growth hormone are immunogenic in that administration to experimental animals of anti-TSH antibody or anti-growth hormone antibody can result in immunological suppression and this can be rectified by the administration preferably of both growth hormone and either T_3 or T_4.

J. M. McKenzie: We have not examined specifically the suppression of T lymphocytes. However, looking at peripheral blood responsiveness to phytohemagglutinin, and making comparisons with data from controls strictly matched for age and sex, we find that the Graves' disease patients are different (M. Bonnyns, P. Cano, R. Osterland, and J. M. McKenzie, unpublished).

R. Volpé: The sex difference does not change before puberty and after the menopause, which has the effect of causing the increased female incidence in Graves' disease.

R. S. Yalow: A short-term period of therapy with steroids can break insulin resistance, probably within a week or two in about 20–50% of the cases. Does such short-term therapy affect the concentration of LATS or the degree of hyperthyroidism? Has one been able to immunize animals with the thyroid receptor and induce hyperthyroidism in some animal model?

J. M. McKenzie: Others have used corticosteroids in patients with Graves' disease and also looked at LATS levels as well as the clinical course, and indeed there is evidence that you can both suppress LATS and influence the clinical course; I do not think there is that experience. It would be very difficult to see a change in LATS, which has a half-life of 7 to 21 days. The effect on thyroid function might be equally difficult to document as rapidly as that, but in any case I do not think the experiments have been carried out.

Immunization by ourselves and a few others, to date, has been either with thyroid gland particulate material or isolated membranes, and one can produce antibodies that will, for instance, stimulate thyroid adenyl cyclase *in vitro*, or the mouse thyroid *in vivo*, but the production of hyperthyroidism in the experimental animal has not been achieved so far.

L. Bullock: In the dog, hypothyroidism is a relatively common endocrine abnormality and is often associated with a syndrome similar to Hashimoto's thyroiditis. Thus one end of the spectrum of thyroid antibodies you referred to has been recognized in another species. By contrast, hyperthyroidism, which is much less common in the dog, is not associated with an immune phenomena but rather with a thyroid tumor. I was wondering if a LATS like substance has been detected in any species other than human?

J. M. McKenzie: I know that hyperthyroidism, with diffuse hyperplastic thyroid, has been described in one horse and one dog; it certainly does not seem to be very frequent.

F. G. Sulman: I should like to refer to your difficulty in producing a certain type of stress that could induce hyperthyroidism by extreme heat or crowding. There is a very simple possibility of attaining this effect in man. You have in the Mediterranean region the hot Sirocco wind and in Israel we have the Sharav, both producing stress. Now, a stress provoked by hot, dry desert winds, or anywhere by temperature changes or weather front changes, produces also a strong reaction of the thyroid in everyone sensitive to it. We have found out that only 15% of heat-sensitive people react via the thyroid (latent hyperthyroeosis). This reaction could easily be traced in the following way: the patients deliver their 24-hour urine every day, and we assay T_4 in the urine, which allows a daily horizontal follow-up. This is a method that has also been applied in London by Chan and in Ulm by Rothenbuchner. It may show that at certain days of heat stress, the thyroid can "explode." Where normal T_4 excretion may be between 10 and 20 µg/day, it may go up to 25, 30, or 40 µg/day, thus attaining values higher than in Graves' disease. With this test, we can easily follow up stress reactions of the thyroid. Interestingly enough, one can treat such cases with lithium; it suppresses the thyroid right away on the same day, a well known fact to all who give lithium for other purposes.

The question of whether the female is really more sensitive or more prone to suffer from Graves' disease should be considered in relation to two facts: (1) When we subjected rats to heat stress, it turned out that the male rat was immediately able to produce androgenic stress hormones in the adrenal glands, and in this way the cortical steroids suppressed the thyroid's reaction in the males. (2) In the females, the opposite was true: the thyroid hormones went up, and the stress hormones of the adrenal gland went down. Receptors for certain hormones and especially for serotonin are sensitive to estrogen, and the same happens also with the receptor for T_4. For this reason, the exact effect of estrogenic hormone on the T_4 receptor should be studied. We have found two steroids that can prevent the heat death of rats. If you breed rats at 37°C all of them die, but if you keep them on dihydrotachysterol (0.1–0.5 mg/kg per day or 5–50 mg/kg of dehydroepiandrosterone) you can salvage all of them and get a survival of 100%. This shows that death in heat stress may be connected with the presence of high amounts of T_4, because when we suppressed T_4, we could let the animals survive.

J. Kowal: Have you studied the effect of TSH on LATS stimulation of cAMP in tissue slices, and vice versa? Do you have data in terms of total cAMP response, and also the time course of the response?

J. M. McKenzie: We have not yet done the experiments required to give you an unequivocal answer on that point.

J. Kowal: Did you or did you not find antibodies to TSH receptors, and are you stating that LATS is an antibody to a TSH receptor, or is this just a hypothesis?

J. M. McKenzie: The hypothesis is that the thyroid-stimulating IgG is an antibody to the; TSH receptor; proof of that is not yet available.

M. M. Martin: The natural history of Graves' disease is to burn itself out. You showed that human thyroid stimulator is present in almost 100% of cases at the time of diagnosis. Does it disappear with treatment, and is there any correlation with the duration of treatment? Will it tend to disappear if treatment is kept up longer? Since thyroid stimulator is present in all patients with hyperthyroidism, why is it that so relatively few of the babies born to thyrotoxic mothers under treatment do show thyrotoxicosis of the newborn? Is there any way in which you can relate the incidence and variation in the severity of this condition with what you find in the mother?

J. M. McKenzie: The duration of the thyroid disease in our experience, distinct from the duration of the treatment, seems to be an important factor in that the majority of patients with high levels of thyroid stimulator; those that tend to cross-react with nonhuman species, come from patients with many years of disease, either unrecognized or relapsing after various forms of therapy.

Regarding the therapy, we have no evidence that the duration of therapy will really influence the disappearance of the thyroid stimulator. To take it from the other point of view, we tend to think that if the stimulator is going to disappear, it will do so most often within a few months of therapy. These are impressions at this stage, not accurate observations.

Regarding neonatal Graves' disease, we are of the opinion, with some supporting data, that the important variable is the concentration of stimulating IgG in the mother; if it is high enough, the child will probably have the transient syndrome.

B. M. Dobyns: I have two questions. The first concerns the patient with very long-standing Graves' disease who had the disease 20 or 30 years ago and had presumably been cured, yet when he dies still has hyperplasia of the thyroid or has a burned out gland replaced by scar tissue and lymphocytes. You have now assayed for LATS over a good many years and no doubt followed some of the individuals. What has happened to their LATS?

Will you give us your recent impressions regarding the relationship between LATS and the severe progressive form of exophthalmos. We have observed that among patients with progressive exophthalmos the LATS-positive response (using the mouse assay) is not found very often. When it is positive, however, it is almost always found in those who have a positive response for the exophthalmos-producing substance as tested in the fundulus. Do you feel that LATS is related to severe progressive exophthalmos?

J. M. McKenzie: The first point has a fairly straightforward answer in that we have followed many patients who at one time had a high LATS, and the majority fall in time, but that time may be 5 or 6 years. We have one patient currently documented with high levels of LATS for 10 years, and the titer remains high. One other patient I know had her hyperthyroidism treated by subtotal thyroidectomy in 1929, and we assayed a high level of LATS in her blood in 1972.

We have no detailed systematic study on the relationship to exophthalmos; we certainly have patients with very bad eyes, who do not have LATS or thyroid stimulator in the blood, although, on the other hand, most of the patients with high levels of stimulator have some degree of ophthalmopathy.

S. Korenman: You appear to claim that there is only one thyroid-stimulating molecule present per patient with Graves' disease. Is this correct?

J. M. McKenzie: We are thinking of a polyclonal response to a single antigen, and the resultant spectrum of immunoglobulins seems to have some similar characteristics in the different patients.

S. Korenman: Do you believe that the various biological effects that you have shown, and those shown by others, are all due to a very narrow range of molecule or a single type of molecule per patient, even though it may be somewhat different from person to person?

J. M. McKenzie: We have difficulty with the nomenclature too, but believe that if we took all the assays that are currently available, then, with optimal sensitivity of those assays (and that is a big drawback since many of them are quite insensitive), a single serum would be active in them all, due to a single species of IgG, at least as defined by isoelectric point.

S. Korenman: I think that is an area of disagreement with Dave Solomon in terms of specificities. With regard to the assay of stimulation of cAMP, is most of the cAMP that you measure in the medium or in the tissue; and does the cAMP have a short or long half-life? Are we looking at a stable substance that has been generated over 2 hours or a rapidly turning over substance that requires continuing high-level activity of the enzymic process?

J. M. McKenzie: We are measuring cAMP in the tissue, and presumably it is "turning over" during the 2 hours of incubation.

J. Geller: How do you reconcile your very good correlation of the thyroid stimulator cAMP assay and the clinical state of Graves' disease after antithyroid drugs with the fact that thyroid suppressibility with ^{131}I uptake after triiodothyronine is accurate in predicting remission or relapse after antithyroid drugs in only about 80% of patients. What accounts for the autonomous thyroid stimulation in the 20% of patients who are in remission and still are not suppressible?

J. M. McKenzie: I think the stimulator is probably there in the patients that you are describing. I should reemphasize though that we have so far studied only a small number of patients. We had some who were clearly euthyroid for a variable time and had suppressible radioiodine uptake, and these patients did not have the stimulator. We have others, as I have shown, who are still abnormal, as defined by thyroid suppressibility, although clinically euthyroid, and they do have the stimulator, but the total series that I referred to is certainly less than 20 patients. I think we must build up a much larger series to give a clear answer to your question.

Ontogenesis of Hypothalamic–Pituitary–Thyroid Function and Metabolism in Man, Sheep, and Rat[1]

DELBERT A. FISHER, JEAN H. DUSSAULT,
JOSEPH SACK,[2] AND INDER J. CHOPRA

*Departments of Pediatrics and Medicine, UCLA School of Medicine, Los Angeles, California
and Department of Medicine, Université Laval, Quebec, Canada*

I. Introduction

During the past three decades our understanding of fetal endocrinology has expanded rapidly. The pioneering work of Dr. Alfred Jost has been extended by numerous investigators, aided more recently by the availability of sensitive and specific hormone radioimmunoassay systems and highly purified synthetic and biosynthesized hormone preparations. These techniques have been applied to studies of aborted human fetuses and to newborn infants, as well as to acute and chronic studies of the fetus and newborn of several mammalian species. The results have confirmed the general concept of fetal autonomy proposed by Dr. Jost and collaborators (Jost, 1966, 1971; Jost and Picon, 1970). Fetal autonomy is particularly clear for the pituitary thyroid system, since the mammalian placenta is essentially impermeable to the iodothyronines as well as thyrotropin (TSH).

It has long been clear that thyroid hormones play an important role in development; this is strikingly evident in amphibian metamorphosis (Etkin, 1968). Although their role in mammalian development is not quite as remarkable, thyroid hormones clearly stimulate somatic growth and are essential for optimal central nervous system growth and development during infancy and childhood. They play an important role in thermogenesis and appear to significantly augment the increase in nonshivering thermogenesis characteristic of

---

[1] Supported in part by grants from the National Institutes of Health, Bethesda, Maryland; HD-04270 from the National Institute of Child Health and Human Development AM-16155 and AM-70225 from the National Institute of Arthritis, Metabolic and Digestive Diseases; Grant 6-41 from The National Foundation–March of Dimes; and grant MA-5730 from the Medical Research Council of Canada.

[2] Present address: Department of Pediatrics, The Chaim Sheba Medical Center, Tel Hashomer, Israel, affiliated to the Tel Aviv University Sackler School of Medicine.

neonatal adaptation to extrauterine life. Finally, thyroid hormones may play a role in lung maturation (Cuestas *et al.*, 1976). Thus the orderly maturation of the hypothalamic–pituitary–thyroid system and of hypothalamic–pituitary control of thyroid hormone production are significant factors conditioning fetal development near term as well as normal extrauterine adaptation and normal growth and development in the extrauterine environment.

II. Placental Transfer of Iodothyronines, TSH, and Hypothalamic Thyrotropin-Releasing Hormone (TRH)

Courrier and Aron in 1929 were the first to study placental transfer of thyroid hormones. They observed that large amounts of animal thyroid failed to alter fetal thyroid gland morphology in pregnant bitches and guinea pigs, and concluded that thyroid hormones did not cross the placental barrier. Subsequent studies by numerous investigators in several species have generally supported this conclusion. In the pregnant guinea pig, rabbit, or rat the administration of radiolabeled thyroxine (T_4) or triiodothyronine (T_3) to the mother results in a fetal to maternal (F/M) serum concentration ratio of labeled hormones of 0.05 or less (Hall and Myant, 1956; Hirvonen and Lybeck, 1956; Postel, 1957; Geloso *et al.*, 1968; London *et al.*, 1963). Thyroidectomy of the fetal rat (decapitation) results in approximately a 70% reduction of mean fetal serum butanol-extractable iodine (BEI) concentration without altering the maternal concentration (Geloso and Bernard, 1967; Geloso *et al.*, 1968) (Fig. 1). The blood of euthyroid rat fetuses of thyroidectomized mothers given radioiodine near term contains labeled T_4, whereas the blood of thyroidectomized fetal littermates does not (Geloso, 1964). Finally the administration of physiological doses of T_4 to pregnant rats or guinea pigs will not prevent drug-induced fetal goiter (Postel, 1957; DiAngelo, 1967) or suppress drug-induced elevations in fetal serum TSH levels (DiAngelo, 1967).

These results indicate that placental permeability to thyroid hormones in these species is limited. However, some transport may occur. The studies of Myant (1958a) indicated some transfer of T_4 in the rabbit and rat. Geloso and Bernard (1967) demonstrated measurable BEI levels and some maternal radiothyroxine in the blood of decapitated rat fetuses. Gray and Galton (1974) showed partial suppression of thyroid gland activity in rat fetuses from mothers given 2 μg of exogenous T_4 per 100 gm body weight per day. Studies of labeled hormone transfer in some instances have shown detectable, albeit small, fractions of apparently organic radioactivity transported to fetal serum (Hall and Myant, 1956; Hirvonen and Lybeck, 1956; Hoskins *et al.*, 1958). Finally, large doses of exogenous T_4 or T_3 given to the mother will partially suppress drug-induced fetal goiter in small mammals (Peterson and Young, 1952; Knobel and Josimovich, 1959). Thus, some small fraction of circulating fetal thyroid hormones may

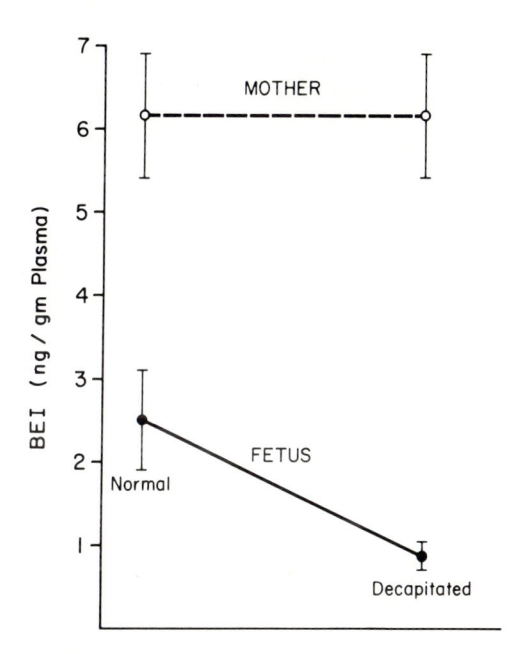

FIG. 1. Serum butanol-extractable iodine (BEI) in normal fetal rats and those thyroidectomized by decapitation. Redrawn from Geloso et al. (1968).

derive from maternal serum, especially in the rat. There are few data quantifying this fraction, but results of Geloso and Bernard (1967), indicating that fetal decapitation reduces mean fetal serum BEI by 65% whereas combined fetal decapitation and maternal thyroidectomy results in an 88% reduction, suggest that some 23% of fetal serum thyroxine in the rat may be derived via placental transfer of maternal hormone.

Considerable data now are available in sheep indicating that placental thyroid hormone transfer is limited. First there is little or no correlation between maternal and fetal serum thyroid hormone concentrations during gestation (Fisher et al., 1972; Thorburn and Hopkins, 1973; Nathanielsz et al., 1973a). This is clearly the case during the last trimester of gestation, when fetal serum free-T_4 concentrations are relatively higher and free-T_3 levels lower than respective maternal concentrations (Table I). Thyroidectomy of the fetal sheep leads to a rapid reduction in fetal serum T_4 concentrations followed by an increase in fetal serum TSH levels; maternal blood thyroid hormone and TSH concentrations remain unchanged (Hopkins and Thorburn, 1971; Thorburn and Hopkins, 1973; Erenberg et al., 1973; Erenberg and Fisher, 1973) (see Fig. 2).

Kinetic studies with labeled thyroid hormones and with labeled TSH also support the view that the sheep placenta is impermeable to thyroid hormones

TABLE I
Total and Free Thyroid Hormone Concentrations in Maternal and Fetal Sheep

	Serum T_3[a]			Serum T_4[a]	
	Total (ng/dl)	Free (pg/dl)	TBG (μg/dl BC)[b]	Total (μg/dl)	Free (ng/dl)
Maternal ($n = 6$)	74 (13)	180 (30)	16.6 (2.2)	5.4 (0.8)	2.4 (0.5)
Fetal ($n = 6$)	<18 (<2.7)	<90 (<10)	7.9 (0.7)	7.5 (1.1)	5.1 (1.1)

[a] Values recorded as mean (SEM)—animals 110–130 days of gestation. T_3, triiodothyronine; T_4, thyroxine.
[b] Maximal binding capacity (BC). TBG, thyroxine-binding globulin.

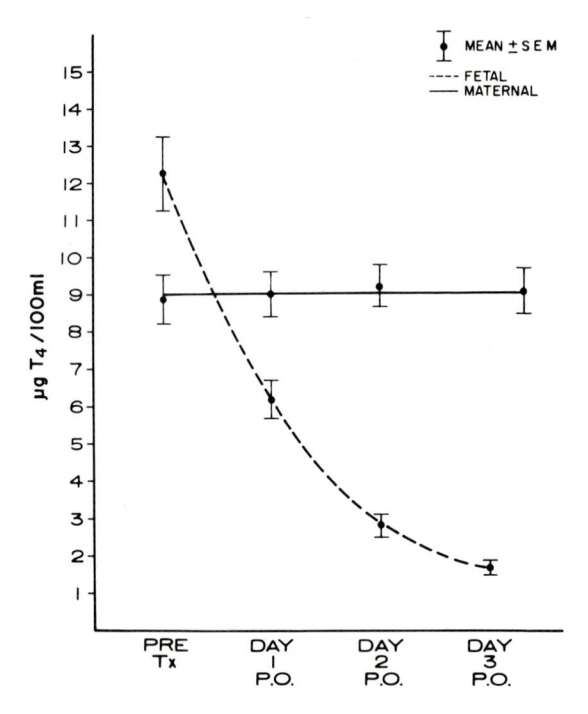

FIG. 2. The fall in fetal serum thyroxine concentration after fetal thyroidectomy (Tx) in the sheep during the last trimester of gestation (nine animals). Maternal serum thyroxine concentrations remain unchanged. From Erenberg *et al.* (1973). Reprinted with permission.

and to TSH. Studies in euthyroid maternal–fetal pairs have shown nearly total impermeability to labeled T_4 or labeled T_3 in both the maternal–fetal (M–F) and fetal–maternal (F–M) directions (Dussault *et al.*, 1971, 1972; Fisher *et al.*, 1972; Nathanielsz *et al.*, 1973a). Similar studies conducted in maternal-fetal sheep pairs after mid-gestation fetal thyroidectomy also have indicated markedly limited M–F and F–M transfer of labeled T_4 and T_3 (Figs. 3 and 4). In these dual-label experiments ^{125}I-labeled hormone injected into the fetus was detectable in maternal serum, and ^{131}I-labeled hormone injected into the mother was detectable in fetal serum, but the transferred hormone reached concentrations less than 0.2% of the dose per liter of plasma. The mean fractional transfer

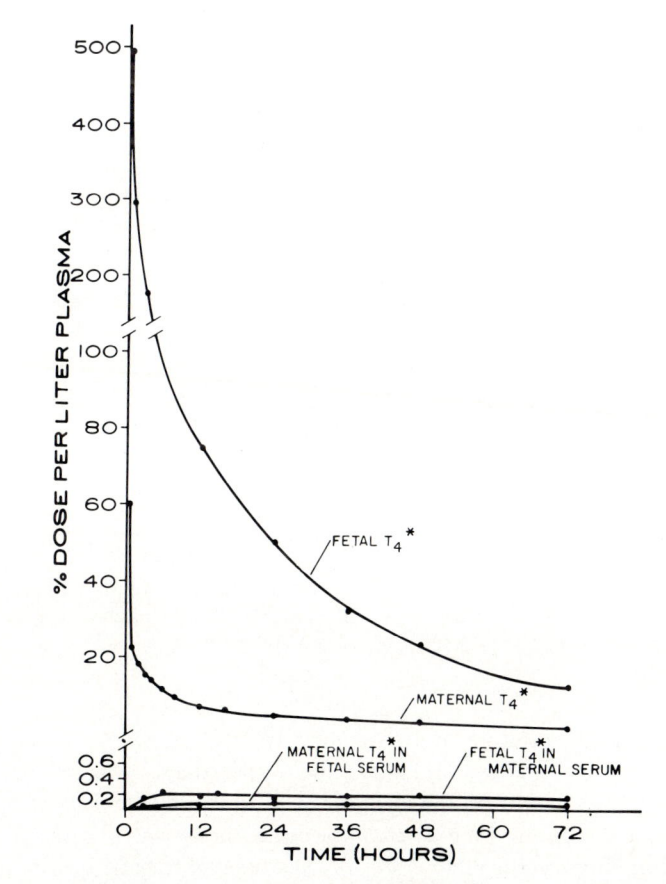

FIG. 3. Placental transfer of labeled thyroxine (T_4) in maternal and hypothyroid fetal sheep pairs. [^{125}I]T_4 was injected into fetal serum and [^{131}I]T_4 into maternal serum. Thereafter measurements of butanol-extractable radioactivity were conducted serially in both mother and fetus. Placental transfer of T_4 was markedly limited in both directions. From Erenberg *et al.* (1973). Reprinted with permission.

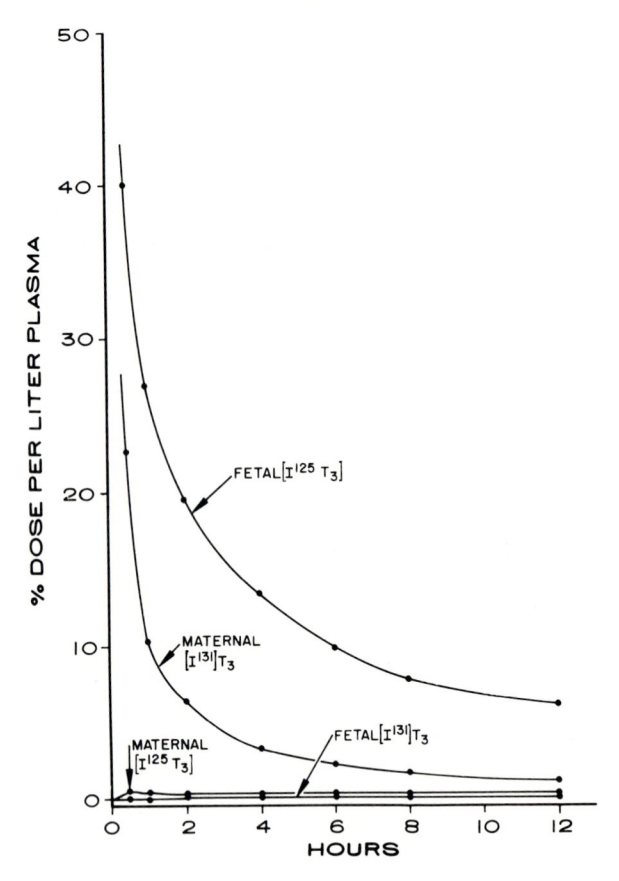

FIG. 4. Placental transfer of labeled triiodothyronine (T$_3$) in maternal and hypothyroid fetal sheep pairs. [^{125}I]T$_3$ was injected into fetal serum, and [^{131}I]T$_3$ into maternal serum. Thereafter measurements of butanol-extractable radioactivity were conducted serially in both mother and fetus. Placental transfer of T$_3$ was markedly limited in both directions. From Erenberg et al. (1973). Reprinted with permission.

rate constants for T$_4$ were 9×10^{-4} and 3×10^{-5}/hr^{-1}, and for T$_3$ 7×10^{-3} and 2×10^{-3}/hr^{-1} in the F–M and M–F directions, respectively. Estimated net transfer of T$_4$ was 0.6 μg and of T$_3$ 0.7 μg in the M–F direction (Erenberg et al., 1973). Fetal T$_4$ plus T$_3$ turnover measured in the same animals was less than 7% of that of euthyroid fetuses (Erenberg et al., 1973). Placental transfer of labeled bovine TSH also was studied in these animals. Results are shown in Figs. 5 and 6. [^{131}I]bTSH injected into fetal serum did not appear in maternal blood (Fig. 5) and [^{125}I]bTSH injected into maternal serum did not appear in fetal blood (Erenberg and Fisher, 1973).

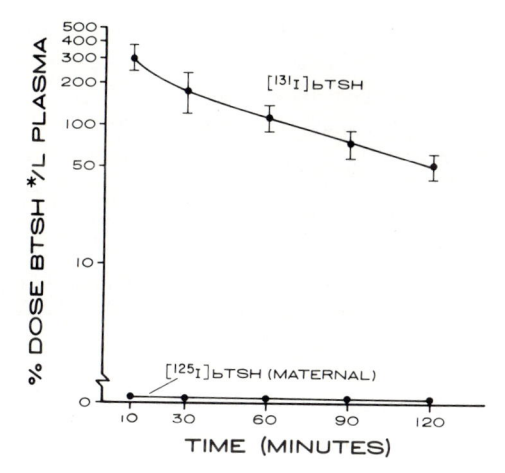

FIG. 5. Fetal–maternal placental transfer of labeled bovine thyrotropin (bTSH) in six sheep. [^{131}I]bTSH was injected into the serum of thyroidectomized fetal sheep. bTSH radioactivity was precipitated from maternal and fetal serum with excess bTSH antiserum. None of the label appeared in maternal serum as immunoprecipitable hormone during a 2-hour period. From Erenberg and Fisher (1973). Reprinted with permission.

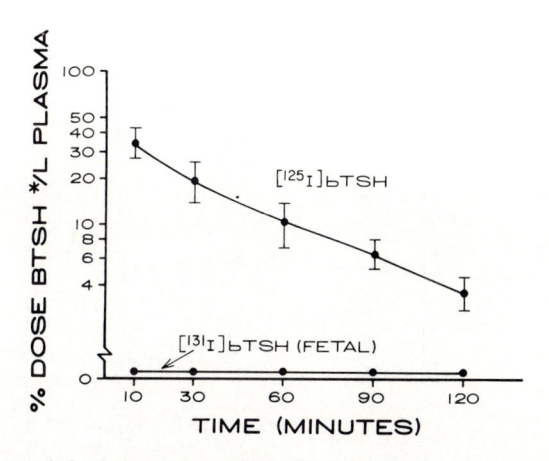

FIG. 6. Maternal–fetal-placental transfer of labeled bovine thyrotropin (bTSH) in six sheep. [^{125}I]bTSH was injected into maternal serum 1–2 weeks after fetal thyroidectomy. bTSH radioactivity was precipitated from maternal and fetal serum with excess bTSH antiserum. None of the label appeared in fetal serum as immunoprecipitable hormone during a 2-hour period. From Erenberg and Fisher (1973). Reprinted with permission.

In the pregnant monkey, Pickering (1968) has shown that the specific activity (SA) of T_4 after maternal injection decreases progressively from maternal serum to placenta to fetal serum to fetal tissues. Moreover, the T_4 SA in fetal serum amounted to only about 4% of that in maternal serum 24 hours after injection. Administration of exogenous T_4 to the pregnant rhesus monkey in quantities adequate to increase maternal serum T_4 concentrations 300% did not alter fetal serum T_4 levels or inhibit T_4 synthesis by the fetal thyroid gland (Pickering, 1968).

The data in man are essentially similar. There are marked base-line maternal–fetal serum concentration gradients of free T_4 during the first 35–38 weeks of gestation, at which time the gradient of mean concentrations tends to reverse (Fig. 7). Fetal serum free T_3 levels are low throughout gestation (Fisher *et al.*, 1973; Abuid *et al.*, 1973; Lieblich and Utiger, 1973; Montalvo *et al.*, 1973).

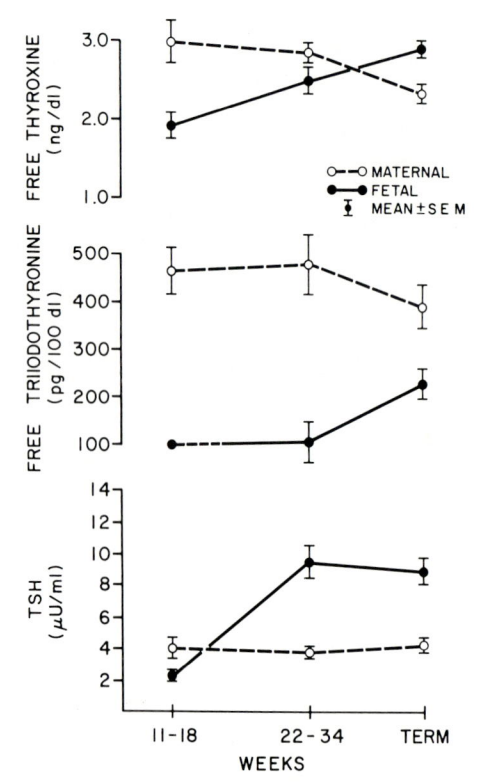

FIG. 7. Maternal and fetal serum free thyroxine, free triiodothyronine and thyrotropin (TSH) concentrations during human gestation. There are no correlations between maternal and fetal levels at any time during gestation. Data were derived from Fisher *et al.* (1970) and Fisher *et al.* (1973).

Fetal serum TSH concentrations, initially low, increase at mid-gestation to values significantly exceeding maternal levels (Fisher et al., 1970, 1973). Placental transfer of labeled T_4 in the M–F direction is minimal during the first trimester of human pregnancy (Myant, 1958b) and markedly limited, although detectable, near term (Grumbach and Werner, 1956; Kearns and Hutson, 1963). Loading of pregnant women at term with large quantities of T_4 (4000 or 8000 µg) leads to significant M–F hormone transfer, but again, the extent is limited; only 1% of the 4000-µg dose of T_4 was transferred during a mean diffusion time of 6 hours, and only 2.8% of the 8000-µg dose was transferred during a mean diffusion time of 16 hours (Table II) (Fisher et al., 1964). Administration of T_3 to pregnant women near term in a dose of 300 µg daily for several weeks produces a modest but variable reduction in cord blood T_4 concentration (Raiti et al., 1967; Dussault et al., 1969) and an increase in the dialyzable fraction of serum T3 (Dussault et al., 1969).

Finally, it is now clear that thyroid agenesis or dysgenesis results in fetal hypothyroidism in man. Screening programs for congenital hypothyroidism have detected several dozen infants with hypothyroidism at birth; all have had low serum T_4 concentrations (< 7 µg/dl) and high serum TSH concentrations (> 100 µU/ml) (Dussault et al., 1976; Klein et al., 1976). Serum thyroxine concentrations in the mothers of these infants are not abnormal (Fig. 8).

These now extensive data in many mammalian species provide convincing evidence that the placenta is impermeable to TSH. In addition, placental transfer of T_4 and T_3 is severely limited in both the M–F and F–M directions. Little or no hormone traverses the placenta at physiological serum concentrations, although at high blood levels some M–F transfer may occur. Even under these circumstances, however, transfer is limited in man. Keynes (1952) reported that chronic administration of 500 µg T_4 daily to a pregnant woman treated with

TABLE II
Maternal and Newborn Butanol-Extractable Iodine (BEI) and Placental Thyroxine (T_4) Transfer Rates After Intravenous Thyroxine Loading of Pregnant Women at Term[a]

Maternal thyroxine dose (µg)	No. of patients	Mean BEI (µg/dl) Maternal	Cord	Mean diffusion time (hr)	Placental T_4 transfer rate (µg/hr)	Total maternal–fetal T_4 transfer (fraction of dose)
0	15	6.4	4.2	–	–	–
4000	5	31.1	7.6	6	6.7	0.010
8000	3	48.7	12.8	16	14.2	0.028

[a]Mean values recorded–from Fisher et al. (1964).

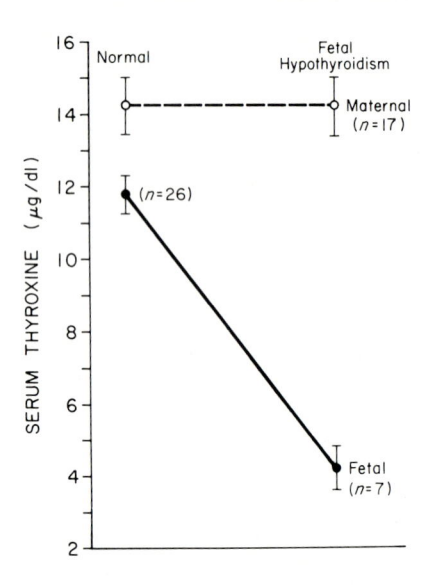

FIG. 8. Maternal and cord blood thyroxine in normal pregnancy and pregnancy associated with congenital thyroid dysgenesis.

propylthiouracil did not prevent fetal goiter. Carr *et al.* (1959) after administering some 1400 mg of desiccated thyroid daily for several months to a pregnant woman at high risk of delivering a hypothyroid infant, observed a cord blood BEI of only 4.5 µg/100 ml in the hypothyroid infant, who was shown to have a small residual of hyperfunctioning thyroid tissue.

There is some evidence that hypothalamic thyrotropin-releasing hormone (TRH) will cross the placenta in the rat (DiAngelo and Wall, 1972; Kajihara *et al.*, 1972), but not in the sheep (Thomas *et al.*, 1975). The significance of TRH in peripheral blood is not clear, but it is doubtful that it is an important factor in maternal–fetal hypothalamic–pituitary–thyroid interaction.

III. Development of Thyroid Function in Man, Sheep, and Rat

A. GENERAL CONSIDERATIONS

The fetal hypothalamic–pituitary–thyroid system develops and functions autonomously of maternal control. The general pattern of ontogenesis of the system is similar in the human fetus, the sheep fetus, and in the fetal-infant rat and can be classified in four stages:

Stage I—Embryogenesis
Stage II—Maturation of the hypothalamus

Stage III—Development of neuroendocrine control
Stage IV—Maturation of thyroid hormone metabolism in peripheral tissues
These maturational events occur during intrauterine development in man and sheep, whereas in the rat stages III and IV are largely extrauterine events. Data will be reviewed separately for the three species, then summarized and compared.

B. DEVELOPMENT OF HUMAN THYROID FUNCTION

1. Embryogenesis

a. The Thyroid Gland. The thyroid of the human embryo derives as a midline outpouching of the endoderm in the floor of the primitive buccal cavity. It is first visible at 16–17 days of gestation in contact with the endothelium of the developing heart. Lateral contributions are derived from the ultimobranchial portions of the fourth pharyngeal pouches. These ultimobranchial anlagen contribute the calcitonin-producing parafollicular C cells (Fisher and Dussault, 1974; Pearse and Carvalheira, 1967). The endodermal thyroid develops as a flasklike vesicle with a narrow neck which at 24 days still connects with the buccal cavity. The vesicle gradually enlarges and becomes bilobed. The stalk finally ruptures by 38–40 days as the thyroid becomes a solid mass of laterally expanding tissue. By 45–50 days the gland has descended to its definitive location in the anterior lower neck. At this stage, the lateral anlagen, having grown downward and forward, contact the median anlage and become incorporated within the developing lateral lobes. By the end of the seventh week the gland has assumed its definitive shape and position and weighs 1–2 mg.

Histologically, the gland develops in three phases: the precolloid, beginning colloid, and follicular-growth phases (Shepard *et al.,* 1964). In man these occur at 47–72, 73–80, and beyond 80 days, respectively. During the precolloid phase small intracellular canaliculi develop as accumulations of colloid material. These gradually enlarge until they disgorge their contents and the colloid material becomes organized into extracellular colloid spaces. During the final period of this follicle organization, by 74 days, iodide concentration and thyroid hormone synthesis can be demonstrated (Shepard, 1967).

Growth and development of the fetal rat thyroid gland does not appear to be thyrotropin (TSH) dependent (Jost, 1966). The gland will develop histologically and store colloid as well as small amounts of hormone in the absence of the pituitary gland. Thyroglobulin synthesis also seems to occur in the absence of TSH. This protein has been detected in human thyroid cells by fluorescent antibody techniques as early as 29 days of gestation (Gitlin and Biasucci, 1969a), whereas TSH is not detected in fetal serum until 10 weeks (Fisher *et al.,* 1970; Greenberg *et al.,*1970). In contrast, optimal concentration of iodine and efficient hormone synthesis by the developing rat thyroid gland do seem to be TSH dependent (Jost, 1966). These functions appear at the time of development of TSH secretory capacity in the human fetus. Moreover, iodine metabolism and T_4 synthesis in nonfunctioning fetal rat thyroid in tissue culture can be stimulated with TSH (Nataf, 1968).

b. The Pituitary Gland. The anterior pituitary gland also is a derivative of the primitive buccal cavity evolving as a dorsel (Rathke's) pouch. By the fifth week Rathke's pouch makes contact with a funnel-shaped derivative of the third cerebral ventricle. This derivative, the infundibular process, is destined to become the neurohypophysis or posterior pituitary gland (Fisher and Dussault, 1974). The buccal connection of the anterior lobe is obliterated by 12 weeks by the developing sphenoid bone, and the pituitary gland becomes partially encapsulated within the bony sella turcica. Cell differentiation can be observed within the anterior lobe by 7–8 weeks; by 8 weeks basophiles can be identified, and eosinophiles are visible by 9–10 weeks (Falin, 1961). The majority of the cells, however, remain "indifferent" (chromophobes). TSH is identifiable by bioassay and immunoassay by 10–12 weeks (Fisher, 1972; Giltin and Biasucci, 1969b; Fukuchi *et al.,* 1970).

2. Maturation of the Hypothalamus

a. The Hypothalamus. The hypothalamus develops as an appendage of the forebrain, which is identifiable in a relatively undifferentiated state by 22 days. The forebrain differentiates into the telencephalon and diencephalon by 34 days. By 5 weeks the third ventricle is visible and the primordium of the neurohypophysis appears. The first hypothalamic nuclei and the fibers of the supraoptic tract can be identified by 12–14 weeks, and the pars tuberalis, median eminence and the remainder of the hypothalamic nuclei are apparent by 16 weeks of gestation (Raiha and Hjelt, 1957). Hyppa (1972) reported the appearance of monoamine fluorescence in the hypothalamus by 10 weeks and in the median eminence by 13 weeks; the hypothalamic hormones TRH, luteotropin-releasing hormone (LHRH), and somatostatin (SRIF) are present by 8–10 weeks (Winters *et al.,* 1974; Kaplan *et al.,* 1976) (Fig. 9). Neurosecretory material can be identified in the neurosecretory hypothalamic nuclei, in the hypothalamic–neurohypophysial tract, and in the infundibular process by 16 weeks (Rinne *et al.,* 1962). Arginine vasopressin (AVT) and vasotocin (AVT) are detectable by radioimmunoassay in the fetal pituitary by 12 weeks (Skowsky and Fisher, 1976).

b. The Pituitary Portal System. There is a paucity of information regarding ontogenesis of the pituitary–portal blood vascular system in the human fetus. It is clear from studies in the rat that a system of capillaries, the supratuberal plexus, develops in association with the pituitary gland. Such capillaries have been identified in the mesenchymal tissue adjoining Rathke's pouch and the

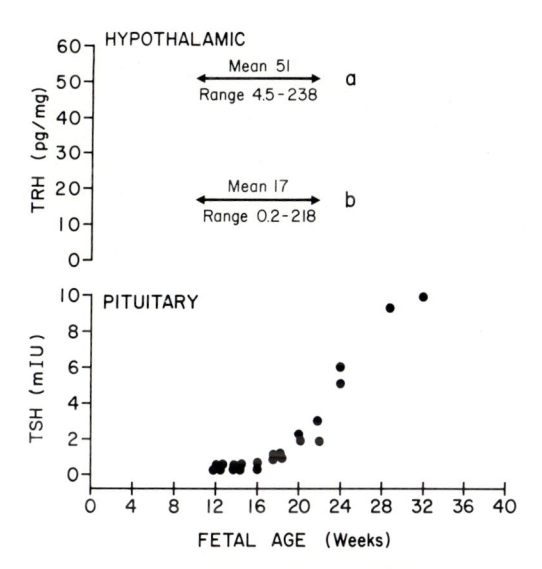

FIG. 9. Hypothalamic thyrotopin-releasing hormone (TRH) concentration and pituitary thyrotropin (TSH) content in the human fetus. TRH is present by 10 weeks. Pituitary TSH content is low before 20 weeks and increases progressively thereafter. TRH data from (a) Winters et al. (1974) and (b) Kaplan et al. (1976).

diencephalon in the human fetus as early as 7–8 weeks of gestation (Falin, 1961). The tufted capillaries of the primary or hypothalamic plexus of the portal vascular system are visible by 15–16 weeks (Niemineva, 1950a,b; Raiha and Hjelt, 1957). There is a progressive maturation of the primary plexus between 15 and 30 weeks with increasing tortuosity and looping of the tufts and a progressive increase in the volume of median eminence capillaries (Niemineva, 1950a).

3. Development of Neuroendocrine Control

a. Pituitary and Serum TSH Concentrations. Pituitary TSH is detectable at 8–10 weeks in the human fetus (Fukuchi et al., 1970; Fisher, 1972). The concentration remains low, however, until 16 weeks, when an increase is evident that continues to 28 weeks. Figure 9 shows pituitary TSH content plotted versus gestational age. Pituitary TSH concentration shows a similar pattern of increase, but it plateaus at 28 weeks whereas content increases progressively thereafter.

Serum TSH is detectable in human fetal serum as early as 10 weeks of gestation, but concentrations remain relatively low until 20 weeks (Fig. 10). The mean concentration of fetal serum TSH after 20 weeks is significantly higher than that before 20 weeks. Moreover the slope of the regression of serum TSH versus age between 10 and 40 weeks is positive whereas the regression of TSH

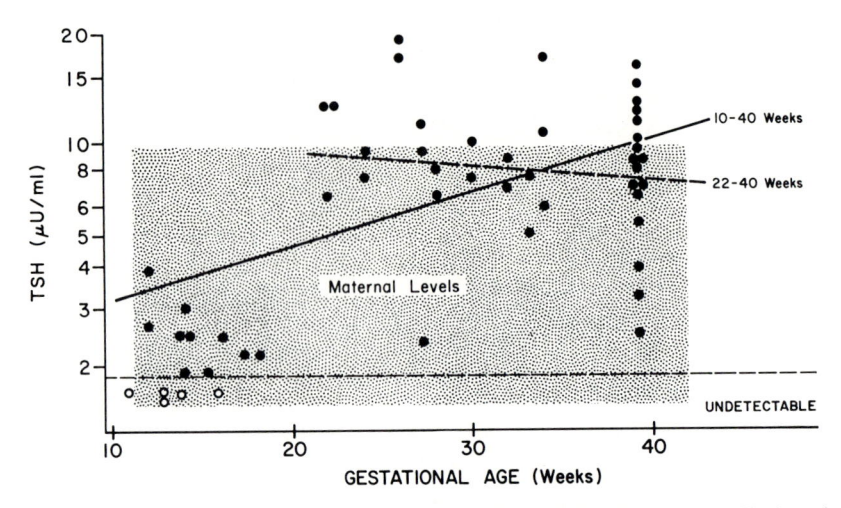

FIG. 10. Serum thyrotropin (TSH) concentration in the human fetus. Each point represents a single fetus or newborn. The range of maternal concentrations is shown as stippled background. The solid line shows the regression of TSH versus gestational age over the range of 10–40 weeks. The interrupted line is the regression of TSH versus gestational age between 22 and 40 weeks. The slopes of the two lines differ statistically. Data from Fisher et al. (1970).

versus age between 22 and 40 weeks has a negative slope (Fig. 10). The slopes are statistically different ($p < 0.01$). This change in slope of the regression of serum TSH on gestational age is compatible with the view that the TSH secretion rate increases at mid-gestation. This view is further supported by the observation that fetal thyroidal radioiodine uptake (after maternal injection) increases at this time. Mean uptake increases from a value of 1.4% per gram of thyroid at 17–21 weeks to 4.6% per gram at 22 weeks (Fisher and Dussault, 1974).

The maturation of the hypothalamus and pituitary portal system are graphically displayed in Fig. 11 for comparison with the changes in pituitary and serum TSH content and concentration. These correlative data suggest the hypothesis that the level of fetal hypothalamic–pituitary function increases markedly at mid-gestation (between 16 and 22 weeks). Since the hypothalamus contains significant concentrations of TRH for several weeks prior to this time (Kaplan et al., 1976; Winters et al., 1974), it would seem likely that histological maturation of the hypothalamus and/or maturation of the pituitary portal system are the critical factors conditioning increasing function of the pituitary–thyroid axis.

b. *Serum T_4 and Negative Feedback.* In response to the thyroid stimulation of the mid-gestation fetus, there is a progressive increase in fetal serum T_4 and free T_4 concentrations during the last half of gestation (Figs. 12 and 13). Most of the increase in serum T_4 levels occurs between 20 and 30 weeks during the

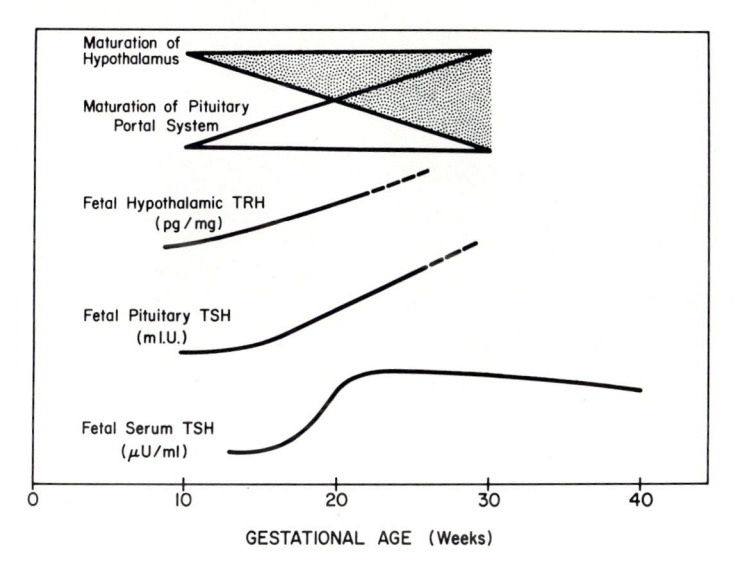

FIG. 11. Maturation of hypothalamic function in the human fetus. The pattern of maturation or change in the hypothalamus, the pituitary portal system, and pituitary and serum thyrotropin (TSH) are graphically summarized. Before 16–20 weeks the hypothalamus and pituitary portal system develop in parallel and the primary plexus of the portal system remains superficial. After this time there is a progressive increase in volume of median eminence capillaries. TRH, thyrotropin-releasing factor.

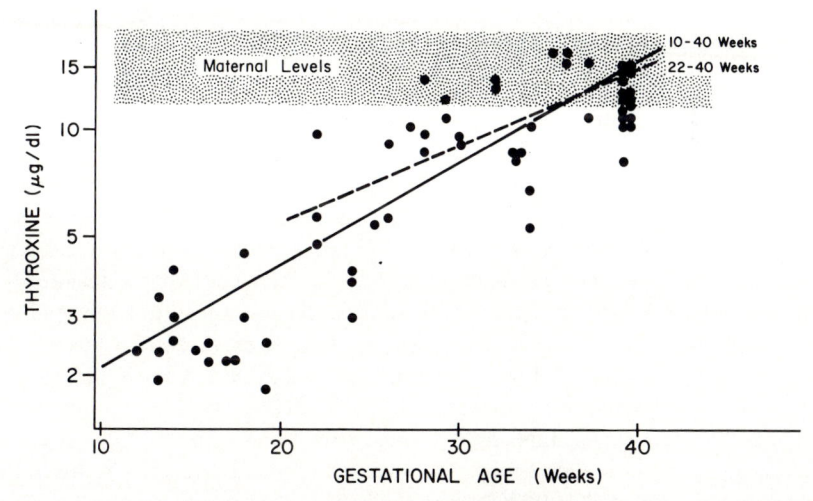

FIG. 12. Serum thyroxine (T_4) concentrations in the human fetus. Each point represents a single fetus or newborn. The ordinate is a log scale. The range of maternal concentrations is shown as stippled background. The solid line shows the regression of T_4 versus gestational age between 10 and 40 weeks. The interrupted line shows the regression of T_4 versus gestational age over the range of 22–40 weeks. The slopes of the lines are statistically similar. Data from Fisher et al. (1970, 1973).

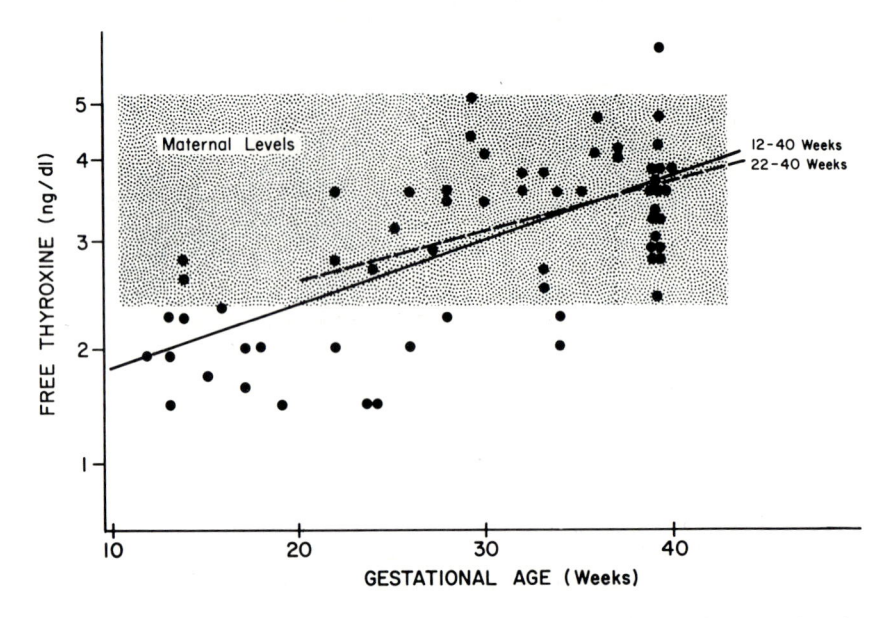

FIG. 13. Serum free thyroxine (T_4) concentration in the human fetus. Each point represents a single fetus or newborn. The ordinate is a log scale. The range of maternal concentrations is shown as stippled background. The solid line shows the regression of free T_4 versus gestational age between 10 and 40 weeks. The interrupted line shows the regression of free T_4 versus gestational age between 22 and 40 weeks. The slopes of the lines are statistically similar. Data from Fisher *et al.* (1970, 1973).

period of hypothalamic and portal system maturation. However, both T_4 and free T_4 values continue to increase slowly after 30 weeks of gestation. This change was more carefully evaluated by measurement of serum T_4, thyroxine-binding globulin (TBG), and TSH concentrations in cord blood of infants delivered at gestational ages between 27 and 48 weeks. These results are summarized in Figs. 14–16. As noted, serum T_4 increased from a mean of 9.3 µg/dl at 30 weeks to 11.0 µg/dl at 45 weeks. Since mean serum TBG did not change between 30 and 45 weeks (Fig. 15), mean free T_4 concentrations also increased as indicated in Fig. 13. During this same period mean serum TSH decreased from a value of 15 to a value of 7 µU/ml (Fig. 16).

These results suggest that fetal serum TSH levels can be suppressed by increasing serum T_4 and perhaps T_3 concentrations. To further evaluate this possibility we injected 200–750 µg of sodium-thyroxine into amniotic fluid of 10 pregnant women 24 hours before elective repeat cesarean section. All women were healthy, with uncomplicated term gestations. The study was approved by the Harbor General Hospital human investigation committee, and written consent was obtained from each mother and father. Serum T_4, and T_3,

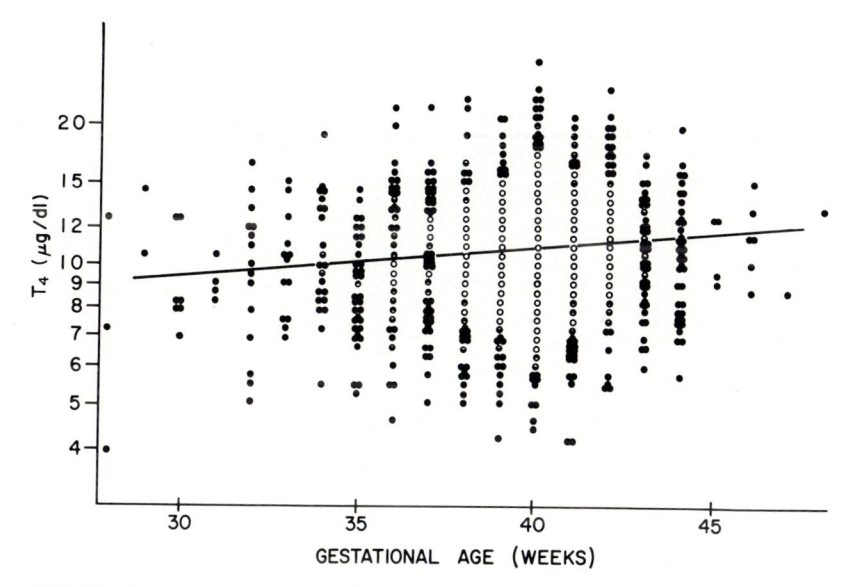

FIG. 14. Cord serum thyroxine (T$_4$) concentrations in newborn infants delivered between 27 and 48 weeks of gestation. The mean T$_4$ level increases from 9.3 μg/dl at 30 weeks to 11.0 μg/dl at 45 weeks. 2683 infants are represented. The regression of T$_4$ versus gestational age during this period is shown as a solid line. From D. A. Fisher *et al.*, unpublished data. •, Single values; ◐, 5 or 6 values; ◖, 7 or 8 values; ○, 9 or more values.

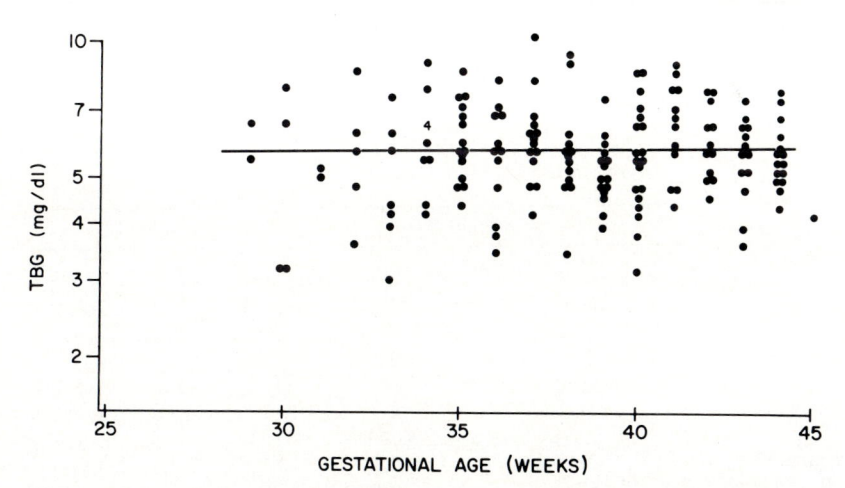

FIG. 15. Cord serum thyroxine-binding globulin (TBG) concentrations measured by radioimmunoassay in newborn infants delivered between 27 and 48 weeks of gestation; 180 infants are represented. Mean TBG did not change during this period. The regression of T$_4$ versus gestational age during this period is shown as a solid line. From D. A. Fisher *et al.*, unpublished data.

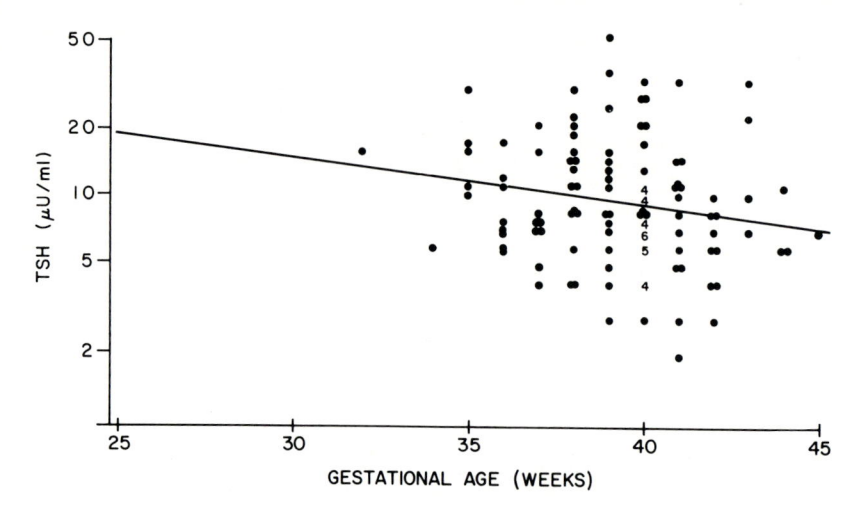

FIG. 16. Cord serum thyrotropin (TSH) concentrations in newborn infants delivered between 30 and 45 weeks of gestation; 125 infants are represented. The regression of TSH versus gestational age during this period is shown as a solid line. Mean TSH decreased from 15 µU/ml at 30 weeks to 7 µU/ml at 45 weeks. From D. A. Fisher et al., unpublished data.

and TSH were measured in maternal blood and cord blood at delivery, and serially during the first 4 hours in the newborn infants. Maternal T_4, T_3, and TSH concentrations were not influenced by the T_4 injection. Data for the infants are shown in Table III. It can be seen that the T_4 injection increased cord blood T_4 concentrations and reduced TSH levels in proportion to the

TABLE III

Effect of Intraamniotic Fluid Injection of Thyroxine (T_4) on Thyrotropin (TSH) and Thyroid Hormone Secretion in the Newborn

Group	No. of infants	Cord blood[a] T_4 (µg/dl)	T_3[b] (ng/dl)	TSH (µU/ml)	Neonatal blood[a] Peak TSH (µU/ml)	4-hr T_4 (µg/dl)	4-hr T_3 (ng/dl)
Control	8	15.3 (1.2)	47 (5)	12 (2)	67 (7)	17.9 (2.6)	173 (27)
Intraamniotic T_4 (200 µg)	6	16.2 (2.9)	46 (9)	11 (1.8)	26 (1.8)	17.9 (1.8)	99 (7)
Intraamniotic T_4 (300–750 µg)	4	26.2 (2.4)	37 (9)	5.5 (1.0)	10.5 (0.9)	33.6 (3.3)	173 (7)

[a]Values are recorded as mean and (SEM).
[b]T_3, triiodothyronine.

dose injected. The T_4 injection also markedly blunted the neonatal TSH surge. These results suggest that serum TSH is suppressible in the term fetus by increasing serum T_4 concentrations. Moreover, this effect occurs without change in serum T_3 levels.

By analogy with data in the rat (Krulich et al., 1976), it is likely that these changes reflect maturation of neuroendocrine control systems for regulation of thyroid function. In the rat this maturation includes a progressive decrease in the TSH response to TRH as well as increasing sensitivity to feedback inhibition of TSH by thyroid hormones. The period during which this maturation occurs in the human fetus and newborn can be estimated by assessing the change in mean serum free T_4/TSH and free T_3/TSH ratios with development. These are shown in Table IV for the 30-week fetus, the term fetus, the 1 month-old infant, and the adult. The early neonatal period was not included because of the alteration in serum thyroid hormone levels conditioned by parturition. As noted there is a progressive increase in both free T_4/TSH and free T_3/TSH ratios between 30 weeks of gestation and 1 month of postnatal life. At this time values were comparable to those in the adult. These data suggest that maturation of hypothalamic–pituitary control of thyroid function in the human infant occurs between 20–30 weeks of gestation and 1 month of postnatal life.

4. Maturation of Tissue Metabolism

It is now clear that the major thyroid hormone produced by the fetus is T_4. Serum T_3 and free T_3 (FT_3) concentrations are unmeasurable throughout most of gestation but increase somewhat during the last 10 weeks (Figs. 17 and 18). Serum triiodothyronine (3,5,3′-iodothyronine) concentrations in cord blood at term average approximately 50 ng/dl whereas the mean maternal concentration approximates 170 ng/dl (Erenberg et al., 1974; Abuid et al., 1973; Lieblich and Utiger, 1973; Montalvo et al., 1973). In contrast, serum concentrations of 3,3′,5′-iodothyronine, or reverse T_3 (RT_3) are high in cord blood (Chopra,

TABLE IV
Maturation of Neuroendocrine Control of Thyrotropin (TSH) Secretion in the Human Fetus and Newborn

Age	Serum free T_4/TSH[a]	Serum free T_3/TSH[b]
30 Weeks of pregnancy	0.17	8
Term	0.25	16
1 Month postnatal	1.3	200
Adult	1.0	160

[a] T_4, thyroxine; ng/ml per µU/ml.
[b] T_3, triiodothyronine; pg/ml per µU/ml.

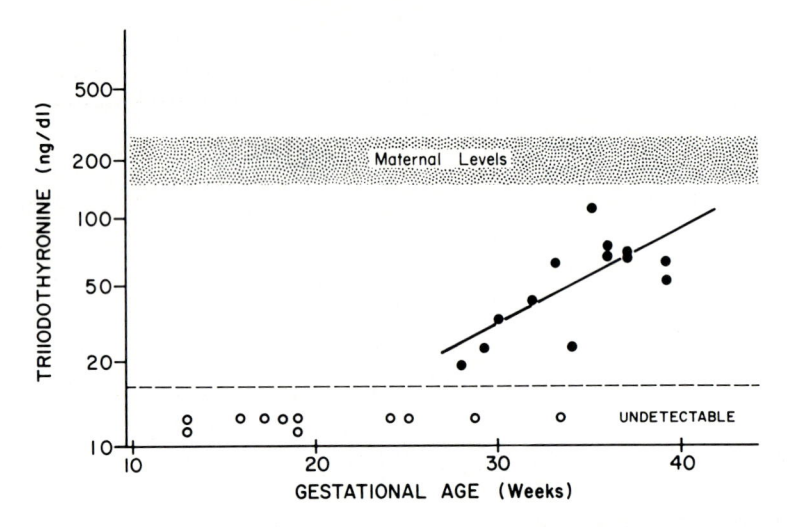

FIG. 17. Serum triiodothyronine concentrations in the human fetus. Each point represents one fetus. The ordinate is a log scale. Open circles represent levels below the limit of sensitivity of our radioimmunoassay system (approximately 15 ng/dl). The range of maternal values is shown as a stippled background. Data from Fisher *et al.* (1973).

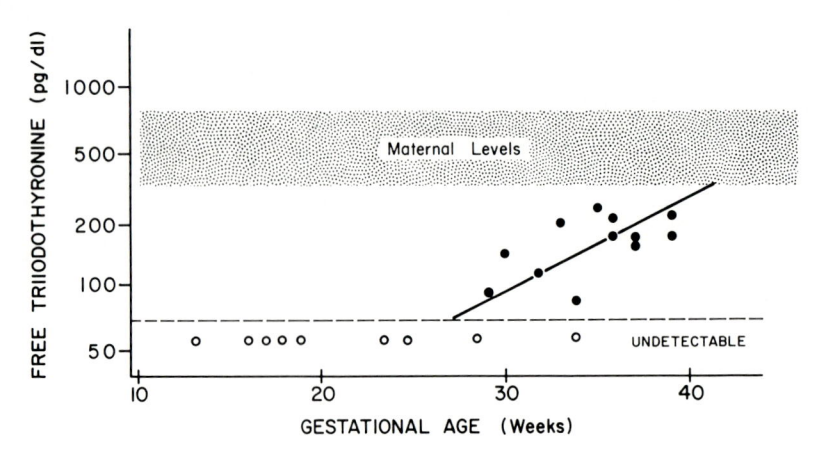

FIG. 18. Serum free triiodothyronine concentrations in the human fetus. The ordinate is a log scale. Each point represents one fetus. Open circles represent levels below the limit of detectability. The range of maternal values is shown as stippled background. Data from Fisher *et al.* (1973).

1974; Chopra et al., 1975a). Both T_3 and RT_3 are synthesized and secreted by the thyroid gland, but the thyroidal concentrations of both hormones are small relative to T_4 (Chopra, 1974; Chopra et al., 1975b). In the euthyroid adult about 75% of the T_3 and 97% of the RT_3 circulating in blood are derived from peripheral deiodination of T_4 (Chopra, 1976). T_3 is derived in tissues by monodeiodination of the outer, or β, benzene ring of the T_4 molecule (containing the hydroxyl side chain); RT_3 is produced by monodeiodination of the inner, or α ring containing the alanine side chain. RT_3 has little biological activity whereas T_3 is 3–4 times more active than thyroxine in most bioassay systems.

From studies in the fetal sheep, it seems likely that fetal serum T_3 levels are low because peripheral production from T_4 is reduced whereas RT_3 production from T_4 is relatively increased (Chopra et al., 1975b). In this species fetal serum T_4 and RT_3 concentrations increase progressively from 50–70 days to term (150 days); fetal serum T_3 levels are unmeasurable until near term, as in the human fetus.

We can construct the pattern of development of thyroid hormone production and metabolism in the human fetus as shown in Fig. 19. In response to

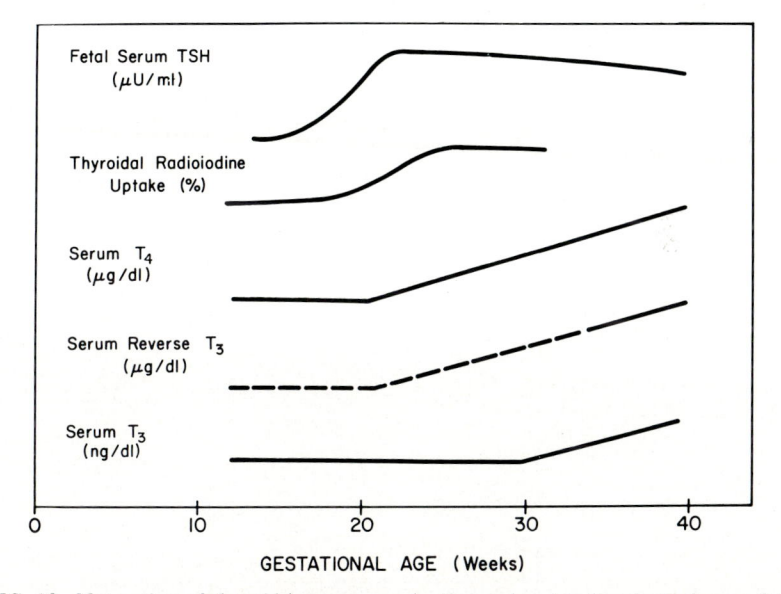

FIG. 19. Maturation of thyroid hormone production and metabolism in the human fetus. Fetal serum thyrotropin (TSH) concentrations increase near mid-gestation followed by an increase in fetal thyroidal radioiodine uptake. Serum thyroxine (T_4) levels, initially low before this time, increase progressively thereafter to term. The increase in fetal serum reverse triiodothyronine concentrations probably parallels the increase in serum T_4 levels. Serum T_3 concentrations increase only near term.

activation of the hypothalamic–pituitary system at mid-gestation, the thyroid gland is stimulated and there is a progressive increase in serum T_4 concentrations to term. Serum RT_3 levels probably increase in parallel and reach high concentrations at term. Serum T_3 values increase much later, and concentrations in cord serum are low. The pattern of high serum RT_3 and low T_3 concentrations in the fetus appears to reflect decreased monodeiodination of T_4 to T_3 and relatively increased monodeiodination of T_4 to RT_3 in peripheral tissues.

During the neonatal period there is a marked increase in serum TSH, peaking at 30 minutes (Fig 20). Serum T_3 concentrations also increase with an early

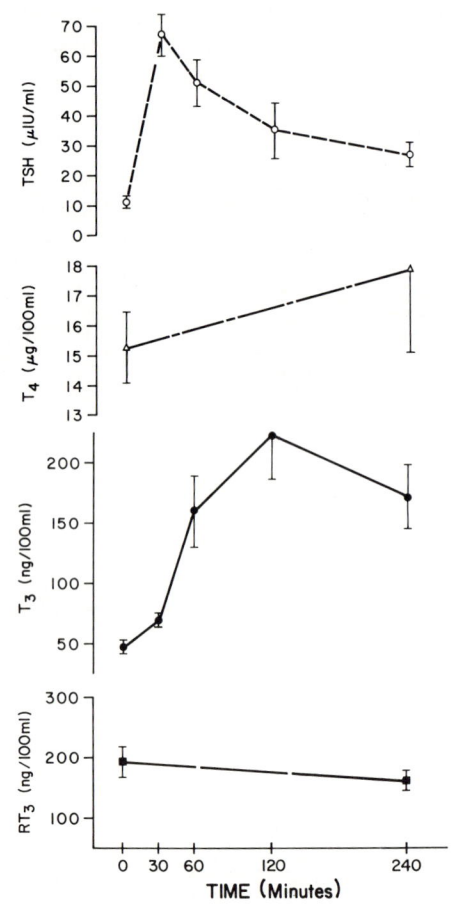

FIG. 20. The changes in serum thyrotropin (TSH) and thyroid hormone concentrations in the newborn period. These data are derived from 8 newborn infants. Values are plotted as mean and SEM. The early TSH surge is followed by increases in serum triiodothyronine (T_3) and thyroxine (T_4) concentrations. Serum reverse T_3 (RT_3) levels remain unchanged or fall slightly.

peak at 120 minutes. Serum T_4 levels increase more gradually and peak at 24 hours in association with a second peak of serum T_3 (Fisher and Odell, 1969; Erenberg et al., 1974; Abuid et al., 1974; Czernichow et al., 1971). Serum reverse T_3 (RT_3) concentrations, already high do not increase further in the neonatal period (Chopra et al., 1975a). Data derived from warming studies in the newborn period have suggested that the neonatal TSH surge may be, at least in part, a response to neonatal cooling (Fisher et al., 1966; Fisher and Odell, 1969). The increase in serum T_4 concentrations during the first 24 hours must occur in response to the TSH surge.

The increase in serum T_3 levels, however, seems to be only partly accountable on the basis of increased T_3 secretion. As noted in Table III, an increase in serum T_3 concentrations occurs within the first 4 hours in spite of marked inhibition of the TSH surge. In the newborn sheep, the TSH surge and the T_3 surge can be dissociated in time by delayed umbilical cord cutting (Sack et al., 1976). Exposure to the extrauterine environment seems to stimulate, in addition to a TSH surge, an increase in the rate of production of T_3 from nonthyroidal sources, presumably via monodeiodination of T_4 in peripheral tissues. Serum T_3 levels do not fall again to fetal levels after birth. Although levels fall somewhat after the 24-hour peak, they equilibrate within a few days to concentrations characteristic of infancy. Serum RT_3 concentrations, initially high, fall within 5–7 days to infant levels. Serum T_3/T_4, RT_3/T_4 and T_3/RT_3 concentration ratios in the fetus and newborn are summarized in Table V between 30 weeks of gestation and 1 month of postnatal life. These data suggest that the T_4 β-ring monodeiodinative pathway(s) in fetal tissues begins to mature *in utero*. In addition, exposure to the extrauterine environment in some way further stimulates this maturation. By 1 month of postnatal life, serum T_3/T_4 and RT_3/T_4 concentration ratios approximate adult values.

TABLE V
Maturation of Tissue Thyroid Hormone Metabolic Systems in the Human Fetus and Newborn[a]

Age	Serum T_3/T_4[b]	Serum RT_3/T_4[c]	Serum T_3/RT_3[d]
30 Weeks of pregnancy	0.002	–	–
Term	0.004	0.014	0.30
1 Month postnatal	0.015	0.004	4.0
Adult	0.016	0.005	3.2

[a] T_3, triiodothyronine; T_4, thyroxine; RT_3, reverse T_3.
[b] Ng/dl per μg/dl.
[c] Ng/dl per μg/dl.
[d] Ng/dl per ng/dl.

C. DEVELOPMENT OF THYROID FUNCTION IN SHEEP

1. Embryogenesis

a. The Thyroid Gland. Barnes and colleagues (1957) studied the development of the thyroid gland in fetal sheep and correlated histological development with function. At 46–48 days of gestation (term is 150 days) the gland is well formed. Histologically there are no visible follicles on day 50. By day 52 small but distinct follicles are visible, and between 50 and 60 days these are limited in number and largely confined to the periphery of the gland. By 70 days the gland resembles the adult gland with extensive colloid filled follicles. Radioautographs of fetal thyroid tissue after administration of radioiodine to the ewe show no accumulation at 46–48 days. On day 50 radioiodine is uniformly distributed through the gland parenchyma. By 52 days focal accumulation reflects the early follicle organization, and this follicular concentrating pattern progresses through 70 days. We have shown that the 70-day fetal ovine thyroid contains T_4 and T_3 by radioimmunoassay.

b. The Pituitary Gland. Alexander et al. (1973) investigated development of the adenohypophysis of fetal sheep, correlating ultrastructure with functional activity. Thyrotrophs, gonadotrophs, and somatotrophs were detectable at 54 days. Prolactin cells were identified at 75 days, and corticotrophs at 95 days. Chromophobe cells predominated at all ages. TSH could be identified in serum at 50 days (Alexander *et al.,* 1973; Thorburn and Hopkins, 1973).

2. Maturation of the Hypothalamus

The Hypothalamus and Portal System. There is little histological information regarding development of the sheep hypothalamus or the pituitary–portal blood vascular system. We have assessed fetal ovine hypothalamic maturation using somatostatin (SRIF) as a marker. SRIF was measured by radioimmunoassay by Drs. Wylie Vale and Marvin Brown of The Salk Institute in La Jolla, California. Tissue was extracted in hot acetic acid (Vale *et al*, 1976). The radioimmunoassay was run with iodinated tyrosine-1-somatostatin using either antiserum A 101 supplied by Dr. Arimura or R 149 supplied by Drs. Reichlin and Patel. This method is not reliable for extraction of TRH, so that this information is not yet available. The SRIF results are shown in Fig. 21. The data are sparse but suggest that SRIF concentrations in fetal hypothalamic tissue increase from low levels at mid-gestation to near term concentrations that approximate adult values.

3. Development of Neuroendocrine Control

a. Pituitary and Serum TSH Concentrations. Pituitary TSH content in fetal sheep between 50 and 132 days of gestation is shown in Fig 21. TSH was

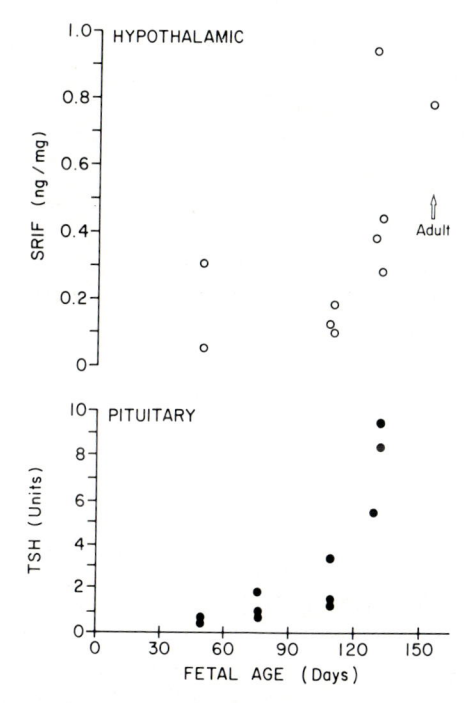

FIG. 21. Hypothalamic somatostatin (SRIF) and pituitary thyrotropin (TSH) in fetal sheep. SRIF and TSH were measured by radioimmunoassay. There is a progressive increase in both hypothalamic SRIF concentration and pituitary TSH content with gestational age.

measured by radioimmunoassay using an antiserum generated against bovine TSH. Bovine TSH β chain was labeled to eliminate LH cross-reaction. NIH ovine TSH (S-6) was used as reference standard. There is a progressive increase in pituitary TSH content (in units) with increasing gestational age, as expected. Pituitary TSH concentration showed no correlation with gestational age varying from 100 to 490 μU/mg wet tissue.

Serum TSH concentrations have been reported for fetal sheep between 50 and 150 days by Thorburn and Hopkins (1973). They observed mean concentrations approximating 3 ng/ml throughout this period. We have measured a mean serum TSH concentration of 5.4 μU/ml (NIH-S6) between 110 and 130 days. It has not been possible to date to show an increase in serum TSH in the fetal sheep comparable to that measured in the human fetus and in the rat.

Earlier data suggest parallel development of TRH and SRIF concentrations in human fetal hypothalamus (Kaplan et al., 1976). Thus the present incomplete data are compatible with the view that fetal hypothalamic development in the sheep occurs between 60 days and term (150 days). The approximately 50% fall in fetal serum T_4 and TSH concentrations after pituitary stalk section of the fetal lamb during the last trimester would support this conclusion (Thorburn and Hopkins, 1973). However, the availability of further information will allow more precise timing of this process.

b. Serum T_4 and Negative Feedback. In the sheep fetus as in the human fetus there is a marked and progressive increase in serum T_4 concentrations beginning at mid-gestation (Fig. 22). Serum T_4 levels increase from concentrations of about 3 µg/dl at 73–77 days to mean values approximating 10 µg/dl at term. Thorburn and Hopkins (1973) and Nathanielsz *et al.* (1973a) have reported similar results.

Data regarding maturation of the TRH response and of negative feedback control in the sheep are sparse. Thorburn and Hopkins (1973) and Erenberg *et al.* (1973) have shown that the sheep fetus during the last trimester responds to thyroidectomy with a rapid decrease in serum T_4 (Fig. 2) and a marked increase in serum TSH concentrations. Hypophysectomy during the same period results in rapid decreases in both serum TSH and T_4 levels (Thorburn and Hopkins, 1973). Administration of exogenous T_4 to the fetal sheep results in suppression

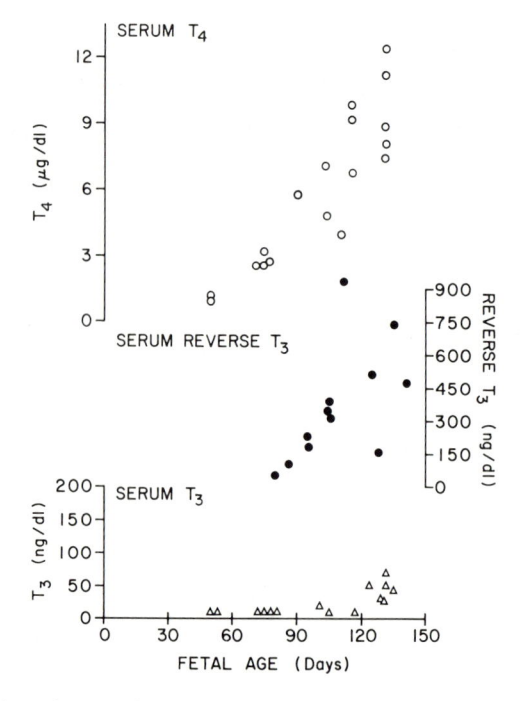

FIG. 22. Maturation of serum thyroid hormone concentrations in the fetal sheep. Serum throxine (T_4), reverse triiodothyronine (T_3), and T_3 are plotted versus gestational age in days.

of fetal serum TSH concentrations (Hopkins, 1972), whereas administration of methylthiouracil to the pregnant ewe will provoke fetal goiter (Lascelles and Setchell, 1959) during the last trimester. Finally, the administration of TRH to the catheterized near-term sheep fetus will provke both TSH and prolactin secretion by the fetal pituitary (Thomas *et al.*, 1975). These data substantiate that TRH responsiveness and negative feedback control do in fact exist during the last trimester of pregnancy in the fetal sheep, but data regarding timing of maturation of these responses are not available.

4. Maturation of Tissue Metabolism

This phase of development of thyroid function has been most completely mapped in the ovine species. The parallel increases in fetal serum T_4 and reverse T_3 (RT_3) concentrations are shown in Fig. 22. Serum T_3 concentrations, by contrast, increase only after 120 days and remain quite low at term. Studies of production rates of T_4, and RT_3, and T_3 in maternal and fetal sheep have been conducted using noncompartmental kinetic methods between 115 and 130 days of gestation (Dussault *et al.*, 1971, 1972; Fisher *et al.*, 1972; Chopra *et al.*, 1975b). These results, summarized in Table VI, have shown that the sheep fetus is producing 2–3 times more T_4 on a microgram per square meter of body surface area basis than are maternal sheep. Nathanielsz *et al.* (1973a) have reported a similar difference. In addition, fetal RT_3 production exceeds T_3 production 4 or more times. Nearly all the T_3 produced is accountable on the basis of thyroid gland secretion, whereas less than 5% of the RT_3 is secreted. T_4 in the fetus is metabolized predominantly via alpha ring monodeiodination in peripheral tissues; β-ring monodeiodination is minimal.

There is a marked increase in serum T_3 concentrations in the newborn sheep (Sack *et al.*, 1976; Nathanielsz *et al.*, 1973b); within 4 hours of birth serum T_3 increases from values approximating 50 ng/dl to levels in excess of 200 ng/dl (Fig. 23). As in the human newborn these changes are preceded by a TSH surge

TABLE VI
Serum Levels and Production Rates of Thyroid Hormones in Maternal and Fetal Sheep[a]

Sheep	T_4 Conc. (μg/dl)	T_4 Prod. (μg/M² /d)	T_3 Conc. (ng/dl)	T_3 Prod. (μg/M² /d)	Reverse T_3 Conc. (ng/dl)	Reverse T_3 Prod. (μg/M² /d)
Maternal	5.7	146	88	62	61	38
	(0.65)	(30)	(23)	(34)	(16)	(5)
Fetal	8.9	335	<32	<27	560	102
	(0.98)	(15)	(9)	(8)	(92)	(8)

[a]Mean and (SEM) values for four animals. T_4, thyroxine; T_3, triiodothyronine.

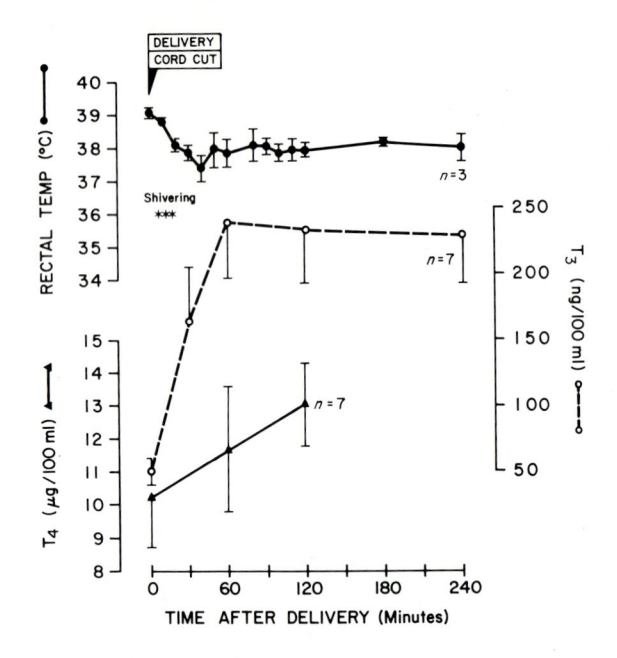

FIG. 23. Rectal temperature, serum triiodothyronine (T_3) and serum thyroxine (T_4) responses to parturition in eight newborn sheep. The transient fall in rectal temperature is associated with a marked rise in serum T_3 and a more modest increase in T_4 concentrations during the first 4 hours. From Sack et al. (1976).

(Fig. 24). However, the extent of the TSH increase (2–3 fold) is less in the sheep than in the human infant, where 8- to 12-fold increases in serum TSH are observed. The neonatal TSH surge in the sheep as in the human infant appears to be provoked by extrauterine cooling. The TSH surge and the increases in serum thyroid hormone concentrations can be pervented by delivering the newborn into a 37°–39°C water bath (Fig. 25 and 26). In this circumstance the umbilical cord remains intact.

As shown in Fig. 27, when umbilical cord cutting was delayed to control for noncutting in the water bath experiments, we observed that increases in serum T_3 and T_4 also were delayed. Delaying umbilical cord cutting did not alter timing of the TSH surge (Fig. 28). We could dissociate the T_3 and TSH responses by delaying cutting of the umbilical cord (Fig 28). The fact that serum T_4 concentrations did not increase during the period of delayed cord cutting (Fig. 27) probably relates to the small number of animals and the variability of this response over the short term. Figure 29 shows that RT_3 levels remained unchanged whether the umbilical cord was cut early or late.

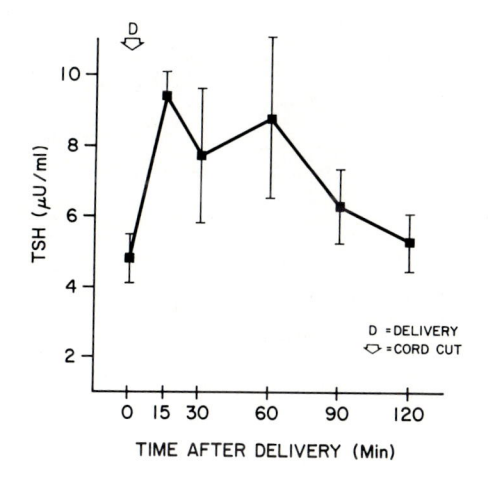

FIG. 24. The serum thyrotropin (TSH) response to parturition in the newborn sheep. Serum TSH concentrations increase within 15 minutes after cutting the umbilical cord and exposure to the extrauterine environment.

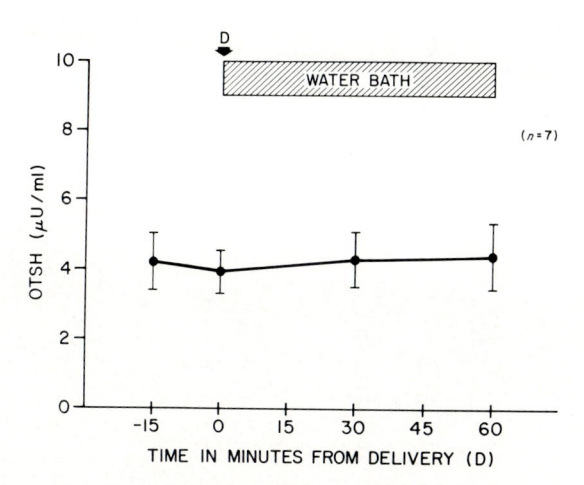

FIG. 25. Serum ovine thyrotropin (OTSH) in seven newborn sheep delivered into a 37°–39°C water bath. The umbilical cord is not cut.

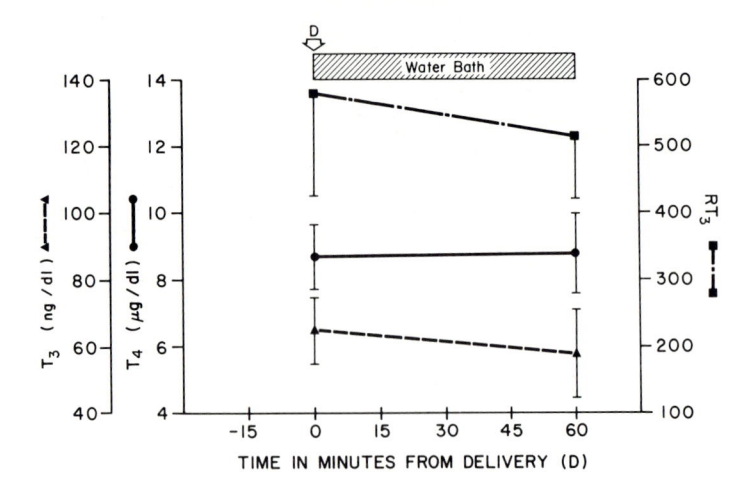

FIG. 26. Serum thyroxine (T_4), triiodothyronine (T_3), and reverse T_3 (RT_3) concentrations in seven newborn sheep delivered into a 37°–39°C water bath. The umbilical cord is not cut.

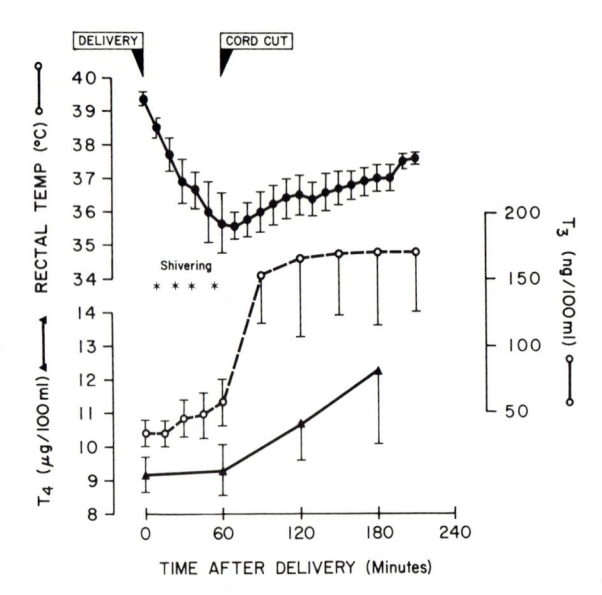

FIG. 27. Rectal temperature, serum triiodothyronine (T_3) and serum thyroxine (T_4) responses to parturition and delayed umbilical cord cutting in four newborn sheep. With delayed cord cutting, hypothermia is marked and the increments in serum T_3 and T_4 are delayed until the time of cord cutting at 60 minutes. From Sack *et al.* (1976).

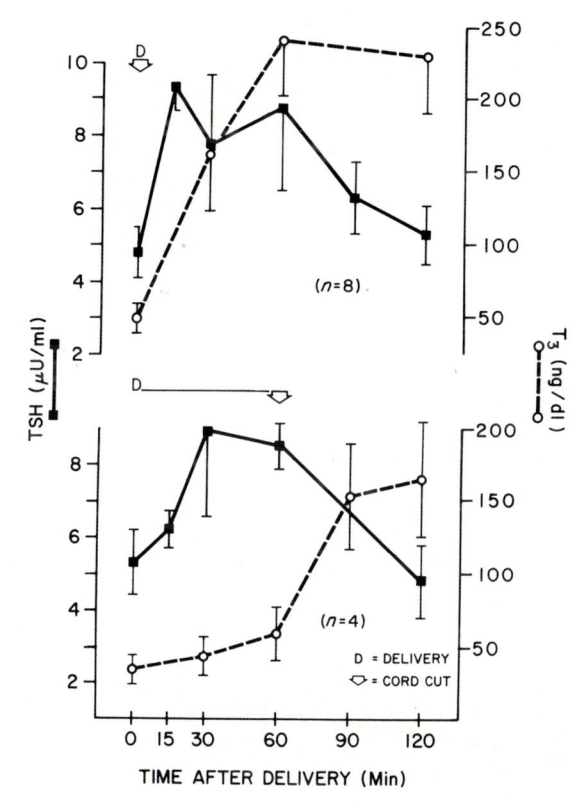

FIG. 28. Serum thyrotropin (TSH) and triiodothyronine (T$_3$) responses in newborn sheep delivered with immediate and (60 minute) delayed umbilical cord cutting Delayed cord cutting delays the T$_3$ surge without altering the timing of the TSH surge.

These studies suggested that the TSH surge was not the important determinant of the early rise in neonatal serum T$_3$ concentrations. Since plasma free fatty acid (FFA) concentrations (measured to reflect the catecholamine response) in the newborn tend to parallel the T$_3$ response (Sack et al., 1976), and since it has recently been suggested that tyrosine hydroxylase (TH), the rate-limiting enzyme in catecholamine biosynthesis, might function as a T$_4$ β-ring monodeiodinase (Dratman, 1974; Dratman et al., 1976), we delivered newborn sheep after premedication with α-methyl-p-tyrosine (MPT), a potent competitive inhibitor of TH (Melmon, 1974). Five fetuses (135–145 days of gestation) with chronic indwelling jugular vein catheters were infused over a 30-minute period with 600 mg to 1 gm of MPT (estimated 200 mg of DL-α-methyl-p-tyrosine methyl ester per kilogram; Aldrich Chemical Co.). The fetuses then were de-

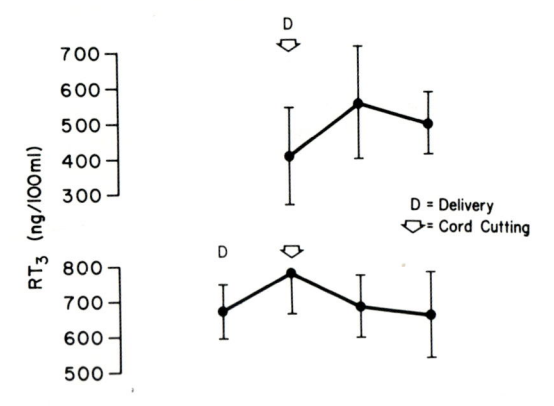

FIG. 29. Serum reverse triiodothyronine (RT$_3$) concentrations in newborn lambs delivered with early and delayed (60 minute) cord cutting. There were no significant short-term changes in serum RT$_3$ levels during these studies.

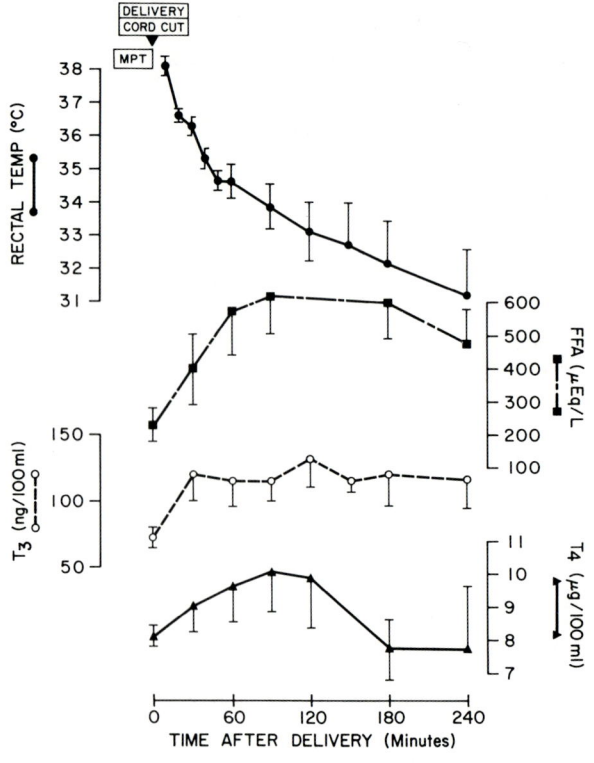

FIG. 30. Rectal temperature and serum free fatty acids (FFA), triiodothyronine (T$_3$), and thyroxine (T$_4$) concentrations in five newborn lambs delivered after premedication with α-methyl-tyrosine (MPT). The umbilical cord was cut at the time of delivery.

livered by cesarean section, and serial measurements of rectal temperature and serum concentrations of TSH, FFA, T_3, T_4, and RT_3 were conducted. Results are shown in Fig. 30. As can be observed. the T_3 surge was markedly inhibited (compare with Fig. 28), whereas the TSH surge (Fig. 31) was essentially similar to that in nonpremedicated animals (Fig 24). The modest increase in T_4 presumably is due to the TSH surge. The FFA response probably is due to the fact that stored catecholamines were not depleted during the 30 minutes of MPT infusion. Reverse T_3 concentrations were not altered by the MPT infusion (Fig. 32).

As a control for the MPT infusion, we conducted a second series of studies premedicating four fetuses (135–145 days gestation) with an estimated 300 mg/kg α-methyl-*m*-tyrosine (MMT; ICN Pharmaceutical Co.). This compound is a noncompetitive inhibitor of the second (L-amino acid decarboxylase) step in the catecholamine synthetic pathway. Its administration provided procedural con-

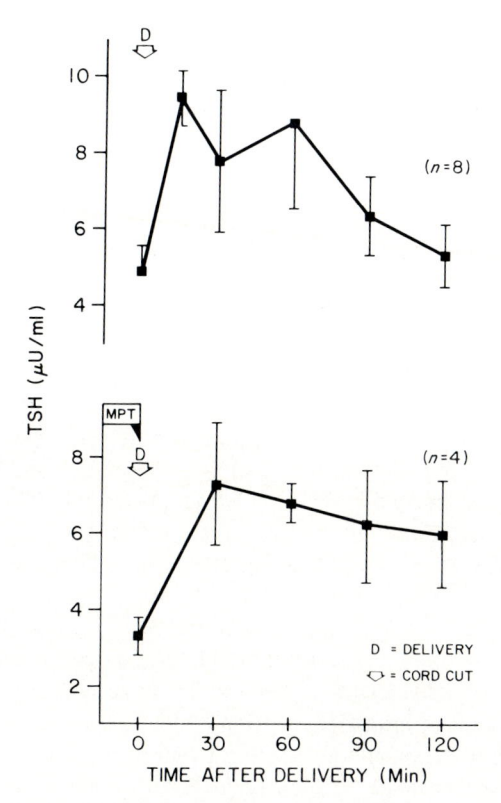

FIG. 31. The serum thyrotropin (TSH) response in eight normal and four α-methyl-tyrosine (MPT) premedicated newborn lambs. The responses are statistically similar.

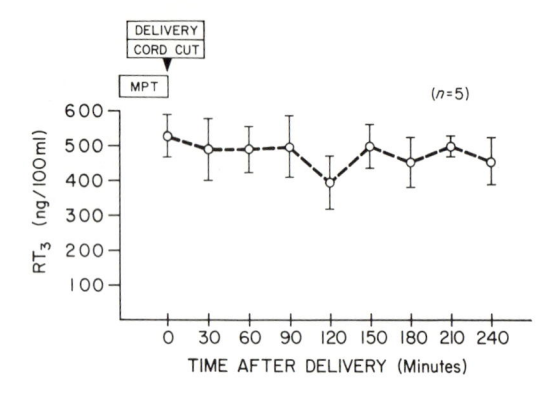

FIG. 32. Reverse triiodothyronine (RT_3) concentrations in five lambs delivered after premedication with α-methyl-tyrosine (MPT).

trol for the MPT experiments. Measurements were carried out as before, with results shown in Fig. 33. In these animals there was a marked hypothermia, although less marked than with MPT premedication. The FFA, T_3, and T_4 responses, however were similar to those of control, nonpremedicated animals. The TSH and RT_3 responses (not shown) also were similar to those of control animals.

These results suggest the possibility that TH in the newborn can act as a T_4 β-ring monodeiodinase. This role for the enzyme might be unique to the newborn because of the large mass of extramedullary, TH containing, chromaffin tissue and the intense stimulation of the enzyme coincident with umbilical cord cutting and extrauterine exposure. The mechanism by which umbilical cord cutting stimulates catecholamine synthesis and release is not clear. The stimulation is very rapid and cannot involve synthesis of enzyme protein. Several factors are known to influence TH activity and/or substrate (tyrosine and oxygen) availability. Oxygen also increases the affinity of the enzyme protein—pteridine cofactor interaction (Melmon, 1974). The marked increase in pO_2 occasioned by cutting the umbilical cord and stimulating respiration may be an important stimulus to the adrenergic system in the newborn (Sack et al., 1976).

In the newborn sheep as in the human neonate, parturition is associated with a marked increase in serum T_3 concentrations and this T_3 surge can be dissociated from the TSH surge. This suggests that the T_3 surge is accountable, to large extent, by stimulation of monodeiodination of T_4 to T_3. The rate of T_4 to RT_3 conversion probably is not altered, since RT_3 levels are stable for several days after birth. The mechanism of the increase in T_4 to T_3 conversion is not entirely clear. The MPT and MMT experiments suggest that TH might play a role in the newborn as a T_4 β-ring monodeiodinase. Other factors, however, could be

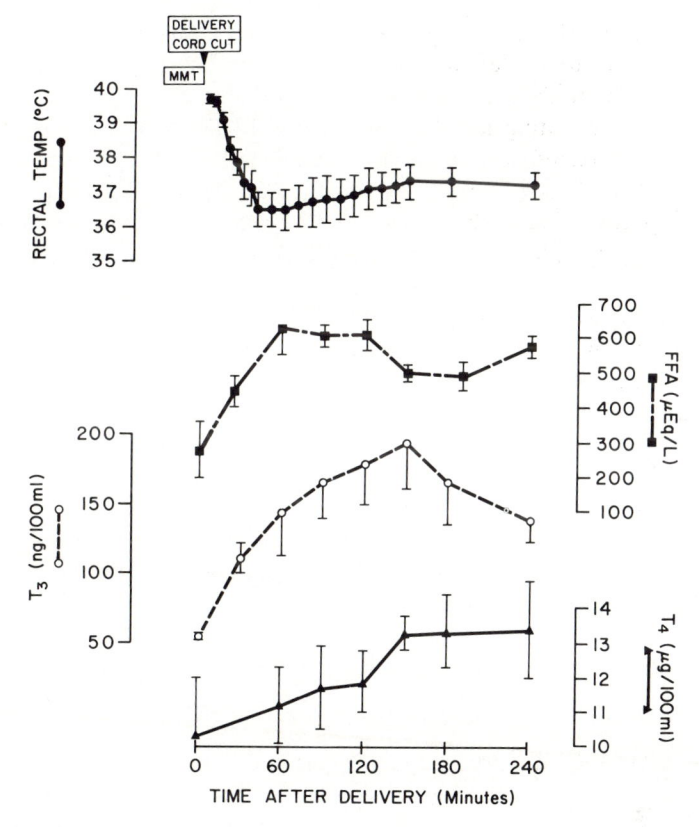

FIG. 33. Rectal temperature and serum free fatty acids (FFA), triiodothyronine (T_3), and thyroxine (T_4) concentrations in four newborn lambs delivered after premedication with α-methyl-tyrosine (MMT). The umbilical cord was cut at the time of delivery.

involved. Whatever the mechanism(s), the maturation of the peripheral systems for T_4 monodeiodination to the adult state appears to occur in the sheep between 120 days of gestation and 1–2 weeks postpartum at which time serum T_3/T_4, and T_3/RT_3 concentration ratios approximate adult values.

D. MATURATION OF THYROID FUNCTION IN THE RAT

1. Embryogenesis

a. The Thyroid Gland. In the rat embryogenesis extends throughout the period of intrauterine development (23 days). The thyroid gland is well formed and in the adult location in the neck by 17 days. Between 17 and 19 days, using fluorescent thyroglobulin labeling techniques, early accumulations of this protein are visible within and between cells (Feldman et al., 1961). This period would correspond to the late precolloid—early colloid stages of human thyroid embryogensis. By 20 days thyroglobulin is identifiable in extracellular follicles. Radioiodine concentrating ability is observed by 17 days, and there is a progressive increase in radioiodine content and concentration between 17 and 20 days. The 20-day thyroid can synthesize T_3 and T_4 (Shepard. 1967). As in the human or the sheep fetus, radioiodine concentrating ability and hormone synthetic capacity correlate with the period of follicular organization (Feldman et al., 1961).

b. *The Pituitary Gland.* There is parallel development of the pituitary gland. The buccal component (Rathke's pouch) is visible by 14 days in intimate contact with the diencephalon. The neurohypophysial component also is visible, albeit smaller, at this time (Glydon, 1957). The pituitary gland, like the thyroid gland, is well developed by 17 days (Glydon 1957) and secretory granules are visible at this time; agranular cells predominate (Fink and Smith, 1971). By 18 days Rathke's pouch and stalk have disappeared and two types of granules (by size) can be distinguished. At 19 days the floor of the sella turcica is closed with cartilage and bone. Between this period and 6 days of postnatal age, there is a progressive increase in number of granular cells as well as numbers of granules per cell in the adenohypophysis. The posterior pituitary develops in parallel and is of adult shape with detectable granularity by 18 days (Fink and Smith, 1971).

2. *Maturation of the Hypothalamus*

a. *The Hypothalamus.* The hypothalamus and median eminence are easily identifiable by 14—15 days of gestation. At this time the median eminence is devoid of cells. Identifiable axon profiles make their appearance by 16 days (Monroe and Paull, 1974). By 17—18 days nuclear condensations are visible in the hypothalamus and acetylcholinesterase can be detected (Pilgrim. 1974). Granular vesicles suggestive of neurosecretory material are visible at this time (Fink and Smith, 1971). Also, at 17—18 days there is a visible thickening of the ventral zone of the median eminence due to an ingrowth of axons (Monroe and Paull, 1974).

By the first postnatal day the number of granular vesicles per fiber in the median eminence has increased considerably, and there is a palisade appearance of the glial and nerve fibers that begins to resemble the adult external layer. Gomori-positive neurosecretory material also is identifiable at this time (Pilgrim, 1974). Monoamine fluorescence is observed in the median eminence during the first postnatal week (Fink and Smith. 1971). There is continuing histological development of the hypothalamus and median eminence during the second week, after which time development is largely complete by morphological criteria (Pilgrim, 1974). The typical adult enzyme pattern is not complete until 4 weeks (Pilgrim, 1974).

TRH is detectable in the hypothalamus at birth in low concentrations (1 pg/μg protein), and increases markedly during the first 2 weeks (Fig 34) (Dussault and Labrie, 1975). Concentrations between 16 and 28 days approximate 50 pg/μg protein, fall transiently at 40 days, and increase again to the adult level.

b. The Pituitary Portal System. Maturation of the neurovascular communication system between the hypothalamus and pituitary is a progressive process continuing throughout the period of hypothalamic maturation. By 17 days of gestation the pituitary gland is surrounded by a capillary network and separated from the brain by a thin layer of mesenchyme and capillaries; this is referred to as the supratuberal plexus, or the secondary plexus of the portal system. Capillaries are present along the ventral surface of the median eminence very early. These have the characteristics of primary portal vessels with an attenuated, fenestrated endothelium, a basal lamina, and a surrounding perivascular space containing connective tissue elements (Monroe and Paull, 1974). During late fetal life, although hypophysial portal vessels are visible, the primary plexus remains superficial. Capillary loops are not seen to penetrate the median eminence until late in the first postnatal week. Thereafter until 5–6 weeks there

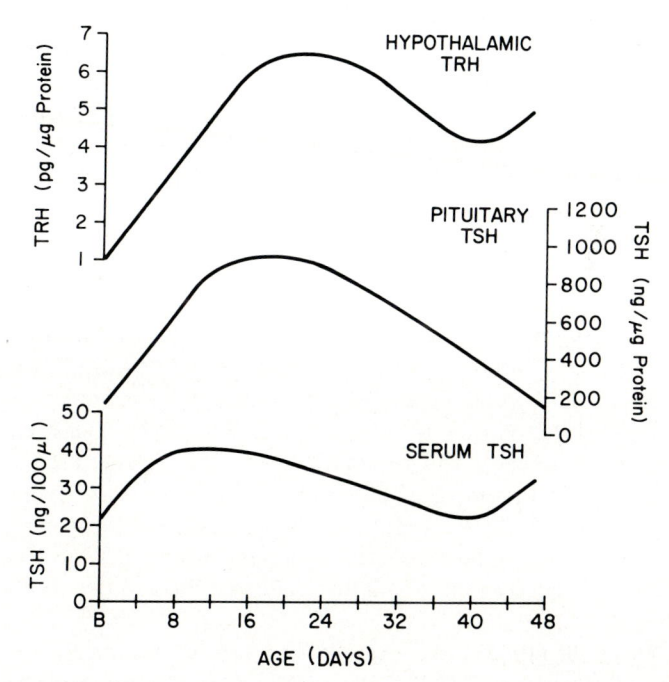

FIG. 34. Maturation of hypothalamic–pituitary function in the newborn rat. Hypothalamic thyrotropin-releasing hormone (TRH), and pituitary and serum thyrotropin (TSH) were measured by specific radioimmunoassay. From Dussault and Labrie (1975).

is a progressive increase in the number and tortuosity of the loops and a progressive increase in the mass of median eminence portal vessels.

The difference between the fetus or early newborn and the adult would seem to be a quantitative one; there are parallel progressive increases in numbers of axon terminals and volume of vesicular inclusions which parallel the increase in complexity and mass of the vascular network.

3. *Development of Neuroendocrine Control*

a. Pituitary and Serum TSH. We have measured pituitary and serum TSH in the infant rat (Dussault and Labrie 1975) and the pattern of change is shown diagrammatically in Fig. 34. Mean pituitary TSH concentration is low at birth (approximately 170 ng/μg protein) and increases rapidly to peak concentrations approximating 1000 ng per microgram of protein by 10–12 days. The mean serum TSH concentration already approximates 20 ng/100 μl at birth but continues to increase to nearly 40 ng/100 μl by 7–8 days. Both pituitary and serum TSH concentrations fall progressively between 14–16 and 40 days.

b. Serum T_4 and Negative Feedback. Serum T_4 concentrations in the newborn rat are low (< 1 μg/dl) and decrease transiently in association with extrauterine adaptation (Dussault and Labrie. 1975; Kieffer *et al.*, 1976). After a nadir at 2 days, serum T_4 concentrations increase rapidly to peak levels approximating 6 μg/dl at 15–20 days and fall during the subsequent 4 weeks to adult levels (Fig. 35). Free T_4 concentrations increase in parallel from a mean value of 0.79 ng/dl at birth to levels of 2.55–2.56 ng/dl at 14–22 days (Table VII). Values fall somewhat thereafter (to 2.23 ng/dl at 26 days), increase transiently at 32 days, and then fall to the adult level by 40 days.

Figure 36 shows the serum free T_4 (FT_4)/TSH concentration ratio plotted versus postnatal age in days. The ratio increases rapidly to the adult level by 15–20 days.

These data suggest that, in the rat as in the human, parallel maturation of the hypothalamus and pituitary portal vascular system is associated with an increase in activity of the thyroid gland. This presumably is due to a progressive increase in serum TSH concentration, although thyroidal TSH responsiveness has not been measured serially. The thyroid stimulation is followed by a progressive increase in serum T_4 and FT_4 concentrations that continues beyond the period of plateau in serum TSH concentrations. Moreover, the increase in serum T_4 and FT_4 relatively exceed the increment in serum TSH so that the FT_4/TSH ratio increases progressively reflecting a maturation of the negative feedback control system.

Krulich *et al.* (1976) have carefully studied the postnatal development of pituitary–thyroid control in the infant rat by measuring the serum TSH response to exogenous TRH, the serum TSH response to exogenous T_3, and the TSH response to methimazole blockade. They observed progressive change in these

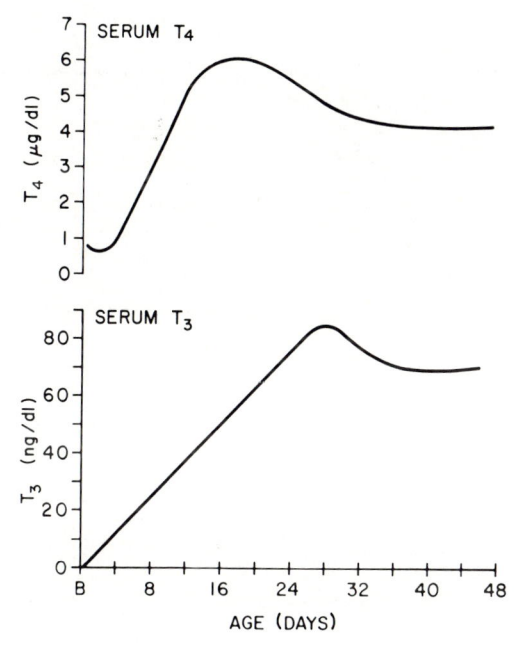

FIG. 35. Maturation of thyroid function in the infant rat. Serum thyroxine (T_4) and triiodothyronine (T_3) concentrations are displayed graphically versus postnatal age in days. From Dussault and Labrie (1975).

TABLE VII
Free Thyroxine and Free Triiodothyronine Concentrations in Infant Rats during the First 40 Days of Life[a]

	Free thyroxine		Free triiodothyronine	
Age in days	% Dialyzable	Ng/dl	% Dialyzable	Ng/dl
5	0.057	0.79	0.22	16
7	0.062	1.25	0.23	42
12	0.068	2.47	0.28	154
14	0.045	2.55	0.28	152
22	0.061	2.56	0.30	271
26	0.046	2.23	0.29	256
32	0.056	2.87	0.27	237
40	0.039	1.81	0.25	206
Adult	0.040	2.21	0.23	208

[a]Each value represents the mean of 6–8 separate animals.

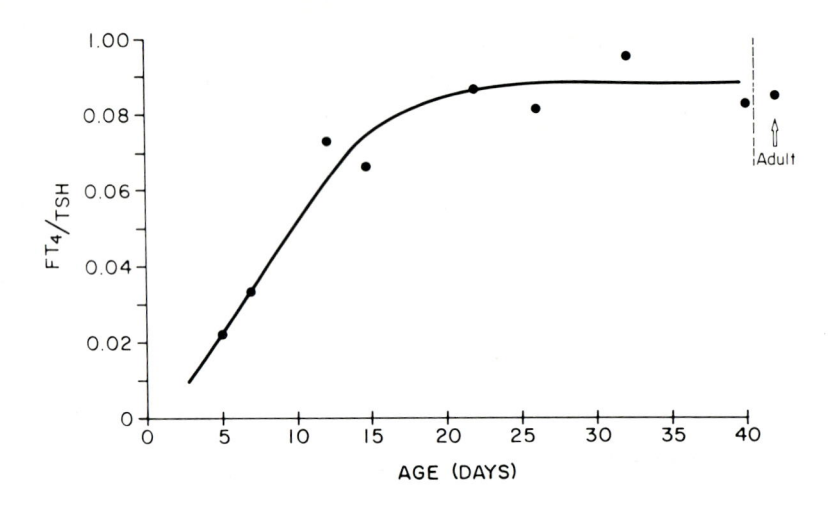

FIG. 36. Maturation of concentration ratio of serum free thyroxine to thyrotropin (FT$_4$'/TSH) in the infant rat. Values (ng/dl per ng/100 μl) are plotted versus postnatal age in days.

responses between 3 and 17 days of age as summarized in Table VIII. Their results indicate a progressive maturation of pituitary sensitivity to TRH and a progressive maturation of negative-feedback sensitivity to thyroid hormones during the first 2 weeks of postnatal life.

4. Maturation of Tissue Metabolism

The pattern of change in serum T$_3$ concentration in the infant rat is shown in Fig. 35. T$_3$ is not measurable at the time of birth and increases progressively during the first month to a mean peak concentration approximating 90 ng/dl.

TABLE VIII
Maturation of Neuroendocrine Control of Thyroid Function in Infant Rats[a,b]

Changes between 3 and 25 days
1. TSH response to TRH—↓ progressively
2. T3 suppression of TSH—↑ sensitivity
3. TSH response to methimazole—↑ progressively

[a]From Krulich et al. (1976).
[b]TSH, thyrotropin; T$_3$, triiodothyronine; TRH, thyrotropin releasing hormone.

Values fall thereafter to a mean of 70–80 ng/dl at 40 and 48 days. Free T_3 concentrations in serum increase in parallel, from a mean value of 16 pg/dl at birth to a peak level of 271 pg/dl at 22 days. There is a subsequent fall to the adult concentration by 40 days (Table VII). The rate of increase in serum T_3 and free T_3 concentrations is much less than the rate of increase in serum T_4 levels in the infant rat.

In order to more carefully assess the significance of the changes in serum T_4 and T_3 concentrations in the infant rat we have conducted measurements of T_4 and T_3 production rates. Kinetic measurements of metabolic clearance were made using radiolabeled T_4 in 6–8 animals for each hormone at ages 5, 7, 12, 14, 22, 26, 32, and 40 days and in adults. Similar studies were conducted using radiolabeled T_3 at ages 5, 12, 22, 26, 32, and 40 days and in adults. Serum concentrations of T_4 or T_3 were measured by radioimmunoassay at the time of the kinetic studies. Production rates were calculated as the product of metabolic clearance rate (ml/hr per 100 gm) and serum hormone concentration for each animal.

Mean and SEM results are shown in Figs. 37 and 38. T_4 production increased from a mean of 16.6 ± 1.37 ng/hr per 100 gm at 5 days to 72.0 ± 5.51 ng/hr per 100 gm at 14 days, and fell gradually thereafter to the adult level of 37.7 ng/hr per 100 gm (Fig. 37). Triiodothyronine production increased from 0.93 ± 0.10 ng/hr per 100 gm at 5 days to 12.6 ± 1.04 ng/hr per 100 gm at 26 days. Mean values approximated those of the adult by 32 days (Fig. 38).

These results indicate that the changes in serum T_4 and T_3 concentration shown in Fig. 35 do in fact represent progressive increases in hormone produc-

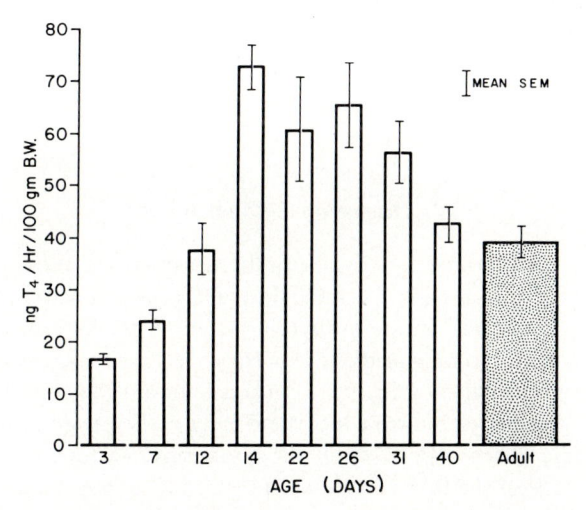

FIG. 37. Thyroxine (T_4) production rates in rats at different times after birth. T_4 production rates, per 100 gm body weight are plotted as mean and SEM.

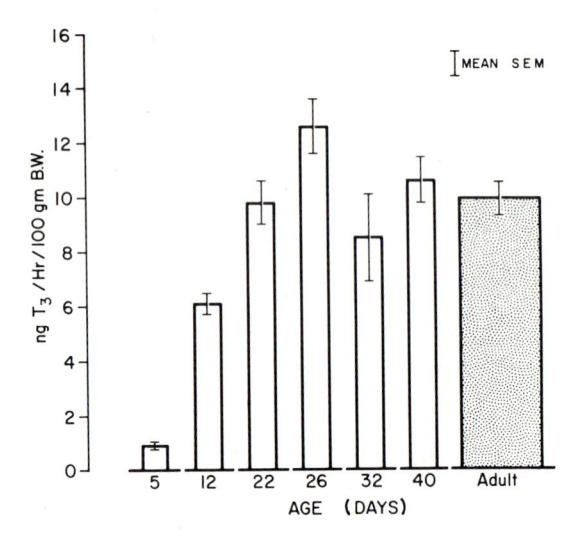

FIG. 38. Triiodothyronine (T_3) production rates in rats at different times after birth. T_3 production rates, per 100 body weight are plotted as mean and SEM.

tion. They also indicate that the fetal and newborn rat, like the fetal human and sheep is relatively T_3 deficient, presumably because of a relatively low rate of production of T_3 in nonthyroidal tissues. The production rate studies do not differentiate, however, between thyroidal and tissue T_3 production. Figures 39 and 40 show T_4/T_3 serum concentration ratios and T_4/T_3 production ratios plotted versus postnatal age in days. Both plots confirm the early T_3 deficiency and suggest that this is rapidly corrected during the first 25–30 days. There may be some continuing maturation of the T_4 tissue metabolic system(s) throughout the first 40 days of postnatal life.

IV. Discussion and Conclusions

The results of our studies and those of others as discussed in the present paper can be summarized as in Table IX. As indicated, the general pattern of ontogenesis of the hypothalamic– pituitary–thyroid system in the human, sheep, and rat species is similar. The timing of the several stages relative to parturition differs in the rat. Human and sheep fetuses are delivered toward the end of the stage of development of neuroendocrine control and early in the period of maturation of tissue metabolic systems. In contrast, the rat is delivered soon after completion of embryogenesis of the thyroid and pituitary glands early in stage II maturation of the hypothalamus. The newborn rat provides a convenient model for study of hypothalamic–pituitary–thyroid maturation since it is readily accessible during this time. The sheep serves as a model for third trimester and neonatal thyroid

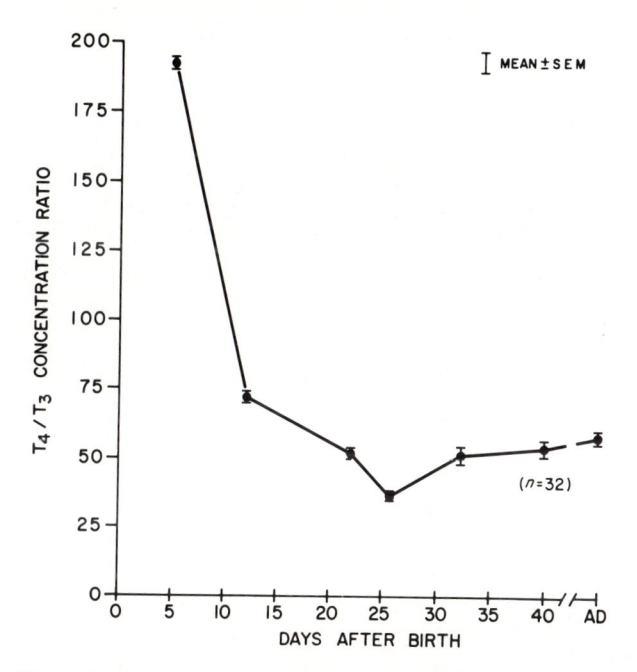

FIG. 39. Thyroxine to triiodothyronine (T_4/T_3) concentration ratios plotted versus postnatal age in days. Values are plotted as mean and SEM. Each point represents 32 measurements.

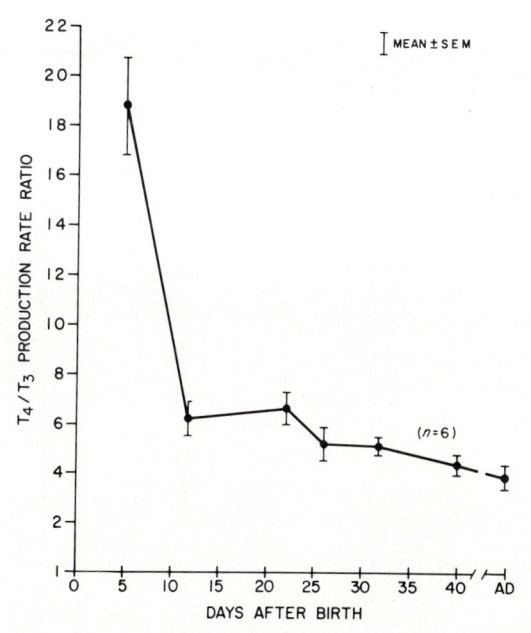

FIG. 40. Thyroxine to triiodothyronine (T_4/T_3) production ratios plotted versus postnatal age in days. Values are plotted as mean and SEM. Each point represents 6 measurements.

TABLE IX
Ontogenesis of Thyroid Function in Man, Sheep, and Rat[a]

Species	Embryogenesis	Maturation of hypothalamus	Development of neuroendocrine control	Maturation of peripheral metabolism
	a. Development of thyroid to mature follicles b. Development of pituitary incl. hormone synthesis and 2° portal plexus	a. Histological maturation b. Increase in TRH concentration c. Maturation of the primary portal plexus	a. Increase in pituitary and serum TSH b. Maturation of negative feedback c. Maturation of TRH response	a. Increase in serum T_3 concentration b. Decrease in serum reverse T_3 (RT3) level c. Equilibration of T_3 and RT_3 production rates
Human (40-week gestation)	0 to 12 g. weeks	10 to 30 g. weeks	16 g. weeks to 1 pp month	30 g. weeks to 1 pp month
Sheep (150-day gestation)	0 to 60 g. days	60 to 150 g. days	70 g. days to 1 pp week	120 g. days to 1 pp week
Rat (23-day gestation)	0 to 20 g. days	16 g. days to 3 pp week	4 pp days to 18 pp days	Birth to 26 pp days

[a] g. = gestational; pp = postpartum.

development and metabolism; it is difficult to use this model to study stage II hypothalamic maturation.

The major developmental events characterizing the four stages of thyroid system development are summarized at the top of Table IX. The periods of hypothalamic maturation (stage II) and development of neuroendocrine control (stage III) are somewhat arbitrarily defined, since both are related intimately to hypothalamic maturation. Also there are too few data in the sheep or in man to temporally characterize stage II precisely. As in the rat, histological maturation of the hypothalamus may be continuing beyond the periods indicated in Table IX. Some data are available in the rat to clarify events during stage II–stage III transition (Fig. 41). In this species hypothalamic nuclei become histologically identifiable by 17–18 days, axonal ingrowth into the median eminence occurs between 16 gestational days and 1–2 postnatal days, vascular loops of the primary portal plexus appear at 4–6 postnatal days, and continued parallel maturation of the hypothalamus and primary portal plexus occurs over several weeks. The increase in hypothalamic TRH concentration tends to parallel this development during the perinatal period (Fig. 41). The fetal thyroid begins to secrete T_4 on day 18 (Geloso and Bernard 1967), but serum T_4 concentrations remain low at birth, fall in the neonatal period, and do not begin to increase rapidly until 2–4 days at the time of beginning vascular looping of the primary plexus of the pituitary portal system. The available serum TSH data are conflict-

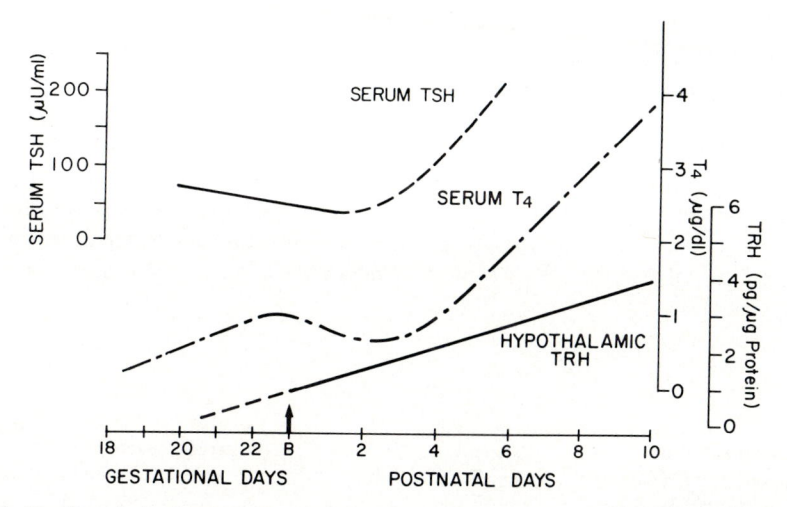

FIG. 41. Hypothalamic–pituitary–thyroid maturational events in the rat between 18 days of gestational age and 10 postnatal days. The thyrotropin-releasing hormone (TRH) data are from Dussault and Labrie (1975); the thyrotropin (TSH) data are from Greer (1975). Thyroxine (T_4) results are a composite of both studies. TSH data are extrapolated from 1 to 6 postnatal days. TRH data are extrapolated from 23 to 20 gestational days.

ing; Dussault and Labrie (1975) measured relatively high serum concentrations at birth (Fig. 34), whereas Greer (1975) measured relatively low concentrations in animals on a high iodine diet throughout the period −4 to +1 perinatal days (Fig. 41). Serum TSH concentrations increase in the human fetus between 16 and 20 weeks, well after the appearance of hypothalamic TRH and approximately at the time of appearance of capillary looping of the primary portal plexus.

Whether a functional median eminence neurovascular link exists *in utero* in the rat is not clear. Antithyroid drugs (Jost *et al.*, 1974) or a low iodine diet (Greer, 1975) administered to the pregnant rat will produce fetal goiter and increase fetal serum TSH concentrations (Greer, 1975) during the last 4 days of fetal life. Decapitation will prevent drug-induced goiter; however, encephalectomy will not (Jost, 1974; Mitskevich and Rumyantseva, 1972). Moreover, encephalectomy produces only a slight reduction in fetal thyroidal radioiodine concentration and in fetal serum T_4 concentration (Jost, 1974). These data confirm the existence of early *pituitary* negative feedback *in utero,* but the functional role of the hypothalamus appears minimal. Clearly maturation of TRH responsiveness and negative feedback sensitivity are largely postnatal events (Krulich *et al.*, 1976) (Fig. 36). The event that appears to correlate most closely with onset of these events in the rat and man is the appearance of early capillary loops of the primary portal plexus.

Functionally, the fetus during stage I development is primarily hypothyroid. During the early phase of stage II development before the midgestation increment in serum T_4 begins the hypothalamic–pituitary–thyroid system functions with features of both secondary and tertiary hypothyroidism; the pituitary response to TRH remains relatively insensitive, and hypothalamic control is minimal. During this period the system operationally resembles that of the adult animal with a transplanted pituitary gland. During stage III, hypothalamic–pituitary control progresses to the adult level of function. Nearly simultaneously, the animal develops the capacity (stage IV) to manufacture active thyroid hormone in peripheral tissues. The factors controlling the latter maturation event(s) are not yet clear.

REFERENCES

Abuid, J.L., Stinson, A., and Larsen, P.R. (1973). *J. Clin. Invest.* **52,** 1195.

Abuid, J.L., Klein, A.H., Foley, T.P., Jr., and Larsen, P.R. (1974). *J. Clin. Endocrinol. Metab.* **39,** 263.

Alexander, D.P., Britton, H.G., Cameron, E., Foster, C.L., and Nixon, D.A. (1973). *J. Physiol. (London)* **230,** 10P–12P.

Barnes, C.M., Warner, D.E., Marks, S., and Bustad, L.K. (1957). *Endocrinology* **60,** 325.

Carr, E.A., Beierwaltes, W.H., Raman, G., Dodson, V.N., Tanton, J., Betts, J., and Stambaugh, R.A. (1959). *J. Clin. Endocrinol. Metab.* **19,** 1.

Chopra, I.J. (1974). *J. Clin. Invest.* **54**, 583.
Chopra, I.J. (1976). *J. Clin. Invest.* **58**, 32.
Chopra, I.J., Sack, J., and Fisher, D.A. (1975a). *J. Clin. Invest.* **55**, 1137.
Chopra, I.J., Sack, J., and Fisher, D.A. (1975b). *Endocrinology* **97**, 1080.
Courrier, R., and Aron, M. (1929). *C. R. Seances Soc. Biol. Ses Fil.* **100**, 839.
Cuestas, R.A., Lindall, A., and Engel, R.R. (1976). *N. Engl. J. Med.* **295**, 297.
Czernichow, P., Greenberg, A.H., Tyson, J., and Blizzard, R.M. (1971). *Pediatr. Res.* **5**, 53.
DiAngelo, S.A. (1967). *Endocrinology* **81**, 132.
DiAngelo, S.A. (1972). *Neuroendocrinology* **9**, 197.
Dratman, M.B. (1974). *J. Theor. Biol.* **46**, 255.
Dratman, M.B., Crutchfield, F.L., Axelrod, J., Colburn, R.W., and Thoa, N. (1976). *Proc. Natl. Acad. Sci. U.S.A.* **73**, 941.
Dussault, J.H., and Labrie, F. (1975). *Endocrinology* **97**, 1321.
Dussault, J.H., Row, V.V., Lickrish, G., and Volpe, R. (1969). *J. Clin. Endocrinol. Metab.* **29**, 595.
Dussault, J.H., Hobel, C.J., and Fisher, D.A. (1971). *Endocrinology* **88**, 47.
Dussault, J.H., Hobel, C.J., DiStefano, J.J., III, Erenberg, A., and Fisher, D.A. (1972). *Endocrinology* **90**, 1301.
Dussault, J.H., Letarte, J., Guyda, H., and Laberge, C. (1976). *J. Pediatr.* (in press).
Erenberg, A., and Fisher, D. A. (1973). *In* "Foetal and Neonatal Physiology," (Comline, R. S., Cross, K. W., Dawes, G. S., and Nathanielsz, P. W., eds.), p. 508. Cambridge Univ. Press, London, and New York.
Erenberg, A., Omori, K., Oh, W., and Fisher, D.A. (1973). *Pediatr. Res.* **7**, 870.
Erenberg, A., Phelps, D.L., Oh, W., and Fisher, D.A. (1974). *Pediatrics* **53**, 211.
Etkin, W. (1968). *In* "Metamorphosis: A Problem in Developmental Biology" (W. Etkin and L.I. Gilbert, eds.), p. 313. Appleton, New York.
Falin, L. (1961). *Acta Anat.* **44**, 188.
Feldman, J.D., Vazquez, J.J., and Kurtz, S.M. (1961). *J. Biophys. Biochem. Cytol.* **11**, 365.
Fink, G., and Smith, G.C. (1971). *Z. Zellforsch. Mikrosk. Anat.* **119**, 208.
Fisher, D. A. (1972). *Excerpta Med. Found. Int. Congr. Ser.* **273**, 1045.
Fisher, D.A., and Dussault, J.H. (1974). *Handb. Physiol., Sect. 7 Endocrinol.* **3**, 21.
Fisher, D.A., and Odell, W.D. (1969). *J. Clin. Invest.* **48**, 1670.
Fisher, D.A., Lehman, H., and Lackey, C. (1964). *J. Clin. Endocrinol. Metab.* **24**, 393.
Fisher, D.A., Oddie, T.H., and Makoski, E. (1966). *Pediatrics* **37**, 583.
Fisher, D.A., Hobel, C.J., Garza, R., and Pierce, C. (1970). *Pediatrics* **46**, 208.
Fisher, D.A., Dussault, J.H., Erenberg, A., and Lam, R.W. (1972). *Pediatr. Res.* **6**, 894.
Fisher, D.A., Dussault, J.H., and Lam, R.W. (1973). *J. Clin. Endocrinol. Metab.* **36**, 397.
Fukuchi, M., Inoue, T., Abe. H., and Kumahara, Y. (1970). *J. Clin. Endocrinol. Metab.* **31**, 564.
Geloso, J.P. (1964). *J. Physiol. (Paris)* **56**, 358.
Geloso, J.P., and Bernard, A. (1967). *Acta Endocrinol. (Copenhagen)* **56**, 561.
Geloso, J.P., Henon, P., Legrand, J., Legrand, C., and Jost, A. (1968). *Gen. Comp. Endocrinol.* **10**, 191.
Gitlin, D., and Biasucci, A. (1969a). *J. Clin. Endocrinol. Metab.* **28**, 849.
Gitlin, D., and Biasucci, A. (1969b). *J. Clin. Endocrinol. Metab.* **29**, 926.
Glydon, R. St. J. (1957). *J. Anat.* **91**, 237.
Gray, B., and Galton, V. A. (1974). *Acta Endocrinol. (Copenhagen)* **75**, 725.
Greenberg, A.H., Czernichow, P., Reba, R.C., Tyson, J., and Blizzard, R.M. (1970). *J. Clin. Invest.* **49**, 1790.
Greer, M.A. (1975). *In* "Perinatal Thyroid Physiology and Disease" (D.A. Fisher and G.N. Burrow, eds.), p. 40. Raven, New York.

Grumbach, M.M., and Werner, S.C. (1956). *J. Clin. Endocrinol. Metab.* **16**, 1392.
Hall, P.F., and Myant, N.B. (1956). *J. Physiol. (London)* **133**, 181.
Hirvonen, L., and Lybeck, H. (1956). *Acta Physiol. Scand.* **36**, 17.
Hopkins, P.S. (1972). Ph.D. Thesis, Macquarie University, Sydney, Australia.
Hopkins, P.S., and Thorburn, G.D. (1971). *J. Endocrinol.* **49**, 549.
Hoskins, L.C., Van Arsdell, P.P., Jr., and Williams, R.H. (1958). *Am. J. Physiol.* **193**, 509.
Hyppa, M. (1972). *Neuroendocrinology* **9**, 257.
Jost, A. (1966). In "The Pituitary Gland" (G.W. Harris and B.T. Donovan, eds.), Vol. 2, p. 299. Butterworth, London.
Jost, A. (1971). In "Hormones in Development" (M. Hamburgh and E.J.W. Barrington, eds.), p. 1. Appleton, New York.
Jost, A., and Picon, L. (1970). *Adv. Metab. Disord.* **4**, 123.
Jost, A., Dupouy, J.P., and Rieutort, M. (1974). In "Integrative Hypothalamic Activity" (D.R. Swaab and J.P. Schade, eds.), p. 209. Elsevier, Amsterdam.
Kajihara, A., Kojima, A., and Onaya, T. (1972). *Endocrinology* **90**, 592.
Kaplan, S.L., Grumbach, M.M., and Aubert, M.L. (1976). *Recent Prog. Horm. Res.* **32**, 161.
Kearns, J.E., and Hutson, W. (1963). *J. Nucl. Med.* **4**, 453.
Keynes, G. (1952). *J. Obstet. Gynaecol. Br. Emp.* **59**, 173.
Kieffer, J.D., Mover, H., Federico, P., and Maloof, F. (1976). *Endocrinology* **98**, 295.
Klein, A.H., Agustin, A.V., Hopwood, N.J., Larsen, P.R., and Foley, T.P., Jr. (1976). *J. Pediatr.* **89**, 545.
Knobil, E., and Josimovich, J.B. (1959). *Ann. N.Y. Acad. Sci.* **75**, 895.
Krulich, L., Hefco, E., and Ojeda, S.R. (1976). *Program, Endocr. Soc.* Abstract No. 179, p. 146.
Lascelles, A.K., and Setchell, B.P. (1959). *Aust. J. Biol. Sci.* **12**, 445.
Lieblich, J.M., and Utiger, R.D. (1973). *J. Pediatr.* **82**, 290.
London, W.T., Money, W.L., and Rawson, R.W. (1963). *Endocrinology* **73**, 205.
Melmon, K.L. (1974). In "Textbook of Endocrinology" (R.H. Williams, ed.), 5th ed., p. 283. Saunders, Philadelphia, Pennsylvania.
Mitskevich, M.S., and Rumyantseva, O.N. (1972). *Ontogenesis* **3**, 376.
Monroe, B.G., and Paull, W.K. (1974). In "Integrative Hypothalamic Activity" (D.F. Swaab and J.P. Schade, eds.), p. 185. Elsevier, Amsterdam.
Montalvo, J.M., Wahner, J.W., Mayberry, W.E., and Lum, R.K. (1973). *Pediatr. Res.* **7**, 706.
Myant, N.B. (1958a). *J. Physiol. (London)* **142**, 329.
Myant, N.B. (1958b). *Clin. Sci.* **17**, 75.
Nataf, B.M. (1968). *Gen. Comp. Endocrinol.* **10**, 159.
Nathanielsz, P.W., Comline, R.S., Silver, M., and Thomas, A.L. (1973a). *J. Endocrinol.* **58**, 535.
Nathanlielsz, P.W., Silver, M., and Comline, R.S. (1973b). *J. Endocrinol.* **58**, 683.
Niemineva, K. (1950a). *Acta Paediatr. Scand.* **39**, 315.
Niemineva, K. (1950b). *Acta Paediatr. Scand.* **39**, 366.
Pearse, A.G.E., and Carvalheira, A.F. (1967). *Nature (London)* **214**, 929.
Peterson, R.R., and Young, W.C. (1952). *Endocrinology* **50**, 218.
Pickering, D.E. (1968). *Gen. Comp. Endocrinol.* **10**, 182.
Pilgrim, C. (1974). In "Integrative Hypothalamic Activity" (D.F. Swaab and J.P. Schade, eds.), p. 97. Elsevier, Amsterdam.
Postel, S. (1957). *Endocrinology* **60**, 53.
Raiha, N., and Hjelt, L. (1957). *Acta Paediatr. Scand.* **72**, 610.
Raiti, S., Holyman, C.B., Scott, R.L., and Blizzard, R. M. (1967). *N. Engl. J. Med.* **277**, 456.
Rinne, U.K., Kivalo, E., and Talanti, S. (1962). *Biol. Neonate* **4**, 351.

Sack, J., Beaudry, M., DeLamater, P.V., Oh, W., and Fisher, D.A. (1976). *Pediatr. Res.* **10**, 169.
Shepard, T.H. (1967). *J. Clin. Endocrinol. Metab.* **27**, 945.
Shepard, T.H., Andersen, H.J., and Andersen, H. (1964). *Anat. Rec.* **148**, 123.
Skowsky, W.R., and Fisher, D.A. (1976). *Pediatr. Res.* (in press).
Thomas, A.L., Jack, P.M.B., Manus, J.G., and Nathanielsz, P.W. (1975). *Biol. Neonate* **26**, 109.
Thorburn, G.D., and Hopkins, P.S. (1973). *In* "Foetal and Neonatal Physiology" (Comline, R.S., Cross, K.W., Dawes, G.S., and Nathanielsz, P.W., eds.) p. 488. Cambridge Univ. Press, London and New York.
Vale, W., Ling, N., Rivier, J., Villarreal, J., Rivier, C., Douglas, C., and Brown, M. (1976). *Metab. Clin. Exp.* **25**, Suppl. 1, 1491.
Winters, A.J., Eskay, R.L., and Porter, J.D. (1974). *J. Clin. Endocrinol. Metab.* **39**, 960.

DISCUSSION

D. Tulchinsky: Is there any association between steroids and thyroid function during pregnancy, particularly fetal corticosteroids, which presumably arise toward the end of pregnancy, and in view of the well-known effect of corticosteroids on thyroid function in nonpregnant patients.

Is there any association between thyroid function and hyaline membrane disease of the newborn, in view of recent reports of an increased thyroid function in infants who do not develop respiratory distress syndromes?

D. Fisher: We have no data on the effect of cortisol on fetal thyroid function. Peter Nathanielsz is here and does have some information. Dr. Chopra already has published data suggesting that cortisol will reduce T_3 levels and increase reverse T_3 concentrations in both hyperthyroid and euthyroid individuals. Therefore, you might expect cortisol to have a similar effect in the fetus. Dr. Nathanielsz's data suggest that cortisol in the fetus does the opposite; that is, it increases T_3 levels in serum.

With regard to your second question, there are animal data suggesting that thyroid hormones have some influence on lung maturation *in utero*. Lung weight and lung tissue DNA and protein concentrations were reduced in our hypothyroid sheep fetuses. There also is information suggesting that thyroid hormones can accelerate the appearance of lung surfactant. A recent paper in the *New England Journal of Medicine* reports that cord-blood T_4 concentrations are low in infants who develop postnatal hyaline membrane disease. In addition, free T_4 index values are low, and TSH concentrations are slightly but significantly high. What this means is not clear; we are not convinced that there was adequate matching of controls. We have observed that both the mass of the infant and the gestational age of the infant are important determinants of cord-blood T_4 concentrations. There also are data suggesting that postnatal serum T_3 concentrations are low in infants with hyaline membrane disease. In adults a variety of kinds of sickness will lower serum T_3 and raise reverse T_3 levels, and adequate controls for "sickness" have not been included in studies of thyroid function in infants with hyaline membrane disease. I think we must conclude that there are data suggesting a role of thyroid hormones in lung maturation and in hyaline membrane disease, but the situation is not yet clear.

J. M. McKenzie: What is actually controlling the maturations *in vitro*? Have you any data, either spontaneously in man or experimentally, of studies where there has been deficiency of function of the thyroid, hypophysis, or hypothalamus? Does removal of one of these organs affect the maturation rate of either of the other two?

Second, does the experimental animal or human *in utero* with a low level of T_3 provide evidence that T_4 itself is active? Do your data bear on the question whether T_4 is necessarily deiodinated to be active? Or does a fetus develop normally in the absence of thyroid hormone?

D. Fisher: We have not done experiments on selective organ removal to look at effects on the rest of the system. Decapitation to remove the hypothalamus and pituitary has been carried out in rodents as well as encephalectomy to remove the hypothalamus and leave the pituitary and the thyroid intact. These alterations do not modify general somatic development. Decapitation only minimally reduces thyroid size. In the rat *in utero* you can produce goiter by feeding a goitrogen to the mother. Goitrogen-induced fetal goiter is prevented if you take out the pituitary, but it is not prevented by encephalectomy. Dr. Jost and colleagues, who did these studies, suggest in the rat that there is development of pituitary negative feedback at a primitive level *in utero,* but that the hypothalamus is minimally active. This conclusion also follows from the rat data I have shown you.

Concerning the effect of T_4 on the fetus, we have thyroidectomized fetal sheep and observed that they develop reasonably well. After a period of 30–60 days of intrauterine hypothyroidism, body weight is minimally altered and organ weights were normal for gestational age with the exception of the lungs, which were small for gestational age. Bone maturation was retarded, and skin maturation was delayed somewhat. Hypothyroid newborn infants picked up in the screening programs in the United States and Canada appear nearly normal at birth in the majority of instances. These data suggest that you can detect evidence of thyroid hormone deficiency in the hypothyroid fetus, but that these effects are minimal. There is very little clinical evidence in my view to suggest that T_4 is critically important to intrauterine growth and development.

The data I presented that were obtained after T_4 loading of mothers at term suggest that it is T_4 that is controlling TSH release in the fetus, not T_3. When we gave T_4, we raised

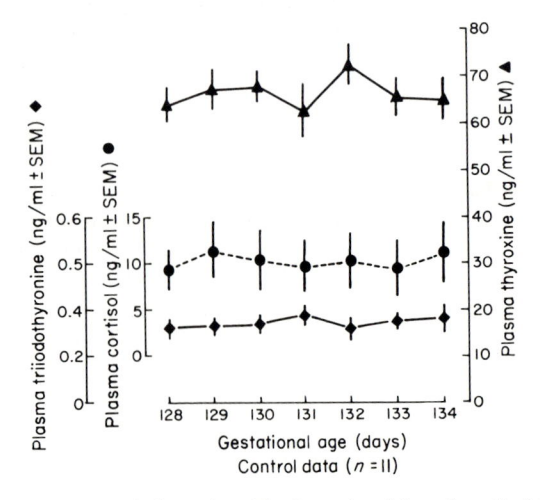

FIG. A. Fetal plasma triiodothyronine (♦), thyroxine (▲), and cortisol (●) concentrations in control chronically catheterized lambs between 128 and 134 days of gestation ($n = 11$; ng/ml; mean ± SEM). Fetal catheterization was performed on day 128. Fetal blood gas and pH measurements were within the normal range throughout. Data in this figure and in Figs. B and C are from A. L. Thomas, F. G. Bass, C. Horn, E. Krane, and P. W. Nathanielsz, *J. Endocrinol.* (in press).

cord-blood T_4 levels and lowered TSH; serum T_3 did not change at all. However, T_3 will suppress TSH if it is given exogenously to the fetus.

R. Volpe: Are you stating that there are no newborn cretins that are clinically cretinous when born?

D. Fisher: I am suggesting that the vast majority of newborn cretins appear normal at birth. It is uncommon that a diagnosis is made in the first 6 weeks on the basis of clinical data.

P. W. Nathanielsz: We have examined the effects of cortisol on T_4 metabolism in the sheep fetus between 128 and 134 days of gestation. Figure A shows the control cortisol, T_4 and T_3 concentrations in 11 fetuses during these 6 days. These fetuses were infused with saline over this period. Figure B shows the changes in fetal plasma cortisol, T_4, and T_3 when gradually incrementing doses of cortisol are infused into the fetus from day 130 in 10 fetuses. Delivery usually occurs in these fetuses within 96 hours of this cortisol infusion. It will be seen that fetal plasma T_4 concentration falls and plasma T_3 concentration rises. Using Fisher's data for metabolic clearance rate (MCR) and volume of distribution of T_4

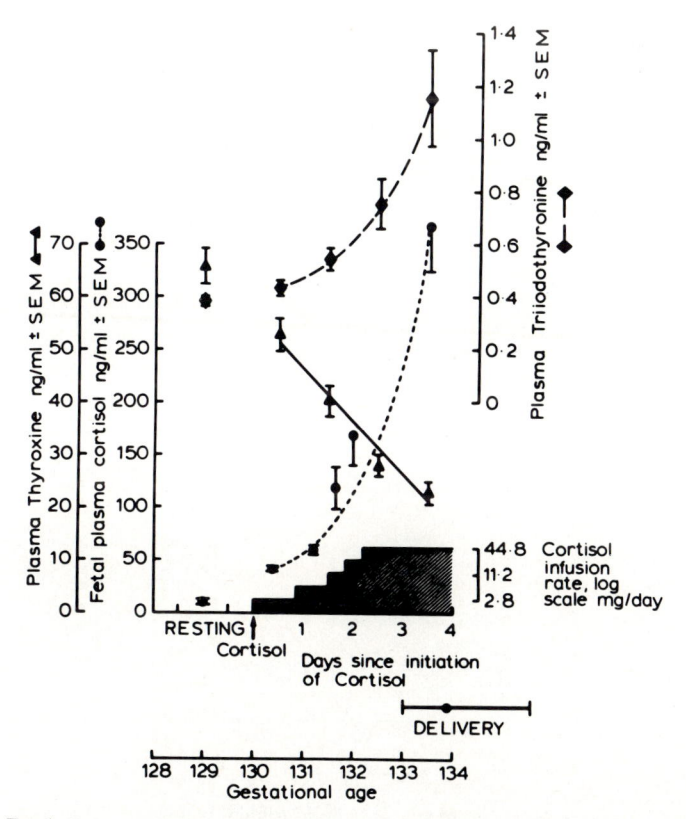

FIG. B. Fetal plasma triiodothyronine (♦), thyroxine (▲), and cortisol (●) concentrations in chronically catheterized lambs between 128 and 134 days of gestation (n = 10; ng/ml; mean ± SEM). Fetal catheterization was performed on day 128. On day 130 an escalating intravascular infusion of cortisol into the fetus was begun.

and T$_3$ in the fetus, we can demonstrate that the fall in plasma T$_4$ exactly equals the amount of T$_4$ that would be needed to generate the T$_3$ produced.

Finally, we have a potentially fascinating observation in seven control animals and one cortisol-infused fetus in which Dr. Inder Chopra has measured fetal reverse T$_3$ concentrations for us. In the seven controls, reverse T$_3$ (Fig. C) remained constant during this period. In the one cortisol-infused fetus reverse T$_3$ fell during the infusion period. The plasma T$_4$ changes in this single fetus are also shown to demonstrate that it behaved like all the other cortisol-infused fetuses in this respect (Fig. B).

Our initial interpretation of these findings is that cortisol is increasing outer-ring deiodination in the sheep fetus and decreasing inner-ring deiodination. Further investigation will be required to assess whether there is a separate action on two separate deiodinases or whether one effect produces a secondary change in the other system. For example, increased outer-ring deiodination may decrease the amount of T$_4$ available for inner-ring deiodination. In this event, reverse T$_3$ concentrations would fall simply as a result of the increased production of T$_3$. I would like to make one observation in relation to the cord-cutting experiments. Not only does respiration begin at this time, but, after cord section, the ductus venosus rapidly closes down. As a result there is a rapid change in blood flow through the liver. There is also a change in the oxygenation of the blood flowing through the liver. Both of these alterations may affect peripheral deiodination in the liver.

D. Fisher: This would suggest that T$_3$ concentrations ought to be high in cord blood of lambs, and we have measured relatively low values. You will recall, however, that we have

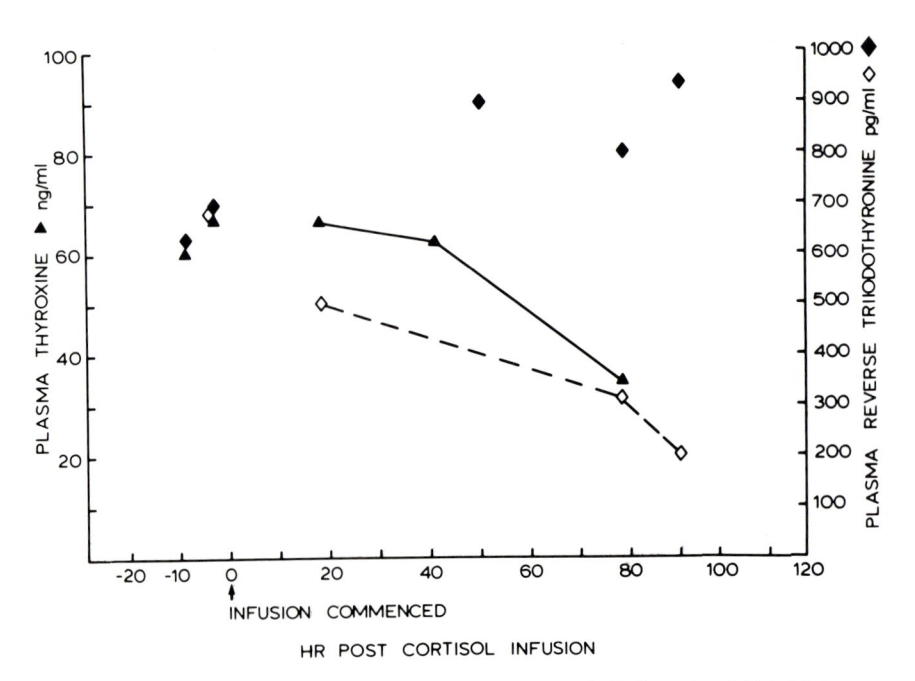

FIG. C. Fetal plasma thyroxine (▲) ng/ml; and reverse triiodothyronine (rT$_3$) (◊) in one fetus in which cortisol was infused intravascularly into the fetus at time 0. For comparison, values are given for the mean plasma rT$_3$ (♦) from seven control fetuses infused only with saline over the same period of gestation.

delivered our animals by cesarian section before they undergo spontaneous labor. We have not looked at cord blood during the last 3 days of normal gestation.

P. W. Nathanielsz: When we are comparing the human and the sheep data, I think it is important that we bear in mind the possible effects of the time scale of the changes in corticosteroid concentrations in the fetus in these two species. The data of both Tulchinsky and Murphy [B.E.P. Murphy, J. Patrick, and R.L. Denton, *J. Clin. Endocrinol. Metab.* **40**, 164 (1975)], admittedly measuring cortisol in amniotic fluid, suggest that the rise in cortisol secretion is slower and occurs over a longer time scale than in fetal plasma in the sheep [J.M. Bassett and G.D. Thorburn, 1969; R.S. Comline, P.W. Nathanielsz, and M. Silver, *J. Physiol. (London)* **207**, 3P–4P (1970)].

R. S. Bernstein: In your data with α-methyl-*p*-tyrosine, you implied that the reason that the rise in T_3 did not take place was because tyrosine hydroxylase might be the enzyme that deidinates the T_4 to T_3. An alternative explanation is that the catecholamine levels would be lower after treatment with the blocker. The control experiment would be to give catecholamines at the same time that one gives the α-methyl-*p*-tyrosine. Have you done that? Have you or anybody studied the deiodination phenomenon *in vitro*, where one could separate out these various factors?

D. Fisher: We have infused epinephrine or norepinephrine during the 60 minutes before cord cutting in delayed-cord cutting experiments. Neither hormone increases T_4 or T_3 concentrations.

A number of people have looked at *in vitro* tissue homogenate systems for an effect of α-methyl-*p*-tyrosine on T_3 production from added T_4 and found no effect.

L. S. Jacobs: I was intrigued by the discrepancy in postnatal TSH rise in the sheep versus the human, especially in light of what I interpreted as a rather more marked fall in rectal temperature in the sheep than what I remember to occur in the human. I wonder whether you might offer some speculation regarding the determinants of this discrepancy, neural or peripheral. You have shown directly that the postnatal T_3 rise in sheep very likely derives from peripheral T_4 deiodination rather than TSH-stimulated thyroidal secretion; one may infer that this is also true in the human neonate. However, in light of the much higher achieved TSH levels, I wonder if it is still possible that thyroidal T_3 secretion may contribute to the neonatal T_3 rise in humans. Information regarding the sensitivity of the neonatal sheep and human thyroid glands to TSH would be helpful in this regard. Are any such data available?

D. Fisher: The reason why the TSH response is greater in the human than in the sheep is not clear. Both species are delivered at a comparable period of development. The hypothermia was relatively exaggerated in our lamb experiments because we were using a cold delivery room. Normally, lambs drop their temperature 1–2 degrees and recover within 30–40 minutes, much as do human infants. However, the TSH response in the newborn lamb is reduced in spite of augmented hypothermia. The pituitary response to TRH is similar in lambs on day 1 and on day 7 in spite of higher levels of T_4 and T_3 on day 1. Also, newborn lambs during the first 2 hours respond to TSH (with elevations of serum T_3) just as they do later. We have not conducted such studies in newborn infants, but I would guess that the lesser TSH surge in the lamb is related to relatively reduced TRH responsiveness to hypothermia.

With regard to your second question, I think that it is likely that the TSH surge contributes to the increase in T_3 levels in the newborn, and the relative TSH contribution in the human infant may be greater because of the greater TSH surge. We have no hard data, however, to quantify the relative contributions of T_3 secretion and T_3 derived from T_4. The human amniotic fluid injection experiments and the sheep α-methyl-tyrosine experiments suggest that a larger proportion of the T_3 surge is due to peripheral conversion than to secretion in both species.

H. Guyda: In response to Dr. Volpe, I would like to differ a bit with Dr. Fisher. Babies diagnosed through the Quebec screening program for hypothyroidism in fact do show clinical evidence of hypothyroidism even as early as 3–4 weeks of age, which is the earliest babies we see. You can even diagnose them at birth, because they are large babies, they are postterm gestationally, and they are usually female. Their anterior and particularly their posterior fontanelles are open, and on clinical grounds their facial features, their mottling, the umbilical hernia, and their lowered body temperature are all very strong evidence that they are hypothyroid right from birth and can be diagnosed if one has a high index of suspicion.

We have been looking at the maturation of the human fetal hypothalamus and pituitary *in vitro* using tissue culture as a means of trying to assess when the maturational events occur. We have studied human fetal pituitaries obtained at the time of abortion, usually between 10 and 20 weeks of gestation. The first observation, and I confine myself to TSH only, is that with increasing gestational age the basal release of TSH from the fetal pituitary increases, so that in the 10–12-week period basal TSH levels are very low, and by the time they get to 18–20 weeks the basal levels are substantially increased over the earlier values.

If one administers TRH to these pituitary cultures one sees a similar pattern; that is, in the younger donors, the magnitude of TSH response to TRH is diminished but increases with increasing gestational age.

In Fig. D, the top panel shows basal release of TSH during days in monolayer culture of a 13-week gestation fetus. The pattern of release is one that is quite uniform throughout all periods of gestation studied; that is, during the first 4 days of culture, the release of TSH is fairly high and diminishes to reach a basal level, usually by the end of the first week in culture. The bottom panel shows the responsiveness to TRH; during the first 3 days in culture the response is quite large and dramatic and decreases with time in culture. With increasing gestational age, the responsiveness in those first 3 days increases as the donor fetal pituitary increases in age. These studies were performed in conjunction with Drs. Cindy Goodyear and Claude Giroud, McGill University.

D. Fisher: It is clear from the work of David Smith in Seattle as well as your own that one can diagnose hypothyroidism early if one establishes an index of several signs and symptoms and uses some critical level of index weight or score as suggestive of hypothyroidism. This only emphasizes that the diagnosis is difficult and the signs subtle.

Your *in vitro* data agree with the *in vivo* results. The fetal pituitary can secrete TSH *in vivo* at 10 weeks, and fetal serum TSH is detectable at low levels until about 20 weeks, when concentrations increase. It would be very interesting to determine whether increasing *in vitro* TRH responsiveness correlates with the apparent increase in pituitary activity between 20 and 35 weeks.

H. Guyda: Those specimens are harder to get.

D. Fisher: Yes, they are.

R. Volpe: I would like to return to the question raised by Dr. Guyda's first remarks. If, as we always thought, the neonatal child is truly hypothyroid at birth, then what is the role of the low T_3 in the last trimester of life? That is, they do need thyroid hormones during the last trimester of life.

D. Fisher: We do not know whether they need it. The striking thing to me is that hypothyroid newborn infants look so normal when their serum TSH concentrations are 350 μU/ml and their thyroxine concentrations are 3–5 μg/dl. As Dr. Guyda pointed out, these infants are large at birth, not small. Dr. Smith has reported that this is due to the fact that their delivery is delayed; they are postmature. This indicates that thyroid hormones are not necessary for somatic growth in the human fetus. Thyroid hormones have some effect on bone maturation and may have some effect on lung maturation. These are of minor significance, however, compared to possible central nervous system effects. It is important

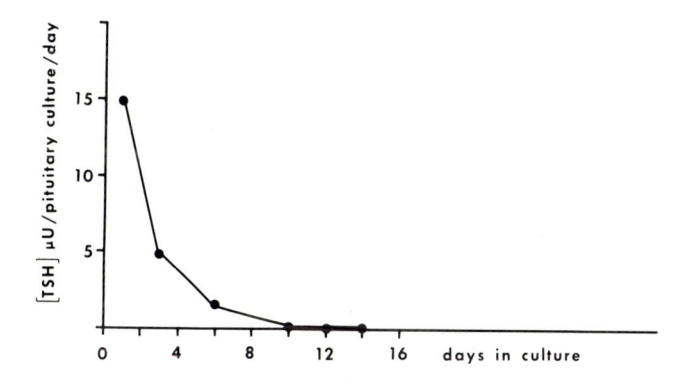

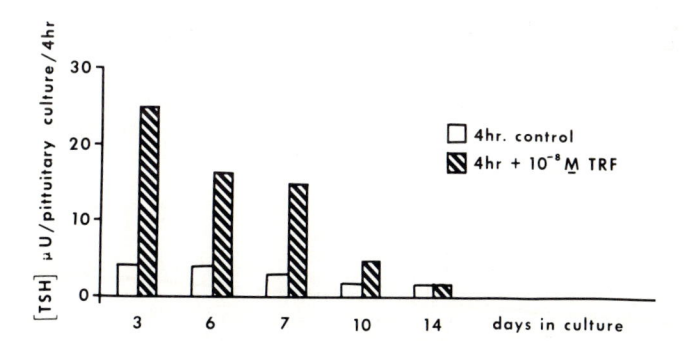

FIG. D. (Top) Release of TSH with time by human fetal pituitary monolayer culture of 13 weeks gestation. (Bottom) Effect of TRF on TSH secretion by human fetal pituitary monolayer cultures of 10 to 14 weeks gestation. Open bars represent 4 hour control period and hatched bars 4 hour treatment period with 10^{-8} M TRF. (Redrawn from Goodyer, C. G., G. Hall, H. Guyda, F. Robert and C.J-P. Giroud, in press).

to determine to what extent any effects of intrauterine thyroid hormone deficiency on the central nervous system are reversible. We know that it is possible to largely prevent the mental retardation of cretinism by treating the infant before 3 months of age. We now need to decide whether treatment before 1 month is better still. These data ought to be available within two or three years.

M. Dratman: I would like to respond to the question raised about the *in vitro* capacity of tyrosine hydroxylase to deiodinate T_4, and about the effect of α-methyl-*p*-tyrosine on the reaction. Studies with partially purified rat adrenal tyrosine hydroxylase, performed in collaboration with Dr. Joseph Coyle, demonstrated that after a 1-hour period of incubation in the presence of all the factors necessary for continued action of the enzyme, but not in their absence, both T_4 and T_3 were deiodinated, and deiodination was inhibited in the presence of α-methyl-*p*-tyrosine. However, we were not able to demonstrate recovery of any labeled products of this deiodination other than iodide, because only outer ring-labeled iodothyronines were available to us at the time these studies were performed. We have not made further efforts to continue with these *in vitro* studies, because we have recognized that a lot of ground work had to be laid before the iodothyronine-deiodinating actions of this only partially purified enzyme could be evaluated properly.

More recent observations are pertinent to the question of tyrosine hydroxylase action and iodothyronine deiodination. Since we have now observed that both T_4 and T_3 are rapidly and selectively taken up into the nerve ending fraction of rat brain *in vivo*, tyrosine hydroxylase and other nerve ending-localized enzymes would have ample opportunity to interact with thyroid hormones. Moreover, we now have evidence for *in situ* generation of T_3 from T_4 within synaptosomes *in vivo*, and have observed that, with time after administration of a single dose of labeled iodothyronine, there is a progressive cytosol to synaptosome concentration gradient for T_3. Therefore an enzyme highly localized in nerve terminals in the brain transforms T_4 to T_3.

Having made these observations in the mature rat, we then studied the ontogeny of this process in the newborn rat from postnatal days 1 through 42. During days 1 through 6 there was no evidence that either T_3 or T_4 were differentially concentrated in any brain particles. Instead, T_4 seemed to be associated with protein fractions of the brain, without apparent reference to their subcellular distribution. However, from day 6 through 18, increasingly selective concentration of T_3 and to a lesser extent T_4 were observed in the synaptosomal fraction of the developing rat brain. After day 18, T_3 was more than 2-fold more concentrated in synaptosomes than in any other subcellular fraction of the rat brain.

D. Fisher: With regard to a possible role of tyrosine hydroxylase in the newborn, we should also remember that the neonate is much different from the adult animal. The newborn has a large mass of paraaortic chromaffin tissue, which contains tyrosine hydroxylase, and the total amount of tyrosine hydroxylase in the newborn per unit weight of total body is very high. There is also a large volume of brown fat, which may contain a high concentration of autonomic nerve endings. Tryosine hydroxylase may under normal circumstances be a weak T_4 β-ring monodeiodinase, but when present in large amounts and intensively stimulated, it might exert a significant effect.

K. Sterling: The first question concerns some of your human neonatal data that seem to contradict some of Inder Chopra's findings with rat liver homogenate suggesting that reverse T_3 formed from T_4 actually inhibits the monodeiodination of T_4 to orthodox T_3. Would you care to comment on that first?

D. Fisher: Yes, Dr. Chopra has suggested that one role of reverse T_3 in the fetus may be the inhibition of β-ring deiodination of T_4 to T_3. I do not believe that this is the only explanation for the low levels of T_3 in fetal serum because there is a dramatic T_3 surge in the newborn period without any change in serum reverse T_3. However, in our intraamniotic fluid T_4 injection experiments, amniotic fluid reverse T_3 concentrations increase dramatically 24 hours after giving the 750-μg intraamniotic fluid T_4 dose. Preliminary data in several other infants suggests that cord blood T_3 concentrations are suppressed. This suggests that reverse T_3 in high concentrations may suppress intrauterine T_4–T_3 monodeiodination.

K. Sterling: The second question is relevant to studies we are conducting on neonatal rats. Barker and Klitgard showed many years ago that hypothyroid rats given replacement therapy have increased Qo_2 in virtually all tissues except the brain, spleen, and testis, and we found an exactly similar distribution of the specific mitochondrial receptor that is present in rat liver, kidney, myocardium skeletal muscle, lung, and small intestine, but not in adult rat brain, spleen, or testis. However, neonatal rats at various ages such as 2, 5, 7, 10, and 12 days do indeed possess the mitochondrial receptor in their brain mitochondria, but it is gone by 14 and 17 days; it is just like the adult brain in our studies. I noticed that you had measured almost all parameters including T_4 to T_3 conversion. Have you done, or do you intend to do, studies of the protein synthetic effects in the neonatal rat brain, and when it turns off, because I gather that the newborn rat is about equivalent to a midpregnancy human or sheep embryo.

D. Fisher: These are very important issues. One obvious area of deficiency in our data is the ontogenesis of receptors: pituitary TRH receptors, thyroid TSH receptors, nuclear receptors for T_3 and T_4, and the mitochondrial receptors. We have not measured these. As you know, others have shown that T_4 stimulates protein synthesis in newborn rat brain, but not in adult rat brain. We have no data on protein synthesis as related to developmental age.

P. W. Nathanielsz: Plasma corticosterone concentrations are rising between 12 and 24 days of postnatal life in the neonatal rat [V.G. Daniels, R.N. Hardy, K.W. Malinowska, and P.W. Nathanielsz, *J. Endocrinol.* **52**, 405–406 (1972)]. It is clearly possible that the adrenal cortex plays a role in the induction of tyrosine hydroxylase.

The kinetics of T_4 production in the fetal sheep have been examined by Dr. Fisher's group and by us [P.W. Nathanielsz, R.S. Comline, M. Silver, and A.L. Thomas, *J. Endocrinol.* **58**, 535–546 (1973); J.H. Dussault, C.J. Hobel, and D.A. Fisher, *ibid.* **88**, 47–51 (1971); A. Erenberg and D.A. Fisher, "Foetal and Neonatal Physiology" (R.S. Comline *et al.*, eds.), pp. 508–526. Cambridge Univ. Press, London and New York, 1973]. As we have seen, the fetal sheep is producing 5–8 times the amount of T_4 per day relative to the pregnant ewe. If reverse T_3 does indeed antagonize the action of T_4, the high levels of reverse T_3 may be the factor that makes it necessary for the fetal thyroid axis to operate at this high level of activity. Immediately after delivery, although the concentration of reverse T_3 remains high for a while owing to the low MCR, as you have demonstrated [I.J. Chopra, J. Sack, and D.A. Fisher, *Endocrinology* **97**, 1080–1088 (1975)], this inhibitory activity may be offset by the activation of other systems that enhance the activity of T_4. As you suggest, the sympathetic nervous system may be one of the most important.

One small point regarding the expression of turnover rates in the fetus. I see that you calculate T_4 production rate relative to the surface area (i.e., proportional to weight 2/3). We have continued to express our data related to body weight since we are undecided regarding what is the most appropriate measure. M. Kleiber [*Annu. Rev. Physiol.* **29**, 1–20 (1967)] suggests that between-species comparisons should be made on the basis of basal metabolic rate. This would give a relationship proportional to body weight of 3/4.

D. Fisher: I will answer the last question first. It was Dr. Sack's idea to express the data relative to surface area. The relative differences in production rates of thyroid hormones were clearer on this basis than on a kilogram body weight basis. We have recalculated your data on a surface area basis and find the same 3- or 4-fold difference that we have measured. Conversely, the 3- or 4-fold difference on a surface area basis in our data, when calculated on a kilogram body weight basis, becomes an 8-fold difference, as we have reported in earlier studies.

With regard to your first question, I agree that the high levels of reverse T_3 in fetal serum probably are exerting some inhibitory effect on monodeiodination of T_4 to T_3. We still have no explanation, however, for the initial elevation of reverse T_3 to inhibitory levels. At birth something could be overriding the inhibition of reverse T_3, or a second β-ring monodeiodinative pathway might be stimulated such as tyrosine hydroxylase.

B.E.P. Murphy: I wonder whether you would comment on thyroidal hormones in amniotic fluid, and on what can be learned from amniocentesis in late pregnancy.

D. Fisher: We have published on this, and I have only limited new information to add. As you know, T_4 is measurable in amniotic fluid; the average level during the last trimester approximates 0.4 μg/dl of fluid. Triiodothyronine concentrations are below our limits of detectability. Reverse T_3 concentrations are high throughout pregnancy, but are higher early in pregnancy and decrease progressively, as Drs. Chopra and Crandall have reported. In the sheep, reverse T_3 levels are low in amniotic fluid. I can not explain this difference, but it means the sheep does not serve as a good model for human amniotic fluid metabolism of thyroid hormones. Thus, we must look to human patients for data. We have the normal data which suggest that reverse T_3 may be the best indicator of fetal thyroid status; now we need information from hypothyroid and hyperthyroid fetuses. These samples are difficult to obtain, but we have had occasion to look at amniotic fluid from one infant treated with intraamniotic T_4 by her physicians, who began treatment at 32 weeks because the mother had been treated with radioiodine at 16 weeks of gestation. Reverse T_3 concentrations in amniotic fluid increased to very high levels after the amniotic fluid injections of T_4 were given. This was the first direct evidence that the human fetus converts T_4 to reverse T_3. And I have mentioned similar increases in reverse T_3 in amniotic fluid of women during the T_4-loading experiments that we are doing. These results suggest the possibility that amniotic fluid might be useful in the diagnosis of fetal hyperthyroidism. We have not had the opportunity of examining amniotic fluid of an infant that turned out to be hypothyroid. Thus, whether amniotic fluid study will help in diagnosis of fetal *hypo*thyrodism is not clear.

R. Guillemin: There is one more system the maturation of which I think would also be of interest; that is the maturation of the enzymic system which in plasma inactivates TRF. As I know you know, Maloof's group recently reported, much to everyone's surprise, that newborn rat blood has no enzymic activity to inactivate TRF for as long as 16 days postnatally.

D. Fisher: I have been intrigued by that observation since you pointed it out to me earlier in the meeting. I wondered whether you had some speculation about what it meant. You commented that you did not think TRF circulating in plasma had very much role. There is information to suggest that TRH crosses the placenta; the only data refuting this is Dr. Nathanielsz's results in the sheep. If TRF does circulate in peripheral blood and does cross the placenta, presumably it could influence the fetus. However, I, like you, feel that circulating TRH probably has little influence on maternal–fetal interaction with regard to the hypothamamic–pituitary–thyroid system.

R. Guillemin: If we do not usually find any measurable TRF in the peripheral blood it might well be because the plasma inactivation is so very powerful. So that, evidence that there may be an absence of this plasma inactivation of the tripeptide at one stage or another of gestational or postnatal age may indeed be of importance, because then it would not be surprising, but possibly expected to find TRF circulating in peripheral blood at that age. This is not the case in the adult.

M. Dratman: Perhaps your comment, Dr. Fisher, about the special susceptibility of the adrenergic nervous system in the fetus and the newborn to metabolic changes, may also be relevant to the susceptibility of this system to the action of glucocorticoids. During the newborn period, intra- and extramedullary chromaffin cells and SIF cells respond to the action of glucocorticoids in a manner that is later lost: these cells are induced by glucocorticoids to produce PNMT, the enzyme that converts noradrenaline to adrenaline, and the inductive process is coupled with induction of tyrosine hydroxylase. In the adult animal, inductive effects of glucocorticoids on catecholamine enzymes are probably confined to the adrenal medulla. There is little information available regarding response of peripheral or central adrenergic nerve-terminal enzymes to glucocorticoids.

In this discussion, a number of references have been made to the importance of the liver in the peripheral metabolism of the thyroid hormones. Falck and Hillarp have recently demonstrated that the hepatocytes receive direct adrenergic innervation. Even if this innervation is not as rich, per cell, as that in other tissues (such as atrium or vas deferens), nevertheless the comparatively large number of liver cells, each receiving even a minimal number of nerve terminals, could then account for a relatively large amount of adrenergic enzyme activity.

The Antimüllerian Hormone

NATHALIE JOSSO, JEAN-YVES PICARD,
AND DIEN TRAN

Unité de Recherches de Génétique Médicale (INSERM), Hôpital des Enfants-Malades, Paris, France

I. Introduction

Müllerian ducts, the primordia for the vertebrate oviduct, were first described by J. Müller in 1930, in young embryos of either sex. The cause of their regression in males was analyzed by Jost in 1947, who reviewed his findings at the 1952 Laurentian hormone conference (Jost, 1953). In the same way that pure mathematical speculation allowed J. C. Adams and Le Verrier to predict the existence of Neptune, although the planet was actually seen only later that year, Jost concluded that the results of his surgical experiments on rabbit fetuses could be explained only by the existence of a distinct testicular morphogenetic secretion responsible for the regression of müllerian ducts in male fetuses. This concept was rapidly accepted by clinicians confronted with disorders of human sexual development. A nomenclature explosion resulted, and the mysterious testicular product was called indifferently müllerian inhibitor, müllerian-inhibiting factor, duct organizer, X-factor or even antifeminine substance (Jost et al., 1972). The term antimüllerian hormone (AMH) is preferred in our laboratory although one might question the appropriateness of designating as "hormone" an endocrine product whose blood concentration is usually not high enough to induce long-distance effects. In intersex patients in whom testicular differentiation has occurred only on one side of the body, müllerian regression is also asymmetrical. However, the inhibition of müllerian ducts of bovine freemartin fetuses, genetic females united to male cotwin(s) by chorionic vascular anastomoses, is an indication that physiological concentrations of AMH may be reached in the blood stream.

Methodological difficulties have been responsible for the slow progression of scientific knowledge concerning AMH. Because the müllerian duct is sensitive to AMH only during a short and early period of fetal development, *in vivo* experiments represent technical feats. The demonstration by Picon (1969) of antimüllerian activity of testicular tissue maintained in organ culture opened a more accessible way of approach, and all the recent data have been obtained with the use of an *in vitro* bioassay for AMH based upon her work. The investigations reported in this paper have been carried out at the Enfants-

Malades Hospital with the support of INSERM, DGRST, and René-Descartes University. We also wish to express our gratitude to Professor Jost for the teaching received by the senior author during her stay in his laboratory and for his continued interest in the progress of our work. Although many questions remain unanswered, the antimüllerian hormone has ceased to be a pleiad of names and has assumed a definite, if still hazy, reality.

II. Embryology of Müllerian Ducts

A. INDIFFERENT OR AMBISEXUAL STAGE

In most higher vertebrates, the first steps of organogenesis of gonaducts are similar in both sexes. In the human embryo, which will serve as our first model, a cleft lined by coelomic epithelium appears between the gonadal and mesonephric parts of the urogenital ridge, at approximately 6 weeks of embryonic development (10 mm). At this stage, the wolffian or mesonephric ducts have already reached the urogenital sinus. This cleft, which persists in females as the abdominal ostium of the fallopian tube, is closed caudally by a solid bud of epithelial cells, which burrows in the mesenchyme lateral to and parallel with the wolffian duct. As the bud increases in length, a lumen appears in its cranial part, in continuity with the abdominal ostium, and gradually extends toward the growing tip. At the caudal end of the mesonephros, the müllerian duct crosses the wolffian duct ventrally to reach its medial side and continues to grow posteriorly until it eventually meets and fuses with the duct of the opposite side. The fusion is at first partial, a septum persisting between the two lumina. This septum however, gradually disappears, leading to the formation of a single uterovaginal canal which in fetuses 30–35 mm in crown–rump length, makes contact with the urogenital sinus, causing an elevation of its dorsal wall, the müllerian tubercle.

According to this concept, which reflects the views of most modern embryologists [review by Didier (1973) and Frutiger (1969)], müllerian ducts in mammals and birds are derived from the coelomic epithelium and not, as in lower vertebrates, from the duplication of the wolffian duct. In the chick embryo, the ultrastructural studies of Didier (1973) have shown that the cytological aspect of müllerian epithelial cells allow them to be easily distinguished from wolffian cells, even though the proliferating müllerian tip is contained within the wolffian basal membrane. However, in the human embryo, Frutiger (1969) has recently upheld the view that the lower segment of the müllerian duct is derived from the wolffian duct. Even the tenants of the opposite opinion agree that the wolffian duct is involved in two aspects of müllerian organogenesis: it induces the formation of the müllerian ostium (Didier, 1971) and serves as guide to the caudal progression of the müllerian duct; experimental interruption of the development of the wolffian duct arrests the growth of the müllerian duct at that point (Gruenwald, 1941; Didier, 1973).

B. SEX DIFFERENTIATION OF MÜLLERIAN DUCTS

Early organogenesis of the müllerian ducts, described in the preceding section, is similar in male and female embryos. Except in reptiles, in whom male müllerian ducts never extend beyond the caudal end of the mesonephros (Raynaud *et al.*, 1970), formation of the ostium, followed by growth of the müllerian duct down to the urogenital sinus, occurs independently of the genetic or gonadal constitution of the embryo. At the end of the ambisexual stage, müllerian morphology undergoes no drastic modification in female mammalian embryos. Coiling and acquisition of tubal fimbriae occur later in fetal life (Price *et al.*, 1968). The contribution of müllerian ducts to vaginal organogenesis varies in different species. In the human embryo, the vagina derives essentially from the vaginal plate, an emanation from the urogenital sinus, and not from the müllerian ducts (Forsberg, 1963). In birds, growth of the right müllerian duct ceases at 9 days of incubation, and only a cloacal remnant persists in the adult female. Growth of the left müllerian duct is essentially due to collagen formation by fibroblasts surrounding müllerian epithelium and is reflected by the ratio of ^{3}H-labeled proline to ^{14}C-labeled amino acid incorporation (Yamamoto *et al.*, 1974).

In male embryos the end of the ambisexual stage is marked by the onset of physiological müllerian degeneration, which chronologically precedes the regression of wolffian ducts in females. Müllerian ducts lose their communications with the coelomic cavity, the cranial extremity persists to form the "appendix testis," and the remainder of the duct undergoes anteroposterior degeneration. In the human, regressive changes of müllerian ducts were found in male fetuses 31–35 mm in crown–rump length, and müllerian ducts had totally disappeared in fetuses of 43 mm or more (Jirasek, 1967). The prostatic utricle, or uterus masculinus, a rudiment of the female duct system, arises from the urogenital sinus in the human (Vilas, 1933), but in some species müllerian ducts make a significant contribution to its organogenesis (Alcala and Conaway, 1968).

The histological aspects of müllerian regression have received surprisingly little attention. The first changes visible by light microscopy consist in a narrowing of the duct and a reduction of epithelial height. Then, the lumen disappears and the remaining epithelial cells seem to be stifled by the fibroblastic ring that surrounds the epithelium. Few necrotic cells are seen. Studies at the ultrastructural level in male fetal mice (Dyche, 1976) indicate that degenerating müllerian epithelial cells, containing vacuoles of debris, are phagocytosed by periductal mesenchymal cells. Forsberg and Abro (1973), in a study of cell degeneration in the uterovaginal anlage of the female mouse, had similarly shown that phagocytosis of degenerating cells by their neighbors occurred at a very early stage: condemned cells did not appear particularly different from surrounding healthy cells and were ingested before evidence of their degeneration became apparent.

In the male chick embryo, both müllerian ducts undergo involution at 9 days of incubation and disappear by day 12 or 13. Using light microscopy, Scheib (1965) could not demonstrate cytoplasmic damage in müllerian epithelial cells, but ultrastructural observations revealed autophagic changes. The role of lysosomes in avian müllerian duct degeneration has also been studied by biochemical analysis (Brachet et al., 1958; Scheib-Pfleger and Wattiaux, 1962; Hamilton and Teng, 1965) taking advantage of the fact that the chick müllerian duct can be isolated by dissection from the rest of the reproductive tract. Between 7 and 9 days of incubation in male ducts, lysosome content of acid hydrolases increases sharply. Protein synthesis continues to rise, though at a slightly slower pace than in the female (Sjöquist and Hultin, 1967). The onset of müllerian degeneration at 9 days is marked by a solubilization of the proteolytic enzymes and a precipitous fall in protein synthesis. These changes, however, represent the consequence rather than the cause of cellular demise.

III. Physiological Studies

A. ROLE OF TESTIS IN MÜLLERIAN REGRESSION

A contemplative approach to study of fetal müllerian development indicates that müllerian ducts are maintained in females and regress in males, but gives no hint regarding the mechanism of these events. Other methods are needed to distinguish between genetic and endocrine control of müllerian development and to determine whether separate mechanisms are required to induce müllerian maintenance in females and regression in males.

1. Clinical Data

Analysis of clinical situations in which gonadal development does not conform to the chromosomal constitution has been helpful. In cases of gonadal agenesis, be the karyotype 45, X as in Turner's syndrome, 46, XX or 46, XY as in pure gonadal dysgenesis, uterus and fallopian tubes are always normal. On the contrary, in males who develop testes in spite of an XX constitution, müllerian regression proceeds normally. These observations are consistent with the hypothesis that müllerian maintenance is a consequence of testicular absence and does not require the presence of ovaries or a XX karyotype.

2. In Vivo Experimental Data

Reenactment of the experiments of nature, i.e., attempts at fetal gonadectomy, are particularly difficult to carry out *in vivo* as müllerian development can be influenced only if the castration is performed at the end of the indifferent stage of müllerian development. For instance, in rats, castration of male fetuses at 18–19 days did not prevent continuing müllerian degeneration (Wells et al., 1954). Raynaud (1950) using X-rays to destroy gonads of younger mice fetuses induced radiolysis of müllerian ducts, so that no conclusions could be drawn. The only convincing experiments concerning mammals in the literature were performed in rabbits by Jost (1947). He showed that castration of fetal males resulted in persistence of müllerian ducts, if the operation was carried out at 19 days.

3. Organ-Culture Experiments

Because of the prohibitive technical difficulties of gonadal manipulation in young mammalian fetuses, *in vitro* techniques have gained increasing popularity and are now widely used. After the preliminary experiments of Jost and Bergerard (1949) and Jost and Bozic (1951), Price and Pannabecker (1959) published a detailed study of *in vitro* sex differentiation of the gonaducts in fetal rats, but they were not able to demonstrate an effect of the fetal testis on müllerian development. The first investigator to prove the determining influence of the mammalian fetal testis upon the müllerian duct *in vitro* was Régine Picon in 1969. She showed that rat fetal müllerian ducts were maintained in rat reproductive tracts explanted without gonads but regressed when cocultured with fetal testicular tissue. Similar results were obtained with tracts of either sex, but the time of explantation was of crucial importance: complete müllerian regression could not be obtained in female tracts aged 15.5 days or more.

Sensitivity of the müllerian duct to testicular hormone is a transient phenomenon. The period during which the müllerian duct is capable of responding to AMH is programmed at the end of the ambisexual stage and is called the "critical period" (Fig. 1). Before the "critical period," growth and maintenance of the müllerian duct are not hormone dependent; after the critical period, male müllerian ducts irreversibly degenerate, even in the absence of AMH, and AMH no longer inhibits female müllerian ducts. All segments of the müllerian duct are not responsive to AMH at the same moment. In the rat, sensitivity of the müllerian duct to testicular hormone progresses craniocaudally from 14 to 16 days postcoitum (p.c.). At 14.5 days p.c., the anterior segment is more sensitive to AMH than the middle part, and the posterior tip still growing toward the urogenital sinus is quite unresponsive. In the human, the critical period is certainly terminated in fetuses 30–32 mm in crown–rump length, as müllerian ducts already begin to degenerate in males at this stage. We have been able to confirm this experimentally. Five human female reproductive tracts from fetuses 25–55 mm in crown–rump length were maintained 2 weeks in organ culture. One side was cultured alone; the other was associated with human fetal testicular tissue (Table I). Müllerian regression was obtained only in the youngest, 25-mm, fetus (Fig. 2).

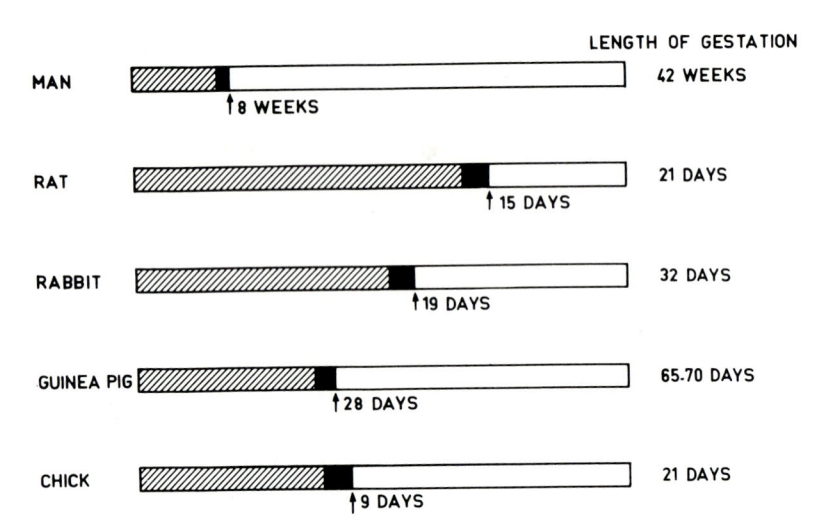

```
                                              LENGTH OF GESTATION
MAN      ▨▨▨▨▨▨▨▨▨▨▨▨▨▨▨▨▨▨▨▨▨▨▨▨▨▨          42 WEEKS
                 ↑8 WEEKS

RAT      ▨▨▨▨▨▨▨▨▨▨▨▨▨▨▨▨▨▨▨▨▨▨▨▨▨▨          21 DAYS
                          ↑15 DAYS

RABBIT   ▨▨▨▨▨▨▨▨▨▨▨▨▨▨▨▨▨▨▨▨                32 DAYS
                    ↑19 DAYS

GUINEA PIG ▨▨▨▨▨▨▨▨▨▨▨▨                      65-70 DAYS
               ↑28 DAYS

CHICK    ▨▨▨▨▨▨▨▨▨▨▨▨                        21 DAYS
               ↑9 DAYS
```

FIG. 1. Sensitivity of müllerian ducts to antimüllerian hormone (AMH) in various species. The hatched zone represents the duration of the ambisexual stage; the black zone, the period of responsiveness of müllerian ducts to AMH. The end of this period is marked by an arrow. Data for this figure were drawn from Josso (1974a) in man, Picon (1969) in the rat, Jost (1947) and Picon (1971) in the rabbit, Price et al. (1975) in the guinea pig and Haffen (1970) in the chick.

TABLE I

Coculture of Reproductive Tracts of Human Female Fetuses with Human Fetal Testicular Tissue[a]

Male fetus		Female fetus		Aspect of female müllerian duct
No.	Size (mm)	No.	Size (mm)	
H 52	90	H 53	53	Maintained
H 76	170	H 75	55	Maintained
H 84	240	H 85	35	Maintained
H 83	150	H 86	38	Maintained
H 108	58	H 109	25	Inhibited

[a] The two sides of a tract were grown separately during 2 weeks; one cultured alone served as control. The müllerian duct on the control side was always maintained. Fetal testicular tissue was able to inhibit the müllerian duct only in the youngest, 25-mm, female fetus; müllerian ducts of the 4 older female fetuses were no longer responsive to antimüllerian hormone.

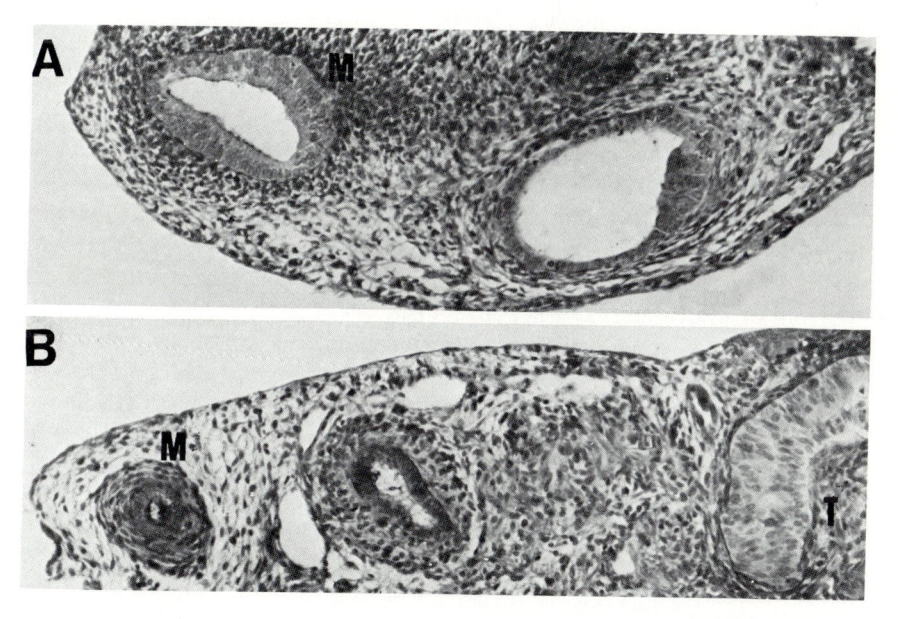

FIG. 2. Response of müllerian duct from a human 25-mm female fetus to human fetal testicular tissue after 15 days in organ culture. (A) Left müllerian duct cultured alone (control). Müllerian duct is maintained. (B) Right müllerian duct after coculture with testicular tissue from 58-mm fetus. Müllerian duct is regressed. M, müllerian duct; T, testicular tissue. Periodic acid–Schiff; × 250. From Josso (1974a) with permission.

B. BIOASSAY FOR ANTIMÜLLERIAN ACTIVITY

The method we use at present for testing antimüllerian activity of tissue or media is derived from the work of Picon (1969). Rat reproductive tracts of Wistar rat fetuses of both sexes aged 14.5 days ± 8 hours are dissected out and their gonads are removed. The tracts are explanted on the grid of a Falcon organ-culture dish, filled with Eagle's minimum essential medium, and placed in an incubator at 37°C in a 95% air–5% CO_2 atmosphere. Antimüllerian activity of tissues can be tested by placing 1-mm^3 explants between the anterior segments of the rat reproductive tracts. Soluble substances can be dissolved in the medium. After 3 days *in vitro*, the tracts are fixed and serially sectioned at 7.5 μm. It is important to obtain a picture of the whole length of the reproductive tract. One section out of ten is stained by PAS-hematoxylin and histologically examined. The wolffian basal membrane stains red with PAS; this feature is a means of identification that may be useful, as wolffian ducts are not yet testosterone dependent at this stage and persist in all tracts.

1. The degree of müllerian duct regression is scored on each individal *section* of coded slides, according to the following criteria, illustrated on Fig. 3: (A) *complete inhibition*—the duct epithelium is replaced by a fibrous whorl; (B) *partial inhibition*—duct epithelium is still visible, but its height is decreased and it rarely contains dividing cells. The lumen is narrow, surrounded by a fibroblastic ring; (C) *no inhibition*—müllerian and wolffian ducts have approximately the same width, müllerian epithelium is normal and shows frequent mitotic figures. Cell density is increased around the duct, but the cells do not have the elongated morphology characteristic of fibroblasts.

2. The scores attributed to the müllerian duct in each section are used to draw a reconstruction of the *müllerian system of a given reproductive tract*. This reconstruction serves to assess the degree of müllerian regression in the reproductive tract, considered as a whole (Fig. 4). Tracts showing no alteration of the müllerian duct on either side are considered normal. Total regression is defined as complete inhibition of the müllerian duct on both sides and on the whole length of the tract, except in the immediate vicinity of the urogenital sinus. Incomplete regression is said to occur in all other cases when müllerian inhibition is heterogeneous, asymmetrical, and/or limited to the anterior segment of the ducts, more sensitive to AMH at this stage of development.

3. The degree of *antimüllerian activity* of a given *tissue* or *medium* can then be expressed as the relative number of rat reproductive tracts in which total, incomplete, or no müllerian inhibition has been elicited after exposure to the tissue or medium during 3 days *in vitro*.

This bioassay has proved to be reliable when performed on coded slides, but care must be exerted to distinguish the hallmarks of AMH action on the müllerian duct from nonspecific changes. For instance, poor culture conditions and toxic substances may alter the appearance of the müllerian duct and produce necrosis or narrowing of the lumen. Usually, however, in these nonspecific situations the epithelium is maintained and lesions are limited to, or maximal in, the posterior segment of the duct, which has differentiated *in vitro* during the 3-day culture period.

The major drawback lies in the semiquantitative nature of the test. The morphological criteria used to score the degree of regression of the müllerian duct do not lend themselves to quantitative analysis, and biochemical criteria cannot be used without prior isolation of the müllerian duct, which we have not

---

FIG. 3. Anterior segment of 14.5-day old rat fetal reproductive tract after 3 days in organ culture. (A) Complete inhibition of müllerian duct: epithelium is replaced by fibroblasts. (B) Incomplete inhibition of müllerian duct: the duct is narrow, and surrounded by a fibroblastic ring. (C) Müllerian duct is normal. M, müllerian duct; W, wolffian duct. Periodic acid–Schiff; × 350.

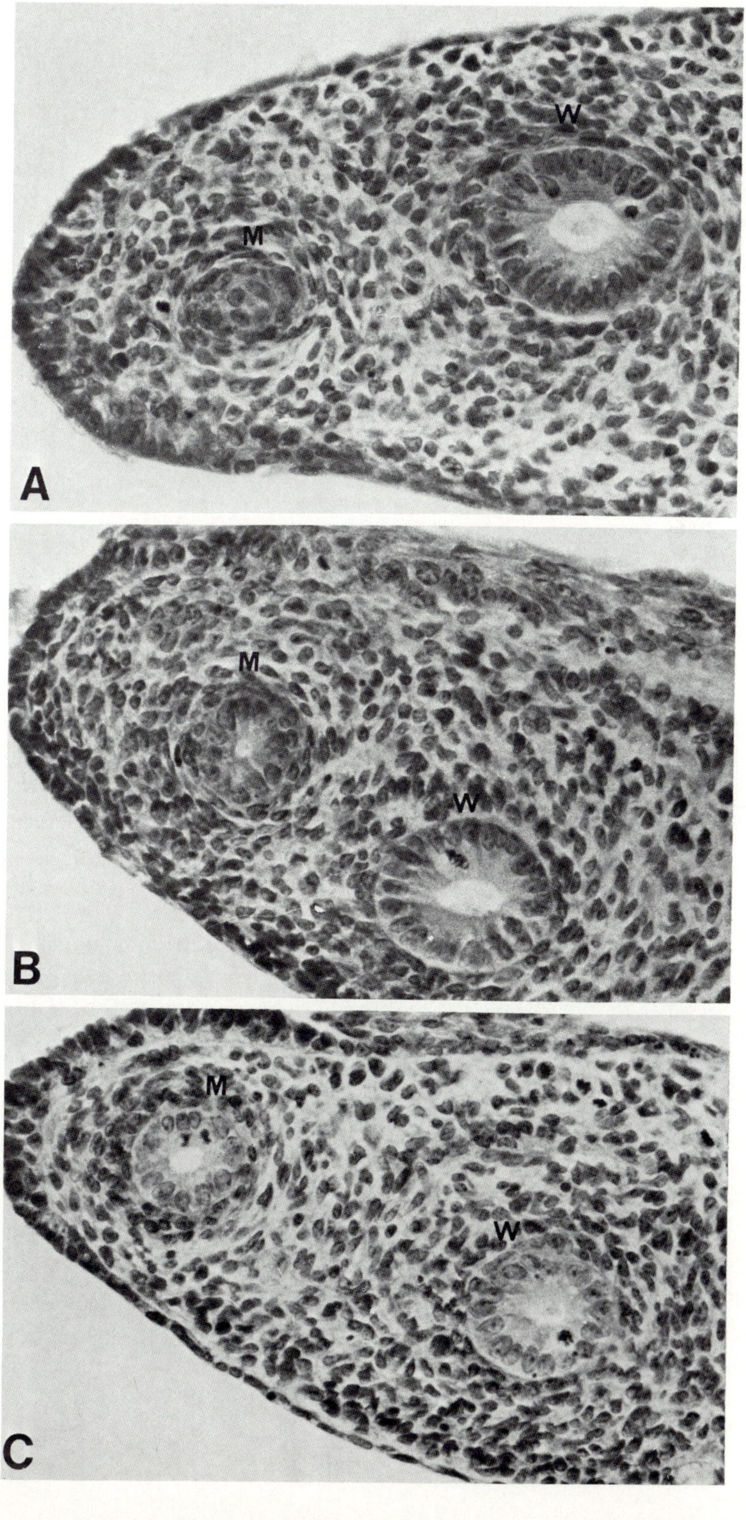
W
M
A
M
W
B
M
W
C

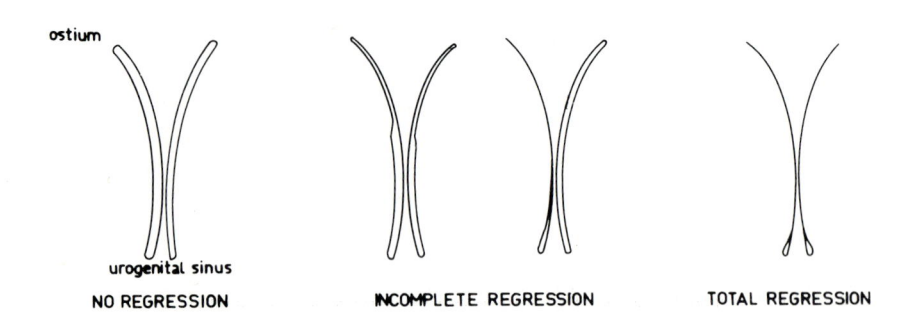

FIG. 4. Criteria used for scoring the degree of regression of the müllerian system in individual rat reproductive tracts.

been able to perform routinely at the critical period, even in species larger than the rat.

C. EVIDENCE FOR A DISCRETE ANTIMÜLLERIAN HORMONE

After the demonstration of the role of the fetal testis in müllerian regression, the question of the nature of the testicular factor involved became open to discussion. Synthesis and secretion of testosterone by the fetal male gonad has been amply demonstrated, and the effect of testosterone on masculinization of the urogenital sinus is now well documented (Goldstein and Wilson, 1975). Is an androgenic steroid also responsible for müllerian regression?

1. Clinical Data

In female fetuses, endowed with a 46, XX karyotype and normal ovaries, intersexuality results from exposure to androgens in early pregnancy. The genitalia of these pseudohermaphroditic children may be so virilized as to mimic the appearance of cryptorchid males, but their fallopian tubes and uterus are always normal. This observation, however, does not unequivocally prove that androgens are incapable of producing müllerian regression unless it can be shown that androgens reached the fetus before the end of the ambisexual stage, i.e., before the 30-mm stage in humans. Most cases of female pseudohermaphroditism are due to congenital adrenal hyperplasia, and in this condition the date of onset of abnormal steroidogenesis by the fetal adrenal is not known. However, in a case reported by Mürset *et al.* (1970) female pseudohermaphroditism resulted from a maternal adrenal tumor, which persisted during the pregnancy. Urethra and labioscrotal folds were completely masculinized, but müllerian duct derivatives were normal.

Sex differentiation of the urogenital sinus and of the müllerian ducts in opposite directions is also seen in XY patients in whom intersexuality is due to fetal testicular failure or to insensitivity of the end organs to androgens. Most male pseudohermaphrodites lack müllerian derivatives, except in cases complicated by unilateral gonadal dysgenesis or by gonadoblastoma. Conversely, müllerian ducts exceptionally persist in normally virilized males, whose external genitalia are normal, apart from a high incidence of cryptorchidism. This syndrome has been reported in siblings (Brook et al., 1973). Failure to produce or to respond to the müllerian inhibiting factor in these patients is not associated with similar disorders in androgen metabolism.

2. In Vivo Experimental Studies

Administration of androgens to pregnant animals results in virilization of their female progeny (Jost, 1953; Schultz and Wilson, 1974), but, as in human androgen-induced female pseudohermaphroditism, müllerian ducts are not affected. This confirms the results of the early experiments of Jost (1947), who showed that müllerian regression in female rabbit fetuses could be produced by a testicular graft, but not by implantation of a testosterone crystal.

Male pseudohermaphroditism can also be experimentally induced by administration to pregnant animals of an antiandrogen, such as cyproterone acetate (Elger, 1967). Cyproterone acetate blocks the cytoplasmic androgen receptor and produces a syndrome of androgen insensitivity similar to that existing in testicular feminization. As in testicular feminization, the genitalia of XY newborns resemble those of their female siblings, but their müllerian ducts are normally regressed.

3. In Vitro Experimental Data

Conflicting results concerning the effect of testosterone upon müllerian development in organ culture have been reported in mammals and in birds. Josso (1971a) found no effect of either testosterone or its 17-hydroxylated derivatives upon the müllerian duct in 14.5-day-old rat reproductive tracts after 3 days in organ culture. On the contrary, Wolff *et al.* (1952), followed recently by Lutz-Ostertag (1974), observed a lysis of avian müllerian ducts exposed to testosterone in organ culture and concluded that testosterone is the müllerian-inhibiting factor in the chick embryo. However, the histological aspect of the testosterone-treated müllerian duct, shown in the Lutz-Ostertag paper, did not support this claim, as extensive vacuolization of the müllerian epithelium is never observed during physiological testicular-induced müllerian regression, but rather suggests a toxic effect. The exact testosterone concentration used in these experiments was not determined, but it was probably high, as müllerian ducts were exposed to testosterone crystals.

Recently Woods *et al.* (1975) have measured testosterone concentration in chick embryo plasma; the peak concentration, 1 nM, was reached at 13.5 days. Obviously, experiments performed using large amounts of testosterone do not reproduce physiological conditions. Tran and Josso (1977) submitted 8-day chick müllerian ducts to various concentrations of testosterone during 4 days in organ culture. Toxic effects on müllerian epithelium were observed at concentrations higher than 15 µM; lower doses did not affect the müllerian duct in any way (Fig. 5). These data support the conclusion that in mammals and birds alike, testosterone is not the physiological antimüllerian hormone.

D. SPECIES SPECIFICITY OF ANTIMÜLLERIAN HORMONE

Interest has been focused recently on the comparative aspects of endocrinology and species-related differences of biological activity and immunoreactivity have been demonstrated for many protein and polypeptide hormones. In view of the accumulating evidence in favor of the nonsteroidal nature of AMH, a study of its species-specificity appeared to be warranted for practical as well as speculative reasons. On one hand, we wished to determine whether our present bioassay allowed us to test antimüllerian activity in testicular tissue of various species. On the other hand, we were interested in testing the responsiveness of the chick müllerian duct to mammalian AMH. The young fetal chick müllerian

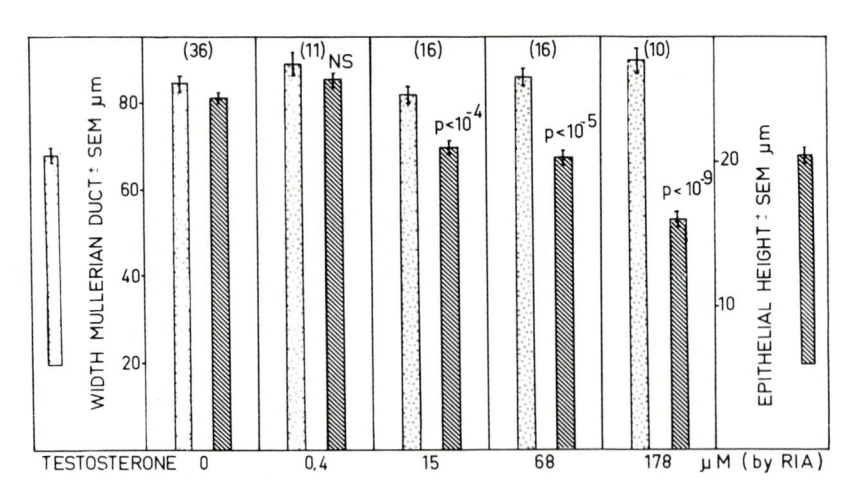

FIG. 5. Effect of testosterone on width and epithelium of müllerian duct of the 8-day chick embryos maintained 4 days in organ culture. Testosterone has no effect upon the width of the müllerian ducts, but high concentrations of the steroid decrease epithelial height. This is considered a toxic effect. Significance was assessed by analysis of variance, experimental groups were compared to controls. Number of experiments is shown in parentheses. RIA, radioimmunoassay.

duct is easily dissected and would be a good starting material for the setting-up of a quantitative bioassay for AMH, based on biochemical criteria.

1. Intramammalian Species Specificity

To this day, no species specificity of antimüllerian biological activity has been shown in mammals. Josso (1970, 1971b) cocultured fetal testicular tissue from 11 human fetuses, aged 7–24 weeks, with 47 14.5-day-old fetal rat reproductive tracts. In all cases in which the human tissue was in good condition, it elicited complete regression of the fetal rat müllerian duct, which had disappeared or was represented only by a whorl of fibrous tissue (Fig. 6) except in the immediate vicinity of the urogenital sinus. Control experiments showed that the rat müllerian duct was maintained after coculture with human fetal ovaries or adrenals.

Similar results were obtained by Picon (1971), who demonstrated reciprocal inhibition of müllerian ducts by fetal testicular tissue in 18-day-old rabbits (Géants des Flandres) and in 14.5-day-old rats. However, complete regression of the rabbit müllerian duct was not obtained, particularly in females, even with homospecific testicular tissue. Apparently, the müllerian duct of the 18-day-old female fetal rabbit has already partly lost its sensitivity to AMH, and explanta-

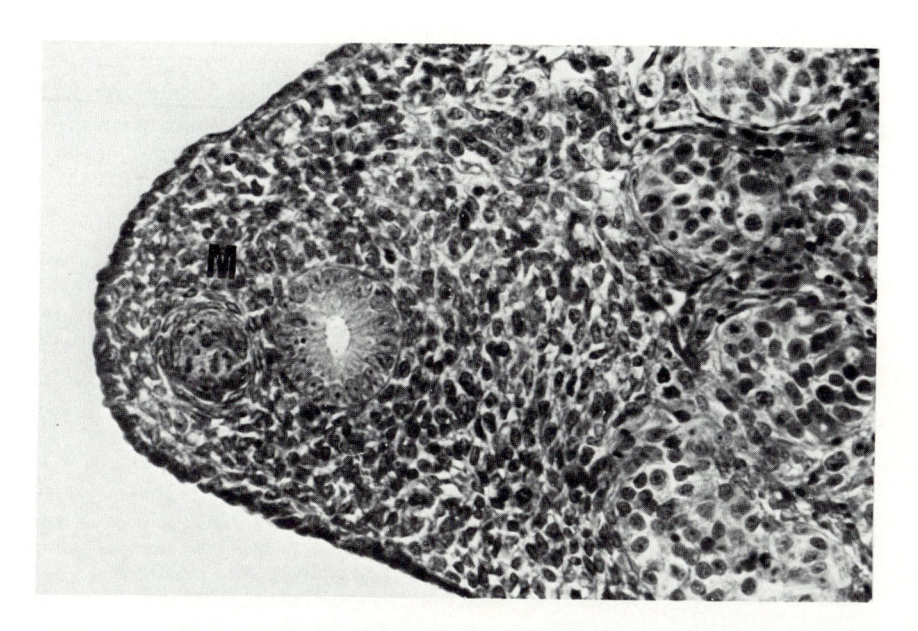

FIG. 6. Association of 14.5-day-old rat fetal reproductive tract to testicular tissue of a 16-week-old human fetus for 3 days in organ culture. Müllerian duct (M) is completely regressed. Hematoxylin and eosin; × 350.

tion should be performed earlier to obtain unequivocal results in this particular strain of rabbits. Calf (Josso, 1973) and pig (Tran *et al.*, 1977) fetal testicular tissue also inhibit the rat müllerian duct *in vitro*.

2. Relationship between Mammalian and Avian AMH

The first attempt to study the biological activity of AMH in different classes of vertebrates was made by Weniger (1963), who reported that mouse fetal testicular tissue has a paradoxical stimulatory effect on the chick müllerian duct. The same author (Weniger, 1965) claimed that chick fetal testes inhibit müllerian ducts of female mice fetuses 15 to 18 days p.c., but, as in the same paper Weniger showed that mouse müllerian ducts do not respond to homospecific testicular tissue after 15 days, the interpretation of these data is not clear. The problem was reinvestigated by Tran and Josso (1977).

a. Effect of Mammalian Testicular Tissue on Chick Müllerian Ducts. Chick müllerian ducts were explanted at 8 days of incubation and exposed to fetal testicular tissue during 4 days in organ culture. Twenty-five pairs of ducts were associated with testicular tissue from 8-day-old fetal chicks, 25 others with testicular tissue from 16- to 20-day-old fetal rats, and 28 with testicular tissue from 2 fetal calves, 18 and 20 cm in vertex–rump length. The antimüllerian activity of these mammalian tissues upon the 14.5-day-old fetal rat müllerian duct has been demonstrated in earlier studies (Picon, 1969; Josso, 1973). Thirty-six pairs of müllerian ducts, cultured alone, served as controls. Chick fetal testicular tissue caused a caudocephalic regression in the chick müllerian ducts associated with it in organ culture. Müllerian regression was complete in 21 cases, in 4 tracts degenerative discontinuous remnants of the müllerian duct were still recognizable (Fig. 7A). Mammalian fetal testicular tissue had a paradoxical effect on chick fetal müllerian ducts: both rat and calf (Fig. 7B) fetal testicular tissue significantly increased the width of the ducts associated to them. The nature of the mammalian fetal testicular factor responsible for this stimulation is not known. Obviously, testosterone can be ruled out, as it does not affect the width of the müllerian duct. Estrogens might be involved. Teng and Teng (1975) have demonstrated the existence of estrogen receptors in the 15-day-old chick müllerian duct; younger stages were not investigated.

FIG. 7. (A) Anterior segment of chick müllerian duct explanted at 8 days and cocultured 4 days with 8-day chick fetal testicular tissue. A degenerating remnant of the müllerian duct is visible. (B) Anterior segment of chick müllerian duct explanted at 8 days and cocultured 4 days with testicular tissue from fetal calf (18 cm in vertex–rump length). Müllerian duct is normal. A mesonephric tubule is seen between testis and müllerian duct. (C) 14.5-day-old rat reproductive tract cocultured 3 days with 13-day-old chick testicular tissue. Note total regression of müllerian duct. M, müllerian duct; W, wolffian duct; m, mesonephric tubule. Periodic acid–Schiff; × 350.

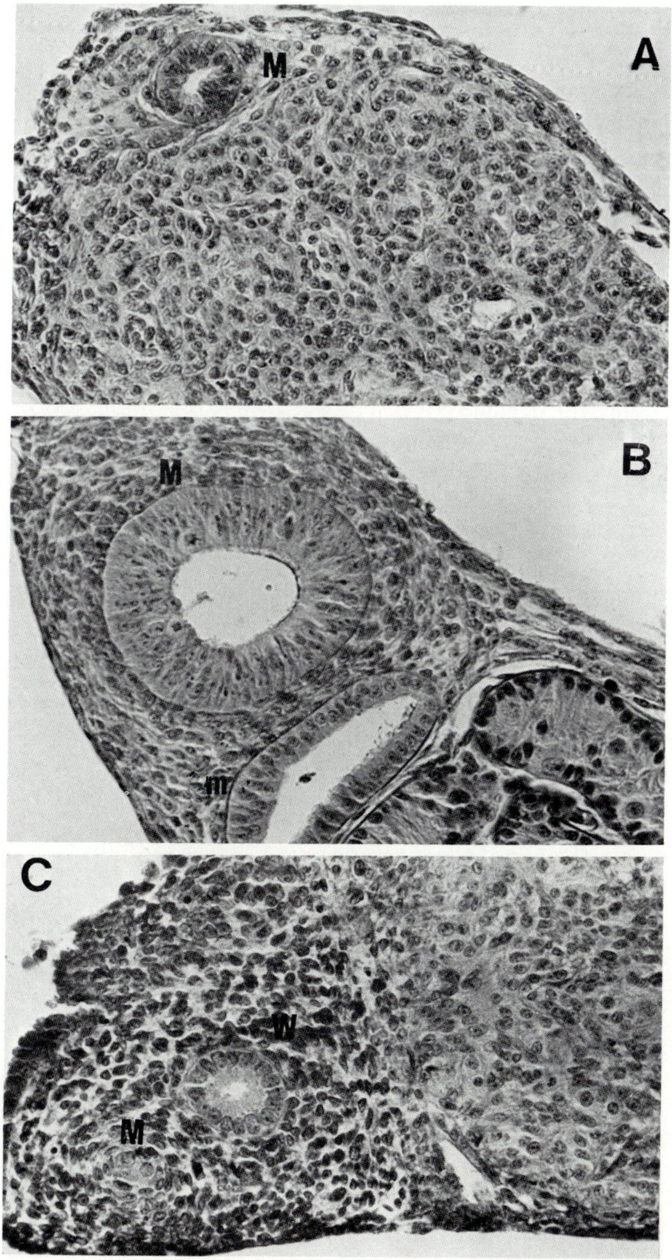
A
M
B
M
m
C
W
M

b. Effect of Chick Fetal Testicular Tissue on Mammalian Müllerian Ducts. Fetal rat reproductive tracts were taken from ten 14.5-day-old fetal rats and cocultured, after removal of their own gonads, with testicular tissue from 13-day-old chick embryos. Stoll *et al.* (1970) have demonstrated a homospecific antimüllerian activity in chick fetal testes up to 15 days of incubation, and this was confirmed by two preliminary experiments associating 13-day-old chick fetal testes with 8-day-old chick müllerian ducts. After a culture period of 3 days, the fetal rat müllerian ducts associated with chick fetal testes were completely inhibited in all cases studied (Fig. 7C). The müllerian epithelium had completely disappeared or was replaced by a fibrous whorl except in the immediate vicinity of the urogenital sinus. Histologically, the effect of chick fetal testes on the rat müllerian duct is indistinguishable from that produced by homospecific testicular tissue under the same conditions.

In conclusion, the demonstration that chick AMH is active in rats prove that the mechanisms for testicular inhibition of the müllerian ducts are very similar in birds and in mammals and that the molecules responsible for the biological effect are closely related. On the other hand, the incapacity of mammalian fetal testicular tissue to inhibit chick müllerian ducts confirms that testosterone is not the antimüllerian factor in birds, as testosterone production by mammalian fetal testes has been amply documented (Goldstein and Wilson, 1975).

E. CHRONOLOGICAL EVOLUTION OF TESTICULAR ANTIMÜLLERIAN ACTIVITY

1. Initiation of AMH Production

Few data concerning the onset of antimüllerian activity in the fetal testis are available in the literature. Using *in vitro* techniques, Picon (1970) showed that antimüllerian activity in the rat fetal testis is low or absent at 13.5 days and appears at 14.5 days. Antimüllerian activity of the guinea pig testis is not detectable before 26 days (Price *et al.*, 1975). Laboratory rodents, however, are not ideally suited to sequential studies of this type, because fetuses in the early stages of development are difficult to manipulate and because sex differentiation occurs within a very short period of time.

Use of larger animals, with a more leisurely pattern of testicular organogenesis would be more appropriate, provided gestational age is accurately determined and provided the antimüllerian activity of testicular tissue is tested on müllerian ducts sensitive to AMH at the time of the bioassay. Otherwise, falsely negative results can be obtained since young homospecific müllerian ducts may not yet be sensitive to AMH or may require a prolonged period of exposure to it before regression is apparent. Tran *et al.* (1977) have recently studied the development of antimüllerian activity of the fetal pig testis, using the 14.5-day-old rat müllerian duct as target organ. The animals were bred in a zootechnical center, and the day of conception was accurately recorded. Fetal sex was determined by cytogenetic analysis of somatic cells (Pelliniemi and Salonius, 1976). Antimüllerian activity was negligible in 27- and 28-day-old fetuses and clearly present at 29 days (Fig. 8).

Leydig cells differentiate at 15 days in the rat (Niemi and Ikonen, 1961), at 27–28 days in the guinea pig (Black and Christensen, 1969), and at 30 days in the pig (Moon and Hardy, 1973). Therefore, in these species, AMH production precedes the appearance of steroidogenetic interstitial cells. This chronological pattern is similar to that found in the human fetus, in whom Leydig cells appear at 30 mm crown–rump length (Jirasek, 1967) whereas antimüllerian activity has been demonstrated in one 20-mm fetus (Josso, 1971b). In contrast, development of antimüllerian activity is synchronized with differentiation of seminiferous tubules. In the pig, although some ultrastructural evidence of testicular organization may be recognized as early as 26 days (Pelliniemi, 1975), by the light microscope, delineation of the tubular basal membrane through silver impregnation was clearly visible only at 29 days (Tran et al., 1977). In the rat, Merchant-Larios (1976) has established a good correlation between ultrastructural differentiation of Sertoli cells and regression of müllerian ducts.

2. *Loss of Antimüllerian Activity*

The chronological evolution of AMH production by the fetal testis has been studied in rats by Picon (1970) and in man by Josso (1972a). The pattern is

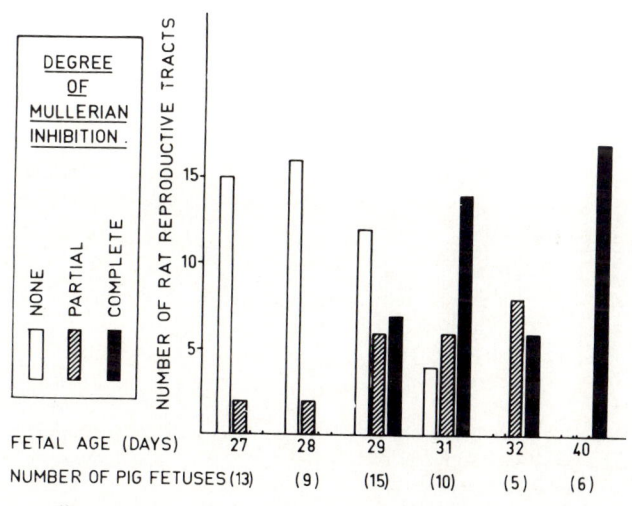

FIG. 8. Antimüllerian activity of the young fetal pig testis expressed as number of 14.5-day-old fetal rat reproductive tracts associated with explants exhibiting totally regressed, incompletely regressed or normal müllerian ducts after 3 days in organ culture.

similar in both species and is characterized by the loss of testicular antimüllerian activity after birth. In the rat, Picon (1970) has measured the length of the müllerian segment inhibited after co-culture with testicular tissue of rats of different ages. She has shown that testicular production of AMH extends into the second postnatal week and is no longer demonstrable 19 days after birth. Donahoe et al. (1976) have arrived at the same conclusions.

Antimüllerian activity of the human testis declines already in late pregnancy and is very low at birth, even in premature infants. The study was carried out on 31 male fetuses (7–32 weeks), obtained after therapeutic, spontaneous or induced abortion, on testicular tissue obtained at autopsy from 11 newborns (27–40 gestational weeks) having survived at most 15 days, and on testicular tissue from 10 postnatal subjects (Table II). The degree of müllerian regression produced by human testicular tissue at various ages is shown on Fig. 9. Fetal testicular tissue exhibited the highest antimüllerian activity, producing total müllerian regression in all tracts except 9. These had been associated with testicular explants from 4 of the oldest fetuses in the series, but no strict correlation between gestational age and antimüllerian activity could be established. The position of the testes did not appear to be critical, as in one fetus, in whom only the right testis was in the scrotum, antimüllerian activity was similar in both testes. Perinatal testicular tissue had low or undetectable antimüllerian activity; here also, no strict correlation could be established with gestational age. Müllerian ducts were normal in all 77 tracts associated with postnatal testicular tissue, even when it was taken from pubertal subjects.

In conclusion, AMH is a fetal hormone whose synthesis begins at the time or immediately after differentiation of seminiferous tubules, before the appearance

TABLE II
Source of Postnatal Testicular Tissue Investigated for Antimüllerian Activity

Age (years)	Diagnosis	Source of tissue
15	Male pseudohermaphroditism	Castration
15	Incomplete testicular feminization syndrome	Biopsy
18	Delayed puberty	Biopsy
20	Hypopituitarism	Biopsy
5/12	Meningitis	Autopsy
19	Testicular feminization syndrome	Castration
15	Cryptorchidism	Biopsy
9	Encephalitis	Autopsy
5	Adrenal hyperplasia	Biopsy
2/12	Transposition of great vessels	Autopsy

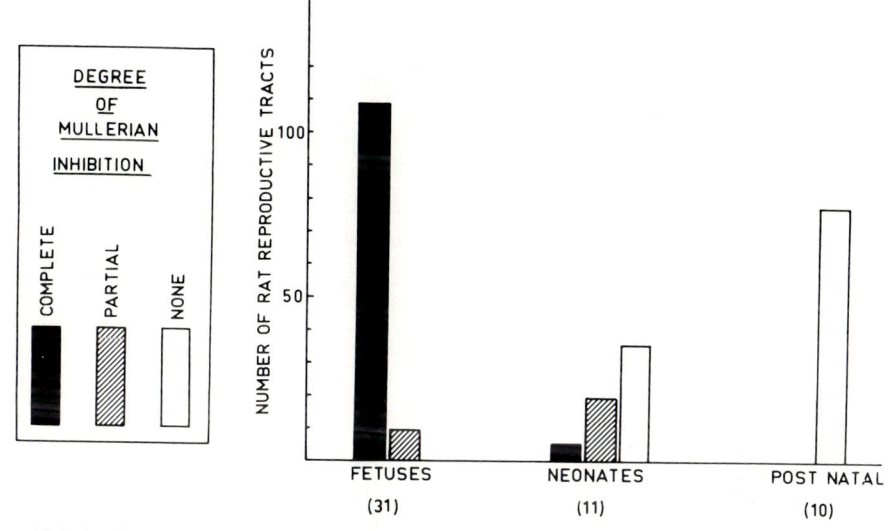

FIG. 9. Chronological evolution of antimüllerian activity of human testicular explants, expressed as in Fig. 8. The number of cases in each group is shown in parentheses. From Josso et al. (1977a) with permission; © 1977 University Park Press, Baltimore.

of Leydig cells. It is no longer demonstrable after birth. There appears to be no correlation between the prolonged antimüllerian activity of the fetal testis and the short period during which the müllerian duct is responsive to AMH.

F. SITE OF AMH PRODUCTION

The lack of correlation between Leydig cell development and antimüllerian activity of the testis suggests that these cells are not involved in AMH synthesis, and that the hormone is probably of tubular origin. Various methods have been used to test and extend this hypothesis.

1. Microdissection of the Calf Fetal Testis (Josso, 1973)

Since the pioneer publication of Christensen and Mason (1965), isolation of seminiferous tubules and interstitial tissue has been widely used in the study of the adult rat testis. Using the modified technique suggested by Hall et al. (1969), Rivarola et al. (1972) also succeeded in dissecting the testis of immature rats, but, as could be expected, with difficulties inversely correlated to the age of the animal. The use of a large species minimizes the problems raised by microdissection in the young animal. We have found the calf fetal testis particularly suited to this purpose. Microdissection of testicular tissue is performed manually, under the dissecting microscope. Testicular tubules are cleaned, one by one, with the aid of insect pins mounted on glass rods, of the clusters of interstitial tissue that adhere to their surface, and are extensively washed until the rinse is clear. The yield of organized interstitial tissue is lower, as the clusters tend to disintegrate in the buffer wash, where cells could eventually be recovered by centrifugation. Attempts to facilitate microdissection by the use of collagenase were not successful. With the method described, reasonably clean testicular tubules can be obtained, but contamination of interstitial tissue by small tubular fragments is difficult to avoid. The best results are obtained with fetal calves 30–40 cm in forehead–rump length, though, with patience, microdissection can be carried out in fetuses only 15 cm long.

Antimüllerian activity of whole fetal testicular tissue and of isolated seminiferous tubules and interstitial tissue was tested in 19 fetal calves by association with 14.5-day-old fetal rat reproductive tracts. No antimüllerian activity was found in the 2 largest fetuses (70 and 75 cm in forehead–rump length). The degree of müllerian inhibition seen in rat reproductive tracts associated with whole or microdissected testicular tissue from the 17 younger fetuses (19–45 cm in forehead–rump length) is shown in Figs. 10 and 11. Whole testicular tissue and isolated seminiferous tubules exhibited marked antimüllerian activity; in contrast, isolated interstitial tissue inhibited müllerian ducts in only 4 out of the 20 fetal reproductive tracts associated to it. In these 4 cases, heavy contamination by tubular fragments was histologically evident. Although our findings do not completely rule out interstitial tissue as an auxiliary source of AMH, seminiferous tubules appear to play a major role in its production during fetal life.

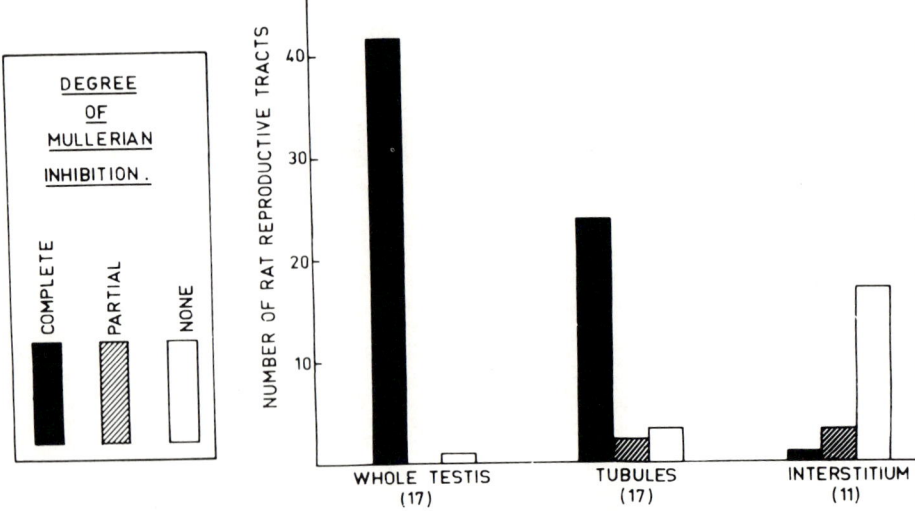

FIG. 10. Site of production of AMH by the calf fetal testis. Comparison of antimüllerian activity of whole testis, seminiferous tubules, and interstitial tissue, expressed as in Fig. 8. From Josso *et al.* (1977b) with permission.

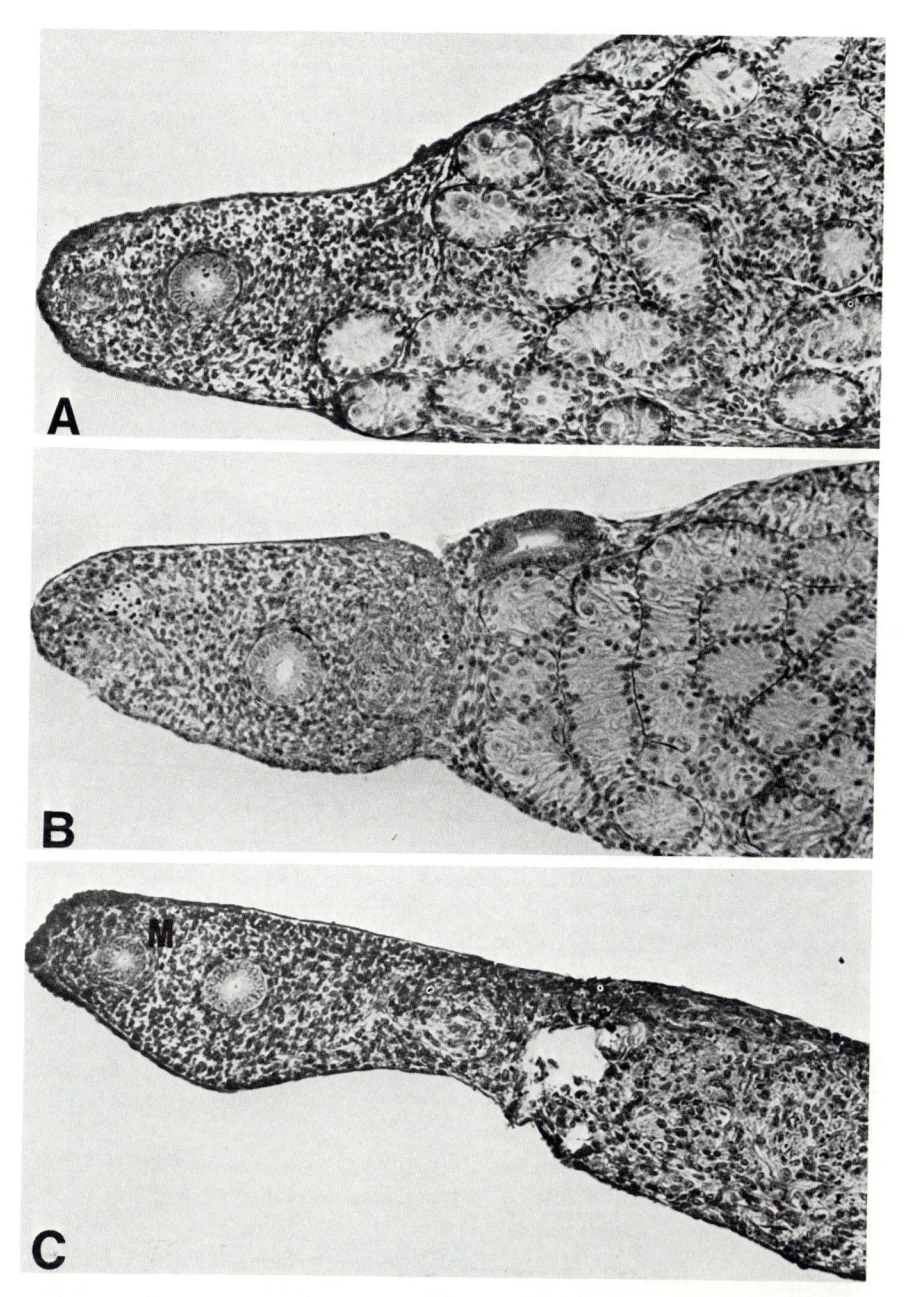

FIG. 11. Effect of components of calf fetal testicular tissue on the 14.5-day-old rat müllerian duct after 3 days in organ culture. (A) Whole testis: total regression of müllerian duct is seen. (B) Isolated seminiferous tubules: total müllerian regression is obtained. (C) Isolated interstitial tissue: müllerian duct (M) is normal. Periodic acid–Schiff; × 350. From Josso et al. (1977a) with permission; © 1977 University Park Press, Baltimore.

2. Irradiation of the Human Fetal Testis (Josso, 1974b)

In order to determine which tubular cells are responsible for AMH synthesis, separation of germ from Sertoli cells is necessary and cannot, obviously, be effected by mechanical means. These cell types, however, differ in their physiological characteristics, and these differences can be used to obtain one type in relatively pure form. For instance, germ cells are extremely radiosensitive and γ-irradiation will selectively destroy the germinal population of seminiferous tubules. This experiment was performed on human fetal testicular tissue, which retains its müllerian-inhibiting activity longer *in vitro* than does the calf fetal testis. Testicular tissue was obtained from seven human fetuses, 60–120 mm in crown–rump length. All except one (H 124) were products of therapeutic interruption of pregnancy. After fixation of control tissue in Bouin's solution, fragments of fresh testicular tissue were explanted in organ culture dishes. Three were prepared for each fetus. One remained unirradiated, and the others were submitted, respectively, to 500 and 700 rads of ^{60}Co γ-rays at a dose rate of 130 rads/minute at room temperature. Irradiation was performed within 2 hours of explantation. After irradiation, the steel grids supporting treated and control explants were transferred to new organ culture dishes and fresh medium was added.

After an interval of 1–2 weeks, depending on the availability of rat fetuses of suitable age, the müllerian-inhibiting activity of the explants was tested by association with 14.5-day-old fetal rat reproductive tracts. Furthermore, control tissue fixed before culture and additional testicular explants were sectioned serially at 5 μm, and one section out of five was stained by periodic acid–Schiff–hematoxylin. The number of normal germ cells present in 50 orthogonal cross sections of seminiferous tubules was determined on a minimum of 10 histologic sections. Necrotic cells or those undergoing nuclear degeneration were not taken into consideration. From these data the average percentage of surviving germ cells was calculated (Table III). In our culture conditions, germinal depletion was already severe in control, unirradiated explants, which after 1–2 weeks *in vitro* contained an average of only 12% of the number of germ cells present at explantation. Exposure of fetal testicular explants to γ-rays increased germ cell damage, but the response to irradiation was variable from one fetus to another.

No correlation between number of surviving germ cells and antimüllerian activity was found. Antimüllerian activity was normal in all explants except those in whom necrosis occurred *in vitro*. In the 14 rat reproductive tracts associated with control fetal testicular explants, extensive müllerian duct inhibition occurred after 3 days. The müllerian duct had either disappeared or was reduced to a fibrous whorl. The same aspect was seen in 21 out of 23 rat reproductive tracts associated with testicular explants treated with 500 rads, and in 20 out of 24 testicular explants treated with 700 rads (Fig. 12). In the

TABLE III
Effect of Organ Culture and γ-Irradiation on Survival of Germ Cells in Human Fetal Testicular Explants

Human fetus	C-R[a] length (mm)	Number of days in culture	Before culture	Control explant	500-Rad explant	700-Rad explant
H 120	120	9	352	49	15	9
H 122	80	9	362	21	16	4
H 124	90	7	140	25	25	18
H 126	110	16	419	94	30	13
H 127	60	4[c]	154	Contamination	9	12
H 128	95	13	334	8	28	7
H 130	120	15	539	80	39	7
Average percentage of surviving cells ($P < 10^{-5}$):				12.9	7.03	3.04

Number of germ cells[b]

[a]CR, crown–rump.

[b]Per 50 cross sections of seminiferous tubules.

[c]The müllerian-inhibiting activity of H 127 was tested the day after irradiation because contamination was noted in control explants, and it was feared that irradiated explants might suffer the same fate.

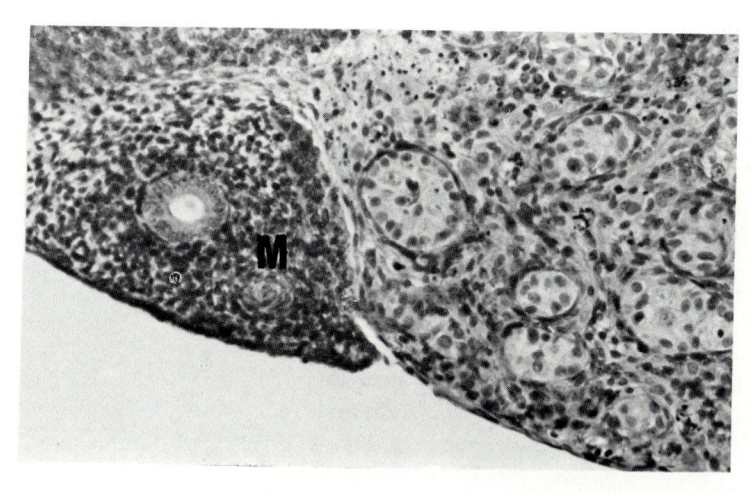

FIG. 12. Inhibition of 14.5-day-old rat fetal müllerian duct (M) by human fetal testicular tissue exposed to 700 R, in organ culture. Antimüllerian activity of testicular tissue is not affected by the near-total destruction of germ cells produced by γ-irradiation.

remaining tracts, associated with necrotic testicular explants, müllerian ducts were normal or only slighty inhibited.

3. *Tissue Culture of Bovine Sertoli Cells (Blanchard and Josso, 1974)*

The effect of microdissection and of irradiation of the fetal testis upon its antimüllerian activity prove that neither interstitial tissue nor germ cells are necessary to AMH synthesis. This leaves the Sertoli cells the most likely candidates for AMH production, and a method that would unequivocally demonstrate their antimüllerian activity appeared desirable.

Steinberger and Steinberger (1966), Kodani and Kodani (1966), and Yamada *et al.* (1972) have shown that in tissue cultures of postnatal testicular cells, germinal elements remain free-floating in the medium, whereas cells of nongerminal origin attach themselves to the surface of the culture vessel and form a monolayer. If, before cell dissociation, seminiferous tubules are isolated from interstitial tissue, the monolayer formed on the flasks plated with tubular cells is formed only by Sertoli cells, free-floating germ cells being eliminated when culture medium is renewed. By application of this method to fetal calf testicular tissue, we have obtained separate cultures of fetal Sertoli and interstitial cells and tested their müllerian-inhibiting activity.

Microdissection was carried out according to a technique previously described (Josso, 1973) in Eagle's minimum essential medium containing 100 U of penicillin and 100 µg of streptomycin per milliliter. To remove adherent interstitial cells, isolated seminiferous tubules were rinsed in phosphate-buffered saline containing 0.25% trypsin. The cleaned tubules were then incubated for 20 minutes at 37°C in a fresh 0.25% trypsin solution, and the cells were dissociated with a Pasteur pipette. After centrifugation at 800 rpm for 5 minutes, a pellet of tubular cells was obtained. Interstitial cells were obtained by aspirating the microdissection medium under control of the dissecting microscope to avoid tubular fragments. The medium was then centrifuged at 800 rpm for 5 minutes. Fragments of undissected testicular tissue were also dissociated in trypsin and the resulting cellular suspension was collected by centrifugation.

Cultures of the cells derived from seminiferous tubules, interstitial tissue, and whole testis were obtained at 37°C by seeding the cellular suspension in plastic tissue culture flasks in Eagle's minimum essential medium supplemented with 10% fetal calf serum. Leighton tubes were also prepared in each case to allow histologic examination of the monolayer. One week after the initiation of the culture, the coverslip was dipped in methanol and the adherent monolayer was stained with Giemsa. At the same time, the cells grown in the plastic culture flasks were harvested by incubation with 0.25% trypsin, washed twice with culture medium, and collected by centrifugation.

Transfer was then effected by pipetting the cellular pellet resuspended in 0.3 ml of culture medium onto a fragment of the vitelline membrane of a hen's egg, which was spread out on the grid of an organ culture dish. The vitelline membrane was then folded back on top of the cellular suspension, and culture medium was added up to the level of the grid. The number of harvested cells did not allow the preparation of more than one culture dish per cell line and per calf fetus. Preliminary experiments showed that in the 3 days that followed the transfer of the cells to organ-culture dishes, most of the cells died and the number of those that survived to produce cell colonies was quite unpredictable. On the other hand, if the cells were left in the vitelline membrane for more than 2 weeks, many cells present in the center of the colonies died.

Müllerian-inhibiting activity of the cells was tested 1 week after the transfer by placing castrated reproductive tracts of 14.5-day-old fetal rats on the surface of the vitelline membrane. Three days later, the cultures were fixed in Bouin's fluid. The excess vitelline membrane was trimmed away and the rat reproductive tracts were serially sectioned, together with the testicular cell colonies wrapped in the vitelline membrane. Altogether, 34 tracts associated with cells derived from whole testis, 28 associated with tubular cells, and 24 associated with interstitial cells were studied. However, in many cases, the number and viability of the cells contained in the folds of the vitelline membrane did not allow the assessment of their müllerian-inhibiting activity. This could be studied in only 16 colonies derived from whole testicular tissue, in 13 tubular cell colonies, and in 9 interstitial cell colonies.

Distinctive morphologic differences were apparent in the various cell lines that formed a monolayer on the coverslip of the Leighton tubes after one week *in vitro*. Tubular cells, with nuclei containing one or more nucleoli, displayed a pavementlike structure and tended to grow in circles reminiscent of seminiferous tubules (Fig. 13A). No cell resembling a gonocyte was seen. Interstitial cell monolayers were formed of elongated cells with a fibroblastic disposition (Fig. 13B).

Morphologic differences were not recognized in the cells after transfer to the egg vitelline membrane. Necrotic cells, numerous in the 2 or 3 days that followed the modification of the culture conditions, had usually disappeared after 10 days. At that time, the folds of the vitelline membrane were covered by one or two layers of testicular cells and in approximately half of the cultures, these developed into colonies of polygonal cells piled on top of one another. Only colonies formed of at least five layers of viable cells will be considered here. Their müllerian-inhibiting activity was reflected by the appearance of the müllerian duct of the rat reproductive tract associated with each for 3 days (Figs. 14 and 15). No müllerian-inhibiting activity was found in the nine interstitial cell colonies studied (Fig. 15A). Müllerian ducts associated with 15 of 16 colonies derived from undissected testicular tissue, and those associated with 10 out of 13 tubular cell colonies (Fig. 15B) were clearly inhibited, although regression was usually not as marked as when organ cultures, but not tissue

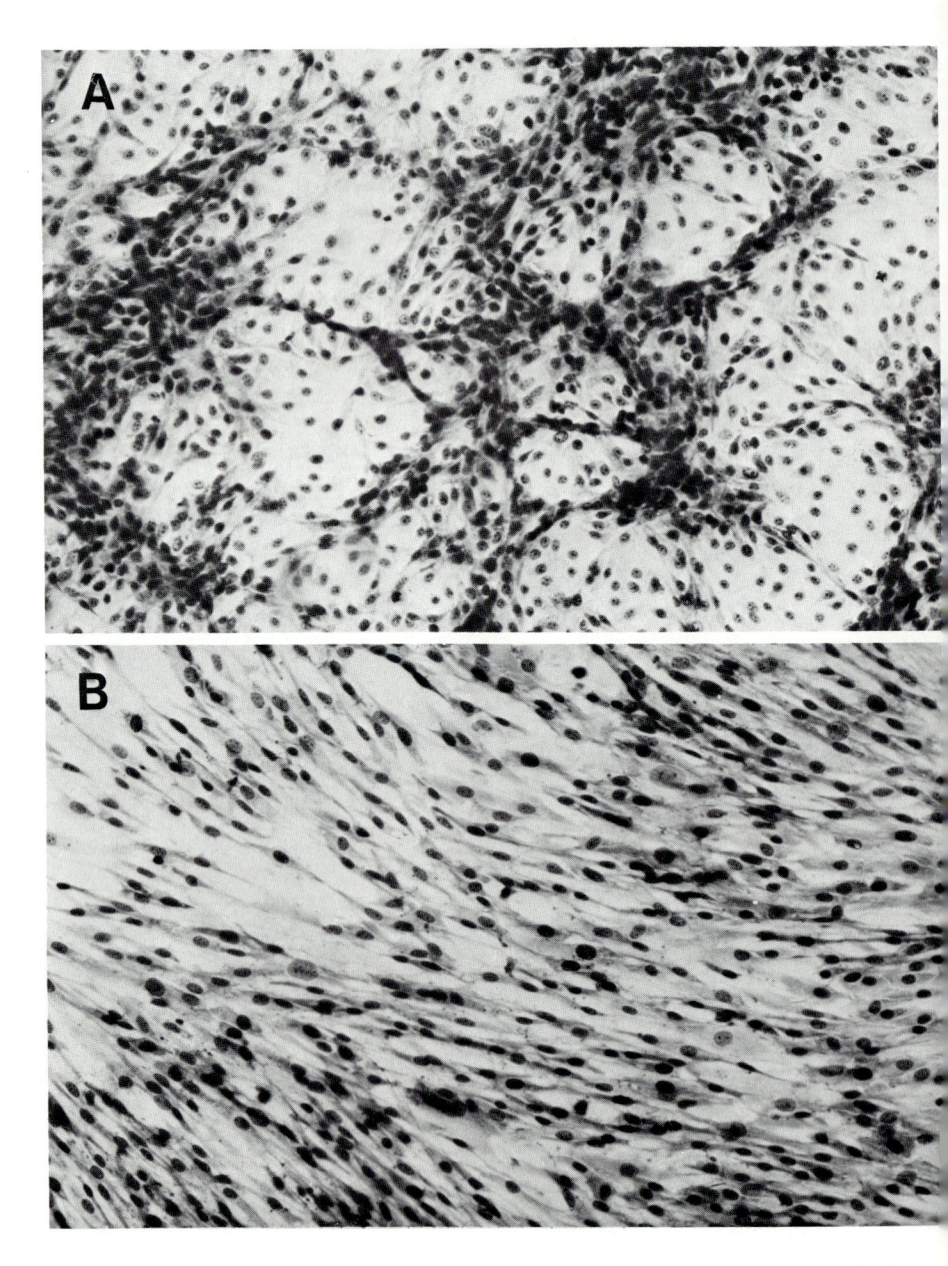

FIG. 13. Aspect of monolayers grown on coverslip of Leighton tubes seeded with fetal bovine testicular cells (Giemsa). (A) Tubular cells: the cells tend to form circular structures reminiscent of seminiferous tubules. (B) Interstitial cells: note elongated fibroblastic pattern of growth.

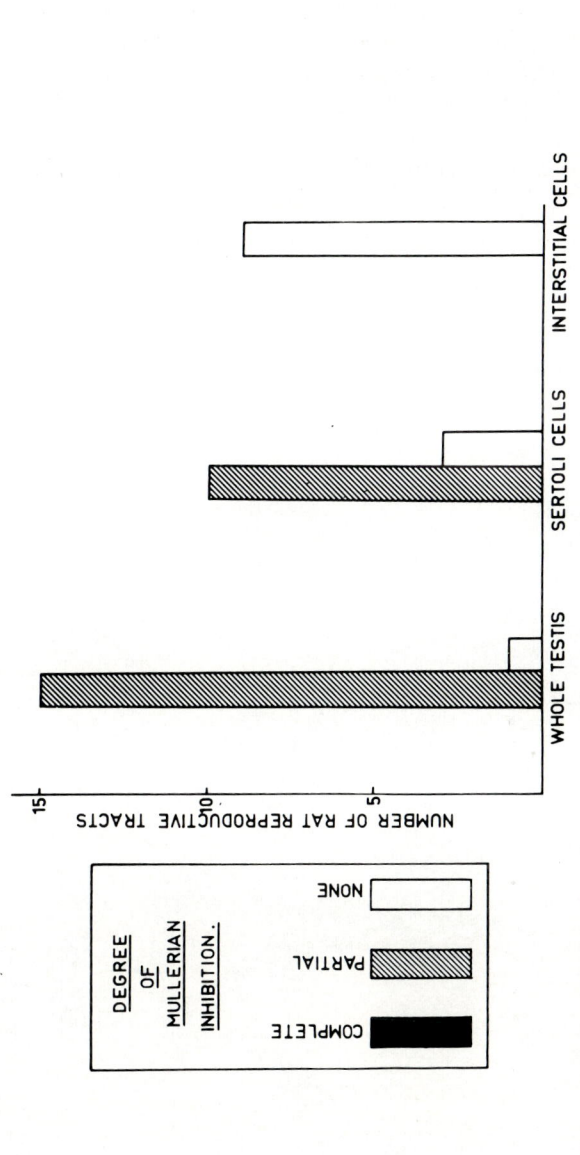

Fig. 14. Effect of bovine testicular cells on 14.5-day-old fetal rat reproductive tracts after 3 days in organ culture, expressed as in Fig. 8. Bioassay is performed after harvesting of monolayers and transfer of the cells to organ-culture conditions. Antimüllerian activity is present only in cells from whole testis or in Sertoli cells, but complete regression of müllerian duct is never obtained. From Josso et al. (1977b) with permission.

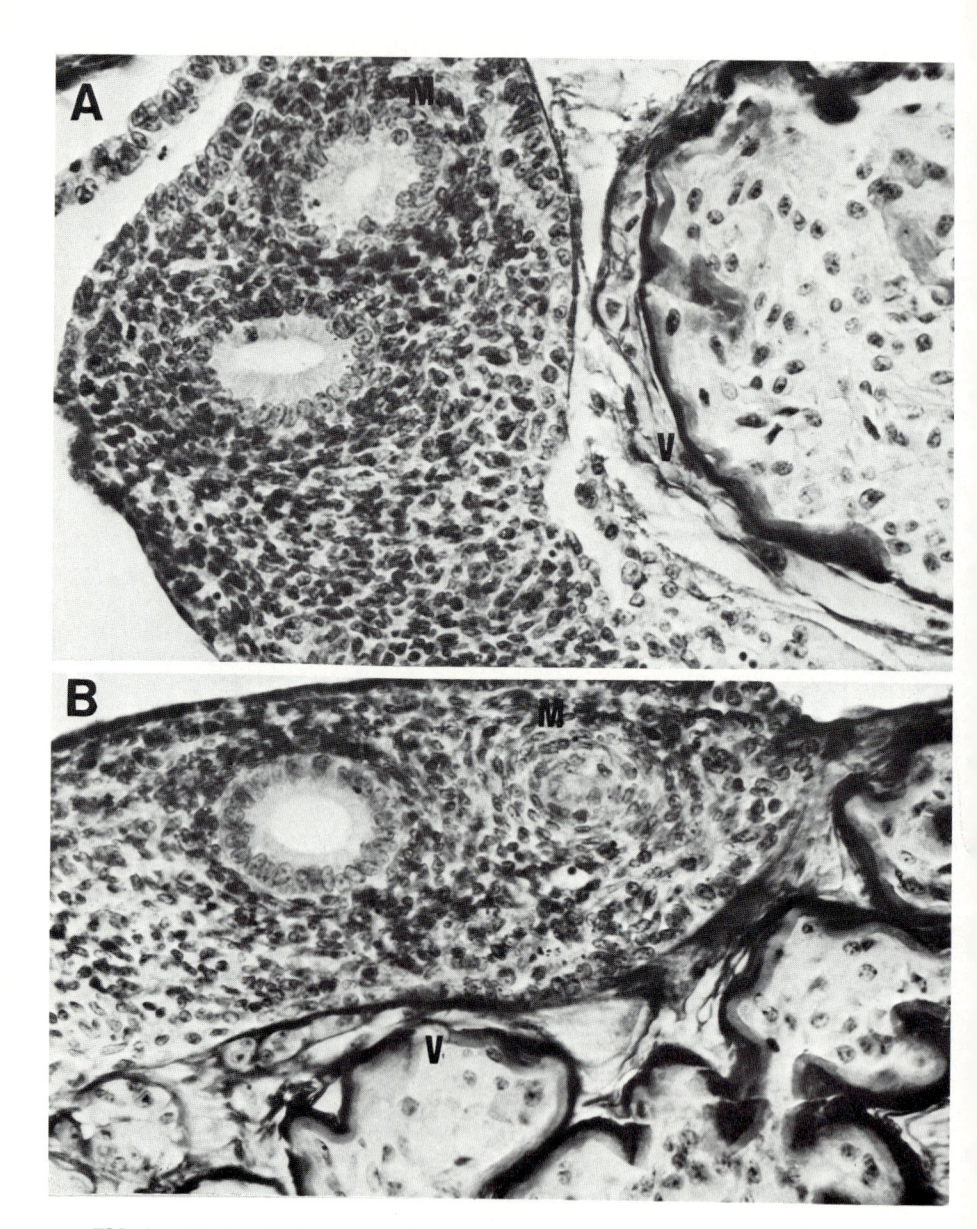

FIG. 15. Effect of cell colonies derived from bovine testicular tissue and wrapped in vitelline membrane of hen's egg on müllerian duct of 14.5-day-old rat reproductive tracts after 3 days in organ culture. (A) Interstitial cell colony: müllerian duct is maintained. (B) Tubular cell colony: müllerian duct is regressed. M, müllerian duct; V, vitelline membrane. Periodic acid–Schiff; × 350. From Josso et al. (1977a) with permission; © 1977 University Park Press, Baltimore.

cultures, are placed next to the müllerian ducts. This is probably due to the unfavorable effect of the switch to organ-culture conditions on cell viability.

In conclusion, data drawn from experimental procedures such as testicular microdissection, irradiation, and cell culture indicate that the fetal Sertoli cell is the source of the antimüllerian hormone. Discrete structures are responsible for the synthesis of the two types of fetal testicular hormones; testosterone is synthesized by Leydig cells under gonadotropin control. Few data concerning the regulation of AMH production by Sertoli cells are presently available; they will be considered in Section III, G.

G. REGULATORY MECHANISMS

1. Role of the Pituitary

The influence of the pituitary on AMH secretion is a matter for controversy. Some experimental evidence suggests depression by the pituitary of testicular antimüllerian activity. In the chick, Maraud et al. (1969) showed that decapitation of chick fetuses allowed their testes to retain antimüllerian activity until day 18 of incubation, at a time when practically none is found in testicular tissue of normal embryos. Growth hormone, TSH, and FSH, but not LH, ACTH, or prolactin, abolished the effect of decapitation. In the rat, Donahoe et al. (1975), in a preliminary communication, indicated that hypophysectomy of weanling rats extended the range of testicular production of AMH well into postnatal life, but were later unable to confirm their findings (Donahoe et al., 1976). However, since hypophysectomy in the first days of life is followed by severe retardation of growth and maturation, these, rather than a specific pituitary factor, might be involved in this process. Initiation of pituitary activity in the human (Gitlin and Biasucci, 1969) and rat (Jost, 1966) fetus is not correlated with a waning of testicular antimüllerian activity. *In vitro*, AMH production by fetal testicular tissue is not pituitary dependent: antimüllerian activity of human fetal testicular tissue is maintained 3 weeks in hormone-free medium (Josso, 1971b).

2. Mediators of Protein Hormone Action

In view of the evidence, which will be reviewed below, of the protein nature of AMH, müllerian ducts have been exposed to mediators of protein hormone action. Prostaglandins E_1, E_2, and $F_2\alpha$ have no effect, even at high concentrations (Jost et al., 1973; Josso, 1974c), but PGA_2 and even more PGB_1 reduce the width of the middle and posterior segment of the müllerian ducts (Table IV). The histological aspect of the ducts exposed to prostaglandins is quite different from that produced by coculture with fetal testicular tissue: narrowing is not

TABLE IV

Width of the Anterior and Middle Segment of the Müllerian Duct in 14.5-Day-Old Fetal Rat Reproductive Tracts Explanted in Vitro with PGE_1, PGA_2, or PGB_1

		\multicolumn{11}{c}{Prostaglandin concentration (µg/ml)}
		PGE_1		PGA_2				PGB_1			
Parameter	Controls	100	100	50	25	10	100	50	25	10
Number of cases	22	17	10	15	9	14	9	15	12	23
Anterior segment										
Mean width (µm)	39.4	38.0	21.5	39.3	41.4	41.1	21.5	33.2	37.4	41.9
Standard error	1.0	1.5	1.8	1.0	1.9	1.3	2.9	2.2	2.3	1.4
Comparison with controls[a] (p)	—	NS	<0.001	NS	NS	NS	<0.001	<0.01	NS	NS
Middle segment										
Mean width (µm)	34.6	36.1	20.7	17.1	36.5	34.2	13.0	22.3	22.6	34.2
Standard error	1.3	1.7	2.2	1.8	2.9	1.6	2.9	1.6	1.9	1.2
Comparison with controls[a] (p)	—	NS	<0.001	<0.001	NS	NS	<0.001	<0.001	<0.001	NS

[a] By Student's t test; NS, not significant.

associated with epithelial damage and fibroblast proliferation. Furthermore, prostaglandins primarily affect the organogenesis of the middle and posterior segments of the ducts, but not the anterior segment, more sensitive to AMH. Physiological AMH-mediated regression is not prevented by addition of indomethacin to the culture medium at ten times the concentration reported to abolish prostaglandin synthesis. We believe, therefore that prostaglandins are not involved in the antimüllerian activity of fetal testicular tissue.

Picon (1976) exposed 14.5-day-old rat reproductive tracts to dibutyryl cAMP and observed no effect on müllerian development. However, dibutyryl cAMP prevented physiological regression of müllerian ducts when these were associated with fetal rat testes in organ culture. Picon did not determine whether dibutyryl cAMP interferes with secretion of AMH by fetal testes or with the response of the müllerian duct to the hormone.

3. Relationship to Other Sertoli Cell Products and to H-Y Antigen

Much attention has recently been focused on the role of Sertoli cells in adult life. It is now well established that testicular androgen-binding protein (ABP) is synthesized by Sertoli cells (Hagenas and Ritzén, 1975). Steinberger and Steinberger (1976) have recently presented evidence in favor of the Sertoli cell origin of inhibin. Both ABP and inhibin play an important role in spermatogenesis and are synthesized during adult life. It is therefore unlikely that either of these proteins is related to AMH, whose synthesis is restricted to the fetal period. Furthermore, the molecular weight reported for inhibin (Keogh *et al.*, 1976) and ABP (Ritzén and French, 1974) is much lower than the value found for AMH, by biochemical studies which will be reported in the next section.

The problem of the relationship of H-Y antigen to AMH may also be discussed. According to Wachtel *et al.* (1975a), in XY individuals, cells of the yet undifferentiated gonad might produce a plasma membrane component concerned in cell–cell recognition and responsible for initiating testicular organization. This compound is recognizable serologically and has been named H-Y antigen, because it is specified by testis-determining gene(s) usually located on the Y chromosome. Ohno *et al.* (1976) ascribed the virilization of the freemartin gonad to secretion of H-Y antigen by gonadal cells transferred from the XY fetus to its female twin. In contrast, Jost *et al.* (1972) suggested that AMH, originating in the testicular tissue of the male fetus and transported in the blood stream, might be responsible both for the regression of müllerian ducts and for the developmental arrest and subsequent virilization of the freemartin ovary. Could H-Y antigen and AMH be related? This is unlikely. AMH production is not detectable in the undifferentiated gonad (Tran *et al.*, 1977) nor is it found after birth. Furthermore, H-Y antigen is always expressed in the heterogametic sex whereas AMH is produced by the male, whatever its sex chromosome constitution (Wachtel *et al.*, 1977b).

IV. Biochemical Studies

The biochemical nature of the hormone responsible for the regression of the müllerian ducts of the male fetus has not yet been elucidated. Since the work of Jost (1947), who showed that the müllerian inhibitor is different from testosterone, progress in this direction has been slow, limited to negative findings indicating that neither the 17-hydroxylated metabolites of testosterone (Josso, 1971a) nor prostaglandins E_1, E_2, $F_2\alpha$ (Jost et al., 1973), A_2 and B_1 (Josso, 1974c) can be identified with AMH. Recently, the macromolecular nature of AMH has been demonstrated and some of its biochemical properties have been studied.

A. MACROMOLECULAR NATURE OF AMH: MEMBRANE PERMEABILITY TO ANTIMÜLLERIAN ACTIVITY

1. Testicular Tissue (Josso, 1972b)

Testicular tissue from 8 human fetuses was maintained in organ culture and associated with 14.5-day-old fetal rat reproductive tracts 0.5–7 days later. Twenty-seven tracts were placed in direct contact with testicular explants, 22 were separated from them by Visking cellulose, and 18 were separated from testicular explants by the vitelline membrane of a hen's egg. The explants were examined after 3 days of association. In all the reproductive tracts in direct contact with the testicular explants, or separated from them by the vitelline membrane, the müllerian ducts were completely regressed, whereas, when Visking cellulose was interposed between testicular tissue and its target organ, no regression was apparent. Control experiments proved that Visking cellulose did not prevent testosterone-mediated maintenance of wolffian ducts by testicular tissue. These findings indicate that Visking cellulose is not permeable to AMH and therefore that the molecular weight of the hormone must exceed the 15,000 permeability limit of the membrane.

2. Incubation Media of Fetal Testes

Further progress was limited by the experimental conditions used up to now, in which hormone secretion and bioassay take place simultaneously. This is suitable for the physiological study of the antimüllerian activity of tissues, but, to investigate the biochemical nature of AMH, it is necessary to obtain the active hormone in a liquid medium. We have found that antimüllerian activity can be detected in incubation media of calf fetal testes (Josso et al., 1975). Attempts to detect AMH in fetal testicular homogenates had failed, probably because, at homogenate concentrations compatible with survival of the explanted müllerian ducts, AMH content was too low. In incubation media, relative concentration of

AMH is probably higher and therefore detectable at total protein concentrations compatible with the bioassay. Also, AMH present in homogenates may have been destroyed by intracellular proteolytic enzymes, but addition of various antiproteases to the homogenate medium did not change the situation.

Testicular tissue was obtained at a local slaughterhouse from 384 fetal calves, less than 65 cm in vertex—rump length, and incubated 4 hours at 37°C in a Dubnoff metabolic shaker, set at 80 oscillations per minute, in a O_2-CO_2 (95:5) atmosphere. [U-^{14}C] L-leucine (sp. act. 297 mCi/mmole) purchased from the Centre de l'Energie Atomique (Saclay, France), was added at a 1 μCi/ml concentration to Eagle's minimum essential medium deprived of leucine, but containing 0.5% fetal calf serum. Five milliliters of this incubation medium were used per gram of tissue.

At the end of the incubation, medium was collected by centrifugation at 12,000 g for 10 minutes at +4°C and frozen. Media were pooled, usually in batches of 50—100 ml, and dialyzed against nonradioactive Eagle's medium. Sample volume was not significantly altered by dialysis. The pools were then concentrated 4-fold by ultrafiltration on an Amicon UM 20 E membrane (molecular cutoff 20,000, according to manufacturer). In each of the 11 pools, before and after dialysis and after concentration, radioactivity was monitored in an Intertechnique SL 30 spectrometer (Plaisir, France) after addition of 5 ml of Instagel emulsifier solution (Packard Instruments Co.) to the sample. Quenching was determined by external standardization, and appropriate corrections were made. Testosterone content was determined by radioimmunoassay (Forest et al., 1973). The specificity of the method was checked by measuring the testosterone content of one pool, before and after dialysis, by the double-isotope derivative method (Rivarola and Migeon, 1966). The ^{14}C:^{3}H ratio obtained at the end of the procedure remained constant throughout 5 successive recrystallizations.

Antimüllerian activity testosterone and ^{14}C concentration was studied before and after dialysis and after concentration. The effect of dialysis and concentration on antimüllerian activity is shown on Fig. 16; mean testosterone and ^{14}C concentration are shown in Fig. 17. The amount of radioactivity remaining after dialysis, a reflection of the amount of newly synthetized proteins released into the medium, correlates well with antimüllerian activity, whereas no correlation between the latter and testosterone concentration can be demonstrated. Heating of the medium (80°C during 30 minutes) abolished its antimüllerian activity. Addition of cycloheximide to the medium during incubation decreased, but did not abolish, its biological activity.

In conclusion, the failure of thorough and prolonged dialysis to influence antimüllerian activity of the medium, the enhancement of this activity in the ultrafiltration retentate, and its correlation with the concentration of newly synthesized proteins but not with testosterone prove the macromolecular nature of AMH. The thermolability of the biological activity and its decrease in

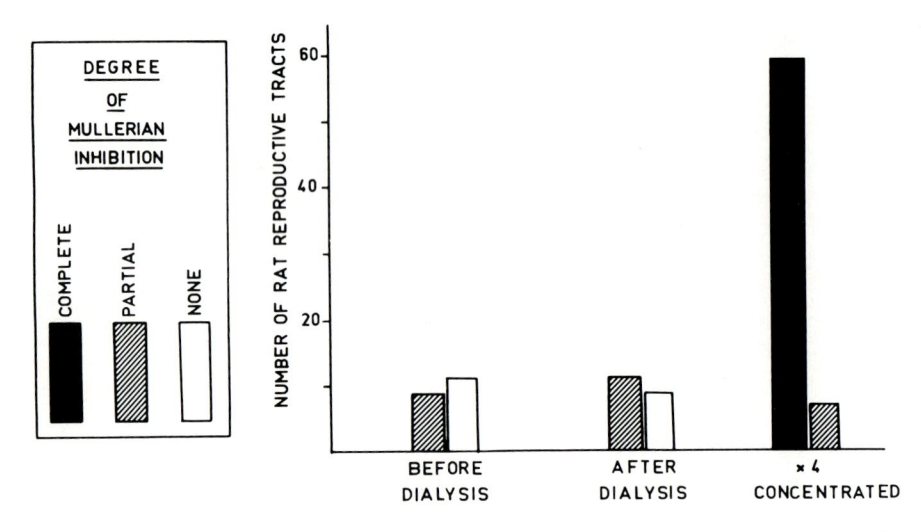

FIG. 16. Effect of dialysis and ultrafiltration on antimüllerian activity of 11 pools of incubation media of calf fetal testes, expressed as in Fig. 8. Antimüllerian activity is not modified by dialysis and is enhanced in the ultrafiltration retentate. From Josso *et al.* (1977b) with permission.

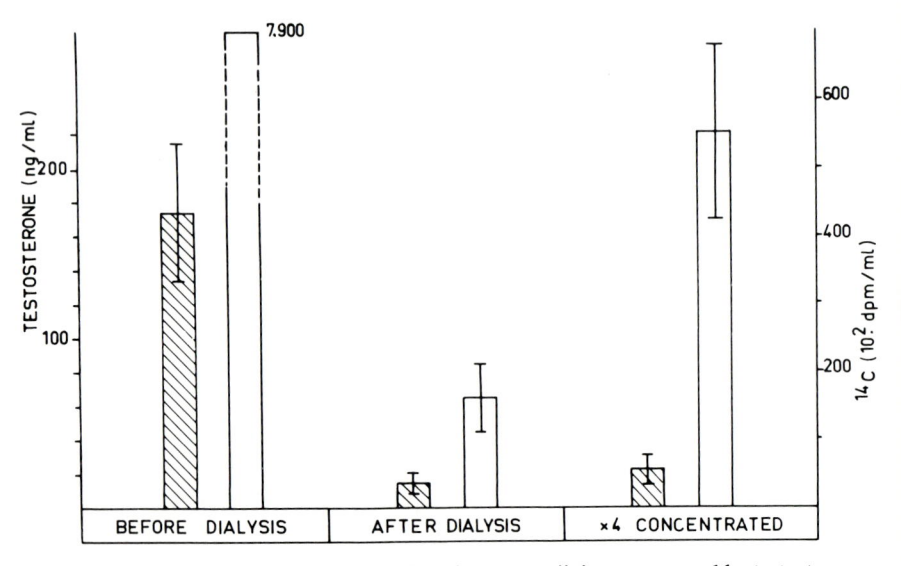

FIG. 17. Effect of dialysis and ultrafiltration on radioimmunoassayable testosterone (hatched columns) and ^{14}C concentration in 11 pools of incubation media. ^{14}C concentration after dialysis reflects neosynthesized protein content.

cycloheximide-treated incubation media indicate that this macromolecule is probably a protein. This hypothesis has been confirmed by the favorable results of application of methods of protein separation to the purification of AMH from the incubation medium.

B. GEL FILTRATION OF AMH (Picard and Josso, 1976)

1. Sephadex G-200

The first attempts at AMH purification were performed using filtration of concentrated incubation media on Sephadex G-200 in Tris-HCl 0.05 M, NaCl 0.12 M, EDTA 0.001 M, dithioerythritol 0.001 M, pH 7.5. Because the sensitivity of the bioassay does not allow the detection of antimüllerian activity in individual fractions, the total eluate was divided into 6–8 pools, which were concentrated, dialyzed against Eagle's medium, and tested for antimüllerian activity in organ culture. The results of 5 individual experiments are shown on Fig. 18. Biological activity was maximal in pools with a mean relative elution volume (V_e/V_o) between 1.13 and 1.37. The excluded fraction was cytotoxic in organ culture and could not be bioassayed. The variations in peak activity seen in the different experiments may be attributed to differences in the amount of AMH present in the sample, since this parameter cannot be accurately measured.

2. Bio-Gel A-5m

Because of the cytotoxicity of the excluded fraction, the position of the elution peak of AMH could not be determined on Sephadex G-200. Therefore, fractions eluted from Sephadex G-200 with a relative elution volume under 1.6, i.e., the excluded fractions and the biologically active following ones, were rechromatographed in the same buffer, on Bio-Gel A-5m, an agarose gel with a higher separation limit. Results of individual experiments are shown on Fig. 19. Maximal biological activity was found in the third and fourth pools, whose mean relative elution volume are, respectively, 1.5 and 1.64; lower activity was found in the second pool (mean V_e/V_o: 1.36).

3. Calibration of the Column

While there is widespread agreement that elution positions upon gel filtration are correlated with general molecular "size," there is no consensus in the literature as to which size parameter is fundamentally responsible for the observed elution behavior of macromolecules. Thus, some workers have presented empirical correlations of elution volume with molecular weight, while others have found the data to correlate best with molecular (Stokes) radius.

a. Molecular Weight. Calibration curves for both columns were constructed

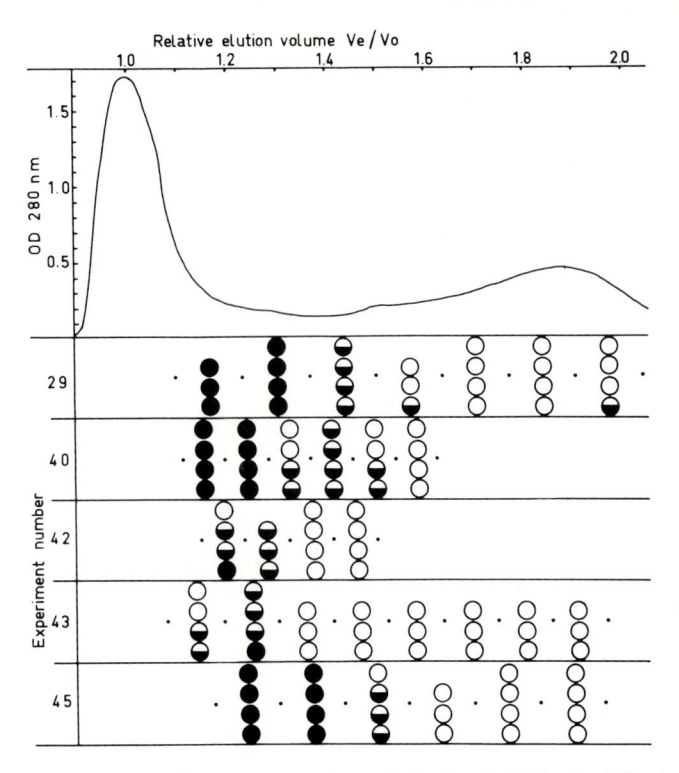

FIG. 18. Elution of antimüllerian activity from Sephadex G-200 in the following buffer: Tris-HCl 0.05 M, NaCl 0.12 M, EDTA 0.001 M, dithioerythritol 0.001 M, pH 7.5. The degree of müllerian inhibition in each individual rat reproductive tract exposed to pooled fractions of column effluent is represented by a symbol: ●, complete inhibition; ◐, incomplete inhibition; ○, no inhibition. Limit relative elution volumes (V_e/V_o) of pools are indicated by dots. The optical density at 280 nm of fractions eluted in a typical experiment (No. 29) is shown at the top. Experiment No. 45 was performed in the presence of dithioerythritol 0.01 M. From Picard and Josso (1976) with permission.

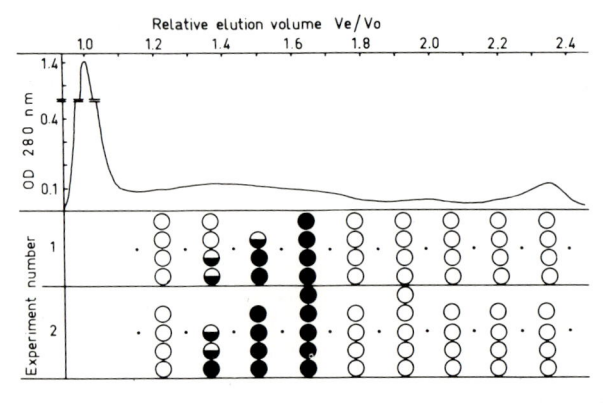

FIG. 19. Elution of antimüllerian activity from Bio-Gel A-5m; same buffer and same symbols as in Fig. 18. Optical density at 280 nm of fractions eluted in experiment No. 1 is shown at the top. From Picard and Josso (1976) with permission.

according to Whitaker (1963) by plotting the relative elution volume against the logarithm of molecular weight of appropriate marker proteins (Fig. 20). Proteins eluted from Sephadex G-200 in the zone of maximum biological activity are considered to have a molecular weight of 295,000–185,000; however, the upper limit is not meaningful, as the biological activity of the excluded fraction could not be tested. The molecular weight of proteins eluted from Bio-Gel A-5m with a relative elution volume corresponding to that of maximal antimüllerian activity, is 320,000–200,000.

  b. *Stokes Radius.* Calibration of the Sephadex G-200 column was performed by plotting the Stokes radius of marker proteins, as given by Siegel and Monty (1966), against a function of their relative elution volume according to the equation of Squire (1964): $(V_e/V_o)^{1/3} = \alpha - \beta a$, where a represents the Stokes radius, V_e/V_o the relative elution volume, and α and β numerical constants. With this representation, the Stokes radius of AMH appears to be in the range from 48 to 74 Å, the usual uncertainty applying to the upper limit. Lack of reliable data concerning the Stokes radius of the high-molecular-weight markers prevented the calibration of the Bio-Gel A-5m column.

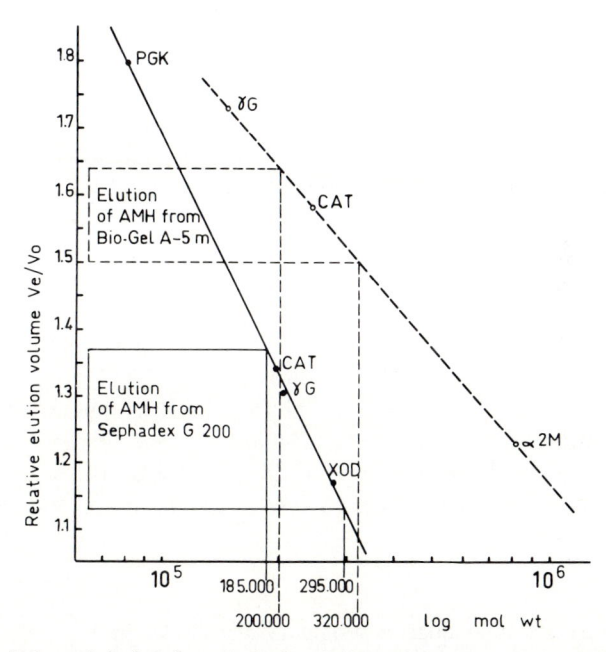

FIG. 20. Correlation of elution from Sephadex G-200 (solid line) and from Bio-Gel A-5m (dashed line) with logarithm of molecular weight of marker proteins. PGK, phosphoglycerate kinase; γG, gamma-G-globulin; CAT, catalase; XOD, xanthine oxidase; α^2M, alpha-2-macroglobulin. Molecular weight data were taken from Schönenberger (1958) for α^2M and from Andrews (1965) for all other markers. From Picard and Josso (1976) with permission.

These figures are surprisingly high, and the question of possible artifacts naturally arises. Rapid elution might result from formation of polymers or aggregates, it would therefore be desirable to repeat the experiments under dissociating conditions. Unfortunately, preliminary investigations showed that drastic methods of depolymerization, such as 8 M urea or 0.1% sodium dodecyl sulfate, irreversibly destroy antimüllerian activity. Dithioerythritol, 0.01 M, did not alter the elution pattern of AMH, but a reducing agent would not be expected to release subunits held together by bonds other than disulfide bridges.

Defective correlation between the elution behavior and molecular weight of AMH may also reflect our lack of information concerning the molecular structure of the hormone. Carbohydrate content, presence of other prosthetic groups, or nonglobular shape of proteins cause a significant deviation from standard behavior (Andrews, 1965). The problem can be solved only by the obtention of AMH in a purified form and its study by other physical methods, such as sucrose density gradient centrifugation.

C. ULTRACENTRIFUGATION IN A DENSITY GRADIENT
(Picard and Josso, 1977a)

1. Methods

Ultracentrifugation of incubation medium purified by 2 filtrations on Sephadex G-200 and one chromatography on DEAE Sephadex (Picard and Josso, 1977b) was performed according to Martin and Ames (1961) with bovine serum albumin (Armour), γ-G-globulin (Miles), and catalase (Boehringer) as reference standards. Samples of 200–300 µl were layered on the top of 5–20% sucrose gradients prepared in 0.05 M Tris-HCl buffer (pH 7.5), EDTA 0.001 M, dithioerythritol 0.001 M, and deposited in 12 ml tubes, using a Densiflow IIC apparatus (Büchler). In one experiment NaCl 0.5 M was added to the buffer. The gradients were spun at 4°C in a Beckman L2-65B ultracentrifuge, equipped with a SW 41 rotor, at 40,000 rpm for 18 hours. The tubes were emptied from the top downward using the Densiflow apparatus and fractions of 0.46 ml were collected. Optical density was read at 280 nm in a Gilford spectrophotometer. For AMH bioassay, fractions were pooled by groups of four and dialyzed against Eagle's medium.

2. Results

Results of six individual experiments are shown in Fig. 21 and were not modified by the addition of 0.5 M NaCl to the buffer in experiment 7. Cumulative results were calculated for those four experiments (Nos. 4, 5, 7, 8) in which the relative position of pool limits and protein markers were identical. These are shown on Fig. 22, as percentage of the total number of tracts exposed

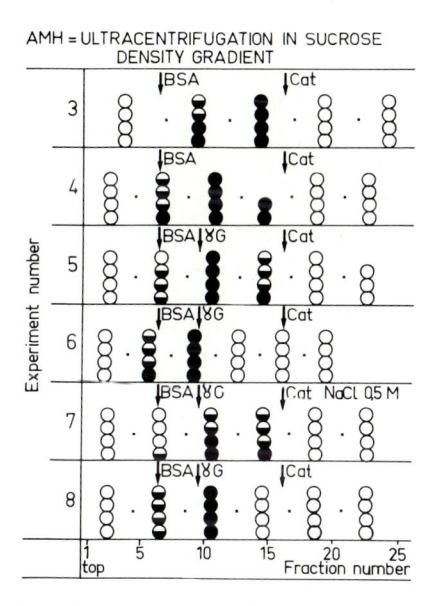

FIG. 21. Sedimentation of AMH after ultracentrifugation in a 5–20% sucrose density gradient, expressed as in Fig. 18. Fractions were collected from top downward, and pooled for determination of antimüllerian activity. Dots indicate pool limits. Position of protein markers is indicated by arrows. Experiment 7 was performed with 0.5 M NaCl added to the buffer.

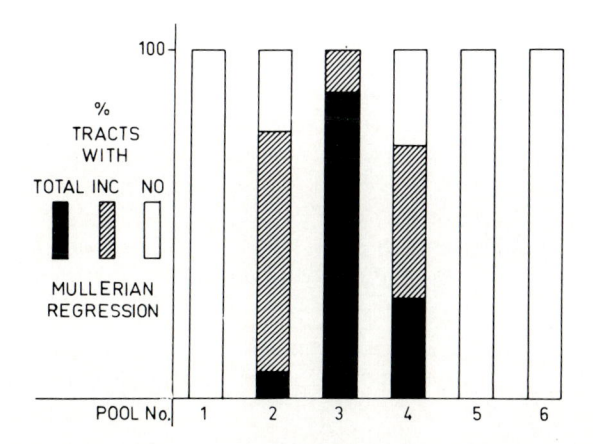

FIG. 22. Cumulative results of four density-gradient ultracentrifugation experiments (4, 5, 7, 8). Each column represents the total number of rat reproductive tracts exposed to a given pool in the four experiments, the percentage of tracts with total, incomplete, or no müllerian regression is indicated on each column.

to a given pool in the course of the four experiments, with either complete, incomplete, or absent müllerian regression. Statistical evaluation of the data was performed using ridit analysis (Bross, 1958; Ben-David et al., 1969), a method designed for the study of nonparametric data. Briefly, the mean ridit for a given pool of fractions expresses the probability that a tract exposed to it will have a degree of müllerian inhibition greater than average. The reference or "average" population here is composed of all the tracts used in the study. After the mean ridit and its SEM had been calculated, according to the method described in detail by Ben-David et al. (1969), the data relative to the pools in which biological activity had been demonstrated were subjected to analysis of variance, and differences between mean ridits of individual pools were tested for significance by Student's t test, using the appropriate residual error term. Results are shown in Table V. Antimüllerian activity of pool 3 differs very significantly from that of pools 2 and 4, which do not differ from each other. No activity was found in pools 1, 5, and 6.

3. Calculation of Sedimentation Constant and Molecular Weight

The sedimentation constant of AMH was calculated by constructing a calibration curve with the three protein markers. According to this curve, the first and last fraction composing the pool containing the peak antimüllerian activity have a sedimentation constant of 6.3 and 9.5, respectively (Fig. 23). A similar curve (Fig. 24) was drawn using the molecular weights of markers. The fit, however, is not as good, since a linear relationship between sedimentation and molecular weight is found only for proteins with similar frictional ratios and partial specific volumes. With this restriction in mind, we find, for the first and last fraction composing the pool of maximal antimüllerian activity, molecular weights of 120,000 and 195,000 respectively. This bracket of values is significantly lower than the 200,000 to 300,000 limits found using gel filtration. The discrepancy is probably due to an anomalous behavior of AMH during the latter procedure—such as one would expect if AMH were, for instance, a glycoprotein.

TABLE V
Statistical Analysis of Results of AMH Bioassay in Pooled Fractions Obtained by Sucrose Gradient Ultracentrifugation

Pool No.	1	2	3	4	5	6
Mean ridit	0.302	0.607	0.865	0.638	0.302	0.302
SEM	0	0.047	0.017	0.063	0	0
Statistical significance of differences between:	\multicolumn{6}{l}{Pools 2 and 3 : $p < .0001$ \\ Pools 3 and 4 : $p < .001$ \\ Pools 2 and 4 : NS}

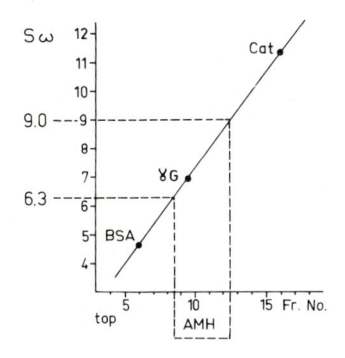

FIG. 23. Sedimentation of AMH in density gradient plotted against sedimentation constant of markers (mean of six experiments).

D. ION-EXCHANGE CHROMATOGRAPHY

Ion-exchange chromatography was used as the second step in the purification of AMH. After two successive filtrations on Sephadex G-200, fractions in the biologically active zone of the eluate were concentrated, dialyzed overnight against the starting buffer, and applied to a column filled with a Sephadex-linked ion exchanger in a ratio of 0.1 gm per milligram of protein.

1. Anion Exchangers

DEAE-Sephadex A-25 equilibrated overnight with Tris-HCl 0.05 M, EDTA 0.001 M, dithioerythritol 0.001 M, pH 8, was used. After emergence of the pass-through fraction, the column was eluted with a NaCl linear gradient, 0 to

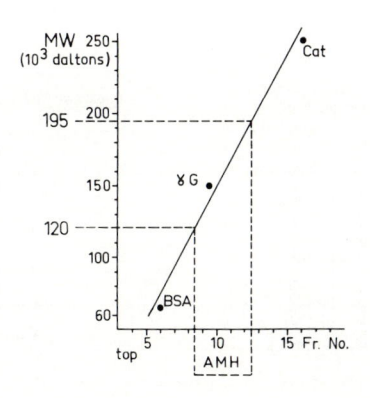

FIG. 24. Sedimentation of AMH in density gradient ploted against molecular weight of markers (mean of six experiments).

0.4 M, at a flow rate of 13 cm/hour. Linearity of the gradient was checked by conductivity measurements, and the gradient eluate up to NaCl 0.3 M was divided into 4 equal pools identified by their conductivity limits. Antimüllerian activity was studied in each of these, in the pass-through fraction, and also, in three instances, after application of NaCl 3 M to the column at the end of the gradient. Results of individual experiments are represented on Fig. 25 and cumulative results in Fig. 26. The variation in peak activity from one experiment to another is similar to that already seen with gel filtration and is probably due to differences in AMH content of the sample. Maximal antimüllerian activity was found in pool 2 (first pool of gradient eluate). Lower and inconstant activity was present in pool 1 (pass-through fraction) and 3 (second pool of gradient eluate). Statistical evaluation, by the ridit method, of the data presented in Fig. 26 is shown in Table VI.

2. Cation Exchangers

CM-Sephadex C-25, equilibrated with sodium phosphate 0.035 M, EDTA 0.001 M, dithioerythritol 0.001 M, pH 6.35, was used in the conditions described above. Biologically active medium eluted either from two Sephadex columns, or from an additional DEAE-Sephadex chromatography was used as starting material. After emergence of the pass-through fraction, a linear gradient of sodium phosphate 0.035 to 0.5 M was applied to the column. In three out of four experiments, antimüllerian activity was found only in the pass-through fraction. When pH was lowered to 5.5 and molarity of the starting buffer to 0.01

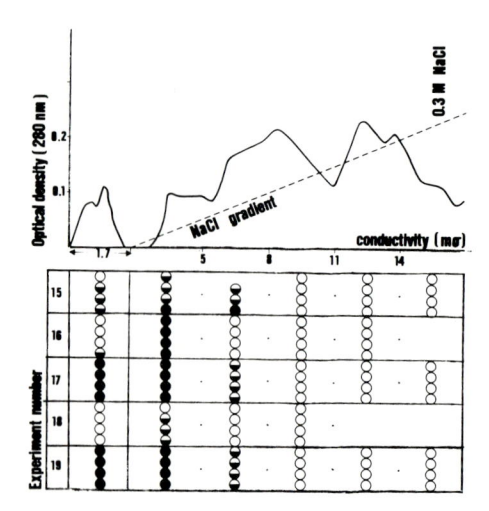

FIG. 25. Elution of antimüllerian activity from DEAE-Sephadex, by NaCl gradient represented as in Fig. 18.

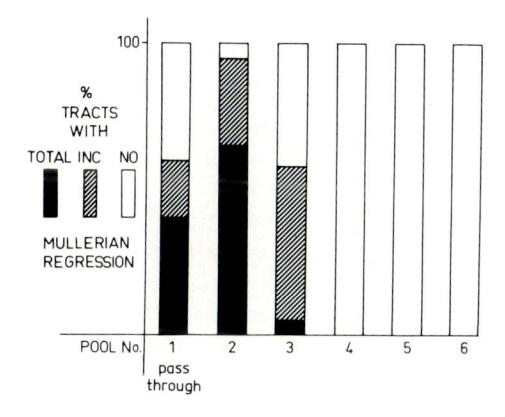

FIG. 26. Cumulative results of chromatography of AMH on DEAE Sephadex represented as in Fig. 22.

M, antimüllerian activity was partially retained on the column. However, these conditions resulted in precipitation of a large amount of antimüllerian activity (Picard and Josso, 1977b).

V. Conclusions

Since Jost (1947) first introduced the concept of the dual nature of the secretion of the fetal testis, the antimüllerian hormone has gained recognition, and some of its physiological characteristics are now well established. Much of the initial progress was due to the development of an *in vitro* bioassay for antimüllerian activity (Picon, 1969) allowing testing of the biological activity of various tissues of different species at different ages on a single, responsive target organ, the müllerian duct of the 14.5-day-old rat fetus.

TABLE VI
Statistical Analysis of Results of AMH Bioassay in Pooled Fractions Obtained by DEAE-Sephadex Chromatography

Pool No.	1	2	3	4, 5, 6
Mean ridit	0.620	0.803	0.543	0.303
SEM	0.061	0.034	0.049	0
Statistical significance of differences between:	\multicolumn{4}{l}{Pools 1 and 2 : $p < .001$ Pools 2 and 3 : $p < .001$ Pools 1 and 3 : NS}

The antimüllerian hormone is synthesized by the fetal Sertoli cells very early in fetal life, as soon as seminiferous tubules are recognizable under the light microscope and before the appearance of fetal Leydig cells. AMH production ceases in the perinatal period and is not resumed at puberty. No intramammalian species specificity has yet been demonstrated by asymmetrical species specificity between mammals and birds has been shown: the mammalian müllerian duct is sensitive to chick AMH, but the reverse is not true.

The biochemical structure of AMH is not yet elucidated, mainly because the bioassay used up to now is not quantitative and therefore not ideally suited to techniques of protein purification. Nevertheless, the macromolecular nature of the hormone has been firmly established and its protein nature is highly probable. A molecular weight of 200,000 to 300,000 has been demonstrated by gel filtration, whereas by density gradient ultracentrifugation, lower values, from 120,000 to 200,000, are found. This discrepancy is compatible with the hypothesis that AMH is a glycoprotein. At conditions imposed by the stability of biological activity (pH 6–8), AMH is not strongly retained either by anion or by cation exchangers, which suggests that its isoelectric point is not far removed from neutrality.

Further characterization of the antimüllerian hormone depends now on the progress of its purification. Usually, the degree of purification of the component of a protein mixture is measured by the increase in its specific activity. Precise quantitation of antimüllerian activity is not possible at the present time, and therefore the degree of purification of AMH, achieved by two Sephadex-G-200 filtrations and one DEAE-Sephadex, cannot be accurately determined. Crude approximations (Picard and Josso, 1977b) indicate that a 36-fold purification has been obtained at the expense of 65% of the hormone initially present in the starting material. Obviously, many more fetal testes, and a great amount of time and effort will be required to isolate pure antimüllerian hormone.

REFERENCES

Alcala, J. R., and Conaway, C. H. (1968). *Folia Primatol.* 9, 216.
Andrews, P. (1965). *Biochem. J.* 96, 595.
Ben-David, M., Heston, W. E., and Rodbard, D. (1969). *J. Natl. Cancer Inst.* 42, 207.
Black, V. H., and Christensen, A. K. (1969). *Am. J. Anat.* 124, 211.
Blanchard, M. G., and Josso, N. (1974). *Pediatr. Res.* 8, 968.
Brachet, J., Decroly-Briers, M., and Hoyez, J. (1958). *Bull. Soc. Chim. Biol.* 40, 2039.
Brook, C. G. D., Wagner, H., Zachmann, M., Prader, A., Armendares, S., Frenk, S., Aleman, P., Najjar, S. S., Slim, M. S., Genton, N., and Bozic, C. (1973). *Br. Med. J.* 1, 771.
Bross, I. D. (1958). *Biometrics* 14, 18.
Christensen, A. K., and Mason, N. R. (1965). *Endocrinology* 76, 646.
Didier, E. (1971). *J. Embryol. Exp. Morphol.* 25, 115.
Didier, E. (1973). *Wilhelm Roux' Arch. Entwicklungsmech. Org.* 172, 287.

Donahoe, P. K., Ito, Y., Marfatia, S. R., and Hendren, W. H. (1975). *Pediatr. Res.* **9**, 289 (abstr.).
Donahoe, P. K., Ito, Y., Marfatia, S., and Hendren, W. H. III. (1976). *Biol. Reprod.* **15**, 329.
Dyche, W. H. (1976). Ph.D. Thesis, University of Pennsylvania, Philadelphia.
Elger, W. (1967). *Arch. Anat. Microsc. Morphol. Exp.* **55**, 657.
Forest, M. G., Cathiard, A. M., and Bertrand, J. A. (1973). *J. Clin. Endocrinol.* **36**, 1132.
Forsberg, J. G. (1963). Ph.D. Thesis, Ed Hakon Ohlssons Boktryckeri, Lund.
Forsberg, J. G., and Abro, A. (1973). *Acta Anat.* **85**, 353.
Frutiger, P. (1969). *Acta Anat.* **72**, 233.
Gitlin, D., and Biasucci, A. (1969). *J. Clin. Endocrinol. Metab.* **29**, 926.
Goldstein, J. L., and Wilson, J. D. (1975). *J. Cell. Physiol.* **85**, 365.
Gruenwald, P. (1941). *Anat. Rec.* **81**, 1.
Haffen, K. (1970). In "Organ Culture" (J. A. Thomas, ed.), p. 121. Academic Press, New York.
Hagenas, L., and Ritzén, E. M. (1975). *Mol. Cell Endocrinol.* **2**, 339.
Hall, P. F., Irby, D. C., and de Kretser, D. M. (1969). *Endocrinology* **84**, 488.
Hamilton, T. H., and Teng, C. S. (1965). In "Organogenesis" (R. L. DeHaan and H. Ursprung, eds.), p. 681. Holt, New York.
Jirasek, J. E. (1967). In "The Testis" (G. E. W. Wolstenholme and M. O'Connor, eds.), p. 3. Ciba Found. Colloq. Endocrinol., Churchill, London.
Josso, N. (1970). *C.R. Hebd. Seances Acad. Sci.* **271**, 2149.
Josso, N. (1971a). *Rev. Eur. Etud. Clin. Biol.* **16**, 694.
Josso, N. (1971b). *J. Clin. Endocrinol. Metab.* **32**, 404.
Josso, N. (1972a). *Biol. Neonate* **20**, 368.
Josso, N. (1972b). *J. Clin. Endocrinol. Metab.* **34**, 265.
Josso, N. (1973). *Endocrinology* **93**, 829.
Josso, N. (1974a). *Pediatr. Ann.* **3**, 67.
Josso, N. (1974b). *Pediatr. Res.* **8**, 755.
Josso, N. (1974c). *Biomedicine* **21**, 225.
Josso, N., Forest, M. G., and Picard, J. Y. (1975). *Biol. Reprod.* **13**, 163.
Josso, N., Picard, J. Y., and Tran, D. (1977a). In "Treatment of Congenital Adrenal Hyperplasia" (P. A. Lee, L. P. Plotnick, A. A. Kowarski, and C. J. Migeon, eds.). University Park Press, Baltimore.
Josso, N., Picard, J. Y., and Tran, D. (1977b). In "Morphogenesis and Malformation of the Genital System" (R. J. Blandau, ed.). A. R. Liss, New York.
Jost, A. (1947). *Arch. Anat. Microsc.. Morphol. Exp.* **36**, 271.
Jost, A. (1953). *Recent Prog. Horm. Res.* **8**, 379.
Jost, A. (1966). In "The Pituitary Gland" (G. W. Harris and B. T. Donovan, eds.), p. 299. Butterworth, London.
Jost, A., and Bergerard, Y. (1949). *C.R. Seances Soc. Biol. Ses Fil.* **143**, 608.
Jost, A., and Bozic, B. (1951). *C.R. Seances Soc. Biol. Ses Fil.* **145**, 647.
Jost, A., Vigier, B., and Prepin, J. (1972). *J. Reprod. Fertil.* **29**, 349.
Jost, A., Vigier, B., Prepin, J., and Perchellet, J. P. (1973). *Recent Prog. Horm. Res.* **29**, 1.
Keogh, E. J., Lee, V. W. K., Rennie, G. C., Burger, H. G., Hudson, B., and de Kretser, D. M. (1976). *Endocrinology* **98**, 997.
Kodani, M., and Kodani, K. (1966). *Proc. Natl. Acad. Sci. U.S.A.* **56**, 1200.
Lutz-Ostertag, Y. (1974). *C.R. Hebd. Seances Acad. Sci.* **278**, 2351.
Maraud, R., Coulaud, H., and Stoll, R. (1969). *C.R. Seances Soc. Biol. Ses Fil.* **163**, 2557.
Martin, R. G., and Ames, B. N. (1961). *J. Biol. Chem.* **236**, 1372.
Merchant-Larios, H. (1976). *Am. J. Anat.* **145**, 319.

Moon, Y. S., and Hardy, M. H. (1973). *Am. J. Anat.* **138**, 253.
Müller, J. (1830). "Bildungsgeschichte der Genitalien." Arnz, Düsseldorf.
Mürset, G., Zachmann, M., Prader, A., Fischer, J., and Labhart, A. (1970). *Acta Endocrinol. (Copenhagen)* **65**, 627.
Niemi, M., and Ikonen, M. (1961). *Nature (London)* **189**, 592.
Ohno, S., Christian, L. C., Wachtel, S. S., and Koo, G. C. (1976). *Nature (London)* **261**, 597.
Pelliniemi, L. J. (1975). *Am. J. Anat.* **144**, 89.
Pelliniemi, L. J., and Salonius, A. L. (1976). *Acta Anat.* **95**, 558.
Picard, J. Y., and Josso, N. (1976). *Biomedicine* **25**, 147.
Picard, J. Y., and Josso, N. (1977a). *Proc. Symp. Birth Defects, 7th*, in press.
Picard, J. Y., and Josso, N. (1977b). Submitted for publication.
Picon, R. (1969). *Arch. Anat. Microsc. Morphol. Exp.* **58**, 1.
Picon, R. (1970). *C.R. Hebd. Seances Acad. Sci.* **271**, 2370.
Picon, R. (1971). *C.R. Hebd. Seances Acad. Sci.* **272**, 98.
Picon, R. (1976). *Mol. Cell. Endocrinol.* **4**, 35.
Price, D., and Pannabecker, R. (1959). *Arch. Anat. Microsc. Morphol. Exp.* **48***bis*, 223.
Price, D., Zaaijer, J. J. P., and Ortiz, E. (1968). *In* "The Mammalian Oviduct" (E. S. E. Hafez and R. J. Blandau, eds.), p. 29. Univ. of Chicago Press, Chicago, Illinois.
Price, D., Zaaijer, J. J. P., Ortiz, E., and Brinkmann, A. O. (1975). *Am. Zool.* **15**, Suppl. 1, 173.
Raynaud, A. (1950). *Arch. Anat. Microsc. Morphol. Exp.* **39**, 518.
Raynaud, A., Pieau, C., and Raynaud, J. (1970). *Ann. Embryol. Morphog.* **3**, 21.
Ritzén, E. M., and French, F. S. (1974). *J. Steroid Biochem.* **5**, 151.
Rivarola, M. A., and Migeon, C. J. (1966). *Steroids* **7**, 103.
Rivarola, M. A., Podesta, E. J., and Chemes, H. E. (1972). *Endocrinology* **91**, 537.
Scheib, D. (1965). *C.R. Hebd. Seances Acad. Sci.* **260**, 1252.
Scheib-Pfleger, D., and Wattiaux, R. (1962). *Dev. Biol.* **5**, 205.
Schönenberger, M., Schmidtberger, R., and Schultze, H. E. (1958). *Z. Naturforsch., Teil B* **13**, 761.
Schultz, F. M., and Wilson, J. D. (1974). *Endocrinology* **94**, 979.
Siegel, L. M., and Monty, K. S. (1966). *Biochim. Biophys. Acta* **112**, 346.
Sjöquist, A., and Hultin, T. (1967). *Experientia* **23**, 544.
Squire, P. G. (1964). *Arch. Biochem. Biophys.* **107**, 471.
Steinberger, A., and Steinberger, E. (1966). *Exp. Cell Res.* **44**, 443.
Steinberger, A., and Steinberger, E. (1976). *Endocrinology* **99**, 918.
Stoll, R., Maraud, R., and Coulaud, H. (1970). *C.R. Seances Soc. Biol. Ses Fil.* **164**, 1013.
Teng, C. S., and Teng, C. T. (1975). *Biochem. J.* **150**, 183.
Tran, D., and Josso, N. (1977). *Biol. Reprod.* **16**, 267.
Tran, D., Meusy, N., and Josso, N. (1977). Submitted for publication.
Vilas, E. (1933). *Z. Anat. Entwicklungsgesch.* **99**, 599.
Wachtel, S. S., Ohno, S., Koo, G. C., and Boyse, E. A. (1975a). *Nature (London)* **257**, 235.
Wachtel, S. S., Koo, G. C., and Boyse, E. A. (1975b). *Nature (London)* **254**, 270.
Wells, L. J., Cavanaugh, M. W., and Maxwell, E. L. (1954). *Anat. Rec.* **118**, 109.
Weniger, J. P. (1963). *Arch. Anat. Microsc. Morphol. Exp.* **52**, 497.
Weniger, J. P. (1965). *Arch. Anat. Microsc. Morphol. Exp.* **54**, 909.
Whitaker, J. R. (1963). *Anal. Chem.* **35**, 1950.
Wolff, E., Lutz-Ostertag, Y., and Haffen, K. (1952). *C.R. Seances Soc. Biol. Ses Fil.* **146**, 1793.
Woods, J. E., Simpson, R. M., and Moore, P. L. (1975). *Gen. Comp. Endocrinol.* **27**, 543.

Yamada, M., Yasue, S., and Matsumoto, K. (1972). *Acta Endocrinol. (Copenhagen)* **71**, 393.
Yamamoto, M., Koshihara, H., and Noumura, T. (1974). *J. Exp. Zool.* **187**, 13.

DISCUSSION

J. C. Orr: Did you try any proteolytic enzymes to see whether they could destroy the activity?

N. Josso: Yes, we did. Trypsin decreased the activity, but we did these experiments with incubation medium, not with purified fractions.

C. W. Bardin: Bill Dyche from the Department of Anatomy, Pennsylvania State University, has recently examined the müllerian duct by electron microscopy. His studies suggest that the mechanism of müllerian duct regression is similar to that of other hormonally induced involutions. The first change to occur is a proliferation of the mesenchyme cells around the müllerian duct. These changes in the mesenchyme occur several hours before any alterations occur in the epithelial cells. Some of your figures suggested a similar phenomenon.

N. Josso: This is true. AMH does not produce necrosis of müllerian epithelium; rather the epithelial cells appear to be stifled by a fibroblastic ring. Epithelial necrosis is caused by toxic agents, but not by AMH.

C. W. Bardin: There is an organized rostral to caudal regression of the müllerian duct. Some investigators have suggested that the portion of the duct next to the testis regresses first since it is the first to come in contact with the hormone. Others believe that the rostral portion of the duct is the first to regress owing to its increased sensitivity to the hormone. Would you comment on these two theories and indicate which one your observations support?

N. Josso: The matter of regression I think can be explained by the fact that the whole length of the müllerian duct is not sensitive to the antimüllerian hormone at the same time. At 14 days in the rat, the anterior part is more sensitive, the middle part less, and the posterior part that is very close to the urogenital sinus is not sensitive at all. You can use this to measure antimüllerian activity. A tissue or medium with a very high antimüllerian activity will inhibit not only the sensitive anterior part, but also the whole of the müllerian duct down to the urogenital sinus. A less active tissue or medium will inhibit only the anterior part, and it will do nothing to the middle part. Now, when I talk about responsiveness or sensitivity of the müllerian duct to AMH, I do not know what that means. There has been a recent paper by R. Picon [*Mol. Cell. Endocrinol.* **4**, 35 (1976)] showing that if you coculture fetal testis with a müllerian duct in the presence of dibutyryl cAMP the nucleotide will prevent the regression that is normally elicited by the fetal testis. Regression occurs from the ostium to the urogenital sinus. I think it is a sensitivity phenomenon.

P. W. Nathanielsz: May I ask the obvious question? In view of the unilateral nature of the effects that you have been describing, are we correct in calling this factor antimüllerian hormone? Also, do you think it is important that the female is the heterogametic sex in the bird whereas in mammals the male is the heterogametic sex?

N. Josso: I can make no comment on your second question. As to whether I have the right to call the antimüllerian factor a hormone, the question is certainly open. According to the criteria required by Dr. Guillemin in his presentation (this volume), physiological concentrations have to be present in peripheral blood. Usually the antimüllerian hormone has essentially a local effect; intersex patients' müllerian ducts regress only on the side of the testis, and probably the concentration in the blood is not high enough to cause regression on the contralateral side. However, in the freemartin the hormone is transferred by a placental anastomosis from the male to the female fetus, and in this instance physiological concentrations of the hormone are no doubt reached in the blood stream.

P. K. Donahoe: We, at the Massachusetts General Hospital, have admired Dr. Josso's work of the last few years. We have worked with this rather meticulous assay and found it to be a very demanding one. We have attempted to make the assay more sensitive in order to look forward to biochemical isolation, and, by using only the cephalic end of the duct and by grading it from 1 to 5, we have achieved this end. Using this more sensitive semiquantitative assay, we have found that the rat maintains activity in the postnatal period until day 21, after which activity disappears. Activity is maintained also in the human using this method of assay, until at least 9 months of age, and possibly beyond, although these latter observations are preliminary.

Working with J. Michael Price at the University of Massachusetts, we have studied the ultrastructure of the regressing müllerian duct. This study has shown that the mechanism of regression is mediated by lysosomes. At day 15 *in vivo*, granules that localize acid phosphatase are found, implying the presence of lysosomes. Granule formation is followed by migration of macrophages across the basement membrane. The macrophages then engulf the cells containing granules and digest them until the cells can no longer be detected. This same series of sequential events is seen *in vitro* as well as *in vivo*. We think that the indifferent mesenchyme is stimulated by the presence of the lysosomes to differentiate into macrophages, which then migrate in and engulf the ducts.

N. Josso: What species did you use?

P. K. Donahoe: The electron microscopic studies were done in the rat.

N. Josso: You are no doubt aware that the role of lysosomes in regression of müllerian ducts has been extensively studied in the chick by J. Brachet, M. Decroly-Briers, and J. Hoyez [*Bull. Soc. Chim. Biol.* **40**, 2039 (1958)], D. Scheib-Pfleger and R. Wattiaux [*Dev. Biol.* **5**, 205 (1962)], and T. H. Hamilton and C. S. Teng [*in* "Organogenesis" (R. L. DeHaan and H. Ursprung (eds.), p. 681. Holt, New York, 1965)].

P. K. Donahoe: I would like to make one correction regarding some of our earlier work in which we had postulated that hypophysectomy might prolong the presence of müllerian-inhibiting substance activity in the rat. We found that this was not the case. By hyphophysectomizing at 20 days of age, we were unable to cause persistence of müllerian-inhibiting substance activity at any time thereafter. Looking at this in another way with Dr. Barry Bercu, we found that anti-LRH given both to the embryo and to the postnatal rat did not cause persistence of müllerian-inhibiting substance.

P. Licht: First of all, a comment on your observation of "species specificity." I appreciate your desire to get a model system—one that would be more convenient than the mouse—and also the desire to come up with some explanation for why birds do not respond to mouse testis. However, I would suggest that before we draw any such conclusions about differences between birds and mammals, or before you give up on your search for a model, you might look at another bird or two. Experience described by Dr. McKenzie at this meeting and data I will discuss later indicate that you just do not pick up one chicken and conclude that birds are "this way or that." Try a duck or a pigeon, for example; don't give up quite so easily. In fact, you might try something even with more dramatic müllerian duct development, like an amphibian.

N. Josso: The reason why I have not tried amphibians is that I thought I would have to culture their testes at a lower temperature than the one I use for rat müllerian ducts.

P. Licht: You can culture many amphibian tissues at 30° or 32° C, and some of the southern *Rana* probably require higher temperatures, and I know that you can culture mammalian testis at 32°C, so there should not be any problem. I also would like to know whether it is really necessary to culture these tissues for 3 days and whether you determined that a constant exposure to testis is required; for example, could you culture for 10 minutes and then remove the testis?

N. Josso: I have not done this, but Régine Picon in her 1969 work has. She showed that you do get a regression after 2 days, but it is not as marked as the one obtained after 3 days.

R. Falvo: I am sure you are aware of the work in the 1960s by Dr. C. Donnell Turner in which he transplanted ovaries and testis next to each other in the kidney capsule. He showed that the fetal testis could inhibit normal ovarian development and actually cause the development of an ovotestis. It was suspected that some fetal factor was responsible for this action. Do you think that the müllerian inhibitory hormone you have described could in some way affect the development of the fetal ovary?

N. Josso: This is a subject that is actually being investigated by Professor Jost. He thinks that the antimüllerian hormone is probably responsible for the inhibition of meiosis in the ovary for the freemartin. Jost has developed a system for the demonstration of meiosis in rat fetal ovaries in organ culture. We have given him some of our active medium, and preliminary results indicate that it inhibits meiosis in this system.

E. B. Thompson: I would like to ask you, and perhaps Dr. Donahoe, to state what the position is with regard to the cell or origin of the putative factor. When you apply your cell separation methods, the cells that adhere to the culture dish are those that produce the factor, but that includes a number of cell types, does it not?

N. Josso: Not really. The cells that I have shown that inhibited müllerian ducts had originated from isolated seminiferous tubules. This means that there were no fibroblasts, only the cells of the myoid sheath and the supporting cells. I assume that there were no germ cells because I assume that the conditions in the fetal testis were the same as those found in the adult testis.

E. B. Thompson: Given that the germinal cells are probably absent, are there only two kinds of cell then? No fibroblasts?

N. Josso: No, there are no fibroblasts in the seminiferous tubule. Microdissection is performed under the dissecting microscope, and the tubules really are quite clean. It is relatively easy to obtain clean seminiferous tubules using very small pins to remove the interstitial tissue. It is much more difficult to do the opposite: I showed you that I sometimes had contamination of interstitial tissue by fragments of seminiferous tubules. Anyway, when we started using this technique, we checked the purity of our preparations by histological examinations, and the seminiferous tubules were really quite clean.

E. B. Thompson: A second point, with regard to the same question: What is the nature of the cells of origin of the factor? I was not clear about your work as compared with that of Dr. Donahoe. Is there a very limited time period, or not, during which the active cells produce the factor? I believe you stated that it was produced during a specific fetal time period, and Dr. Donahoe had data indicating continued production for some time postnatally.

N. Josso: Sertoli cells start producing AMH shortly after testicular differentiation, and they continue to produce up to nearly the end of fetal life in humans. Dr. Donahoe says that she has a more sensitive test and can find AMH in the first year of life in humans. In my experience, antimüllerian activity is high until 27 fetal weeks, then decreases. In the rat, the fetal testis starts making AMH at 14 days and produces large amounts until 4 days postnatally. Antimüllerian activity is no longer found after 15 days, according to Régine Picon.

J. M. Mc Kenzie: Would you elaborate a little more, please, on the influence of the pituitary? Did you refer to some experiments in which hypophysectomy had some effect on the antimüllerian factor?

N. Josso: Dr. Jost has not done work of this kind. The work I referred to in my paper was published by R. Maraud, H. Coulaud, and R. Stoll [*C.R. Seances Soc. Biol. Ses Fil.* **163**, 2557 (1969)] and was in the chick embryo.

S. Leeman: Did I understand you to say that, when you made an initial homogenate or extraction of the tissue, you could not find activity?

N. Josso: Yes, I did say that. There are two explanations for this inability to demonstrate activity in homogenate: The first is that the amount of protein that the rat reproductive tracts will tolerate in the culture medium is limited, so that you have to dilute the homogenate very much. Probably the hormone is present in a very minute concentration, and so the maximum amount of hormone that we were able to test without inducing neucrosis was too low. The other possibility is that there was a proteolysis. I favor the first possibility because the amount of antiproteolytic agents I added to this homogenate is fantastic. I tried all the antiproteolytic agents I could lay my hands on, and I never got any effects. What we mean to do now is perhaps to submit the homogenate to procedures that we know will purify the hormone and then seen whether we can get something. We have not done that yet.

J. S. Roberts: Although you call the inhibiting substance a hormone, I think you would agree that it could be called a local hormone. If so, it seems unusually large at a molecular weight of 300,000. You seem to have a globular protein. Could it not be that the substance is merely a carrier for the truly active material, which might be a smaller peptide or perhaps a lipoidal substance of some sort? Do you have a way of excluding these possibilities?

N. Josso: No, maybe someone could give me an idea.

S. Cohen: As I recall, Dr. Jost, when he was here a few years ago, spoke about the testosterone-producing capabilities of the fetal testis. It seems to me this is an odd gland, making both the steroidal hormone and a protein hormone. Do you wish to comment on that? It must be a very confusing set of enzymes that is present.

N. Josso: No, because there are two different kinds of cells. You have Leydig cells making testosterone and Sertoli cells making a protein. In the adult testis, you know that people have been getting very excited over the proteins that are made by Sertoli cells, for instance, ABP.

D. Rodbard: Have you tested any of your fractions for activity as inhibin? I realize that inhibin appears to have a totally different molecular weight and it arises at a different stage in development, but it is conceivable that there might be some relationship between these two products of the Sertoli cell.

N. Josso: I have not tested it for the simple reason that the chronological evolution is quite different for each. I do not find any antimüllerian activity in the postnatal testis at a time when people do find inhibin.

J. R. Pasqualini: Do you know whether estrogens can protect the müllerian duct from the action of the antimüllerian hormone? I ask this because, in studies that we carried out in the presence of estrogen receptors in the müllerian ducts and uterus of guinea pig fetuses, we found that estradiol receptors appear very early during fetal development. The quantities of specifically bound estradiol in the cytosol are 80–100 fmoles per milligram of protein at 30 days of gestation, and they increase to 450–550 fmoles at the end of gestation (55–62 days) [*J. Steroid Biochem.* **7**, 0000 (1976)]. In your culture studies on the regression of müllerian ducts, have you investigated the effect if you incubated in the presence of estrogens?

N. Josso: We have not performed this type of experiment. In the course of normal development of the müllerian duct, insensitivity to AMH and acquisition of estrogen receptors are chronologically correlated, but a causal relationship between the two phenomena has not been established. In Jost's 1947 experiments, the presence of the fetal ovary did not prevent the grafted fetal testis from inhibiting the female müllerian duct.

R. O. Greep: When you got down to the final stages of chemical purification, the hormone in question appeared to have vanished. Have you considered the possibility that perhaps there is not a müllerian-inhibitory factor in the male, but a müllerian-preservation factor in the female?

N. Josso: I disagree. The biological activity has not vanished at all. We only have trouble in correlating the biological activity we find in our fractions with a band in polyacrylamide-gel electrophoresis. The only way that we can recognize the hormone at present is by its biological activity. I would not say that the hormone vanishes at the end of the purification.

R. O. Greep: What does preserve the müllerian duct in the female?

N. Josso: The müllerian duct in female fetuses is never exposed to testicular tissue.

J. Vaitukaitis: Would you like to speculate on whether or not "antimüllerian factor" has a function other than just stimulating regression of the müllerian duct, since AMF need be around for only a very short period of time developmentally in the early first trimester—but yet it persists postnatally for a relatively long time in the human.

N. Josso: In my experience, antimüllerian activity does not persist postnatally in humans. Even so, in human male fetuses müllerian ducts are completely regressed at 9 weeks, and yet the testis goes on making the hormone until at least 27 weeks. Of course, I agree with you, it is tempting to look for another function. According to Dr. Jost, AMH may play a role in testicular differentiation by preventing testicular germ cells from entering the prophase of meiosis, as do fetal ovogonia.

M. M. Martin: Might it be worthwhile to look into the question of whether this particular factor does not protect the testis from the effect of maternal estrogens? This could be tested perhaps by adding estrogens to see whether this material in some way protects the fetal testis from being damaged.

N. Josso: I cannot answer that. I do not know.

Evolution of Gonadotropin Structure and Function[1]

PAUL LICHT,* HAROLD PAPKOFF,†‡
SUSAN W. FARMER,† CHARLES H. MULLER,*§
HING WO TSUI,* AND DAVID CREWS*

*Department of Zoology and §Cancer Research Laboratory, University of California, Berkeley, California; and †Hormone Research Laboratory and ‡Reproductive Endocrinology Center, University of California, San Francisco, California

I. Introduction

It has long been recognized that the control of gonadal function by pituitary hormones (gonadotropins) is a general feature of vertebrate reproductive physiology. Numerous studies on the hormones of eutherian mammals have established the existence of two chemically distinct types of gonadotropin molecules in the pituitary—luteinizing hormone (LH) and follicle-stimulating hormone (FSH). Recent biochemical studies on these molecules have revealed that each is a glycoprotein consisting of two chemically nonidentical subunits, designated α and β. Although the physiological actions of the two gonadotropins in mammals are still subject to intensive investigation, it is agreed that FSH and LH have somewhat different roles in the regulation of gonadal function. In brief, the processes of ovulation and gonadal steroid secretion (especially postovulatory progesterone in ovaries and androgen in testes) are considered to depend primarily upon LH. The unique actions of FSH are somewhat more difficult to summarize, but this hormone is classically associated with the events involved in gonadal growth and maintenance, including "preparation" of the gonad for the subsequent actions of LH.

Until recently, it was tacitly assumed that the nature and actions of the gonadotropins in other vertebrates, or at least in the "higher" tetrapod species, conformed to this mammalian pattern. However, as more detailed comparative studies were undertaken with nonmammalian species, "unexpected" results began to emerge. Many of the physiological data were compatible with the hypothesis that some groups, such as amphibians and reptiles, might possess only

---

[1] Studies reported herein were supported by grants from the National Science Foundation to P. Licht and H. Papkoff (currently BMS-75-16138) and funds from the Committee on Research of the University of California. C. H. Muller was supported by NIH training grant CA05045.

one gonadotropin with broad actions, or if two hormones, without the same duality of actions as FSH and LH in mammals (van Oordt and de Kort, 1969; Licht, 1974a). This idea was especially interesting in light of the high degree of structural homology evident among the subunits of mammalian LH, FSH, and thyrotropin (TSH), which suggests that the three may have arisen from a common ancestral glycoprotein molecule.

Unfortunately, the greater part of the early comparative work by necessity involved mammalian (usually ovine) hormones, frequently of uncertain purity. Consequently, some of the apparently anomalous responses in nonmammalian species may have been due to the use of heterologous hormones. The demonstration of phylogenetic specificity[2] —mammalian hormones were often less potent than the homologous hormones in some nonmammalian species and many nonmammalian hormones were essentially inactive in mammals—led to further concern about the use of heterologous hormones. We became convinced that fundamental questions on the nature of endocrine physiology in nonmammals and the evolution of vertebrate hormone structure could be answered only by direct study of the nonmammalian hormones, with special emphasis on homologous systems.

We decided to focus our attention on the three classes of tetrapods—Amphibia, Reptilia, and Aves—that, in a simplified view of the phylogenetic history of vertebrates, represent the presumptive "intermediate" stages in mammalian evolution (Fig. 1). To facilitate direct comparisons between mammalian and nonmammalian species, we employed many of the techniques for purification and physiological characterization that were used in mammalian studies. Thus, we hoped to examine both the basic hormonal regulation of reproduction in the different nonmammalian vertebrates and the general patterns of gonadotropin structure and function.

II. Basic Requirements for Comparative Studies on Pituitary Hormones

Several problems unique to studies on the hormones of nonmammalian species had to be overcome. These were related to the acquisition of pituitary tissue, methods of bioassay, and fractionation procedures.

---

[2] The term "phylogenetic specificity" refers to the insensitivity of one species to the hormones derived from a different (heterologous) species; some authors have used the terms species specificity, taxonomic specificity, or zoological specificity for the same concept. The term hormonal specificity refers to the ability of a tissue to discriminate among different types of hormones from a single species. In the context of the present report this term is used in connection with the specificity of a response for FSH vs LH.

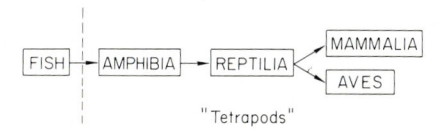

FIG. 1. Simplified scheme of the phylogenetic relationships among vertebrate classes.

A. COLLECTION AND CONDITION OF GLANDS

We considered it essential to have glands from representatives of several major orders within each class before meaningful generalizations about evolutionary patterns could be made. Moreover, a relatively large quantity of glands had to be available before sufficient amounts of hormone could be prepared for the various biochemical and physiological studies we wished to make. The problem of gland acquisition was compounded by principles of conservation: large numbers of wild animals should not be (and were not) sacrificed solely for the purpose of extracting glands.

These problems were eventually resolved by exploiting several industries and projects in which the heads or glands of various animals could be obtained as by-products. For example, our work on reptiles was initiated with the glands from snapping turtles (*Chelydra serpentina*) being slaughtered for the turtle soup industry in the United States. Subsequently, our best supply of turtle pituitaries came from a project involving commercial culture of the green sea turtle (*Chelonia mydas*) on Grand Cayman Island in the Caribbean. Since these two turtle species represent distinct and ancient lineages, they also offered an opportunity to examine evolutionary changes within a single order (Chelonia) of reptiles. Large numbers of snake heads (order Squamata) were acquired from the snake shops of Hong Kong and the rattlesnake hunts of the southwestern United States—heads are a waste product in these activities; about five species representing three snake families (Colubridae, Elapidae, Crotalidae) were obtained in this way. Material representing the third order of reptiles (Crocodilia) was collected in connection with an experimental hunting season on the American Alligator (*Alligator mississipiensis*) in Louisiana; about 500 glands from large adults were obtained with the cooperation of the staff of the Rockefeller Wildlife Refuge. Thus, we had access to representatives of all three major orders of Reptilia. Comparisons among these could allow inferences about the basic reptilian stock since the three orders are thought to have diverged from a common ancestor about 250–300 million years ago.

Avian pituitaries were readily available from the commercial poultry industry which supplied tissues from domesticated chickens, turkeys (order Galliformes), ducks (order Anseriformes), and pigeons (order Columbiformes). Initial studies on a noneutherian mammal were made with glands from the red kangaroo supplied from several projects in Australia.

Studies on amphibians focused on frogs (order Anura), beginning with a relatively small supply of bullfrog (*Rana catesbeiana*) glands obtained in connection with a local frog-leg operation in California. However, we subsequently discovered that vast numbers of bullfrogs were being used in the Japanese frog-leg industry (i.e., most American bullfrog-legs are made in Japan), and we were able to buy almost 2 tons of frozen heads—about 90,000. The glands extracted from these have been a major source of material for the preparation of various pituitary hormones in our laboratories. A small quantity of glands from a second frog (*Rana pipiens*) was also examined for comparative purposes. A representative of a second amphibian order (Urodela), the tiger salamander (*Ambystoma tigrinum*), was obtained from Texas in small numbers but sufficient to get basic information on variability among amphibian hormones. Unfortunately, material representing the third major amphibian order (Apoda) was unavailable.

Methods of processing the pituitaries varied and should be considered in interpreting data; details for individual species are presented in separate studies. Glands were removed from freshly killed specimens and dehydrated in cold acetone, lyophilized or frozen on Dry Ice; sometimes, several batches from a single species were treated differently. Also, not all tissues were in equally good condition. The alligator glands had to be removed from heads taken from animals killed in the field and transported to a central facility, sometimes up to 12 hours after death. In the original study on bullfrogs, glands were flash-frozen from freshly killed specimens, but those from Japan required defrosting of heads and refreezing glands, and we could not be certain of the condition of the heads when originally frozen.

In general, the glands were taken from sexually mature specimens of both sexes. The *Chelonia* glands were primarily from immature subadults, and presumably this was the case for most of the chickens, turkeys, and ducks. In separate tests with glands collected from adult sea turtles and chickens, we were unable to discern any differences related to the age of the animals.

B. BIOASSAY

Identification and purification of gonadotropin obviously depend on the availability of suitably specific bioassays. Early experience in comparative endocrinology and preliminary studies in our laboratories indicated that hormones from many nonmammalian species were relatively inactive in the *in vivo* mammalian bioassays frequently employed to define and quantify FSH and LH. For example, even the most purified nonmammalian preparations tended to be minimally active in the Parlow (1961) ovarian ascorbic acid depletion test for LH in rats (OAAD). Although the potencies of highly purified gonadotropins from several species approached that of the NIH-FSH-S1 in the Steelman–Pohley (1953) ovarian augmentation test in rats, the relatively large quantities of material required for potency estimates in this assay limited its use in most of our fractionation studies. Furthermore, there was some concern that mammalian bioassays might not retain their characteristic FSH/LH specificity when challenged with heterologous nonmammalian hormones (Licht *et al.*, 1976a; Licht and Midgley, 1976b). Thus, while it is of interest to have information on relative potencies of nonmammalian preparations in these standard bioassays for FSH and LH, they were not considered suitable for routine use and were employed only to characterize final preparations.

Ideally, each hormone should be tested in the species from which it was derived; indeed, this was one of our major objectives. However, this was impractical for the initial stages of fractionation before the physiological nature of the hormone was known. Consequently, to facilitate the comparative approach, we first sought to develop a series of bioassays that would allow estimates of potency for the full spectrum of species to be studied, i.e., tests that showed broad phylogenetic cross-reactivity. Furthermore, it was clear that the existence of dual gonadotropins could only be resolved with bioassays that were specific for FSH or LH. With these objectives in mind, we examined a variety of gonadotropin responses in nonmammalian species. Hormonal specificity was tentatively defined with the use of highly purified ovine FSH and LH. Four bioassays were selected for routine use (Table I).

Two of the assay systems were essentially nonspecific for FSH and LH (i.e., the two ovine hormones were about equipotent) and served to measure "total" gonadotropin activity during fractionation. The first was a well known gonadotropin assay involving the uptake of ^{32}P by gonads of day-old cockerels (see Breneman *et al.*, 1962). The second assay was based on spermiation in frogs (the early Galli–Mainini human pregnancy test). Although these two assays were once considered to be more sensitive to LH than FSH, studies with highly purified preparations indicated a lack of specificity (Licht, 1973a,b).

Two additional bioassays showed a much higher degree of hormonal specificity when tested with the mammalian preparations (Table I). Studies with the lizard *Anolis carolinensis* indicated that FSH (from ovine and human origin) was several orders of magnitude more potent than comparably purified LH for the stimulation or maintenance of testicular activity (Licht and Pearson, 1969; Licht and Hartree, 1971; Licht and Papkoff, 1973). The low activity of LH (less than 1% of FSH) was at first attributed to FSH contamination, but subsequent tests confirmed that ovine LH did have intrinsic activity (Licht and Papkoff, 1973); thus, the assay was not entirely FSH-specific, a problem that will be discussed in

TABLE I
Specificities of Nonmammalian Gonadotropin Bioassays for Ovine FSH and LH[a,b]

Bioassay	Minimal effective dosage (μg)[c] FSH	LH
Lizard (*Anolis*) testis weight maintenance	0.01	20
Frog (*Xenopus*) *in vitro* ovulation	>200	0.5
Frog (*Hyla*) spermiation	0.5	1
Chick testis ^{32}P uptake	1	1

[a] Modified from Licht and Papkoff (1974d).
[b] Based on highly purified ovine hormones: luteinizing hormone (LH) = 2 × NIH-LH-S1 in ovarian ascorbic acid depletion assay; follicle-stimulating hormone (FSH) = 50× NIH-FSH-S1 in Steelman–Pohley assay.
[c] Minimal effective doses are representative and may vary with the conditions under which the tests are performed. These are the doses (given in μg/ml or μg/injection) required to obtain significant stimulation over control levels.

greater detail later. In this assay, testicular weight in physiologically or surgically hypophysectomized animals was used to quantify hormonal activity; indices of precision ranged from 0.2 to 0.35.

LH-like activity was quantified by an *in vitro* assay involving the ovulation of follicles from ovarian segments from frogs. Tests were performed with three species—*Hyla, Rana,* and *Xenopus* (and other species are known to be usable). In all cases, the ovulation response was found to be highly specific for LH; the maturation of the oocyte, as indicated by germinal vesicle breakdown prior to ovulation, is equally specific and sensitive for LH. Not all species of frog were equally suitable for comparative studies; e.g., *Rana* ovaries appear to be especially insensitive to mammalian preparations. The African clawed frog *Xenopus laevis* was used most frequently because it showed a broad phylogenetic cross-reactivity, was easy to maintain in the laboratory, and had "ripe" follicles throughout the year, especially if primed with pregnant mare serum gonadotropin (PMSG) (Thornton, 1971). Potency estimates were based on the percentage of follicles ovulating (Licht and Papkoff, 1974b).

Ovine (NIH) reference preparations were employed in each assay; thus, all hormonal activities are expressed in terms of the NIH-LH-S1 or NIH-FSH-S1 standard. An increase in relative potency of a material during fractionation was taken as evidence of increased purification, although some interactions among hormones in crude mixtures may complicate this interpretation. However, we particularly wish to emphasize the need for caution when inferring differential purities among hormones from different species based on results of a single bioassay. Phylogenetic specificity in a bioassay may completely obscure differences in purity. This problem was illustrated by comparisons of relative potencies for a series of purified LH preparations from seven mammalian species in the *Xenopus ovulation* assay and the OAAD rat assay (Licht and Papkoff, 1976). Relative potencies correlated poorly, and there were even marked differences in the rank-order of potencies between the two assays (Table II). Additional examples of this phenomenon will be discussed later in connection with comparisons between *in vivo* and *in vitro* bioassays and the effects of differential clearance rates.

With the above reservations in mind, we hypothesized that if a particular species possessed two discrete gonadotropins whose physiological and biochemical properties paralleled those of mammalian FSH and LH, then it should be possible to isolate two separate fractions with markedly different potency ratios in the *Anolis* testis assay and the *Xenopus* ovulation assay. Both fractions should be approximately equipotent in the chick testis and frog spermiation assays.

TABLE II

*Comparison of Relative Potencies of Various Mammalian Luteinizing Hormone (LH) Preparations in Mammalian (OAAD) and Amphibian (*Xenopus *Ovulation) Bioassays*[a]

Species of LH	Identification	Average potency (× NIH-LH-S1) OAAD[b]	Xenopus[c]	Discrimination ratio OAAD/Xenopus
Sheep	NIH-LH-S18	1	1	1
	G3-222B	2.75	2.8	1
Cow	NIH-LH-B2	1	1.05	1
	G3-78,84B	1	1.3	0.8
Horse	E-24BR	3	0.6	5
Pig	W-49BG	1.2	0.10	12
Rat	NIAMMD-I-1	1	<0.25	>4
Rabbit	EX 130GB	0.8	0.025	32
Human	GI-102	4.1	0.05	82
Monkey	W-244 S	nt	0.06	–
Dog	EX-209BS	nt	0.02	–

[a]From Licht and Papkoff (1976).
[b]Ovarian ascorbic acid depletion assay.
[c]The lowest dose of NIH-LH-S18 required to induce ovulation was 0.2–0.4 µg/ml in the three assays, and the MEDs were all approximately 1 µg/ml.

Although results subsequently obtained with purified nonmammalian hormones indicated that the apparent FSH/LH specificity of each assay might vary with the source of the hormone, this specificity proved sufficiently consistent to allow discrimination between FSH and LH from most of the nonmammalian species studied.

III. Fractionation of Pituitary Hormones

A. PROCEDURES

Several different protocols for purification of pituitary hormones have been employed during the course of our study; all were based on techniques used for mammalian hormones to facilitate comparisons of chemical characteristics of the final products. These procedures combine a series of independent steps that are used, first, to extract the hormones from pituitary tissue; second, to separate the gonadotropins from other pituitary hormones and then to separate the gonadotropins from one another (assuming that more than one can be identified); and, finally, to achieve a high degree of purity of the isolated hormones. The order and number of individual procedures have varied in our studies; Fig. 2 presents a flow diagram of the protocol that we now consider to be the most effective for obtaining high yields and purified products. A brief resumé of these techniques follows:

1. Extraction and separation of gonadotropins from pituitary tissue were accomplished by homogenizing the tissue in cold water with a Waring blender, adjusting the pH to 9.5 with CaO, and extracting with stirring for 3 hours at 4°C (acetone-dried or lyophilized materials were extracted longer). This procedure extracted all the glycoproteins, growth hormone (GH), and probably also prolactin and adrenocorticotropin. After removal of the insoluble residue by centrifugation, the alkaline extract was adjusted to $0.15\ M\ (NH_4)_2SO_4$ and the pH to 4 by addition of freshly prepared $0.2\ M\ HPO_3$; the resulting precipitate contained the GH (Farmer et al., 1976). The gonadotropins and TSH in the supernatant were harvested by precipitation with addition of $(NH_4)_2SO_4$ to a concentration of 0.8 saturation (in earlier studies we showed that the vast majority of these glycoprotein hormones are precipitated at 0.6 saturation). This precipitate was then dissolved in a small volume of $0.2\ M\ K_2HPO_4$, heated to 55–60°C for 2 minutes to destroy proteolytic enzymes (Papkoff et al., 1965), dialyzed, and lyophilized. The lyophilized material was then extracted with 10% NH_4Ac, pH 5.1–40% EtOH to obtain a glycoprotein concentrate (Hartree, 1966).

2. Separation of presumptive LH from FSH and the further purification of each were accomplished by a series of three ion-exchange chromatographies— Amberlite CG-50, sulfoethyl Sephadex (SE) C-50, and DEAE-cellulose. In al-

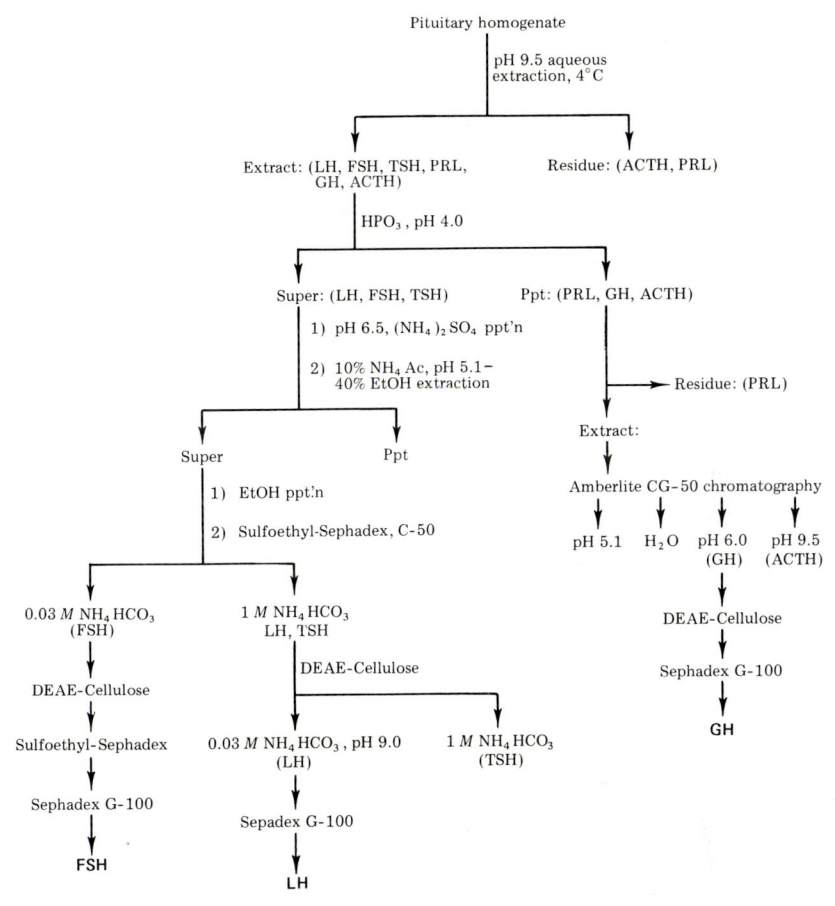

```
                         Pituitary homogenate
                         │
                         │ pH 9.5 aqueous
                         │ extraction, 4°C
          ┌──────────────┴──────────────┐
          ▼                             ▼
Extract: (LH, FSH, TSH, PRL,     Residue: (ACTH, PRL)
         GH, ACTH)
          │
          │ HPO₃, pH 4.0
   ┌──────┴──────┐
   ▼             ▼
Super: (LH, FSH, TSH)    Ppt: (PRL, GH, ACTH)

1) pH 6.5, (NH₄)₂SO₄ ppt'n
2) 10% NH₄ Ac, pH 5.1 –
   40% EtOH extraction          ──▶ Residue: (PRL)
                                 Extract:
   Super        Ppt                  │
                                     ▼
1) EtOH ppt'n                 Amberlite CG-50 chromatography
2) Sulfoethyl-Sephadex, C-50
                              pH 5.1   H₂O   pH 6.0   pH 9.5
                                             (GH)    (ACTH)
0.03 M NH₄HCO₃   1 M NH₄HCO₃
    (FSH)         LH, TSH             DEAE-Cellulose
                                             │
DEAE-Cellulose   DEAE-Cellulose       Sephadex G-100
                                             │
Sulfoethyl-Sephadex  0.03 M NH₄HCO₃, pH 9.0   1 M NH₄HCO₃   GH
                         (LH)                    (TSH)
Sephadex G-100
                     Sepadex G-100
    FSH
                           LH
```

FIG. 2. Protocol for the fractionation of pituitary hormones, with particular reference to gonadotropins and growth hormone (GH), from various nomammalian species. LH, luteinizing hormone; FSH, follicle-stimulating hormone; TSH, tyroid-stimulating hormone; PRL, prolactin, ACTH, adrenocorticotropin; Super, supernatant; Ppt, precipitate. Details on individual methods are given in Licht and Papkoff (1974a, b), Farmer *et al.* (1975), and Papkoff *et al.* (1976a, b).

most every species examined, we found a major separation of the two gonadotropins on each of the three systems, and there was a remarkable similarity in the behavior of both types of hormones among species (Table III). The two notable exceptions (duck and snakes) will be discussed later. By employing all three chromatographic procedures, we were able to minimize the cross-contamination between hormones. Details of each system have been given elsewhere (Papkoff and Li, 1958; Papkoff *et al.*, 1962, 1976a,b; Licht and Papkoff, 1974a,b). In brief, the first system was Amberlite CG-50 equilibrated with pH

TABLE III
Fractionation of Gonadotropins by Ion-Exchange Chromatography[a]

Species	Amberlite CG-50 pH 5.1	Amberlite CG-50 pH 6.0	DEAE, pH 9.0 0.03 M[b]	DEAE, pH 9.0 1 M[b]	Sulfoethyl-Sephadex C-50 0.03 M[b]	Sulfoethyl-Sephadex C-50 1 M[b]
Mammals	FSH	LH	LH	FSH	FSH	LH
Marsupials						
Kangaroo	FSH	LH	LH	FSH	FSH	LH
Birds						
Chicken	FSH	LH	LH	FSH	FSH	LH
Turkey	FSH	LH	LH	FSH	FSH	LH
Duck	FSH/LH	–	nt[c]	nt	–	FSH/LH
Reptiles						
Sea turtle	FSH	LH	LH	FSH	FSH	LH
Snapping turtle	FSH	LH	LH	FSH	FSH	LH
Alligator	FSH	LH	LH	FSH	FSH	LH
Snakes (3 spp.)	–	FSH	FSH	FSH	FSH	FSH
Amphibians						
Bullfrog	FSH	LH	LH	FSH	FSH	LH
Leopard frog	nt	nt	LH	FSH	FSH	LH
Tiger salamander	FSH	LH/FSH	LH/FSH	FSH	LH/FSH	FSH

[a]Shows where most of each type of gonadotropin is recovered in each system. FSH is defined as the material with high activity in *Anolis* lizard testes weight bioassay and low potency in *Xenopus in vitro* ovulation assay; and LH as the material with the reverse biological profile.
[b]NH$_4$HCO$_3$ solutions.
[c]nt, not tested.

5.1 phosphate buffer adjusted to 12% saturation with (NH$_4$)$_2$SO$_4$; under these conditions, FSH was largely unadsorbed. Additional fractions were eluted with H$_2$O, then pH 6.0 buffer, and finally NaOH; LH eluted with the pH 6.0 buffer. SE C-50 was equilibrated with 0.03 M NH$_4$HCO$_3$; FSH was unadsorbed and LH eluted with 1 M NH$_4$HCO$_3$. In the third system, the DEAE was equilibrated with 0.03 M NH$_4$HCO$_3$, pH 9; LH was unadsorbed and FSH was eluted with 0.2 M NH$_4$HCO$_3$.

3. Considerable purification of each gonadotropin was achieved by the steps above. When sufficient material was available after the chromatographic procedures, final purification of each gonadotropin was effected by gel filtration on Sephadex G-100 equilibrated with 0.05 M NH$_4$HCO$_3$. FSH tended to elute at a V_e/V_o of 1.6–1.8 and the LH at 1.8–1.9; in several cases, two distinct FSH fractions with approximately equal biological activity were obtained in this step (Licht *et al.*, 1976b).

B. IDENTIFICATION OF GONADOTROPINS BY BIOASSAY

Of the 12 nonmammalian species studied by the above procedures, all pituitaries, except those from the duck and snakes (3 genera), yielded two separate gonadotropin fractions that were found to differ with respect to both biological profile (Table IV) and biochemical characterization. In six of the species studied, there was sufficient material to obtain a highly purified preparation for detailed biochemical characterization. A summary of the biological activities of the two

TABLE IV
Relative Potencies of Purified Gonadotropins from Various Tetrapod Pituitaries

Source of hormone	Type of hormone	Purity[a]	Anolis (assay[b] (NIH-FSH-S1)	Xenopus assay[c] (× NIH-LH-S1)	Anolis:Xenopus Ratio
Sheep (*Ovis*)	FSH	+	100	<0.005	>20,000
	LH	+	0.04	2.0	0.02
Human (*Homo*)	FSH		>20	nt[f]	–
	LH		<1	0.05	–
Snapping turtle	FSH	+	30	0.002	15,000
(*Chelydra*)	LH	+	0.13	1.8	0.07
Sea turtle	FSH	+	4.0	0.004	1,135
(*Chelonia*)	LH	+	1.0[d]	1.0	1.0[e]
Alligator	FSH	+	2.5	<0.002	>11,250
(*Alligator*)	LH	+	1.0[d]	3.3	0.3[c]
Snake (*Ptyas, Bungarus, Crotalus*)[d]	"FSH"		0.2–0.5	Nil	–
	"LH"		0.2–0.5	Nil	–
Chicken (*Gallus*)	FSH	+	4.3	<0.0015	>2,600
	LH	+	0.85	0.08	10
Turkey (*Meleagris*)	FSH	+	4.0	0.004	1,000
	LH	+	nt	0.33	–
Duck (*Anas*)	"FSH/LH"		3.0	0.05	–
Bullfrog (*Rana catesbeiana*)	FSH	+	7.0	0.004	1,750
	LH	+	0.12	0.33	0.36
Leopard frog	FSH		5.8	0.03	193
(*Rana pipiens*)	LH		0.12	0.30	0.4
Salamander	FSH		0.1	0.002	50
(*Ambystoma*)	LH		1.0[d]	2.0	0.5

[a]+ designates those preparations considered to represent highly purified final products.
[b]Estimates based on the maintenance of testis weight in hypophysectomized lizard injected for 10–12 days.
[c]Estimates based on *in vitro* ovulation from ovarian segments.
[d]Snake fractions tentatively identified as follicle-stimulating hormone (FSH) and luteinizing hormone (LH) on the basis of chromatographic behavior, since no fractions were active in the *Xenopus* ovulation assay (Licht, 1974b).
[e]Dose response slopes in *Anolis* assay were nonparallel to those of FSH, including NIH reference preparation.
[f]nt, not tested.

fractions from each species in our most specific bioassays is provided in Table IV. Considerable interspecific variation was evident among potencies of both hormones, including those considered to be equally purified. For example, in *Anolis* bioassay, none of the nonmammalian hormones were as potent as ovine FSH; snapping turtle (*Chelydra*) FSH was more potent than others, and the remainder were relatively similar to one another. Variations among the potencies of nonmammalian LH preparations were less than observed among a series of highly purified mammalian LH preparations (Table II). Since wide interspecific variations have also been observed in several other physiological tests, we interpreted them as being reflections of phylogenetic specificity of the assay system rather than inequalities in the extent of purification. Thus, comparisons among fractions are best restricted to an individual species. Comparisons of potencies between purified preparations of each gonadotropin and the alkaline extracts of the pituitaries from which they were derived indicated a considerable purification of each hormone: relative potencies of FSH preparations tended to be more than 100-fold greater than the extracts, and LH preparations usually exceeded 20-fold the potency of extracts.

In general, potency estimates of the two gonadotropins in the opposite gonadotropin bioassay revealed a low degree of cross-contamination. In most cases, the LH activity in FSH corresponded to only 1% or less of the respective purified LH. The apparent FSH contamination in the six purified LH preparations was somewhat higher (0.4 to 40%), but may reflect the lack of complete FSH specificity of the *Anolis* assay. With the two most potent LH preparations (*Chelonia* and *Alligator*), dose-response curves in *Anolis* were not parallel to those for FSH (Licht *et al.*, 1976a,b). *Ambystoma* represented a case where the fraction identified as LH was more active than the presumptive FSH; however, the FSH was not as extensively purified as LH.

Results for the three species of snake and the duck presented a special situation in which two biologically distinct gonadotropin fractions could not be identified. In the duck, we purified a single fraction that was highly potent (e.g., by comparison with both chicken FSH and LH) in both *Anolis* and *Xenopus* bioassay (Table IV). The chromatographic behavior in the duck gonadotropin most closely resembled that of an FSH on Amberlite CG-50 but an LH on SEC-50 (Table III). This situation may be like that observed in *Ambystoma*, where LH had high intrinsic activity in *Anolis* and the separate FSH was either relatively weak in potency or low in concentration (Licht *et al.*, 1975).

Results for all snake pituitaries were even more extreme than those for the duck. In snakes (and related squamates, the lizards), the pituitary appears devoid of a hormone that will induce ovulation in any amphibian tested (Licht, 1974b); thus, there is no method for clearly demonstrating LH bioactivity in these reptiles. When the snake fractions were monitored by the *Anolis* bioassay, the gonadotropin activity (FSH?) had a somewhat anomalous chromatographic behavior: the majority was absorbed on Amberlite (LH-like), and this was then about equally distributed between the fractions expected to contain FSH and LH on SE C-50 and DEAE (Table III). Thus, there was evidence for two chemically dissimilar types of gonadotropins from chromatographic studies, but they could not be readily defined as FSH and LH. We were also unable to distinguish between these two materials in a variety of other hormone assays, including steroid production by snake and turtle testes (Tsui, 1976a), radio-ligand studies for FSH binding (Licht and Midgley, 1976b) and immunological measurements with antisera against bird and mammalian LH (Licht et al., 1974). Either the squamate reptiles lack an LH-like gonadotropin in the pituitary or their LH has diverged sufficiently from those of all other tetrapods to render it unrecognizable by various biological and immunochemical systems. It is also surprising that the snake gonadotropin fractions had relatively low potencies in *Anolis* (and other assays) considering the number of purification steps to which they had been subjected.

Studies with a fourth species of bird, the pigeon, demonstrated a further problem for comparative work (S. W. Farmer, H. Papkoff, and P. Licht, unpublished). Preliminary attempts to purify gonadotropins were unsuccessful because of an almost complete lack of biological activity in all our standard assays. It was unclear whether these results reflected another case of extreme phylogenetic specificity in the assays or very low hormone concentrations in the glands (pituitaries were from birds of mixed age and sex). These data illustrate the need for considerable caution in generalizing about the hormones of any vertebrate class based on a single representative. Of the three orders of birds examined, only one (Galliformes) has yielded results that support the existence of a distinct FSH and LH molecule with similarities to those of mammals.

Several of the purified nonmammalian hormones have been examined in two of the standard *in vivo* bioassays for gonadotropins in mammals. Tests of FSH fractions in the Steelman–Pohley bioassay indicated that our chicken, turkey, alligator, and sea turtle preparations ranged from 1 to 4 × NIH-FSH-S1 in potency. Similar values have been reported for chicken and turkey FSH in other laboratories (Furuya and Ishii, 1974; Wentworth, 1971). In the OAAD bioassay for LH in rats, however, we obtained values of only about 0.03 × NIH-LH-S1 for purified chicken and turkey LH, and we were unable to demonstrate significant activity for either alligator, turtle, or frog LH (they were < 0.005 × NIH-LH-S1). Several other laboratories have reported similarly low potencies for chicken LH in this assay (Furr and Cunningham, 1970; Follett et al., 1972; Furuya and Ishii, 1974), but Wentworth (1971), in an abstract, said that turkey LH was about 1 × NIH-LH-S1. Numerous physiological studies and binding analyses to be discussed in detail later provide evidence that these relatively low potencies are reflections of phylogenetic specificity in the mammalian assays rather than lack of purity of the nonmammalian hormones.

IV. Biochemical Characterization of Nonmammalian Gonadotropins

The strong resemblance in chromatographic behavior between the fractions identified as FSH and LH from most nonmammalian species and their mammalian counterparts provided preliminary evidence of chemical similarity among the different species of FSH and of LH. Biochemical analyses on the six highly purified species of LH and FSH (chicken, turkey, alligator, snapping turtle, sea turtle, and bullfrog) yielded additional evidence for the homologies between mammalian and nonmammalian hormones.

A. ASSESSMENT OF PURITY

Purity has been difficult to access in both mammalian and nonmammalian gonadotropins. All gonadotropins studied to date are glycoproteins, and even where a highly purified protein has been prepared, there can be heterogeneity in the carbohydrate moiety of the molecule (Bell et al., 1969). Likewise, gonadotropins have been shown to be composed of two nonidentical subunits, and therefore both amino and carboxy-terminal analyses may reveal at least two amino acids. Furthermore, gonadotropin purity cannot be directly assessed by disc gel electrophoresis since these hormones give diffuse and poorly straining bands. However, we do have data that collectively suggest that the nonmammalian preparations carried through the entire purification procedure were comparable in purity to highly purified mammalian gonadotropin preparations.

The bioassay data indicated that the final gonadotropin preparations were considerably purified over the initial alkaline extracts of the glands. In addition to the very low level of cross-contamination between gonadotropins (Table IV), we observed minimal contamination with growth hormone and prolactin as evidenced by disc gel electrophoresis (Papkoff et al., 1976b; Licht et al., 1976b). The question of possible contamination with TSH is more difficult to resolve at present. Preliminary assays for TSH in our preparations were based on the thyroidal uptake of radioiodine in baby turtles and to some extent in hypophysectomized lizards. These results showed that both FSH and LH from the snapping turtle had marginal TSH activity compared with a third glycoprotein fraction from this species (P. Licht and D. MacKinzie, unpublished). However, in the bullfrog and sea turtle, the FSH was essentially free of TSH, but the LH fraction showed appreciable activity in baby turtles which we have not yet been able to eliminate. Preliminary evidence based on thyroidal uptake of radioiodine in frogs and radiophosphorus uptake in chicks (Licht and Papkoff, 1974b) suggested that the baby turtle assay may not be able to discriminate between bullfrog LH and TSH; in fact, the frog LH may be relatively free of TSH. Further analysis of TSH must await the development of improved bioassays.

Another criterion of purity is the behavior of each fraction on gel filtration. Gonadotropins that were studied on Sephadex G-100 had reasonably symmetrical peaks; published examples are shown in Papkoff et al. (1976b) and Licht et al. (1976b).

B. CARBOHYDRATE ANALYSIS

Colorimetric methods were employed for the determination of carbohydrate content; these data are compared with those obtained for mammalian hormones in Tables V and VI. There was considerable variation among the species in both the amounts and types of sugars present with no clear phylogenetic pattern. The content of sialic acid is of importance because of its effect on biological activity; desialylation with neuraminidase greatly reduced potency of some nonmammalian hormones when tested in *Anolis* (Licht and Papkoff, 1972, 1974c) and other lizards (Burns and Richards, 1974). It is noteworthy that all the FSH preparations possessed significant amounts of sialic acid, whereas some LH preparations did not. Studies with mammalian hormones suggested that FSH had more sialic acid than did LH, but chicken and sea turtle gonadotropins showed the reverse. The high levels of sialic acid in these two species of LH may contribute to their relatively high intrinsic activity in *Anolis* (Table IV), although aligator LH which was also potent has relatively little sialic acid. Surprisingly, neuraminidase failed to inactive the snapping turtle FSH despite its sialic acid content (Licht and Papkoff, 1974c).

C. NH_2-TERMINAL AND AMINO ACID COMPOSITION

Amino-terminal analyses were performed by the dansyl technique (Gray, 1964; Wood and Wang, 1967) and amino acid analysis of hydrolyzed samples

TABLE V
Carbohydrate Composition of Purified Pituitary Luteinizing Hormone (LH)[a]

Species of LH	Hexose	Hexosamine	Sialic acid	Total
Sheep	7.2	9.1	0.4	16.7
Human	5.9–11.3	4.0–5.1	0.7–2.0	12.3–17.3
Turkey	4.9	1.0	Nil	5.9
Chicken	5.2	7.1	1.4	13.7
Alligator	3.9	7.3	0.4	11.6
Sea turtle	6.4	8.9	1.9	17.2
Snapping turtle	5.0	1.8	Nil	6.8
Bullfrog	3.6	5.5	0.2	9.3

[a]Based on Sairam and Papkoff (1974), Farmer et al. (1975), Licht et al. (1976b), Papkoff et al. (1976a,b), and S. W. Farmer and H. Papkoff (unpublished data).

TABLE VI
Carbohydrate Composition of Purified Pituitary FSH[a]

Species of FSH	Hexose	Hexosamine	Sialic acid	Total
Sheep	5.7	4.5	2.8	13.0
Human	3.9–11.6	2.4–9.1	1.4–5.2	8.1–25.9
Turkey	3.0	3.5	0.7	7.2
Chicken	3.1	3.2	0.8	7.1
Sea turtle	4.9	6.6	0.6	12.1
Snapping turtle	4.6	6.6	1.2	12.4
Bullfrog	7.2	9.8	3.0	20.0

Carbohydrates (gm/100 gm glycoprotein)

[a]See Table VI for references.

by the method of Spackman *et al.* (1958) in a Beckman Automatic Amino Acid Analyzer. In most cases, two or more residues were found at the amino terminus of the gonadotropins, and the same residues were found repeatedly in both LH and FSH (phenylalanine, alanine, leucine, and glycine); similar results were obtained with mammalian gonadotropins.

TABLE VII
Amino Acid Composition of Various Species of Luteinizing Hormone (LH)[a]

Amino acid	Human	Sheep	Chicken	Turkey	Alligator	Snapping turtle	Sea turtle	Bullfrog
Lysine	8	12	14	12	15	14	14	13
Histidine	6	6	3	3	5	5	5	6
Arginine	13	11	10	9	10	9	8	9
Aspartic	12	11	16	13	13	17	17	24
Threonine	15	16	14	18	18	17	20	21
Serine	15	14	14	13	15	14	14	17
Glutamic	16	14	21	17	18	16	15	14
Proline	22	27	18	23	20	17	20	16
Glycine	12	11	22	22	14	14	14	8
Alanine	9	15	18	17	17	16	14	10
1/2-Cystine	22	22	16	20	20	19	21	22
Valine	18	13	12	13	12	13	12	12
Methionine	5	7	4	4	4	5	4	7
Isoleucine	6	7	7	8	8	9	8	12
Leucine	12	14	13	11	12	11	10	12
Tyrosine	6	7	3	7	3	11	10	8
Phenylalanine	6	8	8	7	10	9	9	6

[a]Calculated on the basis of 215 residues per mole. See Table V for references.

The amino acid composition of various species of LH are shown in Table VII and of FSH in Table VIII. The compositions of all gonadotropins were calculated on the basis of 215 residues per mole for ease in comparison: this value represents the number of residues known to be present in ovine LH from structural studies (Papkoff et al., 1973) and is close to the number present in human FSH (Shome and Parlow, 1974a,b). All species of LH clearly resembled one another in amino acid composition, particularly with respect to the high content of proline and cystine. An independent analysis of chicken LH by Godden and Scanes (1975) was in agreement with our values, except for a slightly higher proline and tyrosine content. All FSH preparations were characterized by a high content of aspartic and glutamic acids as well as cystine. Nonmammalian species were like mammalian in that LH was distinguished from FSH by higher contents of proline and methionine and a lower content of aspartic plus glutamic acids. Godden and Scanes (1975) showed similar differences between chicken FSH and LH. Among all of the amino acid compositions examined, there were more similarities than differences, and the differences did not show any clear phylogenetic pattern; e.g., the FSH from one mammal was no more like the FSH from another mammal than it was like the FSH from a nonmammalian species.

TABLE VIII
Amino Acid Composition of Various Species of Follicle-Stimulating Hormone (FSH)[a]

Amino acid	Human	Sheep	Chicken	Turkey	Snapping turtle	Sea turtle	Bullfrog
Lysine	13	18	9	4	11	6	10
Histidine	6	6	2	1	4	1	3
Arginine	8	9	7	9	8	12	7
Aspartic	15	22	21	20	22	27	23
Threonine	22	15	11	11	18	12	20
Serine	15	14	14	11	17	16	16
Glutamic	20	26	29	30	22	23	21
Proline	11	12	16	13	16	15	14
Glycine	10	11	24	33	23	23	13
Alanine	10	16	22	22	13	13	11
1/2-Cystine	22	12	14	22	13	24	21
Valine	13	11	11	10	15	8	16
Methionine	4	3	3	2	2	3	4
Isoleucine	7	6	7	4	9	6	8
Leucine	9	19	16	17	12	18	11
Tyrosine	11	7	5	3	6	4	11
Phenylalanine	7	11	5	3	5	4	5

[a] Calculated on the basis of 215 residues per mole except for human FSH, which has 204 (Shome and Parlow, 1974a, b).

D. SUBUNIT STRUCTURE

The first suggestion that gonadotropins were composed of subunits came from the work of Li and Starman (1964). They found that the sedimentation coefficient of ovine LH decreased in acidic solution compared to values obtained in neutral solution. When sedimentation coefficients at pH 7.4 and 1.3 were determined for an avian, reptilian, and amphibian LH, similar evidence was obtained for the presence of subunits (Table IX). Evidence for subunits in gonadotropins from two piscine species has also been found (Burzawa-Gerard *et al.*, 1975; Burzawa-Gerard, 1975). Thus, the subunit nature of gonadotropin appears to represent a primitive and conservative characteristic of this molecule.

The countercurrent distribution technique originally employed in studies with ovine LH (Papkoff and Samy, 1967) enabled us to prepare small quantities of the two subunits from both snapping turtle and sea turtle LH. In an initial study with the snapping turtle, we demonstrated that the amino acid compositions of the two subunits (designated α and β in accordance with mammalian nomenclature) were significantly different from one another and had a strong resemblance to the corresponding subunits of ovine LH (Papkoff *et al.*, 1976b). These and new data for the subunits of sea turtle LH are shown in Table X. Results for the sea turtle materials closely paralleled those previously observed with snap- from the sea turtle.

In preliminary biological studies with the snapping turtle subunits, we showed that the individual subunits, as in mammalian materials, were significantly less active than the intact molecule and that the two subunits could be recombined with substantial recovery of LH bioactivity (Papkoff *et al.*, 1976b). Of special interest, however, was the demonstration that the two subunits of snapping turtle LH could be recombined with the alternate subunits of ovine LH (i.e., turtle α + ovine β and ovine α + turtle β) to generate new "hybrid" molecules that possessed significantly greater activity than any individual subunit compo-

TABLE IX
Sedimentation Coefficients of Several Species of Luteinizing Hormone (LH)[a]

Species	Sedimentation coefficient, $s_{20,w}$ (S)	
	pH 7.4, Tris	pH 1.3, HC1–KC1
Sheep	2.55	1.76
Turkey	2.08	1.54
Snapping turtle	2.76	1.78
Bullfrog	2.48	1.80

TABLE IX
Sedimentation Coefficients of Several Species of Luteinizing Hormone (LH)[a]

| | Alpha subunits ||| Beta subunits |||
Amino acid	Sheep	Snapping turtle	Sea turtle	Sheep	Snapping turtle	Sea turtle
Lysine	10	9	12	2	5	3
Histidine	3	3	4	3	4	2
Arginine	3	4	4	8	8	4
Aspartic	6	9	7	5	7	6
Threonine	9	9	9	7	5	7
Serine	6	6	5	8	7	10
Glutamic	8	7	7	6	6	7
Proline	7	6	7	20	20	13
Glycine	4	6	5	7	7	10
Alanine	7	6	4	8	10	14
1/2-Cystine	10	9	9	12	11	12
Valine	5	4	4	8	10	11
Methionine	4	2	3	3	2	1
Isoleucine	2	4	4	5	3	3
Leucine	2	4	4	12	5	6
Tyrosine	5	4	4	2	4	4
Phenylalanine	5	4	5	3	5	5

[a]Calculated on the basis of 96 residues for alpha; 119 for beta. Data for ovine material from Papkoff *et al.* (1973) and for snapping turtle material from Papkoff *et al.* (1976b).

nent alone. These tests were based on the ovulation assay in *Xenopus,* a system in which both intact molecules are essentially equipotent (cf. Table IV). Unfortunately, there was insufficient material to determine whether the interspecific recombinents were active in a biological system in which one of the intact molecules was relatively inactive.

Further evidence for partial homology among gonadotropin subunits comes from preliminary studies with bovine LH and a fish gonadotropin (LH cannot be defined). Burzawa-Gerard and Fontaine (1976) reported that bovine LH-α combined with a subunit (presumptively β) from carp gonadotropin to yield a hybrid that was more active by frog spermiation test than either the intact bovine or carp molecule alone. This recombinant also showed slight activity in an adenyl cyclase test in frog ovaries, but not in fish ovaries. However, the alternative recombination (carp α + bovine β) had no intrinsic activity in any assay, and neither recombinant was shown to have activity in a mammalian system.

Recent studies in our laboratories on the LH from a sea turtle (*Chelonia mydas*) provide more extensive evidence for the homologies between the reptilian and ovine gonadotropins. Subunits of this turtle LH were prepared as described for the snapping turtle LH (Papkoff et al., 1976b). The individual subunits displayed the expected behavior: they were significantly less active than the intact molecule, and they could be recombined with full regeneration of activity compared to the intact turtle LH (Table XI). Interspecific recombinants were prepared with the individual subunits of ovine LH, and these were tested in three different types of assays designed to elucidate the basis of phylogenetic specificity: (1) radioligand assay for [^{125}I]hCG binding in rat ovarian tissue in which the turtle hormone was inactive; (2) *Xenopus* ovulation in which both species of intact LH were about equipotent; and (3) *in vitro* steroid production by turtle testis in which the ovine LH was relatively inactive. Results for the three assays are summarized in Table XI. These data demonstrated that biologically active interspecific recombinants occurred between each α and β subunit as shown by the significant increase in potency of each recombinant when compared to the potencies of its individual components. More important, these

TABLE XI
Biological Activities of Sea Turtle Luteinizing Hormone (LH), Ovine LH, and Their Subunits and Recombinants in Various Biological Systems

	Relative potency in three assays[a]		
	[^{125}I]hCG binding to rat ovary	*In vitro* androgen production by turtle testis	*In vitro* ovulation in *Xenopus*
Ovine (O)			
LH	100	1.4	100
LH-α	0.2	0.4	0.3
LH-β	1.5	<0.06	0.3
LH-α + LH-β	39.8 (0.85)	0.1 (0.23)	37 (0.3)
Chelonia (Ch)			
LH	0.07	100	25
LH-α	0.06	3	<0.3
LH-β	0.03	27	2.0
LH-α + STLH-β	0.096 (0.045)	300 (15)	42 (<1.2)
Interspecific recombinants			
OLH-α + *Ch* LH-β	1.5 (0.76)	280 (13.7)	32
ChLH-α + OLH-β	29.1 (0.13)	2.5 (1.5)	12

[a]In each assay, the activity of the native (intact) molecule showing the highest potency was taken as 100%, and others were computed as a percentage of this value; each value represents the mean of duplicate assays in each system. The values in parentheses indicate the expected activity of recombinants based on the relative potencies of the individual subunit. All assays were performed on aliquots of the same solutions. All hormones and subunit mixtures incubated for 24 hours at 37°C at a concentration of 1 mg/ml in buffer before assay.

hybrid molecules sometimes had significantly greater activity than either of the two intact molecules. The potency of these hybrids varied with the type of assay employed. In the mammalian test, where only ovine LH was active, the recombinant possessing the ovine β was the most active, whereas in the turtle testosterone assay, where only turtle LH tended to be active, just the recombinant with turtle β showed appreciable activity. In *Xenopus,* where both intact molecules were active, there was only a slight difference between the potencies of the two hybrid recombinants. Thus, the α subunit was clearly important for the full manifestation of biological activity of the LH molecule, but the major expression of activity was related to the characteristics of the β subunit. The two species of α subunit were able to substitute for one another, but the β subunit conferred the species uniqueness to the molecule; thus, phylogenetic specificity of assays derives primarily from the β subunit.

E. IMMUNOCHEMICAL CHARACTERISTICS

Immunological studies have provided valuable evidence for the discrete nature of the FSH and LH molecules from various nonmammalian species as well as information on immunochemical relatedness among the hormones from different tetrapod classes. Hormonally specific antisera have been reported for chicken FSH and LH (Follett *et al.,* 1972; Godden *et al.,* 1975), turkey LH (Wentworth *et al.,* 1976), and we have generated specific antisera against bullfrog FSH and LH (Daniels *et al.,* 1977) and sea turtle and snapping turtle FSH and LH (P. Licht, D. MacKenzie, and E. Daniels, unpublished). All of these showed that the opposite antigen (homologous FSH or LH) had little cross-contamination. We have also raised antiserum in rabbits against the β-subunit of sea turtle LH. The highest specificity for turtle LH in RIA is obtained with this antiserum or when the anti-LH serum is used in conjunction with radioiodinated LH-β as tracer. As expected, specificity is reduced when LH-α is used as a tracer.

Several of the above antisera have been shown to be capable of neutralizing the biological activity of the homologous antigen, further verifying the specificity of the serum. In addition, similar biological tests indicated significant phylogenetic relatedness with heterologous species. For example, the antichicken LH serum blocked the biological activity of turtle and crocodilian LH, as well as showing a high degree of cross-reactivity by radioimmunoassay (RIA): snapping turtle LH was about 70% as immunoreactive as the homologous chicken LH (Licht *et al.,* 1974). This antiserum has been used to measure immunoreactive LH in numerous bird species, although the specificity has not been fully tested in these cases. Specificity of heterologous hormones may be a significant problem since, in studies with the reptilian hormones it was found that some fractions without biological activity showed almost as much immunoreactivity as the best LH. It is of interest, however, that the antichicken LH serum did not show appreciable cross reaction with hormones from reptiles, amphibians, or mammals (Scanes *et al.*, 1972; Licht *et al.*, 1974). Alternatively, we have studied several antisera against mammalian LH that cross-reacted with turtle but not snake, crocodilian, or bird hormones.

An antiserum against the β subunit of bovine LH neutralized the activity of snapping turtle LH, and this turtle LH showed about 10% the potency of mammalian LH in RIA. A second mammalian antiserum, raised against ovine LH showed slightly lower cross-reaction with turtle LH by RIA, but inhibition curves were parallel. However, as was the case with the antichicken LH serum, some cross-reaction was also observed with turtle fractions that lacked LH bioactivity. In contrast, biological and RIA tests with these two mammalian antisera failed to show any appreciable cross-reaction with chicken, snake, or crocodilian hormones (Licht *et al.*, 1974). Thus, the turtle LH appears to be immunologically intermediate to those of mammals and birds/crocodilians. The squamate (lizard and snake) hormones do not show a clear immunological relationship to any other group.

Information on immunorelatedness among FSH preparations is less complete than for LH. Croix *et al.* (1974) showed a cross-reaction of duck pituitary extract in RIA for mammalian FSH (utilizing antiovine FSH serum and iodinated rat FSH), but the specificity of the assay for avain hormones was not validated in any way. Follett (1976) subsequently reported that chicken FSH, but not LH, cross-reacted with a parallel inhibition curve in this same heterologous RIA. Immunoactivity was also observed with plasma samples from a variety of other avian species—quail, ducks, canaries, and sparrows. The antisera that we have generated against bullfrog FSH and LH and turtle FSH do not appear to cross-react with pituitaries or purified hormones from any other genera (E. Daniels, D. MacKenzie, and P. Licht, unpublished).

These biochemical and immunochemical data are consistent with the hypothesis that two separate gonadotropins exist in many of the birds, reptiles, and amphibians. Furthermore, the two hormones are chemically related to the FSH and LH of mammals.

V. Physiological Actions of Gonadotropins

The availability of purified FSH and LH from diverse nonmammalian species allows us to reexamine hormonal regulation of gonadal functions in the tetrapods. The establishment of the existence of two distinct gonadotropins in each class of tetrapods does not necessarily imply that the hormones have the same duality of function as they do in mammals. Only by comparing the actions of each hormone in homologous and heterologous species can meaningful conclusions be drawn about the evolution of hormone structure and function.

The large body of information dealing with the effects elicited by mammalian hormones in nonmammalian species makes an important contribution to our understanding of gonadotropin evolution, and we will rely upon several recent reviews of this work (Redshaw, 1972; Lofts, 1974; Lofts and Murton, 1973; Licht, 1974a; Nalbandov, 1976). However, present discussion focuses primarily on current studies with purified nonmammalian hormones, with attention to the newest data on the actions of purified mammalian hormones. These studies have been organized under the regulation of reproduction in the two sexes with separate discussions of data available for each class.

A. REPRODUCTION IN THE MALE

1. Testicular Growth, Spermatogenesis, and Spermiation

REPTILIA: Early investigations on the effects of mammalian gonadotropins in lizards provided the initial impetus for our comparative work. These studies demonstrated that ovine and human FSH were capable of restoring and maintaining full spermatogenic activity (from early divisions of primary spermatocytes to liberation of mature spermatozoa) in intact or hypophysectomized lizards (reviewed in Licht, 1974a). Similar results were obtained with snakes (Licht, 1972a; Tsui and Licht, 1974) and turtles (Licht, 1972b; Callard et al., 1975). It was once thought that the reptilian gonadal system was rather sluggish in its response to gonadotropins, but this view probably resulted from the failure to recognize the importance of temperature on reptilian endocrine responses. Despite their poikilothermal nature, reptiles (and amphibians) may exercise considerable control over body temperature by behavioral means, and knowledge of the characteristic preferred temperature of each species is critical to understanding its physiology. For example, at low body temperature the testes (and ovaries) of lizards may respond very slowly to hypophysectomy or exogenous hormone treatment; virtually no stimulation of spermatogenesis occurred even after several months of gonadotropin injection if the lizards were kept at moderate temperatures, 20–25°C or room temperature. In contrast, if warmed to their preferred levels (e.g., 32°C in *Anolis*) rapid testicular growth with spermatogenesis and androgen secretion was evident within 2 weeks (Licht and Pearson, 1969; Licht, 1975; Fig. 3). It must also be recognized that different aspects of gonadal activity may show different temperature sensitivities; e.g., steroid secretion tended to be less sensitive to low temperatures than was gonadal growth (Licht and Pearson, 1969; Licht, 1972c, 1975). Thus, inadequate control of temperature may lead to erroneous conclusions about the full spectrum of activities of a hormone. Temperature dependence was evident with both mammalian and reptilian hormones; thus, it is a characteristic of the reptilian gonad, not the source of the hormone (Licht, 1975).

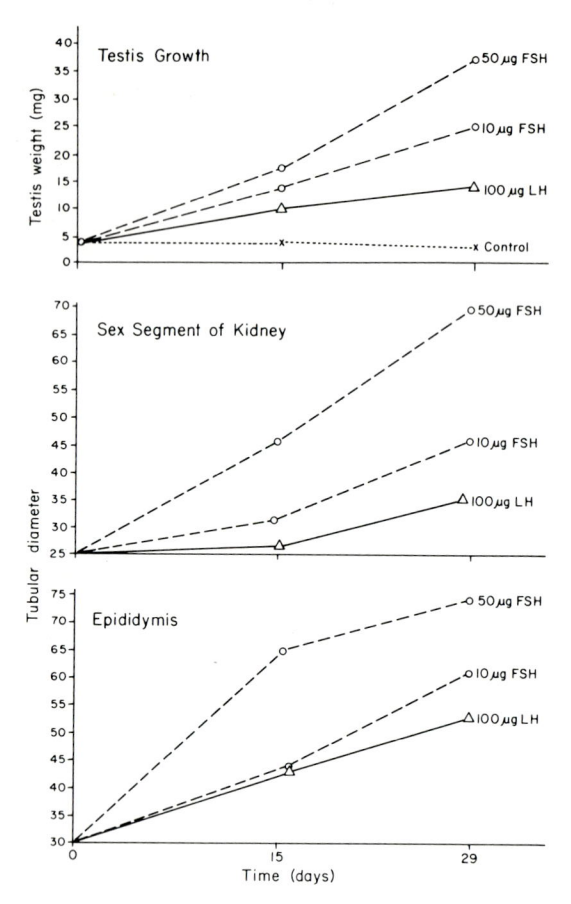

FIG. 3. Effects of ovine (NIH) follicle-stimulating hormone (FSH) and luteinizing hormone (LH) on the growth of the testis and accessory sexual structures (androgen production) in "physiologically" hypophysectomized lizards, *Anolis carolinensis*. Hormones were injected every other day at the doses indicated, and animals were autopsied at 15 and 29 days. Animals, starting with fully regressed gonads in September, were kept at 30°C with a short photoperiod to suppress endogenous gonadotropin during treatment: saline-injected controls showed no development. Each point represents the mean for 5–7 animals. Based on Licht and Pearson (1969).

Male reptiles are highly sensitive to mammalian FSH; injection of only 10 ng of highly purified ovine FSH (50 × NIH-FSH-S1) on alternate days for 2 weeks was sufficient to stimulate full gonadal development (Licht and Pearson, 1969; Licht and Tsui, 1975). The sensitivity of the testis varied with its condition: about 10-fold higher doses were required to stimulate the regressed testis than to maintain one that was already enlarged and active (Licht and Pearson, 1969; Reddy and Prasad, 1971). Treatment of highly purified ovine FSH with anti-LH serum did not affect its potency, confirming that the FSH was effective alone (Licht and Tsui, 1975) (Table XII). In fact, it appeared that only the β subunit of FSH was required for stimulation of testicular activity in the lizard (Licht and Papkoff, 1971).

Mammalian (ovine and human) LH appeared to be relatively inactive (e.g., about 0.05% as potent as purified ovine FSH) in the stimulation of testis growth and spermatogenesis (Table I and Fig. 3). When low doses of ovine LH were given, they had little effect on spermatogenesis (e.g., Jalali *et al.*, 1975; Callard *et al.*, 1975). Consequently, it was initially suggested that the minimal activity might be due largely to trace FSH contamination. However, there is now good evidence to support the conclusion that LH-like gonadotropins have intrinsic activity in testicular development in reptiles. For example, immunological studies demonstrated that anti-FSH serum fully neutralized the activity of ovine FSH without affecting the activity of the LH in the lizard (Fig. 4). In hypophysectomized turtles, hCG stimulated a marked increase in testis weight and spermatogenic activity (Combescot, 1955, 1958), but these studies did not compare the potency of an FSH. The potential importance of pituitary LH for testicular growth in reptiles was further substantiated by studies with hormones from the nonmammalian species.

Early studies indicated that both avian (chicken) FSH and LH stimulated spermatogenesis in the lizard and that avian LH was more potent than mammalian LH (Licht and Hartree, 1971). Extensive comparative data on the effects of avian and other gonadotropins on the lizard testis were obtained in connection with the routine use of *Anolis* bioassay for pituitary fractionations in our laboratories (Table IV). Maintenance of testis growth and spermatogenesis in hypophysectomized lizards have been observed with all species of amphibian,

TABLE XII
Effects of Chronically Injected Highly Purified Ovine Follicle-Stimulating Hormone (FSH) on Testis Weight and Testosterone Production in Hypophysectomized Lizards, Anolis carolinensis[a]

Treatment[b]	N	Testis weight (mg)	Plasma testosterone (ng/ml)
Saline	6	3.8 ± 1.2	<0.6 ± 0.2
FSH + NSS	7	35.3 ± 2.5	146.9 ± 33.3
FSH + anti-LH	7	30.0 ± 2.1	186.9 ± 15.0

[a]Data from Licht and Tsui (1975).
[b]Lizards were injected six times every other day starting 48 hours after hypophysectomy and sacrificed 2 hours after the last injection. FSH was given at 0.7 μg per injection; this was pretreated with either normal sheep serum (NSS) or an antiserum against the β subunit of bovine LH (anti-LH).

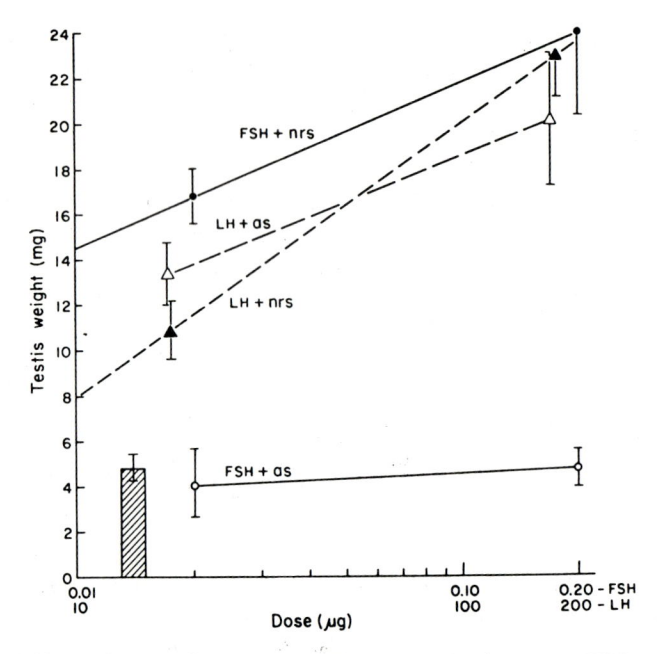

FIG. 4. Effects of a specific antiovine follicle-stimulating hormone (FSH) serum on the activities of highly purified ovine FSH (50 × NIH-FSH-S1) and luteinizing hormone (LH) (2 × NIH-LH-S1) on testis weight in hypophysectomized lizards, *Anolis carolinensis*. Each hormone was treated with either the antiserum (as) or normal rabbit serum (nrs) and injected 5 times on alternate days; controls (hatched bar at lower left) were injected with saline. Means and standard errors are shown for the 4–6 animals per dose. Note the difference in scales for the doses of the two hormones. From Licht and Papkoff (1973), with permission.

reptilian, and avian FSH and LH examined. In most cases, the potency of LH was considerably less than that of the respective FSH, but several factors suggested that LH had intrinsic activity. Some of the LH preparations, e.g., from *Alligator* and *Chelonia*, were almost equipotent to the FSH; and the tiger salamander (*Ambystoma*) fraction considered to be LH was more potent than the FSH fraction. As already discussed we could find no evidence of an FSH separate from LH from the duck. It is noteworthy that the dose-response slopes for these relatively potent LH preparations were not parallel to those for FSH (Licht et al., 1976b). Furthermore, independent estimates of FSH activity in these LH preparations of RIA (D. MacKenzie and P. Licht, unpublished) and radioreceptor assay (Licht and Midgley, 1976b) indicated a low FSH contamination. We have also shown that an antiserum against bullfrog FSH inactivated the bullfrog FSH, but not the LH, when tested in *Anolis* (Daniels et al., 1977).

Preliminary observations on the clearance rates of FSH and LH in the lizard may help explain the differential potency of these two hormones when tested *in vivo*. When the disappearance of rat FSH and ovine LH from the lizards' plasma were followed by RIA, the FSH persisted for days compared with a few hours for LH (P. Licht and B. Goldman, unpublished). In this connection, hCG was shown to be more effective than ovine LH in the lizard (Jones, 1973); as in mammals, hCG appears to have a longer half-life than LH in nonmammalian species (Roos and Jørgensen, 1974). Thus, it seems likely that LH may have substantial effects on testicular growth and spermatogenesis in reptiles, but the potency of some LH preparations (notably ovine) have been underestimated in studies involving infrequent injections over long periods. This problem must be borne in mind throughout the further discussion of relative potencies of different hormonal preparations when tested *in vivo*.

AVES: Although the hypothesis that avian spermatogenesis is primarily under the control of FSH and steroidogenesis under the control of LH is prevalent (e.g., Godden and Scanes, 1975), there is considerable evidence to indicate that spermatogenesis may not be specifically dependent on either gonadotropin. Numerous studies with young chicks have shown that both ovine FSH and LH stimulate enlargement of gonads as well as the uptake of radiophosphorus (Breneman *et al.*, 1962; Licht, 1973b). In general, attempts to stimulate full development of spermatogenesis in avian gonads with mammalian hormones have met with poor success compared with the high potency of avian pituitary extracts (Lofts and Murton, 1973). While mammalian FSH can be used to stimulate birds in some cases, ovine LH was reported to be more effective than FSH for promoting enlargement of seminiferous tubules (e.g., Lofts *et al.*, 1973).

Recent observation on the effects of purified chicken and turkey hormones also indicated that the stimulation of ^{32}P uptake by chick testis is about equally sensitive to FSH and LH (Follett *et al.*, 1972; Furuya and Ishii, 1974; Wentworth, 1971; Farmer *et al.*, 1975; Godden and Scanes, 1975). Similar nonspecificity for FSH and LH was observed with hormones from reptiles and frogs (Licht and Papkoff, 1974a,b). In more complete studies on the chick testis, Ishii and Furuya (1975) concluded that chicken FSH was more potent than chicken LH for promoting testis weight, but their bioassay data indicated a significant LH contamination in the FSH, and they did not use hypophysectomized animals to rule out interaction with endogenous gonadotropins.

The most thorough study of the effects of avian hormones on the bird testis was that of Brown *et al.* (1975) using chicken hormones in intact and hypophysectomized quail. In the intact, sexually immature bird, both gonadotropins stimulated testis growth and development, although complete spermatogenesis was not observed with either hormone. Some of the actions of LH were attributed to interactions with endogenous hormones, since the difference between FSH and LH was more pronounced in hypophysectomized birds. FSH appeared to have its primary effect on the Sertoli cells, causing both hypertrophy and cell division. Although some of the effects of LH were lost after hypophysectomy (e.g., nuclear migration in Sertoli cells was absent), the LH continued to exert distinctive effects on the seminiferous tubules.

AMPHIBIA: Seasonal histological changes, effects of hypophysectomy and injection of pituitary homogenates indicated that spermatogenesis was regulated by pituitary hormones in amphibians (reviewed by Dodd and Wiebe, 1968; Basu, 1969; Lofts, 1974). Results of numerous experiments using relatively impure ovine and porcine pituitary gonadotropins (from NIH and Armour) in normal and hypophysectomized amphibians (both anurans and urodeles) suggested that initial stages of spermatogenesis were primarily dependent on FSH (Burgos and Ladman, 1957; Lofts, 1961; Andrieux and Collenot, 1970; Andrieux et al., 1973). Vellano et al. (1974), using the newt *Triturus cristatus carnifex*, reported that the combination of ovine LH + FSH was more potent than FSH alone in stimulating spermatogonial proliferation and formation of primary spermatocytes. However, either LH or FSH alone was capable of stimulating further primary spermatocyte maturation (after LH + FSH pretreatment), but not complete spermatogenesis. Combined LH + FSH injections stimulated complete spermatogenesis only in July at relatively warm temperature (22°–23°C). In contrast, in intact or hypophysectomized larval salamanders (*Ambystoma tigrinum*) PMSG or FSH stimulated spermatogenesis up to secondary spermatogonia but hCG and LH were inactive (Moore and Norris, 1973; Moore, 1974). At warmer temperatures, FSH stimulated production of secondary spermatocytes (Moore, 1975).

Comparison of various hCG and PMSG preparations in the 1950s demonstrated that only PMSG stimulated spermatogenesis (see Lofts, 1974). More recently, however, other workers have shown hCG to be equal to or more potent than PMSG in stimulating spermatogenesis in amphibians (Iwasawa et al., 1973; Kasinathan and Basu, 1973). Additionally, hCG was shown to stimulate spermatogenesis to some degree by Basu (1964) and Guha (1976); but in one study it was less potent than PMSG (Basu, 1964).

Interpretation of the combined results is virtually impossible owing to the differences in hormone preparations, doses, species tested, criteria examined, and, perhaps most important, the different environmental conditions and seasons in which the experiments were conducted (see van Oordt, 1960; Basu and Nandi, 1965; Werner, 1969; Vellano et al., 1974; Rastogi et al., 1976). Furthermore, in many cases complete spermatogenesis could not be elicited by administration of exogenous gonadotropins. Nevertheless, the positive results obtained both with FSH-like preparations and with hCG suggested that the spermatogenic response may have less gonadotropin specificity than initially believed.

Preliminary experiments (Muller, 1976a) have shown both bullfrog LH and FSH to affect spermatogenesis in the bullfrog. Hypophysectomized male bullfrogs injected daily with 25–50 μg of bullfrog LH or FSH for 3 weeks showed an increase in the number of germ-cell cysts. Additionally, *in vitro* incubation of testis slices with [^{3}H] thymidine revealed increased incorporation of counts into acid-precipitable material in testes from frogs treated with either gonadotropin *in vivo* (Table XIII).

Injection of pituitary preparations into tadpoles or newly metamorphosed male anurans resulted in a dialation and partial evacuation of seminiferous tubules within 48 hours. Ovine FSH, injected over a 3-week period, induced a similar response in immature *Bufo arenarum* and *Cerotaphrys ornata*, but LH was not studied (Pizarro and Burgos, 1963; Pisanó and Burgos, 1971, respectively); similar treatment of *Pleurodema cinerea* stimulated complete spermatogenesis (Pisanó and Burgos, 1962). The short-term response in larval *Xenopus* and *Alytes obstetricans* has been used as a bioassay for mammalian and nonmammalian gonadotropins (see Simon and Reinboth, 1966; Delsol *et al.*, 1970). As in the spermiation assay in adult anurans, this assay does not discriminate between mammalian LH and FSH preparations. Although the exact mechanism for the response is not known, it closely resembles spermiation, which involves hydration of the tubules (Russo and Burgos, 1969).

Spermiation, the release of spermatozoa from the Sertoli cells, was once thought to be an LH-specific response (see Atz and Pickford, 1954). However, Barr (1969) found that the relative potencies of LH and FSH in this test were

TABLE XIII
Incorporation of [^{3}H] Thymidine by Bullfrog Testis[a]

		Cpm adjusted for μg protein	
In vivo treatment[b]	N	Acid precipitable[c]	Acid soluble[c]
Saline	4	210.2 ± 38.0	2500.6 ± 315.2
Rana FSH	3	960.8 ± 224.8[d]	2019.2 ± 498.9
Rana LH	3	621.4 ± 173.2[e]	2095.3 ± 511.2

[a]Data from C. H. Muller (unpublished). Testis slices (20–30) were incubated for 18 hours in Krebs–Ringer bicarbonate buffer with 0.1 μCi of [^{3}H] thymidine (6 μCi/mmole).

[b]Hypophysectomized bullfrogs were injected 21 days with saline or 25–50 μg of bullfrog FSH or LH.

[c]Mean ± SE.

[d]Different from saline: $p < 0.02$.

[e]Two animals were clearly stimulated over control levels, but the third was not.

dependent upon both the source of gonadotropins and the species tested. Recently, tests in three anuran species demonstrated that potency ratios between NIH ovine LH and FSH were quite different depending on the test species (Licht, 1973a); in fact, ovine (NIH) FSH was more potent than LH in two species (*Hyla* and *Eleuthrodactylus*). While of the two NIH hormones, only the LH was active in the third frog (*Rana pipiens*), a highly purified ovine FSH was found to be more potent than a highly purified LH in this species.

Spermiation responses were also observed in *Rana pipiens* and *Hyla regilla* with both FSH and LH from snapping turtles and bullfrogs (Licht and Papkoff, 1974a,b). Most important was the demonstration that bullfrog LH and FSH were about equipotent on spermiation in the bullfrog (Muller, 1976a, and unpublished observations). Information on the control of spermiation in urodele and caecilian amphibians is lacking.

2. Testicular Steroid Production

REPTILIA: Initial evidence for the role of the two gonadotropins in the regulation of testicular steroid production in reptiles was based on indirect histological criteria of androgenic activities in chronically treated animals. Observations made in conjunction with studies of mammalian hormones *in vivo* indicated that androgen was stimulated by both FSH and LH, but that FSH was usually considerably more potent (Fig. 3; reviewed in Licht, 1974a). The criteria for stimulation included hypertrophy of the androgen-dependent epididymis and renal sexual segment in hypophysectomized snakes and lizards and of conspicuous hypertrophy of the Leydig cells in snakes, lizards, and turtles (Figs. 3 and 5). In some cases LH stimulated the interstitium, but not the seminiferous tubules (e.g., Eyeson, 1971; Jalali *et al.*, 1975; Callard *et al.*, 1975), but only a single dose level was tested; it must be recognized that spermatogenesis and androgen production may have different hormonal sensitivities. As with spermatogenesis, the relatively low potency of mammalian LH (e.g., Figs. 3 and 4) raised the possibility that the activity of LH was due largely to FSH contamination. However, intrinsic steroid-stimulating activity of ovine LH was demonstrated with the aid of anti-FSH serum (Licht and Papkoff, 1973) and by the fact that hCG stimulated enlargement of interstitial cell size in intact lizards (Jones, 1973) and hypophysectomized turtles (Combescot, 1955, 1958). The observation that hCG was more potent than ovine LH in the lizard (Jones, 1973) again suggested that the low activity of the latter might be related to its relatively short half-life.

Stimulation of accessory sexual structures was also observed with both purified FSH and LH from all the nonmammalian species studied in the *Anolis* assay. In general, the steroid-stimulating activities of the preparations tended to parallel

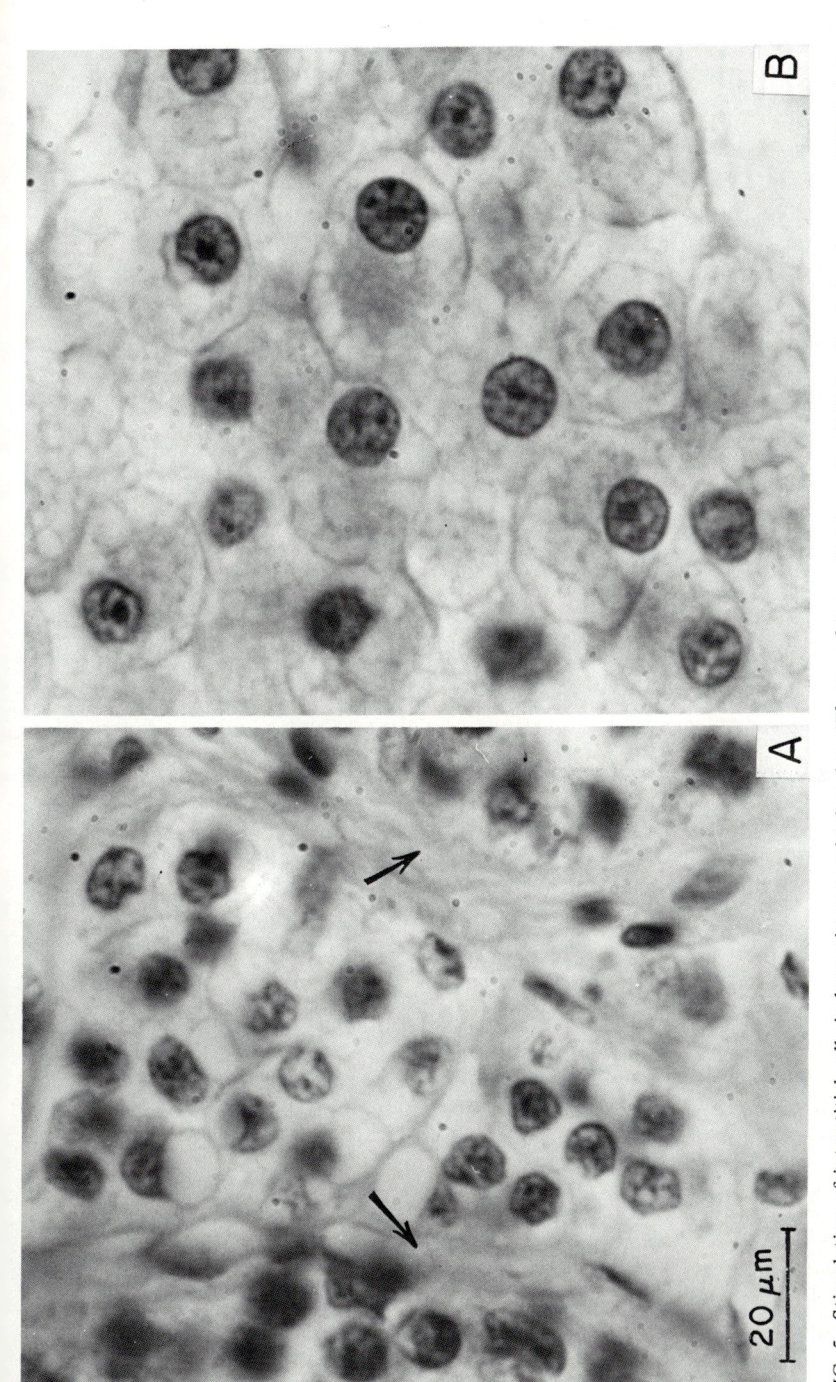

FIG. 5. Stimulation of interstitial cells in hypophysectomized snakes (*Thamnophis sauritis*) by ovine follicle-stimulating hormone (FSH). Animals were injected for 3 weeks after hypophysectomy with 25 μg of NIH-FSH-S8 on alternate days. The interstitium of saline-injected controls is shown on the left (A); arrows indicate the margins of two seminiferous tubules bordering the clump of Leydig cells. At the right, the Leydig cells in FSH-treated animals have expanded to fill the frame at the same magnification. Reprinted from Licht (1972a), with permission.

their potency as determined by testis weight changes and spermatogenesis (Table IV); i.e., FSH was usually somewhat more potent than LH.

The first direct evidence of androgen stimulation by gonadotropins in reptiles came from RIA analyses of plasma androgen in hypophysectomized *Anolis* that were treated chronically with highly purified ovine FSH (Table XII). In these same studies, the intrinsic activity of the FSH was confirmed by the demonstration that its potency was unaffected by treatment with anti-LH serum. We have subsequently shown similar *in vivo* stimulation of plasma androgen levels by both ovine FSH and LH (from NIH). Results with these hormones demonstrated that the responsiveness of the gonad varied greatly, depending on season and the physiological state; e.g., sensitivity to exogenous gonadotropin decreased during the nonbreeding season or after hypophysectomy (Table XIV). Interestingly, the relative potencies of the two ovine gonadotropins were essentially the same when studied by acute injection; human FSH and hCG also showed the same potency in *Anolis* if plasma androgen was examined 1–2 hours after a single injection (P. Licht, unpublished). Similar data were obtained for hypophysectomized snakes (Tsui, 1976a). These latter observations were in sharp contrast to those made in chronically injected lizards and snakes, probably because the LH failed to stimulate or maintain testis activity. Measurements of plasma androgen made in conjunction with our routine bioassay of different species of nonmammalian hormones during fractionation also showed consistent stimulation of androgen by both FSH and LH from all species examined (Table XV). In these studies, FSH tended to appear more potent than the corresponding LH, but this was probably because of the problems with chronic treatment, since testis weight was not equally maintained by all hormone treatments (Table XV). In the few cases where nonmammalian hormones were compared by acute injection (without differential pretreatment), there was little difference between their potencies (P. Licht, unpublished).

Callard *et al.* (1975) reported that ovine (NIH) FSH stimulated an elevation in plasma androgen and an increase in Leydig cell number in intact and hypophysectomized turtles. In contrast, ovine LH at the same single dose was only weakly active in intact animals and had little effect on plasma androgen (despite an increase in Leydig cell number) in hypophysectomized turtles. In view of the experiments described for lizards, this lack of response to ovine LH was possibly due in part to the chronic treatment schedule since the gonads were more stimulated by FSH. Moreover, it was probably also related to the turtles' insensitivity (phylogenetic specificty) toward mammalian LH as indicated by *in vitro* studies. Owens *et al.* (1976) used a relatively large injection of bovine (NIH) FSH to stimulate plasma testosterone as an aid to sexing immature sea turtles: females had no measureable testosterone, whereas males had measurable levels, which were further increased about 10-fold by FSH (LH was not tested in this study).

TABLE XIV

Variation in the Responsiveness of Testicular Androgen Production to Ovine (NIH) Gonadotropins in the Lizard Anolis carolinensis[a]

Expt. no.	Month	Condition of lizards	10-Day hormone pretreatment	Final testis weight (mg)[b]	Hormone tested	Plasma androgen (ng/ml)[b]
1	February	Hypophysectomized	None	9.5 ± 0.75	FSH, 3 µg	<0.8
			FSH, 3 µg	49.5 ± 0.66	FSH, 3 µg	45.4 ± 8.3
2	September	Intact	None	3.5 ± 0.2	FSH or LH, 1–5 µg	<0.8
					FSH, 50 µg	6.7 ± 3.7
					LH, 50 µg	4.5 ± 2.1
		Intact, 28 days photostimulation	None	31.1 ± 6.2	FSH, 0.5 µg	39.0 ± 10.7
3	September	Intact	FSH, 1 µg	12.5 ± 1.35	LH, 2.5 µg	14.8 ± 3.2
			FSH, 5 µg	21.9 ± 2.2	LH, 2.5 µg	53.7 ± 13.3
			FSH, 5 µg	27.0 ± 2.45	FSH, 0.5 µg	12.2 ± 3.5
			FSH, 25 µg	35.1 ± 4.2	FSH, 0.5 µg	28.9 ± 3.6

[a] Animals were either tested for androgen stimulation immediately after collection (Expt. 2), after laboratory photostimulation with long days (14L:10D) at 30°C (Expt. 2), or after 10 days of hormone treatment (Expt. 1 and 3). This pretreatment consisted of 5 injections on alternate days. In each case, testicular androgen responsiveness was tested with a single intraperitoneal injection of either follicle-stimulating hormone (FSH) or luteinizing hormone (LH) (see Hormone tested) 1.5–2 hours before final plasma sampling. Without the final injection of hormone, plasma androgen was nondetectable (<0.8 ng/ml) except in photostimulated animals (2–5 ng/ml).

[b] Mean ± standard error. All values differed significantly ($p<0.02$) from saline-treated controls except the one androgen value indicated as nondetectable in Expt. 2.

TABLE XV
Stimulation of in Vivo Increases in Plasma Testosterone in Hypophysectomized Lizards with Diverse Pituitary Gonadotropins[a]

Expt. no.	Hormonal pretreatment (a)	N	Final hormone injection (b)	Testis weight (c)	Plasma androgen (ng/ml) (d)
	Saline		None	5–10	0.05
I	Bullfrog FSH, 0.75 µg	5	Bullfrog FSH, 0.5 µg	49.2 ± 2.2	9.0 ± 2.5
	1.5 µg	5	Bullfrog FSH, 0.5 µg	66.3 ± 2.6	23.9 ± 5.0
	Chicken LH, 4 µg	5	Chick LH, 0.25 µg	57.6 ± 2.6	28.6 ± 4.6
	Chicken FSH, 0.25 µg	5	Chick FSH, 0.5 µg	35.9 ± 1.1	10.0 ± 3.0
	Chicken FSH, 1.5 µg	5	Chick FSH, 0.25 µg	70.6 ± 7.0	24.2 ± 3.3
II	Chicken FSH, 1.5 µg	4	Chick FSH, 1.5 µg	58.2 ± 4.0	32.7 ± 11.6
	Turkey FSH, 1.5 µg	4	Turkey FSH, 1 µg	56.3 ± 5.5	58.7 ± 7.0
	Snake "FSH," 40 µg	5	Snake "FSH," 15 µg	43.6 ± 6.8	39.9 ± 4.6
	Snake "LH," 30 µg	5	Snake "LH," 15 µg	47.8 ± 1.8	41 ± 9.1
	Bullfrog LH, 4 µg		Same as (a)	25.2 ± 4.0	13.2 ± 3.4
III	Bullfrog LH, 20 µg	5	Same as (a)	34.3 ± 4.9	14.7 ± 2.2
	Bullfrog FSH, 0.12 µg	6	Same as (a)	19 ± 4	9.7 ± 3.3
	Bullfrog FSH, 0.6 µg	5	Same as (a)	41.5 ± 4.5	30.9 ± 7.1
	Alligator FSH, 0.8 µg	6	Same as (a)	32.0 ± 3.6	19.2 ± 4.4
	Alligator FSH, 4 µg	6	Same as (a)	51.7 ± 5.4	48.3 ± 11.3
IV	Sea turtle FSH, 15 µg	5	Sea turtle FSH, 2 µg	46.2 ± 4.1	92.4 ± 10
	Sea turtle LH, 15µg	5	Sea turtle FSH, 2 µg	17.8 ± 1.6	62.6 ± 16

[a] Androgen responses (column d) were obtained as part of the routine bioassay of hormones in *Anolis* testis weight-maintenance assay. Hypophysectomized lizards were injected four times on alternate days starting 48 hours after hypophysectomy, and killed 48 hours after the last regular injection (column a). Two hours before autopsy, they were injected with hormone to test for androgen (column b), and a plasma sample was taken for radioimmunoassay analysis of androgen. Without this final injection, plasma testosterone would be at control levels. Although testes were in different conditions at the start of various assays (reflected in final testis weight—column c), all controls were similar. FSH, follicle-stimulating hormone; LH, luteinizing hormone.

We have shown that plasma androgen in hypophysectomized turtles (*Kinosternon*) was very responsive to both FSH and LH derived from turtle pituitaries (Fig. 6 and Table XVI). The most highly purified turtle FSH tested was somewhat more potent than the best LH preparation (Table XVI).

A series of *in vitro* studies on testicular androgen production have allowed broader comparisons to be made among the numerous species of purified FSH and LH; these data provide evidence for the direct effects of the two gonadotropins on the testis. To begin, we examined the gonadotropin specificity and phylogenetic specificity of two mammalian tissues: minced rabbit testis and isolated rat Leydig cells, both of which were previously shown to be highly sensitive and specific for LH (e.g., Hall and Eik-Nes, 1962; Connell and Eik-Nes, 1968; Ewing *et al.*, 1975; Dufau *et al.*, 1972; Ramachandran and Sairam, 1975).

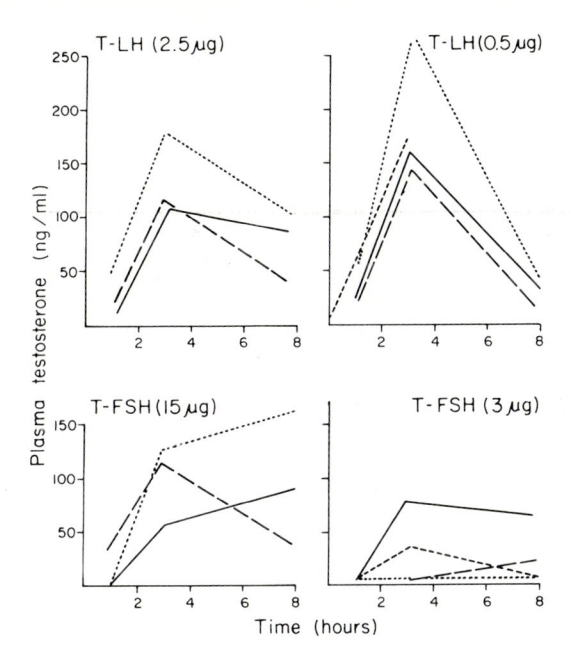

FIG. 6. Stimulation of *in vivo* plasma testosterone levels in hypophysectomized turtles (*Kinosternon subrubrum*) after a single injection of snapping turtle (*Chelydra*) follicle-stimulating hormone (FSH) or luteinizing hormone (LH). Each hormone was tested at two doses in 3 or 4 individual turtles. Hormones were administered by intracardiac injection and plasma samples were taken by cardiac puncture at 1, 3, and 8 hours. Each curve shows the response for an individual animal. The turtle FSH used here was not considered to be as highly purified as the LH preparation. Data from Tsui (1976b).

TABLE XVI
Plasma Levels of Testosterone in Hypophysectomized Turtles
(Kinosternum) *after a Single Injection of Snapping Turtle* (Chelydra) *Gonadotropin*[a]

Hormone treatment	Dose (ng)	N	Initial level (ng/ml)	Final level[b] (ng/ml)	Significance[c]
Chelydra FSH[e] (T-48C)	40	6	4.1 ± 0.8[d]	7.9 ± 2.1[d]	$p < 0.05$
	200	6	2.7 ± 0.6	27.9 ± 11.8	$p < 0.01$
Chelydra LH[e] (T-28B)	20	6	2.6 ± 0.4	3.6 ± 0.5	$p > 0.1$
	100	5	3.6 ± 0.9	20.2 ± 3.0	$p < 0.01$

[a]Modified from Tsui (1976b).
[b]Blood was sampled by heart puncture at 1, 3, and 8 hours after a single intracardiac injection of hormone; the maximal levels observed among all samples for each individual were used to compute these values.
[c]Matched-paired *t* test comparing final and initial levels.
[d]Mean ± standard error.
[e]FSH, follicle-stimulating hormone; LH, luteinizing hormone.

Our results with mammalian hormones were in accord with these earlier studies. In the rabbit, several preparations of mammalian LH were active between 1 and 50 ng/ml, whereas FSH (rat and ovine) was only about 1% as potent and the action of ovine FSH could be blocked by anti-LH serum (Licht *et al.*, 1976a) (Fig. 7). Similarly, *in vitro* production of testosterone by rat Leydig cells (according to the methods of Ramachandran and Sairam, 1975) was stimulated by 1–10 ng of ovine LH, and other species of mammalian LH were 0.1–10 times as active as ovine LH. Highly purified ovine FSH gave a parallel response curve with a similar maximal response, but it was only about 0.05% as potent as the LH. However, this low activity may be intrinsic to FSH in this system since it could not be blocked fully by anti-LH serum (S. W. Farmer and H. Papkoff, unpublished).

All nonmammalian hormones had very low potencies when tested in the two mammalian systems. The nonmammalian preparations produced parallel dose-response slopes with equivalent (or greater) maximal responses, but they were active only above 1 µg/ml (Fig. 7); i.e., they were less than 1% as potent as mammalian LH (Licht *et al.*, 1975; Papkoff *et al.*, 1976a). There was no clear phylogenetic pattern in relative potencies. Furthermore, in both assay systems,

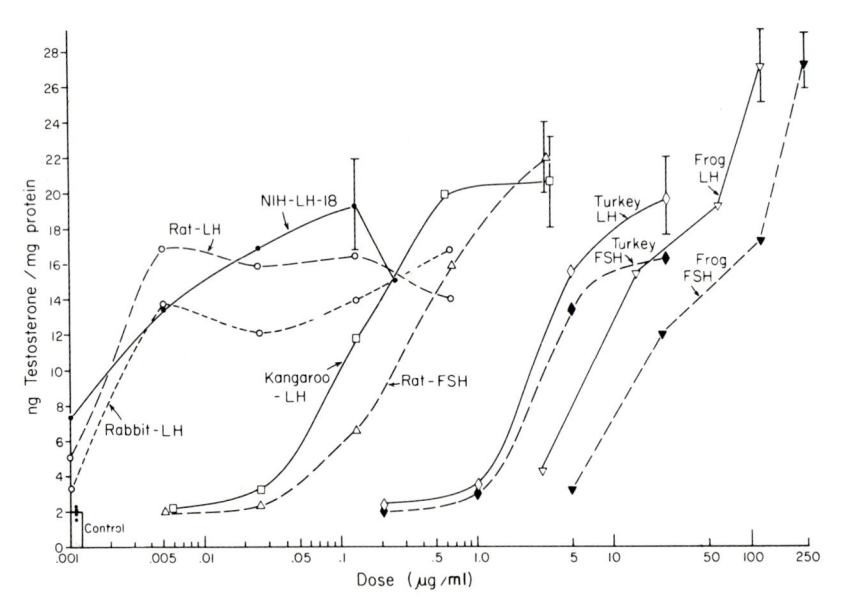

FIG. 7. Stimulation of *in vitro* androgen production by minced rabbit testis with follicle-stimulating hormone (FSH) and luteinizing hormone (LH) preparations derived from a variety of tetrapod species. Testis minces were incubated at 36.5°C for 2 hours in the presence of hormone at the doses shown: each point represents the mean of duplicate tubes (in a few cases, the range of responses are also shown). Modified from Licht *et al.* (1976a).

the difference between each species of nonmammalian FSH and LH was not as great as between the two hormones from mammals. In the rabbit, the potency of nonmammalian FSH sometimes approached that of the respective LH (e.g., Fig. 7), and in rat Leydig cells, the potencies of FSH varied from 3% (alligator) to about 30% (turkey and bullfrog) (S. W. Farmer and H. Papkoff, unpublished). These activities of FSH were much too great to be accounted for by LH contamination. Thus, the mammalian steroidogenic tissues showed a high degree of phylogenetic specificity toward nonmammalian hormones as a group, and their LH specificity was greatly reduced when they were challenged with heterologous gonadotropins.

In vitro studies with a wide variety of reptilian testes (from all three orders) revealed a more general lack of FSH/LH specificity than was apparent in mammals (Figs. 8–13). Thus far, we have examined responses in four species of

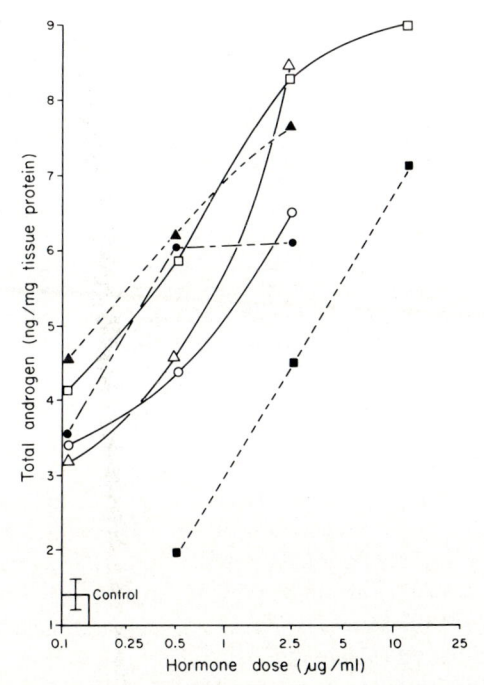

FIG. 8. Stimulation of *in vitro* androgen production by minced testis from an adult snapping turtle (*Chelydra serpentina*). Tissues were incubated for 2 hours at 30°C in various doses of follicle-stimulating hormone (FSH) or luteinizing hormone (LH) derived from sheep (□, ■), snapping turtle (*Chelydra*) (○, ●), or sea turtle (*Chelonia mydas*) (△, ▲). Each point represents the mean of 2–3 replicates; values for tubes containing saline only (controls) are shown at the lower left. For each species of hormone, values for FSH are indicated by the *opened* symbols and solid lines, and LH by the *filled* symbols with dashed curves. Data from Tsui (1976b).

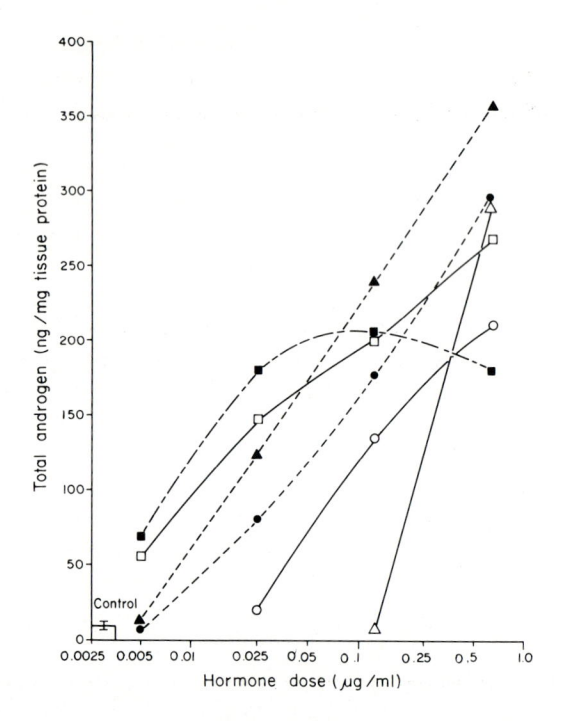

FIG. 9. Stimulation of *in vitro* androgen production by minced testis from the snake *Thamnophis sirtalis* in response to gonadotropins from sheep (□, ■) snake (*Ptyas*) (○, ●), or snapping turtle (*Chelydra*) (△, ▲). Follicle-stimulating hormone (FSH) is indicated by open symbols with solid curves; luteinizing hormine (LH) by filled symbols and dashed curves. See Fig. 8 for details. Data from Tsui (1976b).

turtles, two species of snakes, two species of lizards, and an alligator. Space does not permit detailed review of these results, but several important generalizations can be made. First, every preparation of both FSH and LH tested had some steroid-stimulating activity in all the reptilian testes. The relative potency of a particular gonadotropin varied among reptiles, indicating the existence of phylogenetic specificity in the different testes; e.g., frog and snake hormones tended to be more potent in snakes than in other species. However, in general, variations in relative potencies were less than those observed between all mammalian and nonmammalian hormones in the mammals (cf. Fig. 7).

Second, the FSH/LH potency ratios varied with both the source of hormones and the source of testes (Fig. 12). Such variability was most notable in the case of mammalian hormones. Comparison of the two standard NIH preparations indicated that turtles were insensitive to ovine LH (e.g., Fig. 8), alligators were slightly more sensitive to LH than to FSH (Fig. 10), and snakes and lizards were about equally responsive to both ovine gonadotropins (e.g., Figs. 9 and 13). Differences between turtles and snakes were also observed in FSH/LH ratios for human hormones (including hCG) and bovine hormones (e.g., Fig. 11). However, FSH/LH potency ratios did not vary with all species of hormones; e.g., snapping turtle (*Chelydra*) FSH was consistently more potent than LH, whereas the reverse was true for sea turtle (*Chelonia*) hormones (Fig. 12). Thus, any conclusion regarding the relative importance of FSH and LH for steroid secretion in reptiles based on a single species of gonadotropin may be misleading; it is only clear that the reptiles as a group lack a strict FSH:LH specificity.

The unexpectedly high potencies of FSH preparations for *in vitro* steroid secretion in reptiles compared to mammals again raised concern about possible LH contaminations. However, several lines of evidence support the conclusion

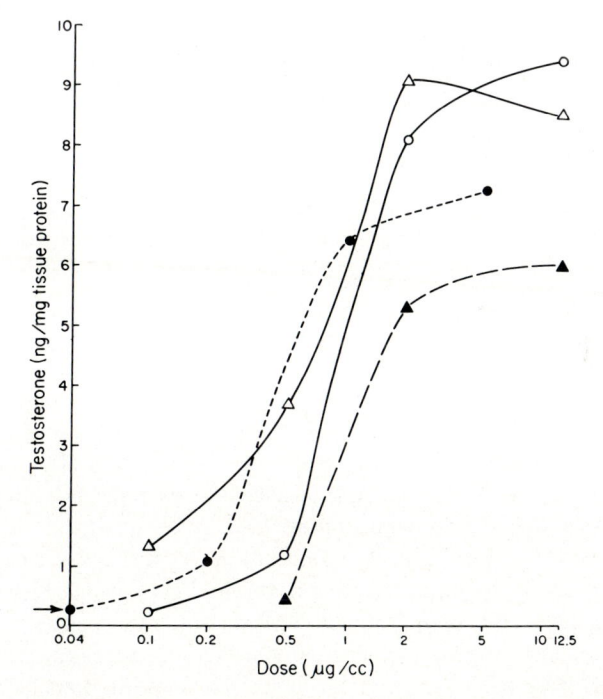

FIG. 10. Stimulation of *in vitro* androgen production by minced testis from an alligator with ovine (△, ▲) or alligator (○, ●) gonadotropins. Responses to follicle-stimulating hormone (FSH) are shown by open symbols and solid curves, and to luteinizing hormone (LH) by filled symbols and dashed curves. See Figs. 8 and 9 for further details. Data from Tsui (1976b).

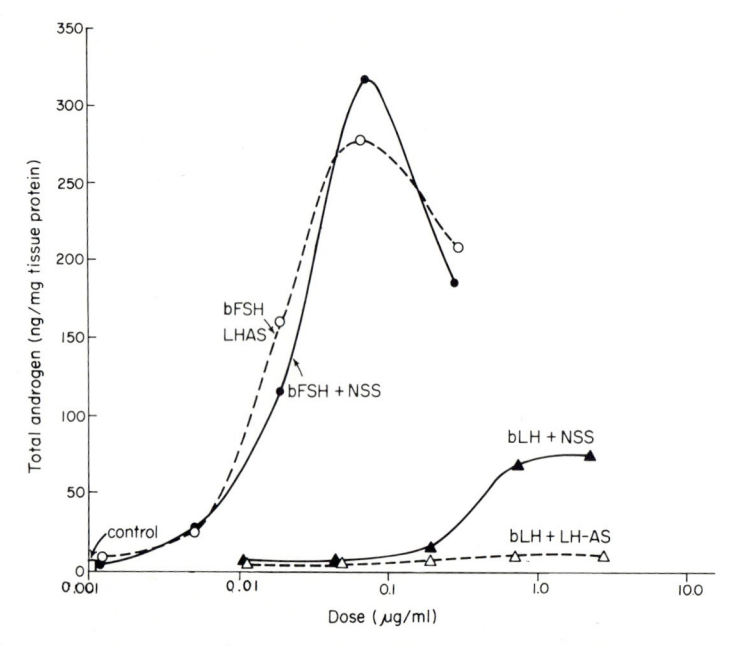

FIG. 11. Stimulation of *in vitro* androgen production by minced testis from the painted turtle (*Chrysemys picta*) by bovine follicle-stimulating hormone (FSH) and luteinizing hormone (LH) in conjunction with immunological purification of hormones. Each hormone was treated with either normal sheep serum (NSS) or an antiserum (LH-AS) raised in sheep to the β subunit of bovine LH. Incubations as described in Fig. 8. Data from Tsui and Licht (1977).

that the activities of FSH were largely intrinsic. The same nonmammalian FSH preparations had low LH contamination in a variety of other studies, including similar *in vitro* steroidogenic analyses in amphibians. Further studies on bovine and frog hormones made with the aid of specific antisera also confirmed the intrinsic activities of FSH from these species. For example, not only did bovine LH have a very low activity compared to FSH in the turtle, but anti-LH serum selectively neutralized the LH (Fig. 11). For the frog hormones, an antifrog FSH serum selectively neutralized the FSH but not the LH (Tsui and Licht, 1977).

In view of the overlap of physiological effects of the two gonadotropins, it was of special interest to ascertain whether the steroid-stimulating actions of the FSH and LH were related to common sites of action in the gonad. This question was resolved by taking advantage of a rather unique anatomical arrangement of Leydig cells in the lizard *Cnemidophorus* (Fig. 19B). In addition to interstitial Leydig cells sparsely scattered throughout the testis, in *Cnemidophorus* there is also a discrete circumtesticular capsule of Leydig cells (Currie and Taylor, 1970).

We found that this subtunica layer could be mechanically peeled off with the tunica, thus cleanly separating Leydig cells from seminiferous tubules. When incubated with ovine gonadotropins, this Leydig cell preparation showed high levels of androgen production—about 25 times greater than an equal weight of tissue from the middle of the testis (Tsui, 1976b). Moreover, ovine FSH and LH (NIH) were equipotent in stimulating increased androgen production from these Leydig cells (Fig. 13).

AVES: Data on the regulation of testicular steroidogenesis in birds are less extensive than for reptiles, but indicate that the gonadotropin dependence of this process is somewhat different in the two classes. Several early *in vivo* studies with ovine hormones indicated that the Leydig cells of some avian species were stimulated by LH and only slightly by FSH (Lofts and Murton, 1973). However, in some cases, the best results were obtained with a combination of the two gonadotropins (Lofts *et al.*, 1973). Preliminary *in vivo* studies with avian

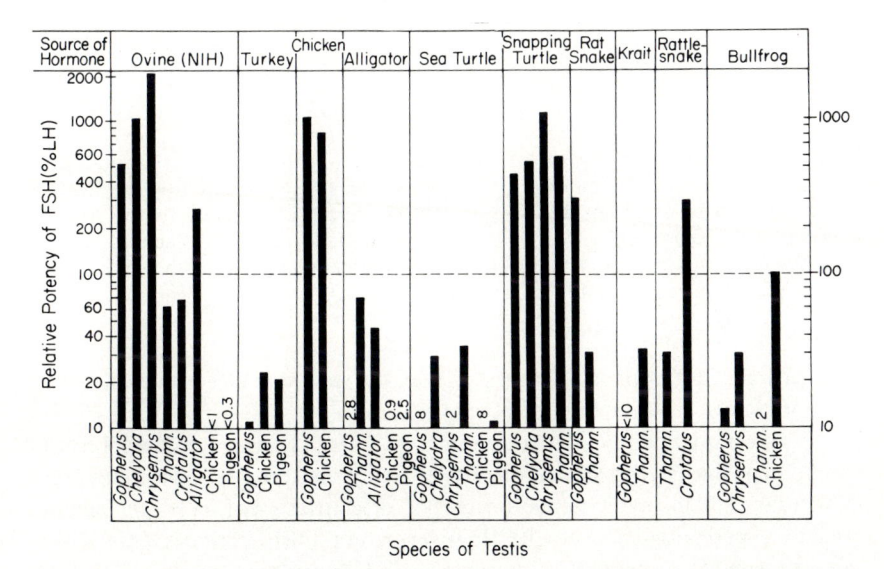

FIG. 12. Summary of relative potencies of follicle-stimulating hormone (FSH) preparations from diverse tetrapod species (listed at the top of graph) based on their ability to stimulate *in vitro* androgen production by minced testis from various species of reptile and bird (listed at the bottom of graph; *Thamn-Thamnophis*). In each experiment, the potency of FSH was computed in terms of the LH from the same species (i.e., FSH:LH potency ratios); thus, 100% indicates that the two types of gonadotropin were equipotent. Figures 8–11, 13, and 14 illustrate the experiments from which the values were derived. Based on data from Tsui (1976b) and Tsui and Licht (1977).

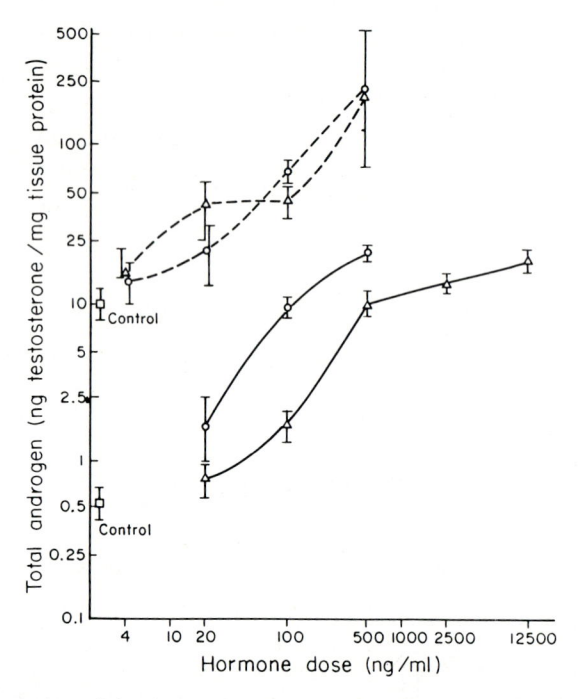

FIG. 13. Stimulation of *in vitro* androgen production by NIH follicle-stimulation hormone (△) and luteinizing hormone (○) in isolated Leydig cells or a mixture of Leydig cells and seminiferous tubules from the lizard *Cnemidophorus tigris*. – – –, data for the "tunica" preparation (this represents a layer of Leydig cells only); ———, data for the remainder of testis (tubules with some interstitium); control tubes without hormone are shown to the left of each. Reprinted from Tsui (1976a), with permission.

(chicken) gonadotropins also suggested that LH was more effective than FSH in the morphological stimulation of Leydig cell.

Ishii and Furuya (1975) reported that chicken LH was considerably more potent than the FSH in stimulating hypertrophy and proliferation of testicular interstitial cells in intact immature chicks; in contrast, LH had only minimal effects on spermatogonial activity. However, several FSH fractions caused significant increases in interstitial cell frequency and/or nuclear diameter, and it was not clear whether this action could be attributed to LH contamination. Studies in sexually immature Japanese quail (Brown *et al.*, 1975) also indicated that chicken FSH fractions were more potent than LH for stimulation of seminiferous tubules, whereas the reverse was true for Leydig cell stimulation. Ultrastructural observations indicated that LH stimulated the development of mature Leydig cells containing the full complement of organelles typical of steroid secretion; in FSH-treated birds, Leydig cell differentiation was minimal, and it was less pronounced in hypophysectomized than in intact birds. In both of these studies with chicken hormones, the authors suggested that the FSH effects on Leydig cells were due primarily to contamination with LH, but an intrinsic action of FSH could not be ruled out. Unfortunately, direct measurements of androgen production, especially after acute injection, were not performed in either of these studies.

The few available data for *in vitro* androgen production by bird testes—similar to those discussed for mammals and reptiles—are unfortunately somewhat conflicting. In our studies, minced chicken and pigeon testes showed very different responses to some FSH and LH preparations from those observed with reptilian tissues (Figs. 12 and 14). In particular, the avian testes were highly specific for ovine LH—the opposite of the response observed in turtles; in this regard, there is good agreement between *in vivo* and *in vitro* data for birds. Reptilian LH (turtle and alligator) was also more potent than FSH in the birds, but the FSH/LH potency ratios were higher than expected from LH contamination in FSH. Also, although frog hormones had only low activities (about 1% of NIH-LH-Sl), the FSH and LH were equipotent. Most important were results for the homologous hormones: turkey FSH was about 25% as potent as LH, and

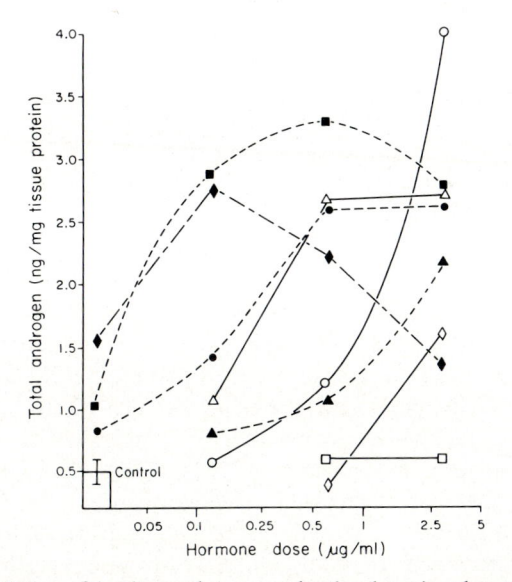

FIG. 14. Stimulation of *in vitro* androgen production by minced testis from the chicken (*Gallus domesticus*) with gonadotropins from various mammalian, avian, or reptilian sources: Values for follicle-stimulating hormone preparations are shown by open symbols and solid curves; for luteinizing hormine, by filled symbols and dashed curves. Tissues were incubated at 36°C (see Fig. 8 for further details): □, ■, ovine; ○, ●, turkey; ◊, ♦, alligator; △, ▲, chicken. Data from Tsui (1976b).

chicken FSH was slightly more potent than LH when tested in minced chicken testis (Figs. 12 and 14). Connell et al. (1966) found stimulation of in vitro testosterone biosynthesis from labeled precursor, [1-^{14}C] acetate, in chick testes incubated with either hCG, PMSG, or ovine (NIH) LH or FSH. At the single dose tested, the responses to the FSH and LH were the same. In contrast, Maung (1976) reported that NIH-FSH was only 0.5% as potent as NIH-LH, and chicken FSH was only about 5% as potent as chicken LH in stimulating in vitro androgen production by chicken Leydig cells.

At present, we cannot account for the discrepancies among the various studies in birds. It should be noted that the purities of the various preparations used by different laboratories have not been confirmed by comparable criteria or bioassays, and in vivo studies were based on young birds in contrast to the mature animals used for some in vitro studies. Thus, Follett's (1976) conclusion that "There is little doubt that LH is the hormone responsible for androgen secretion in birds" seems somewhat premature.

AMPHIBIA: The comparative effects of mammalian pituitary LH and FSH have been studied in hypophysectomized mature male amphibians: the anurans *Rana pipiens* (Burgos and Ladman, 1957) and *Rana temporaria* (Lofts, 1961), and the urodele *Pleurodeles waltli* (Andrieux et al., 1973). In addition, mammalian LH, FSH, hCG, and PMSG have been used in other studies of amphibian testicular development, histology, histochemistry, and biochemistry (see Lofts, 1974). Most evidence from such studies demonstrated that only LH and hCG induced changes in Leydig cells indicative of steroidogenesis. Stimulation of androgen-dependent tissues, such as thumb pad or Wolffian duct, was evident only after treatment with LH-like preparations. However, some indirect evidence of FSH-induced steroidogenesis was suggested by the histochemical results of Lofts (1961) and the 3 β-HSD assays reported by Wiebe (1970). Lofts (1961) demonstrated that mammalian FSH, but not LH, induced a depletion of cholesterol-positive lipid from Sertoli cells of hypophysectomized *R. temporaria*, concomitant with the resumption of spermatogenesis. In the same experiment, LH had no effect on Sertoli lipid or spermatogenesis, but did affect Leydig cell lipid accumulation. In amphibians, these sustentacular (Sertoli) cells have been ascribed a possible steroidogenic function by some workers (van Oordt and Brands, 1970; Saidapur and Nadkarni, 1973, 1975; Lofts, 1972, 1974). From work involving other nonmammalian vertebrates, Lofts (1972) suggested that FSH might stimulate Sertoli steroidogenesis. Other evidence for FSH-induced steroidogenesis came from studies of 3 β-HSD activity in subcellular fractions of *Xenopus laevis* testis. Wiebe (1970) injected mature male *X. laevis* with PMSG, hCG, NIH-LH, or NIH-FSH for 10 days and found that all of these gonadotropins stimulated 3 β-HSD activity; if anything, FSH and PMSG were more potent than LH and hCG. In later work, Wiebe (1972a,b) did not report effects of FSH, but did verify stimulation of 3 β-HSD by mammalian LH in testis homogenates from

R. pipiens. Interestingly, Pesonen and Rapola (1962), using histochemical techniques, found that both hCG and PMSG stimulated Leydig 3 β-HSD in *X. laevis*, but neither gonadotropin affected 3 β-HSD activity in testes of *Bufo bufo*. As Wiebe (1972b) pointed out, further studies of gonadotropin control of steroidogenesis in isolated Leydig and Sertoli cells are required.

We have studied purified bullfrog gonadotropins with respect to both *in vivo* androgen production in mature male bullfrogs, and *in vitro* steroidogenesis by sliced and minced bullfrog testis (Muller, 1975, 1976a,b); plasma or incubation medium was chromatographed and analyzed by radioimmunoassay. Injection of bullfrog LH (0.002–0.1 μg/gm body weight) into intact bullfrogs stimulated up to a 10-fold increase in plasma 5 α-dihydrotestosterone (DHT) and a 6.5-fold increase in plasma testosterone. Injection of bullfrog FSH produced little or no change in these plasma steroids. In another experiment, hypophysectomized bullfrogs were injected with 20 μg of bullfrog LH, FSH, 100 μg of ovine prolactin, or combinations of the hormones; plasma samples were collected by cardiac puncture 2 hours after injection (Fig. 15). Groups receiving LH had significantly higher ($p < 0.01$) plasma testosterone and DHT than either controls or FSH-treated groups; FSH or FSH + prolactin had small but statistically significant effects. Consistent with these observations was the finding that hypophysectomized bullfrogs injected with bullfrog LH for 1 month showed a significant increase in Wolffian duct weight, and hyperplasia of the duct epithelium, whereas no stimulation occurred in frogs receiving FSH.

In vitro incubation of bullfrog testis with bullfrog gonadotropins further confirmed that androgen secretion in this system was an LH-specific response (Muller, 1975, 1976a,b). Both testosterone and DHT were stimulated in a dose-related manner *in vitro* in response to bullfrog LH, but not to FSH. Results of one experiment are illustrated in Fig. 16. No interaction of bullfrog LH with bullfrog FSH, ovine prolactin, or bovine growth hormone was observed. In other incubation experiments, bullfrog FSH had only 0.1–0.3% LH contamination or intrinsic activity. In this system bullfrog LH is 3–10 times more potent than other amphibian (*Ambystoma*) or reptilian (*Alligator* and *Chelydra*) LH, and 30–50 times more potent than NIH ovine LH; NIH ovine FSH and *Alligator* FSH were inactive (Muller, 1976a, and unpublished observations).

A few experiments were conducted to test the effects of amphibian LH and FSH in urodele testes *in vitro* (C. H. Muller, unpublished). Results of several incubations of testes from plethodontid urodeles (*Bolitoglossa* and *Pseudoeurycea*) with bullfrog LH or FSH suggested a slight loss of gonadotropin specificity for testosterone production in these heterologous systems, although LH was still clearly more potent than FSH. *Ambystoma* LH was about 3 times more potent than bullfrog LH, and about 50 times more potent than *Ambystoma* FSH when tested in *Ambystoma*. However, using this salamander testis, bullfrog LH was only 15 times more potent than bullfrog FSH—a result pointing

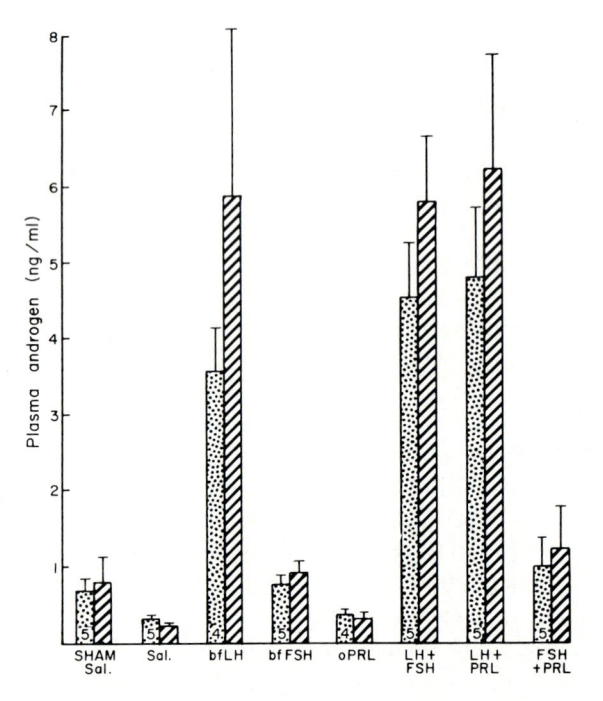

FIG. 15. Stimulation of *in vivo* androgen production by adult bullfrogs with homologous (bullfrog) gonadotropins and ovine prolactin (PRL). Animals were either sham-operated or hypophysectomized and injected with gonadotropins on alternate days with saline, 20 μg of bullfrog luteinizing hormone (bfLH) and/or follicle-stimulating hormone (bfFSH), 100 μg of PRL, or a mixture of gonadotropins and PRL for 14 days. Plasma was sampled by cardiac puncture 2 hours after the last injection and analyzed for testosterone (stippled bars) or dihydrotestorone (crosshatched bars) by radioimmunoassay following chromatography. Bars and vertical lines show the means and standard errors; sample sizes are shown at base of each set of data. Only the values for LH treatment (LH, LH + FSH, and LH + PRL) differ significantly from controls, and there were no differences among the three groups receiving LH. Data from Muller (1976a).

to the importance of using homologous hormones in studying gonadotropin specificity of responses.

B. REPRODUCTION IN THE FEMALE

1. Ovarian Growth and Ovulation

In considering the potential roles of the two gonadotropins on the ovary of nonmammalian tetrapods, a number of different events need to be recognized, not all of which are present in mammals. The production of an egg may involve

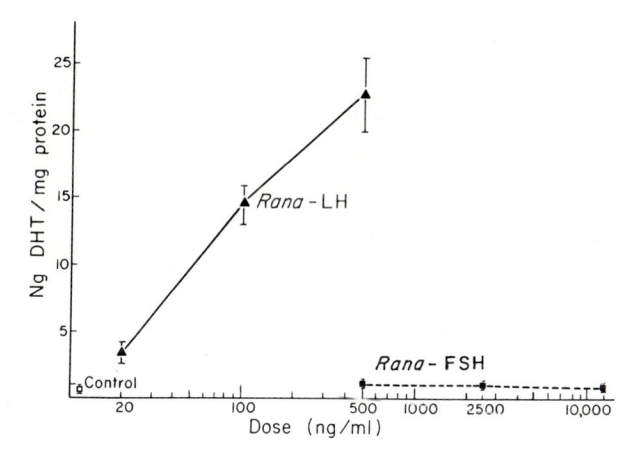

FIG. 16. Stimulation of *in vitro* androgen production by minced testis from bullfrogs (*Rana catesbeiana*) with follicle-stimulating hormone (FSH) and luteinizing hormone (LH) from the same species. Tissues were incubated 3 hours at 30°C. Values show mean and standard errors for 4 replicates. Data from Muller (1975).

oogonial division (this may occur into adulthood in amphibians and reptiles), oogenesis (differentiation of oocytes and their recruitment into the follicular pool), follicular formation (investment of oocytes with associated layers such as theca externa, theca interna, and granulosa) progressive differentiation of the follicular structure (e.g., the number and types of granulosa cells may change), follicular growth (vitellogenesis and yolk deposition), oocyte maturation (final meiotic division or germinal vesicle breakdown) and, finally, ovulation. Steroid production may be involved in the mediation of some of these events, but this will be considered as a separate topic. Not all these events are the same in all species; the process of vitellogenesis when the follicle becomes conspicuously enlarged (and ovarian weight increases dramatically) is perhaps one of the most distinctive features of ovarian development in all nonmammalian tetrapods. Unfortunately, few studies have attempted to elucidate the relative importance of FSH and LH for each process in any of the nonmammalian tetrapods; in fact, it is not clear to what extent some are even dependent on gonadotropins. For example, in amphibians, previtellogenic follicular growth is considered to be gonadotropin-independent (Jørgensen, 1973), but gonadotropins appear to have important effects on many aspects of early oocyte development in some reptiles (Jones *et al.*, 1975).

REPTILIA: Although data for female reptiles are less complete than for males, the ovary appears to show the same broad responsiveness to FSH as does the testis. Initial studies with ovine and human FSH and LH in our laboratory indicated that FSH alone was capable of stimulating and supporting full development of the ovary up to and including ovulation in intact or hypophysectomized lizards (reviewed in Licht, 1974a). Recently, the intrinsic activity of ovine FSH was confirmed by the demonstration that the broad actions of highly purified ovine FSH in female lizards were unaffected by treatment with anti-LH serum (Licht and Tsui, 1975). Ovine and human LH showed only weak activities in connection with ovarian growth, and there was no evidence of synergism between FSH and LH (Licht, 1970), but hCG stimulated vitellogenesis and oocyte maturation in intact lizards (Panigel, 1956).

Callard and Ziegler (1970) found that in the lizard *Dipsosaurus* the gonadotropin (PMSG) stimulation of vitellogenesis was dependent on growth hormone, but, in *Anolis*, FSH was fully active when tested alone. Attempts to demonstrate gonadotropin stimulation of ovarian growth in other reptiles (snakes and turtles) have been largely unsuccessful, and the presence of growth hormone did not improve results (Licht, 1972b). However, these reptiles have not been as thoroughly studied as lizards, and experiments may have been complicated by seasonal cycles or age effects.

Jones and co-workers have utilized ovine (NIH) FSH to study many aspects of ovarian development and function in the lizard *Anolis carolinensis*. Their investigations on immature females (Jones et al., 1975) demonstrated that FSH affected very early stages of differentiation and enlargement of oocytes (oogenesis), the subsequent formation of follicles, and later differentiation of follicular granulosa cells. FSH appeared to maintain the size hierarchy among follicles by differentially increasing vascularity of the largest follicles (Jones et al., 1973; Gerrard et al., 1973); this latter effect may be mediated through estrogen production (Jones, 1975). FSH treatment was recently shown to stimulate increased incorporation of [^{3}H] thymidine into ovarian follicular cells in *Anolis*, including the ovarian medullary stroma, the basal granulosa in larger follicles, and the theca of all follicles (R. Tokarz, personal communication). These effects were apparently not mediated through estrogen, since, despite estrogen-induced increases in ovarian weight (due to vitellogenesis), the steroid had no significant effect on thymidine uptake. It is noteworthy that some of these actions of FSH on the ovary are similar to those that were attributed to LH in studies with mammals (e.g., Ryle, 1972). Unfortunately, the extent to which LH might affect the various ovarian processes discussed above has not been examined.

The very limited data for the actions of nonmammalian gonadotropins in female reptiles closely parallel those observed with mammalian preparations. In an early investigation with chicken gonadotropin (Licht and Hartree, 1971), the avian FSH was more potent than LH for ovarian growth and ovulation in the lizard, but the relative cross-contamination between the two preparations was not well established. In any case, the avian LH appeared to be more potent than mammalian LH for the stimulation of vitellogenesis in adult lizards. More recent data for purified reptilian (turtle) hormones also indicated that FSH was more effective than LH for stimulating both ovarian growth and ovulation in lizards

(Licht and Crews, 1975). In this case, the turtle FSH was more potent than LH in chronic studies involving stimulation of the regressed ovary as well as in acute experiments to induce ovulation of already "ripe" follicles. An anti-LH serum did not reduce the effectiveness of the turtle FSH in any of these respects (Licht and Crews, 1975).

These data indicated that the reptilian ovary may respond to both FSH and LH, but there was no clear evidence for a differential action of the gonadotropins. In general, FSH tended to be more potent than LH in these *in vivo* studies. Additional information on the relative actions of the two types of hormone will be considered later in connection with the regulation of steroid production.

AVES: Many of the data for gonadotropin actions on the avian ovary deal with a few domesticated species, especially the chicken. Nalbandov (1976) has recently reviewed much of this literature and provided an interesting account of his own extensive efforts in this area. Many of these data point to the fact that the heterologous mammalian gonadotropins were inferior to those of the birds' own pituitary in stimulating the avian ovary; hence, information on characterized homologous avian hormone preparations was critical. Nevertheless, there was evidence for a stimulation of gonadal growth by ovine FSH in a variety of avian species such as chicken, duck, and pigeon (Lofts and Murton, 1973); the potential actions of LH on ovarian growth have not been well studied. Limited data for avian hormones suggested that ovarian growth in birds might not be specific for FSH or LH. Mitchell (1970) tested a partially purified chicken FSH; this was prepared according to the methods of Hartree and Cunningham (1969), but was not characterized for purity or potency. This gonadotropin fraction was essentially without effect in hypophysectomized hens if treatment was started after ovarian regression occurred, 10–15 days after the operation. Some maintenance of ovarian and oviducal weight and prolongation of ovulation were observed if hormone injections were started within an hour of hypophysectomy, but the response to the presumptive FSH preparation was poor compared with crude pituitary extracts from chickens. The ovulatory action of the preparation was ascribed to LH contamination (based on the assumption that only LH could induce ovulation), but this contamination was not confirmed by independent assay for LH. Imai (1973) reported that both FSH and LH-enriched fractions derived from chicken glands were about equipotent in stimulating follicular development of the regressed ovaries of intact hens treated with Methallibure, but again there was little information on the degree of cross-contamination in the two preparations.

The term "ovulation-inducing hormone" has been adopted by Nalbandov and co-workers to refer to a hypophysial factor in birds, because of some difficulties in explaining patterns of ovarian function with mammalian hormones (Nalbandov, 1976). In any case, ovulation in birds shows a somewhat different relationship to mammalian gonadotropins from that described in reptiles. In particular, several workers—using birds in varying physiological states—have shown that ovine LH could induce ovulation, whereas ovine FSH was largely inactive (Lofts and Murton, 1973; Imai, 1973; Nalbandov, 1976). The results did not rule out the possibility that ovulation may be induced by appropriate doses of purified FSH, since only the relatively impure NIH preparations were tested. Interestingly, hCG did not induce ovulation in the hen (Nalbandov, 1976).

Studies with hormones derived from chicken pituitaries also suggested that LH was more potent than FSH for ovulation, but FSH preparations induced some ovulation; it was not know whether this resulted from LH contamination (Imai, 1973).

AMPHIBIA: Information on the effects of gonadotropins on ovarian development in amphibians is based almost exclusively on mammalian hormonal preparations, and mostly on nonpituitary materials such as hCG and PMSG. These data indicated that the growth of the anuran ovary, including recruitment of new follicles and vitellogenesis, was relatively nonspecific toward FSH and LH (Lofts, 1974). FSH was generally more effective than LH in hypophysectomized animals, but hCG tended to be more potent than both pituitary hormones. The difference between the potencies of hCG and pituitary (ovine) LH has been attributed to the faster clearance rate of the latter from the amphibian circulation (Roos and Jørgensen, 1974).

A voluminous literature has recently developed on the process of vitellogenesis in the amphibian ovary. Although such studies were generally not oriented toward elucidating the relative importance of the two pituitary hormones, they supply valuable insights into the possible modes of actions of these gonadotropins. The recent review of the physiology of vitellogenesis by Follett and Redshaw (1974) provides a useful framework for further studies into the specific roles of FSH and LH. They concluded that gonadotropins did not act directly on vitellogenin formation, but rather that vitellogenesis occurred in the liver and was mediated by ovarian estrogens that were stimulated by gonadotropins. The action of pituitary hormones on estrogen production has not been defined, but PMSG greatly increased estrogen biosynthesis from the amphibian ovary *in vitro* (Redshaw and Nicholls, 1971). Estrogens stimulated hepatic yolk protein synthesis which led to a large increase in plasma vitellogenin; gonadotropins then played a primary role in the selective translocation of this material into the growing oocyte. For example, when ovine FSH was injected into normal intact female *Xenopus*, there was a significant decrease in several protein and lipid components of the plasma. When frogs were given estradiol and FSH together, there was a highly significant increase in the incorporation of vitellogenin into yolk platelet lipid and the protein component of the oocyte when compared with untreated controls receiving estradiol alone (Follett and Redshaw, 1974). Emmerson and Kjaer (1974) showed that hCG had similar actions in the toad *Bufo bufo*.

Unlike the various stages of ovarian growth, final oocyte maturation (germinal vesicle breakdown) and ovulation in amphibians appear to be highly specific for LH-like hormones. In contrast to the high potencies of FSH preparations for ovulation in reptiles, only LH appeared capable of inducing ovulation in both anuran and urodele amphibians. The LH specificity of amphibian oocyte maturation and ovulation has been suggested for a long time from tests with mammalian hormones, and this specificity was confirmed when the homologous amphibian hormones were studied (Licht and Papkoff, 1974b; Licht *et al.*, 1975). Furthermore, the same high LH specificity was observed when hormones derived from a variety of reptilian, avian, and amphibian species were examined in both the anurans and urodeles (see Table I). It was this broad phylogenetic cross-reactivity and consistent LH specificity of the *Xenopus* ovary that made it particularly valuable for identifying and quantifying LH in our comparative studies. In fact, this situation represented a somewhat unique example of a "receptor" that could discriminate consistently between the two molecules, since, in other mammalian and nonmammalian assays examined, FSH frequently showed appreciable activity when compared to LH (e.g., Licht and Midgley, 1976a,b; Licht *et al.*, 1976a). Thus, a pronounced dichotomy is evident between amphibians and reptiles in the gonadotropin regulation of ovulation, and this difference is a characteristic of the tissue rather than of the gonadotropins.

2. Ovarian Steroid Production

REPTILIA: A consistent marked hypertrophy of the oviducts of lizards treated with mammalian gonadotropins indicates a stimulation of estrogen secretion. In all the *in vivo* studies involving chronic hormonal treatment, mammalian FSH appeared to be considerably more potent than LH for stimulation of the female accessory sexual structures, although both hormones were effective (Licht, 1970, 1974a). Furthermore, as in studies on the testis, the treatment of the ovine FSH with a specific antiserum against LH did not reduce its apparent estrogen-stimulating activity (Licht and Tsui, 1975). Similar results were obtained when female lizards were injected with hormones derived from a reptile, the snapping turtle (Licht and Crews, 1975). These *in vivo* studies suggested that estrogen production might be stimulated by both FSH and LH from several sources, but no evidence was obtained for progesterone production.

Data for the gonadotropin regulation of progesterone production by reptilian ovarian tissues came primarily from *in vitro* studies in several species of turtle and crocodilian, using mammalian, avian, reptilian, and frog hormones. In our laboratory, endogenous *in vitro* estrogen and progesterone secretion were measured by RIA of chromatographed incubation media and tissues. In the three species of turtle studied, both mammalian FSH and LH were active in stimulating pronounced increases in progesterone production, but the ovine LH was considerably less potent than ovine FSH (Fig. 17) and only FSH stimulated estrogen. In the crocodilian (*Caiman*) ovary, the two mammalian hormones did not differ appreciably in activity. More important, when homologous hormones were tested (e.g., alligator hormones in the crocodilian and snapping turtle hormones in the snapping turtle and other turtles), there was again no clear specificity in the stimulation of progesterone, although in the homologous turtle system, purified FSH was again more potent than LH.

Results for preovulatory follicular tissues were the same as for luteal tissues in both the amount of progesterone produced and the sensitivities to FSH and LH (Crews and Licht, 1975a; cf. Fig. 17). In a separate experiment, we demonstrated that the source of this progesterone and the site of action of both gonadotropins in the preovulatory follicle wall were the theca interna and granulosa—theca externa was inactive (Crews and Licht, 1975b).

In studies of *in vitro* steroid biosynthesis of turtle follicular tissues based on conversion of labeled precursors, [^{3}H]cholesterol and [^{3}H]pregnenolone, the results of Chan and Callard (1974) and Callard *et al.* (1976) for FSH and LH actions differed substantially from our observations on steroid production. With preovulatory tissues (Chan and Callard, 1974), ovine LH enhanced the synthesis of pregnenolone, progesterone, dehydroepiandosterone, and androstenedione, whereas ovine FSH was essentially without effect. Subsequent studies with ovine and chicken hormones (Callard *et al.*, 1976) also showed that LH stimulated

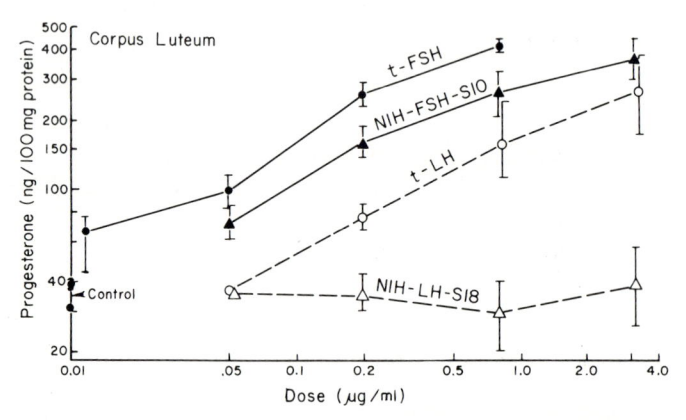

FIG. 17. Stimulation of *in vitro* progesterone production in luteal tissue from snapping turtle (*Chelydra serpentina*) with ovine (NIH) or snapping turtle (t) gonadotropins. Mean and standard error for three replicates at each dose are shown. Tissues were incubated for 5 hours at 30°C, and the medium was assayed by radioimmunoassay after chromatography. FSH, follicle-stimulating hormone; LH, luteinizing hormone. Modified from Licht and Crews (1976).

estrogen biosynthesis from tritiated precursors in preovulatory tissues and yields of progestin in postovulatory tissues of turtles; FSH affects were minimal, especially in postovulatory tissues. Reptilian gonadotropins were not tested in this investigation. At present, we cannot account for the discrepancy between the two types of *in vitro* studies, since the latter studies examined only a single dose of each gonadotropin at a single time interval.

AVES: Data for the gonadotropin regulation of steroids in the bird ovary are limited. Shahabi *et al.* (1975) recently demonstrated that systemic injection of 25 µg ovine (NIH) LH into intact hens caused a premature rise in plasma progesterone and follicle wall progesterone concentration corresponding to those normally observed before ovulation. The LH injection did not duplicate the normal rise in estrogen (it was suggested that FSH might be required for this). Effects of ovine FSH on progesterone were not examined in this study, although based on results for testicular androgen secretion, one would not expect ovine FSH to be active (cf. Fig. 14). However, the potential importance of avian FSH cannot be ruled out.

AMPHIBIA: Information on steroid secretions (especially progesterone) by the amphibian ovary indicates that these are closely linked to the processes of oocyte maturation (e.g., germinal vesicle breakdown) and ovulation. The limited data for estrogen stimulation were discussed under ovarian growth. While it was clear that oocyte maturation and ovulation were under the control of pituitary gonadotropins, much evidence indicated that these gonadotropin effects were mediated by progestational steroids produced in the ovaries, including the follicular membranes (e.g., Redshaw, 1972; Schuetz, 1974). In view of the specificity for LH observed in maturation and ovulation, it would be expected that progesterone was primarily LH dependent. Indeed, direct evidence for such specificity was obtained by measurement of *in vitro* progesterone production by ovarian segments from an anuran, bullfrog (Fig. 18), and urodele, tiger salamander (Licht and Crews, 1976). In both cases, homologous hormones were used to show a high LH specificity. In the bullfrog, the same LH specificity was also shown for ovine hormones, although the ovine LH was much less potent than the frog LH (Fig. 18).

Gonadotropins may have several different actions on the follicle that lead to the progesterone-mediated effects. For example, ovarian follicles from the frog *Rana pipiens* showed greatly increased uptake of isotopically labeled pregnenolone when treated with extracts of frog pituitary (Snyder and Schuetz, 1973). Mammalian LH had the same effect, although this hormone was much less potent than crude frog pituitary extract (Snyder and Biggers, 1975). Other mammalian hormones, including ovine FSH and hCG were found to be inactive; the inactivity of hCG was probably a reflection of species specificity (*Rana* females are unusually insensitive to this human hormone). The mechanism of the enhanced follicular steroid uptake is unknown, but it may be related to the

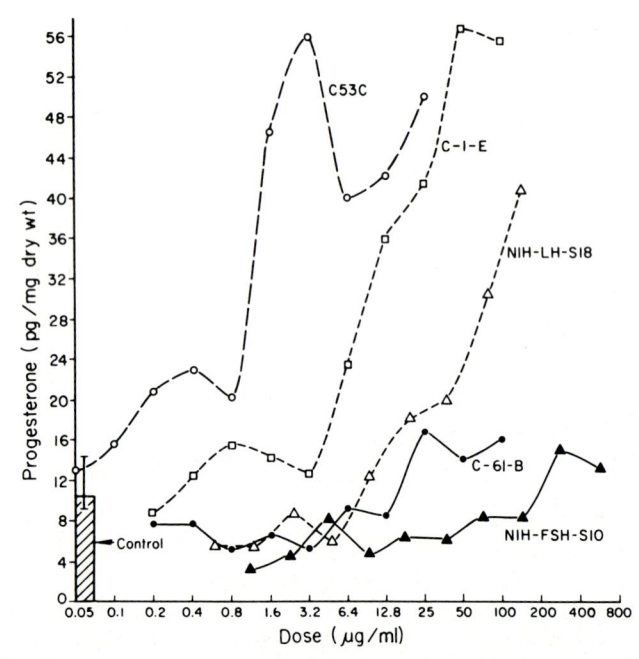

FIG. 18. Stimulation of *in vitro* production of progesterone in ovarian fragments from a bullfrog *Rana catesbeiana* by ovine (NIH) gonadotropins or homologous bullfrog gonadotropins; C53C is purified bullfrog lutenizing hormone (LH), C-61-B is bullfrog follicle-stimulating hormone (FSH), and C-1-E is crude extract of bullfrog pituitary. Tissues were incubated for 24 hours at 19°C and the medium was assayed by radioimmunoassay after chromatography. Reprinted from Licht and Crews (1976), with permission.

many changes in the follicular layers and vitelline membrane (Schuetz, 1972). Several characteristics of the LH-induced effects on the ovarian follicle suggested that this hormone might induce the synthesis of steroids in the follicular envelope by activation of RNA and possibly protein synthesis (Snyder and Biggers, 1975).

VI. Binding of Gonadotropins

A wealth of data has recently been accumulated on the binding of gonadotropins to various mammalian testicular and ovarian tissues. These have shown that the gonads possess distinct binding sites or "receptors" for FSH and LH which appear to be linked to the physiological actions of the hormones. It was thus of interest to determine how the broad overlap in physiological actions of FSH and LH in some nonmammalian species was related to gonadotropin binding sites. For example, do the two gonadotropins share the same receptors in reptiles? Collaborative studies were undertaken with Dr. A. R. Midgley, Jr. to determine whether techniques developed for examining gonadotropin receptors in mammals might be applicable to amphibians, reptiles, and birds. The techniques included topical autoradiography to localize binding sites and competitive inhibition analyses to quantify the hormonal specificity of these sites.

A. BINDING FOR RADIOIODINATED HUMAN FOLLICLE-STIMULATING HORMONE ([^{125}I]hFSH)

Our investigations were initially based on using [^{125}I]hFSH as radioligand to define FSH binding sites. We were encouraged to find that this radioligand not only bound to homogenates of reptilian and avian gonadal tissues (testes and ovaries), but also showed a higher level of binding than was observed in some mammalian gonads. The binding was specific to the gonads—[^{125}I]hFSH showed no specific binding to muscle, liver, or thyroid—and it exhibited many of the characteristics, such as dependence on temperature and tissue concentration, that have been reported in mammalian studies (Licht and Midgley, 1976a). We were unable to demonstrate any specific binding of this ligand to amphibian ovarian or testicular tissues.

Topical autoradiography on frozen sections allowed us to localize the binding of [^{125}I]hFSH in turtle, lizard, and snake gonads; again, amphibian (*Rana*) ovaries and testes did not show specific binding (Licht and Midgley, 1977). Results of these studies were consistent with previous physiological data in showing a broad distribution of the FSH binding within the reptilian testis (Fig. 19). Whereas similar methodology demonstrated that the interstitial tissues (Leydig cells) of the mammalian testis were highly specific for LH (e.g., Desjardins *et al*., 1975; Castro *et al*., 1972), conspicuous labeling of interstitial tissues with [^{125}I]hFSH was observed in all the reptiles; binding to Leydig cells was particularly striking in the lizard *Cnemidophorus* with its thick, subtunic, circumtesticular layer of Leydig tissue. Binding also occurred to elements within the seminiferous tubules corresponding to the situation in mammals with [^{125}I]hFSH (Fig. 19).

In ovaries from mammalian species studied to date, FSH binding sites were restricted to granulosa cells (Midgley, 1973; Zeleznik *et al*., 1974; Ranjaniemi and Midgley, 1975). A comparable localization was observed in reptilian ovaries, but binding was not restricted to granulosa. In addition, [^{125}I]hFSH bound to the vitelline membrane, the theca interna, and especially the epithelial layer surrounding the follicles in reptiles (Fig. 19C).

Competitive inhibition studies with mammalian and nonmammalian hormones were performed with testes and ovaries from five species of turtles, a snake, a lizard, and two birds to quantify the hormonal specificity of [^{125}I]hFSH binding; a mammalian tissue, porcine granulosa, served as a control (Licht and

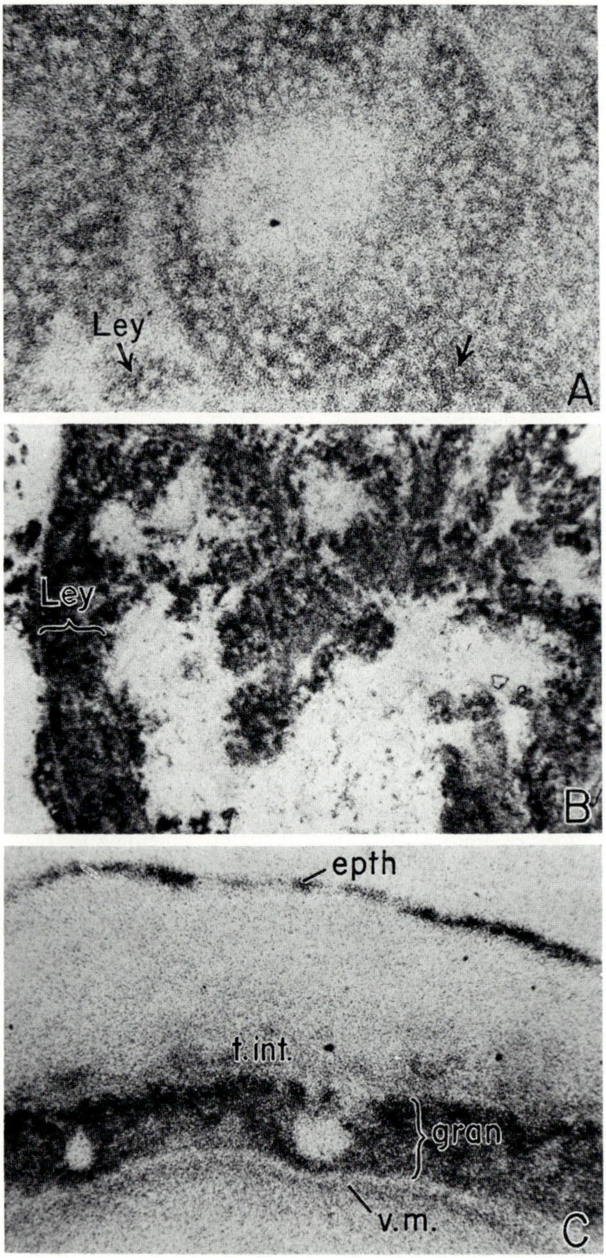
Ley
A
Ley
B
epth
t. int.
gran
v.m.
C

Midgley, 1976a,b). We first examined a series of purified mammalian pituitary preparations. In all cases, the binding of [^{125}I]hFSH was unaffected by TSH, GH, or prolactin; i.e., it was specific for gonadotropins (e.g., Figs. 20 and 21). Purified mammalian FSH from four species—rat, ovine, bovine, and human—effectively competed with only minor variation in activities; their activities in reptiles and birds were similar to those exhibited in radioligand assays with mammalian tissue (Fig. 22). However, notable interspecific variations were observed among the various gonadal tissues in their sensitivities to the LH preparations from these mammals. The porcine granulosa cells, along with gonads from all five turtles, and both birds were highly specific for mammalian FSH; i.e., none of the LH preparations showed appreciable activity in competing for binding of [^{125}I]hFSH (e.g., Figs. 20 and 24). In contrast, all mammalian LH preparations exhibited considerable activity when tested in snakes and lizards. In these squamate tissues, the activities of the four species of mammalian LH differed, but some (especially bovine) were almost as active as the corresponding FSH in competing for [^{125}I]hFSH binding (e.g., Figs. 21 and 23). Comparisons of FSH/LH potency ratios (Fig. 23) illustrate the variations in the specificity of binding. Data support the hypothesis that the physiological actions of FSH and LH may be mediated through the same receptors in squamate reptiles. The situation was less clear for birds and for turtles; if LH did interact with FSH binding sites, its affinity must have been very low.

Better evidence for nonspecificity of binding sites in reptiles and birds came from studies on the nonmammalian gonadotropins (Licht and Midgley, 1976b). Considerable phylogenetic specificity in hormonal activity was evident as illustrated by the variations in activity of a particular hormone when tested in different tissues (Fig. 22). For example, [^{125}I]hFSH binding to the porcine

---

FIG. 19. Unstained autoradiograph of lizard and snake gonads showing the distribution of specifically bound ^{125}I-labeled human follicle-stimulating hormone, [^{125}I]hFSH. Frozen sections were incubated with [^{125}I]hFSH for 1 hour at room temperature, washed, and exposed for about 2 months. When the initial incubation was performed in the presence of 50 ng of unlabeled hFSH, virtually no deposition of silver grains was seen (see area outside of follicle in C for example of background). Thus, all darkly stained areas represent deposition of silver grains corresponding to [^{125}I]hFSH (Licht and Midgley, 1977). (A) Internal region of garter snake (*Thamnophis sirtalis*) testis showing the large rounded seminiferous tubules and clumps of labeled Leydig cells (Ley). (B) Middle and edge of testis from lizard *Cnemidophorus tigris* showing intense labeling over the spermatogonia within the seminiferous tubules as well as over the thick subtunic layer of Leydig cells (Ley). (C) Portion of the follicle wall of a small yolking follicle from the ovary of garter snake (*Thamnophis sirtalis*). Concentration of silver grains indicating [^{125}I]hFSH binding are localized over the external epithelial (epth) layer around the follicle, on scattered cells in the theca interna (t. int.), over most of the granulosa (gran), and on the vitelline membrane (v.m.) around the oocyte.

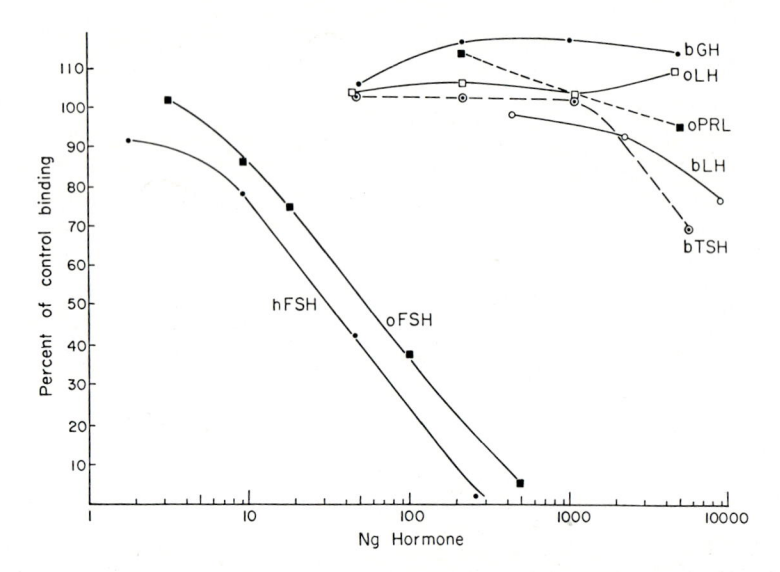

FIG. 20. Competition for the binding of [125] I-labeled human follicle-stimulating hormone ([125] I]hFSH) to turtle (*Chrysemys*) testis *in vitro* by various pituitary hormones derived from human (h), ovine (o), or bovine (b) sources. Points represent the mean of duplicate determinations: binding in the absence of unlabeled hormone was 6.2% of tracer. GH, growth hormone; LH, luteinizing hormone; PRL, prolactin; TSH, thyroid-stimulating hormone. Reprinted from Licht and Midgley (1976a), with permission.

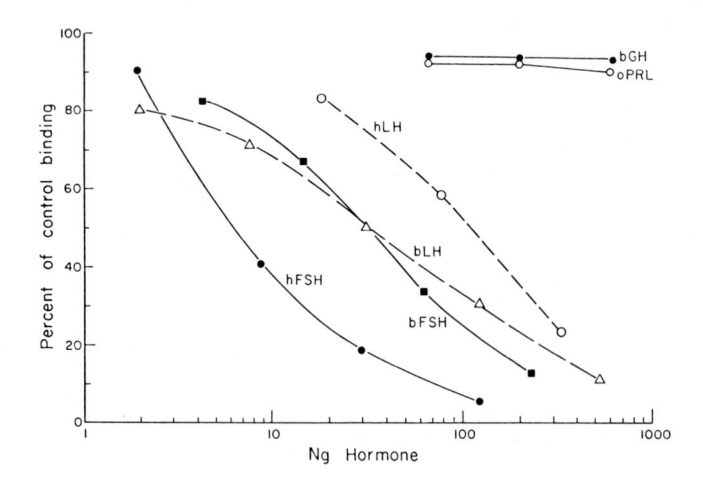

FIG. 21. Competition for the binding of [125] I]hFSH to snake (*Thamnophis*) testis *in vitro* by various mammalian pituitary hormones (see Fig. 20 for details); binding was 14.2% of tracer in absence of unlabeled hormone. Reprinted from Licht and Midgley (1976a), with permission.

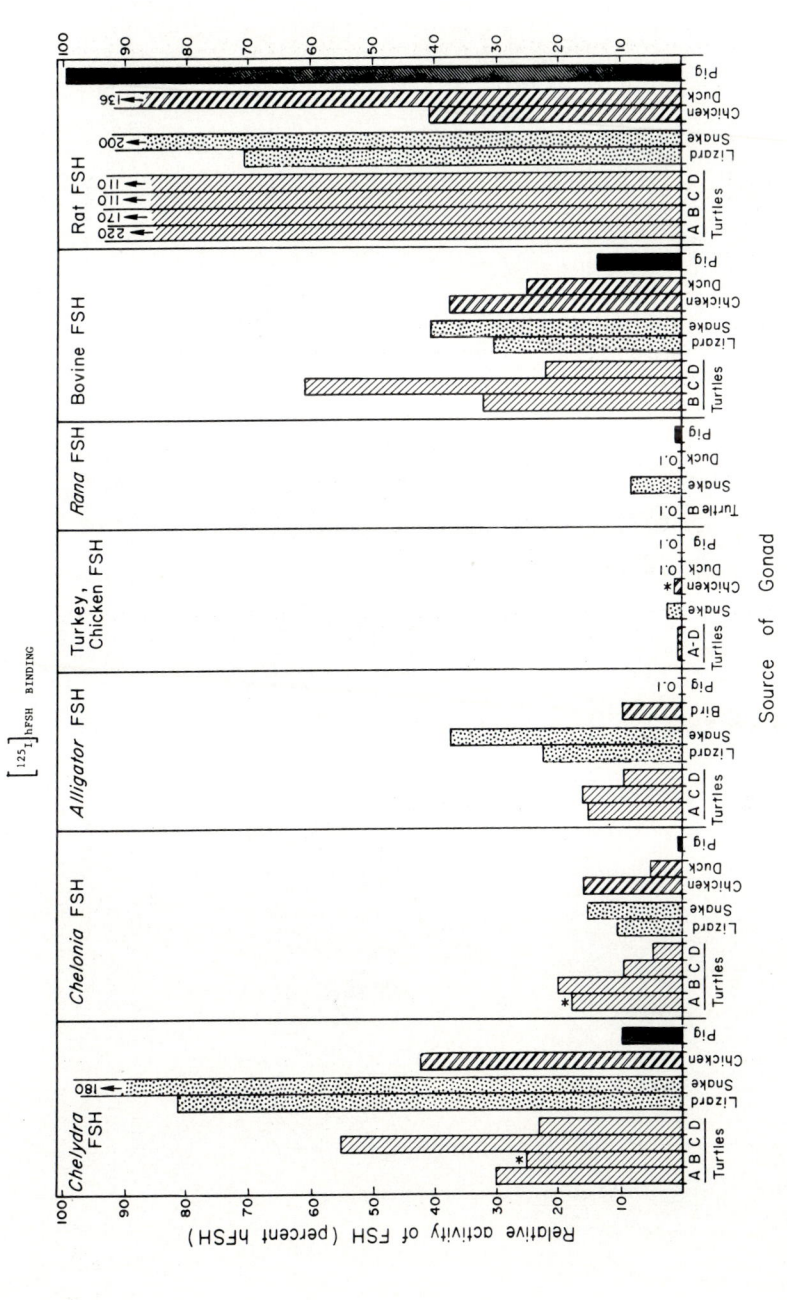

FIG. 22. Relative activities of various species of follicle-stimulating hormone (FSH) (shown at top of graph) in the competition for [^{125}I]hFSH binding to gonadal homogenates (shown at bottom of graph). In each case, the realtive activity of each preparation was computed in terms of a highly purified hFSH standard (AFP-574C). See Figs. 20, 21, and 24 for examples of individual experiments. Based on data from Licht and Midgley (1976a, b).

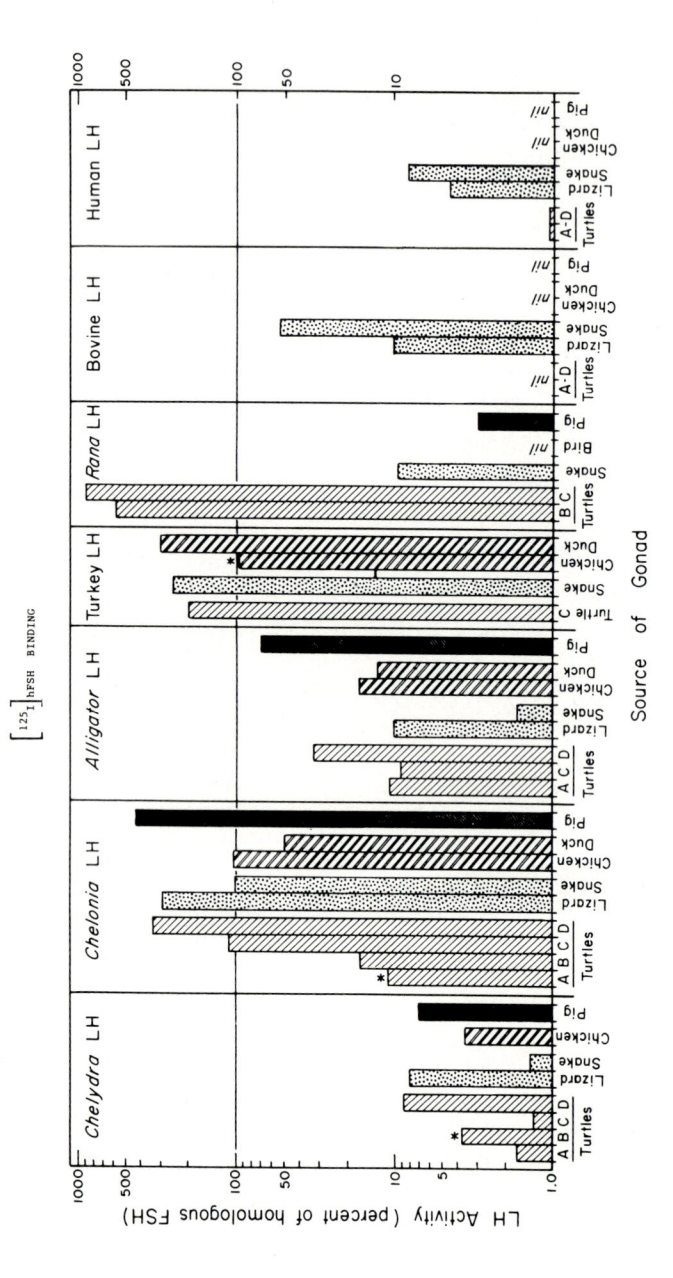

FIG. 23. Activities of luteinizing hormone (LH) preparations relative to the homologous follicle-stimulating hormone (FSH) (source shown at top) in competing for the binding of [^{125}I]hFSH to various species of reptilian, avian, and mammalian gonads (shown at bottom of graph). A value of 100% (see horizontal line on graph) indicates that an LH was equipotent to the corresponding FSH. See Figs. 20, 21, and 24 for examples of individual experiments. Based on data in Licht and Midgley (1976a, b).

granulosa was relatively insensitive to all nonmammalian hormones (Figs. 22 and 24), consistent with our earlier observations of the low potency of the nonmammalian hormones in mammalian bioassays. Phylogenetic specificity was also exhibited among nonmammalian tissues. Frog and snake hormones were essentially inactive in competing for [^{125}I]hFSH binding in birds and turtles; in fact, snake hormones were inactive in all tissues, except snakes (Fig. 22).

Perhaps the most interesting aspect of phylogenetic specificity was in the FSH/LH potency ratios (Fig. 23). Some LH preparations showed unexpectedly high activity, sometimes up to 10-fold greater than the corresponding FSH, in competing for [^{125}I]hFSH binding. The potency of a particular LH varied with the source of the gonad: e.g., *Rana* LH was highly potent in turtles but not snakes or pig; *Chelonia* LH was relatively potent in all tissues except its homologous one. A variety of evidence supported the conclusion that these activities represented intrinsic properties of the LH preparations and were not due to FSH contamination (see Licht and Midgley, 1976b).

B. BINDING OF A RADIOIODINATED NONMAMMALIAN GONADOTROPIN, [^{125}I] *Chelonia* FSH

A major concern in interpreting data based on the use of [^{125}I]hFSH was that the binding of a heterologous ligand might create artifacts in hormonal specificity by binding to sites that normally differentiated between homologous FSH

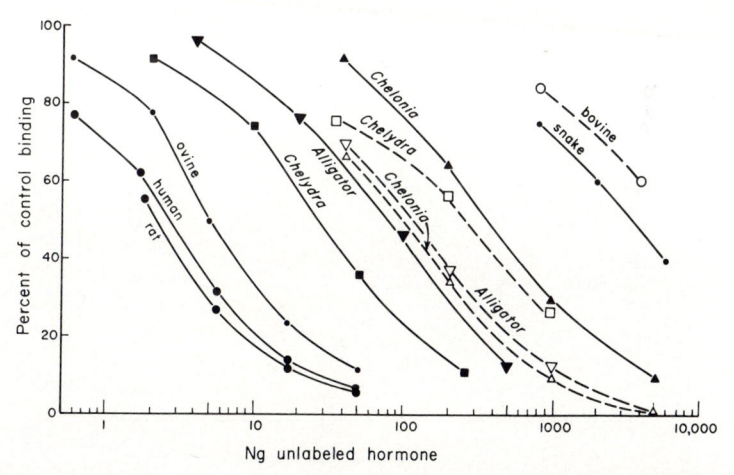

FIG. 24. Activities of various mammalian and reptilian gonadotropin preparations in competition for the binding of [^{125}I]hFSH to a mammalian tissue, porcine granulosa cells. Solid curves and filled symbols show inhibition by follicle-stimulating hormone (FSH) preparations; and dashed curves and open symbols for luteinizing hormone (LH) preparations.

and LH. We were especially concerned with extremes such as the sea turtle hormones. However, preliminary studies with iodinated *Chelonia* FSH did not support the above hypothesis that nonspecificity was a result of heterologous ligand.

When sea turtle FSH was iodinated by the chloramine T method as described by Goldfine *et al.* (1974), it showed 5–10% specific binding to gonadal homogenates from snake, lizard, and several turtle species, but no specific binding could be demonstrated in bird or mammalian gonads. When the ^{125}I-labeled *Chelonia* FSH was used as a ligand in binding tests with the reptilian tissues (including the homologous testis and ovary; e.g., Figs. 25 and 26), estimates of activities of mammalian and *Chelonia* gonadotropins were remarkably similar to those obtained with [^{125}I]hFSH (Table XVII).

C. BINDING OF LUTEINIZING HORMONES (HUMAN LH AND hCG)

The use of LH as a radioligand would clearly be of value in resolving the question of whether separate binding sites exist for the two gonadotropins. A series of such experiments have now been performed in collaboration with Dr. A. R. Midgley, Jr. Studies began with [^{125}I]hCG since this material had been widely used to demonstrate binding sites in mammalian systems and was generally easier to radioiodinate without loss of binding ability than was pituitary

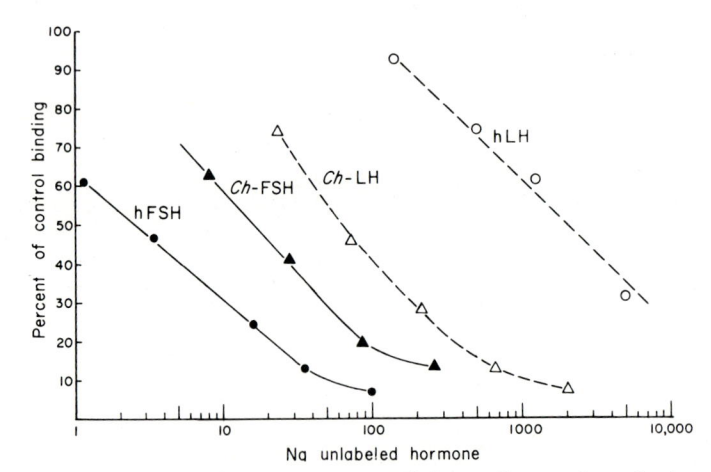

FIG. 25. Activities of human (h) or sea turtle (*Chelonia*, *Ch*) gonadotropins in competition for the binding of ^{125}I-labeled *Chelonia* follicle stimulating hormone (FSH) to homogenates of *Chelonia* ovary. LH, luteinizing hormone. Data from P. Licht (unpublished).

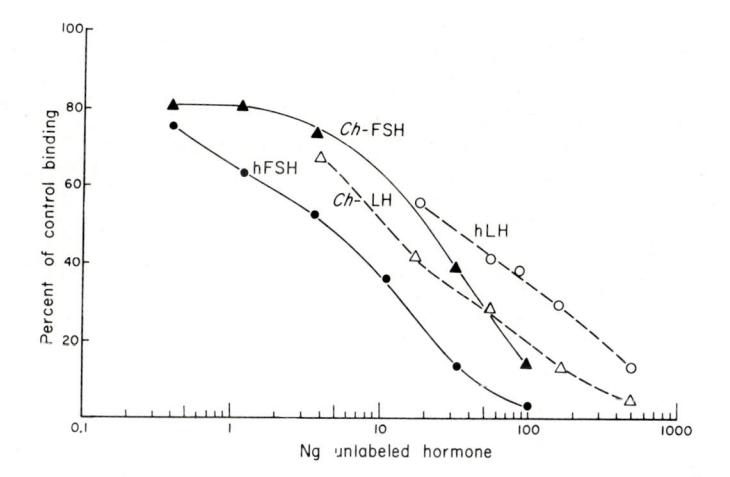

FIG. 26. Activities of human (h) or sea turtle (*Chelonia*, *Ch*) gonadotropins in competition for the binding of ^{125}I-labeled *Chelonia* follicle-stimulating hormone (FSH) to homogenates of snake (*Thamnophis*) testis. LH, luteinizing hormone. Data from P. Licht (unpublished).

TABLE XVII
Comparison of Hormonal Activities in Binding Studies Using Radioiodinated Human FSH (*hFSH) and Chelonia FSH (*ChFSH) as Radioligand with Reptilian Gonadal Homogenates

Hormone preparation	Lizard testis *hFSH	Lizard testis *ChFSH	Snake testis *hFSH	Snake testis *ChFSH	*Trionyx* ovary *hFSH	*Trionyx* ovary *ChFSH	*Chelonia* ovary *hFSH	*Chelonia* ovary *ChFSH
Ovine FSH	0.54	0.55	0.72	0.40	0.61	0.43	0.35	0.33
Ovine LH	0.02	0.035	0.13	0.13	0.002	0.001	0.002	0.003
Chelonia FSH	0.11	0.16	0.15	0.25	0.045	0.035	0.10	0.20
Chelonia LH	0.36	0.30	0.15	0.33	0.11	0.10	0.03	0.05

Relative potency (XhFSH)[a] in four gonadal preparations

[a]In each binding assay, a highly purified preparation of human FSH (AFP-574C) was used as a reference. Values for activities with [^{125}I]hFSH as ligand are based on Licht and Midgley (1976a, b). Examples of data for binding with ^{125}I-labeled *Chelonia* FSH are shown in Figs. 25 and 26.

LH. Also, tests in our laboratory, as well as several others, showed that the binding of [^{125}I]hCG to mammalian testicular and ovarian tissues was highly specific for pituitary LH; i.e., the very low levels of competitive inhibition shown by FSH preparations were consistent with their trace LH contamination. We obtained 30–60% specific binding of [^{125}I]hCG to homogenates of ovarian tissue from hyper-pseudopregnant rats (primed with PMSG and then hCG).

We first examined the activities of various nonmammalian gonadotropins in competitive inhibition studies with rat ovarian tissues (Fig. 27). These and parallel studies made with radioiodinated human LH ([^{125}I]hLH) confirmed earlier observations in mammalian bioassay: namely, the nonmammalian hormones showed extremely low affinities for the LH binding sites in the mammal. Nonparallelism in competitive inhibition slopes for the nonmammalian and mammalian hormones prevented accurate quantification of activities, but frog, bird, and turtle hormones were clearly far less active than ovine LH (Fig. 27). When sufficiently high doses were tested, it was possible to demonstrate a significant (10-fold) difference between LH and FSH from each nonmammalian species (Fig. 27).

Much to our surprise, we were unable to demonstrate specific binding of [^{125}I]hCG to any of the nonmammalian tissues by either topical autoradiography or tests with homogenates. The total counts associated with tissue pellets represented only a few percent of the total counts added (the typical nonspecific

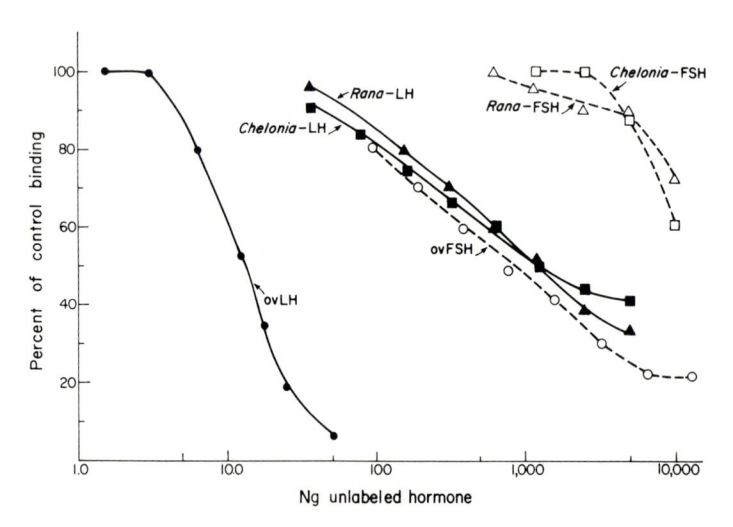

FIG. 27. Activities of ovine (ov), sea turtle (*Chelonia*), and bullfrog (*Rana*) gonadotropins in competition for the binding of ^{125}I-labeled human chorionic gonadotropin to ovarian homogenates from pseudopregnant rats. Data from P. Licht and A. R. Midgley, Jr. (unpublished).

level), and these could not be displaced by the addition of excess amounts of unlabeled gonadotropin. Although hCG is known to be physiologically active in reptiles and amphibians, we considered the possibility that it might behave abnormally in heterologous systems because of its nonpituitary origin. However, we were equally unsuccessful in demonstrating binding of [^{125}I]hLH to nonmammalian tissues by methods that demonstrated 5–10% specific binding to rat ovaries.

In an effort to see whether the lack of LH binding might have resulted from inadequate experimental design, we tested the [^{125}I]hCG and [^{125}I]hLH preparations for binding under a variety of conditions: tissues were minced instead of homogenized; incubations were performed in Krebs–bicarbonate Ringer solution in which hormonal stimulation of testosterone production was studied; temperature was varied over a wide range; homogenates were subjected to a variable number of washes in case an inhibitor were present; and *in vivo* binding was explored. Thus far, none of these modifications has resulted in the demonstration of specific binding. To test whether the reptilian tissues might have a high-affinity, low-capacity binding system for hCG, we experimented with an [^{125}I]hCG that was iodinated to a level about five times that normally used. This radioligand continued to show high binding with mammalian tissues, but still no appreciable binding was detected with the reptilian or amphibian tissues.

Even though the iodinated hLH and hCG bound to mammalian tissues, the possibility was considered that iodination might selectively damage the molecule at the sites involved in binding to nonmammalian tissues. Accordingly, several experiments were performed to determine whether binding of unlabeled mammalian LH could be demonstrated with reptilian or amphibian tissues. We utilized hCG and ovine LH with their known high biological potency in nonmammalian systems and measured binding by RIA techniques. Tissues were incubated with small amounts of the unlabeled hormone, and RIA was then used to determine whether the amount of hormone in the medium had decreased owing to uptake by tissue, and whether a comparable amount of bound hormone could be eluted from the tissue by moderate heat treatment. This protocol had proved effective in demonstrating binding of unlabeled hCG and human and rat LH to rat ovarian tissues (A. R. Midgley, Jr., unpublished). However, we again failed to show any appreciable binding of LH to the homogenates or minced segments of nonmammalian tissues.

Thus far, instead of providing further insights into the specificity and nature of FSH and LH binding sites in nonmammalian species, our studies with LH-like hormones have presented an enigma for understanding gonadotropin binding per se. We know that the various LH preparations tested are biologically active and, in some cases, almost equipotent to FSH. We know that they can express their biological effects under at least some of the *in vitro* and *in vivo* conditions where we could not demonstrate specific binding. Furthermore, we know that the failure to show binding is not due to the enzymic destruction of the [^{125}I]hCG by the reptilian tissues; [^{125}I]hCG continued to have normal levels of binding to rat ovary after preincubation with reptilian gonadal homogenates. We can only suggest that the physiological actions of LH in the nonmammalian tetrapods may not be associated with the same kind of binding characteristics that are observed with the mammalian tissues studied to date.

VII. Concluding Remarks

When we began our studies, a key issue in the evolution of vertebrate reproduction concerned the possibility that nonmammalian vertebrates might possess only a single pituitary gonadotropin. This question now appears resolved for the tetrapods: two chemically distinct gonadotropin molecules have been isolated from the pituitaries of representatives of each class of tetrapods. Thus, the existence of two separate gonadotropins may be considered a primitive characteristic of tetrapod vertebrates. The possibility still exists that certain groups (orders) may, in fact, possess only one functional gonadotropin; in particular, evidence for two is still lacking for squamate reptiles (lizards and snakes). The situation in birds is also not fully resolved; too few avian orders have been examined to evaluate the significance of findings for ducks. However, we are inclined to view these few exceptions with caution, since the apparent absence of a hormone may reflect a low pituitary content or biochemical changes in the molecule which render it undetectable by the few available immunological and physiological tests. If a second gonadotropin should definitely be found lacking in some tetrapods, it seems reasonable to interpret this as a secondary evolutionary loss rather than as a primitive characteristic.

The information accumulating on fish gonadotropins also has interesting evolutionary implications. Detailed discussion of these data is beyond the scope of the present report, but evidence derived from fractionation and physiological studies on several species of teleost is conflicting. Several fractionation studies point to the existence of a single gonadotropin with broad physiological actions (e.g., Donaldson *et al.*, 1972; Burzawa-Gerard and Fontaine, 1972; Burzawa-Gerard *et al.*, 1975; Breton *et al.*, 1976). However, other fractionation studies (Idler *et al.*, 1975a,b; Haider and Blüm, 1976) and numerous histochemical and cytological studies on the fish pituitary (reviewed by Reinboth, 1972; Schreibman *et al.*, 1973; Olivereau, 1976) indicate the presence of two chemically distinct gonadotropins in fish. Again, even if only one gonadotropin is found in teleosts, the possibility must be considered that this situation represents a derived rather than a primitive characteristic.

In addition to establishing the existence of two gonadotropins in tetrapods, a variety of chemical and physiological data support the contention that the two hormones are homologous to the FSH and LH of mammals. This conclusion is based on the parallelism between nonmammalian and mammalian hormones in biochemical characteristics (e.g., chromatographic behavior, amino acid composition, carbohydrate composition, and immunochemical properties) and biological profiles (e.g., *Anolis* and *Xenopus* assays). Preliminary studies on the subunits of LH provide direct evidence of a commonality of structure in reptilian and mammalian hormones. Unfortunately, the relationship between fish gonadotropins and the FSH and LH of tetrapods is still unclear (Papkoff *et al.*, 1977).

While the biochemical data suggest a degree of structural conservatism among different species of FSH and of LH, physiological studies point to important variations in the aspects of the molecules that are related to their biological activities. Such divergence is evidenced by the numerous examples of phylogenetic specificity; in fact, structural variations may be sufficiently great to render the hormone of one species virtually inactive in a heterologous one. A striking example of this type of specificity is provided by the general lack of activity exhibited by nonmammalian hormones in mammalian tissues. Similar situations also exist among the nonmammalian species (even between members of the same class), and mammalian hormones are not always fully potent when tested in nonmammalian species. Data from radioligand studies indicate that phylogenetic specificty in hormonal actions is related, in part, to differential affinities for gonadotropin binding sites on the target tissue.

Knowledge of this "incompatibility" among different species of tissue and hormone may ultimately provide a valuable tool for elucidating the relationships between hormone structure and function. Hopefully, as more detailed information becomes available on hormone structure (e.g., amino acid sequence), we will be able to relate the structural differences in FSH or LH from different species to their biological cross-reactivity. We already know from subunit studies that the biologically relevant differences reside in the structure of the β subunit. Since α subunits may show a high degree of interchangeability, structural similarities should exist at the sites of the subunit associated with binding to the β subunit. Sturctural overlap in the biologically active sites of the FSH and LH molecules is also indicated by several of the physiological and binding studies (in particular, the interchangeability of heterologous FSH and LH in a system that is otherwise specific for its own hormones). Thus, detailed comparisons of common structural features of the two types of gonadotropin should provide additional insights into features of the molecules that are important for expression of biological activity.

Another fundamental concern for understanding gonadotropin evolution is the question of which type of hormone (i.e., FSH or LH) regulates each aspect of gonadal function. The use of heterologous hormones alone in this type of research may lead to erroneous conclusions, since processes do not always show the same hormonal specificity for gonadotropins from all species; thus, it is especially important to have information on the actions of homologous hormones. Data now available on this subject strongly suggest that pronounced evolutionary changes have occurred in the gonadotropin specificity of several gonadal processes; i.e., each gonadotropin may not play the same physiological role in all tetrapods. In general, differences are evident at the level of the class, the Reptilia being the most notably different from all other classes.

There is little evidence for strict gonadotropin specificity in the early stages of either ovarian growth or spermatogenesis in amphibians, reptiles, or birds. Both FSH- and LH-like hormones (including PMSG and hCG) from heterologous and homologous sources appear to be capable of stimulating some development of gonadal growth in these classes. In contrast, ovulation, ovarian progesterone secretion and testicular androgen secretion may show a high degree of specificity. In amphibians, these three processes exhibit definite specificity for LH, of either heterologous or homologous origin. Thus, in this regard, the physiology of gonadotropins in amphibians resembles closely that seen in the few mammals studied. In marked contrast to this pattern, both *in vivo* and *in vitro* studies indicate that the reptilian gonads show little or no difference in their responsiveness to FSH and LH. If anything, when certain heterologous hormones were tested in some reptiles (e.g., mammalian hormones in turtles), FSH was considerably more potent than LH. Thus, although each class may possess two distinct pituitary gonadotropins, a clear duality of hormone action may not always be evident in gonadal regulation. In amphibians, we have little evidence for a unique action of FSH; whereas in reptiles, there are few data to demonstrate a unique action for either gonadotropin. Hopefully, current work related to endogenous hormone levels (e.g., RIA studies) will provide additional insights into whether the two gonadotropins might play different roles under normal *in vivo* conditions.

Thus far, the variations in gonadotropin specificity of ovulation and gonadal steroid production present the most striking example of evolutionary divergence in gonadotropin function. The difference between amphibians and reptiles in this regard appears to be a function of their gonadal receptors rather than of the biochemical characteristics of their gonadotropins, since the characteristic response of each class is essentially independent of the source of the hormone tested. The situation is less clear for birds and mammals since heterologous hormones often show more interchangeability in action than do the homologous ones. In connection with lack of LH specificity in reptiles, the broad actions of FSH on steroid secretion appear to be related to the modification of receptors on tissues normally associated with LH effects, rather than with a change in the tissue site of steroid secretion. This conclusion is supported for the male by the demonstration of both FSH binding to Leydig cells and by the FSH stimulation of testosterone secretion by Leydig cells; in the reptilian ovary, FSH and LH also stimulate progesterone by the same general type of follicular tissue. It is still unclear whether these tissues have one or two separate "receptors" for the two gonadotropins, but preliminary evidence from binding studies supports the possibility of a common receptor for FSH and LH in some reptiles. Resolution of this problem must await demonstration of LH binding.

Can the interspecific differences in the responses to gonadotropins be reconciled with the phylogeny of tetrapods to provide insight into the general pattern of gonadotropin evolution? We began our evolutionary considerations with a simplified scheme of vertebrate phylogeny indicating that reptiles are "intermediate" between amphibians and mammals or amphibians and birds (Fig. 1). However, this pattern certainly does not emerge from the comparative physiological data now available on the actions of FSH and LH in representatives from each class. The discrepancy is due, in part, to the failure of such a simplified scheme to recognize that considerable evolutionary change has occurred within each class as it diverged into the living orders now available for study (Fig. 28).

The similarity between urodele and anuran amphibians in LH specificity argues that this aspect of dual hormone control was probably a characteristic of the ancestral amphibian stock and hence a primitive tetrapod trait. Since the mammals still appear to reflect this same "primitive" form of gonadotropin regulation, it was presumably present in the early reptiles. An important feature

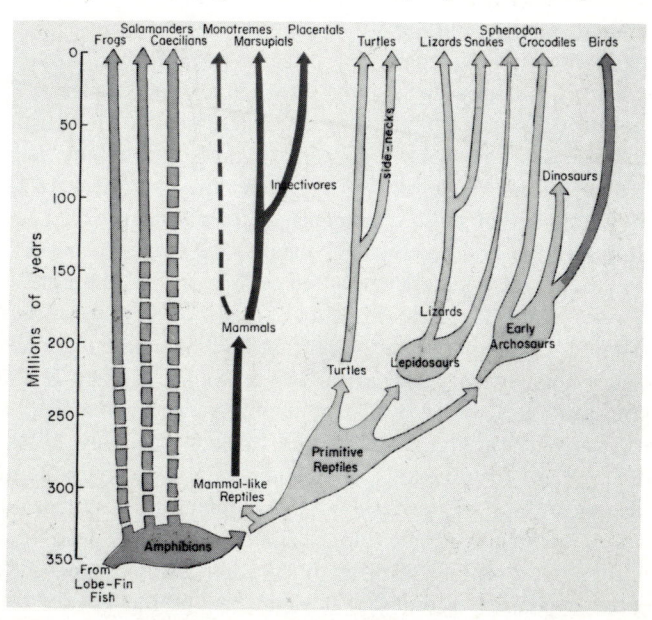

FIG. 28. Phylogeny of the living orders of amphibians, reptiles, birds, and mammals. Approximate times of divergence of stocks may be estimated from the scale on the vertical axis. Dashed portions of lineages indicate major gaps in the fossil record. Graph prepared by Dr. John Ruben.

of the phylogenetic scheme presented in Fig. 28 is that the mammalian lineage diverged from the reptilian stock very early in reptilian evolution, probably before the divergence into the modern orders of reptiles. Since the three orders of reptiles examined appear to be essentially alike in lacking LH specificity, it seems likely that their change from the primitive condition occurred relatively early, perhaps only shortly after their divergence from the common stock from which the mammalian lineage arose. The present reptilian condition can be explained by a single event that occurred in their common ancestor, e.g., in the captorhinids.

The birds would be expected to show a predominantly reptilian (especially crocodilian) pattern of gonadotropin specificity (Fig. 28); i.e., FSH should exhibit broad physiological actions. Unfortunately, the limited data for birds do not allow conclusions about their LH specificity. Birds differ from reptiles in their responses to mammalian hormones (FSH is less potent for steroid production in birds), but data for homologous avian gonadotropins are conflicting. It is possible that gonadal receptors in birds have diverged from those of their reptilian ancestors; in fact, they may have secondarily converged with the more primitive pattern. Unfortunately, we still have little basis on which to speculate about the adaptive significance of the various evolutionary changes that may have occurred in gonadotropin function.

Several problems and directions for future comparative studies on gonadotropin physiology are suggested by the information now available. First, more comparative information is required for mammals, including both eutherian and noneutherian species. The problem of reduced gonadotropin specificity that was frequently observed with heterologous hormones may be of special importance in view of the fact that most mammalian studies have been based on a few experimental species. Broad comparative data now available on gonadotropin function may be useful when evaluating the actions of FSH and LH in mammals; these data give us an idea of the potential breadth of actions of each gonadotropin. It should not be surprising to find that FSH may have intrinsic actions in mammals once thought to be characteristic of only LH, when it is recognized that the two hormones may show almost complete overlap in activity in another group that is derived from the same ancestral stock as mammals.

A second problem for all comparative studies relates to the potential importance of clearance rates for the different gonadotropins. As has been recognized in studies on mammals, differential clearance rates may lead to erroneous conclusions about the actions of different hormones when tested by chronic *in vivo* injection. Evidence suggests that this may be a problem for comparing FSH and LH in nonmammals and that the magnitude of the problem may vary with the source of the hormone. Finally, as more laboratories independently begin to prepare and study nonmammalian hormones, it is essential that some form of standardization of preparations be accomplished.

ACKNOWLEDGMENTS

The authors wish to express their gratitude to the large number of persons who assisted in many ways during the course of the investigations presented herein. Barbara Licht provided invaluable advice in preparing and editing this manuscript. Figures were prepared by E. Reid; H. Meakin, C. Hopkins, J. D. Nelson, K. Hoey and J. Knorr provided technical assistance in the laboratory. The acquisition of the numerous species of pituitary gland on which our studies were based could not have been accomplished without the help of L. McNease, T. Joanen, K. Chiu, F. Rose, G. Megdon, S. D. Bradshaw, W. H. Burke, C. S. Nicoll, A. V. Nalbandov, J. Reeves, T. Hayashida, and B. Wentworth. Additional purified hormones were generously supplied by A. F. Parlow, J. Pierce, L. E. Reichert, Jr., and the hormone distribution program of the NIH. A. R. Midgley, Jr., D. MacKenzie, E. Daniels, A. Bona, B. Goldman, and R. Tokarz allowed us to refer to unpublished data. We thank C. H. Li for his support and encouragement.

REFERENCES

Andrieux, B., and Collenot, A. (1970). *Ann. Endocrinol.* **31**, 531.
Andrieux, B., Collenot, A., Collenot, G., and Pergrale, C. (1973). *Ann. Endocrinol.* **34**, 711.
Atz, E. H., and Pickford, G. E. (1954). *Zoologica (N.Y.)* **39**, 117.
Barr, W. A. (1969). *In* "Perspectives in Endocrinology: Hormones in the Lines of Lower Vertebrates" (E. J. W. Barrington and C. A. Jorgensen, eds.), pp. 164–238. Academic Press, New York.
Basu, S. L. (1964). *Folia Biol. (Krakow)* **12**, 203.
Basu, S. L. (1969). *Gen. Comp. Endocrinol., Suppl.* **2**, 203.
Basu, S. L., and Nandi, J. (1965). *J. Exp. Zool.* **159**, 93.
Bell, J., Canfield, R. E., and Sciarra, J. J. (1969). *Endocrinology* **84**, 298.
Breneman, W. R., Zeller, F. J., and Creek, R. O. (1962). *Endocrinology* **71**, 790.
Breton, B., Jalabert, B., and Reinaud, P. (1976). *Ann. Biol. Anim., Biochim., Biophys.* **16**, 25.
Brown, N. L., Boyle, J.-D., Scanes, C. G., and Follett, B. K. (1975). *Cell Tissue Res.* **156**, 499.
Burgos, M. H., and Ladman, A. J. (1957). *Endocrinology* **61**, 20.
Burns, J. M., and Richards, J. S. (1974). *Comp. Biochem. Physiol. A* **47**, 655.
Burzawa-Gerard, E. (1975). *C. R. Hebd. Seances Acad. Sci., Ser. D* **279**, 1681.
Burzawa-Gerard, E., and Fontaine, Y. A. (1972). *Gen. Comp. Endocrinol., Suppl.* **3**, 715.
Burzawa-Gerard, E., and Fontaine, Y. A. (1976). *C. R. Seances Soc. Biol. Ses Fil.* **282**, 97.
Burzawa-Gerard, E., Goncharov, B. F., and Fontaine, Y. A. (1975). *Gen. Comp. Endocrinol.* **27**, 296.
Callard, I. P., and Ziegler, H. (1970). *J. Endocrinol.* **47**, 131.
Callard, I. P., Callard, G. V., Anderson, C. K., and Eccles, S. S. (1975). *J. Endocrinol.* **65**, 9P (abstr.).
Callard, I. P., McChesney, I., Scanes, C., and Callard, G. V. (1976). *Gen. Comp. Endocrinol.* **28**, 2.
Castro, A. F., Alonso, A., and Mancini, R. E. (1972). *J. Endocrinol.* **52**, 129.
Chan, S. W. C., and Callard, I. P. (1974). *J. Endocrinol.* **62**, 267.
Combescot, C. (1955). *C. R. Seances Soc. Biol. Ses. Fil.* **149**, 1969.
Combescot, C. (1958). *C. R. Seances Soc. Biol. Ses Fil.* **152**, 1077.
Connell, G. M., and Eik-Nes, K. B. (1968). *Steroids* **12**, 507.
Connell, G. M., Connell, C. J., and Eik-Nes, K. B. (1966). *Gen. Comp. Endocrinol.* **7**, 158.
Crews, D., and Licht, P. (1975a). *Gen. Comp. Endocrinol.* **27**, 71.

Crews, D., and Licht, P. (1975b). *Gen. Comp. Endocrinol.* **27**, 553.
Croix, D., Hendrick, J. C., Ballhazart, J., and Franchimont, P. (1974). *C. R. Seances Soc. Biol. Ses Fil.* **168**, 136.
Currie, C., and Taylor, H. L. (1970). *J. Morphol.* **132**, 101.
Daniels, E., Licht, P., Farmer, S. W., and Papkoff, H. (1977). *Gen. Comp. Endocrinol.* (in press).
Delsol, M., Burzawa-Gerard, E., Flatin, J., Fontaine, Y. A., and Leray, C. (1970). *Ann. Endocrinol.* **31**, 481.
Desjardins, C., Zeleznik, A. J., Midgley, A. R., Jr., and Reichert, L. E., Jr. (1975). *In* "Hormone Binding and Target Cell Activation in the Testes" (M. L. Dufau and A. R. Means, ed.), pp. 221–235. Plenum, New York.
Dodd, J. M., and Wiebe, J. P. (1968). *Arch. Anat., Histol. Embryol.* **51**, 157.
Donaldson, E. M., Yamazaki, F., Dye, H. M., and Philleo, W. W. (1972). *Gen. Comp. Endocrinol.* **18**, 469.
Dufau, M. L., Catt, K. J., and Tsuruhara, T. (1972). *Endocrinology* **90**, 1032.
Emmerson, B., and Kjaer, K. (1974). *Gen. Comp. Endocrinol.* **22**, 261.
Ewing, L. L., Brown, B., Irby, D. C., and Jardine, I. (1975). *Endocrinology* **96**, 610.
Eyeson, K. N. (1971). *Gen. Comp. Endocrinol.* **16**, 342.
Farmer, S. W., Papkoff, H., and Licht, P. (1975). *Biol. Reprod.* **12**, 415.
Farmer, S. W., Papkoff, H., and Hayashida, T. (1976). *Endocrinology* **99**, 692.
Follett, B. K. (1976). *J. Endocrinol.* **69**, 117.
Follett, B. K., and Redshaw, M. R. (1974). *In* "Physiology of the Amphibia" (B. Lofts, ed.), Vol. 2, pp. 219–308. Academic Press, New York.
Follett, B. K., Scanes, C. G., and Cunningham, F. J. (1972). *J. Endocrinol.* **52**, 359.
Furr, B. J. A., and Cunningham, F. J. (1970). *Br. Poult. Sci.* **11**, 7.
Furuya, T., and Ishii, S. (1974). *Endocrinol. Jpn.* **21**, 329.
Gerrard, A. M., Jones, R. E., and Roth, J. J. (1973). *J. Morphol.* **141**, 227.
Godden, P., Scanes, C. G., and Sharp, P. J. (1975). *J. Endocrinol.* **67**, 20P (abstr.).
Godden, P. M. M., and Scanes, C. G. (1975). *Gen. Comp. Endocrinol.* **27**, 538.
Goldfine, I. D., Amir, S. M., Peterson, A. W., and Ingbar, S. H. (1974). *Endocrinology* **95**, 1228.
Gray, W. R. (1964). *In* "Methods in Enzymology" (C. H. W. Hirs, ed.), Vol. 11, p. 469. Academic Press, New York.
Guha, K. K. (1976). *Gen. Comp. Endocrinol.* **29**, 278 (abstr.).
Haider, S. G., and Blüm, V. (1976). *Gen. Comp. Endocrinol.* **29**, 251 (abstr.).
Hall, P. F., and Eik-Nes, K. B. (1962). *Biochim. Biophys. Acta* **63**, 411.
Hartree, A. S. (1966). *Biochem. J.* **100**, 754.
Hartree, A. S., and Cunningham, F. J. (1969). *J. Endocrinol.* **43**, 609.
Idler, D. R., Bazar, L. S., and Hwang, S. J. (1975a). *Endocr. Res. Commun.* **2**, 215.
Idler, D. R., Hwang, S. J., and Bazar, L. S. (1975b). *Endocr. Res. Commun.* **2**, 237.
Imai, K. (1973). *J. Reprod. Fertil.* **33**, 91.
Ishii, S., and Furuya, T. (1975). *Gen. Comp. Endocrinol.* **25**, 1.
Iwasawa, H., Michibata, H., and Satoh, N. (1973). *Sci. Rep. Niigata Univ., Ser. D* **10**, 71.
Jalali, S., Arslan, M., and Qazi, M. H. (1975). *Islamabad J. Sci.* **2**, 10.
Jones, R. E. (1973). *Gen. Comp. Endocrinol.* **20**, 567.
Jones, R. E. (1975). *Gen. Comp. Endocrinol.* **25**, 211.
Jones, R. E., Roth, J. J. Gerrard, A. M., and Kiely, R. G. (1973). *Gen. Comp. Endocrinol.* **20**, 190.
Jones, R. E., Tokarz, R. R., and LaGreek, F. T. (1975). *Gen. Comp. Endocrinol.* **26**, 354.

Jørgensen, C. B. (1973). In "The Development and Maturation of the Ovary and its Functions" (H. Peters, ed.), pp. 132–151. Excerpta Med. Found., Amsterdam.
Kasinathan, S., and Basu, S. L. (1973). *Acta Morphol. Acad. Sci. Hung.* 21, 249.
Li, C. H., and Starman, B. (1964). *Nature (London)* 202, 291.
Licht, P. (1970). *Gen. Comp. Endocrinol.* 14, 98.
Licht, P. (1972a). *Gen. Comp. Endocrinol.* 19, 273.
Licht, P. (1972b). *Gen. Comp. Endocrinol.* 19, 282.
Licht, P. (1972c). *Gen. Comp. Endocrinol., Suppl.* 3, 477.
Licht, P. (1973a). *Gen. Comp. Endocrinol.* 20, 522.
Licht, P. (1973b). *Gen. Comp. Endocrinol.* 20, 592.
Licht, P. (1974a). *Chem. Zool.* 9, 399.
Licht, P. (1974b). *Gen. Comp. Endocrinol.* 22, 463.
Licht, P. (1975). *Comp. Biochem. Physiol. A* 50, 221.
Licht, P., and Crews, D. (1975). *Gen. Comp. Endocrinol.* 25, 467.
Licht, P., and Crews, D. (1976). *Gen. Comp. Endocrinol.* 29, 141.
Licht, P., and Hartree, A. S. (1971). *J. Endocrinol.* 53, 329.
Licht, P., and Midgley, A. R., Jr. (1976a). *Biol. Reprod.* 15, 195.
Licht, P., and Midgley, A. R., Jr. (1976b). *Gen. Comp. Endocrinol.* 30, 364.
Licht, P., and Midgley, A. R., Jr. (1977). *Biol. Reprod.* 16, 117.
Licht, P., and Papkoff, H. (1971). *Gen. Comp. Endocrinol.* 16, 586.
Licht, P., and Papkoff, H. (1972). *Gen. Comp. Endocrinol.* 19, 102.
Licht, P., and Papkoff, H. (1973). *Gen. Comp. Endocrinol.* 20, 172.
Licht, P., and Papkoff, H. (1974a). *Gen. Comp. Endocrinol.* 22, 218.
Licht, P., and Papkoff, H. (1974b). *Endocrinology* 94, 477.
Licht, P., and Papkoff, H. (1974c). *Gen. Comp. Endocrinol.* 23, 415.
Licht, P., and Papkoff, H. (1974d). In "Gonadotropins and Gonadal Function" (N. R. Mougdal, ed.), pp. 101–114. Academic Press, New York.
Licht, P., and Papkoff, H. (1976). *Gen. Comp. Endocrinol.* 29, 552.
Licht, P., and Pearson, A. K. (1969). *Gen. Comp. Endocrinol.* 13, 367.
Licht, P., and Tsui, H. W. (1975). *Biol. Reprod.* 12, 346.
Licht, P., Papkoff, H., Goldman, B. D., Follett, B. K., and Scanes, C. G. (1974). *Gen. Comp. Endocrinol.* 24, 168.
Licht, P., Farmer, S. W., and Papkoff, H. (1975). *Life Sci.* 17, 1049.
Licht, P., Muller, C., and Tsui, H. W. (1976a). *Biol. Reprod.* 14, 194.
Licht, P., Farmer, S. W., and Papkoff, H. (1976b). *Biol. Reprod.* 14, 222.
Lofts, B. (1961). *Gen. Comp. Endocrinol.* 1, 179.
Lofts, B. (1972). *Gen. Comp. Endocrinol., Suppl.* 3, 636.
Lofts, B. (1974). In "Physiology of the Amphibia" (B. Lofts, ed.), Vol. 2, pp. 107–218. Academic Press, New York.
Lofts, B., and Murton, R. K. (1973). In "Avian Biology" (D. S. Farmer and J. R. King, eds.), Vol. 3, pp. 1–108. Academic Press, New York.
Lofts, B., Murton, R. K., and Thearle, R. J. P. (1973). *Gen. Comp. Endocrinol.* 21, 202.
Maung, Z. W. (1976). *Gen. Comp. Endocrinol.* 29, 254. (Abstr. No. 37).
Midgley, A. R., Jr. (1973). In "Receptors for Reproductive Hormones" (B. W. O'Malley and A. R. Means, eds.), pp. 365–378. Plenum, New York.
Mitchell, M. E. (1970). *J. Reprod. Fertil.* 22, 233.
Moore, F. L. (1974). Ph.D. Thesis, University of Colorado, Boulder.
Moore, F. L. (1975). *Gen. Comp. Endocrinol.* 26, 525.
Moore, F. L., and Norris, D. O. (1973). *J. Colo.-Wyo. Acad. Sci.* 7, 39.

Muller, C. H. (1975). *Soc. Study Reprod., Annu. Meet., 8th* Abstract, p. 95.
Muller, C. H. (1976a). *Am. Zool.* 16, 259(abstr.).
Muller, C. H. (1976b). *Program Abstr., Endocrinol. Soc. Annu. Meet., 58th* Abstract, p. 268.
Nalbandov, A. V. (1976). "Reproductive Physiology of Mammals and Birds," 3rd ed. Freeman, San Francisco, California.
Olivereau, M. (1976). *Gen. Comp. Endocrinol.* 28, 82.
Owens, D. W., Hendrickson, J. R., Lance, V., and Callard, I. P. (1976). *Am. Zool.* 16, 258 (Abstr. No. 438).
Panigel, M. (1956). *Ann. Sci. Nat., Zool., Biol. Anim.* [11] 18, 569.
Papkoff, H., and Li, C. H. (1958). *Biochim. Biophys. Acta* 29, 145.
Papkoff, H., and Samy, T. S. A. (1967). *Biochim. Biophys. Acta* 147, 175.
Papkoff, H., Li, C. H., and Liu, W.-K. (1962). *Arch. Biochem. Biophys.* 96, 216.
Papkoff, H., Gospodarowicz, D., Candiotti, A., and Li, C. H. (1965). *Arch. Biochem. Biophys.* 111, 431.
Papkoff, H., Sairam, M. R., Farmer, S. W., and Li, C. H. (1973). *Recent Prog. Horm. Res.* 29, 563.
Papkoff, H., Farmer, S. W., and Licht, P. (1976a). *Endocrinology* 98, 767.
Papkoff, H., Farmer, S. W., and Licht, P. (1976b). *Life Sci.* 18, 245.
Papkoff, H., Farmer, S. W., and Licht, P. (1977). *Excerpta Med. Found. Int. Congr. Ser.* 403, 77.
Parlow, A. F. (1961). *In* "Human Pituitary Gonadotropins" (A. Albert, ed.), p. 300. Thomas, Springfield, Illinois.
Pesonen, S., and Rapola, J. (1962). *Gen. Comp. Endocrinol.* 2, 425.
Pierce, J. G., Liao, T., Howard, S. M., Shome, B., and Cornell, J. S. (1971). *Recent Prog. Horm. Res.* 27, 165.
Pisanó, A., and Burgos, M. H. (1962). *Acta Physiol. Lat. Am.* 12, 51.
Pisanó, A., and Burgos, M. H. (1971). *Gen. Comp. Endocrinol.* 16, 176.
Pizarro, N., and Burgos, M. H. (1963). *Gen. Comp. Endocrinol.* 3, 644.
Rajaniemi, H. J., and Midgley, A. R., Jr. (1975). *In* "Methods in Enzymology" (J. G. Hardman and B. W. O'Malley, eds.), Vol. 37, Part B, p. 145. Academic Press, New York.
Ramachandran, J., and Sairam, M. R. (1975). *Arch. Biochem. Biophys.* 167, 294.
Rastogi, R. K., Iela, L., Saxena, P. K., and Chieffi, S. (1976). *J. Exp. Zool.* 196, 151.
Reddy, P. R. K., and Prasad, M. R. N. (1971). *Gen. Comp. Endocrinol.* 16, 288.
Redshaw, M. R. (1972). *Am. Zool.* 12, 289.
Redshaw, M. R., and Nicholls, T. J. (1971). *Gen. Comp. Endocrinol.* 16, 85.
Reinboth, R. (1972). *Am. Zool.* 12, 307.
Roos, J., and Jørgensen, C. B. (1974). *Gen. Comp. Endocrinol.* 23, 432.
Russo, J., and Burgos, M. H. (1969). *Gen. Comp. Endocrinol.* 13, 185.
Ryle, M. (1972). *J. Reprod. Fertil.* 30, 395.
Saidapur, S. K., and Nadkarni, V. B. (1973). *Gen. Comp. Endocrinol.* 21, 225.
Saidapur, S. K., and Nadkarni, V. B. (1975). *Indian J. Exp. Biol.* 13, 432.
Sairam, M. R., and Papkoff, H. (1974). *Hand. Physiol., Sect. 7: Endocrinol.* 4, Part 2, 111.
Scanes, C. G., Follett, B. K., and Goos, H. J. (1972). *Gen. Comp. Endocrinol.* 19, 596.
Schreibman, M. P., Leatherland, J. F., and McKeown, B. A. (1973). *Am. Zool.* 13, 719.
Schuetz, A. W. (1972). *In* "Oogenesis" (J. D. Biggers and A. W. Schuetz, eds.), pp. 479–511. Univ. Park Press, Baltimore, Maryland.
Schuetz, A. W. (1974). *Biol. Reprod.* 10, 150.
Shahabi, N. A., Bahr, J. M., and Nalbandov, A. V. (1975). *Endocrinology* 96, 969.

Shome, B., and Parlow, A. F. (1974a). *J. Clin. Endocrinol. Metab.* **39**, 199.
Shome, B., and Parlow, A. F. (1974b). *J. Clin. Endocrinol. Metab.* **39**, 203.
Simon, N., and Reinboth, R. (1966). *Verh. Dtsch. Zool. Ges.* **30**, Suppl., 254.
Snyder, B. W., and Biggers, J. D. (1975). *Gen. Comp. Endocrinol.* **27**, 465.
Snyder, B. W., and Schuetz, A. W. (1973). *J. Exp. Zool.* **188**, 333.
Spackman, D. H., Stein, W. H., and Moore, S. (1958). *Anal. Chem.* **30**, 1190.
Steelman, S. L., and Pohley, F. M. (1953). *Endocrinology* **53**, 604.
Thornton, V. F. (1971). *Gen. Comp. Endocrinol.* **16**, 599.
Tsui, H. W. (1976a). Ph.D. Thesis, University of California, Berkeley.
Tsui, H. W. (1976b). *Gen. Comp. Endocrinol.* **28**, 386.
Tsui, H. W., and Licht, P. (1974). *Gen. Comp. Endocrinol.* **22**, 277.
Tsui, H. W., and Licht, P. (1977). *Gen. Comp. Endocrinol.* (in press).
van Oordt, P. G. W. J. (1960). *Symp. Zool. Soc. London* **2**, 29.
van Oordt, P. G. W. J., and Brands, F. (1970). *J. Endocrinol.* **48**, 1 (abstr.).
van Oordt, P. G. W. J., and de Kort, E. J. (1969). *Colloq. Int. C. N. R. S.* **177**, 345.
Vellano, C., Sacerdote, M., and Mazzi, V. (1974). *Monit. Zool. Ital.* **8**, 177.
Wentworth, B. C. (1971). *Biol. Reprod.* **5**, 107 (abstr.).
Wentworth, B. C., Burke, W., and Birrenkott, G. P. (1976). *Gen. Comp. Endocrinol.* **29**, 119.
Werner, J. K. (1969). *Copeia* p. 592.
Wiebe, J. P. (1970). *J. Endocrinol.* **47**, 439.
Wiebe, J. P. (1972a). *Science* **175**, 67.
Wiebe, J. P. (1972b). *Gen. Comp. Endocrinol., Suppl.* **3**, 626.
Wood, K. R., and Wang, K. T. (1967). *Biochim. Biophys. Acta* **133**, 369.
Zeleznik, A. J., Midgley, A. R., Jr., and Reichert, L. E., Jr. (1974). *Endocrinology* **95**, 818.

DISCUSSION

A. Segaloff: I wonder how extensively you have tested both pH specificity and temperature specificity for the receptor assay, because the alligator has the most tremendous changes in pH of almost any animal. The "alkaline tide" in an alligator is absolutely unbelievable, and I wonder how much this might affect apparent receptor levels.

P. Licht: We have not yet been able to do much receptor work on the alligator itself, but we have examined all these parameters in the other reptiles, and the reptilian tissue behaves just like the mammalian tissue. For example, if you change the temperature you get reduced receptor binding and so on. But I do not think that this could account for our inability to demonstrate LH binding because we have even tried it *in vivo*, where presumably the animal was taking care of itself, and we could not show appreciable binding of LH.

J. Vaitukaitis: In your hybridization studies with complementary turtle and sheep subunits, did the lack of biological activity observed in some assay systems reflect the fact that the subunits did not recombine, or that they recombined but did not interact with the receptor? Second, would you comment on whether, in those nonmammalian species in which you found enhanced biological activity with human gonadotropins, the activity was due to increased plasma half-lives of those hormones in the reptilian species?

P. Licht: It was the LH subunits with which we were working. I think we can conclude that the two kinds of hybrids molecules did both combine, because if you examine the three different kinds of assays, you see that one of the hybrids was fully potent in one assay whereas the other hybrid was highly potent in the other assay, so both types of interspecific mixtures must have recombined. The fact that they recombined is also shown by the observation that if you add the two subunits separately to the tissue without previous incubation under conditions to give recombination, the subunit mixture is inactive.

As to your second question: There are enormous differences in the plasma half-life of these different hormones in the reptile. We have directly measured the half-life of several FSHs and LHs, and indeed we find as a rule that FSHs have very long half-lives relative to LHs. I think this difference explains why, from most of our *in vivo* studies, we concluded that LH was so inactive. LH is quite potent in most reptiles (especially squamates) in terms of its ability to interact with the receptor, but when you use ordinary *in vivo* methods with chronic injections, given once every 24 or 48 hours, the LH appears to be inactive because it has such a short half-life. Also the difference species of FSH have different half-lives, and that is probably why one is more potent than another. I think, therefore, that many cases of what appears to be species specificity *in vivo* are artifacts of differences in half-life. Nonetheless, I think the concept of species specificity is still very valid, because as you see from the *in vitro* studies, where half-life is presumably less a problem, that it still shows up quite dramatically.

J. Vaitukaitis: Did you subject the hybridized subunits to gel filtration?

P. Licht: No, we have not yet tried to chemically characterize the subunits in this manner.

J. Vaitukaitis: What proof do you have that the subunits hybridized?

P. Licht: I think the proof is the fact that we created a combination that showed enormous increases in biological activity that did not exist if we just took the two subunits and put them in the test tube at the time of the assay. After incubating the two subunits overnight at 37°C at 1 mg/ml, we ended up with a material that was far more potent than the two subunits individually. Unfortunately, we have only very minute amounts of these subunits at the moment. But, we now have a new batch of sea turtle glands and hope to prepare more for further chemical studies.

J. Furth: Regarding your evolutionary discussion of gonadotropins with other hormones, you mentioned growth hormone. The relationship of gonadotropins–in mammals most evident with thyrotopins that you did not mention–TSH has built-in gonadotropin property, which is biologically and immunologically demonstrable. This relationship was well studied; the TSH subunits were separated and the reactive group is commonly named as alpha, and the specific group as beta. Thyrotropic tumors of mice have invariably marked gonadotropic activity. In your use, does the term alpha refer to the group-reacting subunit and a beta to the specific subunit?

P. Licht: I did not mention TSH simply because I did not have time. We are very interested in the TSH. One of the reasons why we have not been able to make as much progress in working with TSH is that of the confusion with gonadotropin. For example, for the last year we have been trying to bioassay for TSH in a test I have in turtles (thyroidal ^{125}I-uptake), but we kept finding that our LH from the frog, for example, is an extremely potent TSH; we even thought that maybe there are not even two separate molecules. However, in the last few weeks, we have started some assays of LH and TSH in the frog itself, and frog LH is not thyrotropic in the frog. We are again dealing with a problem heterologous hormones–the LH of one species may act like a TSH when injected into another species. Alternatively, is the TSH like LH? Is it gonadotropic? So far in our studies I have injected purified mammalian TSH into many animals and it does not act like an LH, it acts only like a TSH. We hope ultimately to get to this subunit question you are talking about when we can purify some TSHs, but we have just not come that far.

J. Furth: The Leydig cell tumors are induced by estrogens. Upon transplantation, at first they masculinize, and later as carried on, they can feminize. So the evidence is that the same cell can produce both types of hormones, androgens and estrogens. This is much in line with your concept.

P. Licht: I no longer really understand what LH and FSH do in mammals, and the reason is that every time I pick up a journal like *Endocrinology*, someone is reporting that FSH is doing something it is not supposed to do. Of course, this pleases me no end, because here we have a whole group of really close relatives—the reptiles—which are doing all kinds of remarkable things with FSH, and maybe we should not be so surprised when a mammal uses FSH to influence ovulation or to influence steroid production. It could be that it is a "dormant" character in many of the vertebrates, and what these groups differ in is simply how they exaggerate each one of these functions.

R. E. Frisch: I wonder whether around the normal time of first ovulation in these animals you noted changes in the deposition of carcass fat or gonadal fat. Also, are there any differences among groups in this?

P. Licht: There are very dramatic seasonal cycles in the carcass fat related to the reproductive cycle. It was actually one of the things that I was studying before I began working on the hormones. In both sexes and in most reptiles that are studied, there is a very nice inverse correlation between reproductive activity and carcass fat. When the testis or ovaries are enlarged, fat bodies tend to disappear.

P. G. Condliffe: Dr. Furth has already brought up the gonadotropic effects of TSH. Many years ago Fontaine discovered the bovine heterotropic factor (HTF) while he was working in my laboratory. In just this fraction, which turned out to be FSH, it had a strong thyrotropic action. Have you any further data regarding TSH from your reptiles? Are there any particular problems other than the bioassay in isolating them? When Fontaine and I isolated TSH we found it to be free of LH by the frog spermiation test. So at least one teleost, the eel, makes a quite specific TSH without LH-like activity.

P. Licht: We have had many problems with the TSHs, as I mentioned earlier. This is not new to the nonmammalian species. Even in the mammals, it is difficult to separate LH from TSH because of their chemical similarity and chemical behavior. We have some evidence now that there is species specificity. Many of the standard bioassays, for example, ^{32}P-uptake by the chick thyroid is very insensitive to amphibian TSH, and I meant to refer to your work with Fontaine with the heterothyrotropin. I think we also have cases of heterothyrotropin now; for example, frog LH may be heterothyrotropic; when injected into a turtle it acts like a TSH. This is what has really plagued us in fractionation, since you may need a bioassay for every species of TSH, and you have to rigorously show that it is a specific TSH assay before you begin.

S. Korenman: Receptor concentration is probably regulated, and in embryological development it appears that hormone and receptor fortuitously turn up at the same time. For the first time I heard from you a suggestion that there was an independent evolution of receptor and hormone. I wonder whether you would comment on the evolutionary consequences of such a concept.

P. Licht: I think there is remarkably little information on which to speculate wisely about this. I am not sure how independent hormone and receptor can be. Obviously the receptor cannot get away from its hormone or it will no longer be a receptor for that hormone. First of all, many of our data indicate that there is a considerable structural similarity between an FSH and an LH. In other words, the two molecules share many chemical properties, and what the receptor might be doing is simply enlarging the area of the molecule that it is responding to. The broader that area, the more likely it is, then, to be responding to parts of the alternate molecule. I do not see why there cannot be changes in the receptor without the hormone ever changing as long as the "fit" does not break down too much.

B. F. Rice: Perhaps I might be able to add some support to Dr. Licht's concepts and answer Dr. Korenman's question. I would also like to ask Dr. Licht for his comments. I find the concept of the evolution of the receptor very intriguing and have some data to present to you. After many years of experience with incubating fresh human corpeus luteum, one dictum has always stood up, that the human corpus luteum will respond only to human gonadotropins, human LH and human chorionic gonadotropin, and bind these two gonadotropins at the LH (hCG) receptor. However, over the past couple of years, because of certain logistical problems, we have collected human corpora lutea and have frozen them in buffer. If one now takes these long-term frozen luteal tissues and studies receptor binding, these preparations that in the past bound only human gonadotropins, will now bind ovine bovine LH, and porcine LH, and even rat LH. Although species specificity was lost, hormone specificity was not—only other species of LH bound; FSH, TSH, growth hormone, prolactin (human and nonhuman) did not. This suggests that there is retention of certain common basic aspects of receptor binding and that the human corpus luteum is only slightly modified from lower species. Our first impression was that this was an artifact, and reiterated the dictum that if one wants to study the human corpus luteum it should be studied in the fresh state. On the basis of your studies, I think this may very well have much more importance on an evolutionary basis.

P. Licht: What this intriguing observation may be telling us is that this binding site or receptor is not a fixed static structure, but can actually be labile, depending on its environmental conditions. This would make it much easier I think to imagine it evolving. If it is capable even in a human being of existing in different states, with regard to "recognizing" other species of hormone, then it could be a fairly simple step in evolution to just "open it up" a little bit, or perhaps we are seeing the removal of some inhibitor or part of the receptor.

N. Schwartz: You have commented on some redundancies in the mammal between FSH and LH with respect to certain functions. There is good evidence now that FSH can cause ovulation, and progesterone secretion, even in the presence of anti-β-LH. It may seem a waste to have two hormones doing the same thing, although a certain amount of redundancy can be useful. But I think the recent data in mammals that indicate that *control* of FSH and LH is independent suggest that one should look at control of rate of secretion, not just tissue or organ action, in order to put the system back together and understand it. FSH is not always secreted at the same time as LH, so that it is not truly redundant, but is performing other kinds of things because different structures are present at different times, in either the ovary or the testis. Do you have any information in the nonmammalian species on control of secretion rates of FSH and LH that may help put together some of this apparently discrepant data on action?

P. Licht: I think that is an extremely critical point. I probably should have prefaced my remarks to the effect that what we have talked about here is the potential capabilities of each hormone. Whether or not a hormone does something in our *in vitro* system or, when we inject it, may have nothing to do with its normal physiological role. I think you are absolutely correct that we should not conclude that LH and FSH are necessarily doing the same thing in a reptile, for example, simply because they are capable of doing it. Perhaps some answers to your questions will be forthcoming soon. A number of groups working with bird hormones have developed nice radioimmunoassays and are following FSH and LH, but as you might expect, so far, when you get one hormone you get the other. We have some very useful antisera against the frog and turtle hormones, and we are now in the process of studying this very aspect. I somehow doubt, if I had to make a prediction, that we are going to find the answer so simply. In any case, what is wrong with redundancy? It, in fact, is the rule in biological systems. Every neurophysiological, every physiological system studied has some redundancy built in; it is a protective device in a way, so perhaps it is real.

S. Farmer: Because of the precious nature of the pituitaries that we have been able to obtain, we have used fractionation procedures designed to purify all the mammalian hormones. We have had considerable success in purifying and characterizing growth hormone as well as the gonadotropins from these species and have begun studies with prolactin and ACTH. These studies [*Endocrinology* **95,** 1560 (1974); **99,** 692 (1976)] have shown that there has also been considerable conservation of the growth hormone structure in a variety of nonmammalian species, and that the growth hormones are clearly distinct from the gonadotropins in each species.

I. Callard: Your hypothesis that mammals have retained a primitive tetrapod condition of LH specificity for steroidogenesis is attractive, but essentially it hinges on observations or assumptions that LH is steroid specific in amphibians and mammals. However, there is a good deal of evidence from hamster and rat granulosa cell preparations and the rodent testis that both FSH and LH are steroidogenic. I wonder whether, when we look more carefully at the amphibians, the "LH specificity" will continue to hold, or whether in this group also both hormones will be steroidogenic. Possibly we will ultimately conclude that both hormones are steroidogenic throughout the vertebrates.

I am also concerned about the *in vivo* versus *in vitro* situation. We, as you know, have used *in vitro* systems a great deal. Using ovine hormones to study progesterone production from turtle follicular tissue *in vitro*, our observations are similar to your own; that is, ovine FSH is very potent, but ovine LH is relatively inactive. In contrast, if we inject ovine FSH or LH *in vivo* we may find them to be equally active or equally inactive, depending on the time of the year in which the study is done.

This brings up the problem of seasonal cycles in receptors in an animal that breeds only once and has a long follicular cycle, which may last 6 months, as in the turtle. We have evidence of a cycle of follicular maturation of response to ovine FSH, since, if we measure three different steroids (progesterone, testosterone, estradiol-17-β) produced by suspended ovarian cells *in vitro* at different times of the year, a different picture of hormone action emerges depending on which one of the three steroids one looks at.

P. Licht: May I comment on your first statement about all these roles of FSH now appearing to exist in mammals. We have already alluded to this, and I agree; that is why I think that rather than regarding LH specificity as an all-or-nothing phenomenon, we are really dealing with a continuum. The data at this moment seem simply to indicate that mammals and amphibians show a relatively greater LH specificty than do reptiles. As to whether we will find exceptions if we look at enough amphibians, I would be the first to agree that, if you look at enough of any of these groups, you will find exceptions. We have, in fact, looked at a large number of salamanders and anurans, and if one uses heterologous hormones again, one begins to see the breakdown in specificity. However, I do not think that this breakdown is the important factor; before we draw a conclusion about how specific a species is, we must study it with its own gonadotropins—that is really the critical issue here. In terms of your comments on seasonality, it is obviously a very important observation, and we have examined a number of aspects of this using mammalian as well as nonmammalian hormones. As we reported at this conference, although the potency of gonadotropins may change in every season, when one hormone works, the other hormone works.

I. Callard: I know you have not yet looked at any Prototheria, and theoretically they should fall in line with the Metatheria and Eutheria. Do you have any information on this group?

P. Licht: When we say that most endocrinology has been done on mammals, we really should say that it has been done on a very few species of eutherian mammals. There is very little information on LH and FSH in noneutherians. That is why we are interested in kangaroo hormones. Obviously this evolutionary theory needs further testing, but I have a way out: if a group does not fit this hypothesis, all I have to propose is that it has secondarily evolved in a slightly different direction.

I. Callard: One more specific comment regarding the male reptile and the site of action of

FSH. You indicated that FSH and LH act on the Leydig cells in the lizard testis, but obtained binding in the tubular area also. Would you care to comment about the possibility that the action of FSH in the mammalian (rodent) testis on the Sertoli cell is a more primitive vertebrate characteristic, which is possibly of greater importance in reptiles or other nonmammalian vertebrates?

P. Licht: I do not think we have enough direct evidence for Sertoli cell regulation to speculate at this time.

B. M. Dobyns: In years past, the pituitary preparations were far less pure than they are today, but, in those early days marked exophthalmos resulted from the administration of LH from certain mammalian pituitaries. These preparations had in them no TSH, or very little. I am curious to know whether, in your wide use of various pituitary preparations from lower animals, you observed exophthalmos; if so, how long did it take to occur?

P. Licht: I cannot say that I have ever seen it, and if we were to see it, our first response would be to assume that we have TSH contamination until proved otherwise.

M. Sairam: I wonder whether you have laid your hands on some nonmammalian FSH β subunit. Some years ago you and Harold Papkoff showed that ovine FSH β subunit was as good as the native hormone molecule in the lizard.

P. Licht: Yes, this is of special interest now that we have far more sensitive binding and physiological tests in reptiles. The observation that Sairam is referring to, and one of the reasons why Papkoff and I first got together, was that we found that not only is mammalian FSH potent in the lizard, but the α subunit is not necessary. β-FSH was as effective as intact FSH, and we were also able to go on to show that in spermiation in the frogs the β subunit of FSH was far more potent than it should be by mammalian assay, indicating that perhaps in the nonmammalian forms the relationship of FSH function and structure may be very different from that in mammals. I recently acquired some subunits of human FSH from the pituitary distribution program; we tested them for their ability to bind to reptilian testis, and I was very disappointed that they were essentially inactive, but we have not yet tested other physiological potencies.

A. Nureddin: Upon recombination of the subunits within the same species, what percent recovery of hormonal activity do you obtain?

P. Licht: It varies from experiment to experiment, even with the ovine subunits. In this particular test, we got something like 40% recombination for ovine subunits and full restoration of activity for turtle subunits.

A. Nureddin: How difficult is it to separate them?

P. Licht: It is essentially the same method as that used for the ovine LH subunits.

Biosynthesis of Parathyroid Hormone

JOEL F. HABENER,* BYRON W. KEMPER,[1]
ALEXANDER RICH,† AND JOHN T. POTTS, JR.*

*Endocrine Unit and Department of Medicine, Harvard Medical School,
and Massachusetts General Hospital, Boston, Massachusetts; and
†Department of Biology, Massachusetts Institute of Technology, Cambridge Massachusetts

I. Introduction

Recent progress in research on parathyroid hormone (PTH) biosynthesis has been marked by the identification, isolation, and structural characterization of two distinct biosynthetic precursors of the hormone (Habener, 1976; Habener and Potts, 1976a). Earlier studies of hormone biosynthesis in parathyroid gland slices resulted in the identification of a proparathyroid hormone (ProPTH) (Kemper et al., 1972; Habener et al., 1972; Cohn et al., 1972b), analogous in many respects to proinsulin, the precursor of insulin described originally by Steiner and Oyer (1967). Subsequently, studies involving the translation of parathyroid messenger RNA in heterologous cell-free systems led to the identification of pre-proparathyroid hormone (Pre-ProPTH), a biosynthetic precursor, preceding ProPTH (Kemper et al., 1974b, 1976a,b; Habener et al., 1975b,c). Present evidence suggests that Pre-ProPTH is the initial product synthesized by the parathyroid cell and represents the total structural information encoded in the gene for PTH. ProPTH appears to be an intermediate product in the pathway of intracellular transport of the hormone. The two precursors of PTH are sequentially cleaved by specific proteases during the cellular translocation of the hormonal polypeptide from its site of synthesis, in the rough endoplasmic reticulum, to the Golgi apparatus, where the final product, PTH, is incorporated into secretory vesicles. The specific precursor peptides themselves may serve important functions in the sequential segregation and transportation of the hormone within cellular organelles.

The discovery and structural analyses of Pre-ProPTH, in addition to parallel development in studies of the biosynthesis of other hormones and secretory proteins (Blobel and Dobberstein, 1975; Blobel and Campbell, 1976; Steiner, 1976), have contributed two concepts that are now generally recognized. One is that posttranslational proteolytic modification of polypeptides and proteins

---

[1] Present address: Department of Physiology and Biophysics, University of Illinois at Urbana, Urbana, Illinois.

destined for secretion from the cell is a process that occurs in most, if not all, secretory tissues. The second is the recognition that the products of this proteolysis form two classes of biosynthetic precursors, the pre-hormones and the prohormones, distinguished on the basis of (a) the chemical nature of their amino acid sequences, (b) the rapidity of proteolytic cleavage after initial translation, (c) the substrate specificity of the enzymic activity, and (d) the site in the subcellular organelles where the proteolysis occurs.

In addition to the elucidation of the biosynthetic pathway for PTH involving discrete proteolytic processing of precursor peptides, new insights have been gained concerning the role of calcium in the regulation of biosynthetic and cellular processes in the parathyroid gland. Recent evidence points to the existence of a calcium-regulated degradative pathway in the parathyroid cell (Chu et al., 1973; Habener et al., 1975a). Changes in concentrations of extracellular calcium appear to rapidly modify the amount of hormone available for secretion by altering the rates of intracellular degradation of hormone. It is now possible to determine whether the enzymic steps in the processing of the PTH precursors may themselves represent points for the regulation of hormone production and secretion.

Important questions arise regarding the possibility that the precursors or the precursor-specific peptides that are split from the precursors during this processing may have biological functions outside the parathyroid gland that are separate and distinct from the biological functions of the precursors or of PTH. Pertinent to this question is information about whether the precursor or split precursor peptides are released from the parathyroid gland into the circulation under either normal or abnormal states of parathyroid function, such as neoplasia or hyperplasia. To answer these questions, sensitive radioimmunoassays for ProPTH using antisera produced by immunization with synthetic ProPTH peptides were developed and applied to analysis of prohormone levels in tissues and blood (Habener et al., 1974a, 1975a, 1976a).

II. Primary Structures of Parathyroid Hormone

The rapid progress in the field of PTH biosynthesis can be attributed in large part to the continued development of more sensitive and effective techniques for the chemical characterization of parathyroid polypeptides. Since the last discussion of PTH at these conferences (Aurbach et al., 1972), work has progressed on the sequence determination of the human hormone (Fig. 1). Initially, the sequence of the amino-terminal 37 residues was established using a high-sensitivity sequencing procedure with ^{35}S-labeled phenylisothiocyanate of high specific activity as the coupling agent (Niall et al., 1974). Subsequently, the complete sequence from residue 44 through the carboxyl-terminal residue, glutamine at position 84, was determined using this procedure (Keutmann et al., 1975). The

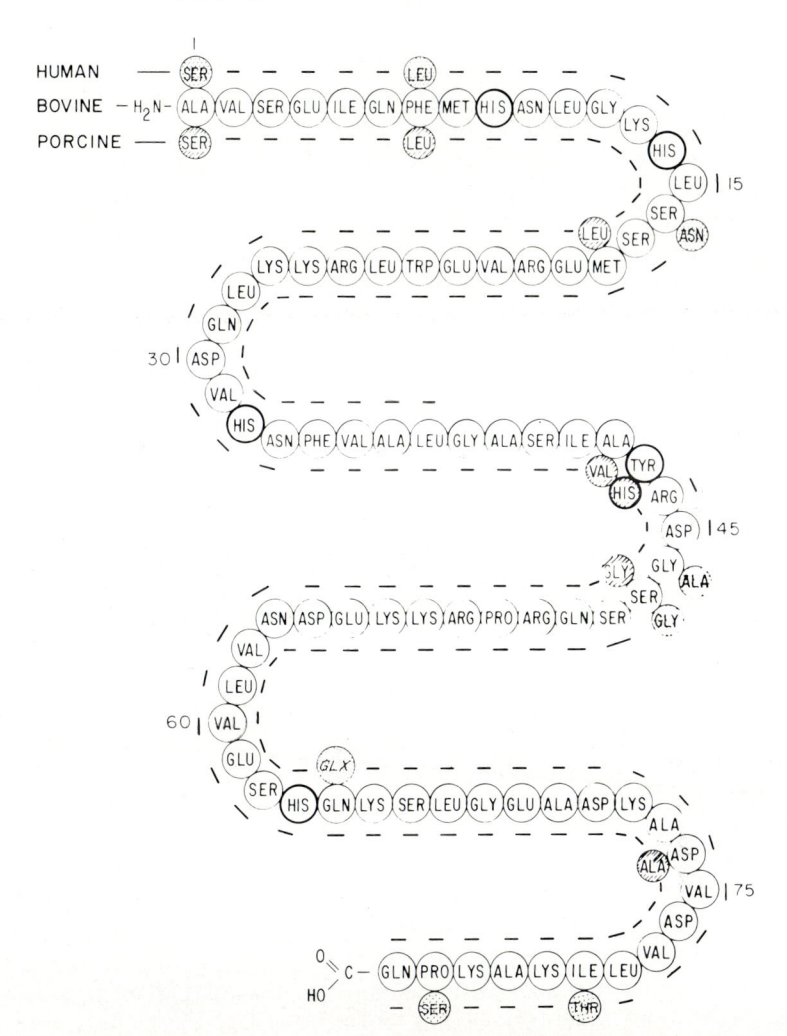

FIG. 1. Amino acid sequences of bovine, porcine, and human parathyroid hormones (PTH). The continuous structure shown in open circles is that of bovine PTH. The appended residues indicate differences in amino acids in the sequence of porcine (hatched circles) and human (stippled circles); a bar indicates an identical residue. Note that the sequence of human PTH is not yet completed. Modified from Keutmann *et al.* (1975).

exact sequence of amino acids from positions 38 through 43 are being studied. The sequence of the human hormone determined thus far differs from bovine PTH in 8 positions and from that of porcine PTH in 4 or possibly 5 positions. It is uncertain whether the amino acid at position 64 is glutamine or glutamic acid. As was found with the bovine and porcine hormones, complete biological activity of the human hormone appears to reside in a limited amino-terminal sequence (Tregear *et al.*, 1974). The chemically synthesized peptide 1–34 of the human hormone expresses the full biological activity of the native hormone in bioassays performed both *in vitro* and *in vivo*. The sequence of human PTH that we have obtained differs at three positions (22, 28, 30) from the structure reported by Brewer *et al.* (1972) and synthesized by Andreatta *et al.* (1973). Recently, Keutmann *et al.* (1975) employed a sensitive microsequencing technique using radioactively labeled PTH synthesized by and extracted from human parathyroid adenomas to confirm the validity of the sequence of the first 37 amino acids as proposed originally by Niall *et al.* (1974). The explanation for the discrepancies in results between our group and Brewer's (1972) has not been found.

The determination of the amino acid sequence of human PTH has made available, through peptide synthesis, large amounts of a biologically active fragment for detailed pharmacological evaluation and exploration of potential therapeutic value (Parsons *et al.*, 1975) as well as specific amino and carboxyl fragments for preparations of fragment-specific assays useful in characterizing hormone metabolism and the heterogeneity of circulating hormone (Segre *et al.*, 1975).

III. Criteria for the Identification of a Biosynthetic Precursor

Before beginning a discussion of the studies that led to the identification of ProPTH, it is helpful to describe the criteria required to establish with certainty the existence of a biosynthetic precursor. Many of the polypeptide and protein hormones are prone to self-aggregation and at times will bind to larger macromolecules, such as plasma proteins, thus giving rise to larger forms of the hormones that may be detected in blood and extracts of tissue by gel filtration and immunological assays or by other analytical techniques that do not employ denaturing conditions, e.g., 0.1% sodium dodecyl sulfate (SDS) or 8 M urea (Antoniades *et al.*, 1965; Zanini *et al.*, 1974; Benveniste *et al.*, 1975). Conversion of such larger forms of the hormones to the authentic hormone by treatment with proteolytic enzymes is sometimes taken as evidence that the larger form contains the hormone within a larger, single-chain polypeptide; i.e., it is a biosynthetic precursor. Such findings alone do not constitute unequivocal evidence for the presence of a biosynthetic precursor, inasmuch as treatment of such aggregates during enzymic digestion may result in the release of hormone that is noncovalently bound to a larger protein, owing to physicochemical changes and/or digestion of the carrier protein. Several criteria should be met before it is concluded that a larger hormonal form is indeed a true biosynthetic precursor. The most convincing evidence for the identification of a biosynthetic precursor consists of a combination of (1) a careful kinetic demonstration of a precursor-to-product relationship by pulse-labeling and pulse-chase experiments and (2) the finding of an additional peptide sequence covalently linked to the hormone either by analyses of chemical or enzymically produced peptides or by analysis of the amino acid sequence of the precursor. Either one of these findings taken alone constitutes suggestive, but not conclusive, evidence for a precursor. The first observation alone may also be a result of differential rates of synthesis and cellular degradation of two unrelated proteins. The second finding alone may be explained alternatively by the existence of a protein that is structurally related to the hormone but arises on the basis of gene duplication and diversification rather than as a biosynthetic precursor of the hormone.

IV. Identification of a Proparathyroid Hormone during Studies of Hormone Biosynthesis in Intact Parathyroid Gland Slices

The existence of a prohormone for parathyroid hormone was suspected before the initiation of our studies on the biosynthesis of the hormone. The earlier discovery of a prohormone for insulin by Steiner and Oyer (1967) made it reasonable to presume that prohormones existed for other polypeptide hormones. Credit for initial evidence of a ProPTH belongs to Cohn and his co-workers (1972a,b). Hamilton *et al.* (1971a) had reported the finding of a polypeptide in extracts of parathyroid glands that differed from PTH in its elution from columns of carboxymethyl cellulose. This substance, termed calcemic fraction A (CFA), appeared to be related to PTH because it produced hypercalcemia when administered to rats (Hamilton *et al.,* 1971b). We confirmed the earlier observations of Hamilton and Cohn (1969) and Sherwood *et al.* (1970) that slices of parathyroid glands would actively incorporate radioactive amino acids into parathyroid hormone *in vitro* and that they retained their responsivity to changes in the concentrations of extracellular calcium (Fig. 2). Thus, slices of bovine parathyroid glands appeared to be a suitable system with which to initiate studies of hormone biosynthesis (Habener and Potts, 1975).

Our attempts to identify a ProPTH were concentrated on an analysis of the radioactive proteins synthesized during incubations of parathyroid gland slices in ^{14}C-labeled amino acids for relatively short periods of 10–150 minutes (pulse-labeling and pulse-chase experiments). Electrophoretic analyses of the radioactively labeled protein extracts prepared from the slices indicated that, at the earliest times of incubation (10 minutes), the principal polypeptide labeled was not PTH, but rather was a more basic polypeptide that migrated more rapidly than PTH in the electric field (Fig. 3) (Kemper *et al.,* 1972). Radioactive PTH did not appear until about 20 minutes of incubation time and then progressively increased in amount relative to the more rapidly migrating peptide during longer

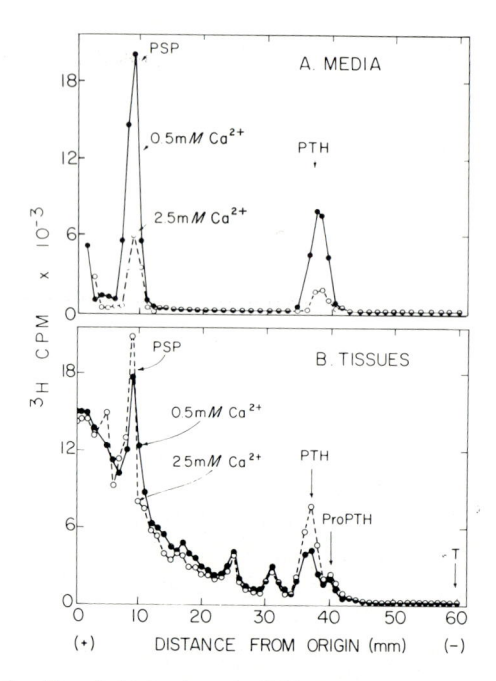

FIG. 2. Polyacrylamide gel electrophoresis (8 M acetate–urea gels, pH 4.4) of extracts of (A) media and (B) tissue, after 4 hours of incubation of parathyroid slices in the presence of low (0.5 mM, ●) and high (2.5 mM, ○) concentrations of calcium. Radioactive protein on the gels has been corrected to counts per minute per milligram of protein in extracts of the tissue slices. T, position of the methyl-green tracker dye; PSP, parathyroid secretory protein, an additional protein of unknown function (not a biosynthetic precursor) that accompanies the secretion of parathyroid hormone (PTH). Direction of migration is from left to right. From Habener *et al.* (1975a).

periods of incubation. A pulse-chase experiment conducted by removing the radioactive amino acids after a 20-minute incubation and replacing them with unlabeled amino acids demonstrated the rapid disappearance of the earlier polypeptide (ProPTH) and nearly quantitative appearance of radioactive PTH (Fig. 4).

The kinetics of the conversion of ProPTH to PTH in these early experiments (Kemper *et al.*, 1972) were somewhat slower than those determined later under conditions in which slices were preincubated for 30–60 minutes before addition of radioactive amino acids to the incubation media (see, for example, Figs. 29 and 32) (Habener *et al.*, 1974b, 1975a, 1976b; Kemper *et al.*, 1975). Confirmation of the identification of ProPTH, which was only tentative based on the pulse-labeling studies, was obtained by finding that the ProPTH isolated from the polyacrylamide gels (1) had a molecular weight higher than that of PTH, by

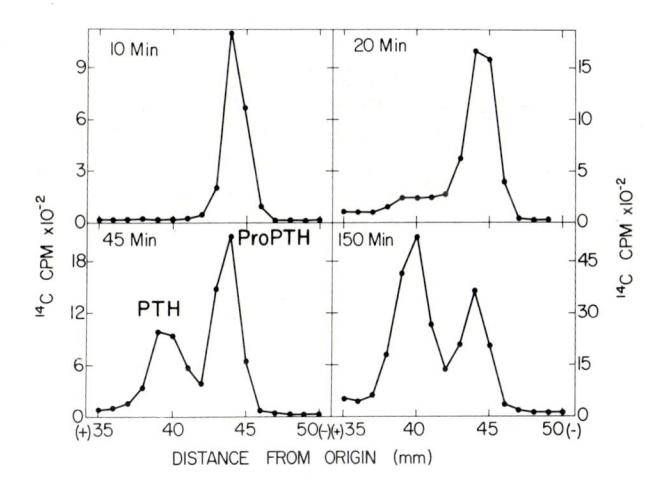

FIG. 3. Polyacrylamide gel electrophoresis of extracts of parathyroid gland slices that had been incubated for times indicated with ^{14}C-labeled amino acids. Only the regions of the gels containing the parathyroid hormone (PTH) polypeptides are shown (see legend to Fig. 2). The more rapidly migrating protein ProPTH was subsequently shown to be an intermediate biosynthetic precursor of PTH. Modified from Kemper *et al.* (1972).

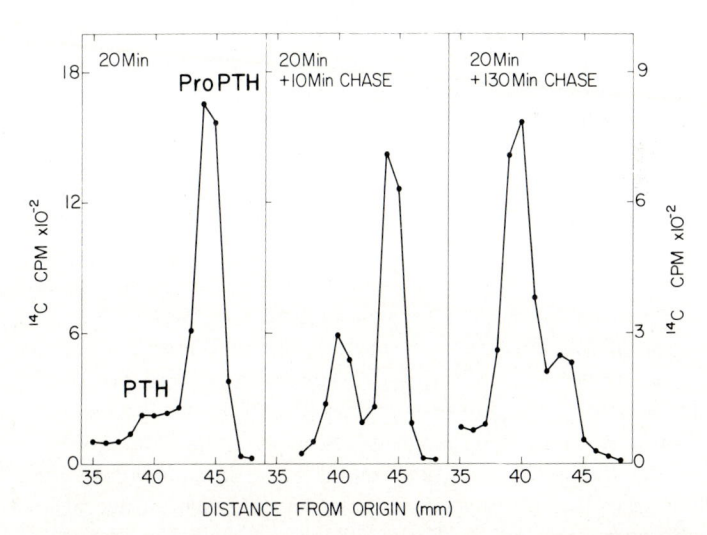

FIG. 4. Conversion of radioactively labeled proparathyroid hormone (ProPTH) to parathyroid hormone (PTH) in the absence of additional incorporation of radioactive amino acids. After a 20-minute incubation with radioactive amino acids, the medium was removed and fresh medium containing unlabeled amino acids was added. Results were analyzed at 20 minutes, 20 minutes + 10 minute chase, and 20 minutes + 130 minute chase. Modified from Kemper *et al.* (1972).

analysis by SDS—acrylamide gel electrophoresis, (2) contained all the tryptic peptides of PTH plus several extra peptides, (3) was bound specifically by antisera to PTH, and (4) could be converted to PTH by treatment with dilute trypsin or by incubation in cell-free extracts of parathyroid tissue (Cohn *et al.*, 1972b; Kemper *et al.*, 1972; Habener *et al.*, 1973). Thus, the evidence from these studies essentially proved the existence of a ProPTH, but final confirmation awaited determination of the amino acid sequence of the prohormone.

V. Primary Structure of Proparathyroid Hormone

During the course of our studies leading to the identification and characterization of ProPTH, we suspected that the ProPTH and the CFA described by Hamilton *et al.* (1971a,b) were probably one and the same polypeptide. This suspicion was confirmed by the report of Cohn *et al.* (1972b) showing, by pulse-chase labeling studies, that CFA was a precursor of PTH, indistinguishable from ProPTH (Kemper *et al.*, 1972). A collaboration was established between our group and Drs. Hamilton and Cohn in an effort to determine the amino acid sequence of ProPTH. Approximately 1—2 mg of what appeared, by electrophoretic analyses, to be relatively homogeneous prohormone had been isolated by successively subjecting acid—urea extracts, prepared from several thousand bovine parathyroid glands, to gel filtration followed by chromatography on CM-cellulose. Evidence derived from earlier analyses of the peptide fragments from radioactively labeled ProPTH produced by cleavages of the prohormone with cyanogen bromide and with dilute acid had indicated that the prohormone contained an additional peptide sequence at the amino terminus of PTH (Habener *et al.*, 1973). This fortuitous situation made it possible to determine the amino-terminal amino acid sequence of ProPTH using the method of sequential degradation and identification of amino acids described by Edman and Begg (1967).

Two independent degradations of the ProPTH were carried out, each for 40 cycles, using the Beckman 890 Sequanator (Fig. 5). A peptide sequence consisting of six amino acids was found covalently linked to the amino-terminal alanine of PTH (Hamilton *et al.*, 1974). Four of the six amino acids of the prohormone are basic (3 lysines, 1 arginine), a finding that explained the rapid mobility of the prohormone on polyacrylamide-gel electrophoresis, late elution from CM-cellulose columns and, in addition, anomalously slow migration on SDS—polyacrylamide gels leading to an interpretation of a spuriously high molecular weight (Kemper *et al.*, 1972). The finding of only six additional amino acids in the sequence of the ProPTH was surprising, inasmuch as repeated analyses of the total amino-acid composition of the same preparations of ProPTH that were subjected to the sequential degradation had shown the presence of 19 extra

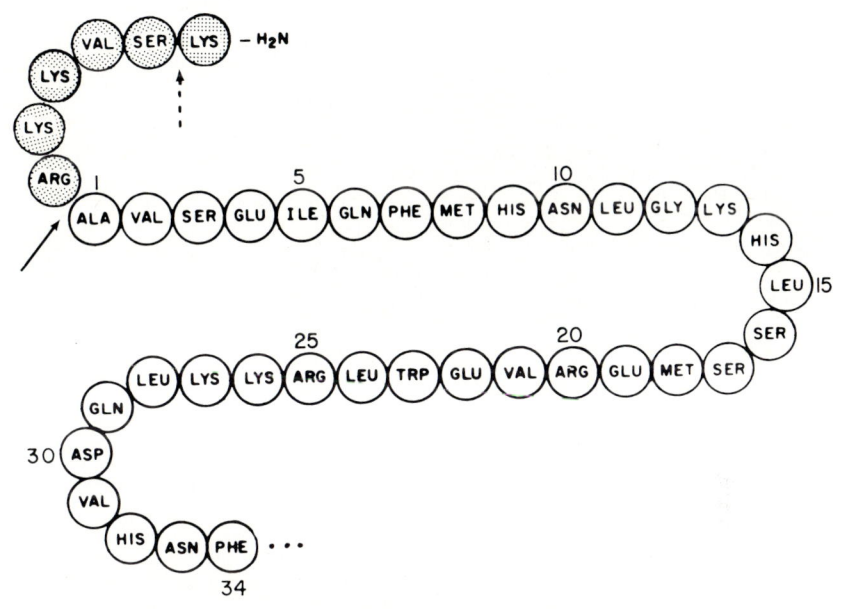

FIG. 5. Amino-terminal structure of bovine parathyroid hormone (PTH). Unshaded residues 1–34 represent the amino-terminal sequence of bovine PTH. The additional six amino-terminal residues found in the prohormone are shown as stippled circles. The solid arrow shows the cleavage point for the conversion of prohormone to hormone; the dashed arrow represents a minor cleavage found in some prohormone preparations. From Potts et al. (1973).

amino acids in addition to those that could be accounted for in the sequence of PTH (Cohn et al., 1972b).

Two possible explanations have been considered that might account for this discrepancy in results. (1) The preparations of ProPTH were contaminated by a small amount of another protein, perhaps a histone, that has a blocked amino terminus (acylated serine, histone IV) that prevented its degradation in the sequenator; and/or, alternatively, (2) the prohormone contains additional peptide sequence at the carboxyl terminus, as well as at the amino terminus of PTH. As yet, neither of these two explanations has been confirmed or excluded with certainty. Our attempts, however, to analyze and to compare the carboxyl-terminal tetrapeptides of ProPTH and PTH produced by digestion of the hormones with trypsin have failed to show any additional peptide sequence at the carboxyl terminus of the prohormone (Fig. 6) (Habener et al., 1973). Cohn et al. (1974a, 1975) have continued to analyze the problem and have presented preliminary data, based on limited degradation of the prohormone with carboxyl-terminal-specific

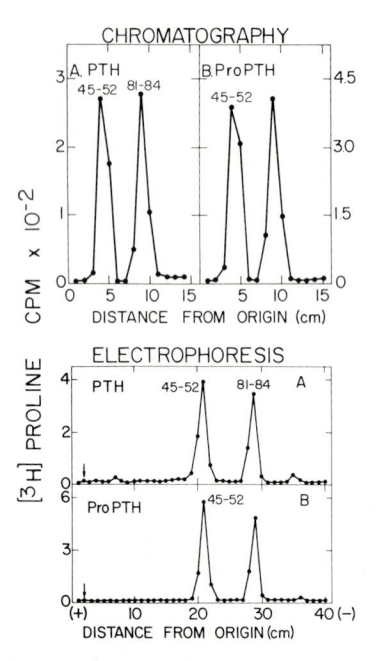

FIG. 6. Identity of carboxyl-terminal sequences of proparathyroid hormone (ProPTH) and parathyroid hormone (PTH) as shown by paper chromatography (upper panel) and electrophoresis (lower panel) of the [^{3}H]proline-labeled tryptic peptides. The mobilities of the carboxyl-terminal tryptic tetrapeptide 81–84 of PTH and the corresponding peptide of ProPTH are identical.

exopeptidases, that there may be additional amino acids at the carboxyl terminus of the ProPTH.

Since the original determination of the amino acid sequence of the bovine ProPTH, the amino acid sequences of the analogous prohormones from human (Jacobs et al., 1974; Huang et al., 1975) and porcine (Chu et al., 1975) species have been analyzed. In all species, a prohormone-specific hexapeptide sequence has been found at the amino terminus of PTH. The amino acid sequence of the hexapeptide of the human prohormone is identical to that of the bovine prohormone. The sequence of the porcine prohormone differs only in the substitution of a proline for serine and isoleucine for valine.

Soon after the primary structures of the bovine and human prohormones were determined, chemical syntheses of numerous fragments and analogs of the prohormones were carried out (Habener et al., 1974a, 1976a). These chemically synthesized peptides consisted of the hexapeptide alone and of the hexapeptide linked to varying lengths of the sequence of PTH through residue 34 (Fig. 5). These fragments have proved to be of great use in characterization of the biological properties of proparathyroid hormone, in investigations of the mechanism of prohormone-to-hormone conversion, and in the development of immunoassays specific for detection of the prohormone in blood and tissue fluids.

VI. Enzymic Conversion of Proparathyroid Hormone to Parathyroid Hormone

Analyses of the tryptic peptides prepared from radioactively labeled ProPTH before the determination of the amino acid sequence of the prohormone indicated that the peptide bond connecting the prohormone-specific hexapeptide to the sequence of PTH was highly sensitive to cleavage by pancreatic trypsin (Habener et al., 1973; Goltzman et al., 1976). Identification of an arginine–alanine bond (arginine–serine in the human prohormone) in the amino acid sequence of ProPTH confirmed the presence of a tryptic-sensitive substrate (Hamilton et al., 1974; Jacobs et al., 1974; Huang et al., 1975). Subsequently, more-detailed investigations of the specificity of the enzymic cleavage of ProPTH were undertaken. Incubations of either the biosynthetically labeled native prohormone or the chemically synthesized fragment of ProPTH consisting of the prohormone hexapeptide and the first 34 amino acids of PTH with dilute trypsin resulted in selective cleavage of the arginine–alanine bond without cleavage elsewhere in the prohormone sequence; higher concentrations of trypsin degraded the prohormone at multiple sites in the polypeptide chain (Fig. 7) (Habener and Potts, 1976a; Goltzman et al., 1976). These observations indicated that the enzymic cleavage of ProPTH to PTH could be accomplished by the

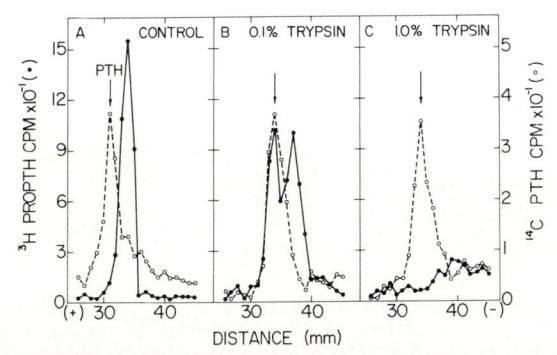

FIG. 7. Conversion of bovine proparathyroid hormone (ProPTH) to parathyroid hormone (PTH) by limited incubation with dilute trypsin. [^{3}H] ProPTH was incubated for 10 minutes at 37°C with either 0.1 or 1.0% (w/w) pancreatic trypsin. Proteolysis was stopped by acidification, and a marker of ^{14}C-labeled bovine parathyroid hormone was added. The digest was analyzed by polyacrylamide-gel electrophoresis; 50% of ProPTH is specifically converted to PTH by 0.1% trypsin, whereas more concentrated trypsin (1.0%) destroys the entire hormonal molecule. From Habener and Potts (1976a).

action of trypsin alone and does not require a specialized enzyme that is unique to the parathyroid gland. In fact, trypsinlike activity that readily converts ProPTH to PTH has been found in a wide variety of tissues including plasma (see Section VIII), renal cortical membranes (see Section VII), and cell-free homogenates of parathyroid glands (Fig. 8).

Recent studies on the enzymic processing of ProPTH in the parathyroid gland have indicated that, in addition to tryptic cleavage of the prohormone, which yields PTH, the hexapeptide sequence is further cleaved at its carboxyl terminus by carboxypeptidase B (Fig. 9) (Habener *et al.*, 1977). Such a combined enzymic processing involving trypsin and carboxypeptidase B was shown previously to occur in the conversion of proinsulin to insulin (Kemmler *et al.*, 1971, 1973), and examination of the primary structures of other known prohormones suggests that this dual enzymic proteolysis may be a universal occurrence in the cellular processing of prohormones (Steiner, 1976). The action of the combined enzymes is most evident when the basic residues usually found at the site of attachment of the extension peptide to the hormone are present on the carboxyl terminus of the hormone itself, as with insulin, and less apparent when the basic residues, as in ProPTH, are at the carboxyl terminus of the prohormone-specific peptide (Goltzman *et al.*, 1976). In the latter case, evidence of carboxypeptidase activity is found with analyses of the prohormone-specific extension peptide from which the basic residues are removed. Several lines of

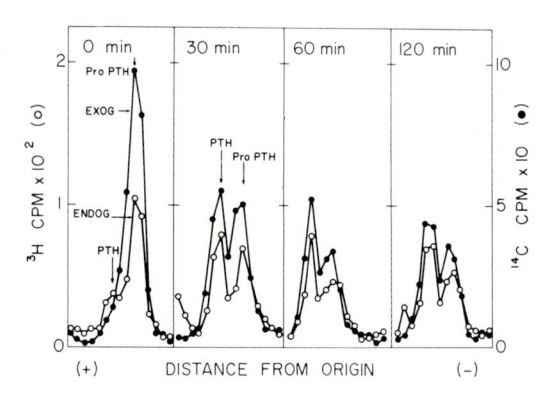

FIG. 8. Time-dependent conversion of proparathyroid hormone (ProPTH) to parathyroid hormone (PTH) during the incubation of radioactively labeled prohormone in homogenates of bovine parathyroid glands. Gland slices were incubated with [^{3}H]leucine for 20 minutes to label ProPTH synthesized endogenously (ENDOG). The slices were homogenized in 0.25 M sucrose–50 mM Tris–10 mM MgCl$_2$ (pH 7.2). [^{14}C]ProPTH, isolated previously from parathyroid slices by electrophoresis, was added exogenously (EXOG) to the supernatant obtained after centrifugation of the homogenate for 5 minutes at 1000 g. The mixtures were incubated at 37°C for the indicated times, after which extracts of the homogenates were prepared and analyzed by electrophoresis on acetate–urea gels.

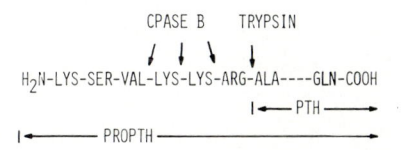

```
                    CPASE B    TRYPSIN
                     / ↓ \  ↓
        H₂N-LYS-SER-VAL-LYS-LYS-ARG-ALA----GLN-COOH
                              |←——— PTH ———→|
        |←————————— PROPTH ——————————————→|
```

FIG. 9. Specific sites in the sequence of proparathyroid hormone (ProPTH) that are enzymically cleaved during its proteolytic conversion to parathyroid hormone (PTH) in the parathyroid gland. Two enzyme activities are involved: trypsin and carboxypeptidase B (cpase B). The amino acids in the sequence of PTH 2–83 are not shown (dashed line).

evidence suggest that the enzymic cleavage of ProPTH occurs in the Golgi apparatus of the parathyroid cell (see Section XIII). Attempts have been made to localize specific enzymic activity in subcellular fraction prepared by differential centrifugation of extracts of parathyroid glands, but they have been only partially successful (Habener *et al.*, 1976c). Some enzymic activity appears to be present in all fractions analyzed, a finding that is presumably attributable to release and redistribution of enzymic activity among the fractions during the process of disruption of the tissue.

VII. Biological and Immunological Properties of Proparathyroid Hormone

The availability of large quantities of the ProPTH tetradecapeptide fragment produced by chemical synthesis (Fig. 5) has allowed extensive comparisons to be made of the biological and immunological activities of the prohormone and of PTH. Early tests of the activity of both native ProPTH and chemically prepared tetradecapeptide in *in vivo* and *in vitro* assay systems indicated that the prohormone was as much as 50% as active as was PTH in production of hypercalcemia in the rat (Hamilton *et al.*, 1971a) and 20% in the chick (Habener and Potts, 1976a), and demonstrated 10% of the activity of PTH in the bone citrate decarboxylase assay (Hamilton *et al.*, 1971a) and 3% in the renal adenylyl cyclase assay (Habener and Potts, 1976a). However, these earlier studies of the biological activity of ProPTH did not take into account the existence of substantial levels of proteolytic (tryptic) activity in the assay systems that convert the ProPTH to PTH. Subsequently, analyses of the stability of ProPTH in blood and extracts of tissues indicated that tryptic activities present in blood and tissues readily and specifically convert ProPTH to PTH (Cohn *et al.*, 1972b; Habener *et al.*, 1974a, 1976a). Thus, results obtained from bioassays in which prohormone is exposed to blood *in vivo* or to preparations of tissues *in vitro* became open to question.

Further improvements in the *in vitro* renal adenylyl cyclase assay involving more-extensive purification of the renal cortical membranes made it possible to test the biological activity of ProPTH under conditions in which proteolytic activity in the assay is very low and in which the small residual amount of tryptic activity remaining could be inhibited by benzamidine (Petreymann *et al.*, 1975). Comparisons of the biological activities of ProPTH and PTH in this modified *in vitro* bioassay have shown that ProPTH exerts less than 0.2% of the activity of a comparable amount of PTH (Fig. 10). At present, it is not certain whether the small degree of biological activity observed in this assay is a true reflection of binding to and activation of the renal membrane receptors for PTH by the ProPTH or whether the activity simply represents a small, chemically undetectable degree of conversion of ProPTH to PTH. The findings of very low biological activity of ProPTH were consistent with earlier studies of synthetic analogs of PTH in which it was shown that extension of PTH at the amino terminus by addition of short sequences of various amino acids resulted in a drastic reduction in the activity of the hormone (Potts *et al.*, 1972). It is apparent that lengthening, as well as shortening, of the amino acid sequence at the amino-terminal end of PTH lowers its biological activity.

In addition to inhibitory effects on the biological activity of PTH, the presence of the prohormone-specific hexapeptide appears to reduce the immunological reactivity of the prohormone with antisera that recognize antigenic determinants within the amino-terminal, but not the carboxyl-terminal sequence of PTH (Fig. 11). This inhibition of immunological activity might be a result of changes in the

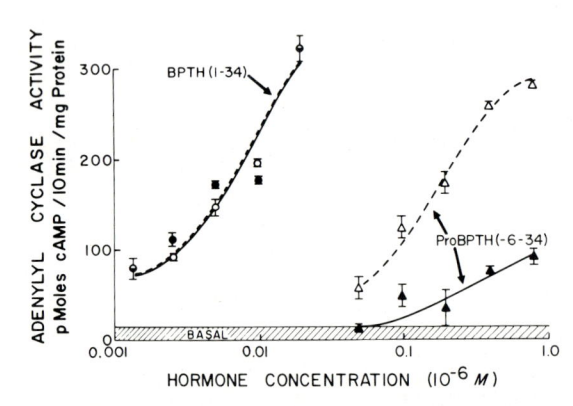

FIG. 10. Stimulation of dog renal cortex adenylyl cyclase activity with bovine parathyroid hormone bPTH-(1–34) or Pro-bPTH-(–6–+34) in the absence or in the presence of 25 mM (7B) benzamidine. The samples were incubated for 10 minutes at 22°C with renal cortical membranes. ○– – – –○, bPTH-(1–34) without benzamidine; ●———●, bPTH-(1–34) with benzamidine; △– – – –△, Pro-bPTH-(–6–+34) without benzamidine; ▲———▲, Pro-bPTH-(–6–+34) with benzamidine. From Petreymann *et al.* (1975).

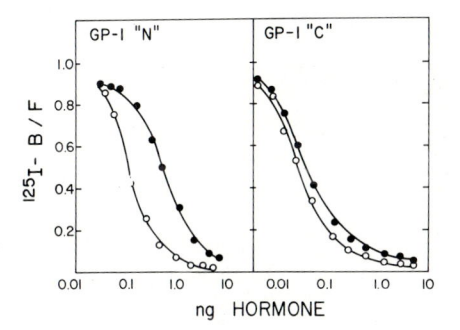

FIG. 11. Relative cross-reactivity of (●) bovine proparathyroid hormone (ProPTH) and (○) parathyroid hormone (PTH) in two radioimmunoassay systems that react specifically with either the amino-terminal (GP-1 "N") or the carboxyl-terminal (GP-1 "C") sequence of PTH. The assay measures the relative competitive displacement of ^{125}I-labeled bovine PTH from specific antisera to PTH. The ratio of the molecular weights of ProPTH to PTH is small (1.08) and is not corrected for in the plots. From Habener and Potts (1976a).

conformation of antigenic determinants in the sequence of PTH induced by the presence of the prohormone hexapeptide. With present information available, we can only infer that secondary structure of the prohormone or hormone may influence immunological reactivity. Cohn *et al.* (1974b) presented evidence based on gel-filtration studies that the amino-terminal regions of ProPTH and PTH may be in a more compact globular form compared with a less compact, more extended configuration in the carboxyl-terminal regions.

VIII. Radioimmunoassays for Proparathyroid Hormone

It has been possible to produce antisera to chemically synthesized peptide fragments of ProPTH that incorporate in their sequence the prohormone hexapeptide and varying lengths of the amino-terminal sequence of PTH (Habener *et al.*, 1974a, 1975a, 1976a). These antisera have been used successfully in competitive-binding radioimmunoassays that in turn have permitted, for the first time, direct measurements to be made of ProPTH levels in extracts of parathyroid glands and in blood. The octadecapeptide consisting of a sequence extending from the amino-terminal lysine of the hexapeptide to the glycine at position 12 in PTH proved to be highly antigenic when injected without conjugation into rabbits. Antisera specific for the recognition of ProPTH and fragments of ProPTH that included the covalent bond arginine–alanine between the hexapeptide and the hormone were obtained (Fig. 12). These antisera showed little or no cross-reactivity with PTH and no reactivity with the hexapeptide. Recently, antisera were produced by immunization of animals with a synthetic polymer of the hexapeptide (three hexapeptide sequences covalently linked to give an

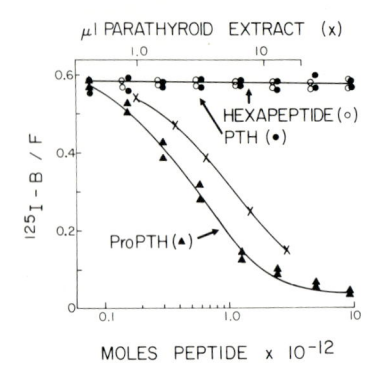

FIG. 12. Radioimmunoassay showing competitive displacement curves produced by native bovine proparathyroid hormone (ProPTH) and synthetic fragments of the prohormone (▲) and by extracts of bovine parathyroid glands (X). Native parathyroid hormone (PTH) and synthetic PTH peptides (●), as well as the prohormone-specific hexapeptide (○), do not react in the assay. From Habener *et al.* (1974a).

octadecapeptide). These antisera react with free hexapeptide, as well as with ProPTH and fragments of ProPTH (Habener *et al.*, 1977).

Application of the assays to an evaluation of the amounts of ProPTH contained in bovine and human parathyroid tissues has shown that ProPTH in bovine glands comprises approximately 7% of the total immunoreactive hormone (93% is PTH) (Fig. 13). Assays, however, of extracts of human parathyroid adenomas revealed levels of prohormone similar to those found in the bovine glands, but the ProPTH comprised 50% of the total immunoreactive hormone, owing to the fact that hormone levels in the adenomas were found to be much lower than levels found in the bovine glands (Habener *et al.*, 1976a). Amounts of PTH in the tissues were also determined by radioimmunoassay. The value of the PTH content in the bovine glands that we obtained is in good agreement with the figure reported by Hamilton *et al.* (1971b) of approximately 200 µg per gram wet weight of tissue. The finding of approximately 20 times more hormone than prohormone in the bovine tissue suggests that little, if any, of the ProPTH is stored in the gland. On a theoretical basis, the amounts of prohormone present could be accounted for entirely by the fraction of ProPTH in transit from its site of synthesis in the rough endoplasmic reticulum to the Golgi apparatus, where proteolytic conversion to PTH is presumed to occur. The surprisingly small amounts of PTH found in the parathyroid adenomas in comparison with the bovine glands may reflect inherent differences in the absolute quantities of cellular organelles for the storage of hormone. Alternatively, the differences in hormone content may in part reflect the relative states of metabolic activities of the tissues—a higher turnover of hormone in the

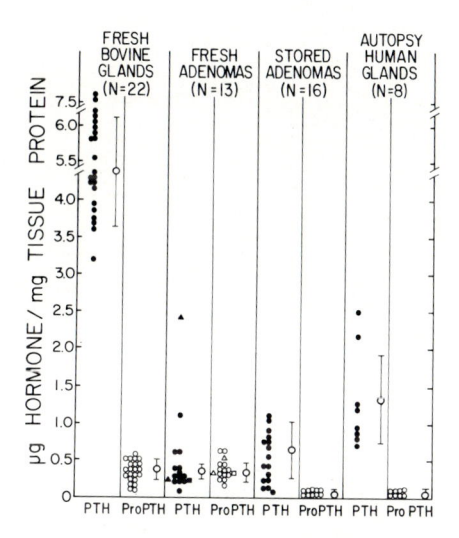

FIG. 13. Amounts of proparathyroid hormone (ProPTH, ○) and parathyroid hormone (PTH, ●) measured by radioimmunoassays in extracts of bovine and human parathyroid tissues. Fresh adenomas were human parathyroid adenomas that were extracted immediately after their surgical removal. Stored adenomas had been frozen (−70°C) for several weeks before extraction. Larger circles in each column represent means ± SEM for the values. Triangular and square symbols denote primary (chief-cell) hyperplasia and parathyroid carcinoma, respectively.

neoplastic adenoma tissue may lead to the storage of a smaller fraction of the hormone synthesized. Additional experience with the radioimmunoassays for ProPTH has brought to light the marked instability of ProPTH in biological fluids such as blood plasma or sera, owing to the presence of trypsinlike activity that readily converts ProPTH to PTH (Fig. 14). It was possible, however, to inhibit the proteolysis of ProPTH by addition of EDTA to samples of blood at the time of collection and by the use of heat-inactivated serum in the buffers used in the immunoassays.

We have been interested in ascertaining whether ProPTH is secreted into the circulation with the expectation that assessments of ProPTH levels in blood might provide a means for the detection of abnormally functioning parathyroid glands. We had expected that, as had been found with neoplastic pancreatic islet tissue, neoplastic parathyroid glands (adenomas) might secrete excessive amounts of the ProPTH. Our earlier studies of the biosynthesis of ProPTH by human parathyroid adenomas suggested that rates and efficiencies of conversion of ProPTH to PTH were impaired in some adenomas (Habener et al., 1972). Studies thus far, however, have failed to detect immunoreactive ProPTH in

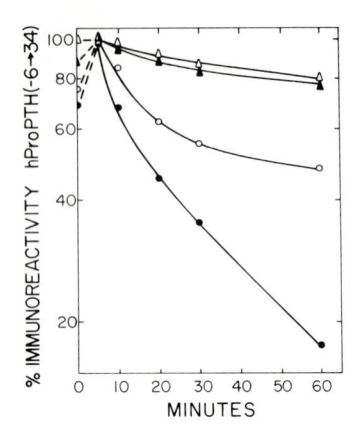

FIG. 14. Effects of temperature and of EDTA on the stability of immunoreactive synthetic human proparathyroid hormone, ProPTH-(−6→+34), in human plasma *in vitro*. Initial concentrations of immunoreactivity ProPTH added to plasma were 20 ng/ml. Aliquots of plasma incubated in an ice bath (0°C) with EDTA (10 m*M*) (△) and without EDTA (▲); and incubated at 37°C with EDTA (10 m*M*) (○) and without EDTA (●). From Habener *et al.* (1976a).

either peripheral blood or venous effluent blood of bovine and human parathyroid glands, where high concentrations of PTH are readily measured (up to 190 ng/ml) (Habener *et al.*, 1976a).

The assays of ProPTH in parathyroid effluents were performed under conditions in which ProPTH was shown to remain immunologically stable. Thus, we can conclude that ProPTH, if released, can only comprise less than 1–2% of the secreted immunoreactive hormone. It does not appear likely, therefore, that assays for ProPTH will be useful in assessments of parathyroid gland secretory activity. However, application of the antisera to analyses, by cellular fractionation and immunofluorescence–electron microscopical techniques, of the subcellular distribution of ProPTH and use of the assay in measurements of ProPTH in parathyroid tissues under stimulated and suppressed conditions will provide important information about the cellular pathways and the regulation of hormone biosynthesis in the parathyroid gland (Habener *et al.*, 1975a, 1976c).

IX. Identification of Pre-Proparathyroid Hormone by Translation of Parathyroid mRNA in a Cell-Free System Derived from Wheat Germ

Several observations made during the studies of ProPTH stimulated us to search for a yet larger biosynthetic precursor of PTH. First was the finding, by sequence analysis, of only six additional amino acids at the amino terminus of PTH (Hamilton *et al.*, 1974), whereas analysis of the total amino acid composition of ProPTH indicated that there were 19–20 amino acids in addition to those in PTH (Cohn *et al.*, 1972b; Hamilton *et al.*, 1974). A second observation was the finding by SDS–gel electrophoresis that the ProPTH had an apparent molecular weight of 11,500–12,000, much higher than the 10,200 for a 90-amino acid ProPTH (Cohn *et al.*, 1972b; Kemper *et al.*, 1972). These observations raised the possibility (yet unresolved) that additional sequence may have been present at the carboxyl terminus of PTH. In addition, some heterogeneity was observed in the ProPTH that was sequenced. Approximately 12% of the sample lacked the amino-terminal lysine, suggesting that more than one species of prohormone differing at the amino terminus might exist (Hamilton *et al.*, 1974).

The opportunity became available for us to examine directly the possibility of the existence of a precursor of PTH larger than ProPTH. The development by Roberts and Paterson (1973) of a highly sensitive cell-free system derived from wheat germ made it possible to translate relatively crude preparations of mRNA under conditions in which virtually no proteolytic activity was present to modify the products of translation. We could therefore analyze the protein products synthesized in response to parathyroid mRNA without the interference of posttranslational cleavages of the polypeptide chains (Kemper *et al.*, 1974a, 1976a,b; Habener *et al.*, 1975c). Parathyroid RNA and a cell-free extract of wheat germ were prepared as outlined in Fig. 15. The fraction of RNA sedimenting between 8 S and 15 S on a sucrose gradient and extracts of wheat germ were combined and were incubated in a cell-free system containing radioactive amino acids that label the newly synthesized proteins (Fig. 16).

The acid-insoluble radioactive proteins that were synthesized in response to the parathyroid RNA were analyzed by electrophoresis on polyacrylamide gels (Fig. 17). The significant finding was that a single band of radioactivity predominated in the electrophoretic patterns, and this protein migrated more slowly than either ProPTH or PTH, indicating that it was of a higher molecular weight than either of the two other forms of the hormone. We termed this major cell-free product (CFP) pre-proparathyroid hormone (Pre-ProPTH) because of the likelihood, not proved at the time, that it was a biosynthetic precursor of ProPTH (Kemper *et al.*, 1974a). Electrophoretic and chromatographic analyses of the radioactive peptide fragments found after cleavages of the protein isolated from the polyacrylamide gels indicated that the protein contained peptides that were indistinguishable from those of ProPTH and PTH and also that it contained additional peptides not found in ProPTH or PTH (Fig. 18). More-detailed analyses of the peptide fragments of the Pre-ProPTH produced by cleavage with cyanogen bromide revealed that the additional sequence of amino acids of the

1. PREP. OF mRNA

 PARATHYROID TISSUE
 | HOMOGENIZE, 0°
 | SUCROSE-TRIS-**SDS**-EDTA
 ↓ CENT. 1000   G, 5 MIN

 PARATHYROID EXTRACT
 | DEPROTEINIZE
 | PHENOL-CHLOROFORM
 ↓ ETHANOL PPTN
RNA
 | SEDIMENTATION
 ↓ SUCROSE GRADIENT
mRNA (8-15 S)

2. PREP. OF WHEAT GERM EXTRACT

 COMMERCIAL WHEAT GERM
 | GRIND-HEPES
 | K+, Mg^{2+}, Ca^{2+}, 2-ME
 ↓ CENT 30 K G, 10 MIN

 S-30 FRACTION
 | PREINCUBATE
 | 30°, 15 MIN
 | ATP, GTP, CP, CPK
 ↓ SEPHADEX G-10
 FROZEN EXTRACT

HEPES, Mg^{2+}, K+, DTT
ATP, GTP, CP, CPK
AMINO ACIDS
[^{35}S] MET, [^{3}H] LYS

PROTEIN SYNTHESIS ASSAY
25-100 μL
23°, 3 HR

↓

CELL-FREE PRODUCTS

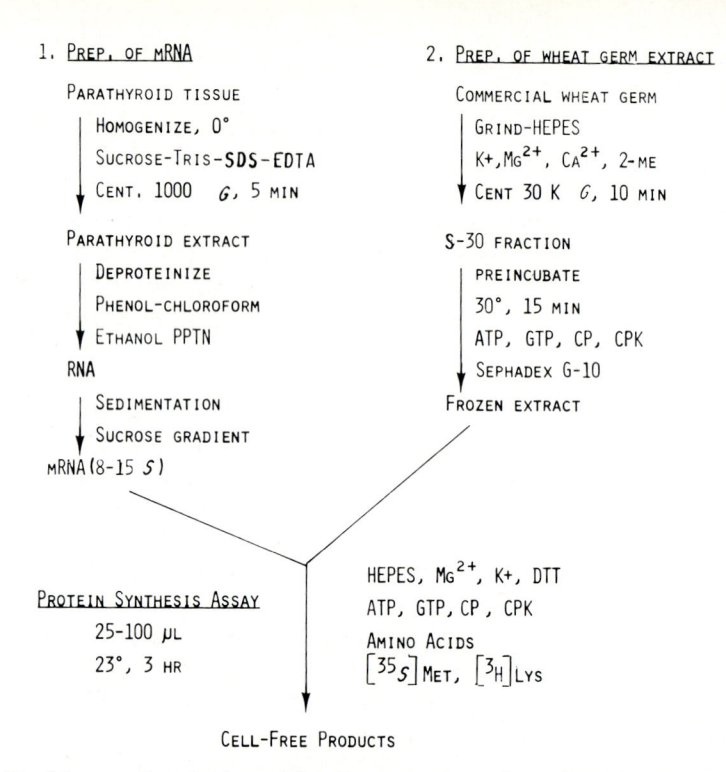

FIG. 15. Schemas of methods used for the preparations of parathyroid messenger RNA and extracts of wheat germ and for the protein-synthesis assay. Cent, centrifugation; PPTN, precipitation.

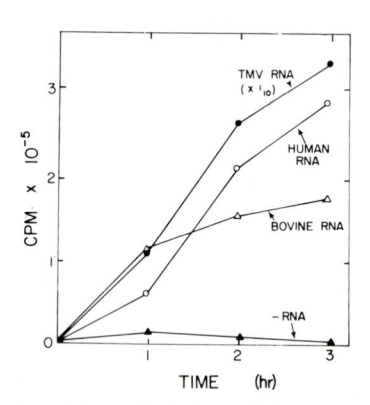

FIG. 16. Stimulation of protein synthesis in a cell-free extract of wheat germ by parathyroid RNA. Four micrograms of 8–15 S parathyroid RNA and 1.3 μg of tobacco mosaic virus (TMV) RNA (as a control) were incubated in 25-μl reaction mixtures for 3 hours at 23°C; 5 μCi of [^{35}S]methionine (132 Ci/mmole) were added to the reaction mixtures. Reagents used in the cell-free system are those described by Kemper et al. (1974a). Values shown are total trichloroacetic acid-insoluble radioactivity. Note the extremely low protein synthesis in the absence of added mRNA (−RNA), characteristic of the wheat-germ extract.

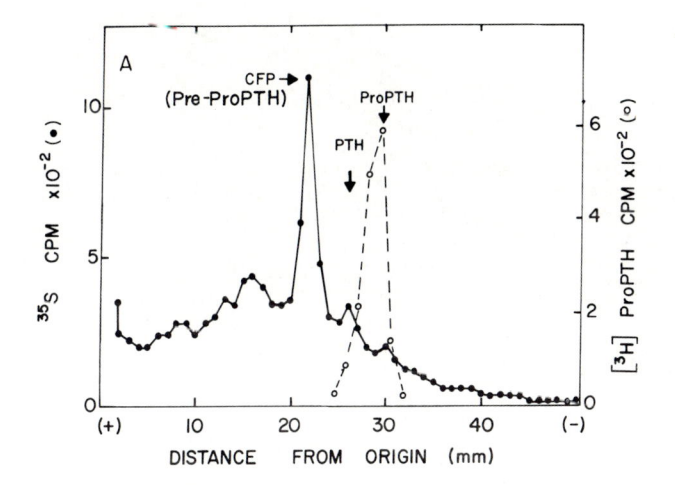

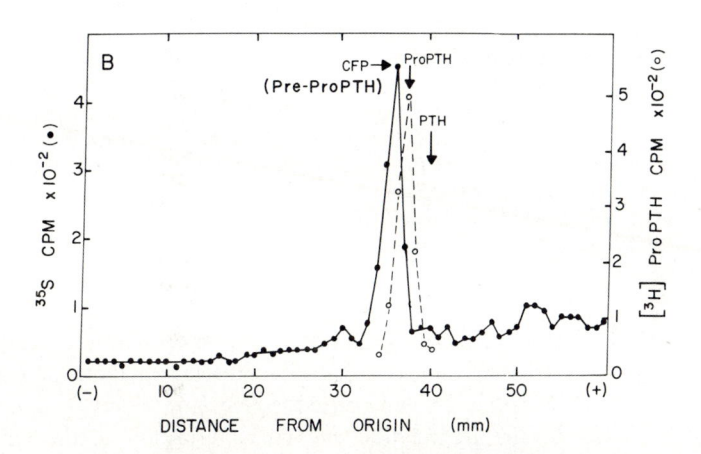

FIG. 17. Analysis by acrylamide gel electrophoresis of the cell-free products of protein synthesis directed by a 8–15 S fraction of RNA from bovine parathyroid glands. The proteins labeled with [^{35}S] methionine from a cell-free incubation were prepared for gel electrophoresis as described by Kemper et al. (1974a). ^{3}H-labeled proparathyroid hormone ([^{3}H] ProPTH), isolated as described previously (Kemper et al., 1972), was added as a marker. The positions of ProPTH and of PTH relative to the major cell-free product (CFP) was determined on separate gels. The CFP was subsequently shown to be a biosynthesis precursor of ProPTH termed pre-proparathyroid hormone (Pre-ProPTH). (A) The equivalent of 10 μl of a 25-μl incubation was analyzed in 10% acrylamide gels at pH 4 in 8 M urea. (B) A protein sample corresponding to 2.5 μl of the incubation mixture was adjusted to 0.1 M sodium phosphate, pH 7.2, 0.5% SDS, 4 M urea, and 1% 2-mercaptoethanol, and heated 100°C for 30 seconds before SDS–acrylamide gel electrophoresis. ●——●, [^{35}S] methionine-labeled cell-free proteins; o— —o [^{3}H] ProPTH. Modified from Kemper et al. (1974a).

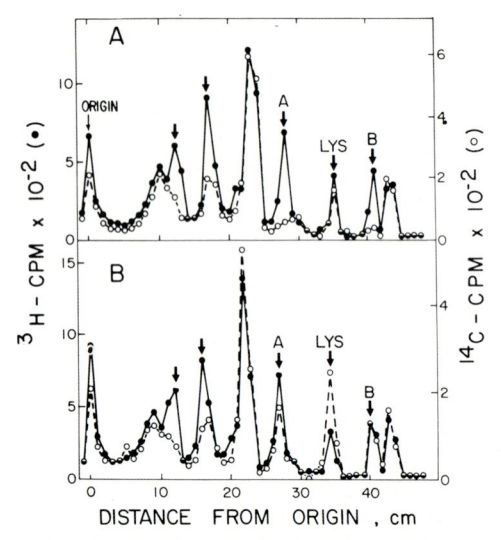

FIG. 18. Coelectrophoresis on paper of the lysine-containing tryptic peptides of (A) parathyroid hormone (PTH) and pre-proparathyroid-hormone (Pre-ProPTH) and (B) ProPTH and Pre-ProPTH. To obtain [^{14}C]lysine-labeled PTH and ProPTH, slices of parathyroid tissue were incubated with 20 μCi/mole of [^{14}C]lysine (0.30 Ci/mmole) in Earle's balanced salt solution with 5% fetal bovine serum for 25 minutes (ProPTH) or 75 minutes (PTH). Pre-ProPTH, ProPTH, and PTH were isolated from sodium dodecyl sulfate–acrylamide gels. Arrows A and B indicate the peaks present in Pre-ProPTH and ProPTH but not PTH, and the unlabeled arrows indicate the extra peaks in Pre-ProPTH compared with ProPTH. ●———●, [^{3}H] Lysine-labeled peptides of Pre-ProPTH; ○– – – –○, [^{14}C]lysine-labeled peptides of PTH (upper panel) or ProPTH (lower panel). From Kemper et al. (1976a).

Pre-ProPTH is at the amino-terminal, not the carboxyl-terminal, end of the polypeptide. The existence of a polypeptide extension at the amino terminus made it feasible to determine the amino acid sequence of the prohormone.

Further studies of the purification and characterization of the mRNA for Pre-ProPTH are in progress. The molecular weight of the parathyroid mRNA is approximately 190,000 as determined by sedimentation analysis on sucrose gradients containing formamide (Kemper and Stolarsky, 1976). This molecular weight corresponds to 550 nucleotides, approximately 200 more than the 345 that are necessary to encode for the 115 amino acids of Pre-ProPTH (3 nucleotides/codon/amino acid). Some, but not all, of these additional nucleotides can be accounted for by the presence of a sequence of polyadenylate at the 3' end of the mRNA. Excess polynucleotide sequence in the mRNAs with respect to the size of the proteins for which they code has been a common finding (Swan et al., 1972; Matthews, 1973; Brawerman, 1974; Marotta et al., 1974). The presence of a polyadenylate sequence in the mRNA for Pre-ProPTH, a finding common to most mRNAs, has facilitated its purification by selective adsorption to and elution from oligo(2′-deoxythymidylic acid)-cellulose columns (Matthews, 1973; Brawerman, 1974) (see B. W. Kemper in the Discussion). Studies are under way using the reverse transcriptase to prepare deoxyribonucleic-acid transcripts complementary to the mRNA for Pre-ProPTH. These DNA copies of the mRNA can then be used for sequence analyses and for construction of hybridization assays for the detection of mRNA levels in parathyroid tissues under various physiological conditions.

X. Primary Structure of Pre-Proparathyroid Hormone

Because of the extremely small absolute amounts of Pre-ProPTH synthesized in the wheat germ system it was not possible, using conventional chemical methods, to identify the amino acids released by the sequential Edman degradation. We therefore resorted to a determination of the amino acid sequence of the radioactively labeled Pre-ProPTH prepared in the wheat germ cell-free system (Kemper et al., 1976a). This microsequencing technique (McKean et al., 1974) simply involves isolation of several different preparations of the Pre-ProPTH synthesized in separate incubations in the cell-free system in the presence of different combinations of radioactive amino acids. Previously, the method had been employed successfully for the determination of the amino-terminal sequence of human ProPTH (Jacobs et al., 1974; Huang et al., 1975). The radioactively labeled protein, plus 2 mg of apomyoglobin added as a carrier, was sequentially degraded in the Beckman 890 Sequenator (Fig. 19). The radioactivity released at each cycle of the degradation was measured by counting an aliquot of the butyl chloride extract containing the phenylthiohydantoin derivatives of the amino acids; in each case, the phenylthiohydantoin derivative of the residue released from myoglobin and, if present, a radioactive residue from the hormone were analyzed on thin-layer chromatography (Jacobs et al., 1974). Verification was thereby obtained that the radioactivity detected at any cycle corresponded to a specific amino acid. The results of approximately 20 degradations carried out to date have revealed the tentative primary structure for Pre-ProPTH shown in Fig. 20. The Pre-ProPTH synthesized in the wheat germ cell-free system is a polypeptide of at least 115 amino acids (Kemper et al., 1976a). The "pre"-peptide consists of a sequence of 25 amino acids covalently attached by a glycyl–lysyl bond to the sequence of ProPTH. One of the amino acids in the "pre" sequence (position −14) has not been identified at this time. Somewhat unusual features of the "pre" sequence are that 5 of the 25 amino acids are methionines, and the overall topography indicates that the sequence is quite hydrophobic.

During the course of these structural analyses by the microsequence technique, a number of issues relating to the reliability of the methods used and the

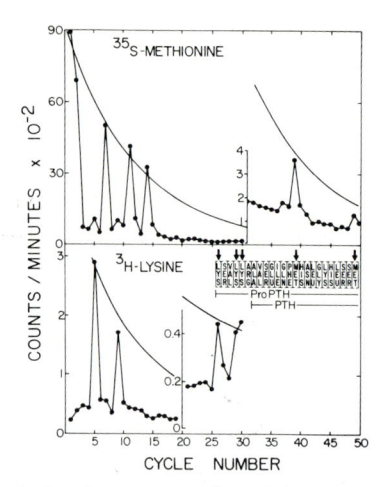

FIG. 19. Sequence analysis of pre-proparathyroid hormone (Pre-ProPTH) labeled with methionine or lysine. The methionine data are from Pre-ProPTH labeled with 400,000 cpm of [^{35}S]methionine (293 Ci/mmole). The lysine data are from Pre-ProPTH labeled only with 50,000 cpm of [^{3}H]lysine. An aliquot of one-third of the total radioactivity released at each degradation cycle of automated Edman degradation was assayed for radioactivity. The identity of the amino acid was confirmed by thin-layer chromatography. The continuous curved lines represent the theoretical recoveries of methionine or lysine at each cycle determined from a least-squares analysis of the log of the [^{35}S]methionine or [^{3}H]lysine radioactivity recovered plotted against the number of cycles. The continuation of the sequential degradation into the amino acid sequences of ProPTH and PTH is indicated by the inset. From Kemper *et al.* (1976a).

correctness of the structure deduced have been systematically evaluated and analyzed. The deductions reached have wider applicability, presumably, to studies in other systems involving study of prehormones or early biosynthetic precursors of other secretory proteins. All the early gene products (as discussed in Section XIII) are present in extremely minute amounts, picograms to nanograms, at most. In one sense, structural analysis of these compounds involves study of a substance that is simply not present if the sensitivity limits available by conventional sequencing techniques are considered as the criteria for detection. Only the great sensitivity afforded by the radioactivity present in the amino acids incorporated permits analysis. Great care must be taken to avoid errors in sequence analysis involving biosynthetic incorporation of trace quantities of amino acids; the issues include (1) biotransformation of one amino acid to another or to an altered form (such as transamidation of glutamic acid and deamidation of glutamine to glutamic acid) before incorporation; (2) failure to incorporate amino acids adequately in certain incubations because of interfering substances in the radioactive amino acids used or other problems affecting the

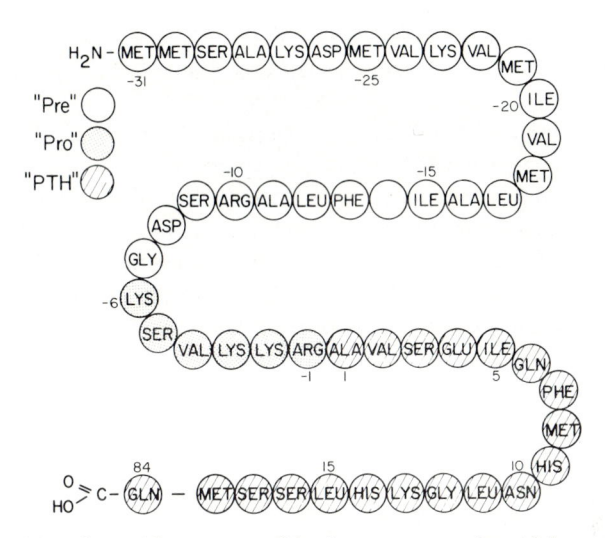

FIG. 20. Partial amino acid sequence of bovine pre-proparathyroid hormone as determined by microsequencing technique. The radiolabeled prehormone was synthesized in the cell-free extract of wheat germ by addition of parathyroid messenger RNA and radioactive amino acids. The NH$_2$-terminal methionine (residue −31) is the initiator amino acid not removed in the wheat germ system.

incorporation efficiency; (3) faulty chemical degradation or errors in analysis of the residues released. We have found that certain techniques are extremely helpful in guarding against these problems. Carrier-protein sequence is carefully analyzed to monitor the chemical degradation and thereby avoid errors in assignment of radioactive residues due to faulty repetitive degradation, such as assigning a given residue to two adjacent positions because of incomplete removal of the residue at the first position. All incubations should be performed several times to avoid errors due to mixing of amino acids; the success of incorporation should be monitored by including residues known to be present in the hormone itself. Degradations should be carried out for many cycles, preferably deep into the hormone region itself (Fig. 19). Detection of a given residue at, and only at, positions in the sequence of the hormone itself already established by conventional sequence studies provides assurance against biotransformation-induced errors. Amino acids can be confidently eliminated if they are not present within the prehormone-specific region but are detected within the prohormone or hormone sequence; such a finding ensures that the residue was incorporated into the nascent chain with sufficient efficiency to be detected at a number of cycles of degradation beyond the region of unknown sequence. Multiple controls of this type, coupled with careful chemical identification of radioactive residues, seem to provide the microsequence approach with the same degree of confidence concerning sequence deductions as do conventional Edman degradations.

As we had done with the ProPTH (Section V), analysis of the carboxyl-terminal region of the Pre-ProPTH was approached by examination of the carboxyl-terminal tetrapeptide labeled with [^{3}H]proline (Fig. 21). The chromatographic and electrophoretic mobilities of the tetrapeptide Ala–Lys–[^{3}H]Pro–Gln were identical to those of the comparable [^{14}C]proline-labeled tetrapeptide of PTH that was analyzed simultaneously with the tetrapeptide of Pre-ProPTH. We interpret these findings to indicate that there is probably no additional sequence of amino acids at the carboxyl terminus of the Pre-ProPTH. Furthermore, the findings suggest that the wheat germ cell-free system has faithfully transcribed the entire mRNA and terminated translation at a natural

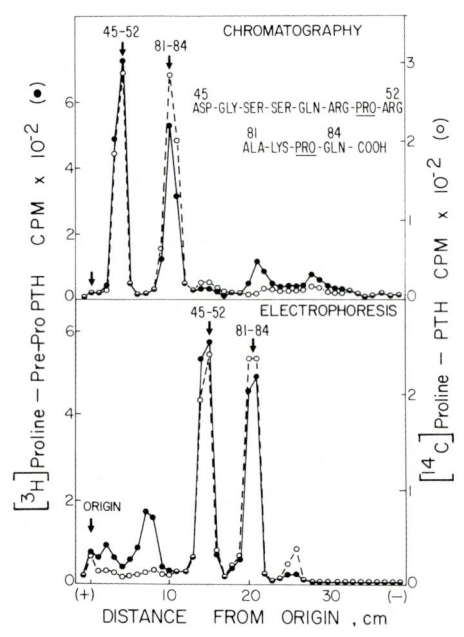

FIG. 21. Paper electrophoresis and chromatography of [^{3}H]proline-labeled pre-proparathyroid hormone (Pre-ProPTH) and [^{14}C]proline-labeled parathyroid hormone (PTH). PTH was isolated from parathyroid slices incubated for 60 minutes with [^{14}C]proline. Pre-ProPTH was isolated from a 50-μl wheat-germ reaction containing 20 μCi of [3,4-^{3}H]proline (34 Ci/mmole) and about 4 μg of 8–15 S parathyroid RNA. The relative migrations of the internal peptide, 45–52, and the carboxyl-terminal peptide, 81–84, are in agreement with previous studies (Potts *et al.*, 1968). Samples were electrophoresed for 2 hours at 3000 kV and chromatographed for 36 hours. ○– – – –○, [^{14}C]PTH; ●———●, [^{3}H]Pre-ProPTH. From Kemper *et al.* (1976b).

termination codon located adjacent to the codon for glutamine, the carboxyl-terminal amino acid of PTH.

We next turned our attention to the question of the fidelity of the wheat-germ cell-free system in the initiation of the translation of the parathyroid messenger RNA. The finding of a methionine as the amino-terminal residue of the Pre-ProPTH (position −31) suggested that it might be the initial amino acid incorporated into the nascent polypeptide chain in response to the initiator AUG codon. All proteins synthesized in eukaryotes are thought to begin with methionine, which is normally removed *in vivo* when the growing nascent chain is approximately 25−30 amino acids long (Wilson and Dintzis, 1970; Jackson and Hunter, 1970; Yoshida *et al.*, 1970; Koffer-Gutmann and Arnstein, 1973). Presumably, this is the length of the nascent chain required for the amino terminus to first emerge from the large ribosomal subunit. Cells contain two distinct transfer RNAs for methionine; one incorporates methionine only at the amino terminus of proteins in response to the specific initiator AUG codon; the other incorporates methionine into the internal positions of the polypeptide in response to internal AUG codons (Lengyel and Sall, 1969; Schreiber, 1971; Lucas-Lenard and Lipmann, 1971).

The procedure that we followed was to carry out a cell-free synthesis in the presence of initiator methionyl tRNA charged with [^{35}S]methionine, excess unlabeled methionine for amino acylation of the methionyl tRNA that recognizes internal AUG codons, and [^{3}H]lysine to serve as an internal label of the Pre-ProPTH (Fig. 22). After the cell-free synthesis, Pre-ProPTH containing both [^{35}S]methionine and [^{3}H]lysine was identified by electrophoresis on SDS–acrylamide gels (Fig. 23A). An aliquot of the radioactive proteins was then subjected to a single-step Edman degradation that removes only the amino-terminal amino acid. After the Edman reaction, the proteins were again analyzed by electrophoresis (Fig. 23B). The electrophoretic analyses showed that there was a marked reduction in the [^{35}S]methionine, but not the [^{3}H]lysine, content of

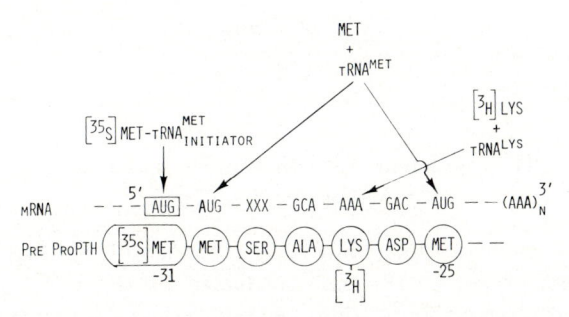

FIG. 22. Schema showing experimental approach used to demonstrate that the amino-terminal methionine (position −31) is the initiator amino acid of pre-proparathyroid hormone (Pre-ProPTH).

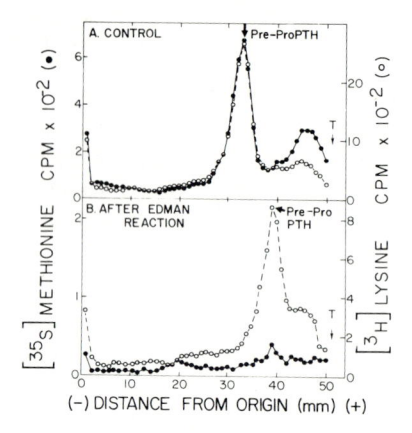

FIG. 23. Sodium dodecyl sulfate (SDS)-gel electrophoresis of protein labeled with initiator [^{35}S]Met-tRNA^{fMet} and [^{3}H]lysine. A wheat-germ reaction was incubated for 3 hours, and the proteins were analyzed by SDS–acrylamide gel electrophoresis. Panel A (control) shows the products of the reaction, whereas panel B shows proteins after a single-step Edman degradation. The [^{3}H]lysine-labeled material serves as an internal marker and a monitor of recovery. ●———●, [^{35}S]Met-tRNA^{fMet}; o— — —o, [^{3}H]lysine. From Kemper et al. (1976b).

the Pre-ProPTH. Thus, we were able to conclude that the amino-terminal methionine of Pre-ProPTH was indeed the initial amino acid that begins the primary sequence of Pre-ProPTH (Kemper et al., 1976b). These observations indicated that initiation of the translation of the parathyroid mRNA was occurring specifically in response to an initiator codon and further indicated that the wheat-germ system apparently lacks the proteolytic activity that removes the initiator methionines from the Pre-ProPTH.

In summary, we believe that these data on the primary structure of Pre-ProPTH strongly indicate that the complete translation of the parathyroid mRNA, including initiation, chain elongation, and termination, is executed faithfully in the wheat-germ system.

XI. Translation of Parathyroid Messenger RNA in an Ascites Cell-Free System

In addition to studies of the products of translation of parathyroid mRNA in the wheat-germ system, we also have conducted less extensive studies of the mRNA-directed products formed in a cell-free system using extracts of Krebs II-ascites tumor cells (Habener et al., 1975b; Kemper et al., 1976b). Information derived from these studies has supplemented our understanding of the processes involved in the biosynthesis of parathyroid hormone. First, they have provided the opportunity to analyze the products of synthesis independently in a second heterologous cell-free system, and, second, it seemed possible that the ascites system contains proteolytic activity that converts larger translation products to smaller ones; that is, it could convert Pre-ProPTH to ProPTH. We found that the major product seen in the Krebs-ascites system was ProPTH (Fig. 24), whereas the only form of the hormone synthesized in the wheat-germ cell-free system was Pre-ProPTH. Furthermore, the small amount of Pre-ProPTH in the ascites system identified by gel electrophoresis was found by analyses of the partial amino acid sequence of [^{35}S] methionine-labeled material to contain two forms of the Pre-ProPTH: one identical to the prehormone synthesized in the wheat-germ system, and another that was lacking the two amino-terminal methionines (Fig. 25). It became apparent from these studies that the ascites system contains specific proteolytic activity for both the selective removal of amino-terminal methionines and the conversion of Pre-ProPTH to ProPTH.

A third finding during studies of PTH synthesis in the ascites system was that proteolytic conversion of the Pre-ProPTH to ProPTH is dependent on translation; only those Pre-ProPTH molecules undergoing synthesis were proteolytically cleaved to give ProPTH. Pre-ProPTH isolated previously from a wheat-germ

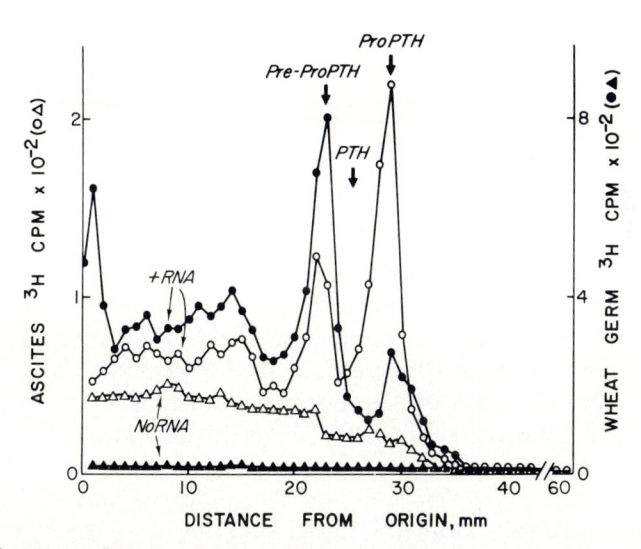

FIG. 24. Electrophoretic profiles (acetate–urea polyacrylamide gels) of [^{3}H]lysine-labeled products synthesized in response to 8–15 S fraction of parathyroid RNA in a Krebs II-ascites cell-free system (open symbols) and in a wheat-germ cell-free system (filled symbols). Arrows indicate electrophoretic positions of pre-proparathyroid hormone (Pre-ProPTH), proparathyroid hormone (ProPTH), and parathyroid hormone (PTH). Migration is from left to right.

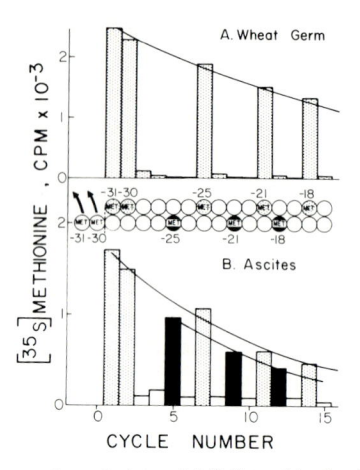

FIG. 25. Sequential Edman degradation of [^{35}S] methionine-labeled pre-proparathyroid hormone (Pre-ProPTH) isolated from parathyroid RNA translation products in cell-free systems derived from wheat germ (A) and Krebs II-ascites tumor cells (B). Bars denote ^{35}S radioactivity released at each cycle of the degradation. Repetitive yields of amino acids released from apomyoglobin carrier at each step were determined by gas chromatography, and the expected yield of radioactivity at each cycle relative to the initial peak is indicated by the curved lines. Note [^{35}S] methionine (B) released at cycles 5, 9, 12 (solid bars) consistent with lower-sequence alignment (terminal methionines removed via cleavage) in addition to pattern of radioactivity seen at cycles 1, 2, 7, 11, and 14 with alternate upper-sequence alignment representing intact Pre-ProPTH also present in ascites system and solely present (A) in wheat-germ system (stippled bars). From Habener *et al.* (1975b).

system and added exogenously to an active protein-synthesizing ascites system remained intact and did not undergo proteolysis (Habener *et al.*, 1975b). Similar results were reported from studies of the translation products synthesized in the ascites system and other cell-free systems containing membranes in response to messenger RNAs from other tissues. The principal translation products of the mRNAs for growth hormone (Bancroft *et al.*, 1973), placental lactogen (Boime *et al.*, 1975), and L chains of immunoglobulins were shown to be the final authentic protein (Milstein *et al.*, 1972; Blobel and Dobberstein, 1975), whereas translations of these same mRNAs in the wheat-germ cell-free system yielded predominantly larger precursor forms of the proteins (Swan *et al.*, 1972; Schechter, 1973; Boime *et al.*, 1975; Blobel and Dobberstein, 1975; Cox *et al.*, 1976). The proteolytic activity in the ascites system, like that in the normal cells of origin of these secreted proteins, is thought to reside in the endoplasmic reticular membranes in the system; proteolytic cleavages of preplacental lactogen (Szczesna and Boime, 1976) and of presecretory immunoglobulin L chains (Blobel and Dobberstein, 1975) have been brought about by addition of membranes to cell-free systems that ordinarily synthesize only the larger pre-forms of the proteins. Apparently, most preparations of wheat-germ extract used in cell-free syntheses do not contain proteolytically active membranes. Continued studies of the properties of the membrane fractions of cell-free systems such as ascites-cell extracts, as well as the cells from which secretory proteins originate, should lead to important new information regarding (1) the nature and location of the proteolytic enzymes involved in the posttranslational modifications of such proteins and (2) the general biological role and significance of the sequential steps of biosynthesis (see Section XIV).

XII. Identification of Pre-Proparathyroid Hormone upon Reexamination of Hormone Biosynthesis in Parathyroid Gland Slices

The discovery of Pre-ProPTH as a major product of the translation of parathyroid mRNA in heterologous cell-free systems strongly suggested that the Pre-ProPTH was a true biosynthetic precursor of ProPTH and of PTH. That is, Pre-ProPTH is the initial product of the parathyroid genome in the course of biosynthesis of the hormone in the parathyroid glands *in vivo*. The distinct possibility remained, however, that Pre-ProPTH could arise from an artifactual synthesis in the cell-free systems. For example, initiation of translation could have occurred at an AUG codon within a region of the mRNA located either proximal or distal to the true initiation site. For this reason, and also to analyze the properties and fate of Pre-ProPTH within the parathyroid cell, we attempted an identification of the Pre-ProPTH in intact parathyroid gland slices utilizing the information derived from the studies in the cell-free systems (Habener *et al.*, 1976b).

Bovine parathyroid gland slices were incubated *in vitro* with [^{35}H] methionine for periods of 1–10 minutes. These times were much shorter than those used in earlier pulse and pulse-chase incubations in studies of ProPTH synthesis (Kemper *et al.*, 1972).

The proteins synthesized by the parathyroid slices during these short periods of incubation were analyzed by electrophoresis on acetate–urea polyacrylamide gels (Fig. 26). A small peak of ^{35}S radioactivity could be detected on the gel that comigrated with ^{3}H-labeled Pre-ProPTH prepared in the wheat-germ cell-free system. The major protein visible by 1 minute of incubation was ProPTH. Pulse-chase and immunoprecipitation studies, however, suggested that the ^{35}S-labeled radioactive material in the region of the [^{3}H] Pre-ProPTH was heterogeneous. The ^{35}S-labeled proteins isolated from the 3–4 mm of the acetate–urea gels where [^{3}H] Pre-ProPTH appeared were reanalyzed on SDS–urea gels, thus revealing a rather marked heterogeneity of the material that appeared as a single peak on the acetate–urea gels (Fig. 27). Four distinct

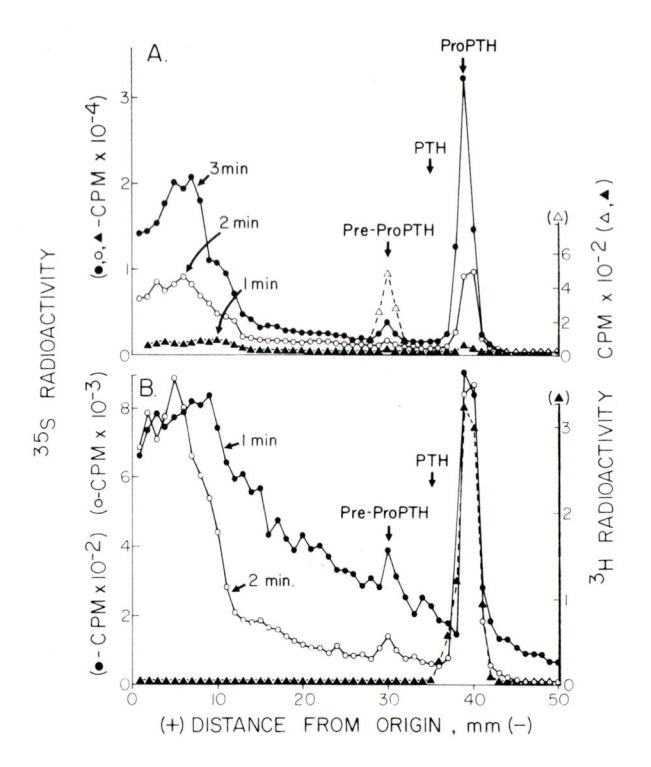

FIG. 26. Electrophoretic patterns on acetate–urea gels of extracts of parathyroid slices incubated with [^{35}S]methionine for 1–3 minutes. Panel A: A marker of [^{3}H]lysine-labeled pre-proparathyroid hormone (Pre-ProPTH), prepared by translation of parathyroid mRNA in a wheat-germ cell-free system, was coelectrophoresed with the [^{35}S]methionine-labeled proteins. Panel B: A marker of [^{3}H]leucine-labeled proparathyroid hormone (ProPTH) was coelectrophoresed with the [^{35}S]methionine-labeled proteins. Peaks of radioactivity corresponding to ProPTH have been coplotted for ease of comparison. T, position of migration of methyl-green tracker dye. Modified from Habener *et al.* (1976b).

proteins were evident on the SDS–urea gels (peaks I–IV). Only one of the four proteins (peak II) corresponded in electrophoretic mobility to the [^{3}H]Pre-ProPTH marker prepared in the wheat-germ cell-free system.

The [^{35}S]Pre-ProPTH could be identified on the SDS–urea gels by 2 minutes of incubation, by which time it had reached a relative steady-state concentration in the tissue. A short, 3-minute chase incubation with unlabeled methionine after a 3-minute pulse incubation resulted in a marked reduction in the [^{35}S]Pre-ProPTH (Fig. 27). Direct evidence that the ^{35}S-labeled protein in peak II of the SDS–urea gels was related to Pre-ProPTH was obtained by an analysis of the ^{35}S-labeled tryptic peptides derived from the protein (Fig. 28). A

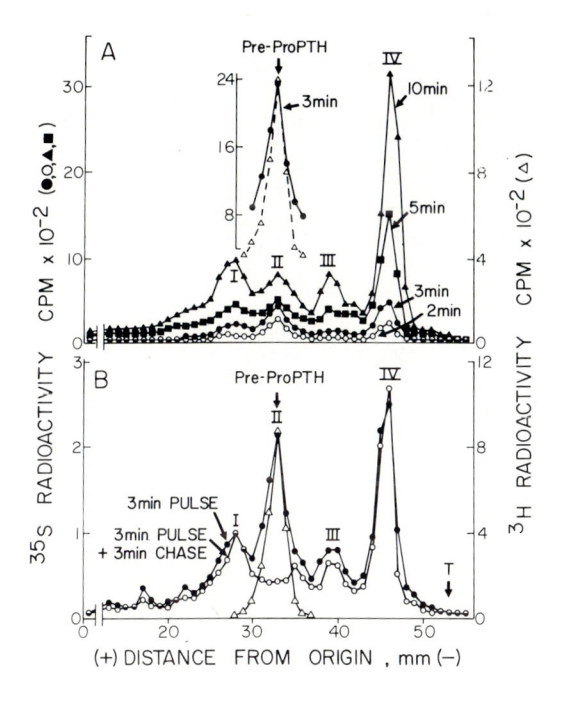

FIG. 27. Reelectrophoresis on sodium dodecyl sulfate–urea polyacrylamide gels of [^{35}S]methionine-labeled proteins in peak corresponding to pre-proparathyroid hormone (Pre-ProPTH) shown in Fig. 26 isolated from acetate–urea gels (Figs. 1 and 2). Tissue extracts that were electrophoresed were prepared from parathyroid slices (panel A) pulse-labeled for 2–10 minutes and (panel B) pulse-labeled for 3 minutes (●) followed by a 3-minute chase incubation (○). [^{3}H]Lysine-labeled Pre-ProPTH, prepared in the wheat-germ cell-free system, was coelectrophoresed with each extract (△). Inset in upper panel (A) shows region of gel from 29 to 36 mm on an expanded scale. I to IV, discrete bands of radioactivity on the gel. From Habener et al. (1976b).

summary of the kinetics of the labeling of Pre-ProPTH, ProPTH, and PTH in the parathyroid gland slices is shown in Fig. 29. Pre-ProPTH and ProPTH are seen at 2 minutes and 1 minute of incubation, respectively. Pre-ProPTH decreases within 3 minutes of a chase incubation. ProPTH, however, does not begin to decrease until approximately 15 minutes of incubation, corresponding in time to the first appearance of PTH (see also Section V). It must be noted, however, that the amount of radioactive Pre-ProPTH relative to the amounts of radioactive ProPTH and PTH is very small; radioactive Pre-ProPTH reaches a relative steady state at a level of 500 cpm per milligram of tissue protein, whereas ProPTH appears at a rate of 25,000 cpm/min per milligram of tissue protein during the first 15 minutes of incubation. Because of this very large discrepancy in the amounts of Pre-ProPTH relative to ProPTH, it has not been possible to quantita-

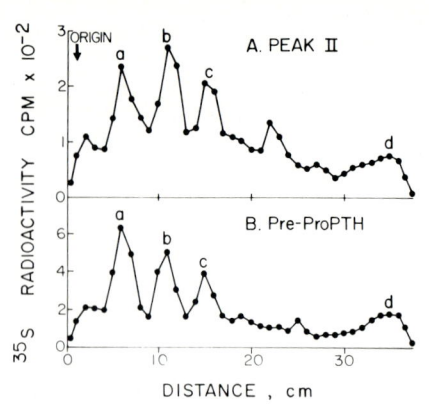

FIG. 28. Chromatographs on paper of [^{35}S]methionine-labeled tryptic peptides prepared from (A) the protein of peak II isolated from a sodium dodecyl sulfate–urea gel (3-minute pulse-label, Fig. 27); and (B) pre-proparathyroid hormone (Pre-ProPTH) synthesized in response to parathyroid mRNA in a wheat-germ cell-free system. From Habener et al. (1976b).

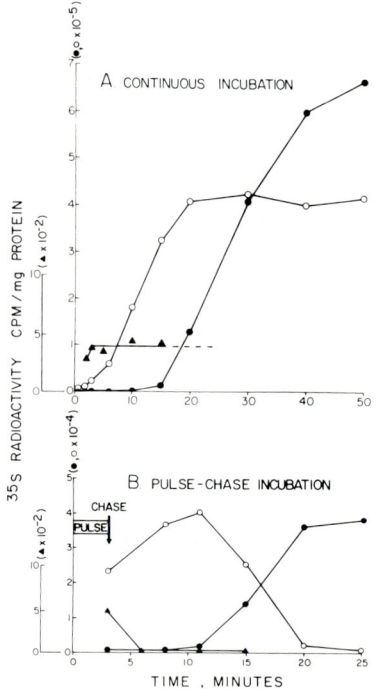

FIG. 29. Kinetics of incorporation by slices of parathyroid glands of [^{35}S]methionine into pre-proparathyroid hormone (▲——▲, proparathyroid hormone (○——○, ProPTH), and parathyroid hormone (●——●, PTH) during (A) a continuous incubation with [^{35}S]methionine and (B) a 3-minute pulse incubation with [^{35}S]methionine followed by a chase incubation with unlabeled methionine. Amounts of radioactive ProPTH and PTH were determined by electrophoresis on acetate–urea gels. Amounts of Pre-ProPTH were determined by reelectrophoresis on urea–sodium dodecyl sulfate gels of the material corresponding to Pre-ProPTH isolated from acetate–urea gels. From Habener et al. (1976b).

tively account for the disappearance of Pre-ProPTH (chase incubation, Fig. 29) in the appearance of ProPTH. These observations indicate that the proteolytic conversion of Pre-ProPTH occurs extremely rapidly (ProPTH is clearly visible within 1 minute of a pulse incubation) and that the conversion must occur at or very close to the site of synthesis of the polypeptide in the rough endoplasmic reticulum (see Section XIII).

XIII. Biosynthetic Precursors in Other Endocrine and Nonendocrine Secretory Tissues

The studies of the biosynthesis of PTH, coupled with studies of the biosynthesis of other secretory proteins that have been conducted over the past several years, have led to the recognition, based on several different criteria, that biosynthetic precursors fall into at least two distinct classes (Table I). These two classes of precursors have been designated (1) prehormones, or preproteins, and (2) prohormones, or proproteins (Campbell and Blobel, 1976). One or both of these precursors have been identified in a number of different endocrine systems and include pre-proinsulin (Chan *et al.*, 1976; Permutt *et al.*, 1976); pre-growth hormone (Sussman *et al.*, 1976); pre-placental lactogen (Boime *et al.*, 1975); pre-prolactin (Evans and Rosenfeld, 1976; Maurer *et al.*, 1976); and procalcitonin (Roos *et al.*, 1974; Moya *et al.*, 1975) (Table II). Interestingly enough, prehormones, but not prohormones, have been found for growth hormone, placental lactogen, and prolactin. It is not clear why these large pituitary hormones appear to lack intermediate prohormonal forms. It is highly likely that prehormonal forms will eventually be found for many if not all other peptide

TABLE I
Two Distinct Classes of Biosynthetic Precursors

I. Prehormones

    A. Identified as the product of mRNA translation in the *wheat-germ* cell-free system
    B. Specific proteolytic cleavage occurs in rough endoplastic reticulum (RER) within seconds before or after completion of synthesis
    C. Amino-terminal "pre"-peptide is hydrophobic
    D. Function? (transport of polypeptide into cisterna of RER)

II. Prohormones

    A. Identified by conventional pulse–chase studies using intact cells
    B. Specific cleavage occurs in Golgi complex 15–20 minutes after completion of synthesis
    C. Cleavage by combined trypsin and carboxypeptidase B
    D. Function? (transport of polypeptide to Golgi complex)

TABLE II
Biosynthetic Precursors of Protein and Polypeptide Hormones

Prohormones	Prehormones
PTH	PTH
Insulin	Insulin
ACTH	? ACTH
Glucagon	Growth hormone
Calcitonin	Placental lactogen
Gastrin[a]	Prolactin
ADH[a]	
γ-LPH[a]	
β-MSH[a]	
α-MSH[a]	

[a]Presumptive—no biosynthetic studies reported.

hormones in addition to PTH and insulin. In fact, recent studies of the biosynthesis of ACTH by mouse pituitary-tumor cells in culture revealed evidence that as many as three biosynthetic precursors may precede the appearance of the final α1-39 ACTH (Mains and Eipper, 1976; Eipper et al., 1976). The limited structural information available strongly suggests that there are prohormonal forms of gastrin (Gregory and Tracy, 1972), glucagon (Tager and Steiner, 1973), β-MSH, derived from β- and γ-lipotropins (Chrétien and Li, 1967) and α-MSH, derived from ACTH (Scott et al., 1973).

Biosynthetic precursors have been found, in addition to those for protein and peptide hormones, for many other secretory proteins that are not hormones (Table III). The pancreatic enzymes (Kassell and Kay, 1973; Segal, 1973; Devillers-Thiery et al., 1975) and bee-venom protein, melittin (Kreil, 1973; Suchanek et al., 1975; Kindås-Mügge et al., 1976) appear to have both pre- and prosecretory precursors, whereas immunoglobulin light chains have only pre-L-

TABLE III
Biosynthetic Precursors of Secretory Proteins Other Than Hormones

Pancreatic enzymes (trypsin, chymotrypsin)
Immunoglobulins (light chains)
Serum albumin
Collagen
Melittin
Vitellogenin
Viral proteins
Diphtheria proteins

chain precursors (Blobel and Dobberstein, 1975; Burstein *et al.,* 1976; Schecter and Burstein, 1976). Other precursors, such as procollagen (Tanzer *et al.,* 1974; Byers *et al.,* 1975; Fessler *et al.,* 1975; Merry *et al.,* 1976), the blood-clotting proteins prothrombin (Shah and Suttie, 1971) and fibrinogen, and angiotensinogen, may be considered to fall into a third class of biosynthetic precursors because certain specific proteolytic cleavages of these proteins occur either in the extracellular matrix or the blood after their secretion from the cell.

XIV. Proposed Roles of Pre-Proparathyroid Hormone and Proparathyroid Hormone in Cellular Transport Processes in the Parathyroid Gland

We have been interested in gaining insights into what, if any, functions the biosynthetic precursors of PTH might serve in cellular processes. Considerable evidence has accumulated that implicates the precursor sequences of Pre-ProPTH and of ProPTH in the processes of segregation and transportation of the polypeptides within the cell. There are data, however, concerning other proposed functions of precursors and how they may or may not apply to the precursors of PTH.

The C peptide of proinsulin apparently permits the proper alignment of and the formation of disulfide linkage between the A and B chains of insulin (Steiner and Clark, 1968). Such a function, however, does not appear to be necessary for PTH, inasmuch as it is a single-chain polypeptide devoid of disulfide bonds (Niall *et al.,* 1970). It has been emphasized that the cell has need for the existence of biologically inactive precursor forms of the digestive enzymes (trypsinogen, chymotrypsinogen) and for collagen (procollagen) that are cleaved after their secretion. In these instances, the presence of active proteins in the cells could have deleterious consequences, such as uncontrolled proteolysis or precipitation of collagen fibrils within the cell itself. Biologically active PTH is the major form of the hormone stored in the parathyroid gland, and therefore there does not appear to be any requirement for the cell to provide an inactive precursor form of the hormone. The possibility that the activities of enzymes that specifically cleave the precursors of PTH might be regulated by changes in extracellular calcium has attracted our attention and may ultimately prove to be an important step in the regulation of the production of the hormone (Habener *et al.,* 1974b, 1974a) (see Section XV). Other explanations for the existence of biosynthetic precursors are more speculative, inasmuch as there is no direct experimental evidence to support them. For example, one might propose that there exists a minimum-size constraint for monocistronic genes and their mRNAs, i.e., a minimum chain length is required for a polypeptide to be synthesized on polyribosomes. Likewise, it is possible, although not proved, that some small biologically active peptides may simply be a product of the evolution of a suitable enzyme that, by proteolytic cleavage, can modify larger, preexisting protein (Adelson, 1971; Dayhoff, 1972; Steiner, 1976). At present, one of the most attractive functions that has been proposed for biosynthetic precursors generally, and for the precursors of PTH in particular, is that of a role for the precursor-specific amino acid sequences in the transport of the polypeptides within their cells of origin.

In 1971, Blobel and Sabatini proposed a provocative hypothesis to explain, on a molecular level, the mechanism by which proteins destined for secretion from cells are able to selectively obtain access to the subcellular organelles involved in the transport and secretion of the proteins (Blobel and Dobberstein, 1975). This hypothesis, known as the "signal hypothesis," appears to be relevant to the Pre-ProPTH that we have found in our studies of PTH biosynthesis. To understand the signal hypothesis, it is first necessary to review briefly the subcellular architecture of secretory cells (Palade, 1975). It has been known for some time that proteins exported from secretory cells are synthesized in the matrix on ribosomes attached to the rough endoplasmic reticulum (RER), whereas most proteins retained by the cell are synthesized on free polyribosomes (Siekevitz and Palade, 1960). Furthermore, exportable proteins synthesized on the RER are transferred from the cytoplasmic matrix across the membrane of the endoplasmic reticulum into the cisternal space as an early event in the sequential segregation of these proteins (Palade, 1975). The proteins subsequently undergo translocation within the cisterna to the Golgi apparatus where they are incorporated into secretory granules and either stored in the cell or further transported to the cytoplasmic membrane within the granules and released into the circulation (exocytosis).

A fundamental question relates to how the polyribosomes synthesizing secretory proteins become preferentially attached to the membrane of the endoplasmic reticulum, a necessary requirement for the discharge of the newly synthesized proteins into the cisternal space and their eventual secretion from the cell. The hypothesis forwarded by Blobel and Sabatini (1971) and Blobel and Dobberstein (1975) attempts to answer this question. Simply stated, they have proposed that the "information" necessary for the attachment of the polyribosome–mRNA–nascent-chain complex resides in the specific sequence of codons at the 5′ region of the RNA for the secretory protein immediately following the initiator AUG codon. This sequence of codons is translated into a sequence of amino acids at the amino terminus of the proteins, and the chemical properties of this "signal" sequence of amino acids result in selective binding of the growing nascent polypeptide chain to the membrane of the endoplasmic reticulum and in some way facilitate the polypeptide transport into the cisterna of the endoplasmic reticulum. Immediately after entry into the cisternal space, the signal sequence is removed, presumably by an endopeptidase ("clipase"), which perhaps is an event required for the protein to assume its native configuration.

Figure 30 depicts a model incorporating our concepts of the biosynthetic pathway including the sequential proteolytic cleavages of the hormone that occur during the biosynthesis of PTH. We envision that Pre-ProPTH of 115 amino acids (Fig. 20) is the initial product of synthesis on the polyribosomes containing the specific mRNA for the hormone. Protein synthesis is initiated on polyribosomes free in the cytoplasmic matrix. The initial amino acid(s), methionine, is removed from the amino terminus of the growing nascent chain by a methionine-specific amino peptidase when the chain is 25–30 amino acids long (about the length of polypeptide necessary to span the distance across the large ribosomal subunit. The hydrophobic "pre" sequence of the Pre-ProPTH then serves as a "signal sequence" to attach the polyribosome–nascent-chain complex to the membrane of the endoplasmic reticulum and to initiate vectorial discharge of the polypeptide into the cisternal space of the endoplasmic reticulum. Immediately after or at some point during the transfer of the polypeptide into the cisterna, the prehormone sequence is removed either by an endopeptidase "clipase" or perhaps by the sequential action of an exopeptidase(s). This proteolytic process occurs extremely rapidly. We have found that the

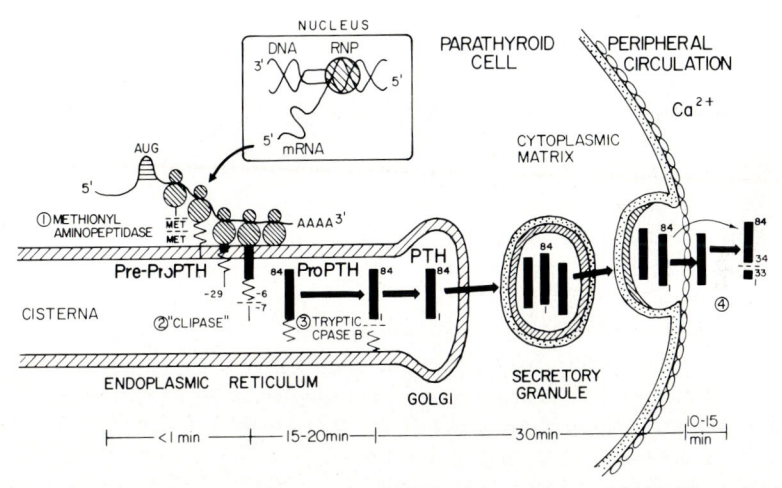

FIG. 30. Schema depicting the proposed intracellular pathway of the biosynthesis of parathyroid hormone. Pre-proparathyroid hormone (Pre-ProPTH), the initial product of synthesis on the ribosomes, is converted into proparathyroid hormone (ProPTH) by removal of (1) the NH_2-terminal methionyl residues and (2) the NH_2-terminal sequence (−29 to −7) of 23 amino acids during and/or within seconds after synthesis, respectively. The conversion of Pre-ProPTH probably occurs during transport of the polypeptide into the cisterna of the rough endoplasmic reticulum. By 20 minutes after synthesis, ProPTH reaches the Golgi region and is converted into PTH by (3) removal of the NH_2-terminal hexapeptide. PTH is stored in the secretory granule until released into the circulation in response to a fall in the blood concentration of calcium. The time needed for these events is given below the schema.

product of this cleavage, ProPTH, is identifiable in parathyroid cells within 1 minute after the introduction of a pulse of [^{35}S]methionine (Habener et al., 1976b) (see Section XII). Because all methionine residues in the Pre-ProPTH are within the amino-terminal one-third of the sequence of the prehormone, only nascent chains that are one-third, or less, completed at the time of exposure to the [^{35}S]methionine could be labeled. Given a rate of polymerization of amino acids during protein synthesis in eukaryotes of from 4–8 amino acids per second (Knopf and Lamfrom, 1965; Hunt et al., 1969), it would take at least 10–20 seconds to complete the synthesis of the carboxyl-terminal two-thirds of the polypeptide. Thus, proteolytic cleavage of the Pre-ProPTH to give ProPTH must occur at least within 1 minute or less of completion of synthesis. In fact, it is possible that proteolytic cleavage of some fraction of the Pre-ProPTH may occur before completion of the synthesis of the nascent polypeptide chain. Evidence was reported by Blobel and Dobberstein (1975) that removal of the amino-terminal signal sequence of amino acids from the pre-L-chain precursor may occur during the elongation of the polypeptide chain and before the chain is completely synthesized.

No direct evidence is available to implicate a transport function for ProPTH. Several lines of evidence, however, indicate that the prohormone-specific hexapeptide sequence may function in some manner in the transport of the polypeptide to the Golgi apparatus of the cell. Clearly, proteolytic conversion of the ProPTH to PTH probably occurs in the Golgi region. Exposure of parathyroid slices *in vitro* to pharmacological agents, such as amine compounds (Chu et al., 1974) and ionophorous antibiotics (Mira-Moser et al., 1976; Ravazzola, 1976), that produce marked cytological disruption of the Golgi apparatus, as well as exposure to inhibitors of microtubular function (vinblastine, colchicine) (Kemper et al., 1975), results in an inhibition of the conversion of ProPTH to PTH and, in some cases, to a marked accumulation of the prohormone in the tissues. Furthermore, analyses of the distribution of autoradiographic grains at the electron microscope level during pulse and pulse-chase incubations of parathyroid slices with [^{3}H]leucine have shown that the grains arrive at the Golgi apparatus at about 15 minutes, which is just the time that proteolytic conversion of ProPTH first begins (J. F. Habener and L. Orci, unpublished).

Thus, present data derived from multiple approaches are consistent with the concept of a highly specialized, vectorially directed transport, via subcellular organelles of hormones and secretory proteins from the sites of synthesis on the RER to sites of storage in secretory vesicles or granules.

XV. Regulation of Parathyroid Hormone Biosynthesis

Calcium is the principal factor known to regulate the activity of the parathyroid glands (Habener and Potts, 1975). The rates of secretion, and ultimately the synthesis, of PTH vary inversely with the concentration of extracellular calcium ion (Patt and Luckhardt, 1942; Copp and Davidson, 1961; Sherwood *et al.*, 1966). Figure 31 schematically summarizes the proposed steps in the biosynthetic pathways of the parathyroid cell at which calcium or other agents may exert a regulatory influence on the synthesis, cleavage, storage, and secretion of PTH. Data concerning actions of calcium at these specific control points are available, but still incomplete.

Calcium does not appear to affect directly the activity of the cleavage enzyme (step 5 of Fig. 31), inasmuch as no alteration in rates of efficiencies of conversion of the prohormone to the hormone is seen throughout a 4-hour period in studies *in vitro* (Habener *et al.*, 1974b, 1975a) (Fig. 32). Chu *et al.* (1973), however, in more chronic studies involving studies of hormone biosynthesis *in vitro* in parathyroids from rats whose blood calcium level has been altered for several days by dietary manipulations, reported the finding of different efficiencies of conversion of ProPTH to PTH.

Potential effects of calcium on synthesis and intracellular storage or turnover of ProPTH and PTH have been more complicated to analyze. Studies *in vitro* (Habener *et al.*, 1974b, 1975a) indicate that there is probably no direct effect of calcium on the regulation of hormone synthesis at the level of translation (step 3 of Fig. 31). Changes in rates of prohormone biosynthesis require hours to become apparent (Habener *et al.*, 1975b), suggesting that transcriptional (step 2, Fig. 31, synthesis of mRNA from the DNA of the genome) rather than translational events may be involved in the regulation of hormone biosynthesis. There is evidence, however, that regulation of intracellular stores of PTH may occur through a pathway of intracellular turnover of PTH or of ProPTH (steps 7 and 8, Fig. 31). High concentrations of extracellular calcium stimulate, and low concentrations inhibit, intracellular degradation of hormone (Fig. 33) (Chu *et al.*,

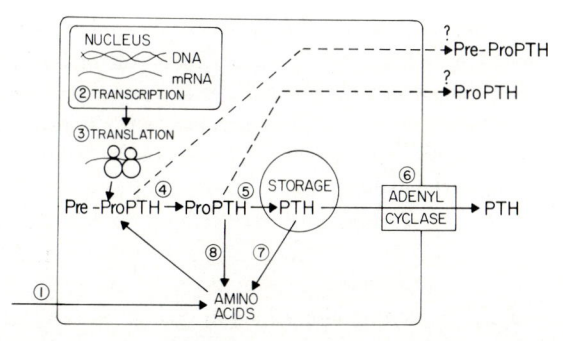

FIG. 31. Model depicting the possible points in the biosynthetic pathways of a parathyroid cell where calcium may exert regulatory effects. PTH, parathyroid hormone; ProPTH, proparathyroid hormone; Pre-ProPTH, pre-proparathyroid hormone. Modified from Habener and Potts (1976a).

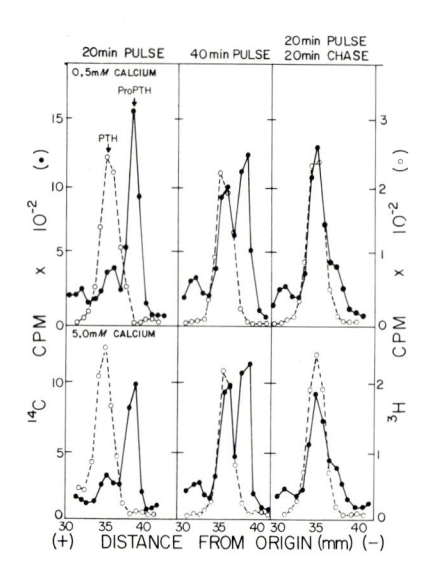

FIG. 32. Study demonstrating lack of an acute effect of calcium in the regulation of conversion of proparathyroid hormone (ProPTH) to parathyroid hormone (PTH). Slices of bovine parathyroids were incubated for 3 hours in media containing high (5.0 mM) and low (0.5 mM) calcium. ^{14}C-labeled amino acids were added for the times indicated, including a chase incubation where radioactive amino acids were removed after 20 minutes, and incubation was continued in presence of unlabeled amino acids for additional 20 minutes. Extracts of tissues were analyzed by polyacrylamide gel electrophoresis. [^{3}H] PTH was added during the initial extraction of tissue to control for small but variable losses of PTH and ProPTH. Synthesis of ProPTH and PTH was stimulated by 15% in low-calcium media, but no change in rate of conversion of ProPTH to PTH was observed. From Habener *et al.* (1974b).

1973; Habener *et al.*, 1975a). It was suggested that inhibition of this degradative pathway, mediated by a lowering of extracellular calcium concentrations, may, in addition to drawing upon preformed stores of hormone, provide a means for a rapid increase of the amounts of hormone available for secretion before the rates of hormone biosynthesis have time to increase to the extent required to meet secretory demands (Habener *et al.*, 1975a). Conversely, stimulation of the degradative pathway by elevations of extracellular calcium may be a mechanism used by the cell to dispose of the excess hormone that is synthesized during the rather long interval of time (hours) required for the suppression of biosynthesis. Observations of the sequential alterations in cellular ultrastructure of the parathyroids that occur after hypercalcemia and suppression of hormone secretion *in vivo* also suggest that calcium effects the degradation of PTH (Roth and Raisz, 1964; Capen, 1971; Oldham *et al.*, 1971). Under conditions of suppression of hormone secretion, cytolysosomal bodies containing degradative enzymes appear in the cell and progressively engulf and digest granular structures thought to contain PTH. Such a role for lysosomes in the parathyroid glands is illustrated in Fig. 34. Similar observations concerning lysosomal bodies and secretory granules were made in the pituitary after suppression of prolactin secretion (Smith and Farquhar, 1966).

More is known about the regulation of the secretion than about the synthesis of PTH. The action of PTH is to increase the concentration of calcium in the extracellular fluid (ECF) through its effects on bone, kidney, and gut. An increase in ECF calcium, in turn, acts on the parathyroid gland to suppress secretion of the hormone (step 6, Fig. 31). This negative-feedback inhibition contributes to the regulation of concentrations of ECF calcium within very narrow limits.

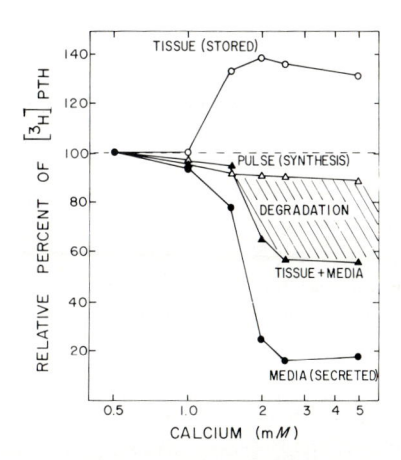

FIG. 33. Effect of extracellular calcium concentrations on synthesis (△), secretion (●), and storage (○) of parathyroid hormone (PTH). The calcium-dependent degradation of PTH (▲) is also shown. Slices of parathyroid glands were incubated *in vitro* for 4 hours in MEM containing [^{3}H]leucine and various concentrations of calcium. Amounts of ^{3}H-labeled PTH were determined by polyacrylamide gel electrophoresis of extracts of tissue (○) and media (●) at the end of the incubation as described previously (Habener *et al.,* 1974b, 1975b). Rates of PTH synthesis (△) were determined by preincubating tissue slices for 3 hours in various concentrations of calcium and then adding [^{3}H]leucine for an additional 35 minutes of incubation (pulse label). Extracts of tissues were analyzed for their content of ^{3}H-labeled ProPTH and PTH by gel electrophoresis. In the higher concentrations of calcium, up to 40% of the [^{3}H]PTH is neither secreted from, nor stored in, the tissue, but, instead, is degraded within the tissue (hatched area). Data are expressed as percentage of [^{3}H]PTH in extracts from tissues and media incubated at lowest concentration of calcium (0.5 mM calcium = 100% [^{3}H]PTH). From Habener (1976).

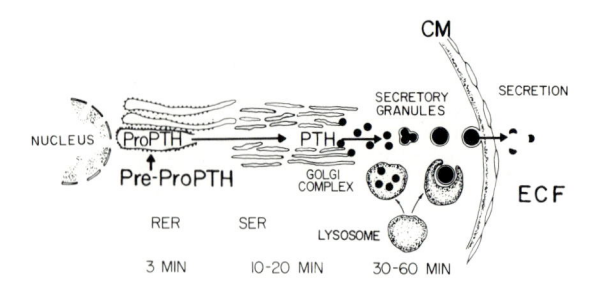

FIG. 34. Schematic depiction of the translocation of parathyroid hormone (PTH) within subcellular organelles and the proposed lysosomal activity that might account for the intracellular degradation of PTH. ProPTH, proparathyroid hormone; Pre-ProPTH, pre-proparathyroid hormone; CM, cell membrane; RER, SER, respectively, rough and smooth endoplasmic reticulum; ECF, extracellular fluid.

Although calcium is the principal ion that influences PTH secretion, magnesium also can change rates of PTH secretion, but only at supraphysiological concentrations, severalfold higher than those found in ECF under normal physiological conditions (Mayer, 1974; Habener and Potts, 1976b). It was shown in studies done both *in vivo* and *in vitro* that, on a molar basis, magnesium is 2–3-times less effective than calcium in suppressing PTH secretion (Mayer, 1974; Habener and Potts, 1976b).

Adenylyl cyclase and the product of the enzyme, cyclic $3':5'$-AMP, are believed to be intermediates in the calcium control of PTH secretion (Abe and Sherwood, 1972; Dufresne and Gitelman, 1972). In addition, studies *in vivo* and *in vitro* have suggested that the stimulation of PTH release might involve β-adrenergic receptors mediated by increases in cAMP (Sherwood and Abe, 1972; Fischer *et al.*, 1973); this latter action, not yet fully understood in physiological terms, involves the response to epinephrine (Mayer, 1976).

XVI. Secretory Products of the Parathyroid Gland

The studies of PTH biosynthesis that have led to the discovery of precursor forms of PTH indicate that products other than PTH may be secreted from the parathyroid gland (Fig. 35). At least one protein distinct from PTH is secreted from parathyroid tissue during studies *in vitro*. This protein was termed parathyroid secretory protein (PSP) (Kemper *et al.*, 1974b). PSP is a high-molecular-weight protein (MW 150,000) that is released from the parathyroid gland in response to changes in calcium concentration in the medium; the fractional stimulation or inhibition of release, respectively, of PSP due to lowering or raising of ambient calcium concentrations shows precisely the same pattern as that of PTH itself. At present, the biological function of PSP is unknown.

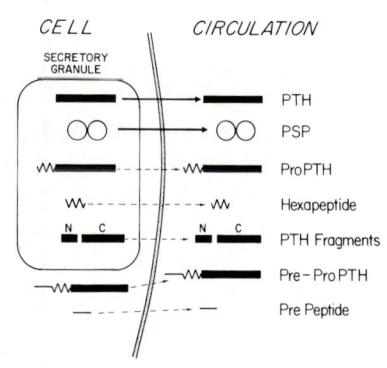

FIG. 35. Schematic representation of products that are either known to be released (solid arrows) or potentially could be released (dashed arrows) from the storage granules in the parathyroid cell. PSP, parathyroid secretory protein; PTH, parathyroid hormone; ProPTH, proparathyroid hormone; Pre-ProPTH, pre-proparathyroid hormone. From Habener (1976).

Clearly, it is not a biosynthetic precursor of PTH, but it may be a transport protein for PTH, analogous to the neurophysins, the proteins involved in the intracellular transport of oxytocin and vasopressin in the posterior pituitary gland (Cheng and Friesen, 1971; Martin *et al.*, 1972). PSP, alternatively, may be important in facilitating concentrations of PTH during packaging of the preformed hormone in secretory granules. The eventual development of an immunoassay specific for the detection of PSP could prove to be particularly useful for monitoring the secretory activity of the parathyroids via detection of PSP, as well as PTH, released from the gland.

Although no direct evidence has yet been obtained to indicate that ProPTH, Pre-ProPTH, or the peptide fragments of the precursors that are removed from ProPTH and PTH in the cell are actually secreted into the circulation, indirect evidence suggests that malignant tumors (not of parathyroid origin) and adenomas of the parathyroids may secrete a precursor along with the hormone. Material of high molecular weight that reacts with antisera to PTH has been observed in the blood of patients with the syndrome of pseudohyperparathyroidism—secretion of PTH-like material by nonparathyroid cancer and hyperparathyroidism due to parathyroid adenomas (Benson *et al.*, 1974). It has not been reported whether this material contains PTH covalently linked in its structure or whether it is noncovalently bound to some larger macromolecule.

Radioimmunoassays for the detection of ProPTH were developed using antisera produced by immunization with synthetic peptides containing the prohormone-specific hexapeptide sequence (Habener *et al.*, 1974a, 1975c). The recognition site of two of the antisera involves the region of hormone sequence at the site of attachment of the prohormone hexapeptide to PTH. Assays based on these antisera readily detect intact, natural ProPTH and synthetic peptides of the prohormone but not PTH or the prohormone-hexapeptide sequence alone. Preliminary application of the prohormone assay has revealed that prohormone is readily measurable in extracts of parathyroid tissue, but not in blood. Initially, it was thought that proteolytic enzymes in blood that rapidly degrade ProPTH prevented the detection of the prohormone (Habener *et al.,* 1974a). Subsequent studies, however, using inhibitors of proteases under conditions in which degradation of ProPTH was prevented, also failed to detect the prohormone, even when searched for directly in parathyroid effluent blood of calves and patients with primary hyperparathyroidism. Within the detection limits of the assays, ProPTH at a concentration of approximately 1–2% of the PTH in the blood samples could have been detected; however, no final conclusions have yet been reached about secretion of prohormone at low levels or in unusual pathophysiological states. Pre-ProPTH has not been isolated in amounts sufficient for immunization of animals. Hence, no immunoassay for its detection is available. If specific and sensitive radioimmunoassays can be developed to detect the presence of ProPTH or of Pre-ProPTH in the circulation, then evidence of the release of the prohormone might serve as a useful marker of abnormal parathyroid gland function or of ectopic hormone production (secretion of PTH by nonparathyroid tumors). It is believed that nonparathyroid tissue, as a consequence of neoplastic transformation, initiates uncontrolled production and release of PTH as a result of what may be abnormalities in the intracellular transport and cleavage of the hormone. Hence, defects such as release of predominantly pre-prohormone or prohormone (lack of specific cleavage enzymes), or predominantly fragments of hormone (excessive proteolytic degradation) might be found in states of uncontrolled parathyroid secretory activity.

XVII. Summary

Much additional information about the nature of the normal controlling processes within the parathyroid cell involved in regulation between initial steps of ProPTH synthesis and eventual release of PTH from storage granules is needed. Such information will be of great fundamental interest in the general area of biosynthesis of proteins destined for secretion from the cell. The parathyroid system may eventually serve as a model with which to evaluate, on a cellular and biochemical level, the failure of homeostatic mechanisms involved in neoplasia or hyperplasia of the parathyroids. The studies may, in turn, facilitate more discriminant diagnoses based on specific immunoassays for the various secretory products defined by present investigations.

ACKNOWLEDGMENTS

The authors wish to acknowledge the important contributions to this work made by Drs. H. T. Keutmann, H. D. Niall, G. W. Tregear, B. Roberts, R. Mulligan, and H. Kronenberg.

We are indebted to T. D. Stevens, J. W. Jacobs, and M. D. Ernst. J.F.H. was supported by a U.S. Public Health Service Research Career Development Award and is currently an investigator for The Howard Hughes Medical Institute. B.W.K. was supported by a fellowship from the Medical Foundation, Boston, Massachusetts.

REFERENCES

Abe, M., and Sherwood, L. M. (1972). *Biochem. Biophys. Res. Commun.* **48**, 396.
Adelson, J. W. (1971). *Nature (London)* **229**, 321.
Andreatta, R. H., Hartmann, A., Jöhl, A., Kamber, B., Maier, R., Riniker, B., Rittel, W., and Sieber, P. (1973). *Helv. Chim. Acta* **56**, 470.
Antoniades, H. N., Huber, A. M., Boshell, B. R., Saravis, C. A., and Gershoff, S. N. (1965). *Endocrinology* **76**, 709.
Aurbach, G. D., Keutmann, H. T., Niall, H. D., Tregear, G. W., O'Riordan, J. L. H., Marcus, R., Marx, S. J., and Potts, J. T., Jr. (1972). *Recent Prog. Horm. Res.* **28**, 353.
Bancroft, F. C., Wu, G. J., and Zubay, G. (1973). *Proc. Natl. Acad. Sci. U.S.A.* **70**, 3646.
Benson, R. C., Jr., Riggs, B. L., Pickard, B. M., and Arnaud, C. D. (1974). *J. Clin. Invest.* **54**, 175.
Benveniste, R., Stachura, M. E., Szabo, M., and Frohman, L. A. (1975). *J. Clin. Endocrinol. Metab.* **41**, 422.
Blobel, G., and Campbell, P. [N.] (1976). Summary of papers presented at the FEBS Advanced Course No. 43: "The Role of Organelles in the Chemical Modification of the Primary Translation Products of Secretory Proteins." The Middlesex Hospital Medical School, London. Organized by P. N. Campbell and G. Blobel. (Privately circulated among participants.)
Blobel, G., and Dobberstein, B. (1975). *J. Cell Biol.* **67**, 835.
Blobel, G., and Sabatini, D. D. (1971). *Biomembranes* **2**, 193.
Boime, I., Boguslowski, S., and Caine, J. (1975). *Biochem. Biophys. Res. Commun.* **62**, 103.
Brawerman, G. (1974). *Annu. Rev. Biochem.* **43**, 621.
Brewer, H. B., Jr., Fairwell, T., Ronan, R., Sizemore, G. W., and Arnaud, C. D. (1972). *Proc. Natl. Acad. Sci. U.S.A.* **69**, 3585.
Burstein, Y., Kantor, F., and Schechter, I. (1976). *Proc. Natl. Acad. Sci. U.S.A.* **73**, 2604.
Byers, P. H., Click, E. M., Harper, E., and Bornstein, P. (1975). *Proc. Natl. Acad. Sci. U.S.A.* **72**, 3009.
Campbell, P. N., and Blobel, G. (1976). *FEBS Lett.* **72**, 215.
Capen, C. C. (1971). *Am. J. Med.* **50**, 598.
Chan, S. J., Keim, P., and Steiner, D. F. (1976). *Proc. Natl. Acad. Sci. U.S.A.* **73**, 1964.
Cheng, K. W., and Friesen, H. G. (1971). *Endocrinology* **88**, 608.
Chrétien, M., and Li, C. H. (1967). *Can. J. Biochem.* **45**, 1163.
Chu, L. L. H., MacGregor, R. R., Anast, C. S., Hamilton, J. W., and Cohn, D. V. (1973). *Endocrinology* **93**, 915.
Chu, L. L. H., MacGregor, R. R., Hamilton, J. W., and Cohn, D. V. (1974). *Endocrinology* **95**, 1431.
Chu, L. L. H., Huang, W. Y., Littledike, E. T., Hamilton, J. W., and Cohn, D. V. (1975). *Biochemistry* **14**, 3631.
Cohn, D. V., MacGregor, R. R., Chu, L. L. H., and Hamilton, J. W. (1972a). *In* "Calcium, Parathyroid Hormone and the Calcitonins" (R. V. Talmage and P. L. Munson, eds.), Int. Congr. Ser. No. 243. pp. 173–182. Excerpta Med. Found. Amsterdam.
Cohn, D. V., MacGregor, R. R., Chu, L. L. H., Kimmel, J. R., and Hamilton, J. W. (1972b). *Proc. Natl. Acad. Sci. U.S.A.* **69**, 1521.

Cohn, D. V., MacGregor, R. R., Chu, L. L. H., Huang, D. W. Y., Anast, C. S., and Hamilton, J. W. (1974a). *Am. J. Med.* **56**, 767.
Cohn, D. V., MacGregor, R. R., Sinha, D., Huang, D. W. Y., Edelhoch, H., and Hamilton, J. W. (1974b). *Arch. Biochem. Biophys.* **164**, 669.
Cohn, D. V., MacGregor, R. R., Chu, L. L. H., and Hamilton, J. W. (1975). *In* "Calcium-Regulating Hormones" (R. V. Talmage, M. Owen, and J. A. Parsons, eds.), Int. Congr. Ser. No. 346, pp. 45–52. Excerpta Med. Found. Amsterdam.
Copp, D. H., and Davidson, A. G. F. (1961). *Proc. Soc. Exp. Biol. Med.* **107**, 342.
Cox, G. S., Weintraub, B. D., Rosen, S. W., and Maxwell, E. S. (1976). *J. Biol. Chem.* **251**, 1723.
Dayhoff, M. O., ed. (1972). "Atlas of Protein Sequence and Structure," Vol. 5. Natl. Biomed. Res. Found., Washington, D.C.
Devillers-Thiery, A., Kindt, T., Scheele, G., and Blobel, G. (1975). *Proc. Natl. Acad. Sci. U.S.A.* **72**, 5016.
Dufresne, L. R., and Gitelman, H. J. (1972). *In* "Calcium, Parathyroid Hormone and the Calcitonins" (R. V. Talmage and P. L. Munson, eds.), Int. Congr. Ser. No. 243, p. 202. Excerpta Med. Found., Amsterdam.
Edman, P., and Begg, G. (1967). *Eur. J. Biochem.* **1**, 80.
Eipper, B. A., Mains, R. E., and Guenzi, D. (1976). *J. Biol. Chem.* **251**, 4121.
Evans, G. A., and Rosenfeld, M. G. (1976). *J. Biol. Chem.* **251**, 2842.
Fessler, L. I., Morris, N. P., and Fessler, J. H. (1975). *Proc. Natl. Acad. Sci. U.S.A.* **72**, 4905.
Fischer, J. A., Blum, J. W., and Binswanger, U. (1973). *J. Clin. Invest.* **52**, 2434.
Goltzman, D., Callahan, E. N., Tregear, G. W., and Potts, J. T., Jr. (1976). *Biochemistry* **15** 5076.
Gregory, R. A., and Tracy, H. J. (1972). *Lancet* **2**, 797.
Habener, J. F. (1976). *Polypeptide Horm.: Mol. Cell. Aspects, Ciba Found. Symp., 1975* No. 41. pp. 197–224.
Habener, J. F., and Potts, J. T., Jr. (1975). *In* "Methods in Enzymology" (B. W. O'Malley and J. G. Hardman, eds.), Vol. 37, p. 345. Academic Press, New York.
Habener, J. F., and Potts, J. T., Jr. (1976a). *Hand. Physiol., Sect. 7: Endocrinol.* **7**, 313–342.
Habener, J. F., and Potts, J. T., Jr. (1976b). *Endocrinology* **98**, 209.
Habener, J. F., Chang, H., and Potts, J. T., Jr. (1977). *Biochemistry* (in press).
Habener, J. F., Kemper, B. [W.], Potts, J. T., Jr., and Rich, A. (1972). *Science* **178**, 630.
Habener, J. F., Kemper, B. [W.], Potts, J. T., Jr., and Rich, A. (1973). *Endocrinology* **92**, 219.
Habener, J. F., Tregear, G. W., Stevens, T. D., Dee, P. C., and Potts, J. T., Jr. (1974a). *Endocr. Res. Commun.* **1**, 1.
Habener, J. F., Kemper, B. [W.], Potts, J. T., Jr., and Rich, A. (1974b). *Endocr. Res. Commun.* **1**, 239.
Habener, J. F., Kemper, B. [W.], and Potts, J. T., Jr. (1975a). *Endocrinology* **97**, 431.
Habener, J. F., Kemper, B. [W.], Potts, J. T., Jr., and Rich, A. (1975b). *Biochem. Biophys. Res. Commun.* **67**, 1114.
Habener, J. F., Kemper, B. [W.], Potts, J. T., Jr., and Rich, A. (1975c). *J. Clin. Invest.* **56**, 1328.
Habener, J. F., Stevens, T. D., Tregear, G. W., and Potts, J. T., Jr. (1976a). *J. Clin. Endocrinol. Metab.* **42**, 520.
Habener, J. F., Potts, J. T., Jr., and Rich, A. (1976b). *J. Biol. Chem.* **251**, 3893.
Habener, J. F., Stevens, T. D., and Potts, J. T., Jr. (1976c). *Program Abstr., 5th Annu. Meet. Endocr. Soc.* Abstract No. 101, p. 107.

Hamilton, J. W., and Cohn, D. V. (1969). *J. Biol. Chem.* **244**, 5421.
Hamilton, J. W., MacGregor, R. R., Chu, L. L. H., and Cohn, D. V. (1971a). *Endocrinology* **89**, 1440.
Hamilton, J. W., Spierto, F. W., MacGregor, R. R., and Cohn, D. V. (1971b). *J. Biol. Chem.* **246**, 3224.
Hamilton, J. W., Niall, H. D., Jacobs, J. W., Keutmann, H. T., Potts, J. T., Jr., and Cohn, D. V. (1974). *Proc. Natl. Acad. Sci. U.S.A.* **71**, 653.
Huang, D. W. Y., Chu, L. L. H., Hamilton, J. W., MacGregor, R. R., and Cohn, D. V. (1975). *Arch. Biochem. Biophys.* **166**, 67.
Hunt, T., Hunter, T., and Munro, A. (1969). *J. Mol. Biol.* **43**, 123.
Jackson, R., and Hunter, T. (1970). *Nature (London)* **227**, 672.
Jacobs, J. W., Kemper, B. [W.], Niall, H. D., Habener, J. F., and Potts, J. T., Jr. (1974). *Nature (London)* **249**, 155.
Kassell, B., and Kay, J. (1973). *Science* **180**, 1022.
Kemmler, W., Peterson, J. D., and Steiner, D. F. (1971). *J. Biol. Chem.* **246**, 6786.
Kemmler, W., Steiner, D. F., and Borg, J. (1973). *J. Biol. Chem.* **248**, 4544.
Kemper, B. [W.], and Stolarsky, L. (1976). *Program Abstr., 58th Annu. Meet. Endocr. Soc.,* Abstract No. 102, p. 107.
Kemper, B. [W.], Habener, J. F., Potts, J. T., Jr., and Rich, A. (1972). *Proc. Natl. Acad. Sci. U.S.A.* **69**, 643.
Kemper, B. [W.], Habener, J. F., Mulligan, R. C., Potts, J. T., Jr., and Rich, A. (1974a). *Proc. Natl. Acad. Sci. U.S.A.* **71**, 3731.
Kemper, B. [W.], Habener, J. F., Rich, A., and Potts, J. T., Jr. (1974b). *Science* **184**, 167.
Kemper, B. [W.], Habener, J. F., Rich, A., and Potts, J. T., Jr. (1975). *Endocrinology* **96**, 903.
Kemper, B. [W.], Habener, J. F., Ernst, M. D., Potts, J. T., Jr., and Rich, A. (1976a). *Biochemistry* **15**, 15.
Kemper, B. [W.], Habener, J. F., Potts, J. T., Jr., and Rich, A. (1976b). *Biochemistry* **15**, 20.
Keutmann, H. T., Niall, H. D., Jacobs, J. W., Barling, P. M., Hendy, G. N., O'Riordan, J. L. H., and Potts, J. T., Jr. (1975). *In* "Calcium-Regulating Hormones" (R. V. Talmage, M. Owen, and J. A. Parsons, eds.), Int. Congr. Ser. No. 346, pp. 9–14. Excerpta Med. Found., Amsterdam.
Kindås-Mügge, I., Frasel, L., and Diggelmann, H. (1976). *J. Mol. Biol.* **105**, 177.
Knopf, P. M., and Lamfrom, H. (1965). *Biochim. Biophys. Acta* **95**, 398.
Koffer-Gutmann, A., and Arnstein, H. R. V. (1973). *Biochem. J.* **134**, 969.
Kreil, G. (1973). *Eur. J. Biochem.* **33**, 558.
Lengyel, P., and Söll, D. (1969). *Bacteriol. Rev.* **33**, 264.
Lucas-Lenard, J., and Lipmann, F. (1971). *Annu. Rev. Biochem.* **40**, 409.
McKean, D. J., Peters, E. H., Waldby, J. I., and Smithies, O. (1974). *Biochemistry* **13**, 3048.
Mains, R. E., and Eipper, B. A. (1976). *J. Biol. Chem.* **251**, 4115.
Marotta, C. A., Forget, B. G., Weissman, S. M., Verma, I. M., McCaffrey, R. P., and Baltimore, D. (1974). *Proc. Natl. Acad. Sci. U.S.A.* **71**, 2300.
Martin, M. J., Chard, T., and Landon, J. (1972). *J. Endocrinol.* **52**, 481.
Matthews, M. B. (1973). *Essays Biochem.* **9**, 59.
Maurer, R. A., Stone, R., and Gorski, J. (1976). *J. Biol. Chem.* **251**, 2801.
Mayer, G. P. (1974). *Program, 56th Annu. Meet. Endocr. Soc.* Abstract No. 252, p. 181.
Mayer, G. P. (1976). *Proc. Int. Congr. Endocrinol., 5th, 1976* (in press).
Merry, A. H., Harwood, R., Woolley, D. E., Grant, M. E., and Jackson, D. S. (1976). *Biochem. Biophys. Res. Commun.* **71**, 83.

Milstein, C., Brownlee, G. G., Harrison, T. M., and Mathews, M. B. (1972). *Nature (London) New Biol.* **239**, 117.
Mira-Moser, F., Schofield, J. G., and Orci, L. (1976). *Eur. J. Clin. Invest.* **6**, 103.
Moya, F., Nieto, A., and Candela, J. L. R.- (1975). *Eur. J. Biochem.* **55**, 407.
Niall, H. D., Keutmann, H. T., Sauer, R., Hogan, M. L., Dawson, B. F., Aurbach, G. D., and Potts, J. T., Jr. (1970). *Hoppe-Seyler's Z. Physiol. Chem.* **351**, 1586.
Niall, H. D., Sauer, R. T., Jacobs, J. W., Keutmann, H. T., Segre, G. V., O'Riordan, J. L. H., Aurbach, G. D., and Potts, J. T., Jr. (1974). *Proc. Natl. Acad. Sci. U.S.A.* **71**, 384.
Oldham, S. B., Fischer, J. A., Capen, C. C., Sizemore, G. W., and Arnaud, C. D. (1971). *Am. J. Med.* **50**, 650.
Palade, G. (1975). *Science* **189**, 347.
Parsons, J. A., Rafferty, B., Gray, D., Reit, B., Keutmann, H. T., Tregear, G. W., and Potts, J. T., Jr. (1975). *In* "Calcium-Regulating Hormones" (R. V. Talmage, M. Owen, and J. A. Parsons, eds.), Int. Congr. Ser. No. 346, pp. 21–26. Excerpta Med. Found., Amsterdam.
Patt, H. M., and Luckhardt, A. B. (1942). *Endocrinology* **31**, 384.
Permutt, M. A., Biesbroeck, J., Chyn, R., Boime, I., Szczesna, E., and McWilliams, D. (1976). *Ciba Found. Symp.* **41**, 97–116.
Peytremann, A., Goltzman, D., Callahan, E. N., Tregear, G. W., and Potts, J. T., Jr. (1975). *Endocrinology* **97**, 1270.
Potts, J. T., Jr., Keutmann, H. T., Niall, H. D., Tregear, G. W., Habener, J. F., O'Riordan, J. L. H., Murray, T. M., Powell, D., and Aurbach, G. D. (1972). *In* "Endocrinology 1971: Proceedings of the Third International Symposium" (S. Taylor, ed.), pp. 333–349. Heinemann, London.
Potts, J. T., Jr., Niall, H. D., Tregear, G. W., van Rietschoten, J., Habener, J. F., Segré, G. V., and Keutmann, H. T. (1973). *Mt. Sinai J. Med.* **40**, 448.
Ravazzola, M. (1976). *Lab Invest.* **35**, 425.
Roberts, B. E., and Paterson, B. M. (1973). *Proc. Natl. Acad. Sci. U.S.A.* **70**, 2320.
Roos, B. A., Okano, K., and Deftos, L. J. (1974). *Biochem. Biophys. Res. Commun.* **60**, 1134.
Roth, S. I., and Raisz, L. G. (1964). *Lab. Invest.* **13**, 331.
Schechter, I. (1973). *Proc. Natl. Acad. Sci. U.S.A.* **70**, 2256.
Schechter, I., and Burstein, Y. (1976). *Biochem. Biophys. Res. Commun.* **68**, 489.
Schreiber, G. (1971). *Angew. Chem., Int. Ed. Engl.* **10**, No. 9, 638.
Scott, A. P., Ratcliffe, J. G., Rees, L. H., Landon, J., Bennett, H. P. J., Lowry, P. J., and McMartin, C. (1973). *Nature (London) New Biol.* **244**, 65.
Segal, H. L. (1973). *Science* **180**, 25.
Segre, G. V., Tregear, G. W., and Potts, J. T., Jr. (1975). *In* "Methods in Enzymology" (B. W. O'Malley and J. G. Hardman, eds.), Vol. 37, p. 38. Academic Press, New York.
Shah, D. V., and Suttie, J. W. (1971). *Proc. Natl. Acad. Sci. U.S.A.* **68**, 1653.
Sherwood, L. M., and Abe, M. (1972). *J. Clin. Invest.* **51**, 88a (Abstr. No. 292).
Sherwood, L. M., Potts, J. T., Jr., Care, A. D., Mayer, G. P., and Aurbach, G. D. (1966). *Nature (London)* **209**, 52.
Sherwood, L. M., Rodman, J. S., and Lundberg, W. B. (1970). *Proc. Natl. Acad. Sci. U.S.A.* **67**, 1631.
Siekevitz, P., and Palade, G. E. (1960). *J. Biophys. Biochem. Cytol.* **7**, 619.
Smith, R. E., and Farquhar, M. G. (1966). *J. Cell Biol.* **31**, 319.
Steiner, D. F. (1976). *In* "Peptide Hormones" (J. A. Parsons, ed.), pp. 49–65. Macmillan, New York.
Steiner, D. F., and Clark, J. L. (1968). *Proc. Natl. Acad. Sci. U.S.A.* **60**, 622.

Steiner, D. F., and Oyer, P. E. (1967). *Proc. Natl. Acad. Sci. U.S.A.* **57**, 473.
Suchanek, G., Kindås-Mügge, I., Kreil, G., and Schreier, M. H. (1975). *Eur. J. Biochem.* **60**, 309.
Sussman, P. M., Tushinski, R. J., and Bancroft, F. C. (1976). *Proc. Natl. Acad. Sci. U.S.A.* **73**, 29.
Swan, D., Aviv, H., and Leder, P. (1972). *Proc. Natl. Acad. Sci. U.S.A.* **69**, 1967.
Szczesna, E., and Boime, I. (1976). *Proc. Natl. Acad. Sci. U.S.A.* **73**, 1179.
Tager, H. S., and Steiner, D. F. (1973). *Proc. Natl. Acad. Sci. U.S.A.* **70**, 2321.
Tanzer, M. L., Church, R. L., Yaeger, J. A., Wampler, D. E., and Park, E. D. (1974). *Proc. Natl. Acad. Sci. U.S.A.* **71**, 3009.
Tregear, G. W., van Rietschoten, J., Greene, E., Niall, H. D., Keutmann, H. T., Parsons, J. A., O'Riordan, J. L. H., and Potts, J. T., Jr. (1974). *Hoppe-Seyler's Z. Physiol. Chem.* **355**, 415.
Wilson, D. B., and Dintzis, H. M. (1970). *Proc. Natl. Acad. Sci. U.S.A.* **66**, 1282.
Yoshida, A., Watanabe, S., and Morris, J. (1970). *Proc. Natl. Acad. Sci. U.S.A.* **67**, 1600.
Zanini, A., Giannattasio, G., and Meldolesi, J. (1974). *Endocrinology* **94**, 104.

DISCUSSION

R. S. Yalow: In a sense we have had two papers today on comparative endocrinology: one compares the same or the same group of hormones among species, which is one aspect of comparative endocrinology; and the other one, an aspect that has interested several groups including our own, transfers from understanding of the biosynthetic and other physiologic mechansims of one hormone, what to look for in terms of the properties of other peptidal hormones. What was learned about proinsulin was transferred into the search for ProPTH. What was learned about the prehormones is now being transferred across the board to many others. My question relates to whether these prehormones are natural products or arise as an artifact of the wheat-germ synthetic system. I am asking this in part because of the evidence you presented suggesting that there is not a universal enzyme converting pattern, a universal cleavage, as there appears to be in the case of the prohormones.

J. Habener: Yes, this was a great concern to us, and initially we had some reservations about interpreting data obtained from translations of mRNAs in the plant system. But we are now certain of the validity of the information. We can identify a small amount of the Pre-ProPTH within intact cells [J. F. Habener, J. T. Potts, Jr., and A. Rich, *J. Biol. Chem.* **251**, 3893 (1976)], which I think is convincing evidence that it is indeed a biosynthetic precursor. Increasing experience with cell-free translation in the wheat-germ system indicates that it is a faithful translation system. Apart from a tendency for premature chain termination under some conditions, there has been no evidence of false initiation or of misinsertion of amino acids into the sequences of the polypeptides. We are becoming more confident that the large products synthesized in cell-free systems are the initial products expressed from the genes *in vivo*.

A. Rubenstein: I enjoyed your talk very much. I would like to ask a question similar to that of Dr. Yalow. If the amino terminus of the prehormone is very rapidly removed, it is difficult to imagine whether you would find the prehormone in tissue extracts, because the amino terminus may be cleaved off before the hormone is completely synthesized. As far as I am aware, Dr. Steiner has never been able to find the prehormone in pancreatic tissue itself, although in the wheat-germ translation system you can obviously find it. I would like to ask also whether you have any additional data about the secretion of the protein associated with parahormone, because its resemblance to the C peptide is extremely fascinating. Also, I wonder whether you have any hypothesis as to what it might be doing.

J. Habener: Studies by G. Blobel and B. Dobberstein [*J. Cell Biol.* **67**, 835 (1975)] indicate that the cleavage of the presequence, or signal sequence, from the pre-L chains occurs during the actual synthesis of the polypeptide chains. Completed presecretory immunoglobulin L chains are found only under conditions where membranes are completely removed. Steiner apparently has not yet identified any pre-proinsulin in his intact cell systems. However, the pulse-labeling times used may not have been sufficiently short and the isolation techniques may not be disciminating enough to actually detect it [S. J. Chan, P. Keim, and D. F. Steiner, *Proc. Natl. Acad. Sci. U.S.A.* **73**, 1964 (1976)]. Bancroft, working with the growth-hormone mRNA, has found a small amount of pre-growth hormone in intact pituitary cells treated with inhibitors of chymotryptic activity (TPCK). The problem appears to be that the cleavages of the prehormones occur so rapidly that it is going to be difficult in many circumstances to identify them in the intact cells. I should emphasize that, in the parathyroid cell, it is still quite likely that the most of the cleavage may occur during the growth of the nascent polypeptide chain and that only a few of the chains may escape cleavage until they are completed. There may be some variability in the time that these nascent chains are processed.

Characterization of the parathyroid secretory protein includes a determination of its molecular weight. It appears to be a protein of MW 150,000 consisting of two subunits presumably of about MW 75,000, and there is reasonably good evidence that it contains carbohydrate [B. [W.] Kemper *et al., Science* **184**, 167 (1974)]. It is synthesized and secreted in parallel with PTH and is distributed in the same fashion as PTH when one disrupts the tissue and analyzes the various subcellular compartments. Its function(s) are unknown. It could be a binding protein analagous to the neurophysins that bind oxytocin and vasopressin. The other thought that had arisen concerns the possibility that it may have some function in the concentration process of the PTH within the secretory granule. Other than these considerations, I have no real speculations as to the function of this protein.

I. D. Raacke: I would like to comment on your proposed possible functions of prehormones and prohormones, and also on Dr. Yalow's question as to whether the prehormone, especially, was possibly an artifact. First, I do not think it is an artifact, since such a precursor is exactly what would be expected from our knowledge of protein synthesis and the fact that all eukaryotes have a methionyl RNA initiator. Since mature eukaryotic proteins do not have methionine end groups, one would expect that there should be preproteins quite generally, not just prehormones that would start with methionine. As to function, I think that the function of pre- and prohormones might be different. Prehormones might simply exist because of the evolutionary necessity of having a single initiator RNA. Processing of the initiator sequence to different end points takes place quite generally in eukaryotes and prokaryotes. In both, this processing is very rapid, occurring while the chain is still being synthesized on the ribosome. (I might add that the corresponding exopeptidases are both species and amino acid specific and have difficulty in detecting prehormones synthesized in homologous systems.) Prohormones, on the other hand, may have several different functions, as you have indicated. But I would question your saying that transport might be the main function, because after all you do get processing in proteins that are not secreted. I would think that perhaps regulation would be a more important function. In some better known systems, the coagulation system, for example, you get both active enzymes and active peptides as products from the processing system. Taking the analogy to hormones, it is interesting to speculate that some or all of the active regulatory peptides, as the releasing factors, are by-products of this posttranslational processing. The existence and functions of prohormones are thus more problematic, but your demonstration of a prehormone starting with methionine, although the first example for a hormone protein, is to be expected as a general phenomenon.

J. Habener: I agree with you that specific proteolytic cleavage is probably a quite general process in the posttranslational modification of proteins. There appear to be several different classes of biosynthetic precursors for proteins, differing in the functions that they serve. Precursors of viral proteins, for example, are a special class of proteins, probably in that several, all of which are used in the assembly of the virus, are cleaved from a single large precursor protein. Proteins involved in the coagulation process, such as prothrombin and fibrinogen, are proteolytically modified in the blood after secretion from their cells of origin. If one looks at protein and polypeptide hormones as a distinct class of proteins destined for export from the cell, there is reasonably good evidence that one function of the biosynthetic precursors may be the transport of the hormones within the subcellular organelles that provide the necessary pathway for hormone secretion.

E. B. Thompson: First a comment about the conversation you are now having. The fact that these peptides must begin with the initiator methionine tRNA codon perhaps implies that there are specific base sequences following, which are evolved for proteins that are to be secreted. The amino acids coded by these codons are those that are hydrophobic and therefore enter the membranes. Now, two questions. The first question: Have you looked at the HnRNA population for parathyroid hormone sequences that can be translated in the wheat-germ system?

J. Habener: No we have not.

E. B. Thompson: The second question concerns the nature of the enzymes and the cleavage enzymes in the peptidase reactions that produce the ultimate active peptide. You suggested, I think, that they may have very general specificity, enzymes such as trypsin, carboxypeptidase, and so forth. Yet there are obviously other sites in the peptide that would be expected to be sensitive to those enzymes. Have you looked to see whether there are special endopeptidases involved, in the Golgi apparatus, for example, that may carry out the reactions that specifically reduce the size of the hormone?

J. Habener: We are attempting to localize and isolate specific cellular enzymes that cleave the biosynthetic precursors of PTH. It has been a very difficult task. When one breaks cells apart and attempts to isolate specific enzymes, there appear to be enzyme activities everywhere. Virtually all the tissue fractions contain both the tryptic and carboxypeptidase activity. I might mention in respect to the comments that you made about the initiator methionine that in a sense all proteins arise from precursors, inasmuch as all proteins are synthesized initially with methionine as the initial amino acid. The amino-terminal methionine is removed from the protein when the nascent polypeptide chain is 25 or 30 amino acids long. The explanation, in functional terms for this process, is, as far as I know, unknown. It is a universal process for all proteins, not just secretory proteins, and presumably has nothing to do with transport.

E. B. Thompson: That was exactly what I was referring to, meaning that those peptides that have evolved to be secreted have taken advantage of the universal process by developing codons inside the initiator codon that specify hydrophobic amino acids.

J. Habener: That is the substance of the "signal hypothesis" that has been proposed by Blobel and Sabatini.

R. S. Yalow: I would like to interject a comment to point out that not all the larger putative prohormones are necessarily biologically inactive. Big gastrin has essentially full biologic activity as measured by an administered dose-response system. Intermediate ACTH, i.e., rat pituitary ACTH, seems to have virtually full biologic activity. Conversion of prohormones by tryptic digestion is not always necessary to release a biologically active form.

S. Korenman: I am curious that you have not mentioned the possibility that the initial sequence in the pre-proparahormone may have a relation to the regulation of transcription.

J. Habener: That is a very nice point and, of course, one that I can only speculate about. As far as I know, there is no evidence that this might be a function of the codons in the message corresponding to the signal peptide. I prefer to think that the mRNA probably contains a polynucleotide sequence that may be involved in the initiation of transcription, but this sequence is located at the 5' end of the initiation codon and is not translated into protein. To translate a sequence of mRNA involved only in regulation of transcriptional processes would seem to be a wasted effort by the cell. The reason that we adhere to the signal hypothesis is that there exists a reasonable amount of experimental evidence in support of a function for the "pre" peptide in the transport of the polypeptide across the membrane of the endoplasmic reticulum.

S. Korenman: I do not think that those are mutually exclusive situations. If a certain amount of information is necessary on DNA to ensure specificity of transcription, then one could allocate 75 nucleotides to that purpose. Those 75 nucleotides could be responsible for the translation of 25 amino acids, which might then be important in the translational process as well. The heterogeneity of these N-terminal peptides in different hormones suggests such a trasncriptional-control derivation.

H. Papkoff: I would like to make a quick observation and maybe elicit a comment. I think it is fair to say that the evidence for precursor hormones is not as well substantiated with respect to the glycoprotein hormones as it is for the polypeptide materials you discussed. In view of the fact that the glycoprotein hormones consist of subunits that somehow must complex together, there would be some advantage to synthesizing a single-chain precursor, which is then split into the alpha and beta subunit. Another point of interest is that the glycoprotein hormones, as you know, contain a very high content of cystine (11 disulfides) and most of the polypeptide hormones you have discussed do not contain any cystine. Do you think this has any relationship to whether or not there might be precursor molecules for the glycoprotein hormones? Do you know of any solid evidence that suggests that there are glycoprotein precursors?

J. Habener: I am not aware of any experimental evidence relating to the existence of biosynthetic precursors in the synthesis of the glycoproteins.

E. M. Brown: A comment and a question with regard to the relationship between calcium, cAMP, and the control of PTH release. In Dr. Aurbach's laboratory we have developed a system of isolated bovine parathyroid cells that has enabled us to look at cellular levels of cAMP in response to several different secretagogs. With β-adrenergic stimulation there is a very nice correlation between cellular levels of cAMP and the release of PTH; likewise there is an inverse relationship between cellular cAMP and extracellular calcium which correlates with PTH release. There are, however, two important differences between these two situations. First, the maximal levels of cAMP associated with maximal β-adrenergic stimulation are about 40-fold higher than those associated with maximal PTH release in response to hypocalcemia. As several groups have shown previously, PTH release with β-adrenergic stimulation is only transitory, and it subsequently becomes refractory to further β-adrenergic stimulation, although the levels of cellular cAMP remain high. So I think it is still a little unclear what the relationship of low-calcium-stimulated PTH release is to cAMP. Have you tried cAMP in your cell-free system to see whether it has any effect?

J. Habener: No, we have not performed any experiments along those lines. I am not certain that adding cAMP to heterologous cell-free systems would provide any useful information regarding regulatory processes *in vivo*.

M. Lippman: In your Fig. 31 you showed a scheme for PTH synthesis and you showed several sites at which this regulation of PTH can potentially be regulated. I wonder whether you could expand on that. Do you have evidence from your cell-free translation system, for example, that mRNA levels are regulated either chronically or acutely by ambient calcium? Is there evidence for regulation at any other stage, either packaging or conversion, or degradation of any of these precursors in a physiologically relevant fashion?

J. Habener: Yes, we believe so. Some evidence exists, albeit indirect, that the synthesis of PTH is regulated at the level of transcriptional, not translational or posttranslational, processes. This is because rates of synthesis of ProPTH, the earliest precursor of PTH for which we have been able to evaluate regulation, change very slowly in response to stimulation or suppression of hormone secretion. If synthesis is regulated at the level of transcription of an RNA, it would appear that the mRNA for the hormone has a relatively long half-life, perhaps on the order of several hours. This situation is in contrast to the findings in studies of insulin biosynthesis, where changes in glucose concentrations result in rapid changes in the rates of proinsulin synthesis. We have also examined rates of conversion of ProPTH to PTH under conditions of stimulation versus suppression using pulse-labeling techniques and have found no differences in the rates of conversion. These studies, however, have been conducted over relatively short periods of time (4–5 hours).

M. Lippman: Rather than using a whole-cell extract, have you attempted to fractionate polysomal mRNA versus whole-cell mRNA in the hope of finding whether or not this putatively stable parathyroid mRNA was present but not on the polysomes? Is there a pool of mRNA for PTH not associated with polysomes?

J. Habener: No, we have not done those studies. We have attempted to isolate mRNA from polyribosomes using immunoprecipitation techniques, but they were unsuccessful owing to very large amounts of contamination of the polyribosomal preparations by hormone presumably bound to membranes. Our efforts have been directed toward purification of the mRNA, followed by preparation of a radioactive complementary DNA to be used in a hybridization assay for detection of mRNA in parathyroid tissues.

J. M. McKenzie: Is there any evidence in the parathyroid for control over synthesis of the hormone involving changes in the ratio of free to membrane-bound ribosomes; similar changes have been recognized in a number of systems. In my own laboratory, Dr. A. V. N. Murthy has shown that thyrotropin will fairly acutely give you relatively more membrane-bound ribosomes, that would be appropriate for the synthesis of a secreted protein, namely thyroglobulin. Does that happen in the parathyroid?

J. Habener: We have no evidence on this point from our own studies. Other workers have examined the morphology of parathyroid tissue from animals in which the parathyroid glands have been either chronically stimulated or suppressed [S. I. Roth and C. C. Capen, *Am. J. Med.* **50**, 598 (1971)]. In the stimulated parathyroid gland, there appears to be an increase in the rough amounts of bound compared to free ribosomes. In fact, there seems to be an increase in the total intracellular content of organelles during chronic stimulation. There is an increase in the amounts of endoplasmic reticulum and Golgi apparatus, and the cells increase in both size and number.

J. E. Rall: I suspect you were a little disappointed to see in these "clipases" the lack of specificity for certain sequences of amino acids so that a single clipase could be identified. It occurs to me that in many prokaryotic systems there is apparently a substantial amount of error correction after protein synthesis. Perhaps in eukaryotes it may be at the level of the prehormones rather than the prohormones, and perhaps a mechanism for error correction could consist of a variety of proteolytics or peptidases present in the cisternae of the endoplasmic reticulum ready to clip off errors. As a second function these proteases might clip off in multiple fragments that pre-portion of the hormone which is probably responsible for its insertion into the membrane.

J. Habener: That is an interesting thought. How would the enzymes recognize an error in the protein, inasmuch as there are so many different amino acid sequences and proteins in general?

J. E. Rall: The pre portion of prehormones is unusual and quite hydrophobic, and you may not find exactly those di- or tripeptides in the rest of the amino acid sequence of normal proteins. Hence, the clipases could recognize both the pre part of the hormone and errors in the rest of the protein.

J. Habener: That is a very provocative suggestion. I would like to elaborate on the nature of the enzymic processes that remove the initial, or "signal", sequences from the presecretory proteins. Thus far, no one has identified the signal sequences that are removed from the polypeptides. This assumption that the process is accomplished by an endopeptidase has not been proved. It is possible that the removal of the sequence may occur by the actions of exopeptidases or dipeptidases during the translocation of the polypeptide through the membrane. Perhaps the basic sequence of amino acids characteristic of the prohormone may act as a stop signal for the exopeptidase activity.

B. Kemper: Since the question of artifacts has come up several times, I thought I might show a couple of slides which relate, I think, to at least one artifact that has not been mentioned so far. That is the possibility that the RNA preparation we are using may be a degraded mRNA preparation and thus that Pre-ProPTH may be actually smaller than the real initial precursor protein made *in vivo*. I have done a couple of experiments to try to determine that our preparation of RNA contains intact PTH mRNA. To do this I have followed the same philosophy we followed with our studies on the precursur proteins and have looked at the ends of the mRNA. It turns out that both termini of mRNAs from eukaryotic cells are unique. Probably the best-studied terminus is the 3' end, which has a string of adenylic acid residues or poly(A) present on it. This has been used to isolate mRNA by utilizing an affinity column that has short chains of deoxythymidylic acid attached to cellulose. The poly(A)-containing RNA binds to the column, and after RNA not containing poly(A) is washed off the column the poly(A)-containing mRNA can be eluted. You can also use this method to show that an mRNA contains poly(A) because if you can purify mRNA by this method it obviously has to contain a poly(A) sequence.

Figure A summarizes results from an oligo(dT) column purification of parathyroid RNA that had been partially purified already by sucrose gradient centrifugation and was equivalent to the 8–15 S fraction that Dr. Habener was just discussing. The right-hand side of the figure simply measures total poly(A) in the samples to show that the column worked. The "total" bar is the total poly(A) that was present. The "unbound" is the RNA that comes through the column and does not bind to it, and it contains very little poly(A). In the bound fraction in this case, we recovered about 50% of the poly(A). The left-hand side shows specific activities (incorporation per absorbance unit of RNA added to the reaction) of these preparations of RNA. The specific activity of the total RNA is shown in the left bar. Although the unbound fraction has a lower specific activity than the total RNA the poly(A)-containing RNA is enriched about 8–10-fold for the mRNA activity that produces Pre-ProPTH. This is presumptive evidence at least that this mRNA has poly(A) at the 3' end, and thus that the 3' end of the mRNA is intact. At the other end of the mRNA, the 5' end, there is also a unique structure, which consists of a methylated nucleotide (a "capped" structure), 7-methylguanosine, which is attached not through its 3' hydroxyl to the 5' hydroxyl of the penultimate base, but is essentially in reverse and is attached with its 5' hydroxyl to the 5' hydroxyl of the penultimate base, thereby leaving a free 3' hydroxyl. This capped structure is required for the activity of eukaryotic mRNAs, and it can be removed chemically by treatment with periodate and aniline. There are a couple of mRNAs that are not "capped" but are nevertheless active, one of which is satellite tobacco necrosis virus RNA.

Figure B shows experiments in which I removed the "caps" from different RNAs by treatment with periodate and aniline, translated the RNAs in wheat-germ system, and

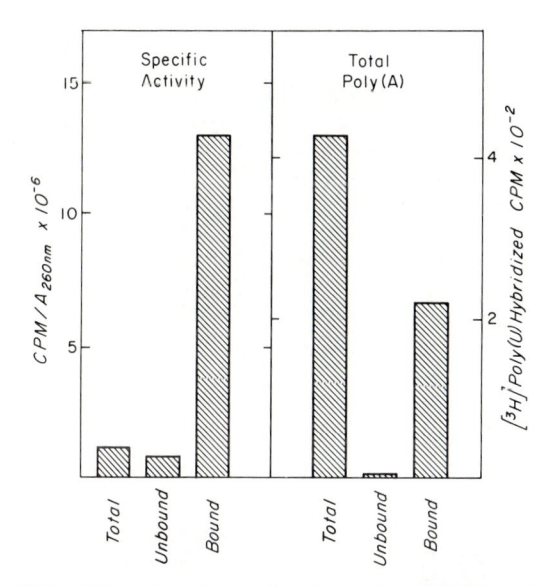

FIG. A. Oligo(dT) cellulose chromatography of parathyroid RNA. A fraction of parathyroid RNA sedimenting from 8 S to 15 S (Total) was dissolved in 0.5 M NaCl, 0.01 M Tris-Cl, pH 7.4, and passed through an oligo(dT) cellulose column. RNA passing through the column was collected (Unbound), and the RNA that hybridized to the oligo(dT) cellulose (Bound) was then eluted with 0.01 M Tris-Cl, pH 7.4. The mRNA activities of the fractions were determined in a wheat-germ-extract protein-synthesizing system, and the content of poly-(A) was determined by hybridization to [^{3}H]poly(U).

analyzed the products by slab-gel electrophoresis. The slide shows an autoradiogram of the slab gel. On the left-hand side, we have translated, in slot a, an mRNA known to have a "cap" on it, hemoglobin mRNA, and the main product, indicated by the H, is hemoglobin. Slot d shows products of the translation of satellite tobacco necrosis virus RNA. In slot b a mixture of those two mRNAs was translated, and then in slot c, which is the critical experiment, a mixture at the two mRNAs was translated after treatment with periodate and aniline to remove the "cap." As you can see, there is very little hemoglobin made, although there is very little effect on the translation of the control noncapped satellite tobacco necrosis virus RNA. The center slot shows the products of a reaction in which mRNA was not added, and it is blank, which gives you an indication of the endogenous protein synthesis in the wheat-germ system.

The right-hand side of the slide is an analogous experiment to the one on the left, but using parathyroid mRNA in this case. Slot f is parathyroid RNA alone. Slot i is a satellite tobacco necrosis virus RNA alone. Slot g is the mixture of the two mRNAs, and h is the mixture after periodate and aniline treatment. Essentially no Pre-ProPTH is produced by the treated RNA. This piece of evidence suggests then that the 5' end of parathyroid mRNA is "capped," and together with the poly(A) data indicates that both the 5' and 3' ends of the mRNA are intact. Our preparation of RNA thus contains intact PTH mRNA, and Pre-ProPTH is not an artifact resulting from the translation of a degraded mRNA.

I could make one comment to Dr. Rubenstein's question about PSP and indicate that it is not present in equal molar concentrations with PTH. It is present in equal ratios at all levels

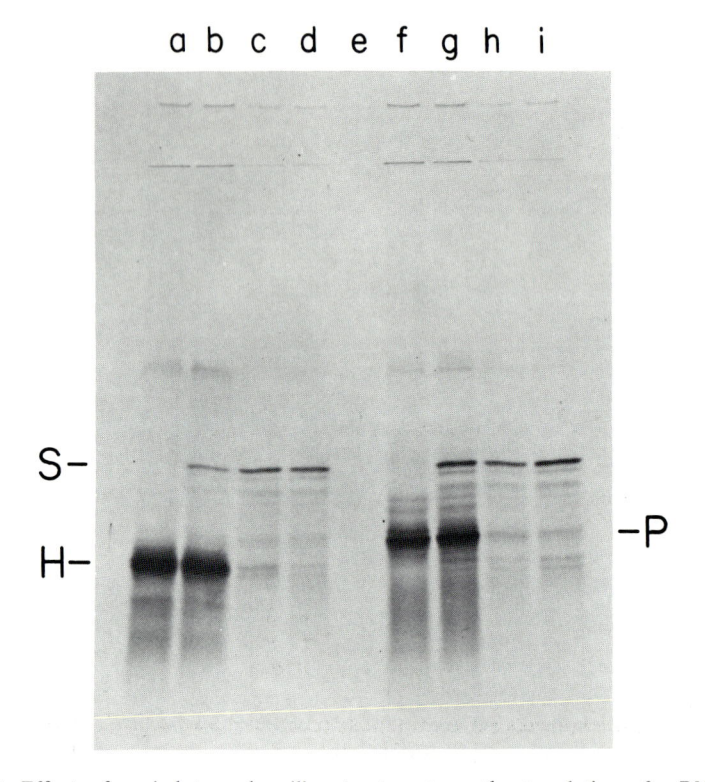

FIG. B. Effect of periodate and aniline treatment on the translation of mRNAs in wheat-germ extracts. Proteins were analyzed by electrophoresis on slab acrylamide gels containing sodium dodecyl sulfate, and the positions of *in vitro* synthesized radioactive proteins were detected by autoradiography. S, H, and P indicate the positions of satellite tobacco necrosis virus (STNV) coat protein, hemoglobin, and Pre-ProPTH, respectively. RNA added to the wheat-germ extract was: slot a, hemoglobin mRNA; b, hemoglobin mRNA + STNV RNA; c, same as b after treatment with periodate and aniline; d, STNV RNA; e, none; f, parathyroid mRNA; g, parathyroid mRNA + STNV RNA; h, same as g after treatment with periodate aniline; i, STNV RNA. From B. Kemper, *Nature (London)* **262**, 321-323 (1976), with permission.

of stimulation, but there are actually about three PTH molecules per each PSP molecule, so it is probably not analogous to the C peptides of proinsulin.

J. Habener: That is a very important achievement. Clearly, the mRNA that we have been using in these studies does contain at least some full-length message.

K. Sterling: I was intrigued by the suggestion by Dr. Rall about the clipase as a possible mode of self-correction of errors in this and other proteins. I believe there is a discrepancy between these observations and Dr. Yalow's observations on "big" PTH in the circulation. Certainly the ProPTH had only 6 amino acids added to the 84 of PTH; when we only knew about this, it seemed much smaller than some of the molecules Dr. Yalow was reporting by Sephadex gel filtration. When I first heard about the Pre-ProPTH, I wondered whether this might not resolve what I gathered to be a discrepancy. However, if it is being degraded while the nascent chain is still growing on polyribosomes, and probably none gets out into the circulation, that would hardly seem to explain or resolve the discrepancy. I wonder if I am right as to whether there is an apparent discrepancy, and do you have some idea as to what the explanation may be.

J. Habener: Your question is: What are my thoughts about the nature of the "big" forms of hormones?

K. Sterling: The biggest ones.

J. Habener: I think this is an open question. Most of these studies, as far as I am aware, have been done by gel filtration and immunoassays under conditions where buffers containing serum were used. These conditions do not eliminate the possibility of aggregation and noncovalent binding of the hormones to larger macromolecules. My feeling is that most of these large forms are aggregates or noncovalent complexes of the hormones. Using the prohormone-specific immunoassay, we have not obtained any convincing evidence that prohormone itself is secreted. Certainly the proinsulin is secreted into the blood stream. One can speculate, however, that neoplastic tissues, those that produce ectopically the hormones, may lack the specific proteolytic enzymes for conversion of the precursors to the hormones, and the precursors themselves may escape from the cell.

K. Sterling: One wild thought, which occurred to me while Dr. Kemper was talking with reference to the parathyroid secretory protein you have called PSP, is that it could function somewhat like a neurophysin with vasopressin. It was a big molecular weight. Could that possibly exist in a linkage with orthodox PTH that may be detected in some of these gel filtrations?

J. Habener: By linkage, you mean a noncovalent linkage, I would presume?

K. Sterling: Yes, however neurophysin and vasopressin are linked.

J. Habener: I would certainly accept that as a possibility. We have no information on that.

R. Yalow: Have you attempted to look for rebinding of PTH to PSP?

J. Habener: Yes, in some very preliminary and not yet complete studies. These were negative. We could demonstrate no binding of PSP to the iodinated PTH or to ProPTH.

J. T. Potts: I wish to amplify the comments that Dr. Habener made earlier about the cleavage process that has been touched on by several speakers. One of the audience asked the question how could trypsin and carboxypeptidase accomplish the conversion of precursor molecules to the hormone given the number of tryptic-sensitive sites that exist within those molecules. I believe it has been the findings of Dr. Steiner's group, as well as those of our own with PTH, that the highly specific grouping of basic residues, which Dr. Habener showed links the prohormone peptide to the hormone sequence for most prohormones, is so favored by sequence alone for cleavage by trypsin that, if we expose these particular polypeptides to dilute solutions of trypsin, then we can get very very efficient conversion of prohormone to hormone—85% or more—by proteolysis of the prohormone-specific tryptic site without cleavage of the tryptic sites within the rest of the molecule. This means, then, that the converting enzyme within the cell need not have any unusual, but only trypsinlike, specificity [D. Goltzman, E. N. Callahan, G. W. Tregear, and J. T. Potts, Jr., *Biochemistry* **15**, 5076 (1976)]. That the issue of prohormone conversion has become more complex, however, is a consequence of the observations of Dr. Steiner's group and various work, including that of Dr. Habener in our group, that with proinsulin a trypsinlike enzyme was not sufficient for conversion to insulin.

An enzyme with specifications like those of carboxypeptidase B may also be involved. If the basic residues that are part of the prohormone extension are added to the amino terminus of the prohormone-specific peptide, as with ProPTH, then a trypsinlike enzyme alone is sufficient, and the carboxypeptidase B activity works on the basic residues left at the carboxyl end of the prohormone-specific peptide. In the case of certain of the other prohormones, such as proinsulin, where some basic residues are at the carboxyl terminus of the hormone portion, then the basic residues are left on the hormone-specific polypeptide after tryptic cleavage. One then appreciates the carboxypeptidase B-like action more readily. Now, what kind of enzymic process actually operates within the cytoskeleton *in vivo* is unknown; the *in vitro* work with enzymes such as trypsin and carboxypeptidase B can only imitate what the cell does. There may be some kind of highly specific, closely spatially linked set of enzymes near the Golgi apparatus unique from, but with overall similarities to, trypsin and carboxypeptidase. On the other hand, like everyone else, I certainly am puzzled by the apparent lack of uniformity of sequence so far identified at the linkage of the prehormone peptide to the prohormone.

The wide variation in sequence seen does not provide any clues as to the specificity of a converting enzyme involved with the conversion of any initial precursor to prohormone or related biosynthetic intermediate, and there is, on the basis of sequences alone, no suggestion that an enzyme of a given type of specificity could be common from one cell to the next, in sharp contrast to the general mechanism that seems evident for prohormone-to-hormone conversion.

Carbon-13 Nuclear Magnetic Resonance Investigations of Hormone Structure and Function

IAN C. P. SMITH AND ROXANNE DESLAURIERS

Division of Biological Sciences, National Research Council of Canada, Ottawa, Canada

I. Introduction

Nuclear magnetic resonance (NMR) of the abundant isotope of hydrogen, ^{1}H, has been of great utility in elucidating the conformations of a wide variety of compounds of biological interest (James, 1975; Dwek, 1973). However, as the complexity and size of the compounds increase, so does the difficulty in distinguishing the resonances of different constituents, and one has to resort to spectrometers operating at very high magnetic fields. The spin–spin couplings between vicinal hydrogens yield conformational information, but often with at least a 2-fold ambiguity. Dynamical information is available in principle from the spin–lattice (T_1) and spin–spin (T_2) relaxation times of ^{1}H. However, owing to the complexities of the relaxation mechanism this information is not readily derived without errors or ambiguities. Carbon-13 NMR offers a resolution of many of these difficulties.

The properties of ^{13}C and ^{1}H are compared in Table I. Both nuclei have spin 1/2, which makes many aspects of ^{13}C interpretable in terms familiar from ^{1}H NMR. Carbon-13 has a lower magnetic moment than ^{1}H, and resonates at 25 MHz in a magnetic field of 23 kilogauss (kG). Owing to its lower magnetic moment, ^{13}C is intrinsically more difficult to observe than ^{1}H. A further

TABLE I
Some Properties of Carbon-13 Nuclei

Property	^{13}C	^{1}H
Spin	1/2	1/2
Frequency (MHz)	25.1	100
Sensitivity	0.016	1.00
Abundance (%)	1.1	99.9
Chemical shift range (ppm)	220	12

difficulty lies in the low natural abundance of ^{13}C; this turns to our advantage, however, in that no coupling between ^{13}C nuclei is observed in spectra taken at natural abundance. In cases where sensitivity becomes a problem, compounds can be enriched in ^{13}C at specific positions at increasingly reasonable cost. The chemical shift range of ^{13}C, that is, the range over which one can detect different types of carbon, is much greater than that for ^{1}H. Finally, the relaxation mechanism for ^{13}C is relatively simple, and the T_1 and T_2 values can be used to study conformational dynamics. Some excellent reference works are available on ^{13}C NMR (Stothers, 1972; Levy and Nelson, 1972; Gray, 1975; Wehrli and Wirthlin, 1976; Levy, 1975, 1976) as well as on its biological applications (Gray, 1973; Smith et al., 1975; Deslauriers and Smith, 1976a; Wenkert et al., 1976; McInnes et al., 1976; Komoroski et al., 1976; Wüthrich, 1976).

II. Use of ^{13}C Chemical Shifts for the Identification of Biological Compounds

The high resolving power of ^{13}C NMR makes it a useful technique for the identification of complex compounds. The relative advantage of ^{13}C over ^{1}H is well demonstrated by the spectra of cholesterol in Fig. 1. Even with a high-frequency ^{1}H spectrometer the methylene and methine protons are difficult to distinguish, whereas the ^{13}C spectrum yields a separate resonance for every carbon atom. Figure 2 shows the assignments of the various ^{13}C resonances of cholesterol and how the chemical shifts vary with solvent. When attempting to identify an unknown compound by reference to ^{13}C spectra of possible compounds, it is imperative to use the same solvent; solvation shifts can act to change dramatically the overall character of a spectrum (Mantsch and Smith, 1973). Carbon-13 NMR has been used recently to characterize 17α-ethyl-5α-estran-17β-ol, an impurity in the anabolic steroid ethylestrenol, 17α-ethylestr-4-en-17β-ol (Hasan et al., 1977).

In some cases ^{13}C NMR may be the only analytical method to detect with certainty the nature and amount of impurities in drug preparations. Such was the case with the ophthalmic drug pilocarpine (I). In most analytical techniques it is extremely difficult to identify the isomer isopilocarpine (II). Neville and co-workers (1976, 1977a,b) showed that ^{13}C can detect and quantitate not only

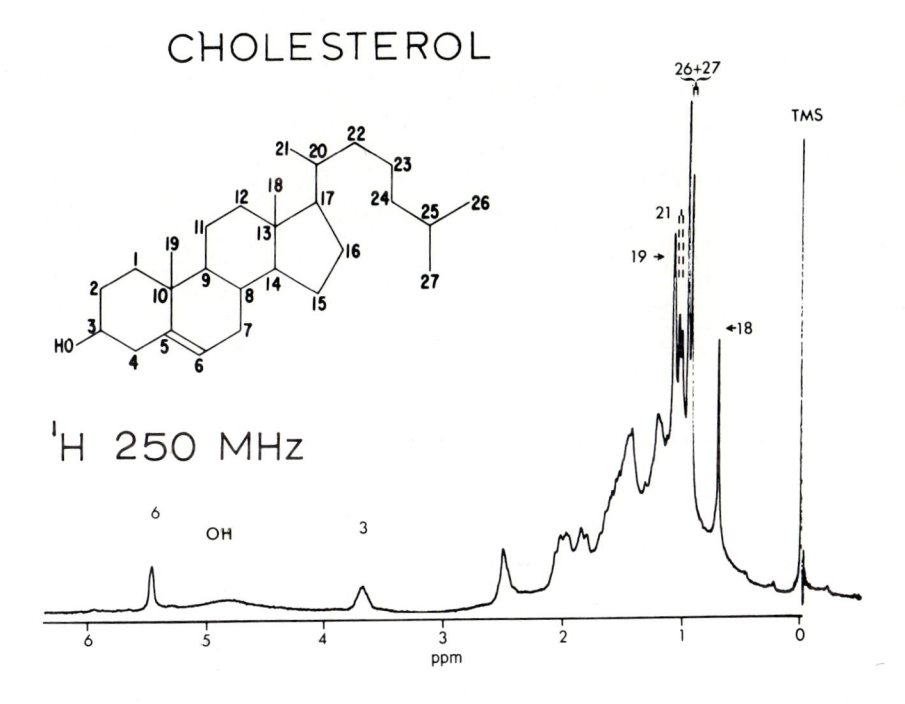

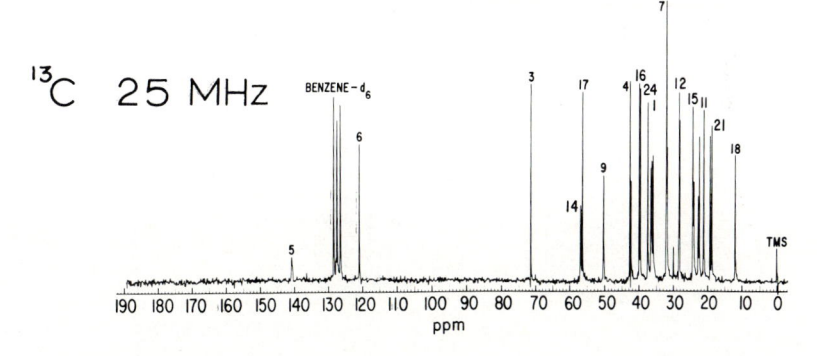

FIG. 1. Comparison of the ¹H NMR (250 MHz) and ¹³C NMR (25 MHz) spectra of cholesterol, 0.2 M in benzene-d_6. The ¹³C spectrum was obtained at natural abundance and required 30,000 accumulations of spectra in a time of 3.3 hours. Adapted with permission from Anet and Levy (1973).

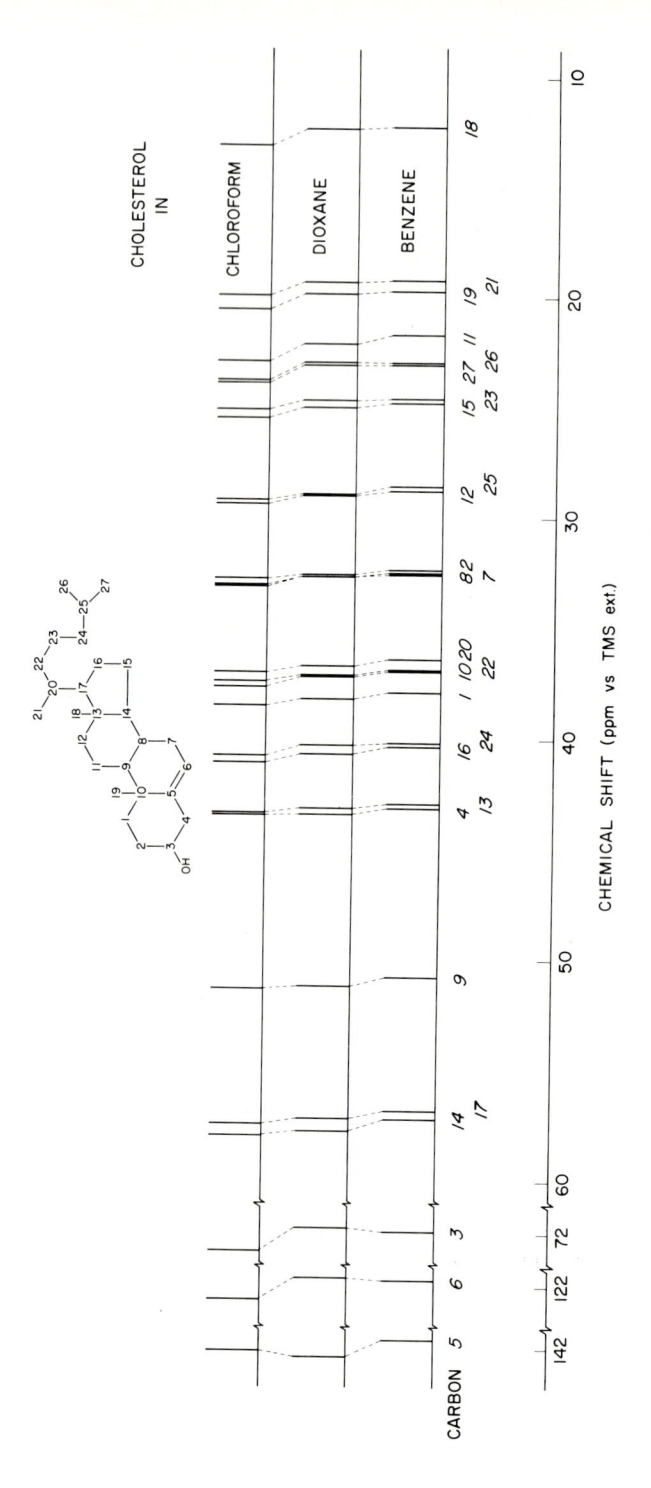

Fig. 2. Schematic representation of the ^{13}C NMR spectra of cholesterol in chloroform, dioxane, and benzene. From Mantsch and Smith (1973).

isopilocarpine, but also the two isomeric products resulting from hydrolysis of the lactone rings of (I) and (II).

III. ^{13}C Chemical Shifts and the Conformations of Peptide Hormones

The ^{13}C chemical shifts of the common amino acids are quite characteristic (Horsley and Sternlicht, 1968; Christl and Roberts, 1972). In addition, they undergo changes upon incorporation into peptides. Figure 3 shows the effect on the α-carbon chemical shift of glycine on incorporation into di- and tripeptides. Note that in Gly-Gly the residues at the carboxyl and amino termini have distinct chemical shifts. In Gly-Gly-Gly this distinction is retained, but the central residue has a chemical shift very similar to that of the unlinked amino acid. This behavior gives us two advantages: the terminal residues in a complex peptide can be identified (in the case of an ambiguity their chemical shifts respond characteristically to pH change); the chemical shifts for aminoacyl residues in a large peptide can be readily predicted from the shifts of the component amino acids (any deviation from the predicted position is indicative of secondary or tertiary structure). In larger peptides, particularly those in which the same amino acid occurs more than once, other assignment techniques must be used. These are: comparison with shorter peptides of homologous sequence;

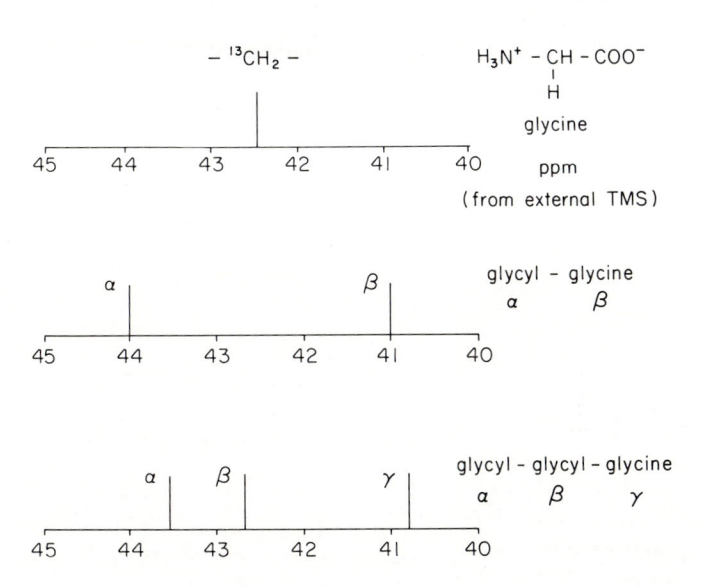

FIG. 3. Schematic representation of the ^{13}C NMR spectra of glycine, glycyl-glycine, and glycyl-glycyl-glycine in peptides. Constructed from the data of Horsley and Sternlicht (1968).

titration of individual residues; study of selectively modified peptides (one or more residues replaced or deleted); ^{13}C or ^{2}H enrichment of specific residues. A comprehensive review of these techniques and their application to peptide hormones has appeared recently (Deslauriers and Smith, 1976).

A. OXYTOCIN

An example of the resolving power of ^{13}C NMR is the neurohypophysial hormone oxytocin, whose entire ^{13}C spectrum is shown in Fig. 4, with an expansion of the aliphatic carbon region in Fig. 5. All resonances have been assigned, although there was some difference of opinion among early studies (Smith *et al.*, 1972; Deslauriers *et al.*, 1972; Lyerla and Freedman, 1972; Brewster *et al.*, 1973).

The correct assignments have now been determined by use of substituted analogs (Walter *et al.*, 1973) and by ^{13}C enrichment (Griffin *et al.*, 1975a,b; Chaiken *et al.*, 1975; Blumenstein and Hruby, 1976). In a study of the oxytocin spectrum as a function of pH, the pH-sensitive resonances were restricted to those of carbon atoms in the cyclic moiety of oxytocin (Walter *et al.*, 1973). As expected, the Cys-1 residue underwent the greatest change in chemical shift with pH. However, unexpected changes were observed in the chemical shifts of carbons far removed from the site of titration. These changes were attributed to conformational alterations in the cyclic moiety, mainly due to a pH-induced change in the dihedral angle of the Cys-1 to Cys-6 disulfide bond.

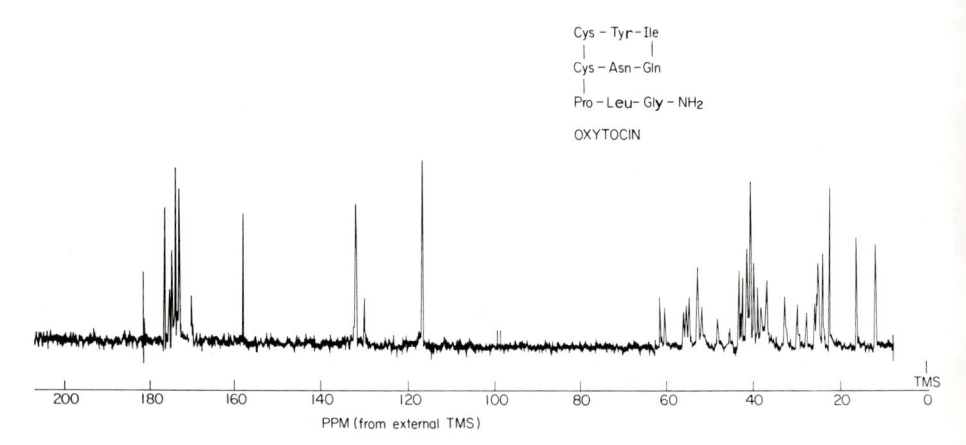

FIG. 4. The 25 MHz ^{13}C NMR spectrum of oxytocin, 40 mg per 0.2 ml of dimethyl sulfoxide (DMSO-d_6), 37°C. The resonances to largest chemical shift are due to the carbonyl carbons, those from 120 to 140 ppm to the tyrosyl aromatic carbons, and those from 15 to 60 ppm to the methyl, methylene, and methine carbons. The spectrum required the averaging of 84,000 accumulations.

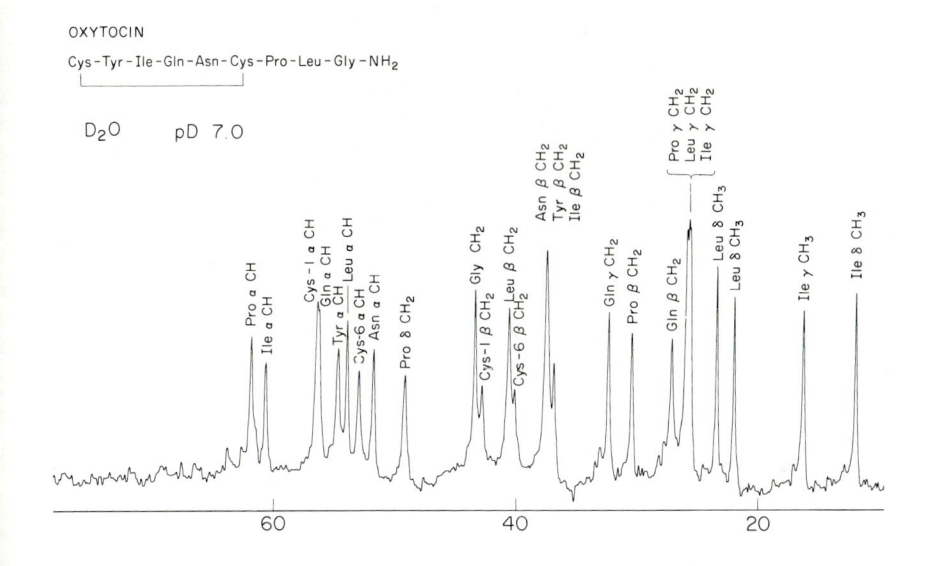

OXYTOCIN

Cys-Tyr-Ile-Gln-Asn-Cys-Pro-Leu-Gly-NH₂

FIG. 5. Expansion of the aliphatic carbon region of the 25 MHz ^{13}C NMR spectrum of oxytocin in D$_2$O, 100 mg/ml, pD 7.0, 37°C, 57,000 accumulations. The assignments of the various carbon resonances are indicated; those for the α-carbons of Cys-1 and Tyr should be interchanged (Deslauriers et al., 1972).

B. CIS AND TRANS ISOMERS OF PROLINE IN PEPTIDE HORMONES

Although most peptide bonds are assumed to prefer the trans conformer with respect to rotation about the —N—C(=O)— bond, the cis conformer has been found to be quite common in peptides containing proline (Fig. 6) (Smith et al., 1972, 1973; Bovey, 1972; Deslauriers et al., 1972; Dorman and Bovey, 1973; Grathwohl et al., 1973; Gratwohl and Wüthrich, 1976a,b). The resonance positions for Pro in the two conformers depend upon the nature of the neighboring amino acids. The chemical shift of the γ-carbon of proline (Dorman and Bovey, 1973) and the difference between the chemical shifts of the β- and γ-carbons (V. Bystrov, personal communication, 1976) seem to be the best monitors of the cis conformer.

Thyrotropin-releasing factor (TRF, <Glu-His-Pro-NH$_2$) was the first biologically active peptide in which a significant population of the cis conformer of proline was found (Smith et al., 1972; Deslauriers et al., 1973a,b). The ^{13}C NMR spectrum of TRF in D$_2$O is shown in Fig. 7; the resonances of the cis conformer of Pro are seen as a shoulder on the resonance of the Pro C$_α$, and

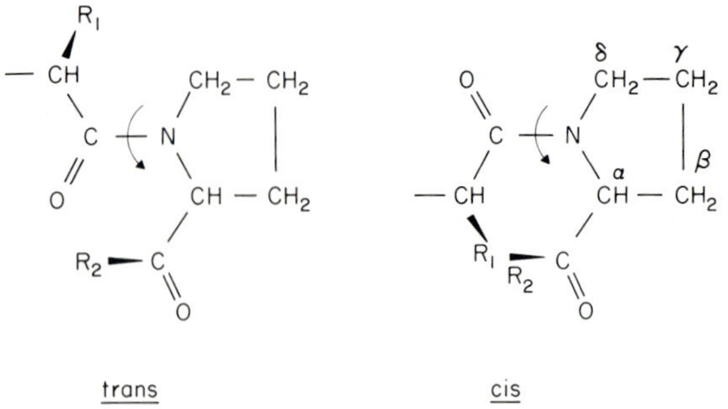

FIG. 6. The cis and trans conformers of proline in an X-Pro amide bond with respect to rotation about the −C(=O)−N− bond.

resonances of lower intensity beside those indicated for the C_β, C_γ, and C_δ of Pro. The population of the cis conformer was found to vary with the nature of the solvent (D_2O, 14%; DMSO, 6%; pyridine, 0%). The accessibility of the cis conformer raises the intriguing possibility that the active conformation bound to the receptor for TRF might contain only the cis conformer.

Angiotensin II (Asp-Arg-Val-Tyr-Val (or Ile) -His-Pro-Phe) has been studied by ^{13}C NMR (Zimmer et al., 1972; Deslauriers et al., 1975a). The prolyl residue was found to be entirely in the trans conformation (Deslauriers et al., 1975a). Recently, it has been found that appreciable amounts of the cis conformer may exist at higher pH values (Galardy et al., 1976). Modification of the peptide to contain three prolyl residues, [Pro^3, Pro^5]-angiotensin II, yielded a ^{13}C NMR spectrum with indications of the presence of cis-proline (Fig. 8a) (Deslauriers et al., 1976a). However, it could not be determined from the spectrum at 25 MHz whether only the two additional prolyl residues, or all three prolyl residues, have some population of cis conformer. The difficulty was resolved by obtaining the spectra at much higher frequency, 90 MHz, which has a resolving power 3.6 times higher (Fig. 8b) (Deslauriers et al., 1976a). Note that the resonance at approximately 21 ppm in Fig. 8a becomes four resonances at the higher frequency in Fig. 8b. The three narrow resonances of almost equal intensity are due to the γ-carbons of the three prolyl residues. The fourth resonance is due to the methyl group of residual acetate from previously buffered solutions of the peptide. Thus, addition of two prolyl residues to angiotensin II causes a sufficiently large conformational change in the peptide to force the original prolyl residue more easily into the cis conformation. From spin−lattice relaxation time measurements (see below) on these systems, it was concluded that the peptide backbone of [Pro^3, Pro^5]-angiotensin II was less rigid than that of the parent hormone.

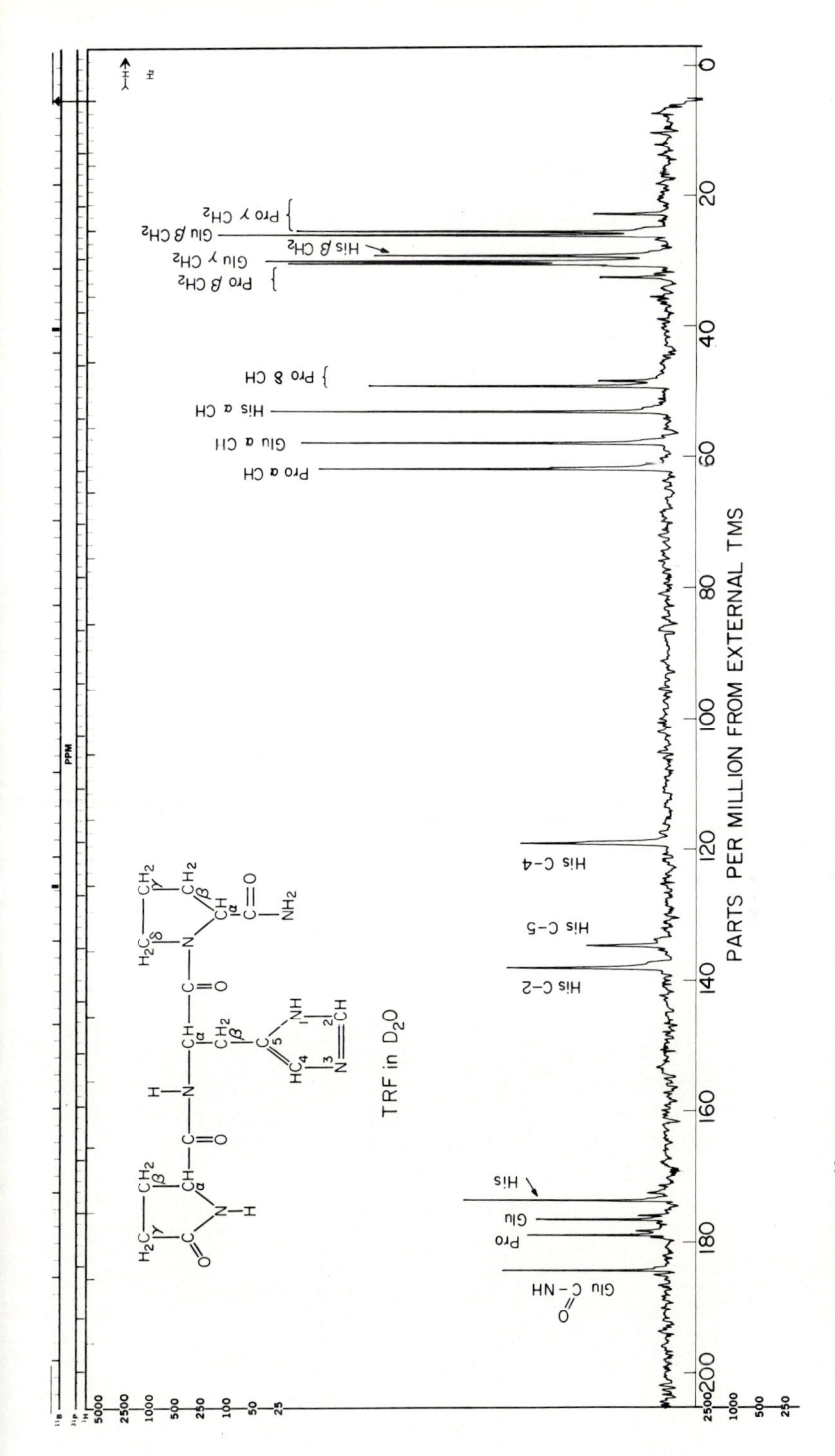

FIG. 7. The 25 MHz ^{13}C NMR spectrum of thyrotropin-releasing factor in D_2O, 80 mg/ml, 37°C, 7400 transients. Resonance assignments are indicated. From Smith et al. (1973).

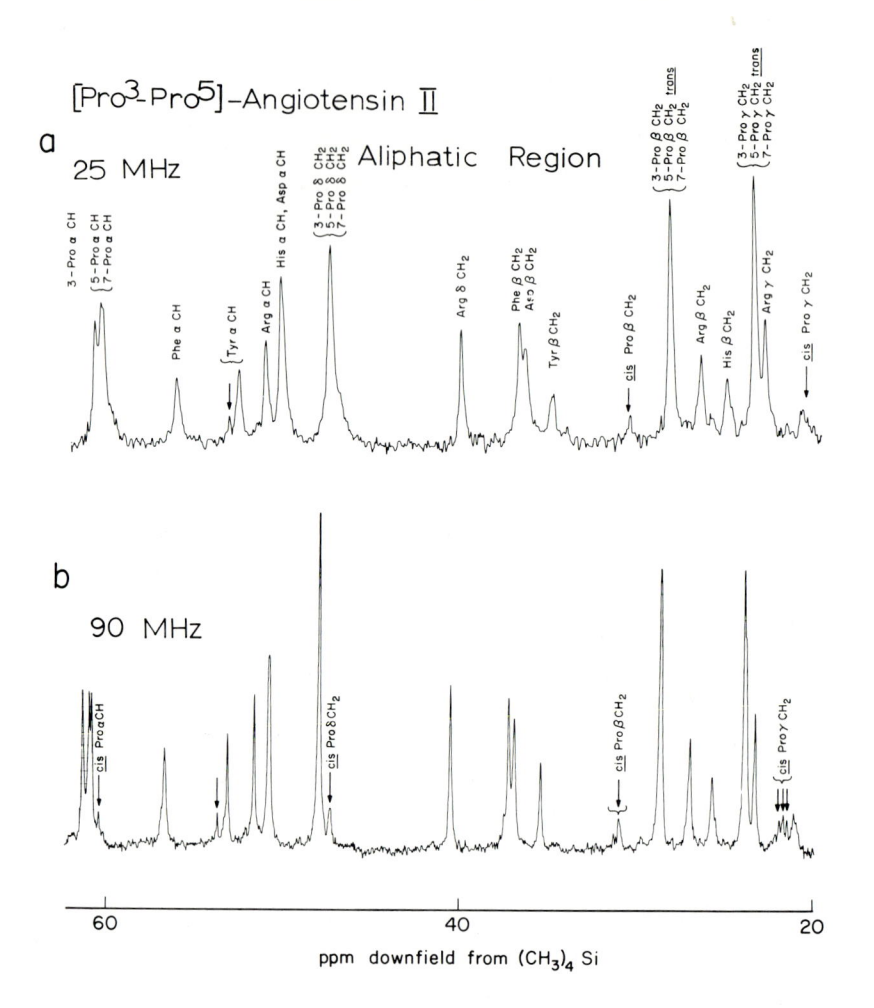

FIG. 8. ^{13}C NMR spectra (aliphatic region) of [Pro^3, Pro^5]-angiotensin II in D$_2$O, 100 mg/ml; (a) 25 MHz, pH observed 5.2, 32°C, 32,000 accumulations; (b) 90 MHz, pH observed 6.0, 20°C, 39,000 accumulations. From Deslauriers *et al.* (1976).

C. THE TAUTOMERIC FORMS OF HISTIDINE IN PEPTIDE HORMONES

The imidazole moiety of neutral histidine can exist in one or both of two tautomeric forms in which a hydrogen atom is attached to the π- or τ-nitrogen. The pH dependences of the ^{13}C chemical shifts of π-methylhistidine and

τ-methylhistidine are dramatically different (Reynolds et al., 1973). It was demonstrated that the tautomeric preference of histidine itself could be determined by measuring the pH dependence of the imidazole chemical shifts. This technique was applied to TRF (Deslauriers et al., 1974a) and angiotensin II (Deslauriers et al., 1975b). In both cases it was concluded that the histidyl residue has a strong preference for the N^τ tautomer. The hyperactivity of the N^τ-methyl derivative of TRF was postulated to be due to the necessity of substitution on N^τ. In TRF itself the preference of the N^τ–H over the N^π–H tautomer is 4:1; in N^τ-methyl TRF no N^π-tautomer is present. This dominance of the N^τ-substitution in the N^τ-methyl TRF, and improved hydrophobic character of the methyl group over the proton, were proposed to enhance interaction of the hormone with the receptor site.

IV. Carbon-13 Spin–Lattice Relaxation Times As Indicators of Molecular Mobility

After excitation to a higher spin state by absorption of radiofrequency energy, a carbon-13 nucleus must return (relax) to the ground state. It does so by a nonradiative transfer of energy to the system in which it finds itself (the lattice). The mechanism of this transfer can vary with the chemical nature of the carbon (Wehrli, 1976), but, for most C–H moieties (sp^3 or sp^2) in compounds of molecular weight greater than 100, relaxation is dominated by the dipole–dipole mechanism (Allerhand et al., 1971). For the case of a molecule rotating rapidly (on the ^{13}C NMR time scale, $>10^{10}$ sec^{-1}) and isotropically, a simple relationship exists between the ^{13}C T_1 values and the correlation time (τ_c) for rotational reorientation of the ^{13}C–^{1}H moiety:

$$1/T_1 = [(N\hbar^2 \gamma_C^2 \gamma_H^2)/r^6]\tau_c \qquad (1)$$

where N is the number of hydrogen atoms at a distance r from the ^{13}C nucleus, $\hbar$ is Planck's constant/2π, and γ_C and γ_H are the magnetogyric ratios for carbon and hydrogen, respectively. Thus, for a series of C—H moieties of equal r value, one can write a convenient rule of thumb:

$$NT_1 \alpha \text{ mobility} \tag{2}$$

the longer the NT_1 value, the more mobile is the carbon atom. Even in the case of rapid motion, two further cases must be considered. It is possible that for a highly organized system the molecular motion is not isotropic—i.e., the molecule can rotate more rapidly about some axes than about others, and one must consider at least two rotational rates. This case and the corresponding relation between the diffusion rates about two nonequivalent axes, and the effective correlation time τ, are shown in Fig. 9. The other, and very common, possibility is that some region of the molecule can rotate rapidly with respect to the rest of the molecule. This is represented in Fig. 10 in terms of the correlation times τ_i and τ_m for internal and overall motion, respectively. For quantitative measurements of correlation times these more complex expressions must be used (Deslauriers and Somorjai, 1976; Somorjai and Deslauriers, 1976; Deslauriers *et al.*, 1977b,c), but often one can gain considerable insight into the dynamics of a system by using the simple expression in Eq. (2).

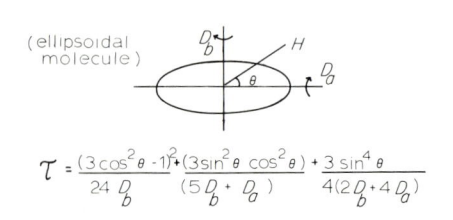

$$\tau = \frac{(3\cos^2\theta - 1)^2}{24 D_b} + \frac{(3\sin^2\theta \cos^2\theta)}{(5 D_b + D_a)} + \frac{3\sin^4\theta}{4(2 D_b + 4 D_a)}$$

D_a = diffusion constant for rotation about the symmetry axis

D_b = diffusion constant for reorientation about any axis perpendicular to the symmetry axis

θ = $\angle$ between C-H and symmetry axis

FIG. 9. Schematic of an ellipsoidal molecule that undergoes anisotropic rotational diffusion. An effective correlation time τ can be defined in terms of the rates of diffusion about the two principal axes, D_a and D_b, and the angle θ between the C—H vector and the major axis a. D_a = diffusion constant for rotation about the symmetry axis; D_b = diffusion constant for reorientation about any axis perpendicular to the symmetry axis.

ISOTROPIC MOLECULAR MOTION WITH INTERNAL MOTION

$$\frac{1}{NT_I} = \frac{\hbar^2 \gamma_H^2 \gamma_C^2 \tau_m}{r^6}\left[\frac{1}{9} + \frac{8}{27}\frac{6\tau_i}{6\tau_i + \tau_m} + \frac{16}{27}\frac{\tau_i}{3\tau_i + \tau_m}\right]$$

FIG. 10. Schematic of a molecule in which motion of a particular group takes place rapidly with respect to the rate of overall motion. The τ_i and τ_m are correlation times for internal and overall motion, respectively. All other terms are as defined previously.

A. POLY-L-LYSINE

A simple example of the use of ^{13}C T_1 values is the mobility of the side chains in oligo- and poly-L-lysine (Saitô and Smith, 1973). Figure 11 shows the T_1 values for the various carbons of (L-Lys)$_{12}$ and (L-Lys)$_n$. Notice that although the molecular weights differ greatly (2850 for (L-Lys)$_{12}$, 17,000 for (L-Lys)$_n$), the magnitudes of the T_1 values measured at low pH are essentially the same.

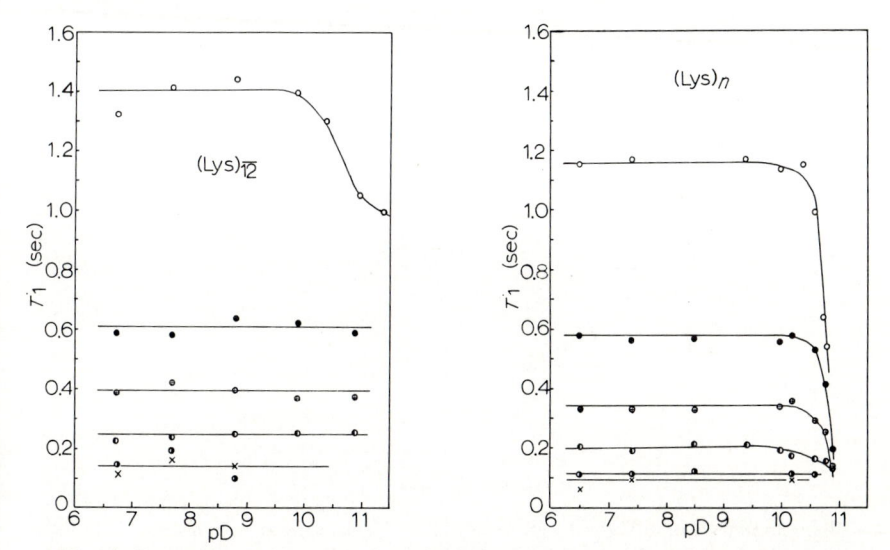

FIG. 11. Dependence of the T_1 values (25 MHz) for individual carbon atoms of poly-L-lysine and (L-lysine)$_{12}$ as a function of the pH of the D$_2$O solution; compounds were 100 mg/ml, 37°C. From top to bottom the various curves represent the behavior of: ○, C=O; ●, C$_\epsilon$; ⊕, C$_\delta$; ◐, C$_\gamma$; ◑, C$_\beta$; X, C$_\alpha$ (Saitô and Smith, 1973).

This is because in the disordered form of (L-Lys)$_n$ the Lys side chains do not interact strongly and their segmental motions are similar in rate and angular amplitude to those of Lys in (L-Lys)$_{12}$. Notice also that the T_1 values decrease steadily from the ε- to the α-carbons in both cases. This reflects the greater restriction of the Lys residues at the point of attachment to the peptide backbone. A considerable difference between the two systems is observed at high pH where (L-Lys)$_n$ forms an ordered helical structure which severely restricts side-chain motions. The helical content of (L-Lys)$_{12}$ has been shown to be very low under these conditions (Yaron *et al.*, 1971). Thus, near pH 11 the T_1 values of (L-Lys)$_n$ drop rapidly with increasing pH whereas those of (L-Lys)$_{12}$ are unchanged.

A further demonstration of the dominance of segmental motion is shown in Fig. 12. On increasing the number of monomers in (L-Lys)$_n$ from 1 to 5, there is a steady decrease in the T_1 values. This is because the rate of overall motion of the peptide is comparable to the rates of segmental motion. For five residues or more, however, the rate of segmental motion is sufficiently greater than that of overall motion, and increasing molecular weight has very little effect on the T_1 values of C$_\beta$ to C$_\epsilon$.

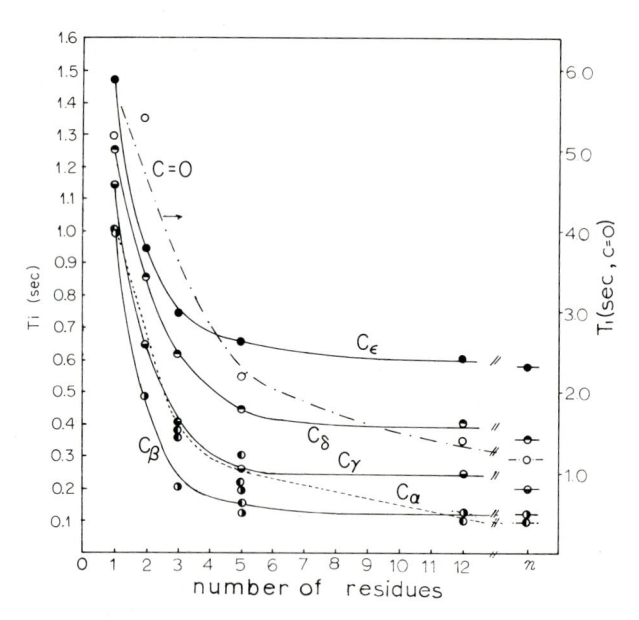

FIG. 12. Dependence of the T_1 values (25 MHz) for the various carbon atoms of lysine in a series of oligolysines of increasing molecular weight. The points for n residues are for poly-L-lysine, molecular weight 17,000. The various curves represent the behavior of: ○, C=O; ●, C$_\epsilon$; ◒, C$_\delta$; ◓, C$_\gamma$; ◐, C$_\beta$; –●–, C$_\alpha$. T = 37°C, pD = 7.5. From Saitô and Smith (1973).

B. OXYTOCIN AND ANGIOTENSIN

An excellent example of the information content of ^{13}C T_1 values is afforded by the neurohypophysial hormone oxytocin, Fig. 13 (Deslauriers et al., 1974b). Within the cyclic portion the α-carbons have NT_1 values that are equal within the experimental error of ± 15%. This indicates that any motion that occurs within the cyclic backbone does so with roughly equal rate and angular amplitude for all α-carbons. Proceeding to the exocyclic tripeptide, we notice a steady increase in the NT_1 values of the α-carbons from Pro to Gly-NH$_2$. The Gly NT_1 value is almost five times greater than that of the Pro. This demonstrates an increasing degree of segmental motion with greater distance from the point of attachment of the tripeptide to the cyclic region and argues against any conformation involving a restriction of the tripeptide by binding to the cyclic region.

Another interesting dynamical feature is the lack of equality of the NT_1 values for the carbon atoms in the prolyl ring. If the five-membered ring were internally rigid, these values should be equal. The large value for the β-carbon with respect to those of the α- and δ-carbons indicates that the prolyl ring is interconverting rapidly between puckered forms at the γ-carbon. This intracyclic mobility has

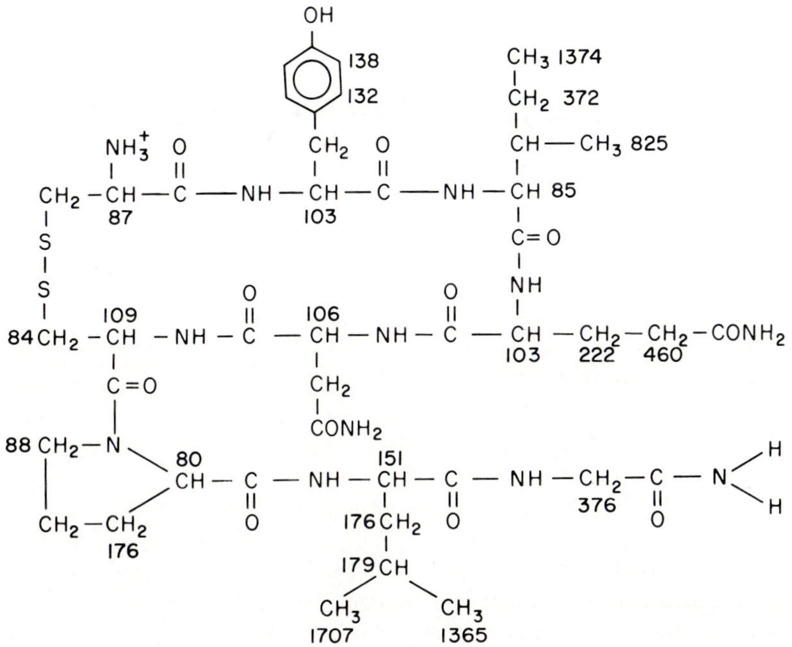

FIG. 13. The NT_1 values (in msec at 25 MHz) for the individual carbon atoms of oxytocin, 100 mg/ml in D$_2$O, pH observed 3.5, 32°C. From Deslauriers et al. (1974b).

been observed in a variety of hormones and peptides containing the pyroglutamyl and prolyl residues (Deslauriers et al., 1973a, 1974a,b,c, 1975a,b, 1976a,b; Deslauriers and Smith, 1976). The position in the ring at which the rapid motion occurs depends upon the nature of the neighboring amino acids.

Segmental motion in the side chains of the amino acids in oxytocin is also manifest in the ^{13}C T_1 values. The tyrosyl residue rotates rapidly about the C_α–C_β bond; this effect is even more pronounced for the tyrosyl and phenylalanyl residues in lysine vasopressin (Deslauriers et al., 1974b). The two methyl groups of each of the leucyl and isoleucyl residues in oxytocin have significantly different mobilities. The increase in NT_1 values for the glutamidyl residue from C_α to C_γ is indicative of an unhindered residue, arguing against any secondary structure involving this residue.

A similar study has been made of the ^{13}C T_1 values of [Ile[5]]-angiotensin II and [Ile[5],Leu[8]]-angiotensin II (Deslauriers et al., 1975a). These linear peptides show properties similar to those of the cyclic portion of oxytocin, suggesting that the backbone is folded to form a relatively rigid conformation.

C. LUTEINIZING HORMONE-RELEASING FACTOR

The decapeptide luteinizing hormone-releasing factor (LRF) and a series of substituted analogs have been studied by the ^{13}C T_1 method (Deslauriers et al., 1975b, 1976b, 1977a). The NT_1 values for LRF in D_2O are shown in Fig. 14. Note that although LRF is a larger molecule than oxytocin, the NT_1 values of its α-carbons are longer than those of oxytocin. Thus, unlike angiotensin (see above), LRF has a significant degree of mobility in the peptide backbone. The NT_1 values for the α-carbons are shown as a function of position in the backbone in Fig. 15. They increase appreciably toward the termini, and show a discontinuous rise for the glycine at position 6. Thus, within a relatively mobile backbone, the glycyl residue shows a particularly high mobility. This is for the most part due to a lack of substituent on the α-carbon; substitution of a D- or L-leucine at position 6 results in an NT_1 value for this position more similar to those of the neighboring residues (Deslauriers et al., 1976b). This demonstrates that the presence of a glycyl residue within the peptide backbone confers upon the backbone a greater degree of conformational freedom. The increased freedom would facilitate any change of conformation required for binding to a receptor site.

A glance at the NT_1 values for the side chains of the LRF yields details of segmental motion. The prolyl and pyroglutamyl rings manifest a high degree of intracyclic motion at C_β and C_γ. No segmental motion of the histidyl and tryptophanyl rings is apparent, whereas some mobility of the tyrosyl aromatic ring is indicated. The rather small degree of segmental motion for the side chain of arginine is characteristic of this amino acid.

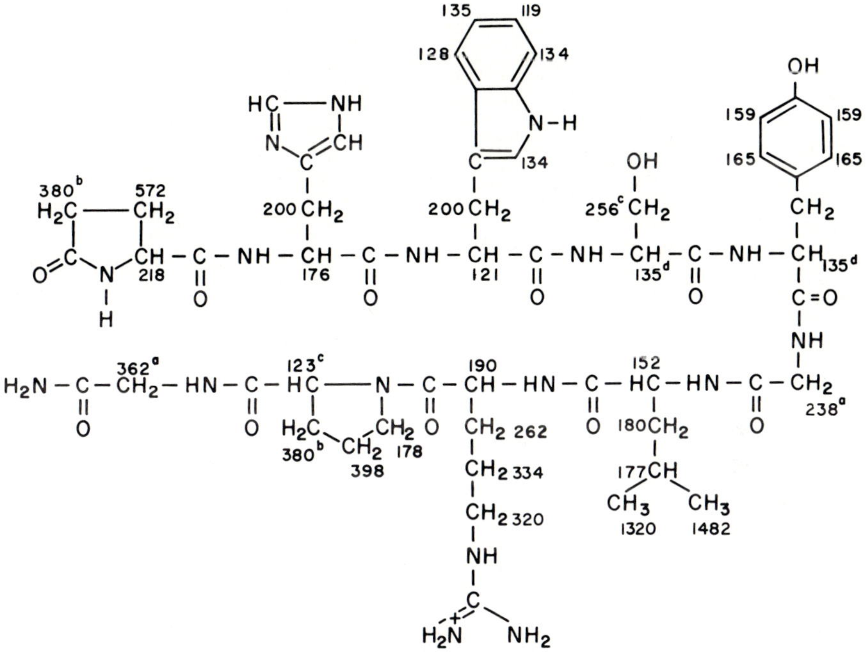

FIG. 14. The NT_1 values (in msec at 25 MHz) for luteinizing hormone-releasing hormone, 100 mg/ml in D_2O, pH observed 6.4, 32°C. The superscript letters indicate carbons whose resonances overlap. From Deslauriers *et al.* (1975b).

Measurement of T_1 values as a function of temperature allows an estimate of the activation energies (Ea) for the motional processes dominating the ^{13}C relaxation (Deslauriers *et al.*, 1977a). The C-terminal glycyl residue has an Ea of only 3.5 kcal whereas the average for the nonterminal α-carbons is 6.0 kcal. In contrast, the Ea for Gly-6 is 5.2 kcal, only slightly lower than for the other nonterminal residues. Looking at the side chains, the arginyl carbons show an average Ea of 7.0 kcal compared to that of the more mobile leucyl residue (5.5 kcal average for C_β and C_γ, 2.3 kcal for the two C_δ methyl groups). The prolyl carbons have an average Ea of 5.0 kcal. The differences between the Ea values for various positions confirm the assumption throughout the discussions above that the T_1 values are determined mainly by local rather than overall motions.

V. Conclusion

The chemical shifts and spin–lattice relaxation times of ^{13}C constitute valuable indicators of the conformations and dynamical properties of hormones and provide a useful complement to conclusions reached by ^1H NMR. An aspect that deserves further study is coupling between ^{13}C and ^1H which has been shown in

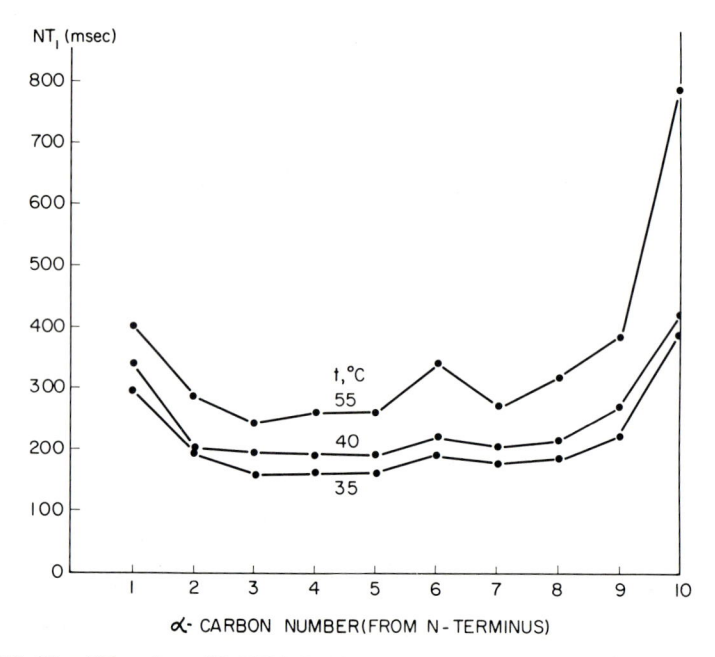

FIG. 15. The NT_1 values (25 MHz) for the α-carbon atoms in the peptide backbone of luteinizing hormone-releasing hormone as a function of position relative to the N-terminus. From Deslauriers et al. (1977b).

a few studies to be sensitive to rotamer preferences in amino acids (Hansen et al., 1975; Rodgers and Roberts, 1973). The binding of small molecules to proteins and receptors is proving to be a fruitful application (Moon and Richards, 1972; Feeney et al., 1973; Griffin et al., 1976a,b; Blumenstein and Hruby, 1975; Chaiken et al., 1975). In these latter cases, enrichment of ^{13}C is a necessity. Studies on the biosynthesis of hormones using ^{13}C-labeled precursors should provide information not easily obtainable by other techniques (Tanaka et al., 1975). With the present availability of sensitive ^{13}C NMR spectrometers operating at a variety of magnetic field strengths, these techniques will rapidly become routine in laboratories concerned with the structure–function relationship in hormones.

ACKNOWLEDGMENT

It is a pleasure to thank our collaborators without whom the studies described here would not have been possible: Drs. C. Garrigou-Lagrange, R. A. Komoroski, G. C. Levy, W. H. McGregor, A. C. M. Paiva, J. H. Seely, K. Schaumburg, H. Saitô, D. Sarantakis, and R. Walter.

REFERENCES

Allerhand, A., Doddrell, D., and Komoroski, R. A. (1971). *J. Chem. Phys.* **55**, 189.
Anet, F. A. L., and Levy, G. C. (1973). *Science* **180**, 141.
Blumenstein, M., and Hruby, V. J. (1976). *Biochem. Biophys. Res. Commun.* **68**, 1052.
Bovey, F. A. (1972). *In* "Chemistry and Biology of Peptides" (J. Meinhofer, ed.), p. 3. Ann Arbor Sci. Publ., Ann Arbor, Michigan.
Brewster, A. I. R., Hruby, V. J., Spatola, A. F., and Bovey, F. A. (1973). *Biochemistry* **12**, 1643.
Chaiken, I. M., Randolph, R. E., and Taylor, H. C. (1975). *Ann. N.Y. Acad. Sci.* **248**, 442.
Christl, M., and Roberts, J. D. (1972). *J. Am. Chem. Soc.* **94**, 4565.
Deslauriers, R., and Smith, I. C. P. (1976). *In* "Topics in Carbon-13 NMR Spectroscopy" (G. C. Levy, ed.), Vol. 2, p. 1. Wiley (Interscience), New York.
Deslauriers, R., and Somorjai, R. L. (1976). *J. Am. Chem. Soc.* **93**, 1931.
Deslauriers, R., Walter, R., and Smith, I. C. P. (1972). *Biochem. Biophys. Res. Commun.* **48**, 854.
Deslauriers, R., Garrigou-Lagrange, C., Bellocq, A.-M., and Smith, I. C. P. (1973a). *FEBS Lett.* **31**, 59.
Deslauriers, R., Walter, R., and Smith, I. C. P. (1973b). *Biochem. Biophys. Res. Commun.* **53**, 244.
Deslauriers, R., McGregor, W. H., Sarantakis, D., and Smith, I. C. P. (1974a). *Biochemistry* **13**, 3443.
Deslauriers, R., Smith, I. C. P., and Walter, R. (1974b). *J. Am. Chem. Soc.* **96**, 2289.
Deslauriers, R., Walter, R., and Smith, I. C. P. (1974c). *J. Biol. Chem.* **249**, 7006.
Deslauriers, R., Paiva, A. C. M., Schaumburg, K., and Smith, I. C. P. (1975a). *Biochemistry* **14**, 878.
Deslauriers, R., Levy, G. C., McGregor, W. H., Sarantakis, D., and Smith, I. C. P. (1975b). *Biochemistry* **14**, 4335.
Deslauriers, R., Komoroski, R. A., Levy, G. C., Paiva, A. C. M., and Smith, I. C. P. (1976a) *FEBS Lett.* **62**, 50.
Deslauriers, R., Komoroski, R. A., Levy, G. C., Seely, J. H., and Smith, I. C. P. (1976b). *Biochemistry* **15**, 4672.
Deslauriers, R., Levy, G. C., McGregor, W. H., Sarantakis, D., and Smith, I. C. P. (1977a). *Eur. J. Biochems.* (in press).
Deslauriers, R., Ralston, E., and Somorjai, R. L. (1977b). *J. Mol. Biol.* (in press).
Deslauriers, R., Ralston, E., and Somorjai, R. L. (1977c). *Nature (London)* (in press).
Dorman, D. G., and Bovey, F. A. (1973). *J. Org. Chem.* **38**, 2379.
Dwek, R. A. (1973). "Nuclear Magnetic Resonance in Biochemistry." Oxford Univ. Press (Clarendon), London and New York.
Feeney, J., Burgen, A. S. V., and Grell, E. (1973). *Eur. J. Biochem.* **34**, 107.
Galardy, R. E., Bleich, H. E., Ziegler, P., and Craig, L. C. (1976). *Biochemistry* **15**, 2303.
Grathwohl, C., and Wüthrich, K. (1976a). *Biopolymers* **15**, 2025.
Grathwohl, C., and Wüthrich, K. (1976b). *Biopolymers* **15**, 2043.
Grathwohl, C., Schwyzer, R., Tun-Kyi, A., and Wüthrich, K. (1973). *FEBS Lett.* **29**, 271.
Gray, G. A. (1973). *Crit. Rev. Biochem.* **1**, 247.
Gray, G. A. (1975). *Anal. Chem.* **47**, 546A.
Griffin, J. H., Alazard, R., DiBello, C., Sala, E., Mermet-Bouvier, R., and Cohen, P. (1975a). *FEBS Lett.* **50**, 168.
Griffin, J. H., DiBello, C., Alazard, R., Sala, E., and Cohen, P. (1975b). *In* "Peptides: Chemistry, Structure and Biology" (R. Walter and J. Meienhofer, eds.), p. 826. Ann. Arbor Sci. Publ., Michigan.

Hansen, P. E., Feeney, J., and Roberts, G. C. K. (1975). *J. Magn. Reson.* **17**, 249.
Hasan, F., Lodge, B. A., and Smith, I. C. P. (1977). *J. Pharm. Sci.* (in press).
Horsley, W. J., and Sternlicht, H. (1968). *J. Am. Chem. Soc.* **90**, 3738.
James, T. L. (1975). "Nuclear Magnetic Resonance in Biochemistry." Academic Press, New York.
Komoroski, R. A., Peat, I. R., and Levy, G. C. (1976). *In* "Topics in Carbon-13 NMR Spectroscopy" (G. C. Levy, ed.), Vol. 2, p. 180. Wiley (Interscience), New York.
Levy, G. C., ed. (1975). "Topics in Carbon-13 NMR Spectroscopy," Vol. 1. Wiley (Interscience), New York.
Levy, G. C., ed. (1976). "Topics in Carbon-13 NMR Spectroscopy," Vol. 2. Wiley (Interscience), New York.
Levy, G. C., and Nelson, G. L. (1972). "Carbon-13 Nuclear Magnetic Resonance for Organic Chemists." Wiley (Interscience), New York.
Lyerla, J. R., and Freedman, M. H. (1972). *J. Biol. Chem.* **247**, 8183.
McInnes, A. G., Walter, J. A., Wright, J. L. C., and Vining, L. C. (1976). *In* "Topics in Carbon-13 NMR Spectroscopy" (G. C. Levy, ed.), Vol. 2, p. 125. Wiley (Interscience), New York.
Mantsch, H. H., and Smith, I. C. P. (1973). *Can. J. Chem.* **51**, 1384.
Moon, R. B., and Richards, J. H. (1972). *J. Am. Chem. Soc.* **94**, 5093.
Neville, G. A., Hasan, F. B., and Smith, I. C. P. (1976). *Can. J. Chem.* **54**, 2094.
Neville, G. A., Hasan, F. B., and Smith, I. C. P. (1977a). *J. Pharm. Sci.* (in press).
Neville, G. A., Hasan, F. B., and Smith, I. C. P. (1977b). *Can. J. Pharm. Sci.* (in press).
Reynolds, W. F., Peat, I. R., Freedman, M. H., and Lyerla, J. R. (1973). *J. Am. Chem. Soc.* **95**, 328.
Rodgers, P., and Roberts, G. C. K. (1973). *FEBS Lett.* **36**, 330.
Saitô, H., and Smith, I. C. P. (1973). *Arch. Biochem. Biophys.* **158**, 154.
Smith, I. C. P., Deslauriers, R., and Walter, R. (1972). *In* "Chemistry and Biology of Peptides" (J. Meienhofer, ed.), p. 29. Ann Arbor Sci. Publ., Ann Arbor, Michigan.
Smith, I. C. P., Deslauriers, R., Saitô, H., Walter, R., Garrigou-Lagrange, C., McGregor, H., and Sarantakis, D. (1973). *Ann. N.Y. Acad. Sci.* **222**, 597.
Smith, I. C. P., Jennings, H. J., and Deslauriers, R. (1975). *Acc. Chem. Res.* **8**, 306.
Somorjai, R. L., and Deslauriers, R. (1976). *J. Am. Chem. Soc.* **98**, 6460.
Stothers, J. B. (1972). "Carbon-13 NMR Spectroscopy." Academic Press, New York.
Tanaka, K., Armitage, I. M., Ramsdell, H. S., Hsia, Y. E., Lipsky, S. R., and Rosenberg, L. E. (1975). *Proc. Natl. Acad. Sci. U.S.A.* **72**, 3692.
Walter, R., Prasad, K. U. M., Deslauriers, R., and Smith, I. C. P. (1973). *Proc. Natl. Acad. Sci. U.S.A.* **70**, 2086.
Wenkert, E., Buckwalter, B. L., Burfitt, I. R., Gasic, M. J., Gottlieb, H. E., Hagaman, E. W., Schell, F. M., Wovkulich, P. M., and Zheleva, A. (1976). *In* "Topics in Carbon-13 NMR Spectroscopy" (G. C. Levy, ed.), Vol. 2, p. 81. Wiley (Interscience), New York.
Wehrli, F. W. (1976). *In* "Topics in Carbon-13 NMR Spectroscopy" (G. C. Levy, ed.), Vol. 2, p. 343. Wiley (Interscience), New York.
Wehrli, F. W., and Wirthlin, T. (1976). "Interpretation of Carbon-13 NMR Spectra." Heyden, London.
Wüthrich, K. (1976). "NMR in Biological Research." Am. Elsevier, New York.
Yaron, A., Katchalski, E., Berger, A., Fasman, G. D., and Sober, H. A. (1971). *Biopolymers* **10**, 1107.
Zimmer, S., Haar, W., Maurer, W., Ruterjans, H., Fermandjian, S., and Fromagoot, P. (1972). *Eur. J. Biochem.* **29**, 80.

DISCUSSION

H. Papkoff: Would you please briefly describe the kind of equipment employed in NMR studies. Also, how are the samples prepared for analysis, and how much steroid or peptide do you need for analysis? Would you then tell how you could determine whether a material had carbohydrate and that carbohydrate affected the conformation. In the case of a glycoprotein hormone, could you determine whether it is dissociating or not under various conditions?

I. Smith: I hope that my relaxation time is sufficiently short that I can remember all those questions. I will give a brief idea of the spectrometer with apologies to those who already know about such things. You would first need a magnet. Iron core magnets weigh 2–4 tons; there are superconducting solenoids that are much smaller. The former operate at about 23 kG and the latter at up to about 100 kG. The price of this kind of system is now of the order of $180,000, including computers and all hardware. The superconducting systems start at about $200,000 and go as high as you can pay, say 0.5 megadollars. The superconducting systems have the advantages of greater dispersion and sensitivity, which is what you want for biological samples. Therefore we are seeing the creation of more and more national NMR centers such as I described. The big component is the computer, which is necessary for the Fourier transform and other data handling. The bigger the computer you have the more versatile manipulation of the data you can do. Then you have radiofrequency sources and amplifiers. Associated hardware for each nucleus, ^{13}C, ^{1}H, ^{2}H, etc., will cost about $10,000 per nucleus. The sample goes in a probe situated in the magnetic field; it is placed in a cylindrical sample tube of diameter 1–30 mm. Thus you can go from very small samples, such as 50 µg of a steroid you obtained from a sea anemone in the 1-mm tube, to the case of a protein where 1 gm/ml exceeds solubility limits, so you use 1 gm in 20 ml of solvent in the 30-mm tube. Thus, you have the whole range of tube sizes optimized for very small samples or, on the other hand, for compounds that are not very soluble or are associated at high concentration. The sample is placed in this tube in solution, usually in water, but it could be any solvent you choose; the volumes required vary from 50 µl up to about 20 ml. The sample can be run at any temperature, the range is continuously variable from well below freezing to 300°C. You can program the computer to change the experiment for you if you wish. You can do series of experiments while you are playing tennis or sleeping or whatever; that is the advantage of computer-controlled instruments. That is the hardware part.

The other question was: If you had a glycoprotein receptor, would you be able to tell whether the glycomoieties affected the protein or the protein affected the glycomoiety, or whether something binding to the receptor would affect either or both? Is that correct?

H. Papkoff: I wondered whether one could determine the contribution of the carbohydrate in a glycoprotein to the conformation of the molecule.

I. Smith: The immediate response is yes. You probably could, but without more detail one cannot guarantee success. If the glyco part interacted with the peptide part so as to change conformation, this would be manifest in both the carbohydrate and the peptide moieties. In fact, this is the region very dear ro our hearts because we are interested in carbohydrates on cell surfaces and we are trying to label these components to look for just such changes when hormones bind to them. As I said, we still have not found the right hormone, but in principle the experiment is a winner. The occasional one might fail, but in principle I would say you have a 50–60% chance.

T. E. Andreoli: Can you look at dispersions of phospholipid vesicles in aqueous solutions? For example, the so-called liposomes? If you can, as a beginning model for receptor type interactions, the red cell binds about 10^2 or more glycoside molecules per cell. That might be a starting point, since you are dealing with a number appreciably larger than the one you were talking about.

I. Smith: In answer to the first part of your question, yes, we are very busily studying ^{13}C and ^{2}H NMR of phospholipid dispersions. The second question was really not a question—it was a very good suggestion. What was the glycoside? I see, a cardiac glycoside. Yes, I think that's a feasible experiment. The T_1's of phospholipids in vesicles are very sensitive to the rate of motion of the acyl chains. Deuterium NMR, which I have not mentioned, is very sensitive to the degree of organization of those chains. So those two experiments are possible; the carbon gives you the dynamics, the deuterium gives you the structure. Those are both very possible experiments with as little as 20 mg of phospholipid in 1 ml of water. Then you add the glycoside to see whether these properties change.

T. E. Andreoli: One other question in that vein: have you looked at the effect of sterols on the mobility of phospholipids and vesicles, vis-à-vis condensing effects?

I. Smith: Yes, indeed. It's a long story, but very interesting because the condensing effect occurs mainly in the top half of the monolayer, not in the bottom half. That is, carbons 2 to 10 of the fatty acid chain are very sensitive to the condensing effect; 11 to 18 are much less sensitive because the cholesterol A to D rings are worth about ten carbon atoms of fatty acyl chain, and the more flexible tail of cholesterol covers the rest of the distance. That has been done mostly by deuterium, but also by carbon, and not only by us. There is quite a lot of activity in that field now.

B. A. Scoggins: Two short questions in relation to the conformation of angiotension: As you all know, a number of species have the valine-5 form as distinct from the isoleucine 5 form of angiotensin II, and I wonder whether this form has the same general conformation as you present for the isoleucine-5 form.

I. Smith: We have not measured the T_1's of the valine-5 form of angiotensin.

B. A. Scoggins: So you do not know whether its conformation is the same sort of hairpin?

I. Smith: No; the proton NMR has been done for valine-5-angiotensin. We have measured many other derivatives of angiotensin Fermandjian and Fromageot are working on valine-5-angiotensin in France. They have not done the ^{13}C T_1's to my knowledge. My guess would be that, based on the analogs we have studied up to now, it makes very little difference. I do not think that amino acid substitution would lead to a large change in overall conformation.

B. A. Scoggins: Second, would you comment on whether you think the 10% of the valine-5 with the *cis*-proline is important from a sort of physiological point of view.

I. Smith: I would hope so, but it would be foolish to say that I believe so. One knows it is there; until one has a receptor that one can bind it to and measure an increase in the percentage of *cis*-proline, it is just a hypothesis—at best a figment of a physical chemist's imagination.

M. B. Vallotton: You mentioned your interest in determining the conformation of the hormone when it is bound to its receptor, but that the problem was that you did not have a sufficiently dense population of receptors in your tube. Now what about determining the conformation of the hormone when it is bound to an antibody? Could you do that?

I. Smith: Yes, that's much easier. One needs to have concentrations of the order to the 10^{-3} or 10^{-4} M in binding site to make it easy; 10^{-6} M if you want to do it the hard way. I think that if you can get enough antibody into the solutions, particularly in one of these larger tubes, the experiment would go.

M. B. Vallotton: Do you have already examples of a change in conformation of a hormone when it is bound by antibody?

I. Smith: No, I do not think that the experiment has been done yet. It is a very interesting experiment, but rather difficult.

S. Korenman: I am curious to know whether you considered estrogen receptor–estrogen interaction. There are some people who have nearly pure estrogen receptor. Further, have you considered electric eel muscarinic receptor and acetylcholine: those are the two that come to mind as being the most probable, where you might be able to get the proper concentration.

I. Smith: Acetylcholine yes, but not by us; people are studying the acetylcholine receptor, as one can make the excitable vesicles. I think the group in London are working on the muscarinic receptor. I do not know what sort of concentrations of receptors one gets in that particular system. What was the other one?

S. Korenman: Estrogen.

I. Smith: I know of nothing along those lines. One of the suggested systems involved was epidermal growth factor, where a slightly higher population of receptor sites is available, but we have not started on that yet; that was the suggestion of Pedro Cuatrecasas.

C. Monder: In the early days of NMR studies a great deal of effort was expended on attempts to study the structure and conformation of polypeptides and proteins using proton magnetic resonance (PMR). However, the thrust of your presentation was focused on ^{13}C magnetic resonance. Has there been equivalent progress in PMR, and what is the relative value of the information obtained using those two techniques?

I. Smith: The progress in PMR has been completely analogous to the progress in carbon NMR. The level at which you can detect protons has been going down all the time, and that means greater and greater sensitivity. The dispersion of the spectrum is going up in a proportional manner. The problem is that the spectral width in hydrogen is still only 10 ppm whereas in carbon it is 200 ppm. So for PMR of a protein for example, it has been calculated by Jardetsky that you would have to go to 1000 MHz or greater to get any sensible resolution out of a protein; that does not correspond to a presently attainable magnetic field. At about one-third that magnetic field for carbon, one begins to resolve individual amino acid resonances in proteins. For example, Allerhand has observed eight separate tryptophans in cytochrome *c* and has identified which tryptophan is which in the spectrum. This level of resolution is not possible by proton yet; I do not mean to downgrade the information you get out of PMR because in exceptional cases like that of histidine the resonances stick nicely out of the proton spectrum. The same applies to some of the aromatic residues and some of the aliphatic residues when they are near aromatic residues. So, there is still progress, but you are starting with a tougher game as far as pulling these resonances apart. It is easier to obtain proton spectra, but harder to do anything with what you get. All progress is parallel in terms of sensitivity and dispersion of the chemical shift. PMR is definitely alive and well, but still has the same old problems.

K. Savard: I want to bring up a historical note in order to allay the concern of our speaker and our chairman about the presentation of highly technical data to this audience. I can remember as a recent postdoctoral fellow, about 100 years ago, in this very room, listening to a man who was to become a prominent National Research Council scientist, Dr. Norman R. Jones, present the application of infrared spectra to the identification of steroid and steroid metabolite. I also recall the highly attentive attitude of the clinical and highly physiological members of the audience—of Dr. Konrad Dobriner, particularly, who referred to the spectra as "viggle-vaggles." I also remember that I helped to run the slide projector at the time.

I. Smith: Could I just say thank you. That's a very nice comment. To hear you mention the name of Norman Jones in the context of my presentation makes me very happy.

L. L. Engel: I am thrilled by what I heard this evening, and I want to suggest to you other model systems that you could use for studying steroid hormone–protein interactions. They are corticosteroid-binding globulin, sex steroid-binding globulin, orosomucoid, and similar compounds; possibly also enzymes that bind steroids may be used. These proteins can be obtained in larger amounts than receptors at the present time.

I. Smith: What sort of concentration of steroids can you get into solution?

L. L. Engel: You can get up to 10^{-4} M.

I. Smith: Yes, that is certainly possible with ^{13}C enrichment.

L. L. Engel: Last year we heard Bill Duax give an elegant presentation of the uses of X-ray crystallography in the study of structure; structure in a somewhat more static form than NMR. Would you like to comment on the relative advantages and disadvantages of X-ray crystallography versus proton and ^{13}C NMR?

I. Smith: When we talk about a family of conformations it is certainly useful to us to have some idea of a static one that is well known, and that is what you determine with the X-ray technique. You usually find the energetically lowest conformation, due either to the intrinsic properties of the molecule or to the packing in the crystal. One hopes that it is not due to the packing in the crystal, but more intrinsic to the molecule itself. If that is true, it is a very valuable input in predicting the most likely conformer in solution, which is the average about which both the conformers would be interconverting. In some cases one observes the same sort of conformation in solution as in the solid state, and in other cases one does not. The X-ray result is always a very good starting point. However, we must know the basic geometry and the distances between atoms from the primary structure before we can interpret our data properly, so I think the two techniques go hand-in-hand very nicely. They do not duplicate each other in any respect I know; they merely complement.

J. C. Orr: If you are so enthusiastic about ^{13}C NMR, why is it that you are working so much with deuterium NMR?

I. Smith: There is a very simple answer to that. When you run into a very difficult problem, you always have to enrich the sample in ^{13}C. If you have to enrich the sample you will notice that the cost of deuterium is one-twentieth that of the corresponding carbon-labeled compound. So we decided to try it, although it is a more difficult nucleus to study in many ways. It turns out that for the membrane problems, where we are trying to get some idea of how a membrane responds to a hormone, deuterium gives the kind of information that carbon cannot give easily; that is, it gives the spacial organization of the membrane, whereas carbon gives the rate of interconversion between conformations. We had to spend a lot of time to develop deuterium because of various difficulties: lack of compounds, insufficient theory, and instrument-related problems. Having achieved this, we are now back to carbon to explore the dynamical aspects of the problem.

C. Monder: I notice that, in order to do studies with peptides and proteins in your ^{13}C NMR system, you dissolved the proteins in D_2O. Does that preclude the use of this technique in the future in biological systems, that is, natural biological systems?

I. Smith: No. The only reason for our use of D_2O in carbon NMR is that one stabilizes the magnetic field by using some other resonance onto which to lock the magnetic field. There is no reason why deuterium has to be used as the solvent. You could have, for example, a coaxial capillary tube containing the deuterium for the lock. When you get to studying carbon NMR *in vivo* you are going to have to use H_2O as a solvent. Deuterium NMR has the advantage that one uses H_2O to lock onto. There is very little deuterium around other than the labeled compound in which one is interested.

The Regulation of Vasopressin Function in Health and Disease[1]

GARY L. ROBERTSON

Department of Endocrinology, Indiana University Medical Center and Veterans Administration Hospital, Indianapolis, Indiana

I. Introduction

Our knowledge concerning control of the antidiuretic hormone arginine vasopressin (AVP), began over 30 years ago with the studies of E. B. Verney and his colleagues in England (Rydin and Verney, 1938; Verney, 1947). In an elegantly conceived and executed series of experiments, they identified most of the major regulatory systems and described their principal features in as much detail as the indirect methods of the day would allow. Over the next 15 to 20 years, much was learned about the chemistry, anatomy, and comparative aspects of the neurohypophysis (Sawyer, 1961). Some progress was also made in confirming and extending our knowledge about the functional and physiologic properties of the system (Sawyer and Mills, 1966), but these efforts were severely hampered by the lack of a simple and reliable method for measuring AVP at the low concentrations normally found in body fluids. Several years ago, this technical obstacle was overcome by the development in our laboratory of a sensitive radioimmunoassay for plasma AVP (Robertson *et al.*, 1973b). This new tool has now made it possible to being to redefine AVP function in more concrete and comprehensive terms. It has also produced some unexpected results that suggest the existence of a potent but previously unrecognized stimulus while challenging certain traditional concepts concerning the importance of others. In the following sections, I would like to review some of these new developments with special emphasis on the physiologic findings and the way they relate to certain well known clinical disturbances of salt and water balance.

II. Methodology

Despite the obvious need, a workable radioimmunoassay for plasma AVP was not achieved until 1970 (Robertson *et al.*, 1970), almost 10 years after the

---

[1] Supported in part by research funds from the Veterans Administration, the Indiana Heart Association, the Indiana Kidney Foundation, and CRC Grant No. PHS M01 RR 750-02 from the National Institutes of Health.

technique had been successfully applied to insulin and many other polypeptide hormones. This delay was due to a number of special problems (Robertson, 1969; Robertson et al., 1973c; Chard, 1973) including the relatively weak antigenicity of native AVP and the exceptional sensitivity required to measure the hormone at physiologic concentrations ($\sim 10^{-12}$ M). The first obstacle, a consequence probably of its small size and/or lack of "foreignness" in the animals commonly used for immunization, was surmounted rather easily by conjugating vasopressin to larger polypeptides, such as albumin (Permutt et al., 1966; Oyama et al., 1971). Enhancing antigenicity in this way did not completely solve the problem, however, because antisera of sufficiently high sensitivity still were difficult to produce and could not be used to full advantage without an immunoreactive tracer of high specific activity. In 1966, Dr. Jesse Roth and his colleagues succeeded in producing such a tracer by modifying the Hunter and Greenwood method (1962) and filtering the radioiodination mixture over a Sephadex column in dilute acetic acid (Roth et al., 1966). This chromatographic system separates radioiodinated AVP from unreacted iodine and other interfering substances as well as from unreacted AVP (Robertson et al., 1973c). Although the tracer thus obtained made possible AVP assays of great sensitivity and specificity, it soon became apparent that they could not be used to measure the hormone in dilute serum or plasma (Robertson, 1969). This unforeseen problem was first encountered during our studies with the Wu-12 antiserum, the best of several produced by immunizing rabbits with unconjugated vasopressin (Wu and Rockey, 1969). We found that even though this assay was specific enough to distinguish between AVP and lysine vasopressin (LVP), incubation with dilute whole plasma yielded unknown values which were several hundredfold higher than expected and did not change appropriately in response to known AVP stimuli. An explanation for these disconcerting results emerged almost by chance from efforts to characterize this abnormal immunoreactivity (Robertson et al., 1970). We found that when plasma from a dehydrated healthy adult was filtered over Sephadex G-25 in dilute ammonium acetate, the immunoreactivity could be recovered in three discrete peaks (Fig. 1). The first and largest peak eluted in the void volume, coincident with the plasma proteins; a second peak of intermediate size eluted somewhat later, just ahead of the plasma salts. Quite unexpectedly, we also found a third, very much smaller peak of immunoreactivity, which eluted last, well beyond the internal volume of the column. This third peak, which almost escaped detection because of its position and small size, proved to be the only one of the immunoreactive fractions that was diminished by water loading or destruction of the neurohypophysis (Robertson et al., 1970). These physiologically appropriate variations in peak III were perplexing at first since it seemed that, by virtue of its size, AVP ought to elute from Sephadex G-25 just ahead of the salt, i.e., coincident with the nonsuppressible peak II. However, subsequent studies in this chromatographic system

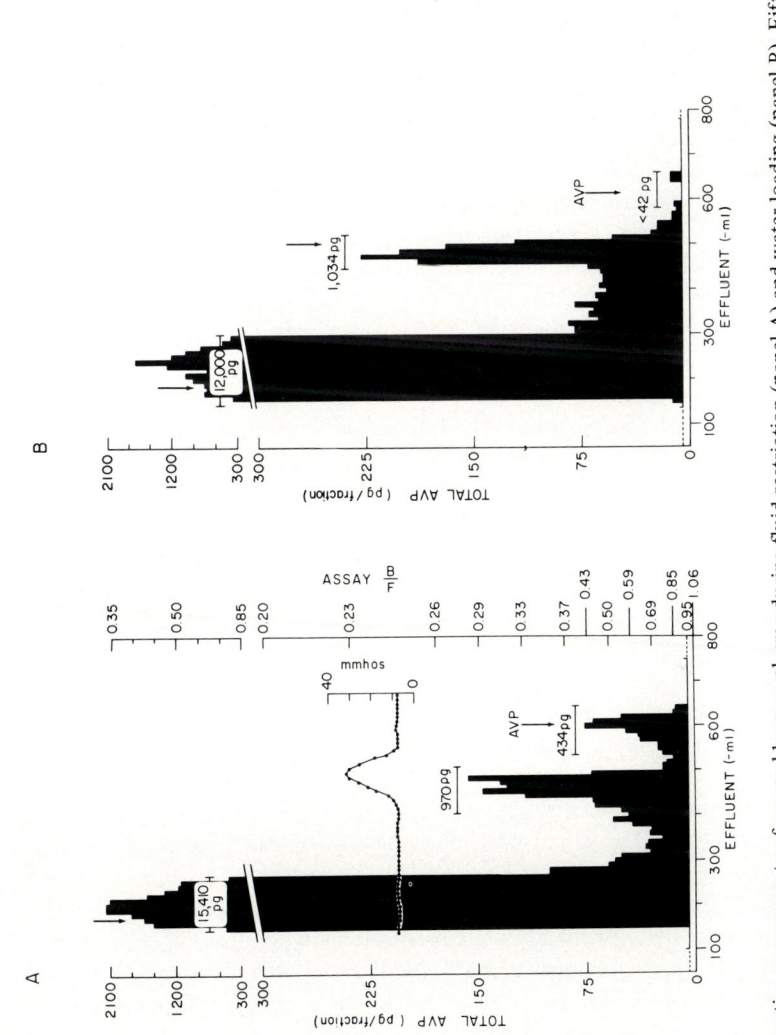

FIG. 1. Immunoreactive components of normal human plasma during fluid restriction (panel A) and water loading (panel B). Fifty milliliters of each plasma was filtered on Sephadex G-25, and the effluent fractions were lyophilized and assayed with the Wu-12 antiserum. From Robertson et al. (1970).

established that purified native or synthetic AVP also eluted after the salt in the same place as peak III (Fig. 1). This result confirmed that peak III was, in fact, identical with endogenous AVP and revealed for the first time the anomalous behavior of the hormone on Sephadex. Further study using more tightly cross-linked gels indicated that this interesting property was due to adsorption of the AVP to Sephadex and was strongly influenced by the pH of the eluting solvent or the coapplication of other peptides or salts (Fig. 2). In retrospect, this unusual chromatographic behavior on Sephadex might have been anticipated, since it has been observed previously with other polypeptides containing a high proportion of aromatic amino acids (Eaker and Porath, 1967; Janson, 1967; Determan and Walter, 1968). Our selection of this particular chromatographic system was truly fortunate since, without its unique retentive properties, the small fraction representing true AVP probably could not have been separated from all the other immunoreactive components detected by the Wu-12 antiserum.

The source, nature, and significance of the immunoreactivity in peaks I and II (Fig. 1) has never been completely established, but it appears to be totally unrelated to vasopressin function. Neither fraction yielded detectable amounts of intact AVP when rechromatographed in the same or different systems or when extracted with hot acetic acid. The most compelling evidence in this regard is that patients with suppressed AVP due to water loading or untreated diabetes insipidus have as much immunoreactivity in peaks I and II as healthy subjects or those with increased AVP due to inappropriate secretion or nephrogenic diabetes insipidus (Robertson et al., 1970, 1973c). Because it did not correlate with either chronic or acute changes in AVP secretion, the activity in peaks I and II could not be attributed to any slowly eliminated by-products of neurosecretion, such as binding proteins, prohormones, or degradation products, and thus could be said to be truly "nonspecific." It is worth noting, therefore, that one of the two nonspecific components cross-reacted in the assay very similarly to the AVP component (peak III), which itself deviated slightly but perceptibly from the binding curve produced by standard AVP (Robertson et al., 1970). Further study showed that the latter deviation probably resulted from subtle changes in the immunoreactivity of the hormone produced by lyophilization (Fig. 3). These disquieting findings provided an early warning that the commonly used cross-reactivity tests were not completely reliable as a means either to negate or confirm the validity of an immunoassay procedure.

Exploiting the unique chromatographic properties of Sephadex permitted us to use the Wu-12 antiserum to measure AVP at physiological concentrations in plasma (Robertson et al., 1970), but the overall procedure proved to be only slightly less cumbersome and imprecise than the bioassay it was supposed to supplant. And despite considerable effort, we were unable to devise a simpler method of extracting AVP from plasma that satisfactorily eliminated all the

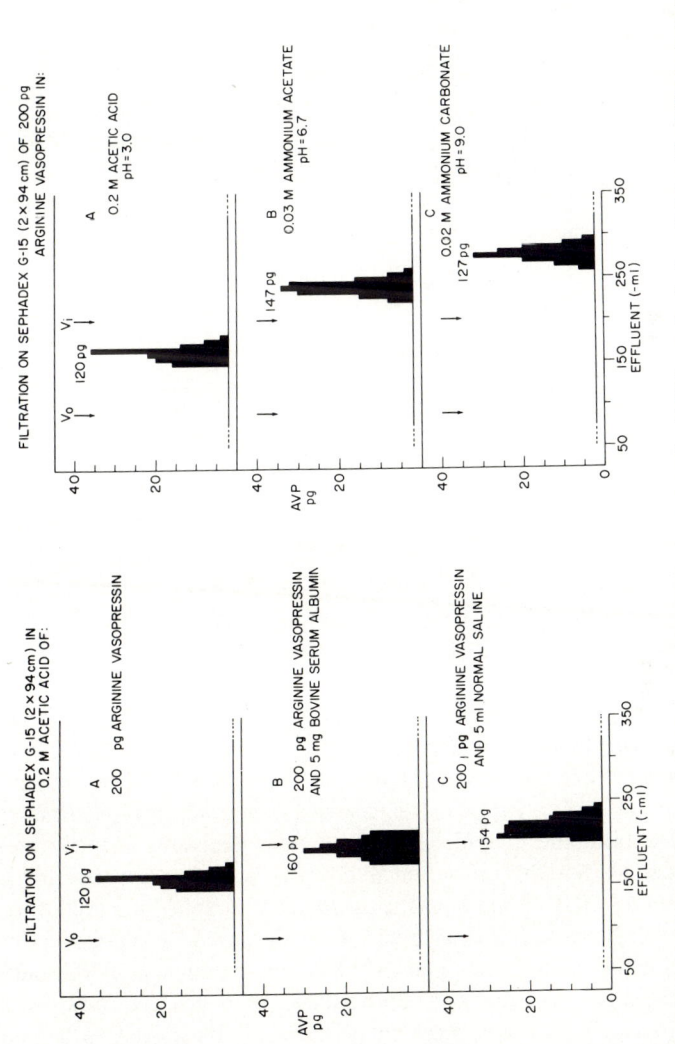

FIG. 2. The chromatographic behavior of purified bovine arginine vasopressin (AVP) on Sephadex G-15 with different eluting solvents or sample composition. Each column fraction was lyophilized prior to assay with the Wu-12 antiserum.

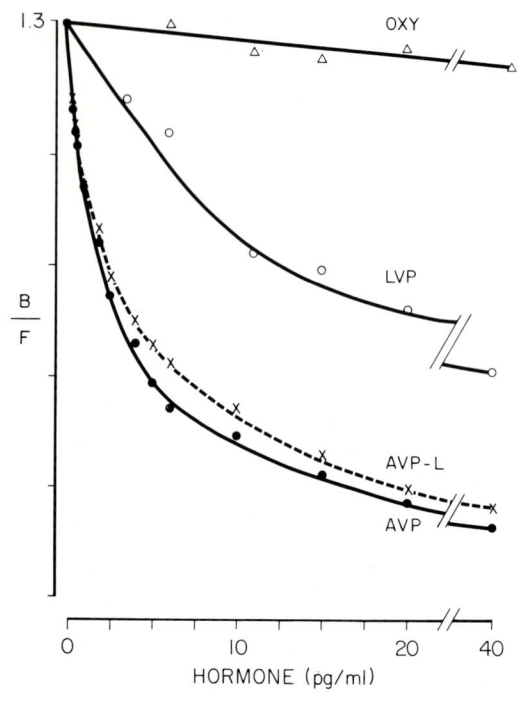

FIG. 3. The cross reactivity of arginine vasopressin (AVP) and its analogs in the Wu-12 assay. AVP-L represents purified bovine AVP after lyophilization in 0.03 M ammonium acetate (pH = 6.7).

nonspecific immunoreactivity to which the Wu-12 antiserum was susceptible. At about the same time, Glick and his colleagues reported the development of a new antiserum that could be used to assay AVP in unextracted human urine (Oyama et al., 1971). Its extraordinary sensitivity and apparent immunity to interference from the myriad of small peptides and salts normally present in urine suggested that this antiserum might also provide a superior method for assaying plasma AVP. When Dr. Glick generously supplied a sample for testing, we were delighted to find that addition of dilute whole plasma to the assay yielded unknown values that approximated those found after Sephadex fractionation and seemed to vary appropriately with changes in water balance (Robertson et al., 1973b). These encouraging results proved to be somewhat misleading, however, since further testing revealed that the immunoactivity of whole plasma did not cross-react exactly like standard AVP and also was heterogeneous when chromatographed on Sephadex (Fig. 4). The total amount of activity found in the excluded, nonhormonal component (3–4 pg/ml) was several hundredfold less than that detected with the Wu-12 antiserum (500

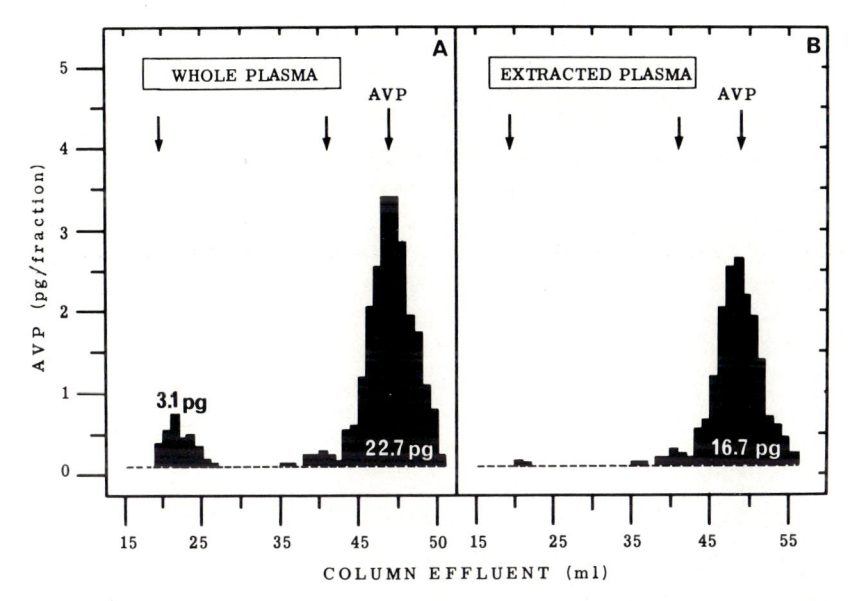

FIG. 4. Immunoreactive components of human plasma as determined by the Gl-I assay before and after acetone extraction. Each sample was chromatographed on Sephadex G-25 in 0.0375 M ammonium acetate, and the eluting fractions were lyophilized prior to assay. From Robertson et al. (1973b), with permission.

pg/ml) but still was sufficient to interfere with accurate measurements of AVP when the latter was present at concentrations within the physiologically effective range (1–5 pg/ml).

Fortunately, however, the nonhormonal activity detected by the Gl-I assay differed qualitatively as well as quantitatively from that detected by the Wu-12 antiserum. Because it was confined to the macromolecular fraction (Fig. 4), it could be easily eliminated by precipitating plasma proteins with a reagent such as 60% acetone, in which AVP is both soluble and stable. The deproteinized plasma extract that remained after evacuation of the acetone contained a single immunoreactive component that was indistinguishable immunologically and chromatographically from pure AVP (Fig. 4). The concentration of the AVP in this fraction was not altered by the extraction procedure and, as discussed in a later section, conformed closely to physiologic expectations in both normal and disordered states of water balance. Thus, with a simple batch extraction procedure and some minor modifications in the assay to increase its sensitivity, we were able for the first time to accurately and efficiently measure plasma AVP at physiologic concentrations in plasma volumes of 1 ml or less (Robertson et al., 1973b).

These experiences and those of others now working in this area illustrate several important lessons about the development and validation of AVP immunoassays. For one thing, it has become clear that nonspecific interference by serum or plasma is not peculiar to the Wu-12 or Gl-I antisera, since it has been found to a greater or lesser extent with all the AVP assays in which it has been carefully sought (Beardwell, 1971; Skowsky *et al.*, 1974; Hayward *et al.*, 1976; Thomas and Lee, 1976). Indeed, our own studies indicate that the small but significant amounts of nonspecific activity detected in plasma with the Gl-I antiserum was due solely to tracer damage and, thus, ought to occur in all plasma assays regardless of the antiserum employed (Robertson *et al.*, 1973b). Little is known about the cause of the tracer damage except that it seems to be greater in serum than in plasma and is not prevented by any of the commonly available protease inhibitors. At least for the present, therefore, some kind of preliminary extraction or deproteinization probably will continue to be needed with all immunoassays of plasma AVP.

A second point worth emphasizing is that individual antisera differ markedly both as to the type and amount of nonspecific interference to which they are susceptible (Figs. 1 and 4). As a consequence, the kind of extraction method employed must be tailored to the characteristics of the individual antiserum. For example, acetone extraction, which eliminated all the nonspecific activity detected by the Gl-I assay (Fig. 4) and seems to have been used successfully with two other antisera (Hayward *et al.*, 1976; Mohring and Mohring, 1975), completely failed to remove the smaller of the two interfering plasma components detected by the Wu-12 antiserum (Robertson *et al.*, 1973c). Other extraction procedures also have been shown to be nontransferable from one antiserum to the next. Thomas and Lee found that when Florisil extracts of human plasma were split and assayed simultaneously with two different antisera, one assay gave values in the expected range and responded appropriately to ordinary changes in water balance, whereas the other assay detected severalfold higher levels of immunoreactivity which showed little or no physiological variation (Thomas and Lee, 1976). Thus, the Florisil extract must have carried over from plasma some nonhormonal substance which interfered with only one of the two antisera.

The studies of Thomas and Lee illustrate a third point, also suggested by our studies with the Gl-I assay. Because AVP normally circulates in plasma at such extremely low concentrations, the persistence of even small amounts of nonspecific immunoreactivity in the extract will suffice to "swamp out" the effects of the hormone and render an assay virtually useless for physiologic studies. For this reason, the developer of a new method must go to extraordinary lengths to ascertain that the extraction procedure eliminates all the interfering substances to which his particular antiserum is susceptible. To this end, several different kinds of tests should be applied since, with the possible exception of Sephadex chromatography, no one criterion is consistently reliable in determining the presence or the absence of nonspecific immunoreactivity. As previously noted, the widely used cross-reactivity test is of limited significance since, on one hand, the interfering substances sometimes produce binding curves nearly parallel to those of the purified standard (Robertson et al., 1970; Thomas and Lee, 1976) whereas on the other, true AVP may be altered immunologically during its extraction from plasma (Fig. 3). Obtaining unknown values "within the expected range" affords no better proof of identity because each immunoassay laboratory now uses a different reference standard and it is not known whether they are all of comparable potency. If some were not, unknown values could be biased in such a way as to falsely suggest either the presence or the absence of nonspecific activity. Comparing immunoassay and bioassay values does not provide a totally satisfactory solution either, because the relative imprecision and insensitivity of the bioassay procedure makes it difficult if not impossible to obtain accurate comparisons at physiologic concentrations, the only condition in which close agreement would be really meaningful. And even without these technical limitations the significance of the results will be open to question until it is firmly established whether or not plasma contains antidiuretic substances other than AVP (Pavel et al., 1973). Even the demonstration that immunoassay values change appropriately in response to known stimuli also may be falsely reassuring since our experience with the Gl-I assay would indicate that such a response, even to mild, physiologic stimuli, does not exclude the presence of small but significant amounts of nonspecific activity (Robertson et al., 1973b). And with more potent stimuli, some kind of response can be observed even in the presence of relatively high levels of interference (Thomas and Lee, 1976).

Because there are significant limitations to each of the more commonly used methods of validating immunoassay results, it is usually advisable to apply additional criteria of a physiochemical nature. Inactivation studies are the simplest, but there is no physical, chemical, or enzymic method available that is known to be specific for the AVP molecule. For this reason, characterization of the unknown immunoreactivity in one or more chromatographic systems is preferable because it provides the least ambiguous identification of unknown immunoreactivity. Sephadex in a neutral or alkaline buffer is particularly effective for this purpose because it separates AVP on the basis of both its size and unusual adsorptive properties (Fig. 2).

The demonstration that immunoassay methods could be used successfully to develop simple and accurate assays for plasma AVP has restimulated much investigative interest in this area. As a consequence, many other laboratories are beginning to develop their own assay procedures (Beardwell, 1971; Skowsky et al., 1974; Hayward et al., 1976; Thomas and Lee, 1976; Mohring and Mohring, 1975; Shimamoto et al., 1976; Morton et al., 1975; Husain et al., 1973). With one or two notable exceptions, the results obtained have been fairly similar (Robertson, 1974) (Table 1), even though the amount and kind of additional validation provided has varied greatly from one laboratory to the next. A greater awareness of the numerous and often subtle defects to which these assays are prone will be needed if these potentially powerful tools are not to become the source of much unnecessary misinformation about AVP function.

III. Physiologic Control

It has been known for almost 30 years that the secretion of AVP can be influenced by a number of factors including changes in blood osmolality, volume, and pressure (Verney, 1947). However, the relative potency of these variables and the way in which they interact in regulating secretion of the hormone under different conditions of salt and water balance has been a subject of some dispute. The advent of the immunoassay technique has now made it possible to begin to define these aspects of AVP physiology in more precise and comprehensive terms.

A. OSMOTIC CONTROL

The osmoregulation of AVP can be readily understood by analyzing the relationship between plasma AVP and plasma osmolality under different conditions of water balance. The existence of a significant positive correlation between these two variables was first observed using our original Wu-12 assay method (Robertson *et al.*, 1970). Subsequent studies with the simpler and more sensitive Gl-I assay confirmed and extended this observation and showed how this relationship could be used to define the major functional properties of the osmoregulatory system (Robertson *et al.*, 1973b; Robertson, 1974; Robertson and Athar, 1976). For example, in samples obtained from a large group of healthy adults in various states of water balance, it is possible to discern a definite relationship between the two variables such that, below a certain level of osmolality, plasma AVP is uniformly suppressed to low or undetactable levels whereas above this point, the hormone increases steeply in proportion to the rise in osmolality (Fig. 5). Appropriate statistical analysis showed that the latter data were significantly correlated ($P < 0.001$) and could be best described in terms of a linear regression function, $y = 0.38 x - 106.4$, in which y represents the plasma AVP concentration in picograms per milliliter and x represents plasma osmolality in milliosmoles per kilogram. This equation can be rearranged to a physiologically more meaningful form, $y = 0.38(x - 280)$, which describes the line in terms of its intercept on the abscissa instead of the ordinate. The value so obtained, 280 mOsm/kg, indicates the level of plasma osmolality at which increases in plasma AVP appear to begin and thereby provides a measure of the threshold or "set" of the osmoreceptor mechanism. The other coordinate, 0.38, indicates the

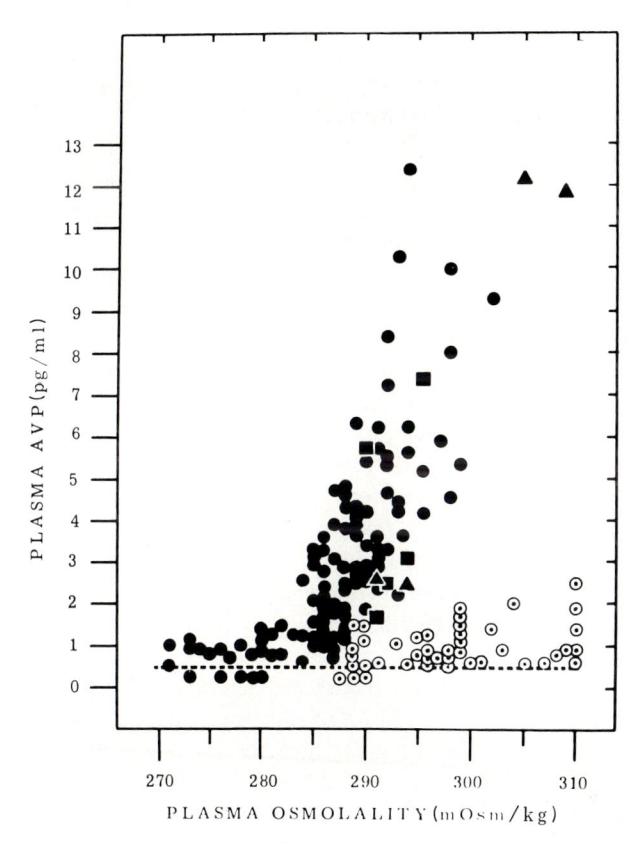

FIG. 5. The relationship of plasma arginine vasopressin (AVP) to plasma osmolality in healthy adults and patients with polyuria of diverse etiologies. Regression analysis of all samples from normal subjects (●; n = 25) in which osmolality was greater than 280 mOsm/kg showed a significant correlation (r = 0.52; P = <0.001) described by a linear function, $y = 0.38 (x - 280)$. In the patients with diabetes insipidus, these two variables also correlated significantly (r = 0.42; P = <0.001) as described by a linear function, $y = 0.03 (x - 275)$. ■, Primary polydipsia (n = 2); ▲, nephrogenic diabetes insipidus (n = 2); pituitary diabetes insipidus (n = 8). From Robertson et al. (1973b), with permission.

slope of the regression line and thereby reflects the sensitivity or gain of the system. Nearly identical values were obtained when the same kind of studies were performed in another large group of healthy adults (Robertson and Athar, 1976). This method of characterizing osmoregulatory function would appear to be widely applicable, since the same kind of relationship between plasma AVP and osmolality has now been observed by several other laboratories using totally different immunoassay procedures and in animals other than man (Robertson et al., 1976a).

This kind of analysis reveals the extraordinary importance of the osmoreceptor system in controlling AVP secretion. In healthy adults, plasma osmolality is normally maintained near 287 mOsm/kg (Robertson *et al.*, 1973b) and, as indicated by the slope of the regression line, a rise or fall of only 1%, or 2.9 mOsm/kg, from this point would suffice to increase or decrease plasma AVP by an average of almost 1 pg/ml. A change of this magnitude is readily detectable by several of the immunoassay procedures (Table I) and also is sufficient to produce marked changes in urinary concentration and flow (Robertson *et al.*, 1973b). This kind of osmoreceptor sensitivity does not depend on a concurrent decline in blood volume since a nearly identical slope is observed during hypertonic saline infusion (Robertson and Athar, 1976). It also is not peculiar to man, since the same kind of analysis in the dog, rat, and monkey yields similarly steep regression coefficients (Robertson *et al.*, 1976a). Thus, these results provide the first direct and unambiguous confirmation of Verney's conclusion, based on indirect studies, that the osmoreceptor is sufficiently sensitive to respond to changes in blood osmolality of 2% or less (Verney, 1947).

The wide scatter that characterizes the relationship between plasma AVP and osmolality in a population of healthy adults (Fig. 5) appears to be due largely to individual differences in osmoreceptor threshold and/or sensitivity rather than to imprecision or instability in its response characteristics. Thus, when multiple determinations of plasma AVP and osmolality were made in 16 healthy adults during hypertonic saline infusion and the data from each were subjected to individual regression analysis, a high degree of correlation consistently was observed even though there were large individual differences in both the slope (0.14 to 0.98) and threshold (276 to 291 mOsm/kg) values (Robertson and Athar, 1976). The basis for these individual differences is unknown, but it must be a relatively stable factor, since we have found the values to be fairly reproducible on repeat testing over periods as long as a year (G. L. Robertson and S. Athar, unpublished observations). Moreover, there seems to be a significant relationship between the set of the osmoreceptor as determined by hypertonic saline infusion and the level at which the individual maintains plasma osmolality under basal conditions (Fig. 6). In each subject, the tonicity of body water appeared to fluctuate over a relatively narrow range (±1.8%) whose boundaries corresponded to the osmotic threshold for AVP secretion below and the threshold for thirst perception above. The close correlation between these three parameters indicates that the individual differences in the set of the osmoreceptor are probably real and have perceptible consequences for the physiology of water balance.

In addition to individual variation in osmoreceptor function, the relationship between plasma AVP and osmolality also can be influenced by the kind of blood solutes present. Under normal conditions sodium and its anions account for

TABLE I
Plasma Arginine Vasopressin (AVP) Concentration in Healthy Adults and Laboratory Animals in Various States of Water Balance[a]

Species	Reference	N	Basal mOsm/kg	Basal Pg/ml	Hydropenic mOsm/kg	Hydropenic Pg/ml	Hydremic mOsm/kg	Hydremic Pg/ml
Human	Robertson et al. (1973b)	12	287 ± 2	2.7 ± 1.4	292 ± 2	5.4 ± 3.4	282 ± 1	1.1 ± 0.8
	Beardwell (1971)	—	—	—	288 ± 2	5.5 ± 0.3	—	—
	Skowsky et al. (1974)	12, 19	—	4.7 ± 1.1	—	23.6 ± 3.1	—	—
	Thomas and Lee (1976)	5	276 ± 2	0.7 ± 0.2	280 ± 1	3.6 ± 1.0	263 ± 3	0.2 ± 0.05
	Shimamoto et al. (1976)	23, 8	—	—	—	7.0 ± 0.4	-	1.7 ± 0.2
	Morton et al. (1975)	6	—	5.4 ± 0.6	—	8.6 ± 0.5	—	3.4 ± 0.8
	Husain et al. (1973)	31, 14	—	—	—	5.6	—	0.7
Monkeys	Hayward et al. (1976)	6	296 ± 0.6	3.0 ± 0.15	307 ± .6	5.8 ± 0.2	282 ± 1	1.5 ± 0.1
	Mohring and Mohring (1975)	—	—	1.6 ± 0.6	—	13.0 ± 3.7	—	—
Rats	Dunn et al. (1973)	16	294 ± 1	2.3 ± 0.9	301 ± 2	13.7 ± 3.0	287 ± 2	0.5 ± 0.1

[a]Samples were obtained during ad libitum fluid intake (basal), after 8–24 hours of fluid deprivation (hydropenic) or 1–2 hours after rapid water loading 2% of total body weight (hydremic). Plasma AVP is expressed as picograms per milliter and plasma osmolality as milliosmoles per kilogram. All values represent the mean ± SE of the mean. All values corrected for extraction losses except Robertson et al. (1973b) and Dunn et al. (1973), where recovery was 100%.

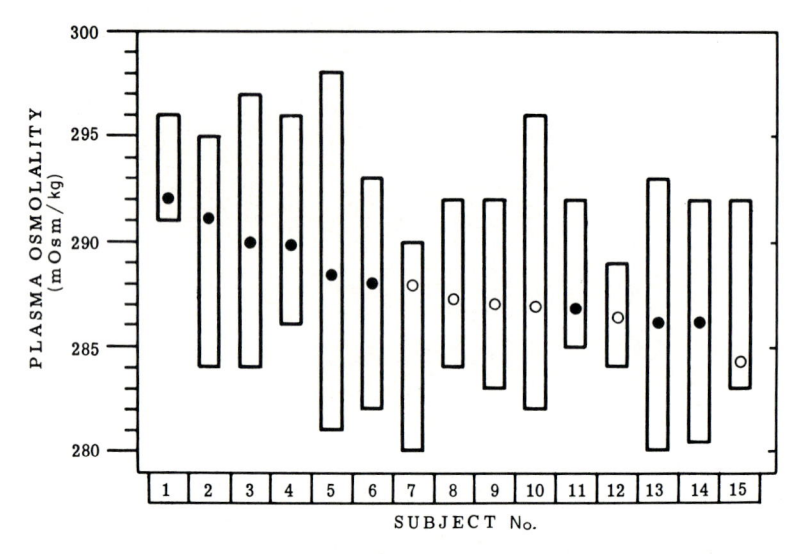

FIG. 6. The relationship of mean basal plasma osmolality to the osmotic threshold for arginine vasopressin (AVP) release and thirst in 15 healthy adults. The circles represent the mean of four separate measurements of plasma osmolality under conditions of ad libitum fluid intake. The bottom of each rectangular box represents the osmotic threshold for AVP release as determined by linear regression analysis of plasma AVP-osmolality data obtained during hypertonic saline infusion. The top of the box represents the level of plasma osmolality at which each subject first reported thirst during the same saline infusion. Data of G. L. Robertson and S. Athar (unpublished observations).

more than 95% of all the osmotically active components of plasma, and it is a change in the concentration of these solutes that is largely if not totally responsible for the AVP response to changes in water balance (see below). However, Verney (1947) showed that sodium salts were not the only solutes capable of stimulating AVP release and concluded, therefore, that the hormone was regulated by an osmoreceptor rather than a specific sodium receptor. Recent studies using our assay and methods of regression analysis have now confirmed Verney's findings and established for the first time the relative potency of nonsodium solutes in evoking AVP release (Athar and Robertson, 1974). For example, we find that the infusion of hypertonic sodium or mannitol both increase plasma AVP even though plasma sodium concentration rises in one and falls in the other. Moreover, the regression line relating plasma AVP to osmolality is indistinguishable during the two infusions (Robertson et al., 1976b; Athar and Robertson, 1974) indicating that, particle for particle, mannitol is just as potent as sodium and its anions in stimulating the osmoreceptor mechanism. However, as Verney (1947) also demonstrated, not all solutes are effective in this regard since raising plasma osmolality by the same amount with hypertonic urea has little or no effect on plasma AVP or sodium concentration (G. L. Robertson and R. Shelton, unpublished observations).

More important, our immunoassay studies have shown that a rise in certain blood solutes can actually cause a fall in plasma AVP. For example, the infusion of hypertonic glucose at a rate sufficient to raise plasma osmolality and lower serum sodium by 2–3% consistently causes a significant fall in plasma AVP, which rapidly reverses as soon as the infusion is stopped and all blood solutes return to basal levels (Athar and Robertson, 1974). This paradoxical response, which would not have been discernible under the water-loaded conditions employed in Verney's indirect studies (Verney, 1947), suggests that the effect of a particular solute on AVP secretion will be a function of its net effect on the total osmotic gradient across some selectively permeable membrane in the brain. Thus, any solute that penetrates this barrier very slowly or not at all would always raise total solute concentration on the systemic side, thereby dehydrating osmosensitive areas in the brain and triggering AVP release. On the other hand, those solutes that penetrate very rapidly would have little or no effect on the total gradient, water flow, or osmoreceptor activity, provided they did not at the same time alter the plasma concentration of some nonpenetrating solute such as sodium. This condition would be fulfilled by a solute like urea, which freely permeates both somatic and brain cells, but would not hold true for glucose, which, because it penetrates somatic cells less readily than brain, causes a shift in free water from the intra- to the extracellular space, thereby lowering plasma sodium concentration. The net effect of this sequence of changes would be to alter the osmotic gradient across the critical brain barrier in such a way as to facilitate migration of free water into osmoreceptor cells, thereby inhibiting AVP secretion. The location of this critical barrier—i.e., whether it corresponds to membranes of hypothalamic capillaries of the brain ventricles or the osmoreceptor cells themselves—and its permeability to other naturally occurring solutes, such as glycerol or acetone, remains to be determined.

Whatever the mechanism for the observed differences in solute effects, their existence has several important physiologic and methodologic implications. First, they indicate that any condition that significantly alters carbohydrate metabolism could affect water balance through osmoreceptor-mediated changes in AVP secretion. For example, the polyuria that characterized the hyperglycemia of uncontrolled diabetes mellitus could be due to suppression of AVP secretion as well as to an increase in solute excretion. Second, they constitute an important limitation to the use of plasma osmolality as a reference point to evaluate AVP function. Thus, in any situation where blood glucose or urea deviate significantly from normal, measurements of plasma sodium may provide a more reliable index of osmoreceptor stimulation.

B. VOLUME CONTROL

Unlike osmolality, there has long been some uncertainty as to relative importance of blood volume in the physiologic control of AVP secretion. Early studies using indirect or bioassay methods showed that reducing blood volume 8% in conscious dogs (Arndt, 1965) or 10% in sheep (Johnson et al., 1970) resulted in a significant increase in antidiuretic activity. However, these studies did not attempt to define the minimum reduction in blood volume required to elicit a hormonal response nor compare it quantitatively to that produced by a comparable osmotic stimulus. Nevertheless, a view emerged that blood volume was at least as potent as blood osmolality in stimulating AVP release and probably played a major role in mediating the hormonal response to physiologic changes in water balance (Gauer, 1968).

Recent studies with our immunoassay procedure have challenged this concept and suggested a somewhat different role for blood volume in regulating AVP secretion. It should be noted that our immunoassay data do not conflict in any way with those obtained previously. Instead, they have only defined more fully the nature of the stimulus–response relationship, and thereby revealed how the earlier experiments may have been misinterpreted. This is illustrated most clearly by our studies in rats (Dunn et al., 1973). We found that when blood volume was reduced isotonically by intraperitoneal injection of polyethylene glycol, plasma AVP increased in a curvilinear rather than a linear fashion (Fig. 7). Thus, little or no increase in plasma AVP was detected until blood volume was reduced by 8–10% even though more severe hypovolemia consistently evoked relatively large hormonal responses not unlike those found in some of the earlier studies. In other words, previous concepts that blood volume is important in the physiologic control of AVP may have been in error because they were based on only one datum point *and* the false assumption that the stimulus response relationship for this variable was linear. By showing that it is not, our results indicate that blood volume is much less important than osmolality in controlling AVP under physiologic conditions, and does not even begin to exert an equivalent influence until the hypovolemia reaches large and unphysiologic proportions.

For obvious reasons, the quantitative effects of hypovolemia on AVP in man have not been as thoroughly characterized but, at least within the physiologic range, appear to be no more potent than in the rat. Immunoassay as well as bioassay studies recently have shown that reducing blood volume 6–9% by phlebotomy has no effect on plasma AVP in healthy, recumbent man (Robertson and Mahr, 1972; Goetz et al., 1974; Robertson, 1976). Similarly, upright posture, which probably reduces central blood volume by 10–15%, has only marginal effects on plasma AVP in normally hydrated adults (Robertson and Athar, 1976). Combining phlebotomy with upright posture results in very large

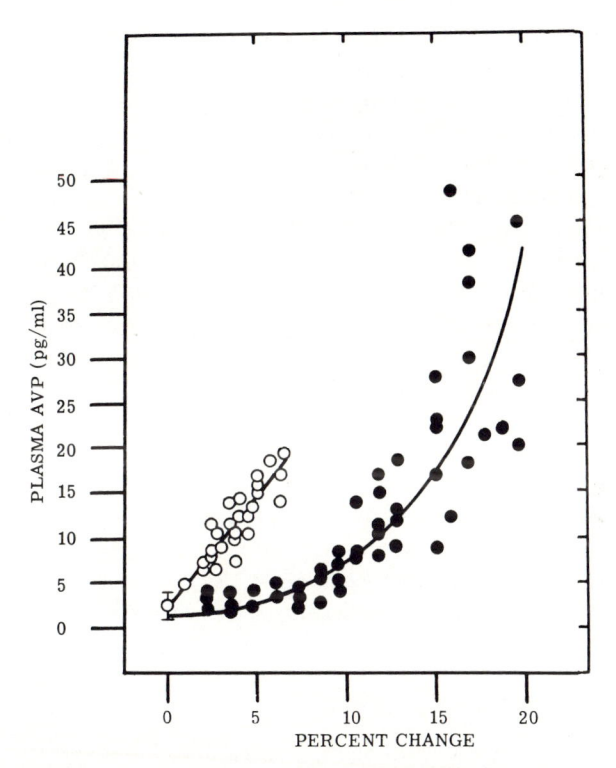

FIG. 7. The relationship of plasma arginine vasopressin (AVP) to percent increase in blood osmolality (○; $P_{AVP} = 2.5\Delta osm + 2.0$) or decrease in blood volume [●; $P_{AVP} = 1.3 \exp(1.7\Delta vol)$] in conscious rats. From Dunn et al. (1973), with permission.

increases in plasma AVP, but is also associated with a sharp fall in blood pressure (Robertson, 1976). In man, therefore, AVP secretion apparently is not affected until blood volume declines by somewhere between 10 and 20%, a "threshold" even higher than in the rat.

The relative insensitivity of AVP to small changes in total blood volume may be explained by the effectiveness with which other compensatory mechanisms maintain central blood volume relatively constant (Gauer et al., 1970). Thus, even though AVP secretion appears to be quite sensitive to changes in left atrial pressure (Gauer, 1968), in conscious animals left atrial pressure may be relatively insensitive to small changes in total blood volume. If this explanation is valid, then anything that impairs the integrity of the primary compensatory mechanisms ought to "sensitize" AVP to the effects of small changes in blood volume. One example may be general anesthesia, which is now known to impair many cardiovascular reflexes (Vatner and Braunwald, 1975). Its use in many early studies may have contributed to the view that both left atrial pressure and AVP secretion were quite sensitive to hypovolemic stimuli.

Whatever the mechanism, the refractoriness of AVP to small changes in blood volume has important implications for our understanding of fluid and electrolyte balance. It is now clear that the hormonal response normally evoked by maneuvers such as water loading (Farber *et al.*, 1975) or fluid deprivation (Robertson and Athar, 1976) is due almost exclusively to the associated changes in blood osmolality instead of volume. This means that the AVP system is geared primarily to maintain the *tonicity* of body fluids and contributes significantly to defending blood volume only when the hypovolemia begins to reach large and threatening proportions. The point at which the primary osmoregulatory mission begins to be compromised probably varies somewhat depending on the effectiveness of the other mechanisms for defending central blood volume. In normal man, however, our data would indicate that it occurs at a total deficit between 10 and 15%. This result agrees remarkably well with a much earlier study by McCance, who showed that a healthy adult slowly depleted of salt by sweating in a heat chamber, did not begin to exhibit a decline in the *tonicity* of body fluid until his extracellular fluid *volume* had been reduced by approximately 12% (McCance, 1936). Below this point, less water than sodium was lost indicating that tonicity was being increasingly compromised in order to conserve volume.

C. PRESSOR CONTROL

Large changes in blood pressure also have long been known to significantly affect AVP secretion (Verney, 1947), and recent evidence indicates that this is the mechanism by which a large number of drugs and procedures effect water excretion (Schrier and Berl, 1975). As with blood volume, however, the relationship between blood pressure and AVP release had not been well defined either in quantitative or qualitative terms. Studies performed in the process of validating our Gl-I assay originally suggested that blood pressure changes as small as 5% could significantly increase plasma AVP (Robertson *et al.*, 1973b). Since healthy adults are known to exhibit fluctuations in arterial pressure much greater than 5% in the course of a normal day (Bevan *et al.*, 1969), our result suggested that this variable could be very important in the moment to moment control of AVP secretion. Accordingly, we performed additional studies to determine the relationship between plasma AVP and the percent fall in blood pressure induced either by the infusion of trimethaphan (Arfonad, Roche Laboratories) or by standing immediately after phlebotomy (Robertson, 1976). Under these conditions, we found a highly significant correlation between these two variables ($n = 0.74$; $P < 0.001$) that was described by an exponential function, $P_{AVP} = e^{0.11\Delta MP + 0.34}$, in which PAVP represents plasma AVP concentration (pg/ml) and ΔMP represents the percentage fall in mean arterial pressure. This result indicates that significant increases in plasma AVP begin to occur in man when mean blood pressure declines by an average of only 5%, an appreciably lower "threshold" than that for volume.

To better define this relationship, we performed similar studies in conscious rats with chronically implanted intraaortic catheters for monitoring blood pressure. When sodium nitroprusside was administered subcutaneously and the rat was sacrificed 10–20 minutes later, a close correlation was found between plasma AVP and the percent fall in mean arterial pressure achieved at the moment of sacrifice (Fig. 8). This relationship again appeared to be curvilinear ($P_{AVP} = e^{0.10 \Delta MP - 0.76}$) and, allowing for the differences in methodology, could not be said to differ quantitatively from that found in man.

The significance of these findings lies in the increased importance they give to blood pressure as a determinant of AVP function. Fluctuations even smaller than those occurring in many healthy individuals can significantly affect the hormone and thus could be responsible for the "episodic" secretion observed by some during normal activity (Katz et al., 1976). It is equally possible that many other clinical or experimental conditions associated with nonosmotic AVP secretion also result from a fall in blood pressure too subtle to be detected except by constant and careful monitoring. This may be particularly relevant to the case of emotional stress, since we have observed several otherwise healthy volunteers in whom the psychological stress of ordinary venepuncture consistently evokes a vasovagal reaction characterized by transient hypotension and marked increases in AVP (see below).

D. INTERACTION OF THE OSMOTIC AND HEMODYNAMIC CONTROL SYSTEMS

The ability of hypotension and, to a lesser extent, hypovolemia to stimulate AVP release raises obvious questions about how these two variables affect the osmoregulatory system. Bioassay studies in conscious sheep had shown previously that reducing blood volume sufficiently to increase AVP secretion did not abolish the effect of a concurrent osmotic stimulus (Johnson et al., 1970). However, these studies did not provide the kind of data needed to determine whether the two stimuli were operating in an independent or an interactive fashion. Accordingly, we undertook to reinvestigate this question in rats by examining the effect of a uniform reduction in blood volume on the relationship between plasma AVP and plasma osmolality over a wide range of values (Dunn et al., 1973). Under these conditions, we found that the hypovolemia did not interfere in any way with either the stimulatory or inhibitory effects of plasma osmolality (Fig. 9A). Instead, it only altered the relationship between these two

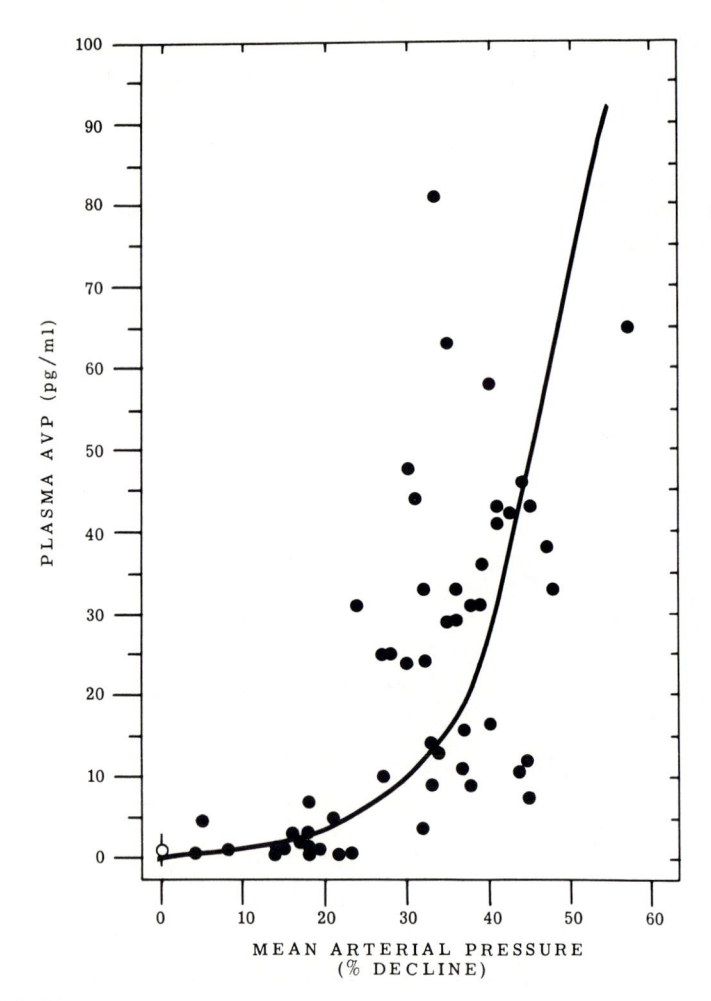

FIG. 8. Plasma AVP as a function of the percent fall in mean arterial pressure in conscious rats ($r = 0.75$; $N = 53$). Each point represents the values from a single animal determined at the time of sacrifice 15 minutes after a subcutaneous injection of sodium nitroprusside. Blood pressure was monitored directly via a cannula chronically implanted in the aorta. Data of R. M. Kinney and G. L. Robertson (unpublished observations).

variables in such a way that they fell along a new regression line which intercepted the osmolality axis slightly below the normal value of 290 mOsm/kg. In other words, hypovolemia seemed to increase AVP not by abolishing or even bypassing the osmoregulatory system, but by lowering the threshold or set of the mechanism. Exactly the same kind of interaction was seen when rats were made moderately hypotensive with isoproterenol (Figure 9B). In man, the

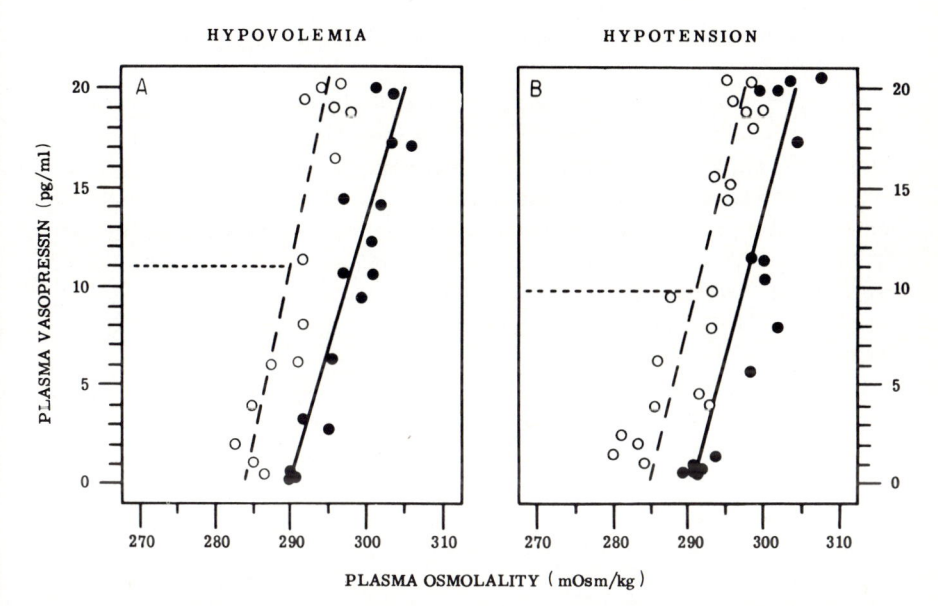

FIG. 9. The effect of hypovolemia or hypotension on the relationship of plasma AVP to plasma osmolality in conscious rats. (A) Blood volume was reduced approximately 15% by intraperitoneal injection of polyethylene glycol (Dunn *et al.*, 1973). (B) Mean arterial pressure was reduced approximately 15% by subcutaneous injection of isoproterenol hydrochloride, and osmolality varied with intraperitoneal injection of hypotonic, isotonic, or hypertonic saline (R. M. Kinney and G. L. Robertson, unpublished observations).

reduction in central blood volume that results from upright posture is associated with a similar shift (Robertson and Athar, 1976).

The demonstration of this kind of interaction between the osmotic and hemodynamic control of vasopressin helps to clarify the way in which the body reconciles its sometimes competing needs to preserve both the tonicity and volume of body fluids. By lowering the "set" of the osmoreceptor instead of inducing a fixed osmotically unresponsive increment in AVP secretion, the organism is able to limit the amount of free water retained and thereby minimize the degree of hypotonicity that might otherwise result. In striking a balance between these two homeostatic needs, the preservation of tonicity clearly seems to be favored, at least initially, since lowering the osmotic threshold from 290 to 286 mOsm/kg would permit an increase in body water of less than 2%, far short of the amount needed to correct the 15% volume depletion that produced the shift. However, as the hypovolemia or hypotension becomes more severe, the osmoregulatory system appears to be increasingly compromised since plasma AVP rises exponentially to very high levels far in excess of those required to produce maximum antidiuresis (Fig. 7). At this point, the primary homeostatic function of the hormone may be to increase vascular tone (Schmid et al., 1974).

The same considerations may also explain why the osmoregulatory system is not more noticeably disturbed by the relatively large fluctuations in blood pressure or central blood volume that occur with normal activity. Even if these hemodynamic variations were great enough to significantly alter AVP secretion, they would do so only by shifting the set of the osmoreceptor by a few percent. Since this kind of change would not permit appreciable changes in free-water balance, the essential features of the osmoregulatory system would be preserved, and body tonicity would be maintained relatively constant even in the face of large variations in water intake.

The existence of this kind of interaction between the osmotic and hemodynamic control systems also provides a valuable insight into the organization of the neurohypophysial unit. Since hypotonicity can completely suppress the effects of hypovolemia or hypotension on AVP secretion, the two kinds of stimuli must interact in some way upon a single, functionally homogeneous population of neurosecretory cells (Robertson et al., 1976a) (Fig. 10).

E. STRESS AND AVP

Since the studies of Verney and his colleagues (Rydin and Verney, 1938), it has been almost axiomatic that emotional or other forms of "nonspecific" stress are effective stimuli to AVP secretion. However, in the course of many studies in both rats and human volunteers, we were impressed with how rarely pain or

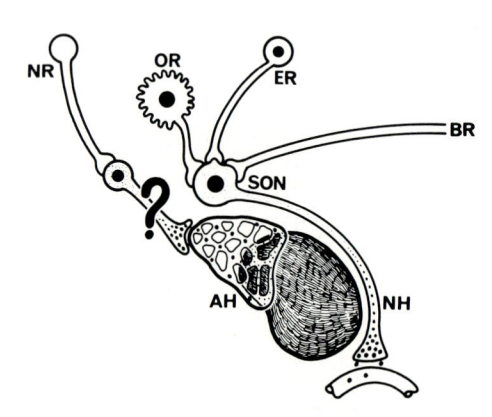

FIG. 10. Schematic representation of hypothetical relationships between the supraoptico(SON)-neurohypophysial (NH) tract and its known regulatory afferents from the osmoreceptors (OR), baroreceptors (BR), and emetic centers (ER). The elements designated by a question mark represent a hypothetical second system linking brain nociceptors (NR) to the adenohypophysis (AH) via some as yet unknown neurosecretory tract.

other obvious stresses incidental to the experiments had any discernible effect on plasma AVP. These observations and the lack of any report directly confirming an effect of stress on AVP in conscious animals prompted us to undertake a systematic study of this question in rats. For this purpose, we chose a variety of stresses known to stimulate the pituitary–adrenal axis, a response thought by some to be mediated or facilitated by secretion of AVP (Nichols, 1961; Yates, 1967). By so doing, we hoped not only to have an objective measure of the effectiveness of the stressful stimulus (plasma corticosterone), but also to obtain some insight into the relationship between these two hormonal systems. To our surprise, we found that none of three stresses employed–light ether anesthesia, water immersion, or pain–had any detectable effect at any time on plasma AVP even though all three resulted in very large increases in plasma corticosterone (Brennan *et al.*, 1975). A typical result from an ether stress experiment is shown in Fig. 11. Exposure for 90 seconds to an atmosphere saturated with diethyl ether produced transient loss of consciousness and a marked rise in plasma corticosterone but had no effect whatsoever on plasma AVP. Similar hormonal results were obtained when rats were immersed in water or given intraperitoneal injections of isotonic saline (Brennan *et al.*, 1975).

The failure of all three types of stress to alter plasma AVP is especially significant in view of the ease with which such effects can be demonstrated in these animals after small changes in blood osmolality (Dunn *et al.*, 1973) (Fig. 7). Although negative results of this kind cannot totally exclude an influence of nonspecific stress on AVP secretion, they do indicate that this variable has a much less important influence than has generally been supposed. Because of its obvious importance for understanding the pathophysiology of AVP function, this whole concept needs to be reexamined in a more critical and systematic fashion. In this regard, it will be particularly important to distinguish between effects that are a direct result of the stress per se and those which are secondary to changes in blood pressure or other recognized stimuli to AVP secretion. For example, we find that exposing rats to high concentrations of ether for longer than 90 seconds results in a distinct fall in blood pressure and an increase in plasma AVP (R. M. Kinney, R. L. Shelton and G. L. Robertson, unpublished observations). In this situation, the ether really is acting not as a "nonspecific" stress, but as just another drug, such as isoproterenol, nitroprusside, or trimethaphan, that can affect AVP secretion by lowering blood pressure. The same might be said for emotional factors since, as noted above, pain or apprehension in some healthy adults evokes a vasovagal reaction that is also associated with hypotension and a rise in AVP. Thus, it is possible that many, if not all, of the AVP effects previously attributed to emotional or nonspecific stress were, in fact, secondary to changes in blood pressure.

Our studies also cast doubt on the role of AVP in mediating or facilitating the effect of nonspecific stress on the pituitary–adrenal axis. Not only do stressful

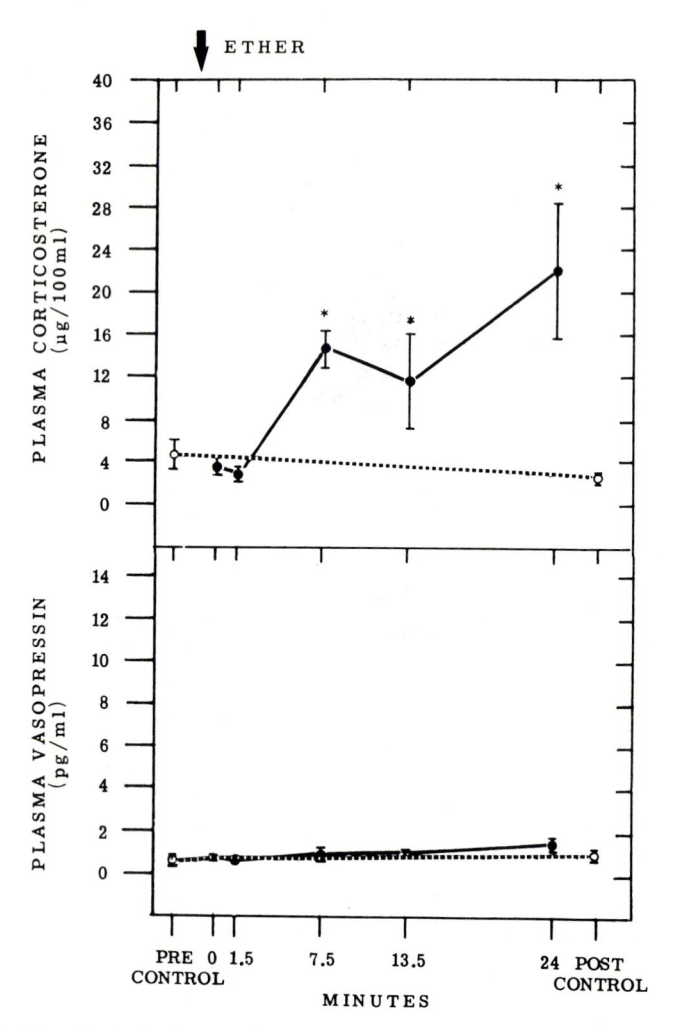

FIG. 11. The effect of ether on plasma corticosterone and arginine vasopressin (AVP) in rats. Filled circles represent the values obtained at frequent intervals after exposure for 90 seconds to an atmosphere saturated with diethyl ether (arrow). Open circles represent control rats. *$P < 0.05$ by unpaired t compared to time 0. Mean ± SEM; $N = 6$. From Brennan *et al.* (1975).

stimuli that increase corticosterone secretion fail to affect plasma AVP (Fig. 11), but osmotic stimuli that increase plasma AVP appear to have no effect on plasma corticosterone (Brennan *et al.*, 1975). This dissociation means either that AVP plays little or no role in pituitary–adrenal activation, as previously has been suggested by studies in Brattleboro rats (McCann *et al.*, 1966; Arimura *et al.*,

1967), or else if AVP does act as a releasing factor, it must be secreted from a distinct group of neurons with functional and anatomic characteristics quite different from those involved in regulating water balance. These hypothetical neurosecretory units would be distinguished by the fact that they do not respond to osmotic stimuli and, when stimulated by afferents from nociceptive areas in the brain, would release AVP at a site from which it does not gain access in significant amounts to the peripheral circulation (Fig. 10). Recent advances in electrophysiologic and immunohistochemical techniques should make it possible to ascertain whether neurosecretory cells fulfilling all these criteria really exist.

F. EMESIS AND AVP

Serendipitous observations made in the course of several other studies have led to the discovery of a new variable that seems to have a potent influence on AVP secretion. During separate investigations conducted in our laboratory and in collaboration with the Gerontology Research Center, we observed that the administration of an oral water load or intravenous ethanol to normal volunteers occasionally resulted in a marked rise in plasma AVP (J. H. Helderman, R. E. Vestal, J. D. Tobin, R. Andres, R. L. Shelton, and G. L. Robertson, unpublished observations). This response was particularly noteworthy because it was contrary to the AVP suppression usually produced by these procedures and, at least in several instances, could not be accounted for by changes in known physiologic stimuli, such as blood osmolality or pressure. Careful review of the clinical records of these studies revealed that in each case the paradoxical rise in plasma AVP occurred in a subject who developed nausea and/or vomiting during the procedure.

To evaluate more systematically the possibility that emesis per se is a specific stimulus to AVP release, we administered varying doses of apomorphine subcutaneously to healthy adult volunteers and closely monitored its effects on plasma AVP, osmolality, arterial pressure, and symptomatology (Shelton *et al.*, 1976c). In standard emetic doses (25 μg/kg), apomorphine produced nausea and vomiting and massive increases in plasma AVP in all subjects tested. More significantly, slightly lower doses of apomorphine (15 μg/kg) which produced nausea without vomiting, also consistently evoked very large increases in plasma AVP. As exemplified by Fig. 12, the AVP response corresponded temporally to the occurrence of nausea and could not be accounted for by changes in blood osmolality or pressure. Thus, it appeared to occur independent of any known AVP stimulus, including a change in intrathoracic pressure and/or blood volume induced by the physical act of vomiting itself. Doses of apomorphine too small to produce nausea (7 μg/kg) had no effect whatsoever on plasma AVP. These findings, as well as the chance observations noted above, suggest the existence of

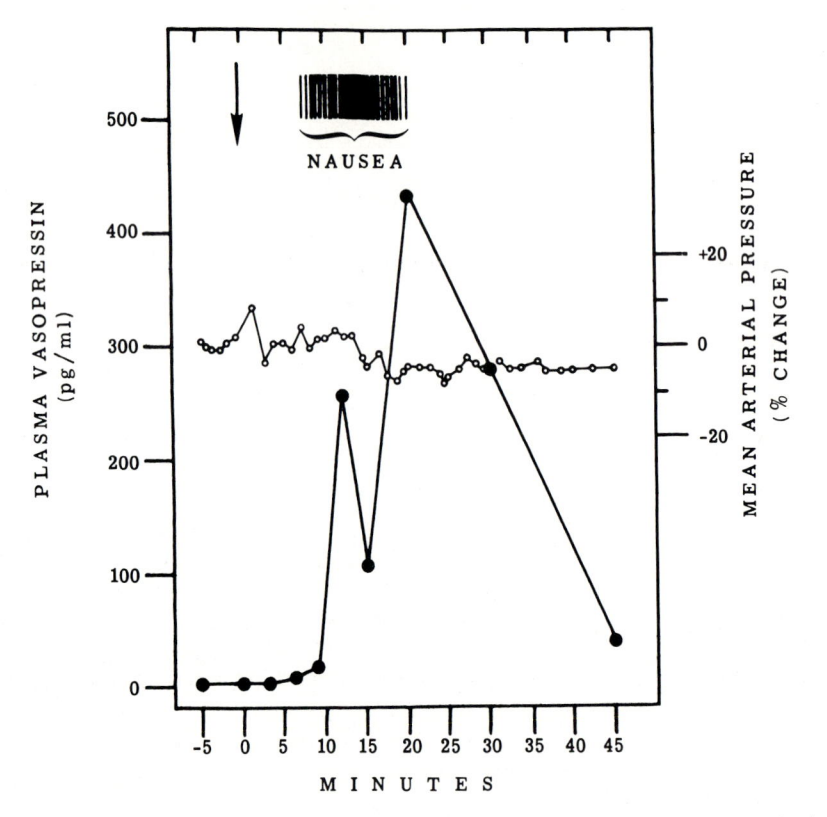

FIG. 12. The effect of apomorphine on plasma arginine vasopressin (AVP) (●) and mean arterial pressure (○) in a healthy adult. Recumbent subject was injected subcutaneously with apomorphine, 14 μg/kg, at time zero (arrow). Duration and intensity of nausea is represented by vertical, spaced lines. From Shelton *et al.* (1976c).

a close and possibly direct relationship between stimulation of the emetic center and AVP release.

It is interesting to note that these are not the first studies to suggest a relationship between the gastrointestinal tract and AVP release. Moran and his colleagues observed that in anesthetized man or experimental animals, traction on the intestines consistently resulted in very large increases in plasma antidiuretic activity (Moran *et al.*, 1964; Ukai *et al.*, 1968). This response, which was said not to be accompanied by "significant" changes in blood pressure and could be abolished by cervical cordotomy but not vagotomy (Ukai *et al.*, 1968), has been widely interpreted as an example of general nociceptive stimulation of AVP secretion. However, in light of our recent negative findings with other noxious stimuli (see above), an alternative interpretation may be that the response evoked by visceral manipulation is in fact a specific effect involving stimulation of the emetic center by afferents arising in the gastrointestinal tract (Wang, 1965). In any case, it is noteworthy that, at present, the only two well documented examples of nonosmotic and nonhemodynamic AVP release both involve some aspect of the nervous control of gastrointestinal function.

The existence of a special relationship between the emetic center and AVP release could have important pathophysiologic implications. Nausea and/or vomiting is a prominent symptom of some of the drugs and clinical disorders associated with inappropriate antidiuresis (Bartter, 1973), and it is conceivable, therefore, that some cases of this syndrome might be effectively treated with antiemetic agents. Clinical investigators should also be especially alert to the occurrence of nausea during a study, since it may be as important as osmotic or hemodynamic variables for interpreting properly the AVP response.

G. DISTRIBUTION AND CLEARANCE

Plasma AVP is determined not only by the rate at which the hormone is secreted, but also by the way in which it is distributed and cleared from body fluids. However, the relative importance of the latter has long been uncertain because, for unknown reasons, bioassay studies have yielded markedly discrepant results. In man, for example, different laboratories have reported that exogenous vasopressin distributes into a space ranging from one-half to more than twice plasma volume and disappears from plasma with a half-time ranging from 1.2 to 42 minutes (Lauson, 1971).

In an attempt to clarify these issues, we have used our radioimmunoassay method and non-steady-state techniques to determine the distribution and clearance of exogenous Pitressin (Parke-Davis Co.) in normally hydrated healthy adults (Maxwell *et al.*, 1976). We found that after intravenous injection the vasopressin immunoreactivity disappeared from plasma as an exponential function composed of two separate phases (Fig. 13). The first phase, which probably represents the time required for the hormone to mix throughout its distribution space, had a half-time of only about 5 minutes. The second phase, which presumably reflects irreversible or metabolic clearance of the hormone, had a much shallower slope with a half-time $(t_{½})$ approximating 18 minutes. Using these data and standard formulas as indicated (Fig. 13), we were able to calculate that the vasopressin had an apparent distribution volume (V_d) of 295 ml/kg and a total metabolic clearance rate (C) of 11.5 ml/kg per minute. Studies in 5 other young adults revealed considerable individual variation in these values, but gave a mean ±SD $t_{½}$, V_d, and C of 16.1 ± 4.6 minutes, 287.6 ± 40.1 ml/kg and 13.5 ± 4.4 ml/kg per minute, respectively.

These findings suggest an explanation for some of the discrepancies noted previously. For example, our demonstration that exogenous vasopressin disap-

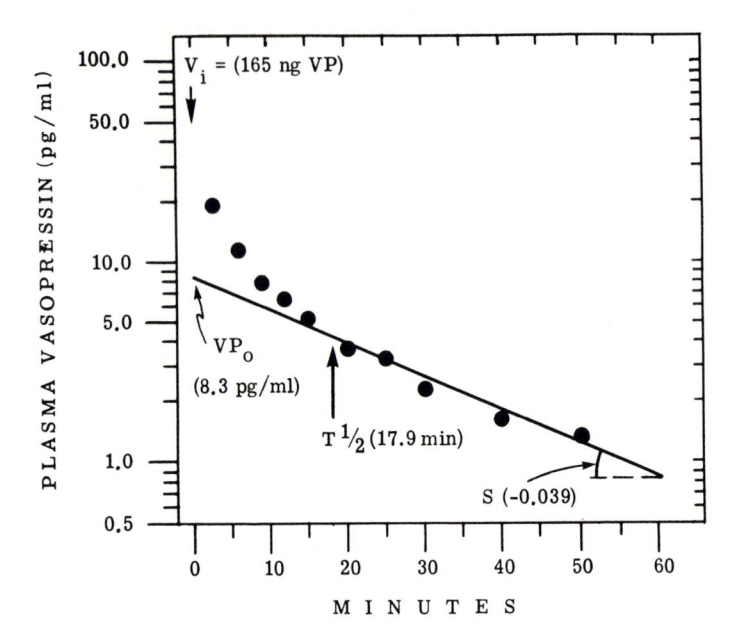

FIG. 13. Plasma vasopressin after an intravenous bolus injection of aqueous Pitressin (Parke-Davis Co.) in a recumbent normally hydrated healthy adult (68.2 kg). $V_d = V_i/VP_o =$ 20,110 ml, or 295 ml/kg; $C = S \cdot V_d = 784$ ml/min, or 11.5 ml/kg per minute. From Maxwell *et al.* (1976).

pears from plasma in two distinct phases suggests that the wide variation in intravascular half-time reported by others may have been due at least partly to a failure to distinguish between the mixing and metabolic phases. As summarized in a recent review (Lauson, 1971), all the very short half-times 1.2–7.3 minutes) were obtained by collecting plasma during the first 10–15 minutes after vasopressin injection (corresponding to our first or mixing phase) whereas all the longer half-times (10–42 minutes) were based on plasma samples collected 10–50 minutes after injection (corresponding to our second, or metabolic, phase). To some extent, the observed differences in intravascular half-time may also be due to individual variation since, in a group of normal adults studied under identical conditions, we have observed second-phase half-times ranging from 8 to 32 minutes. However, the mean of these values, 16.1 minutes, compares favorably to the mean of all the previously reported second-phase values (21.3 minutes) (Lauson, 1971) as well as those obtained previously following prolonged infusion of Pitressin (20.9 minutes) (Robertson *et al.*, 1973b) or the injection of [^{125}I] AVP (24.1 minutes) (Baumann and Dingman, 1976). Thus, all the evidence now available is at least consistent with the view that in healthy, normally hydrated recumbent adults, vasopressin mixes throughout its distribution space with a half-time of approximately 5 minutes and is cleared metabolically with a somewhat more variable half-time that averages 20 minutes.

However, our data do not help to explain or reconcile the relatively large discrepancies in the apparent distribution volume of vasopressin obtained by other laboratories. In fact, our V_d of 287.6 ml/kg is almost 4-fold greater than the highest value obtained by bioassay (Lauson, 1971) and exactly 2-fold greater than the mean value obtained by injecting [^{125}I] AVP (Baumann and Dingman, 1976). The lower values cannot be attributed to erroneous use of the first or mixing phase data since, in both studies, the results were based on plasma samples collected 8 minutes or more after injection. The higher value obtained by immunoassay cannot be explained by measurement of immunologically reactive degradation products since our assay does not detect residual activity after *in vitro* degradation of vasopressin by liver or kidney (Robertson *et al.*, 1973b) and yields an *in vivo* half-time for the metabolic phase that does not differ significantly from that obtained by bioassay (see above).

Despite these unresolved differences, there can no longer be any reasonable doubt that vasopressin is rapidly distributed throughout a space at least as large as the extracellular fluid volume. Not only do both the [^{125}I] AVP method (Baumann and Dingman, 1976) and our immunoassay (Maxwell *et al.*, 1976) give calculated distribution volumes that equal or exceed the thiocyanate space in man, but simultaneous measurements of endogenous vasopressin in the lymph and plasma of dogs (Maxwell *et al.*, 1976; Brook and Share, 1976) have confirmed unequivocally that the hormone rapidly equilibrates between these two compartments. This conclusion is fully consistent with the capillary diffusion characteristics that would be expected from its small size and apparent lack of binding to any of the macromolecular components of plasma (Lauson, 1971). Thus, the major uncertainty remaining is whether AVP is confined to the extracellular space as suggested by the [^{125}I] AVP studies (Baumann and Dingman, 1976), or also distributes reversibly into one or more additional compartments, as suggested by our results (Maxwell *et al.*, 1976).

Owing largely to the difference in calculated distribution volume, our estimate of mean ±SD metabolic clearance rate (13.5 ± 4.4 ml/kg per minute) also is significantly greater than that obtained in the [^{125}I] AVP study (4.1 ± 0.9 ml/kg per minute) (Baumann and Dingman, 1976). It should be noted, however, that our value is very close to the mean metabolic clearance rate of 11.1 ml/kg per minute obtained by Lee and Jones using a steady-state infusion method that does not involve calculating the volume of distribution (Lauson, 1971). Thus, the weight of the evidence would again seem to favor the higher value, although more studies, particularly of the steady-state variety, are clearly needed. The study of Lee and Jones also indicates that the clearance rate can vary considerably from one person to the next, ranging from as low as 5 to as high as 18 ml/kg per minute. This result also agrees well with our own findings and suggests that some of the observed discrepancies may be due simply to normal variation. Metabolic differences of this magnitude would be expected to have a pronounced effect on the regulation of endogenous plasma AVP, and it will be of considerable interest to determine how reproducible they are and how much they contribute to the individual differences in osmoregulatory function noted earlier (Section III, A).

H. URINARY AVP

The measurement of urinary AVP also has been used to assess neurohypophysial function (Oyama *et al.*, 1971; Miller and Moses, 1972; Fressinaud *et al.*, 1974; Merkelbach *et al.*, 1975). This approach is technically attractive because it does not require as much assay sensitivity as the plasma method and also has the advantage, at least in theory, of providing an integrated picture of AVP secretion over a period of time. However, there seems to be at least one problem inherent in the urinary method that seriously limits its utility for studying AVP function. This problem came to light in the course of studies comparing the effect of different kinds of osmotic stimuli on plasma and urinary AVP (Robertson, 1972). In a group of healthy adults we observed that the rate of AVP excretion was increased much less by fluid restriction than by the infusion of hypertonic saline even though the two procedures produced comparable increases in plasma osmolality and AVP concentration. Further studies showed that this disparity was due to the fact that the urinary clearance of vasopressin decreased almost 100% during fluid restriction but increased 300–400% during the infusion of hypertonic saline. These changes in AVP clearance were much larger than the associated changes in creatinine clearance which accompanied the two procedures, but were of approximately the same magnitude and direction as the changes in total solute clearance. In fact, sequential determinations of AVP and solute clearance in 2 healthy recumbent adults during hypertonic saline infusion revealed a very close positive correlation between these two variables (Fig. 14). The slope of the regression line describing this relationship was markedly different in each person, but both lines intercepted the ordinate very near its junction with the abscissa.

These findings reveal several important characteristics of the renal handling of AVP that must be kept in mind when interpreting changes in excretion rate. First, under the usual conditions of hydration and solute excretion, the urinary clearance of AVP averages only 25 ml/min. On a body-weight basis, this is about 0.3 ml/kg per minute, or less than 3% of the total clearance rate as determined by both steady-state and non-steady-state studies (Section III, G). This result agrees well with previous determinations of the fraction of exogenous vasopressin that is excreted in the urine (Lauson, 1971) and points up how easily

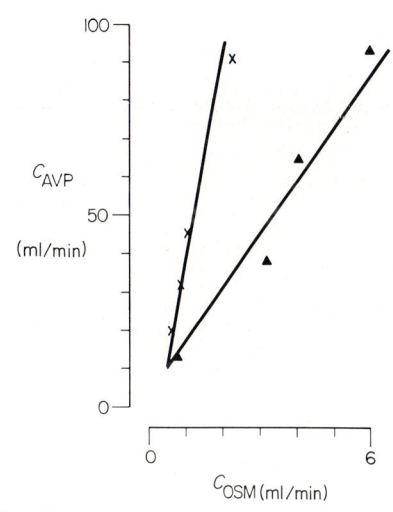

FIG. 14. Relationship of urinary arginine vasopressin (AVP) clearance (C_{AVP}) to total solute clearance (C_{OSM}) in 2 healthy adults (×, ▲) during the infusion of hypertonic saline. From Robertson (1972).

large changes in AVP excretion rate could result from extremely small changes in the total metabolic clearance of the hormone. Second, depending on the solute clearance rate, the urinary clearance of AVP in a given individual can vary markedly from as low as 10 to as high as 100 ml/min (Fig. 14). This means that the rate of AVP excretion can change by as much as 10-fold independent of any change in the secretion or plasma levels of the hormone. This kind of dissociation can result in changes in AVP excretion that differ in direction as well as magnitude from those occurring in plasma. For example, we find that the mild solute diuresis which normally results from a large water load sometimes is accompanied by a transient paradoxical rise in AVP excretion at the very time when plasma osmolality and AVP concentration are beginning to decrease (Robertson, 1972). Thus, changes in AVP excretion can never be assumed to reflect changes in plasma AVP unless they occur in the absence of, or are adjusted for, appropriate changes in solute clearance. Third, the renal handling of AVP also may be significantly influenced by factors other than solute excretion rate. The existence of this as yet undefined factor is indicated by the marked difference in slope of the regression lines in Fig. 14. In practical terms, it means that adjusting AVP excretion values for differences in solute excretion may only partially normalize their relationship to plasma AVP. In certain situations, therefore, it may be necessary to employ other methods, such as urinary recovery of exogenous vasopressin, to validate or rectify excretion values.

It should be noted that the foregoing considerations do not totally invalidate the use of urinary AVP measurements for particular research or clinical purposes. As previously noted (Oyama *et al.*, 1971; Miller and Moses, 1972), they can be of considerable diagnostic help in clinical disorders, such as diabetes insipidus, where one is looking for relatively large changes in AVP function. This applies also to nephrogenic diabetes insipidus, since our studies indicate that this disorder does not result in any change in the renal handling of AVP (unpublished observations). They may also be useful for investigating disorders of abnormal water retention (Miller and Moses, 1972; Merkelbach *et al.*, 1975) although the variable changes in solute excretion that accompany many of these conditions may complicate interpretation of the results.

IV. Pathophysiology

Many clinical disorders of fluid and electrolyte balance now can be characterized in terms of a specific disturbance in one or more aspects of AVP physiology.

A. DIABETES INSIPIDUS

The vasopressin deficiency characteristic of acquired pituitary diabetes insipidus behaves almost uniformly as though it were due to a marked decrease in the gain or sensitivity of the osmoregulatory system. In most of these patients, plasma AVP is low but detectable under basal conditions and tends to increase slightly when water intake is restricted (Fig. 15). However, the increase in vasopressin thus achieved is invariably subnormal relative to the degree of hypertonic dehydration present. Thus, when the AVP values are plotted as a function of simultaneous plasma osmolality, patients with diabetes insipidus almost always can be readily distinguished from normals or those with other types of polyuria (Fig. 5). The demonstration that the relationshiop of plasma AVP to osmolality is normal in patients with primary polydipsia or nephrogenic diabetes insipidus indicates that, unlike the case in some other endocrine systems, the responsiveness of the neurohypophysis is not appreciably altered by chronic suppression or stimulation. This conclusion is consistent with earlier studies (Barlow and de Wardener, 1959) showing that the subnormal antidiuretic response to fluid restriction in primary polydipsia (Fig. 15) is due to impaired renal responsiveness to vasopressin.

When the plasma AVP-osmolality data from patients with diabetes insipidus is subjected to regression analysis, a significant positive correlation can still be demonstrated ($r = 0.42$; $P < 0.001$), but the resultant regression function, $y = 0.03(x - 275)$ has a slope less than 1/10 the normal value. Analysis of the data obtained during hypertonic saline infusion in individual patients yields similar

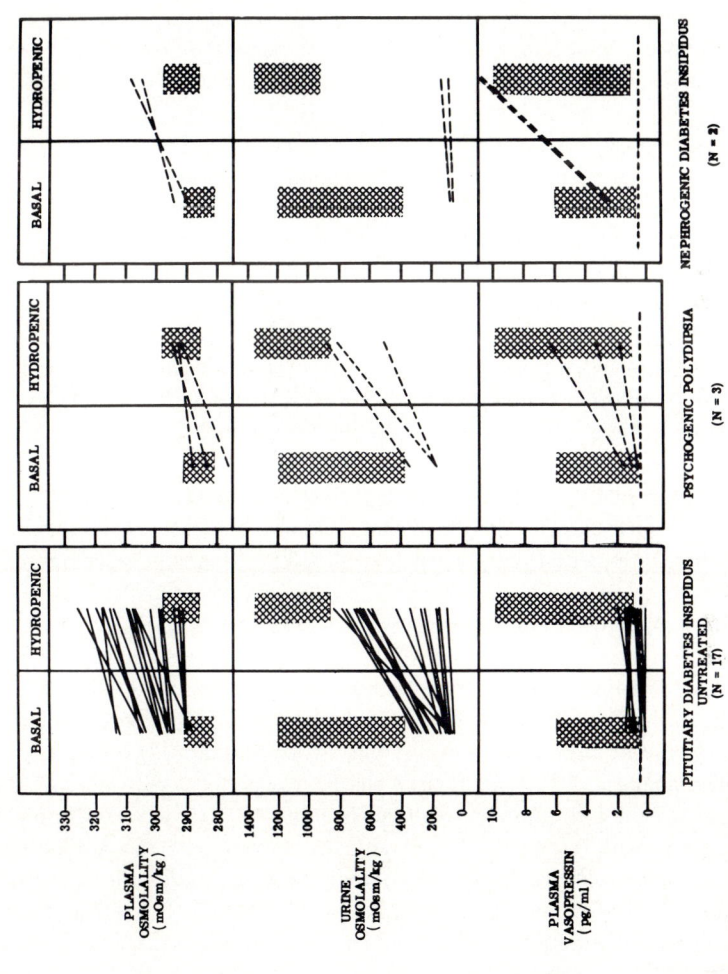

FIG. 15. Plasma osmolality, urine osmolality, and plasma arginine vasopressin (AVP) before and after varying periods of fluid restriction in patients with polyuria of different etiology. Crosshatched zones represent the range of values obtained in 24 healthy adults under similar conditions. Data of G. L. Robertson, E. Mahr, and S. Athar (unpublished observations).

values (Robertson *et al.*, 1976b). This confirms that the osmoregulation of AVP remains qualitatively normal even though quantitatively deficient in most cases of acquired diabetes insipidus. The demonstration that the slope of the regression function is consistently less than 20% of normal in patients with diabetes insipidus indicates that the secretory response of the neurhypophysis must be reduced more than 80% for the disorder to be manifest clinically.

It is now apparent that this kind of decrease in osmotically mediated AVP secretion can occur in at least two ways. Probably the most common and widely recognized cause is a reduction in the amount of AVP available for secretion due to extensive destruction of the neurohypophysis. Such cases classically result from surgical hypophysectomy and, as might be expected, exhibit quantitatively similar reductions in their AVP response to hemodynamic, emetic, and osmotic stimuli (R. L. Shelton, and G. L. Robertson, unpublished observations).

Less commonly, the same kind of hormone deficiency can result from a lesion that selectively abolishes osmotically mediated AVP secretion without damaging the neurohypophysis or diminishing its response to nonosmotic stimuli (Robertson *et al.*, 1976a; DeRobertis *et al.*, 1974; Shelton *et al.*, 1976a; Halter *et al.*, 1976). Since these patients also lack osmotically mediated thirst, they usually present with signs and symptoms atypical of diabetes insipidus. Owing to inadequate water intake, they tend to develop much more severe hypertonic dehydration and, as a consequence, are able to stimulate the release of AVP via the relatively insensitive volume-dependent mechanisms. Thus, polyuria and polydipsia are usually absent altogether, and the only obvious abnormality will be severe chronic hypernatremia, with or without its attendant neuromuscular disabilities.

These unusual patients constitute important experiments of nature because they provide a unique insight into several fundamental aspects of AVP regulation. The observation that osmoregulatory function can be totally ablated without impairing quantitatively or qualitatively the AVP response to hemodynamic or emetic stimuli provides strong evidence that the osmoreceptor mechanism is anatomically separate from the neurosecretory cells where AVP is synthetized, stored, and secreted (Robertson *et al.*, 1976a). This conclusion is fully consistent with other studies in animals showing that an antidiuretic response can be evoked by injecting hypertonic saline or sucrose near, but not directly into, the supraoptic nucleus (Peck and Blass, 1975; Bennett and Pert, 1974) and with electrophysiologic studies in monkeys demonstrating the existence of specific, osmosensitive cells in several areas of the anterior hypothalamus (Hayward and Jennings, 1973; Hayward and Vincent, 1970; Vincent *et al.*, 1972). Our observations in these patients and those relating to the interaction of osmotic and hemodynamic stimuli (Section III, D) (Robertson *et al.*, 1976a) also indicate that the neurohypophysial unit must be organized along lines similar to the scheme depicted in Fig. 10; i.e., that stimuli from the hemodynamic receptors alter osmotically mediated AVP secretion not by acting on the osmoreceptor itself, but by modifying its stimulatory effects on the same population of neurosecretory cells. Finally, these patients also provide proof of the conclusion, originally advanced on the basis of the comparative stimulus response data (Fig. 7) (Dunn et al., 1973; Robertson and Athar, 1976; Robertson, 1976), that the AVP response evoked by ordinary changes in free water balance is due almost exclusively to the associated changes in blood osmolality rather than volume. This is shown by the fact that the loss of osmoreceptor function results in a markedly deficient AVP response to hypertonic dehydration, even though the volume control pathways appear to be completely intact in these patients (Robertson et al., 1976a; DeRobertis et al., 1974; Shelton et al., 1976a; Halter et al., 1976).

B. ANTIDIURETIC EFFECT OF CHLORPROPAMIDE AND CARBAMAZEPINE

In 1966, it was discovered almost by chance that the oral sulfonylurea chlorpropamide had an antidiuretic effect in patients with diabetes insipidus (Arduino et al., 1966). These observations were quickly confirmed (Meinders et al., 1967), and the drug has since come to be widely used in the treatment of this disorder. However, there is still some disagreement as to precisely how chlorpropamide produces its antidiuretic effect (Miller and Moses, 1976) *In vitro* studies with the toad bladder were the first to suggest that it acted at the renal level by potentiating the hydrosmotic effects of small amounts of vasopressin (Ingelfinger and Hayes, 1969; Mendoza, 1969). *In vivo* evidence supporting such a mechanism was obtained shortly thereafter in the Brattleboro rat (Miller and Moses, 1970; Berndt et al., 1970), and many clinical studies employing indirect methods were also interpreted as being consistent with this view (Miller and Moses, 1976). However, immunoassay studies of vasopressin excretion in normal subjects have suggested that chlorpropamide may also increase secretion of the hormone (Moses et al., 1973). These findings have led to speculation that the drug might have a similar effect in patients with diabetes insipidus and thus act by two mechanisms to correct the polyuria.

Our studies using chlorpropamide to treat patients with diabetes insipidus are fully consistent with previous observations regarding its clinical efficacy. In all patients, basal urine output was reduced by 30–70%; and in half, polyuria was abolished completely (G. L. Robertson, E. A. Mahr, and S. Athar, unpublished observations). In each case, the decline in urine output was accompanied by a proportionate increase in urine osmolality and water retention, as indicated by a gain in weight and return of plasma osmolality toward normal (Fig. 16). However, in none of our patients was the antidiuretic response accompanied by an increase in plasma AVP (Robertson and Mahr, 1971) (Fig. 16). In fact,

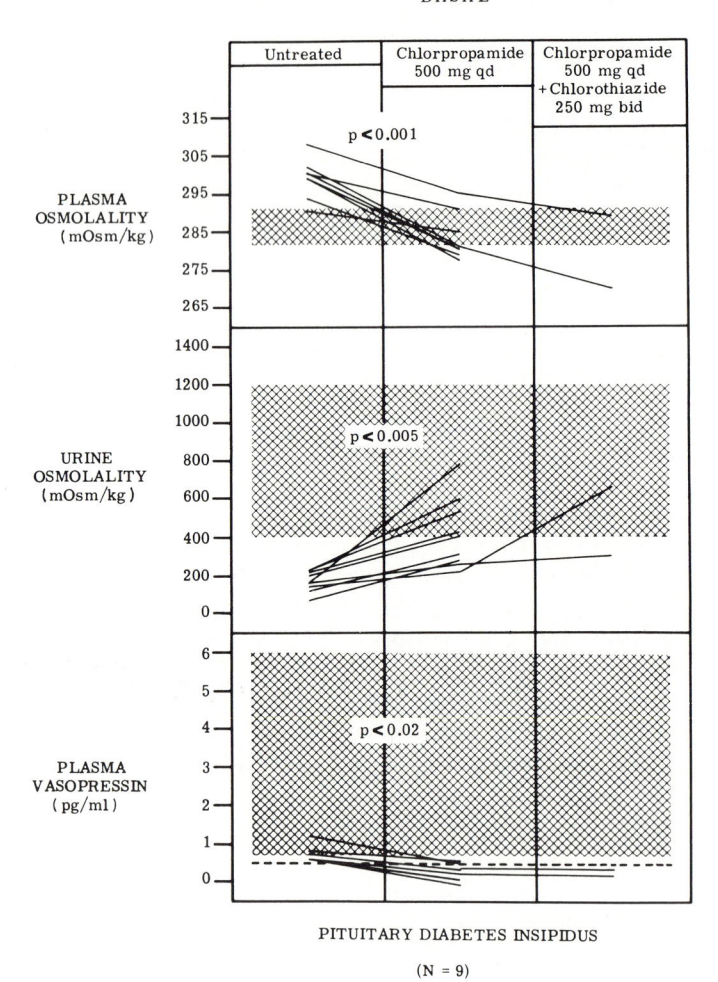

FIG. 16. The effect of chlorpropamide alone or combined with chlorothiazide on plasma osmolality, urine osmolality, and plasma arginine vasopressin in patients with diabetes insipidus. All values were obtained recumbent, on ad libitum fluid intake on the morning immediately before and the third day after institution of therapy. The crosshatched areas represent the range of values found in healthy untreated adults under similar conditions. From Robertson and Mahr (1971).

plasma AVP, which was low but usually detectable under basal conditions, fell significantly during treatment, often to unmeasurable levels. We attribute this fall in plasma AVP to the improved hydration produced by chlorpropamide since, as noted above, secretion of the hormone in these patients is still responsive to osmotic stimuli. Moreover, we find no increase in the distribution or metabolic clearance of AVP during therapy (Robertson and Mahr, 1971). Thus, our results do not support the hypothesis that an increase in AVP secretion is one of the mechanisms by which chlorpropamide induces antidiuresis in diabetes insipidus. Instead, they indicate that its therapeutic action is due largely, if not solely, to a direct renal effect like that originally suggested by the *in vitro* studies (Ingelfinger and Hayes, 1969; Mendoza, 1969).

It should be noted that another drug, carbamazepine, has also been shown to reduce polyuria in patients with diabetes insipidus (Tietze and Finkenwirth, 1970). Since it is known to have important effects on nerve tissue and has been reported to cause an increase in the antidiuretic activity of plasma (Frahn *et al.*, 1969; Kimura *et al.*, 1974), its effects on urine concentration have been attributed to an increase in AVP secretion. However, in studies conducted with collaborators in the Netherlands, we found that administration of carbamazepine to patients with diabetes insipidus or normal subjects caused no increase in plasma AVP as determined by radioimmunoassay (Meinders *et al.*, 1974). If anything, the drug appeared to suppress plasma AVP, much as did chlorpropamide. The discrepancy between the bioassay and immunoassay results is difficult to explain, particularly in view of the very large increases noted by bioassay. However, it again raises questions about the specificity of the bioassay method and whether, in some instances, it may not detect antidiuretic substances other than AVP.

C. SYNDROME OF INAPPROPRIATE ANTIDIURESIS

In 1957, Schwartz and Bartter proposed, on the basis of indirect evidence, that the occurrence of hyponatremia, urinary concentration, and renal sodium wasting in a patient with bronchogenic carcinoma was due to inappropriate secretion of the antidiuretic hormone (Schwartz *et al.*, 1957). This syndrome (SIADH) has now been observed in association with several other types of malignancy as well as a great variety of nontumorous conditions (Bartter, 1973). Both bioassay (Fishman and Bethune, 1968; Baumann *et al.*, 1972) and immunoassay (Beardwell, 1971; Shimamoto *et al.*, 1976; Morton *et al.*, 1975; Miller and Moses, 1972; Merkelbach *et al.*, 1975; Robertson and Mahr, 1973; Robertson *et al.*, 1973a; DeFronzo *et al.*, 1974; Zimbler *et al.*, 1975) studies have confirmed that AVP secretion is inadequately suppressed in patients with SIADH, but the frequency and nature of the hormone dysfunction has not been thoroughly defined.

Accordingly, we began 5 years ago to assess AVP function in every hyponatremic patient referred to our endocrine service who fulfilled the usual clinical criteria for the diagnosis of SIADH (Shelton et al., 1976b). During this period, measurements of plasma AVP have been obtained in 106 such patients before correction of the hyponatremia (Fig. 17). In the vast majority, plasma AVP was inappropriately elevated to a degree that could readily account for the antidiuresis present. In many, however, the absolute values were no greater than those normally found in healthy adults (1–5 pg/ml) and could be recognized as excessive only in relation to the abnormal hydration state. Just as in the polyuric disorders, therefore, the hormone dysfunction may not be readily apparent unless the AVP value is expressed as a function of plasma osmolality.

Further study has revealed that this defect in the osmoregulation of AVP can behave in one of several different ways (Shelton et al., 1976b). When plasma AVP was measured sequentially in 26 of these patients during the therapeutic infusions of hypertonic saline, four different types of response were observed. The first type, found in about 20% of the patients, was characterized by large

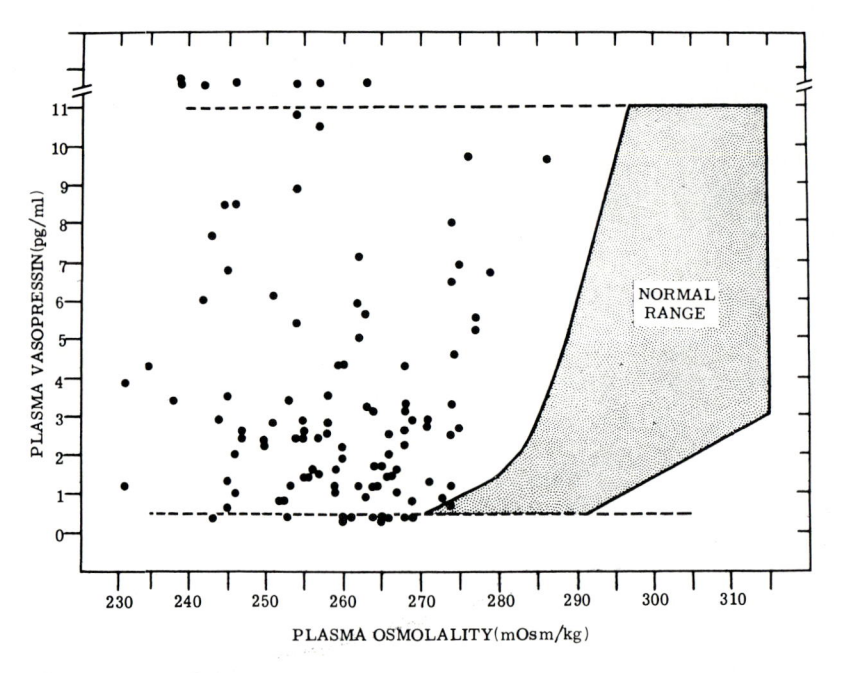

FIG. 17. The relationship of plasma arginine vasopressin to plasma osmolality in the syndrome of inappropriate antidiuresis. Each value represents a single patient. From Shelton et al. (1976b).

erratic changes in plasma AVP that showed no relationship whatsoever to the rise in plasma osmolality. This pattern was observed in association with nontumorous as well as tumorous disease and suggested that AVP release had been completely divorced from osmoreceptor control and was occurring either at random or in response to transient and erratic, nonosmotic stimuli. The existence of this kind of pattern means that one or two measurements of AVP are never sufficient to give a reliable picture either of the kind of defect present or of its responsiveness to a particular experimental maneuver.

The second type, found in about 35% of these patients, was characterized by a prompt and progressive increase in plasma AVP which correlated closely with the rise in plasma osmolality. In each of these patients, the regression line describing this relationship had a slope which paralleled the normal response but intercepted the osmolality axis well below 275 mOsm/kg, the lowest level found in healthy adults. This pattern, which has been observed in association with a wide variety of diseases, indicates that the osmotic control of AVP secretion was qualitatively normal but that the "set" of the system had been reduced to abnormally low levels. The mechanism by which this resetting of the osmoreceptor occurs is still unknown, but, in view of the frequency and diversity of clinical settings in which it occurs, it probably represents the common result of several different pathogenetic processes. Reductions in blood volume and/or pressure produce a similar kind of shift in osmoreceptor function (Section III, D) but cannot play a role in these patients since fairly marked degrees of hypovolemia or hypotension would be required and, by definition (Bartter, 1973), neither abnormality is present in SIADH. However, the same kind of resetting might result from false hemodynamic signals, such as those generated by neuropathic lesions in the afferent limb of the volume and/or baroreceptor system (Robertson et al., 1973a), or from a totally unrelated disturbance, such as a reduction in the osmotic "stuffing" of osmoreceptor neurons (Flear and Singh, 1973). Regardless of the etiology, however, patients with this kind of defect might be expected to exhibit certain clinical characteristics not seen in the other types of SIADH. Because the osmotic control of vasopressin is *qualitatively* normal, increasing water intake should lead eventually to maximum urinary dilution and, from that, a rate of output sufficient to prevent further increases in water balance. Two groups of investigators recently have described several patients who exhibit precisely this behavior (DeFronzo et al., 1976; Michelis et al., 1974) and quite probably are examples of this kind of osmoregulatory defect.

The third type, also found in about 35% of our patients, was characterized by plasma AVP levels that were elevated initially but did not change significantly until plasma osmolality exceeded the normal threshold level. At this point, plasma AVP began to rise appropriately in response to further increases in tonicity. This pattern, which has been observed in several patients with meningitis and/or basilar skull fractures but also occurs in association with other diseases, indicates that the syndrome was due to a constant, nonsuppressible release or "leak" of AVP under hypotonic conditions despite otherwise normal osmoreceptor function. Unlike those with the type two defect, these patients would be expected to exhibit the findings typical of SIADH at all times, since maximum AVP suppression and urinary dilution would not be expected to occur regardless of how severe the water retention and hypotonicity became.

The fourth type, found in about 10% of our patients, was characterized by plasma AVP levels that appeared to be normally suppressed under hypotonic conditions and did not rise until plasma osmolality exceeded the normal threshold level. This interesting and unusual pattern was associated with continuous production of a concentrated urine, suggesting that the diluting defect was due not to abnormal secretion of AVP, but to some other factor, such as a change in renal sensitivity to the hormone or the elaboration of some other as yet unrecognized antidiuretic substance.

REFERENCES

Arduino, F., Ferraz, F. P. J., and Rodrigues, J. (1966). *J. Clin. Endocrinol. Metab.* **26**, 1325.
Arimura, A., Saito, T., Bowers, C. Y., and Schally, A. V. (1967). *Acta Endocrinol. (Copenhagen)* **54**, 155.
Arndt, J. O. (1965). *Pfluegers Arch. Gesamte Physiol. Menschen Tiere* **282**, 313.
Athar, S., and Robertson, G. L. (1974). *Clin. Res.* **22**, 335A.
Barlow, E. D., and de Wardener, H. E. (1959). *Q. J. Med.* **28**, 235.
Bartter, F. C. (1973). *Dis. Mon.* November, pp. 1–47.
Baumann, G., and Dingman, J. F. (1976). *J. Clin. Invest.* **57**, 1109.
Baumann, G., Lopez-Amor, E., and Dingman, J. F. (1972). *Am. J. Med.* **52**, 19.
Beardwell, C. G. (1971). *J. Clin. Endocrinol. Metab.* **33**, 254.
Bennett, C. T., and Pert, A. (1974). *Brain Res.* **78**, 151.
Berndt, W. O., Miller, M., Kettyle, W. M., and Valtin, H. (1970). *Endocrinology* **86**, 1028.
Bevan, A. T., Honour, A. J., and Stott, F. H. (1969). *Clin. Sci.* **36**, 329.
Brennan, T. C., Shelton, R. L., and Robertson, G. L. (1975). *Clin. Res.* **23**, 234A.
Brook, A. H., and Share, L. (1966). *Endocrinology* **78**, 779.
Chard, T. (1973). *J. Endocrinol.* **58**, 143.
DeFronzo, R. A., Colvin, O. M., Braine, H., Robertson, G. L., and Davis, P. J. (1974). *Cancer* **33**, 483.
DeFronzo, R. A., Goldberg, M., and Agus, Z. S. (1976). *Ann. Intern. Med.* **84**, 538.
DeRobertis, F. R., Michelis, M. F., and Davis, B. B. (1974). *Arch. Intern. Med.* **134**, 889.
Determan, H., and Walter, I. (1968). *Nature (London)* **219**, 604.
Dunn, F. L., Brennan, T. J., Nelson, A. E., and Robertson, G. L. (1973). *J. Clin. Invest.* **52**, 3212.
Eaker, D., and Porath, J. (1967). *Sep. Sci.* **2**, 507.
Farber, M. O., Bright, T. P., Strawbridge, R. A., Robertson, G. L., and Manfredi, F. (1975). *J. Lab. Clin. Med.* **85**, 41.
Fishman, M. P., and Bethune, J. E. (1968). *Ann. Intern. Med.* **68**, 806.
Flear, D. T. G., and Singh, C. M. (1973). *Br. J. Anaesth.* **45**, 976.
Frahn, H., Smejkal, E., and Kratzenstein, R. (1969). *Acta Endocrinol. (Copenhagen)* **138**, Suppl., 240.

Fressinaud, P., Corval, P., and Menard, J. (1974). *Kidney Int.* **6,** 184.
Gauer, O. H. (1968). *Fed. Proc., Fed. Am. Soc. Exp. Biol.* **27,** 1132.
Gauer, O. H., Henry, J. P., and Behn, C. (1970). *Annu. Rev. Physiol.* **32,** 547.
Goetz, K. L., Bond, G. C., and Smith, W. E. (1974). *Proc. Soc. Exp. Biol. Med.* **145,** 277.
Halter, J., Goldberg, A., Robertson, G., and Porte, D. (1976). *Clin. Res.* **24,** 99A.
Hayward, J. N., and Jennings, D. P. (1973). *J. Physiol. (London)* **232,** 545.
Hayward, J. N., and Vincent, J. D. (1970). *J. Physiol. (London)* **210,** 947.
Hayward, J. N., Pavasuthipaisit, K., Perez-Lopez, F. R., and Sofroniew, M. W. (1976). *Endocrinology* **98,** 975.
Hunter, W. M., and Greenwood, F. C. (1962). *Nature (London)* **194,** 495.
Husain, M. K., Fernando, N., Shapiro, M., Kagan, A., and Glick, S. M. (1973). *J. Clin. Endocrinol. Metab.* **37,** 616.
Ingelfinger, J. R., and Hayes, R. M. (1969). *J. Clin. Endocrinol. Metab.* **29,** 738.
Janson, J.-C. (1967). *J. Chromatogr.* **28,** 12.
Johnson, J. A., Zehr, J. E., and Moore, W. W. (1970). *Am. J. Physiol.* **218,** 1273.
Katz, F. H., Loeffel, D. E., Roper, E. F., Lock, J. P., and Husain, M. (1976). *58th Annu. Meet. Endocr. Soc.* Abstract No. 322 (supplement to *Endocrinology*).
Kimura, T., Matsui, K., Sato, T., and Yoshinaga, K. (1974). *J. Clin. Endocrinol. Metab.* **38,** 356.
Lauson, H. D. (1971). *Hand. Physiol., Sect. 7: Endocrinol.* **6,** 287.
McCance, R. A. (1936). *Proc. R. Soc. London, Ser. B* **119,** 245.
McCann, S. M., Antunes-Rodrigues, J., Nallar, R., and Valtin, H. (1966). *Endocrinology* **76,** 1058.
Maxwell, D., McMurray, S., Szwed, J., Shelton, R., and Robertson, G. (1976). *Clin. Res.* **24,** 407A.
Meinders, A. E., Touber, J. L., and de Vries, L. A. (1967). *Lancet* **2,** 544.
Meinders, A. E., Cejka, V., and Robertson, G. L. (1974). *Clin. Sci. Mol. Med.* **47,** 289.
Mendoza, S. (1969). *Endocrinology* **84,** 411.
Merkelbach, U., Czernichow, P., Gaillard, R. C., and Vallotton, M. B. (1975). *Acta Endocrinol. (Copenhagen)* **80,** 453.
Michelis, M. F., Fusco, R. D., Bragdon, R. W., and Davis, B. D. (1974). *Am. J. Med. Sci.* **267,** 267.
Miller, M., and Moses, A. M. (1970). *Endocrinology* **86,** 1024.
Miller, M., and Moses, A. M. (1972). *Ann. Intern. Med.* **77,** 715.
Miller, M., and Moses, A. M. (1976). *Kidney Int.* **10,** 96.
Mohring, B., and Mohring, J. (1975). *Life Sci.* **17,** 1307.
Moran, W. H., Jr., Miltenberger, F. W., Shuayb, W. A., and Zimmerman, B. (1964). *Surgery* **56,** 99.
Morton, J. J., Padfield, P. L., and Forsling, M. L. (1975). *J. Endocrinol.* **65,** 411.
Moses, A. M., Numann, P., and Miller, M. (1973). *Metab., Clin. Exp.* **22,** 59.
Nichols, B. L., Jr. (1961). *Yale J. Biol. Med.* **33,** 416.
Oyama, S. N., Kagan, A., and Glick, S. M. (1971). *J. Clin. Endocrinol. Metab.* **33,** 739.
Pavel, S., Dorcescu, M., Petrescu-Holban, R., and Ghinea, E. (1973). *Science* **181,** 1252.
Peck, J. W., and Blass, E. M. (1975). *Am. J. Physiol.* **228,** 1501.
Permutt, M. A., Parker, C. W., and Utiger, R. D. (1966). *Endocrinology* **78,** 809.
Robertson, G. L. (1969). *Br. J. Hosp. Med.* **2,** 1481.
Robertson, G. L. (1972). *Clin. Res.* **20,** 778.
Robertson, G. L. (1974). *Annu. Rev. Med.* **25,** 315.
Robertson, G. L. (1977). *Proc. Int. Congr. Endocrinol., 5th, 1976* (in press).
Robertson, G. L., and Athar, S. (1976). *J. Clin. Endocrinol. Metab.* **42,** 613.

Robertson, G. L., and Mahr, E. (1971). *Proc. 53rd Annu. Meet. Endocr. Soc.* Abstract No. 125 (supplement to *Endocrinology*, Vol. 88).
Robertson, G., and Mahr, E. (1972). *J. Clin. Invest.* 51, Abstract No. 79a.
Robertson, G., and Mahr, E. (1973). *Proc. Int. Congr. Endocrinol., 4th, 1972* Excerpta Med. Found. Int. Congr. Ser. No. 256, Abstract No. 118.
Robertson, G. L., Klein, L. A., Roth, J., and Gorden, P. (1970). *Proc. Natl. Acad. Sci. U.S.A.* 66, 1298.
Robertson, G. L., Bhoopalam, N., and Zelkowitz, L. J. (1973a). *Arch. Intern. Med.* 132, 717.
Robertson, G. L., Mahr, E. A., Athar, S., and Sinha, T. (1973b). *J. Clin. Invest.* 52, 2340.
Robertson, G. L., Roth, J., Beardwell, C., Klein, L. A., Petersen, M. J., and Gorden, P. (1973c). *In* "Methods in Investigative and Diagnostic Endocrinology" (S. A. Berson, ed.), Vol. 2A, p. 656. North-Holland Publ., Amsterdam.
Robertson, G. L., Athar, S., and Shelton, R. (1977). *Physiol. Rev.* (in press).
Robertson, G. L., Shelton, R. L., and Athar, S. (1976b). *Kidney Int.* 10, 25.
Roth, J., Klein, L. A., and Petersen, M. J. (1966). *J. Clin. Invest.* 45, 1064.
Rydin, H., and Verney, E. B. (1938). *Q. J. Exp. Physiol.* 27, 343.
Sawyer, W. H. (1961). *Recent Prog. Horm. Res.* 17, 437.
Sawyer, W. H., and Mills, E. M. (1966). *In* "Neuroendocrinology" (L. Martini and W. F. Ganong, eds.), Vol. 1, p. 187. Academic Press, New York.
Schmid, P. G., Abboud, F. M., Wendling, M. G., Ramberg, E. S., Mark, A. L., Heistad, D. D., and Eckstein, J. W. (1974). *Am. J. Physiol.* 227, 998.
Schrier, R. W., and Berl, T. (1975). *N. Engl. J. Med.* 292, 81.
Schwartz, W. B., Bennett, W., Curelop, S., and Bartter, F. C. (1957). *Am. J. Med.* 23, 529.
Shelton, R., Athar, S., and Robertson, G. (1976a). *Clin. Res.* 24, 101A.
Shelton, R., Athar, S., and Robertson, G. (1976b). *Proc. 58th Annu. Meet. Endocr. Soc.* Abstract No. 323 (supplement to *Endocrinology*).
Shelton, R. L., Kinney, R. M., and Robertson, G. L. (1976c). *Clin. Res.* 24, 531A.
Shimamoto, K., Murase, T., and Yamaji, T. (1976). *J. Lab. Clin. Med.* 87, 338.
Skowsky, W. R., Rosenbloom, A. A., and Fisher, D. A. (1974). *J. Clin. Endocrinol. Metab.* 38, 278.
Thomas, T. H., and Lee, M. R. (1976). *Clin. Sci. Mol. Med.* 51, 525.
Tietze, H. U., and Finkenwirth, H. (1970). *Monatsschr. Kinderheilkd.* 118, 237.
Ukai, M., Moran, W. H., and Zimmermann, B. (1968). *Ann. Surg.* 168, 16.
Vatner, S. F., and Braunwald, E. (1975). *N. Engl. J. Med.* 293, 970.
Verney, E. B. (1947). *Proc. R. Soc. London, Ser. B* 135, 25.
Vincent, J. D., Arnould, E., and Bioulac, B. (1972). *Brain Res.* 44, 371.
Wang, S. C. (1965). *Physiol. Pharmacol.* 2, 255.
Wu, W. H., and Rockey, J. H. (1969). *Biochemistry* 8, 2719.
Yates, F. E. (1967). *In* "The Adrenal Cortex" (A. B. Eisenstein, ed.), p. 133. Little, Brown, Boston, Massachusetts.
Zimbler, H., Robertson, G. L., Bartter, F. C., Delea, C. S., and Pomeroy, T. (1975). *J. Clin. Endocrinol. Metab.* 41, 390.

DISCUSSION

G. Flouret: Are you aware that lysine vasopressin and arginine vasopressin react with acetone? Could that affect your results when you do acetone extractions?

G. Robertson: Yes, we are aware of that. We looked into this question, and under the condition that we do the acetone extractions from plasma, there appears to be no effect on the immunoreactivity of the vasopressin. We have not looked at the bioactivity, but our recovery of added hormone is 100%; as you can see, the immunologic cross-reactivity also is not changed by exposure to the acetone. It would appear that the conditions necessary for the effects to which you refer must be somewhat different from the conditions we use in our extraction procedure.

G. Flouret: You can still have an acetone derivative of vasopressin, but not on the antigenic determinant. So that you might be measuring the analog rather than the hormone, but that might be all right.

G. Robertson: Yes, I think that all we can say is that it does not alter vasopressin either qualitatively or quantitatively as seen by the immunoassay. You are quite right, that does not mean that there is actually not a derivative present. It is simply not detected by the assay.

G. Flouret: It is also possible that your acetone derivative may have different binding properties and that you might be grossly underestimating the concentration of the hormone, if you are using a binding assay. Have you thought about that?

G. Robertson: Why would you think that?

G. Flouret: The acetone derivative is a different compound to begin with. It might bind differently, and, for example, it might displace your labeled hormone differently than the hormone itself would.

G. Robertson: I think the recovery studies answer that point. If we take a water-loaded plasma, for example, with very low levels, add a known amount of hormone to the plasma, and carry it through the extraction procedure, we recover 100% in terms of concentration. If a derivative were formed that had less activity, it should be reflected in the recovery.

W. H. Sawyer: I might comment that, in experiments that we did with the same antiserum several years ago, we found out that the acetone derivative of lysine vasopressin, I believe, cross-reacted very nicely in the assay; so the formation of the derivative would not change your radioimmunoassay, although it would destroy biological activity.

G. Robertson: I might add also that, if an acetone derivative is formed, it must have chromatographic properties identical to those of native vasopressin.

R. S. Bernstein: I am interested in the paradoxical response of vasopressin to glucose infusion. Glucose is somewhat different from the other osmotically active substances because it can enter cells and be metabolized. I wonder whether this is an insulin-sensitive function. For example, for diabetics do you get the same negative slope as for nondiabetics? Also, if this process is not insulin sensitive, it may be one of the mechanisms of production of hyperosmolar nonketotic diabetic coma, because as the patients become hyperosmotic with glucose, they would then excrete more free water and become more hyperosmotic. This would be a self-perpetuating mechanism.

G. Robertson: In answer to the first part of the question, we do not think that insulin is involved, but we are just now doing studies to attempt to determine that. We are repeating these infusions in insulin-dependent diabetics to see whether they exhibit the same paradoxical suppression as do normal individuals. We think that this suppression is really due to the fact that a rise in blood glucose lowers plasma sodium concentrations. Sodium is an effective solute, so what you are doing is replacing an effective solute with one that is ineffective insofar as stimulating the receptor goes. Urea does not have this effect because it penetrates somatic cells relatively freely and therefore does not lower serum sodium concentration. It is only because glucose does not penetrate somatic cells so readily that you get this reduction in plasma sodium concentration, which we think is probably responsible for the suppressive effect. But we have not ruled out the other possibility that you mention.

P. W. Nathanielsz: Speaking probably as one of the few in this room who were actually taught by Verney, I may say that we were always impressed by the meticulous precision to which you referred in your communication. I am sure he would have been delighted to see your data. I would like to ask whether you have any information regarding the biological activity of the various peaks you demonstrated after Sephadex separation of blood? If I may add a second, separate, question arising from evidence that both α- and β-adrenergic receptors influence the action of AVP on the renal tubule: Do you have any information as to whether these systems may affect the release of AVP? For example, do infusions of catecholamines influence the release of AVP in response to any of the stimuli you have tested?

G. Robertson: I'll take your second question first. We have no evidence ourselves, but Dr. Robert Schrier has looked at the effects of several catecholamines on vasopressin and concluded that they are largely if not totally secondary to the changes in blood pressure that they produce. Thus, they probably do not affect vasopressin secretion directly but are mediated through the neurogenic afferents from pressure-sensitive receptors in the large arteries of the thorax and neck. In other words, catecholamines probably shift the osmoreceptor in much the same way as other pressure- or volume-altering stimuli, but I do not think they act as mediators in this regard.

P. W. Nathanielsz: Have you looked at the bioactivity of these various fractions? You have gone into great detail, and you went through in detail the various physical chemical criteria that must be utilized to say that this is AVP and came down very nicely to show that AVP is that third peak, but what about the bioactivity of those three peaks?

G. Robertson: No, we did not look at the bioactivity in any of the three peaks. We have extracted peak 1 and peak 2, by a variety of methods and attempted to recognize vasopressin by physical chemical methods. For example, using hot acetic acid and a number of precipitating agents, we have not been able to recover anything which by immunoassay and chromatography behaves like vasopressin. But we have not looked at the bioassay material.

F. C. Bartter: I would like to return to the question of glucose and sodium. If you were measuring the plasma sodium while you were measuring the glucose, and if you, in fact, did have decreases of sodium, why did you not get *no* delta, *pari passu* with the increments of glucose, of plasma osmolality at all? If indeed you got no delta of osmolality, I wonder why you expected plasma AVP to change. Do you postulate that there is greater concentration of glucose intracellularly than extracellularly in the specific osmoreceptor cells? All your discussion seemed to imply that the changes in plasma AVP were the result of changes in secretion of AVP. I wonder whether you have data concerning the metabolic clearance of AVP, changes of which might influence blood values in an entirely different way.

G. Robertson: In regard to the first question, the rises in blood glucose in these individuals were quite large. We had to infuse 20% of glucose at a maximum rate in normal individuals to produce the observed increase in osmolality. Blood sugars of 300–400 mg/100 ml were attained at the peak, at the end of the infusion. Under this condition, you do obtain a rise in plasma osmolality even though you produce a very significant depression in plasma sodium concentration. However, I think the point that you raise about possibly increasing the intracellular solute content of the osmoreceptor is a very valid one and a possible interpretation of the results also. I believe that we cannot distinguish between these two possibilities at the present time. I would not be surprised if there were not a rise in intracellular solute in the osmoreceptor in association with these marked changes in blood glucose, and maybe that supplements the effects of the hyponatremia on vasopressin secretion.

As for the second question: yes, we do have a fair amount of information about the volume of distribution and metabolic clearance of vasopressin, and I am sorry that I did not get a chance to review it this morning. The hormone does appear to be distributed into a space slightly larger than the extracellular compartment. Variations in the size of this compartment could have some effect on the slope of the individual regression lines, for example, since secreting the same amount of hormone into different volumes would yield a slightly different plasma level. This seems particularly pertinent if you want to study vasopressin in patients who have edema. This is an obvious application of the assay, and it is a factor that you are going to have to take into consideration because, as you know, anybody with generalized edema has an increase in extracellular fluid volume of anywhere between 25 and 50%. A change in distribution volume of this magnitude is going to have a very substantial effect on the slopes of the regression line quite independent of what is happening with secretion rates. We also find, as you imply, that there are significant individual variations in metabolic clearance. The average half-time appears to be about 18 minutes, but it can vary from as low as 6 to as high as 40 minutes in some individuals. We do not really know why there is such a large variation, but I would not be surprised if, in part, it has to do with such factors as cardiac output or the rate of blood flow through organs, such as the kidney or liver, which appear to be the principal sites of vasopressin clearance. It is quite possible, therefore, that some of the changes that we are seeing in the hypotensive hypovolemic state are, in fact, due to changes in metabolic clearance. Theoretically, however, this should change only the slope of the line and should not really change the calculated intercept value. Whether that is true or not I do not know, but I interpret the fact that we change the intercept to mean that we are seeing a real change in vasopressin secretion, not just a change in plasma vasopressin due to differences in metabolic clearance, although it may be present at the same time.

K. Sterling: I think this is interesting, and I should like to spend a few moments pursuing points Drs. Bernstein and Bartter have brought up. If we assume that the real effect on the osmoreceptor reflects the ratio of osmotically active particles in the extracellular fluid to those in the nerve receptor cells themselves, you might be able to rationalize these effects of glucose infusion. You have explained fully, I think, the urea effect, since urea distributes throughout total body water. However, I believe that some nerve cells will take up glucose very rapidly regardless of the presence or the absence of insulin, and it is entirely conceivable that they could take it up disproportionately to the hyperglycemia that they achieve, which, as you mentioned, is countered by the lowering of sodium because glucose is known to pull water out of most of the cells of the body when you get an abrupt 400 mg/100 ml hyperglycemia. I should like to hear your reaction to the question of whether it may ever be technically feasible for you to study the intracellular colligative properties. Also, I wish just briefly to comment on the work of Dan Fine, which occurred after that of Verney but before that of Roth and Glick. I believe that it was in the mid-1950s that Fine was working at Maimonides Hospital and did some venesections on himself to the extent of about 12,000–15,000 ml, showing by inference a rise in aldosterone output which was the main purpose of this work. The findings also seemed compatible with the expected abrupt rise in vasopressin. This work by Dan Fine was, I think, published in the *Journal of Clinical Investigation* in the mid-1950s.

G. Robertson: Thank you, Dr. Sterling; I was not aware of the latter study. In regard to the first question, again I think that the question of intracellular solute content in the osmoreceptor itself is an extremely important one here, but I do not know at present exactly how to get a handle on this question. We are looking at it now, for example, by attempting to characterize vasopressin function in patients with hyperosmolar coma, thinking that if anyone would have an abnormal increase in intracellular content of some unknown solute it might be these patients. If this were the case, we might be able to show a marked displacement of the normal osmoregulatory relationship to the right. But at the moment, I cannot think of a cleaner experimental approach to this question. I certainly would be more than happy to hear any suggestions.

M. M. Martin: Apart from the vasopressin effect on hyperosmolar coma, I wonder whether this may have a bearing on cerebral edema when patients go into coma after being rehydrated rather than when they are in a severe state of dehydration.

Your observation on the effect of hyperglycemia on serum vasopressin is very interesting. For some time I have been intrigued by the lack of correlation between the amount of sugar in the urine of children with diabetes and their 24-hour urine volume. One has always assumed that the large volume passed by a diabetic was the result of an osmotic diuresis. Yet, very often we see children who test all negative and yet bring in volumes of 2 or 3 liters which contain only moderate amounts of sugar. I wonder now whether the explanation might not be in your observation that hyperglycemia inhibits vasopressin secretion and we are seeing a form of water diuresis unrelated to the amount of sugar present. This is an entirely new interpretation and very exciting indeed.

One question I have relates to studies we did some years ago in patients with combined anterior and posterior hypopituitarism. Urine flow was a function of solute load. Steroids were required for the elaboration of a dilute urine, and vasopressin to produce a concentrated urine. There was a lot of argument at the time as to whether vasopressin and the steroids acted, as it were, complementary—each affecting urine osmolality depending on whether the urine should be hypertonic or hypotonic—or whether the steroids in fact had some profound effect on turning on or turning off vasopressin. Have you any information on the effect of steroid on vasopressin levels under various conditions of hyper- and hypotonicity?

G. Robertson: Yes, we agree very much with what you suggest—that is, the polyuria that occurs in hyperglycemia states such as diabetes mellitus may be due not only to the solute diuresis, which is undoubtedly important, but also to the suppression of plasma vasopressin. In support of that, Dr. Robert Zerbe, who is working with me this year, has recently been looking at plasma vasopressin levels in some uncontrolled diabetics who are also mildly hyponatremic and has found in some that the hormone is suppressed to very low or undetectable levels. What appears to happen, however, is that, if the hyperglycemia persists, the solute diuresis that results eventually depletes total body water enough so that sodium is brought back into the normal range. Under these conditions, we are finding plasma vasopressin levels that are completely normal or even in some cases elevated if plasma sodium is also elevated. So it would look as though this vasopressin mechanism plays a role in the polyuria only in the early stages before there is any significant free-water depletion. Once that occurs, sodium is restored more or less to normal level and so is plasma vasopressin.

M. M. Martin: Yes, I was referring to the reasonably well controlled diabetic who nevertheless drinks an awful lot, and yet there is no explanation in the total 24-hour urine sugar to explain it. I think your observation might be the explanation for the thirst and polyuria of the diabetic who is not in ketoacidosis and not dehydrated.

G. Robertson: Are you suggesting that this is due to a change again in intracellular solute content, or is it mediated strictly through changes in extracellular sodium concentration?

M. M. Martin: I do not really know, but the point is that this type of diabetic will run blood sugars around 300, 350, or 400 mg/dl for some hours each day. Now, if under those conditions he turns off vasopressin he could conceivably have a water diuresis, which then will make him thirsty, and explain the 2, 3, and 4 liters of urine that contains only 30 gm of glucose.

G. Robertson: I agree completely that the low plasma vasopressin levels could account for the low urinary concentration, and thus for the relatively low content of solute or glucose in their urine. I do not know what effect hyperglycemia per se has on thirst.

In regard to your second question, the story with adrenal hormones and water metabolism appears to be a relatively complicated one, that is all I can say. We hoped that we could answer it rather simply in a straightforward way, but this does not appear to be the case. One thing I think we can say is that adrenal insufficiency does not protect against the stimulatory effects of hypovolemia or hypotension, so that, whenever you have a patient with adrenal insufficiency who is either hypovolemic or hypotensive, you are going to find elevated levels of plasma vasopressin and an inability to excrete a water load. We have found this in several patients with this condition and, in collaboration with Dr. Schwyer in Denver, also in some adrenalectomized dogs that he has been studying. We find that withdrawing glucocorticoids results in very significant elevations in plasma vasopressin which respond inappropriately to a water loading. There is some disagreement yet as to whether that is mediated through hemodynamic consequences of glucocorticoid deficiency, but it is clear that vasopressin is probably mediating or at least contributing significantly to the water excretion. The problem is that we also have studied at least two patients who have Addison's disease in association with impaired water excretion and hyponatremia, and in whom plasma vasopressin levels are suppressed to very low or barely detectable levels. Unfortunately, this really complicates the picture because it suggests that maybe deficiency of glucocorticoids per se may result in some defect in the kidney such that it cannot form a maximally dilute urine. At the present time, therefore, I cannot give you a clear answer to exactly what the total mechanism is.

M. M. Martin: What you might want to try is to infuse cortisol at a very low rate so as to raise the glomerular filtration rate (GFR) to normal and then observe the effect of that on serum vasopressin and water diuresis.

G. Robertson: We have tried that; unfortunately we found that, in contrast to what others have reported, this does not have a dramatic or immediate effect on water excretion, at least in our patients. We are not quite sure exactly how to interpret this at the present time.

D. A. Fisher: I have a comment and a question. We have been interested in comparing vasopressin responsiveness in the fetus, the newborn, and the adult, and decided to compare plots of serum vasopressin concentration versus serum osmolality. Dr. Richard Weitzman in our laboratory, who has been doing this work, began with adult sheep that were dehydrated, normally hydrated, and water loaded. His results are shown in Fig. A. The points represent mean values of ten 3-minute samples over 30 minutes for both vasopressin concentration and osmolality. When he plotted these integrated data, he obtained a best fit with a log-linear relationship. In this regard, it is interesting to recall that the kidney response to vasopressin is log linear with respect to the dose of exogenous vasopressin.

My question related to Fig. B. Dr. Sharon Siegel in our laboratory has administered furosemide to the newborn lamb to study the response of the renin–angiotensin system. When she injected furosemide, the animals became very thirsty. Because of this she measured their serum vasopressin concentrations and noted a dramatic response. This response did not occur in nephrectomized animals. As you know, there are considerable data suggesting that exogenous angiotensin II will stimulate vasopressin release. Our data suggest that endogenous angiotensin also stimulates vasopressin secretion. This may have some relevance to hemodynamic control of vasopressin. Have you looked at the vasopressin response to angiotensin or measured angiotensin during your hemorrhage experiments?

G. Robertson: In reply to your question; no, we have not looked at angiotensin effects at all. I think, however, that the weight of the evidence from a number of other laboratories would indicate that probably the renin angiotensin system does have significant effects on antidiuretic hormone release. The only two questions in my mind at the present time are, first, whether it ever really does so at what might be called physiologic levels of activity, and, second, whether it does so directly or only by potentiating a concurrent osmotic

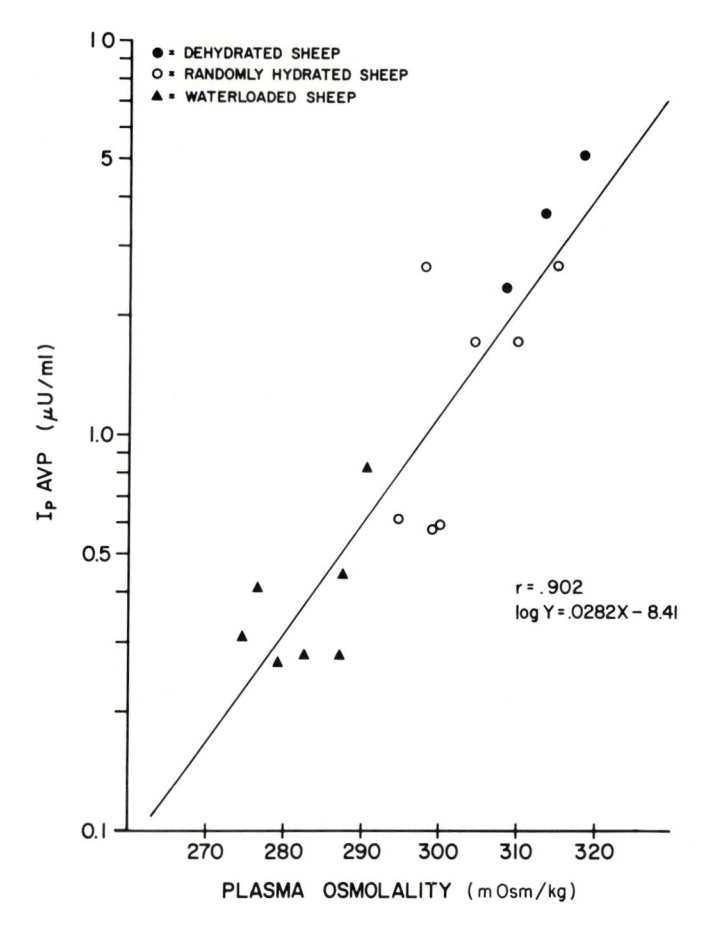

FIG. A. Linear regression of the log of integrated plasma AVP versus plasma osmolality. Each point represents the mean of samples collected at 3 minute intervals. There is no particular osmolality which is associated with an abrupt use in plasma vasopressin.

stimulus. In that regard, I wonder whether you could tell how much volume depletion the furosemide produces in the lamb and what level of plasma renin you were achieving in these experiments.

D. A. Fisher: The response is first detected at 8–10 minutes. The diuresis is maximal in the newborn sheep at about 30 minutes. There are no changes in hematocrit or blood pressure at the time the first increase in plasma renin or vasopressin is observed. The data suggest that the response is triggered by a direct action of furosemide on the kidney rather than by salt and water depletion. The response occurs too early for a depletion effect. This is a pharmacologic response, but it suggests that endogenous angiotensin may be involved in some way in the response of vasopressin to hemodynamic changes.

G. Robertson: In regard to the first point, about whether the osmoregulatory response is really linear or curvilinear: I think you have to be very careful, if you are going to draw those conclusions, to do the study in the same individual. Because there is significant individual variation in the characteristics to the osmoreceptor, you sometimes can obtain data by chance selection of individual subjects that will suggest indeed that the relationship is curvilinear. However, looking at it in greater detail in a single individual, we have not been able to obtain any suggestion whatsoever of a curvilinear response, at least over the range of osmolalities that we have used in human subjects.

M. B. Vallotton: First, I wish to congratulate you on your brilliant work and excellent presentation. As you are aware, we have applied a radioimmunoassay for AVP to determine antidiuretic hormone (ADH) in urine in humans, using an antiserum that is specific for the C-terminal tripeptide of AVP [P. Czernichow, U. Merkelbach, and M. B. Vallotton, *Acta Endocrinol. (Copenhagen)* **80**, 444–452 (1975); P. Czernichow, A. Reinharz, and M. B. Vallotton, *Immunochemistry* **11**, 47–53 (1974)]. We have observed sex difference in daily urinary excretion of AVP, males excreting twice as much AVP as females [U. Merkelbach, P. Czernichow, R. C. Gaillard, and M. B. Vallotton, *Acta Endocrinol. (Copenhagen)* **80**, 453–464 (1975)].

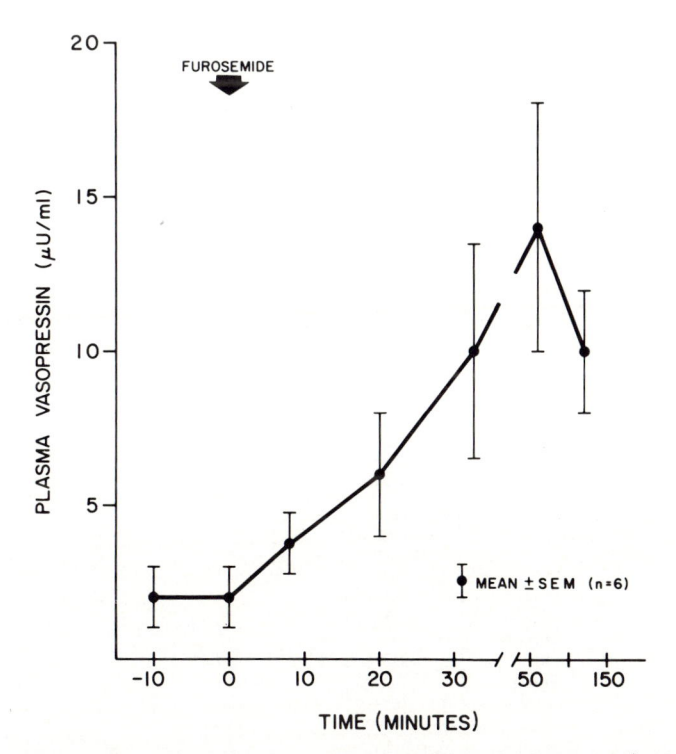

FIG. B. Vasopressin response to furosemide in the newborn lamb. Two mg/kg furosemide was given at the indicated time. The increases in both PRA and plasma vasopressin occurred within 8–10 minutes.

This prompted us to look for a difference in the response to an osmotic stimulus between the two sexes in normal human subjects. As seen on the top part of Fig. C, one observes after infusion of hypertonic saline that the hourly excretion rate of AVP increases in males (top left panel), but that the rise of AVP is much more dramatic in females (top right panel) both in absolute and relative terms. In order to try to find an explanation for this difference between the sexes, we administered to normal human male subjects 0.5 mg of ethynylestradiol daily for 4 days, performing the saline loading test on day 4. As shown on the lower right panel, there is a considerable difference compared to values for the men not receiving estrogen, their response now being not different, as assessed by analysis of variance, from the response of the women.

As seen also on Fig. C (left lower panel), when we infuse in normal male subjects angiotensin II (at the rate of 7 ng/kg per minute), a much greater hourly excretion rate of AVP during the control period and a more marked response to hypertonic saline are induced, thus confirming in man the stimulator effect as well as the potentiating effect of angiotensin II upon AVP release as found in dogs by K. Shimizu, L. Share, and J. R. Claybaugh [*Endocrinology* **93**, 42 (1973)] so far for the response to osmotic stimuli.

In collaboration with Drs. Hemmer, Viquerat, and Sutes, we have studied a series of 8 patients requiring prolonged mechanical ventilation for traumatic chest injury, monitoring the daily urinary excretion rate of AVP under four ventilatory conditions: (Fig. D): continuous positive pressure ventilation (CPPV), intermittent positive pressure ventilation (IPPV), then during continuous positive airway pressure (CPAP), and finally while spontaneously breathing (SB). During CPPV, there is a marked increase in ADH excretion over the normal range during the whole period and a significant decrease toward normal values after transition from CPPV to IPPV, but there is no concomitant change in free-water clearance with osmolal clearance, as shown at the bottom of Fig. D. ADH excretion remained within

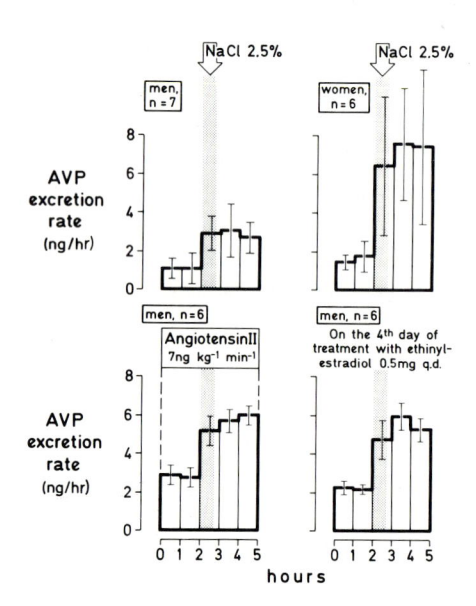

FIG. C. Carter-Robbins test for daily urinary excretion of arginine vasopressin (AVP) in normal men and women and in men receiving ethynglestradiol or angiotensin II.

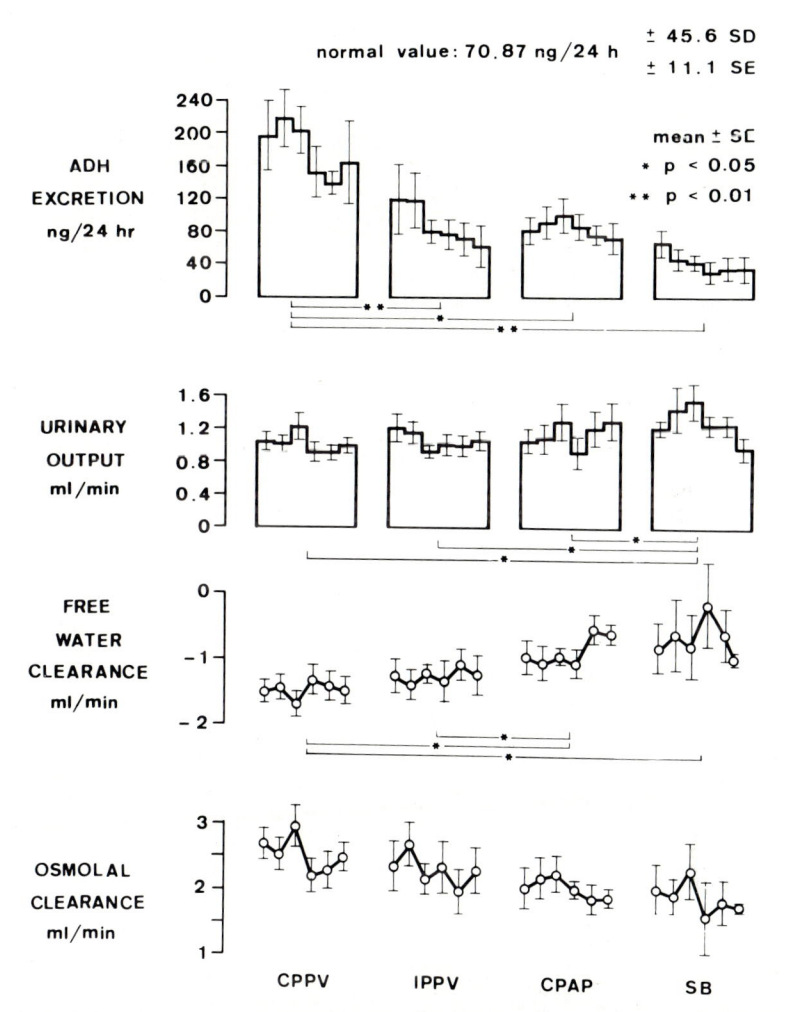

FIG. D. Effects of four different ventilatory conditions on excretion of antidiuretic hormone (ADH) and on water handling. CPPV, continuous positive-pressure ventilation; IPPV, intermittent positive-pressure ventilation; CPAP, continuous positive airway pressure; SB, spontaneous breathing.

the normal range during CPAP, as during IPPV, and decreased further during spontaneous breathing.

Our explanation, therefore, is that the difference in secretion of AVP during CPPV and IPPV is probably due to a volume mechanism activated by cardiocirculatory changes induced by CPPV, which is associated with antidiuresis without change in osmolal clearance. Determination of urinary ADH, granting all the necessary controls and careful validation and taking into account all related parameters, appears to be useful for physiological and clinical studies.

G. Robertson: In my presentation, I did not get a chance to get into the question of the factors that influence the renal clearance of vasopressin, but it is quite clear that they are quite pronounced and that there are very marked changes in renal clearance of the hormone in association with changes in solute excretion rate. I do not think that this invalidates the use of urinary assays of vasopressin for answering certain kinds of questions, but it does place a very heavy obligation on the investigator to be sure that the changes that he is observing are not due simply to change in the renal clearance of the hormone. For example, I think the experiments in which you show the sex difference are quite interesting, because when we use plasma measurements, as I indicated earlier, we cannot find any male–female difference in responsiveness to osmotic stimulation under conditions very similar to those you have used here. There is a lot of individual variation, but when we average the mean responses in the males and females they are identical. I am not sure why there is discrepancy between our results, but I would encourage you to look very carefully to see (1) whether there were differences in solute excretion, and (2) whether there are male–female differences in renal clearance of the hormone independent of solute excretion rate. I think this is still a completely unresolved question.

R. E. Frisch: Even though you do not find differences between male and female, have gonadal hormones been specifically studies in relation to vasopressin or osmotic stimuli–that is, experimentally?

G. Robertson: We have not, but in the studies that Dr. Vallotton just mentioned, he gave ethynlestradiol to males and found increased urinary excretion of vasopressin in response to a standardized stimulus.

R. E. Frisch: At the time of the adolescent growth spurt and puberty, there are large changes in fractional body water and also apparently in blood volume. Have there been studies of osmotic regulation during this period?

G. Robertson: We have not looked at any patients that young. However, we have looked at this question in subjects ranging from 20 to 80 years of age, and there is a definite change in osmoregulatory function with aging. Fortunately, the sensitivity seems to improve with age, which is somewhat the reverse of what you might expect. That is, the amount of hormone release per unit increment and osmolality is greatly increased in elderly subjects over younger individuals, but we have not looked at it in anyone younger than 20.

R. E. Frisch: It would be interesting to know whether there are differences between prepubertal and postpubertal boys and girls.

T. E. Andreoli: Can you tell whether the sensitivity of your assay is sufficiently precise so that you can look at the relationship between urine osmolality and AVP levels when you are in the range of urine osmolalities under 300 and in the range of plasma AVP levels under 1? Your curve begins to tell off and it is a singularly important issue, I think, in trying to evaluate, for example, the Schwartz border syndrome and other disorders.

G. Robertson: In answer to the first question–no. I think immunoassays have many advantages, but precision, particularly at the low end of the standard curve, is not one of them. The precision of our assay in that range is such that we cannot tell the difference between a circumlating level assay of, let us say, 0.7 and one of 0.5, or one of 1.00. In that range, we know from the slope of that line that changes so small must have a significant effect on urinary concentration. Therefore, we have a blind spot, as it were, in terms of urinary osmolality that is fairly substantial, maybe of the order of 100–150 mOsm/kg.

T. E. Andreoli: If you assume that the line you get, the regression line, which is quite tight between, say, 1 and 10, with changes in urine, or 1 and 5 pg/ml, and extrapolate that line down to, let us say, a zero urine osmolality, is there a relationship to be made there only by extrapolating rather than by using the data points themselves? That is, at what point would you expect AVP levels to shut off completely?

G. Robertson: At what point would I expect urine concentration to cease, for maximum urinary dilution to occur? By extrapolating the line that I showed you on Fig. X, it would appear to occur at a plasma hormone level around 0.5 pgs, which is our limit of detectability. However, I think that if you look at that figure, you will see tremendous individual variation and that some who appear to have circulating levels as high as 1.5 pg have maximum urinary dilution. Part of this individual variation may be due, as I say, to the inherent imprecision of the assay in the range, but also, I should not be a bit surprised if there were not other very important renal factors that modulated the effect of a given level of hormone on urinary concentration. For example, I think Dr. Schrier's studies in regard to renal prostaglandins suggest very strongly that there are intrinsic renal mechanisms that may alter that relationship substantially.

Dr. Gill: I was very interested in your findings of high values for plasma vasopressin in adrenalectomized dogs. A number of years ago, we observed that intravenous infusion of normal saline in patients with untreated Addison's disease decreased urinary osmolality to values as low as those observed in normal subjects during a water diuresis [*J. Clin. Invest.* **41**, 1078 (1962)]. Infusion of albumin similarly decreased urinary osmolality and increased the clearance of solute-free water whereas infusion of hypotonic mannitol, which increased sodium excretion and therefore delivery of sodium to the diluting segment of the nephron, did not increase the clearance of free water. This observation, that infusion of normal saline could produce a normally dilute urine, appeared to cast doubt on the hypothesis that, in the absence of cortisol, collecting ducts cannot maintain the characteristics necessary for excretion of a dilute urine. On the basis of our findings, we concluded that circulating antidiuretic hormone was probably increased and that a relationship existed between some function of the volume of extracellular fluid and the secretion of antidiuretic hormone. One would have hoped that the controversy over the mechanism of impaired water excretion in adrenal insufficiency would have been resolved by your very elegant assay, but your findings of hyponatremia in association with a very low value for plasma vasopressin in two patients with Addison's disease will probably serve to keep the argument alive for a while longer.

G. Robertson: I am sorry that they may and am aware of your studies. We were always of the impression that the volume and pressure changes probably were important and that vasopressin was involved. However, I must add a certain disclaimer here, which is pertinent also to Dr. Andreoli's question: when we are confronted with small changes in urinary dilution or free-water excretion, I am not absolutely convinced that the assay is sufficiently sensitive or precise as to exclude absolutely the possibility that small amounts of circulating vasopressin are in fact responsible for the diluting defect. So, even with our assays in their present form, I am not sure that we are going to be able to answer this question with certainty.

B. Little: One last question. Have you done these studies in subjects during pregnancy? For example, are the slope and the intercept the same for hypertension and hypovolemia during pregnancy? Also is the half-life or metabolic clearance rate of AVD changed during pregnancy? There is one very simple observation that has been made recently: in hypertension of pregnancy, blood pressure goes up at night rather than down, as in the nonpregnant woman, just as one example of the difference between pregnant and nonpregnant homeostatic mechanisms.

G. Robertson: No, I have a very simple answer for that. We have not had a chance yet to do any studies during pregnancy. I am not quite sure how much of a problem we would have in regard to the vasopressinases that circulate in this condition. This is a technical problem that I think we would have to surmount before attempting any physiologic studies.

Some Considerations of the Role of Antidiuretic Hormone in Water Homeostasis

THOMAS E. ANDREOLI AND JAMES A. SCHAFER

Division of Nephrology, Department of Medicine, and Department of Physiology and Biophysics, University of Alabama School of Medicine, Birmingham, Alabama

I. Introduction

The purpose of this paper is to review some recent observations on the role of antidiuretic hormone (ADH) in the maintenance of water homeostasis, with particular regard to the physical processes by which ADH modulates the rate of water permeation through apical (or luminal) surfaces of hormone-responsive epithelia. In vertebrate species having widely disparate salt and water intakes, the plasma osmolality is virtually constant, ranging between 285 and 295 mOsm kg^{-1}. Because plasma osmolality expresses the ratio of aqueous solutes to total body water, it is evident that invariance of plasma osmolality in the presence of nonisotonic solute and water ingestion depends on the ability of the kidney to dissociate solute and water excretion: one excretes urine which is either hypotonic to plasma, in the case of relative water excess, or hypertonic to plasma, in the case of relative water deficit.

Since proximal tubular absorption of glomerular ultrafiltrate is an iostonic process, the independent modulation of solute and water excretion is evidently effected by events occurring in more distal nephron sites. The major components of the physiological factors responsible for osmotic homeostasis of the body fluids are illustrated schematically in Fig. 1. Slight increments in plasma osmolality, or reductions in plasma volume, stimulate ADH release from the neurohypophysis. Within collecting duct cells, ADH binding to contraluminal plasma membranes results in activation of the enzyme adenyl cyclase, which catalyzes the formation of cyclic 3′,5′-adenosine monophosphate (cAMP) from ATP. In turn, cAMP increases the water permeability of luminal membranes in collecting duct cells, thereby permitting osmotic equilibration of collecting-duct fluid with a medullary interstitium made hypertonic by the renal countercurrent multiplication and exchange systems. Thus in the presence of circulating ADH, normal individuals excrete urine that is hypertonic to plasma.

It is evident that a detailed discussion of even a fraction of the information available on the processes cited above is well beyond the scope of this presentation. The present inquiry has been restricted to a single set of issues: the

```
                        Collecting Duct Cell
              contraluminal              luminal
               membrane                membrane
  ↑plasma osmolality  ⎫
                      ⎬  → ADH ──→ binding
  ↓plasma volume      ⎭                          increased
                                     ──→ cAMP ──→ water
                              ↓                  permeability
                           adenyl
                           cyclase
                                    ←── ATP

              Interstitium                        Lumen
  outer medulla:  300 mOsm/kg
                                        ⬅         ⬅  H₂O
  papillary tip:  1400 mOsm/kg
                                        ⬅         ⬅  H₂O
```

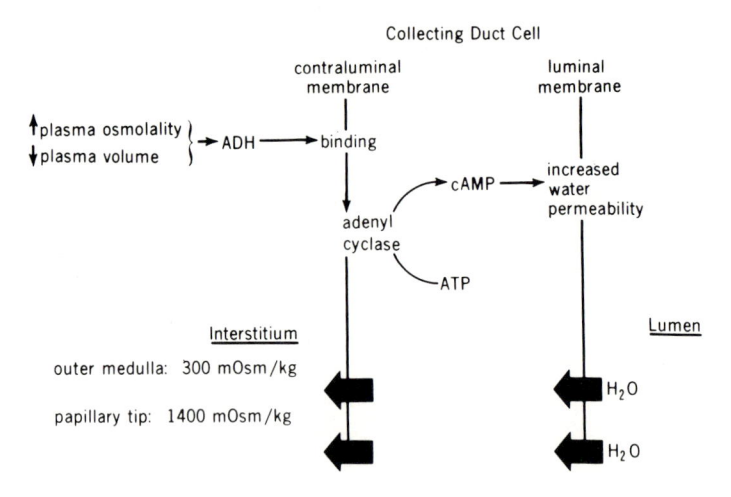

FIG. 1. Components of the water homeostasis system. ADH, antidiuretic hormone.

mechanism of water and solute transport across apical surfaces of hormone-sensitive epithelia, particularly mammalian cortical and outer medullary collecting tubules; and the ways in which ADH, or the second messenger, cAMP, affects these processes.[1]

II. Nomenclature

The effects of ADH on membrane transport processes in hormone-sensitive epithelia have been examined with various neurophypophyseal extracts as well

---

[1] For convenience, the reader is directed to the reviews indicated below for information on aspects of the physiology, biochemistry, and pharmacology of ADH not touched on in this article: the chemistry of ADH, and the relationship of hormone structure to hormone action (du Vigneaud, 1969; Sawyer, 1967; Walter et al., 1967, 1977); structural analogs of ADH, particularly those having clinical efficacy in the treatment of pituitary diabetes insipidus (Cort et al., 1975; Edwards et al., 1973; Sawyer et al., 1974a,b); biosynthesis, storage, and release of ADH from the neurohypophysis (Douglas and Poisner, 1964; Dreifuss, 1975; Dreifuss et al., 1971; Sachs, 1967; Scharrer and Scharrer, 1954; Verney, 1947); the physiological stimuli to ADH release (Berl et al., 1974; Dunn et al., 1973; Goetz et al., 1974; Leaf and Frazier, 1961; Robertson, 1974; Robertson and Athar, 1976; Robertson et al., 1977; Schrier et al., 1977; Verney, 1947); renal concentrating and diluting mechanisms (Berliner, 1976; Berliner and Bennett, 1967; Gottschalk, 1961, 1964; Kokko and Rector, 1972; Marsh and Azen, 1975; Pennell et al., 1974; Rector, 1977; Stephenson, 1972); the effects of cAMP on the kinetics of protein phosphorylation–dephosphorylation reactions in apical plasma membranes (Forte et al., 1972; Handler and Orloff, 1963, 1971; Kinne and Schwartz, 1977; Michelakis, 1970; Schwartz et al., 1974); the effects of cAMP on intracellular microtubule and microfilament assemblies (Carasso et al., 1973; Dousa and Barnes, 1974;

as with synthetic ADH analogs. In this review, the term ADH will be utilized to connote: arginine vasopressin, the naturally occurring antidiuretic hormone in most mammals; Pitressin, a commercially available mammalian neurohypophyseal extract containing predominantly arginine vasopressin; lysine vasopressin; and arginine vasotocin (Sawyer, 1967).

The surfaces of ADH-responsive cells frequently have a different nomenclature, depending on the epithelium studied. We shall use the terms *apical membranes* and *basolateral membranes* to indicate, respectively, epithelial cell membranes continuous only with junctional complexes and those adjacent either to basement membranes or to lateral intercellular spaces.

Finally, it is widely accepted that the effects of ADH on transport processes in hormone-sensitive epithelia are mediated in accord with the second messenger hypothesis (Sutherland, 1961–1962): the hormone binds to receptors on basolateral surfaces of responsive epithelial cells and results in an adenyl cyclase-mediated acceleration of cAMP synthesis from ATP. Subsequently, cAMP effects changes—involving possibly the kinetics of apical membrane-bound protein phosphorylation–dephosphorylation reactions, and/or cellular microtubule and microfilament structures—that modify transport processes in apical membranes. In this paper, the terms *ADH-dependent* or *ADH-mediated* will be used interchangeably to signify those transport processes in apical plasma membranes that are affected by, and/or dependent on, exposure of basolateral membranes of the epithelium to the hormone.

III. Some Biophysical Considerations

A. MEASUREMENT OF WATER AND SOLUTE TRANSPORT ACROSS MEMBRANES

Two methods of assessing water permeation across membranes have been widely utilized. In the first method, net water flow across a membrane is measured when either hydrostatic and/or osmotic pressure gradients exist across the membrane, and J_v (cm^3 sec^{-1} cm^{-2}), the volume flow across the membrane, is given by the Starling equation:

$$J_v = L_p(\Delta\pi - \Delta P) \tag{1}$$

where L_p (cm sec^{-1} atm^{-1}) is the coefficient of hydraulic conductivity, ΔP (atm)

Taylor, 1977; Taylor *et al.*, 1973); and the effects of ADH on the physical and structural characteristics of apical plasma membranes in amphibian epithelia (Grantham, 1970; Kachadorian *et al.*, 1975).

is the hydrostatic pressure gradient, and $\Delta\pi$ (atm), the osmotic pressure gradient, is

$$\Delta\pi = RT \sum_{i=1}^{n} \sigma_i \Delta C_i \qquad (2)$$

where R is the gas constant, T is absolute temperature, σ_i is the reflection coefficient of the ith solute (see below), and ΔC_i is the transmembrane concentration difference of the ith solute. For convenience, L_p may be expressed as P_f, a water permeation coefficient having the dimensions cm sec^{-1}, by the relationship

$$P_f = L_p(RT/\bar{V}_w) \qquad (3)$$

where $\bar{V}_w$ (cm^3 mole^{-1}) is the partial molar volume of water.

A second method of assessing water permeation depends on measuring the diffusion of labeled water (for example, THO) across a membrane. In an ideal water-diffusion experiment, both solutions bathing a membrane are identical except for the tracer water concentration, and net volume flow is zero. Under these conditions, the diffusional water permeability coefficient P_{D_w} (cm sec^{-1}) may be computed from Fick's first law:

$$P_{D_w} = -J^*_{1\to 2}/\Delta C^* \qquad (4)$$

where $J^*_{1\to 2}$ is the flux of tracer water (cpm sec^{-1} cm^{-2}) from solution 1 to solution 2, and ΔC^* is the concentration difference of tracer water (cpm cm^{-3}) between solutions 2 and 1. For convenience, solution 2 ordinarily contains little or no tracer with respect to solution 1. The diffusional permeability coefficient for the ith solute (P_{D_i}, cm sec^{-1}) may likewise be measured from the tracer flux of a labeled solute species, for example, [^{14}C] butyramide.

B. ANALYSIS OF P_f, P_{D_w}, and P_{D_i} VALUES

In principle, P_f will equal P_{D_w} when net water transport across a given membrane occurs exclusively by a solubility-diffusion process: thus, if P_f and P_{D_w} are identical, it may be argued that the rate of water diffusion across a membrane (estimated from the diffusion of tracer water at zero volume flow) is sufficient to account for the net flux of water occurring across a membrane in the presence of osmotic and/or hydrostatic gradients across that membrane. But in the vast majority of natural and synthetic membranes, P_f exceeds P_{D_w}, i.e., the P_f/P_{D_w} ratio exceeds unity (Andreoli and Schafer, 1976). Two classes of explanations have been set forth to account for this discrepancy: one theory argues that, during net volume flow, there occurs bulk, or Poiseuille-type, water flow through membrane pores; the other theory posits the presence of stagnant, or unstirred, layers adjacent to the membrane, which offer an appreciable resistance to water diffusion at zero volume flow, but act as negligible resistances to net volume flow. We now consider these issues in further detail.

1. The Pore Argument

Kocfoed-Johnsen and Ussing (1953) and Pappenheimer (1953) reasoned that membranes might contain aqueous pores sufficiently large to permit laminar net volume flow. In aqueous cylinders, laminar, or Poiseuille, flow is proportional to r^4, where r is the tube radius, while diffusion is proportional to r^2. Accordingly, in a porous membrane, P_f will exceed P_{D_w}. A number of workers have pointed out that, by combining Poiseuille's law for laminar flow with Fick's first law of diffusion, the relationship between P_f and P_{D_w} for a porous membrane may be expressed as (Koefoed-Johnsen and Ussing, 1964; Pappenheimer, 1964; Robbins and Mauro, 1960):

$$P_f/P_{D_w} = 1 + [RT/(8\eta D_w^o \bar{V}_w)] r^2 \qquad (5)$$

where the pore radius r is given in Ångstrom units (10^{-8} cm), η is aqueous viscosity (poise), and D_w^o is the free diffusion coefficient of water in water (cm² sec⁻¹). At 25°C, the term $RT/8\eta D_w^o \bar{V}_w$ for relatively dilute aqueous solutions may be expressed as:

$$RT/(8\eta D_w^o) \simeq 0.08 \text{ Å}^{-2}$$

Thus Eq. (5) may, for practical purposes, be rewritten as

$$P_f/P_{D_w} = 1 + 0.08 \, r^2 \qquad (6)$$

It is clear from Eq. (6) that, for pore radii $\geqslant 3$ Å, the P_f/P_{D_w} ratio will be $\geqslant 1.7$, i.e., a P_f/P_{D_w} discrepancy that ought be discernible experimentally.

Laminar water flow through membrane pores may also affect solute transport. In this regard, consider the thermodynamic term, reflection coefficient (σ_i, for the ith solute): a solute completely excluded, or reflected, from a membrane has a unity reflection coefficient; but if the membrane is coarsely selective (i.e., if the membrane does not distinguish between a solute species and water), the solute is assigned a zero reflection coefficient. Thus the reflection coefficient provides an index of the degree to which a given membrane distinguishes between water and solute molecules, and hence, as indicated in Eq. (2), defines the "effective" osmotic pressure produced by a given solute concentration gradient across a given membrane.

Koefoed-Johnsen and Ussing (1953) pointed out that, in a porous membrane, solutes having reflection coefficients less than unity, i.e., solutes able to enter membrane pores, would exhibit an acceleration of flow in the same direction, and a retardation in the opposite direction, of solvent flow. Koefoed-Johnsen and Ussing (1953) termed this entrainment of solute and solvent flows within membrane pores the "solvent drag" effect. Quantitatively, the contributions of entrainment to net solute flow may be described by the expression (Kedem and Katchalsky, 1958):

$$J_i^e = J_v(1 - \sigma_i)[(C_i^1 + C_i^2)/2] \tag{7}$$

where J_i^e is the entrained component of the net flux of the ith solute, and $[(C_i^1 + C_i^2)/2]$ is the mean of the concentrations of the ith solute in the two aqueous phases bathing the membrane.

It is somewhat surprising, yet at the same time gratifying, to note that Eqs. (5)–(7), which derive from expressions formulated with macroscopic coefficients (i.e., the relationship between bulk-phase viscosity and the pore radius, both in the Fick equation and in Poiseuille's law), describe accurately water and nonelectrolyte transport in pores having molecular dimensions. Thus in ~ 60 Å-thick lipid bilayer membranes, the polyene antibiotic amphotericin B and cholesterol interact to form multimolecular aggregates with the characteristics of pores having ~ 4 Å radii (Andreoli, 1973; Andreoli et al., 1969; Holz and Finkelstein, 1970). And within such porous membranes (after appropriate corrections are made for unstirred layer effects; see below): the experimentally observed P_f/P_{D_w} ratio agrees well with that predicted from Eq. 6 for a 4-Å pore radius (Andreoli, 1973; Holz and Finkelstein, 1970); and likewise, the experimentally observed effect of volume flow on net urea flux [in such porous membranes, σ_{urea} is 0.31 (Andreoli et al., 1971)] may be accounted for quantitatively in terms of Eq. (7). Moreover, in human erythrocytes, both the experimentally observed P_f/P_{D_w} ratio and the degree of molecular seiving for hydrophilic nonelectrolytes (i.e., the variation of σ_i with the molecular radius of a nonelectrolyte), are consistent with the view that plasma membranes of red blood cells contain the equivalent of 4-Å pores (Solomon and Gary-Bobo, 1972).

2. The Unstirred Layer Problem

Alternatively, virtually all transport phenomena presumed to be referable to membrane pores—i.e., P_f/P_{D_w} ratios in excess of unity, the solvent drag effect, and even electrokinetic phenomena, such as streaming potentials or electroosmosis—may be apparent rather than real, if there are unstirred layers adjacent to a membrane. These layers may be viewed picturesquely as unmixed, diffusion-limited regions of water interposed between a membrane surface and a well mixed solution. Defined in this manner, an unstirred layer forms a resistance for diffusion from the well mixed solution to the membrane interface, but is a negligible constraint to laminar flow.

It is interesting to note that the unstirred layer problem is not a newcomer. In 1904, Nernst called attention to the problem of boundary layers, and the possible rate-limiting effects such layers might exert on any process involving interactions between heterogeneous systems (e.g., an aqueous solution and a lipid membrane). And in 1936, Teorell brought the unstirred layer issue into sharp focus: he cautioned, "Ultimately every attempt to study what is hidden behind the terms penetration and permeability has to face the conditions within diffusion layers." Despite these caveats, the unstirred layer problem was largely ignored by transport physiologists until 1963, when Dainty pointed out that unstirred layer effects, rather than pores, might account for P_f/P_{D_w} ratios in excess of unity in biological or synthetic membranes.

It now seems feasible to consider two different types of unstirred layers (Schafer et al., 1974a) depending on the geometric constraints to flow-diffusion processes within the latter. In the first kind, which we (Schafer et al., 1974a) have termed type A, the area available for flow-diffusion processes is approximately equal to the area of the membrane with which the unstirred layer interfaces. This type of unstirred layer occurs, for example, in series with synthetic membranes (Andreoli and Troutman, 1971; Cass and Finkelstein, 1967; Ginzburg and Katchalsky, 1963; Hanai and Haydon, 1966), various epithelia (Dainty and House, 1966a; Diamond, 1966; Green and Otori, 1970; Hays and Franki, 1970), and—parenthetically an issue of major importance to clinicians—in most hemodialysis machines, where approximately 80% of the total resistance to the dialysance of molecules such as urea and NaCl is referable to unstirred layers in the blood adjacent to the hemodialyzer membrane (Colton, 1967). In certain instances, the thickness of a type A unstirred layer may be reduced by more effective stirring of the bulk solutions (Cass and Finkelstein, 1967; Colton, 1967; Dainty and House, 1966a; Ginzburg and Katchalsky, 1963; Green and Otori, 1970; Hays and Franki, 1970). Table I summarizes the effective unstirred thicknesses that have been measured for various type A unstirred layers, either in series with lipid bilayer membranes or with ADH-sensitive epithelia; omitted from Table I is the mammalian cortical collecting tubule,

TABLE I
Type A Unstirred Layers Associated with Synthetic Membranes and with Antidiuretic Hormone Sensitive Epithelia[a]

Preparation	Total	Apical	Basolateral	References[b]
Lipid bilayer membrane	100–125	–	–	1, 2
Frog skin	130–290	30–60	100–230	3, 4
Toad bladder	1055	–	236	5

Unstirred layer thickness (cm sec^{-1} × 10^4)

[a]Adapted from Andreoli and Schafer (1976).
[b]Key to references: (1) Andreoli and Troutman, 1971. (2) Cass and Finkelstein, 1967. (3) Dainty and House, 1966a. (4) Dainty and House, 1966b. (5) Hays and Franki, 1970.

since the possibility of unstirred layers in series with this preparation will be treated in detail in a subsequent section.

A membrane of given area may also be in series with an unstirred layer having a smaller effective area within which transport processes occur: we have termed (Schafer *et al.*, 1974a) such an unstirred layer type B. Clearly, for the steady-state transport of water between two aqueous phases separated by a membrane in series with a type B unstirred layer, the velocity of water flow in the latter will vary with the fractional area available for flow.

It is particularly relevant to note that, in many instances, a type A unstirred layer represents a vexing constraint to experimental design. For example, bulk phase unstirred layers adjacent to a synthetic lipid bilayer membrane are not necessary determinants of transport events across the membrane; rather, because the investigator cannot stir the bulk phases sufficiently well to eliminate the unstirred layers, the latter offer constraints to water and solute transport above and beyond those inherent in the membrane. Evidently, these constraints must be considered explicitly in order to make accurate deductions about transport processes within the membrane.

On the other hand, a type B unstirred layer may be a necessary determinant of transport processes of biological relevance. Consider, for example, the functional organization of transporting epithelia (Farquhar and Palade, 1963): the apical plasma membranes are fused to form junctional complexes; and immediately adjacent to junctional complexes, the basolateral plasma membranes of adjacent cells are separated by intercellular channels which represent, with respect to apical plasma membranes, type B unstirred layers. It is widely believed that transepithelial solute and water transport processes in many epithelia—notably the small intestine and the proximal renal tubule—involve, at least in part, transport through intercellular spaces. Thus it is evident that a detailed understanding of the physiology of epithelia depends on a careful assessment of the diffusion constraints within intercellular spaces. Indeed, as we shall argue below, it may also be that an understanding of the mode of water transfer through ADH-sensitive epithelia requires a consideration of the cytoplasm of apical cells as a type B unstirred layer in series with apical plasma membranes.

We now consider explicitly the ways in which unstirred layers may affect measurements of water and solute transport processes. As stressed originally by Dainty (1963), net water transport across a membrane may be entirely diffusional in nature, but because of unstirred layers, P_f excceds P_{D_w}. To develop this issue, we note that $1/P_{D_w}$ is the total resistance to tracer water diffusion. Thus for a membrane in series with unstirred layers, we have

$$1/P_{D_w} = 1/P_w^m + \alpha/D_w^o \tag{8}$$

where P_w^m is the true permeability coefficient for water diffusion across the membrane, and α is the sum of the effective thicknesses of the unstirred layers on either side of a membrane. Clearly, it is difficult to envision that, in bulk solutions, there exists a discrete boundary between stirred and unstirred regions. Thus the α term in Eq. (8) should be considered as an operational quantity that defines the effective diffusional resistance of stagnant regions adjacent to a membrane by assuming that the frictional constraints to diffusion in stagnant regions are the same as those in bulk solutions.

Dainty (1963) further considered that unstirred layers offered little resistance to net water flow, and that volume flow through unstirred layers would have a negligible effect on the solute concentration profiles within unstirred layers. Given these assumptions, P_f becomes an accurate index of P_w^m in a membrane where net water transport is diffusional, and Eq. (8) may be rewritten as:

$$1/P_{D_w} = 1/P_f + \alpha/D_w^o \qquad (9)$$

Equation (9) indicates that, for entirely diffusional osmotic water flow in a nonporous membrane, P_f exceeds P_{D_w} when the α/D_w^o term becomes large with respect to $1/P_f$. The validity of Eq. (9) has been verified in certain instances. In synthetic lipid bilayer membranes, where the mode of net water tranport is almost certainly diffusional (Andreoli and Troutman, 1971; Cass and Finkelstein, 1967; Hanai and Haydon, 1966), vigorous aqueous phase stirring reduces P_f/P_{D_w} ratios to unity (Cass and Finkelstein, 1967); and, in the presence of unstirred layers, the appropriate value of α (estimated independently) accounts quantitatively for discrepancies between P_f and P_{D_w} (Andreoli and Troutman, 1971).

In epithelia, however, the problem is more complex. Wright *et al.* (1972) recently observed that initial rates of osmotic water flow in rabbit gallbladder were considerably greater than the steady-state values. These workers deduced that the discrepancy between initial and steady-state osmotic flows was the consequence of flow-dependent changes in the concentration of solutes in the lateral intercellular spaces, and indicated the possibility that, as a result of such solute polarization, steady-state osmotic flow measurements in epithelia might result in significant underestimates of P_f. In order to consider the effects of solute polarization within unstirred layers on P_f determinations in epithelia, we have recently analyzed (Schafer *et al.*, 1974a) quantitatively the effects of unstirred layers and unstirred layer geometry on transient and steady-state osmotic volume flows. The relevant equations are given by Schafer *et al.* (1974a); the present paper sets out a qualitative description of the processes.

Consider a membrane separating two solutions, I and II; solution I contains a relatively small unstirred layer of thickness β, and solution II contains a relatively large unstirred layer of thickness γ, thus α in Eqs. (8) and (9) equals (β + γ). Both solutions initially contain water; and at zero time, a solute having a unity reflection coefficient is added either to solution I or to solution II.

Figure 2 illustrates schematically the temporal sequence of osmotic volume flow for these two circumstances. When solute is added to solution I, which contains the thinner unstirred layer β, the flow from II to I peaks rapidly and declines monotonically to a steady-state value. The initial rise in $J_v^{II \to I}$ is referable to a rapid diffusion of solute from solution I through β, which is relatively thin, to the membrane interface. Subsequently, volume flow through the unstirred layers decreases the solute concentration at the interface between the membrane and unstirred layer β; or, stated in another way, the sweeping-away effect of volume flow within the unstirred layer β results in a steady-state solute concentration adjacent to the membrane which is less than that in bulk solution I. As a consequence, $J_v^{II \to I}$ declines to a steady-state value. And the value of P_f, estimated according to Eqs. (1)–(3) from the steady-state $J_v^{II \to I}$, is erroneously low.

The situation is quite different when the solute is added to solution II. In this instance, the time required for solute diffusion from solution II to the membrane interface is considerably greater than from solution I to the interface, since γ is thicker than β. Consequently, an initial peak does not occur, and $J_v^{I \to II}$ increases monotonically to the steady state. But again, because of the sweeping-away effect of volume flow within the unstirred layers, the solute concentration at the interface between the membrane and the unstirred layer γ is less than in bulk solution II, the steady-state value of $J_v^{I \to II}$ is less than the peak value of $J_v^{II \to I}$, and the value of P_f computed according to Eqs. (1)–(3) from the steady-state $J_v^{I \to II}$ is appreciably less than the true P_f.

It is evident from these considerations that an assessment of the mode of water transport across membranes requires not only the traditional assessment of P_f

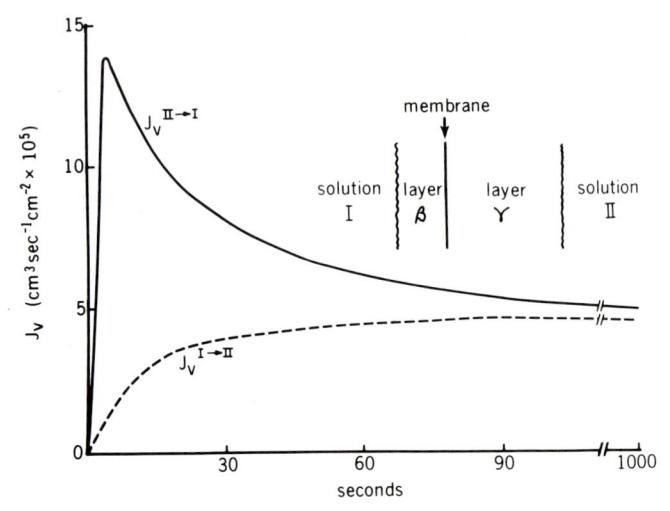

FIG. 2. A schematic illustration of the osmotic transient phenomenon. Adapted from Schafer *et al.* (1974a).

and P_{D_w}, but also an explicit assessment of the effects of unstirred layers on P_{D_w} and a quantitative analysis of the osmotic transient phenomenon (Fig. 2) with respect to the accuracy of P_f determinations. Indeed it is probable that, in rather water-permeable epithelia such as the gallbladder, current estimates of P_f may underestimate the true osmotic water permeability by as much as 10-fold (Schafer *et al.*, 1974a; Wright *et al.*, 1972).

Finally, acceleration of solute flow in the direction of solvent flow may occur when unstirred layers are in series with a nonporous membrane: in such a circumstance, the increment in solute flux produced by solvent flow is not the consequence of coupling solute and solvent flows within membrane pores, but rather the result of flow-dependent changes in solute concentration processes within the unstirred layers.

Figure 3 illustrates such a process schematically. A homogeneous, nonporous membrane is bounded on either side by an unstirred layer having a thickness $\alpha/2$. Solution I and solution II each contain identical concentrations of the ith nonelectrolyte having a unity reflection coefficient; thus from Eq. (7), J_i^e, the entrained component of the ith solute flux, is zero by definition. And since the ith solute concentrations in solutions I and II are equal, the net diffusional flux of solute sould also be zero. Now if a second impermeant solute is added to solution II, thereby producing osmotic volume flow from I to II, solute polarization within the unstirred layers results in a concentration gradient for the ith solute across the membrane interfaces. Accordingly, there will be a net flux of the ith solute in the direction of solvent flow, but the flux is due to diffusion rather than to a real solvent drag effect. A quantitative account of the ways in which one can distinguish between the contributions of unstirred layers and

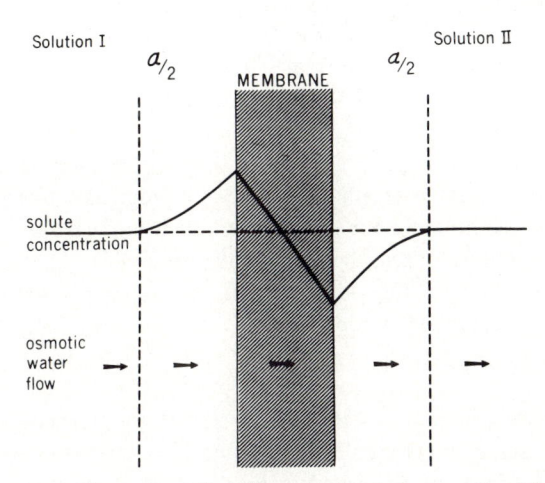

FIG. 3. A schematic illustration of artifactual solvent drag produced by unstirred layer effects.

solvent drag to the fluxes of solutes across membranes in series with unstirred layers has been presented elsewhere (Andreoli and Shafer, 1976; Andreoli et al., 1971).

IV. The Effects of ADH on Water and Solute Permeation Rates in Mammalian Collecting Tubules

We now consider the effects of ADH on water and solute transport coefficients. In 1966, Burg and his co-workers described a technique for perfusing individual nephron segments isolated by freehand dissection of rabbit kidney. Thus it became possible to evaluate the effects of ADH on transport processes in discrete segments of the mammalian nephron. In the same year, Grantham and Burg (1966) documented clearly the ADH-dependent increases in P_f and P_{D_w}, but not in urea permeation, occurring in cortical collecting tubules. This section will be concerned with a summary of the effects of ADH on water and solute transport coefficients in mammalian cortical and outer medullary collecting ducts obtained from rabbit kidney and microperfused according to the Burg technique.

A. THE RATE-LIMITING SURFACE FOR WATER AND SOLUTE PERMEATION

Grantham and Burg (1966) first showed that cortical collecting tubules perfused *in vitro* became progressively impermeable to water in the absence of ADH: after 3 hours of perfusion, both P_f and P_{D_w} reached stable minima; and, when ADH was added to the bathing media, both P_f and P_{D_w} rose promptly. Figure 4 illustrates the morphologic correlates obtained in our laboratory (Schafer and Andreoli, 1972a) for such observations. When cortical collecting tubules were perfused and bathed with, respectively, hypotonic and isotonic Krebs–Ringer buffers (in which NaCl accounts for the majority of solution osmolality) in the absence of ADH, the tubular epithelium gradually became flattened, and the lateral intercellular spaces were progressively less visible. When ADH was added to the bath, the cells swelled, their luminal borders bulged, and the lateral intercellular spaces became distinctly visible. In other words, the results in Fig. 4 illustrate that luminal surfaces (i.e., apical plasma membranes and/or junctional complexes) constitute the ADH-sensitive interface for water transport, and, in the absence of ADH, the rate-limiting site for water permeation.

Luminal surfaces are also the rate-limiting site for the transport of hydrophilic nonelectrolytes across cortical collecting tubules, both in the presence and in the absence of ADH. Thus, as illustrated in Fig. 5, when cortical collecting tubules were perfused and bathed with isosmotic Krebs–Ringer solutions, replacement

Lumen
125 mOsm/liter KRP

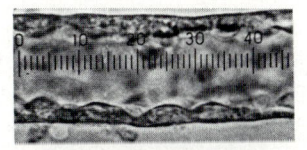

Bath
290 mOsm/liter KRB

76 minutes

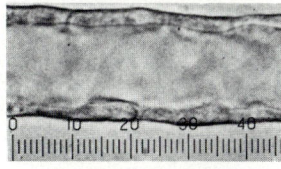

186 minutes

200 minutes: bath ADH = 250 μU/ml

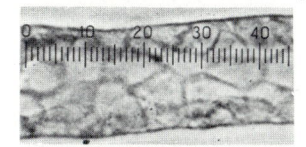

210 minutes

FIG. 4. The rate-limiting site for water permeation. The segment of cortical collecting tubule was perfused with 125 mOsm liter^{-1} Krebs–Ringer phosphate buffer (KRP) and bathed in 290 mOsm liter^{-1} Krebs–Ringer bicarbonate buffer (KRB) at 25°C. The elapsed time after the rabbit was sacrificed is shown below each photomicrograph. Antidiuretic hormone (ADH) was added to the bathing solution at 200 minutes. The interval between small divisions in the photomicrograph scale is 2.13 μm. Adapted from Schafer and Andreoli (1972a).

of a portion of NaCl in the perfusing solutions with urea had no effect on tubular morphology; but when urea replaced a portion of the NaCl in the bathing solutions, the tubular cells swelled rapidly. One may conclude from these observations that, even in the presence of ADH, luminal surfaces of cortical collecting tubules are rather impermeable to urea, especially in comparison to basolateral membranes (Andreoli and Schafer, 1976; Schafer and Andreoli, 1972b).

B. EFFECT OF ADH ON WATER AND HYDROPHILIC SOLUTE TRANSPORT COEFFICIENTS

A summary of quantitative estimates on the effects of ADH on water permeation in cortical collecting tubules is presented in Table II: ADH produced approximately a 10-fold increase in P_f and about a 4-fold increment in P_{D_w};

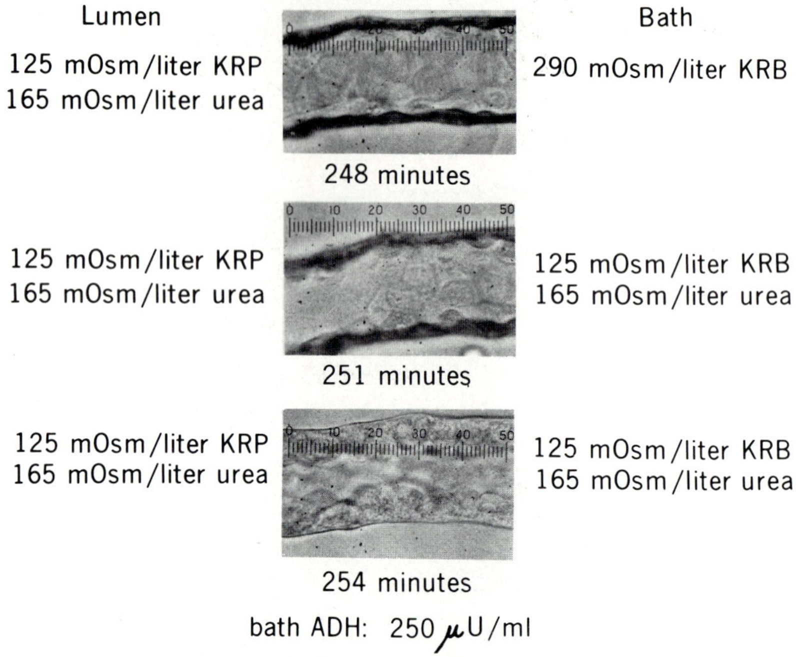

Lumen		Bath
125 mOsm/liter KRP 165 mOsm/liter urea	248 minutes	290 mOsm/liter KRB
125 mOsm/liter KRP 165 mOsm/liter urea	251 minutes	125 mOsm/liter KRB 165 mOsm/liter urea
125 mOsm/liter KRP 165 mOsm/liter urea	254 minutes	125 mOsm/liter KRB 165 mOsm/liter urea

bath ADH: 250 μU/ml

FIG. 5. The rate-limiting site for urea permeation in the cortical collecting tubule. The perfusing solution was 125 mOsm liter^{-1} Krebs–Ringer phosphate buffer (KRP) plus 165 mOsm liter^{-1} urea throughout the experiment. The bathing solution was initially 290 mOsm liter^{-1} Krebs–Ringer bicarbonate buffer (KRB), but was changed to 125 mOsm liter^{-1} KRB plus 165 mOsm liter^{-1} urea at 250 minutes after the rabbit was sacrificed. The bathing solution contained 250 μU ml^{-1} of ADH throughout the experiment. The interval between small divisions in the photomicrograph scale is 2.13 μm. Adapted from Schafer and Andreoli (1972b).

thus the hormone increased the P_f/P_{D_w} ratio from 4.25 to 13.1. As indicated in Table II, P_f was measured from steady-state lumen to bath osmotic volume flows when the lumen and bath were, respectively, either hypotonic and isotonic or isotonic and hypertonic.

In contrast, at least two lines of evidence indicate the hormone has little effect on the transepithelial permeation rates for hydrophilic solutes across these tubules. Burg et al. (1970; Grantham and Burg, 1966) observed that, both in the presence and in the absence of ADH, the permeability coefficients for small hydrophilic species, such as urea, thiourea, and acetamide, were in the range of 0.03–0.05 × 10^{-4} cm sec^{-1} (i.e., vanishingly small with respect to the ADH-dependent values of P_f and P_{D_w}) and not detectably affected by ADH. Our laboratory has confirmed these observations (Table III). We have also found that outer medullary collecting tubules had permeability coefficients for urea and

TABLE II
The Effect of Antidiuretic Hormone (ADH) on P_f and P_{D_w} in Cortical Collecting Tubules[a]

Bath [ADH] (μU ml^{-1})	P_f (cm sec^{-1} × 10^4)	P_{D_w} (cm sec^{-1} × 10^4)
0	20 ± 4.0	4.7 ± 0.5
	(n = 34)	(n = 7)
250	186 ± 8.9	14.2 ± 0.6
	(n = 18)	(n = 7)

[a] P_f, the osmotic water permeability coefficient, was computed from lumen to bath osmotic volume fluxes in isolated cortical collecting tubules under either of two sets of conditions: when the perfusate and bath contained, respectively, 125 mOsm liter^{-1} Krebs–Ringer phosphate buffer (KRP) and 290 mOsm liter^{-1} Krebs–Ringer bicarbonate buffer (KRB); or, when the perfusate and bath contained, respectively, 290 mOsm liter^{-1}; KRP and 455 mOsm liter^{-1} KRB. P_{D_w}, the diffusional water-permeability coefficient, was measured from lumen to bath THO fluxes in isolated cortical collecting tubules at zero volume flow: the perfusate and bath contained, respectively, 290 mOsm liter^{-1} KRP and 290 mOsm liter^{-1} KRB. The results are expressed as mean values ± SEM for the numbers of tubules indicated in parentheses. All experiments were carried out at 25 ± 0.5°C. Adapted from Schafer and Andreoli (1972a) and Schafer et al. (1974b).

thiourea that were both ADH-unresponsive and as small as the comparable values in cortical collecting tubules.[2]

Moreover, the osmotic "effectiveness" of various solutes was assessed by comparing lumen to bath osmotic volume flows produced by a given NaCl concentration (when the perfusate and bath contained, respectively, lower and higher concentrations of Krebs–Ringer buffer, e.g., Fig. 4) with lumen to bath osmotic flows produced when a portion of the NaCl in the Krebs–Ringer bathing solutions was replaced with a test solute. Thus by utilizing Eqs. (1)–(3), it was possible to measure the ratio P_f^a/P_f^b, where a and b refer to the solutes tested.

The results, shown in Table IV, indicate that, in the presence of ADH, urea, sucrose, and NaCl were equally effective in producing transepithelial osmotic

---

[2] It should be noted that not all segments of the mammalian collecting duct have such low transepithelial urea permeation rates. In the *in vitro* papillary collecting duct, Rocha and Kokko (1974) found the transepithelial urea permeability coefficient to be 0.22 × 10^{-4} cm sec^{-1}. Likewise, Morgan et al. (1968) found the reflection coefficient of urea to be 0.4 in the inner medullary collecting duct. At present, no information is available that permits an insight into the reasons for these differences in urea permeation rates in papillary with respect to either cortical or outer medullary collecting ducts.

TABLE III
The Effect of Antidiuretic Hormone (ADH) on Amide Permeability Coefficients in Cortical Collecting Tubules[a]

Solute	ADH (μU ml^{-1})	P_{Di} (cm sec^{-1} $\times$ 10^4)
Urea	0	0.03
	250	0.02
Thiourea	0	0.03
	250	0.02

[a] P_{Di} for each of the test species was measured from unidirectional ^{14}C-tracer fluxes at zero volume flow. Adapted from Schafer and Andreoli (1972b).

water flow. If it is assumed that L_p in Eq. (1) is a constant, these data indicate that the reflection coefficients (Eq. 2) of NaCl, urea, and sucrose were indistinguishable. Since the effective hydrodynamic radii of urea (radius ~ 2.0 Å) and sucrose (radius ~ 5.2 Å) differ by a factor of nearly 3-fold, these observations are consistent with two general conclusions: no detectable degree of molecular sieving for hydrophilic solutes occurred across luminal surfaces (i.e., apical membranes and/or junctional complexes), even in the presence of ADH; and the ADH-dependent reflection coefficients of NaCl, urea, and sucrose across the luminal membranes of these tubules were each unity.

Alternatively, the reflection coefficient for the highly lipophilic solute butanol was zero in these tubules, both with and without ADH (Schafer and Andreoli,

TABLE IV
The Antidiuretic Hormone (ADH)-Dependent P_f^a/P_f^b Ratio in Cortical Collecting Tubules[a]

Solutes a/b	P_f^a/P_f^b
Urea/NaCl	1.01 ± 0.04 (n = 7)
Urea/sucrose	0.97 ± 0.04 (n = 4)
Sucrose/NaCl	1.08 ± 0.05 (n = 5)

[a] P_f was measured from lumen to bath osmotic volume flows at 25 ± 0.5°C. The perfusate contained 125 mOsm liter^{-1} of Krebs–Ringer phosphate buffer. The bath contained either 290 mOsm liter^{-1} of Krebs–Ringer bicarbonate (KRB) buffer, or 125 mOsm liter^{-1} KRB buffer plus 165 mOsm liter^{-1} test solute (either urea or sucrose). The results are expressed as mean values ± SEM for the numbers of tubules indicated in parentheses. Adapted from Schafer and Andreoli (1972b).

1972b). This result is not surprising since, in synthetic lipid bilayer membranes, the permeability coefficients for highly lipophilic species are sufficiently high that tracer diffusion rates for such molecules are entirely limited by bulk-phase unstirred layers, rather than bilayer membranes (Andreoli and Troutman, 1971; Holz and Finkelstein, 1970; Schafer and Andreoli, 1972a). In fact, Läuger et al. (1967) have deduced that the permeability coefficients for highly lipophilic solutes in lipid bilayer membranes may be in excess of 10 cm sec^{-1}, i.e., more than 10^2-fold greater than the ADH-dependent values of P_f in cortical collecting tubules.

C. PATHWAY FOR WATER PERMEATION: APICAL MEMBRANES OR JUNCTIONAL COMPLEXES

From experiments of the type illustrated in Fig. 4, it was possible to infer that luminal surfaces of cortical collecting tubules constituted the rate-limiting site for water transport, and the final locus of action of ADH-mediated effects on transepithelial water permeation (Ganote et al., 1968; Grantham et al., 1969; Schafer and Andreoli, 1972a). However, these data provide no information concerning the fractional area available for osmotic water flow in luminal surfaces, or, in other words, whether osmotic volume flow traversed luminal plasma membranes or junctional complexes. To evaluate this issue, we assessed the effects of luminal hypertonicity on ADH-independent water and solute permeation in cortical collecting tubules.

In this regard, it is now well recognized that apical hypertonicity has differing effects on tissue permeability, depending on the type of epithelium studied. In so-called "leaky" epithelia (Frömter and Diamond, 1972) having relatively low transepithelial electrical resistances (6–100 ohm-cm^2), such as the mammalian proximal tubule, gallbladder, or small intestine, the responses to apical hypertonicity include a collapse of intercellular spaces (Loeschke et al., 1970; Smulders et al., 1972; Wright et al., 1972), an increase in transepithelial electrical resistance (Loeschke et al., 1970; Smulders et al., 1972), and a decrease in the transepithelial permeation rates for hydrophilic solutes (Loeschke et al., 1970; Smulders et al., 1972). In contrast, so-called "tight" epithelia having high electrical resistances (300–2000 ohm-cm^2), such as toad urinary bladder or the frog skin, exhibit lower transepithelial resistances (DiBona and Civan, 1973; Lindley et al., 1964; Ussing, 1966; Ussing and Windhager, 1964) and morphologically detectable transformations—presumed to represent widening of junctional complexes (Erlij and Martinez-Palomo, 1972; Ussing and Windhager, 1964)—in response to apical solution hypertonicity. In short, apical solution hypertonicity decreases the conductance of the extracellular pathway in electrically leaky epithelia and increases the conductance of the extracellular pathway in electrically tight epithelia.

One other effect of apical solution hypertonicity on transport events in electrically tight epithelia is particularly noteworthy. Ussing (1966, 1969) observed that, when apical and basolateral surfaces of amphibian skin were exposed to, respectively, hypertonic and isotonic solutions, net osmotic volume flow from basolateral to apical solutions was accompanied by increases in the passive permeability of the skin to hydrophilic solutes. Flux asymmetry for solutes also occurred; i.e., despite the fact that net osmotic volume flow was in the outward direction, inward sucrose flux exceeded outward sucrose flux when the inner and outer bathing solutions contained identical sucrose solutions. Subsequently, van Bruggen et al. (Franz and van Bruggen, 1967; Franz et al. 1968; Galey and van Bruggen, 1970) and Biber and Curran (1968) confirmed these observations in various amphibian epithelia.

Isolated rabbit cortical collecting tubules have transepithelial resistances of approximately 850 ohm-cm^2 (Helman et al., 1971), and for convenience, may be classed as electrically tight epithelia. Table V summarizes the effects of varying the tonicity of luminal solutions on the water and urea permeability properties of these tubules in the absence of ADH. Two major results are evident. First, there was appreciable rectification of osmotic volume flow: the P_f for bath to lumen osmotic flows (83×10^{-4} cm sec^{-1}) was more than four times greater than the P_f (20×10^{-4} cm sec^{-1}) for lumen to bath osmotic flows. Second, when the luminal solutions were made hypertonic with urea, the permeation coefficient for urea in the lumen to bath direction, 0.22×10^{-4} cm sec^{-1}, was more than six times greater than the permeation coefficient for urea in the bath to lumen direction, 0.033×10^{-4} cm sec^{-1}; and the latter was indistinguishable from 0.03×10^{-4} cm sec^{-1}, the urea permeability coefficient measured from [^{14}C] urea lumen to bath fluxes at zero volume flow, either with or without ADH (Table III). In short, Table V indicates that, in the absence of ADH, bath to lumen osmotic volume flow produced by luminal hypertonicity resulted in an increment in urea permeation rates, but only in a direction opposite to osmotic volume flow; flux asymmetry for urea; and significant osmotic flow rectification due to an increase in P_f, bath to lumen, with respect to P_f, lumen to bath.

Two classes of explanations have been set forth to account for solute flux asymmetry associated with apical solution hypertonicity and basolateral to apical solution volume flow. Ussing (1969) proposed that osmotic flow produced by apical hypertonicity increased the leakiness of junctional complexes and created volume flow from cells to intercellular spaces which drained toward the basolateral surface of the epithelium. Alternatively, others (Biber and Curran, 1968; Franz and van Bruggen, 1967; Franz et al., 1968; Galey and van Bruggen, 1970) have suggested that apical hypertonicity and basolateral to apical solution volume flow results in unsealing of junctional complexes; apical solutes that permeate junctional complexes, and have higher apical than basolateral

TABLE V
Effect of Varying Luminal Osmolality on Antidiuretic Hormone (ADH)-Independent Water and Urea Permeation in Cortical Collecting Tubules[a]

Lumen (mOsm liter^{-1})	Bath (mOsm liter^{-1})	P_f (cm sec^{-1} × 10^4) Lumen → bath	Bath → lumen
125 KRP[b]	290 KRB[b]	20 ± 4 (n = 34)	—
290 KRP + 200 urea	290 KRB	—	83 ± 15 (n = 5)

Lumen (mOsm liter^{-1})	Bath (mOsm liter^{-1})	$P_{D_{urea}}$ (cm sec^{-1} × 10^4) Lumen → bath	Bath → lumen
125 KRP	290 KRB	0.045 ± 0.004 (n = 4)	—
290 KRP + 200 urea	290 KRB	0.22 ± 0.02 (n = 4)	0.033 ± 0.002 (n = 4)

[a] P_f was measured from lumen to bath or bath to lumen volume fluxes and $P_{D_{urea}}$ from bath to lumen or lumen to bath [^{14}C]urea fluxes. The data are expressed as mean values ± SEM for the numbers of tubules indicated in parentheses. The experiments were carried out at 25 ± 0.5°C. Adapted from Schafer *et al.* (1974b).

[b] KRP, Krebs–Ringer phosphate buffer; KRB, Krebs–Ringer bicarbonate buffer.

solution concentrations, then move along their chemical gradients and result in flux asymmetry.

In terms of either of the above explanations, acceleration of solute flow in a direction opposite to volume flow produced by apical hypertonicity requires that the solute used to generate apical hypertonicity penetrate junctional complexes. Thus we may interpret the results in Table V to indicate that, when luminal solutions were made hypertonic with urea, the junctional complexes of these cortical collecting tubules were relatively permeable to urea; and the bath to lumen P_f of 83 × 10^{-4} cm sec^{-1} provides an estimate of the water conductance of junctional complexes that were relatively permeable to urea. But this bath to lumen value of P_f is less than half of 186 × 10^{-4} cm sec^{-1} (Table II), the ADH-dependent P_f obtained from lumen to bath osmotic flows under conditions where junctional complexes were impermeant to urea (Table III).

These observations are consistent with the view that the rate of water transport through junctional complexes that are relatively permeable to urea is inadequate to account for the ADH-dependent rate of water transport from lumen to bath under conditions where junctional complexes are virtually closed to urea. Based on these considerations, we argue that ADH-dependent lumen to bath osmosis in mammalian cortical collecting tubules involves water flow primarily through luminal plasma membranes rather than junctional complexes (Andreoli and Schafer, 1976; Schafer et al., 1974b). Models invoking mainly a transcellular rather than an extracellular route for ADH-dependent apical to basolateral solution osmosis have also been proposed by other workers, both for rabbit cortical collecting tubules (Ganote et al., 1968; Grantham et al., 1969) and for ADH-sensitive amphibian epithelia (Civan, 1970; Jard et al., 1971).

D. THE P_f/P_{D_w} RATIO IN CORTICAL COLLECTING TUBULES

The results in Table II indicate that ADH increases P_f, P_{D_w}, and the P_f/P_{D_w} ratio in cortical collecting tubules, where luminal plasma membranes are the rate-limiting site for water permeation (Fig. 4). Thus in view of the considerations raised in Section III, two major issues need to be addressed: the validity, or accuracy, of P_f determinations; and the origin of the discrepancy between P_f and P_{D_w}.

In order to assess the validity of P_f determinations, we carried out a quantitative analysis of the osmotic transient phenomenon (e.g., Fig. 2) and an experimental assessment of the time course of osmotic volume flows in cortical collecting tubules (Schafer et al., 1974a). The cardinal experimental observations are shown in Fig. 6: a rapid-time resolution of osmotic volume flow in these tubules indicated that no osmotic transients were detectable as early as 20 seconds after initiating osmosis; or, stated in another way, the ADH-dependent values of P_f, computed from either initial or steady-state osmotic flows, were the same. A quantitative analysis of the data in Fig. 6, taking into account the known diffusion constraints of the cell layer (see below), indicated that the P_f values in Fig. 6, which are in close agreement with those listed in Table II, were essentially unaffected by solute polarization within the epithelial cell layer (Schafer et al., 1974a). Thus we may argue that, in cortical collecting tubules, P_f provides an accurate measure of the hydraulic conductivity of luminal membranes.

We now consider the disparity between P_f and P_{D_w} (Table II). If the P_f/P_{D_w} discrepancy in these tubules were due to laminar water flow through membrane pores, the ADH-independent and ADH-dependent P_f/P_{D_w} ratios predict, in terms of Eq. 6, that the hormone increases the radii of luminal membrane pores from 6.4 Å to 12.1 Å. But the ADH-dependent reflection coefficient for urea is unity in these tubules (Table IV), and $P_{D_{urea}}$ is unaffected by the hormone

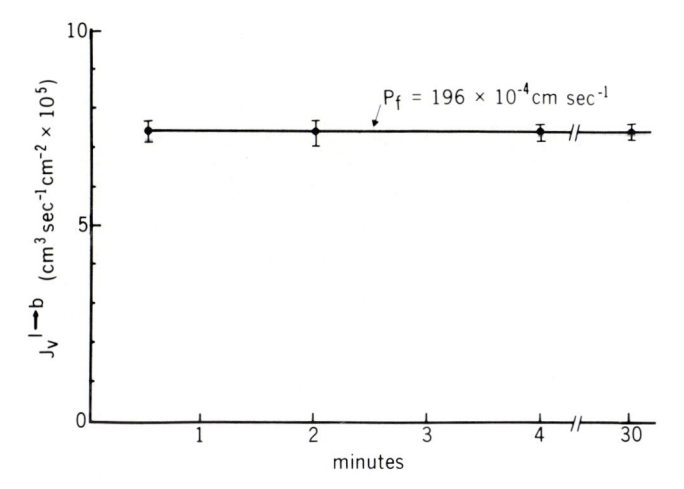

[Figure: plot of $J_v^{l \to b}$ (cm³ sec⁻¹ cm⁻² × 10⁵) vs minutes, showing $P_f = 196 \times 10^{-4}$ cm sec⁻¹]

FIG. 6. Rapid time resolution of lumen to bath osmosis in cortical collecting tubules. The lumen and bath both contained 290 mOsm liter⁻¹ Krebs–Ringer buffer prior to zero time; at zero time osmosis was begun by rapidly raising the bath osmolality. The results are expressed as mean values ± SEM ($n = 4$). Adapted from Schafer et al. (1974a).

(Table III). Since the effective hydrodynamic radius of urea is approximately 2.0 Å, one expects from pore theory (Andreoli and Troutman, 1971; Kedem and Katchalsky, 1958; Kofoed-Johnsen and Ussing, 1953; Pappenheimer, 1953; Robbins and Mauro, 1960; Solomon and Gary-Bobo, 1972) that, if ADH increased luminal membrane pore radii from 6.4 Å to 12.1 Å, the ADH-dependent reflection coefficient for urea would be less than unity and the hormone would have increased $P_{D_{urea}}$ appreciably. Thus pore theory per se is not adequate to account for the effects of ADH on water transport across luminal membranes of cortical collecting tubules.

However, unstirred layer effects, due either to bulk-phase unstirred layers or cellular constraints to diffusion, might account for the disparity between P_f and P_{D_w}. Taking D_w° to be 2.36×10^{-5} cm² sec⁻¹ (Wang et al., 1953), Eq. (9) predicts an α value of 153×10^{-4} cm to rationalize the ADH-dependent disparity between P_f and P_{D_w}, i.e., to argue that the route of ADH-dependent osmotic water flow across luminal membranes was entirely diffusional, and that P_{D_w} was less than P_f because of diffusion resistances in series with luminal membranes.

But the epithelium of these tubules consists of a single layer of cells, approximately 6.0 to 7.0×10^{-4} cm in thickness, surrounded by a single basement membrane (Fig. 4). Thus, in contrast to epithelial preparations such as amphibian urinary bladder, amphibian skin, or gallbladder, a consideration of water and solute flows across cortical collecting tubules is not subject to unstirred layer artifacts referable to stromal tissue elements. And the internal diameter of tubules is only 22 to 25 × 10^{-4} cm (Fig. 4). Moreover, 10-fold increments in the viscosity of perfusing and bathing solutions, a maneuver which predictably increases the diffusion resistance of bulk phase unstirred layers in series with synthetic bilayer membranes (Andreoli and Troutman, 1971; Holz and Finkelstein, 1970; Schafer and Andreoli, 1972a), have no effect on the ADH-dependent values of P_{D_w} in cortical collecting tubules (Schafer and Andreoli, 1972a). So it seems improbable that bulk phase unstirred layers contributed to the disparity between P_f and P_{D_w} in these tubules.

Alternatively, the epithelial cell layer itself, exclusive of luminal membranes, might act as a diffusion resistance sufficient to account for the ADH-dependent P_f/P_{D_w} discrepancy. The thickness of cortical collecting tubule epithelium is 6 to 7 × 10^{-4} cm $\sec^{-1}$, and the value of α required from Eq. (9) to account for the difference between the ADH-dependent values of P_f and P_{D_w} is, as indicated above, approximately 153 × 10^{-4} cm. In other words, in order for cellular diffusion constraints to account for the ADH-dependent P_f/P_{D_w} ratio, one requires that the resistance to water diffusion in the epithelial cell layer be 21–25 times greater than in free solution.

To test this possibility, the permeability coefficients of highly lipophilic solutes, such as butanol, 5-hydroxyindole, and pyridine, were measured in cortical collecting tubules. These solutes were chosen for the following reasons: first, each of these solutes has a relatively high oil:water partition coefficient (β_{oil}), i.e., in excess of 0.5 (Collander and Bärlund, 1933), and hence might be expected to permeate readily the hydrophobic regions of luminal membranes; second, in agreement with this view, solutes such as butanol have zero reflection coefficients in cortical collecting tubules (Schafer and Andreoli, 1972b); third, in synthetic bilayer membranes, the permeability coefficients of these solutes, although too great to be measured accurately by conventional methods (Holz and Finkelstein, 1970; Schafer and Andreoli, 1972a), are substantially in excess of 150 to 200 × 10^{-4} cm $\sec^{-1}$; and, finally, for such lipophilic solutes, P_{D_i} through synthetic bilayer membranes or the hydrophobic core of biological membranes may actually exceed 10 cm $\sec^{-1}$ (Läuger et al., 1967). Based on these considerations, it was reasoned that $1/P_{D_i}$ (R_{D_i}, sec cm^{-1}) for these lipophilic solutes would provide an index to the diffusion resistance of the epithelial cell layer of cortical collecting tubules, exclusive of luminal membranes (Schafer and Andreoli, 1972a).

Table VI summarizes the P_{D_i} values for butanol, 5-hydroxyindole, and pyridine in cortical collecting tubules, and also lists both the observed R_{D_i} values (i.e., $1/P_{D_i}$) and the values of R_{D_i} predicted for the diffusion of these solutes through a 6 × 10^{-4} cm layer of water, i.e., a layer of water having a thickness equivalent to the cell layer of cortical collecting tubules. Two observations are particularly relevant. First, ADH had no effect on the P_{D_i} values for these

TABLE VI
Diffusional Permeability Coefficients of Lipophilic Solutes in Cortical Collecting Tubules[a]

Solute	ADH (μU ml^{-1})	P_{D_i} (cm sec $\times$ 10^4)	R_{D_i} (sec cm^{-1}) Observed	Predicted
Pyridine	0	12.6 ± 0.87 (5)	794	68.2
	250	13.2 ± 0.73 (5)	—	
Butanol	0	12.2 ± 0.30 (4)	820	67.8
	250	12.4 ± 0.30 (4)	—	
5-Hydroxyindole	0	4.03 ± 0.24 (5)	2480	92.8
	250	3.94 ± 0.30 (5)	—	—

[a] P_{D_i} for each of the solutes was measured when the lumen and bath contained, respectively, 290 mOsm liter^{-1} of Krebs–Ringer phosphate buffer and 290 mOsm liter^{-1} of Krebs–Ringer bicarbonate buffer, and net volume flow was zero. All experiments were carried out at 25 ± 0.5°C. The data are expressed as mean values ± SEM for the numbers of tubules listed in parentheses. Adapted from Schafer and Andreoli (1972a).

solutes. Second, the observed R_{D_i} values were 12–26 times greater than the R_{D_i} values predicted for diffusion in free solution. Thus, to the extent that the resistance of the epithelium to the diffusion of these solutes provides an index to the constraints imposed by the cell layer, exclusive of luminal plasma membranes, to transepithelial water diffusion, these data imply that the mode of water transport across luminal membranes during osmosis is diffusional. According to this view, P_f, rather than P_{D_w}, measures the rate of water diffusion across luminal membranes; and, the mechanism of action of ADH is to enhance the rate of water diffusion across luminal membranes.

It is important to note in this connection that this conclusion in no way precludes the possibility that water traverses luminal membranes through hydrophilic sites having the characteristics of narrow aqueous channels, and that ADH increases the water permeability of luminal membranes of collecting tubules by increasing the number of such hydrophilic sites in these membranes. For example, an aqueous channel having an effective radius of 1.8–2.0 Å would virtually exclude a solute such as urea; and, from Eq. (6), pore radii in the range

1.8–2.0 Å would yield P_f/P_{D_w} ratios of 1.25 to 1.32, i.e., indicating that water traverses such channels almost entirely by a diffusion process. This issue will be discussed in further detail in a subsequent section.

E. ORIGIN OF CELLULAR CONSTRAINTS TO DIFFUSION

In principle, it might be argued that the cellular diffusion constraints accounting for both the ADH-dependent P_f/P_{D_w} discrepancy (Table II) and the observed resistance to the diffusion of highly lipophilic solutes (Table VI) might be due either to frictional and/or geometric constraints. It is assumed that the cells of cortical collecting tubules are isotonic with the bathing solutions; the frictional constraints imposed by cytoplasmic viscosity would have to be due primarily to macromolecules. The relationship between diffusion and viscosity may be described by the Stokes–Einstein expression:

$$D_i^\circ = RT/(N 6\pi a_i \eta) \qquad (10)$$

where N = Avogrado's number, a_i is the radius of the diffusing species, and η is the viscosity. The elegant studies of Wang et al. (1954) have shown clearly that alterations in bulk viscosity produced by albumin have relatively small effects on D_w°. Since the number and mobility of macromolecules in aqueous solution are both negligibly small with respect to water molecules, small molecules diffusing in water are "obstructed," in the terminology of Wang et al., to a relatively slight degree by macromolecules (Wang, 1954). On the other hand, the linear reduction in bulk flow with increasing viscosity expressed by Poiseuille's law is relatively independent of the molecular species used to perturb viscosity. Thus cellular constraints referable to viscosity changes would be expected to retard volume flow to a greater extent than water diffusion, in direct contrast to the experimental observations on P_f and P_{D_w} (Table II).

Rather, we have argued (Schafer et al., 1974a) that geometric factors, more specifically a reduction in the area available for water transport in the cell layer, account for cellular constraints to diffusion. In the limit, a 25-fold increase in the diffusion resistance of the epithelial cell layer, with respect to an equivalent thickness of water, would be due either to a tortuosity factor that increases path length for water transport from 7.5×10^{-4} cm (the actual epithelial thickness) to an effective length of 150×10^{-4} cm or to a 25-fold reduction in the area available for water transport. In order to assess these two possibilities, the experimental data in Fig. 6 were analyzed in terms of the equations used to predict osmotic transients of the type illustrated in Fig. 2.

The results, described in detail elsewhere (Schafer et al., 1974a), are shown in Fig. 7. If the effective path length for water transport in the cell layer was increased even 7-fold, to 50×10^{-4} cm, the curves in Fig. 7 indicate that an osmotic transient would have been detected; but as shown in Fig. 6, no osmotic

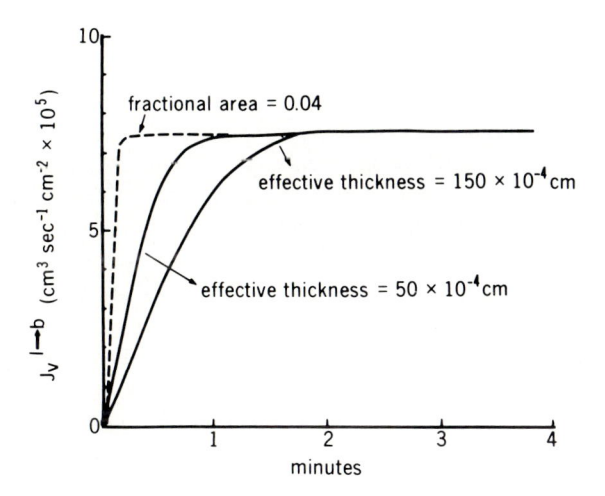

FIG. 7. A theoretical analysis of the osmotic transient phenomenon in cortical collecting tubules. The thoretical curves were drawn from the experimental data and conditions shown in Fig. 6. Note that the steady-state $J_v^{l \to b}$ predicted in Fig. 7 is the same as the experimentally observed $J_v^{l \to b}$ in Fig. 6. Adapted from Schafer et al. (1974a).

transients were evident within this time frame. However, as shown in Fig. 7, a reduction in the fractional area available for water transport to 0.04 (i.e., enough to account for the ADH-dependent P_f/P_{D_w} discrepancy in Table II), did not predict either a significant error in the measurement of P_f or a detectable osmotic transient.

On the basis of these experimental and theoretical observations, we have suggested (Schafer et al., 1974a) that geometric rather than frictional constraints account for the increased resistance of the cell layer of these cortical collecting tubules for the diffusion of lipophilic species and water. The geometric constraints appear to be the consequence of approximately a 25-fold reduction in the area available for water transport, and result in relatively small errors in underestimating P_f. In short, we consider the epithelial cell layer to be a type B unstirred layer.

F. TRANSCELLULAR VERSUS PARACELLULAR WATER FLOW

The fact that osmotic volume flow from lumen to bath widens intercellular spaces in the ADH-treated cortical collecting tubule (e.g., Fig. 4) does not, in our view, indicate that most lumen to bath osmosis traverses a paracellular route. Rather, as discussed in connection with Table V, it seems likely that osmotic volume flow exits the lumen via a cellular rather than a paracellular route.

Moreover, it may be that, once across luminal plasma membranes, the bulk of lumen to bath osmotic flow involves a transcellular route.

The average height and width of cortical collecting tubule cells is approximately 10×10^{-4} cm (Ganote et al., 1962; Grantham et al., 1968) so that the area of a tubular cell facing the lumen is roughly 10^{-6} cm^2. The reported width of intercellular spaces in collecting tubule cells is 75–300 Å (Ganote et al., 1968; Grantham et al., 1969). If the chosen width is 300 Å, the area of two lateral spaces adjacent to a tubular cell is 6×10^{-9} cm^2. Thus the approximate fractional area of intercellular spaces with respect to the luminal cell surface is about 0.001, i.e., 1/40th the area restriction required to account for the disparity between P_f and P_{D_w} (Table II; Fig. 7). These calculations, admittedly rough approximations, imply that, for a 25-fold reduction in the area available for water transport (cf. Fig. 7), a major fraction of volume flow probably involves transcellular pathways. We assume the latter are in part responsible for diffusion constraints to lipophilic species (e.g., Table VI).

G. PERMEATION OF MODERATELY LIPOPHILIC SOLUTES

Recently, Pietras and Wright (1975) and Levine et al. (1976) observed that, in toad urinary bladder, ADH produced small increments in the diffusional permeability coefficients for moderately lipophilic species. In cortical collecting tubules, the choice of moderately lipophilic solutes that may be evaluated for ADH-dependent effects on P_{D_i} is limited by certain technical and conceptual considerations. On the one hand, it is difficult with present techniques for microperfusing single renal tubules to measure small differences, e.g., 0.03×10^{-4} cm sec^{-1}, among solutes having P_{D_i} values in the range $0.02-0.06 \times 10^{-4}$ cm sec^{-1}. Alternatively, as discussed in connection with Table VI, it is likely that, in the case of highly lipophilic solutes, the epithelial cell layer exclusive of luminal membranes constitutes the principal resistance to tracer diffusion. Thus, in order to evaluate the effect of ADH on permeation rates for moderately lipophilic species, solutes were chosen whose P_{D_i} values were in the range 0.1 to 1.0×10^{-4} cm sec^{-1}, i.e., P_{D_i} values that could be determined for solutes where luminal membranes, rather than the epithelial cell layer, might reasonably be construed as constituting the major rate-limiting permeation site (Al-Zahid et al., 1977).

As indicated in Table VII, it is clear that, in paired observations on cortical collecting tubules, ADH produced significant increments in the permeability coefficients for three moderately lipophilic solutes: isobutyramide, n-butyramide, and antipyrine. However, it should be noted, by comparing Tables VI and VII, that both the ADH-independent and the ADH-dependent values of P_f are appreciably greater than the comparable P_{D_i} values listed in Table VII. Thus, either with or without hormone, the rate of water diffusion across these

TABLE VII
Effect of Antidiuretic Hormone (ADH) on the Diffusional Permeability Coefficients for Moderately Lipophilic Solutes in Cortical Collecting Tubules[a]

Solute	β_{oil}	Bath [ADH] (μU ml^{-1})	P_{D_i} (cm sec^{-1} × 10^4)	p (cm sec^{-1} × 10^4)	Tubules (number)
Isobutyramide	0.0095	0	0.12 ± 0.02	–	
	0.0095	250	0.26 ± 0.05	–	
	Mean paired difference:		0.14 ± 0.04	<0.01	6
n-Butyramide	0.0095	0	0.14 ± 0.04	–	
	0.0095	250	0.28 ± 0.02	–	
	Mean paired difference:		0.14 ± 0.02	< 0.01	6
Antipyrine	0.032	0	0.25 ± 0.07	–	
	0.032	250	0.40 ± 0.06	–	
	Mean paired difference:		0.16 ± 0.03	<0.01	5

[a]Paired measurements of P_{D_i}, with and without ADH, were carried out on the same tubule. The data are expressed as mean values ± SEM for the indicated numbers of tubules, and the mean paired differences and p values were computed from these data. The oil:water partition coefficients were obtained from Collander and Bärlund (1933). Adapted from Al-Zahid et al. (1977).

membranes is appreciably greater than the diffusion rates for these moderately lipophilic solutes.

V. Activation Energies for Water and Solute Permeation

In thermodynamic terms, the energy expenditure for a diffusion process, i.e., the free-energy change for a diffusion process, is the sum of two parts: an enthalpic component, representing the heat of activation; and the entropy change. It happens that the enthalpic component of the total free-energy change is very nearly the same as the activation energy, which may be measured experimentally. Thus it seemed reasonable to consider that a comparison of the activation energies for P_f, with respect to P_{D_i} for the solutes listed in Table VII, and an assessment of the effects of ADH on these energies, might provide some insight into the mode of action of ADH on luminal membranes.

The experiments (Al-Zahid et al., 1977) involved a measurement of the activation energies for water and solute permeation in cortical collecting tubules, either in the presence or in the absence of ADH. In the case of solute permeation P_{D_i} at zero volume flow (e.g., Table VII) was taken as an index of solute diffusion across luminal membranes; in the case of water, P_f rather than P_{D_w} (see above) was used to evaluate water diffusion across luminal membranes.

It was possible to make measurements only at two temperatures, 23–25°C and 37°C, rather than over a wide range of temperatures. This choice of conditions was dictated by two factors: (a) It was desirable to evaluate activation energies of P_f and P_{D_i} within the same temperature range. But below 23–25°C, ADH-independent values of P_f are difficult to distinguish reliably from zero; and above 37°C, cortical collecting tubules tend to desquamate rapidly. (b) In the range 23–37°C, the ADH-independent variations in P_f are sufficiently small that temperature variations of smaller magnitude would not have permitted reproducible detection of significant differences in P_f within a given tubule. Thus apparent activation energies were calculated from paired comparisons of either P_f and/or P_{D_i}, within a given tubule, according to the Arrhenius relationship:

$$\ln P_1/P_2 = -(E_A/R)[(1/T_1) - (1/T_2)] \qquad (11)$$

where P_1 and P_2 are either P_f or P_{D_i} measured at, respectively, T_1 and T_2, and E_A (kcal mole^{-1}) is the apparent activation energy.

Table VIII compares the effects of ADH on the apparent activation energies for P_f and n-butyramide permeation, and Table IX compares the ADH-dependent activation energies for permeation of the three solutes listed in Table VII. Table VIII illustrates clearly that the apparent activation energy for n-butyramide permeation was more than twice as great as that for P_f, and that ADH has no significant effect on the activation energies for water or n-butyramide transport. Table IX indicates that, in the presence of ADH, there was no significant difference among the apparent activation energies for n-butyramide, isobutyramide, and antipyrine permeation.

TABLE VIII

Effect of Antidiuretic Hormone (ADH) on the Apparent Activation Energies for Water and n-Butyramide Permeation in Cortical Collecting Tubules[a]

Coefficient	Apparent E_A (kcal mole^{-1})		p
	–ADH	+ADH	
P_f	9.35 ± 0.92 (n = 5)	8.9 ± 1.5 (n = 4)	>0.5
$P_{D_{n\text{-butyramide}}}$	19.40 ± 1.6 (n = 5)	16.6 ± 1.6 (n = 4)	>0.1

[a]The activation energies for P_f and n-butyramide permeation in cortical collecting tubules were computed from simultaneous, paired measurements of these coefficients at two temperatures (23°C and 37°C) in the same tubule. The data are expressed as mean values ± SEM for the number of tubules listed in parentheses. Adapted from Al-Zahid et al. (1977).

TABLE IX

A Comparison of the Activation Energies for Antidiuretic Hormone (ADH)-Dependent Permeation of Moderately Lipophilic Solutes in Cortical Collecting Tubules[a]

Solute	E_A (kcal mole^{-1})	p
n-Butyramide	17.50 ± 1.36 ($n = 4$)	–
Isobutyramide	15.81 ± 1.20 ($n = 5$)	>0.5
Antipyrine	19.61 ± 1.80 ($n = 6$)	>0.2

[a] E_A for each of these solutes was computed from paired measurements of P_{D_i} at 25°C and 37°C for each of the three solutes listed. The data are expressed as mean values ± SEM for the numbers of tubules listed in parentheses. The p column compares the E_A for a given solute with respect to that for n-butyramide. Adapted from Al-Zahid et al. (1977).

VI. Comparison of ADH Effects in Collecting Tubules and Amphibian Epithelia

A vast body of literature has detailed the effects of antidiuretic hormone on transport processes in amphibian epithelia. These observations have not only provided major insights into the molecular mechanism of action of this hormone, but also have formed one of the major cornerstones for our understanding of many other kinds of transport processes in biological membranes. In this section, we summarize some of these observations and compare these with experimental results in mammalian collecting tubules.

A. THE DUAL-BARRIER HYPOTHESIS

Capraro and Bernini (1952) showed that ADH increased dramatically the rate of osmotic water transport from apical to basolateral surfaces of frog skin. Koefoed-Johnsen and Ussing (1953) confirmed and extended these observations by showing that ADH increased both P_f and P_{D_w}, and elevated the P_f/P_{D_w} ratio from 5.3 to 33.5.

A number of critical measurements of the effects of ADH on transport processes in amphibian skin were made by Ussing et al. Ussing and Zerahn (1951) showed that the hormone increased apical to basolateral Na$^+$ flux. Andersen and Ussing (1957) observed that ADH increased the apparent permeability coefficients of acetamide (effective hydrodynamic radius $\simeq$ 2.5 Å) and urea (effective hydrodynamic radius ≃ 2.0 Å), although the hormone-dependent permeation coefficients for these solutes were appreciably lower than the ADH-dependent value of P_{D_w}. These workers (Andersen and Ussing, 1957) also found that the net flux of thiourea and acetamide in ADH-treated frog skin increased in proportion to the rate of osmotic volume flow in the same direction, and attributed the effect to entrainment of solute and solvent flows within aqueous membrane pores (i.e., the solvent drag effect).

Andersen and Ussing (1957) integrated these results into their classical dual barrier hypothesis for the action of ADH on amphibian epithelia. The apical surfaces of frog skin were visualized as a series barrier: a dense outer diffusion barrier which limited solute but not water penetration—thereby accounting for the low ratio of P_{D_i} for urea and acetamide with respect to P_{D_w}—and was unresponsive to ADH; and a porous inner barrier sensitive to ADH. Pores in the inner barrier were presumed to account for the disparity between P_f and P_{D_w} and the apparent solvent drag effect; from the effects of ADH on P_f and P_{D_w}, Koefoed-Johnsen and Ussing (1953) computed that ADH increased the radii of pores in the inner barrier from 6 to 20 Å [Eq. (6)]. Finally, it was reasoned (Andersen and Ussing, 1957) that water, urea, Na$^+$, and other solutes crossed inner barriers through these pores and that hormone-dependent increases in the permeation rates for these molecules were due to the ADH-dependent increases in pore size.

Leaf and co-workers subsequently provided strong evidence that the dual-barrier model might also account for the action of ADH on toad urinary bladder. These workers reported that in that tissue the apical membrane was the rate-limiting, ADH-sensitive site for water and urea permeation (Maffly et al., 1960); ADH increased P_f, P_{D_w}, and the P_f/P_{D_w} ratio sufficiently, in terms of Eq. (6), to increase pore radii in the inner barrier from 8.4 Å to 40 Å (Hays and Leaf, 1962a); and ADH increased the apparent permeability coefficients for urea, other low-molecular-weight amides, and Na$^+$, and produced apparent solvent drag for urea (Leaf and Hays, 1962). As was the case with frog skin, Leaf and Hays (1962) found that the ADH-dependent permeation coefficients for urea, other small amides, and Na$^+$ were smaller than would be expected for molecular sieving through pores with 40 Å radii. Hence these workers deduced that an outer diffusion barrier limited the access of such solutes to the inner, porous barrier.

Hays and Leaf (1962b) also found that ADH reduced the activation energy for THO diffusion across the toad bladder from 9.8 kcal mole^{-1} to 4.1 kcal mole^{-1}. These findings were interpreted to indicate that, without ADH, pores in the inner barrier contained highly structured water; with ADH, pores in the inner barrier were enlarged [as indicated above, from 8.4 to 40 Å, based on the effect of ADH on P_f/P_{D_w} ratios (Hays and Leaf, 1962a)], and water within these pores assumed the characteristics of liquid water.

B. DISSOCIATION OF ADH EFFECTS ON SOLUTE AND WATER PERMEATION

A cardinal requirement of the dual-barrier hypothesis is the assumption that Na^+, urea, and water permeate apical membranes at the same sites, specifically the pores in the inner barrier. But a substantial body of evidence now indicates that the effects of ADH, or cAMP, on Na^+, water, and urea permeation may be dissociated.

Within toad urinary bladder, a number of pharmacologic agents have been shown to have a specific inhibitory effect on the ADH-mediated increment in urea, but not water or Na^+ transport. Both phloretin (Levine et al., 1973a), an aglucone of phlorizin, and tanning agents, such as chromate or tannic acid (Schuchter et al., 1973), inhibited significantly transepithelial urea permeation, either with or without ADH, but had no effect on ADH-mediated net Na^+ transport or water transport. Moreover, by analyzing the kinetics of urea transport in toad urinary bladder, Hays (1972a) and Levine et al. (1973a,b, 1975b) concluded that, in that tissue, urea and other small amides were transported across apical membranes via an ADH-sensitive facilitated diffusion route, which differed from the water transport pathway.

It has also been possible to inhibit selectively ADH-mediated increments in Na^+ and water transport in toad urinary bladder. For example, agents such as cytochalasin B (Carasso et al., 1973; Taylor et al., 1973), colchicine (Taylor et al., 1973), and vinblastine (Taylor et al., 1973), as well as anesthetics such as halothane and methoxyfluorane (Levine et al., 1975a), have been shown to inhibit the hydroosmotic response of toad bladder to ADH without affecting the stimulatory effects of the hormone on Na^+ or urea transport. Alternatively, both the diuretic amiloride (Bentley, 1968) and the coronary vasodilator verapamil (Bentley, 1974) block the stimulation of Na^+, but not water transport, produced in toad bladder by either ADH or cAMP.

The dual-barrier hypothesis also requires that the effects of ADH on Na^+ transport be referable solely to hormone-mediated increments in apical membrane Na^+ permeability. But it now appears likely that, in a variety of amphibian epithelia, including amphibian skin (Morel and Bastide, 1965), frog bladder (Janáček and Ryborá, 1970; Janáček et al., 1972), and toad bladder (Finn, 1971; Finn and Rockoff, 1971), ADH augments both passive Na^+ entry at apical surfaces and active Na^+ efflux at basolateral surfaces. Further, Finn (1975) has proposed that apical and basolateral membranes may "signal" one another, possibly by chemical means, and that the effects of ADH on these two surfaces may be related to the "signal" effect.

Taken together, these observations are consistent with the view that, in amphibian epithelia, Na^+, water, and urea and other small amides permeate apical membranes through parallel, dissociable routes rather than through a common pathway. In the case of urea permeation, the molecular mechanisms involved appear to have many of the characteristics of a saturable, facilitated diffusion system, rather than a simple solubility diffusion process (Hays, 1972a; Levine et al., 1973a,b, 1975b). In the case of Na^+ transport, apical Na^+ entry may also be related to the effects of ADH on Na^+ transport across basolateral membranes (Finn, 1975). In other words, a dual-barrier model, which assumes that Na^+, urea, and water traverse apical membranes via a single transport pathway, and that Na^+, water and urea transport through this pathway may be accounted for quantitatively by molecular sieving theory, is not adequate to rationalize the effects of ADH on the transport of these molecules across apical membranes of amphibian epithelia.

C. ADH EFFECTS IN AMPHIBIAN AND MAMMALIAN EPITHELIA

The effects of ADH on urea and Na^+ transport differ in toad urinary bladder and in cortical collecting tubules. As indicated above, ADH augments urea transport in toad urinary bladder (Leaf and Hays, 1962) and amphibian skin (Andersen and Ussing, 1957), possibly by a facilitated diffusion pathway (Hays, 1972a; Levine et al., 1973a,b, 1975b). But in cortical and outer medullary collecting tubules, the hormone has no augmenting effect on urea permeation (Tables III and IV). Likewise, ADH augments transepithelial Na^+ flux and reduces transepithelial electrical resistance in amphibian epithelia (Helman and Miller, 1974; Leaf et al., 1958; Leaf and Dempsey, 1958; Ussing and Zerahn, 1951), but not in isolated rabbit cortical collecting tubules (Frindt and Burg, 1972; Helman et al., 1971). Thus in cortical collecting tubules, ADH can enhance water permeation without simultaneously affecting either urea or Na^+ transport.

The effects of ADH on water transport across luminal membranes of amphibian epithelia and cortical collecting tubules warrant comparison, especially in regard to unstirred layer effects in amphibian epithelia: (a) In cortical collecting tubules at 23–25°C, the ADH-dependent P_f is approximately 186×10^{-4} cm sec^{-1} (Table II); in toad urinary bladder, the ADH-dependent P_f is in the range 185 to 230×10^{-4} cm sec^{-1} (Hays and Franki, 1970; Hays and Leaf, 1962a). (b) In cortical collecting tubules at 23–25°C, the ADH-dependent P_{D_w} is 14.2×10^{-4} cm sec^{-1} (Table II). In toad bladder, Hays and Franki (1970) found that a significant fraction of the total tissue resistance to water diffusion was due to unstirred layer effects referable to stromal elements; in the epithelial cell layer of that tissue, these workers found the ADH-dependent P_{D_w} to be 11×10^{-4} cm sec^{-1}. (c) As shown in Table VIII, the apparent activation energy for water diffusion across cortical collecting tubules, either with or without ADH, is approximately 9 kcal $mole^{-1}$. In toad urinary bladder, Hays et al. (1971) found that those observations (Hays and Leaf, 1962b) indicating that ADH reduced the activation energy for transepithelial water diffusion were probably due to the fact that, with ADH, water diffusion across the tissue was rate limited by bulk-phase unstirred layers rather than apical plasma membranes; in experiments designed to eliminate unstirred layer artifacts, Hays et al. (1971) found that the activation energy for water diffusion across toad urinary bladder was approximately 10 kcal mole^{-1}, both in the presence and in the absence of ADH.

(d) Finally, Hays (1972b) showed convincingly that the apparent solvent drag effect in toad bladder could be abolished by rigorous stirring of aqueous phases bathing the isolated toad bladder, i.e., that acceleration of solute flux in the direction of solvent flow was due to flow-dependent solute polarization in unstirred layers, rather than to an actual solvent drag effect. We have analyzed (Andreoli and Schafer, 1976) quantitatively the original data of Leaf and Hays (1962) showing apparent solvent drag for urea in terms of the unstirred layer thickness for the experimental conditions under which the apparent solvent drag effect was measured (Table I) (Hays and Franki, 1970; Leaf and Hays, 1962); and Equations (5) and (6), which permit a distinction between the true solvent drag phenomenon, i.e., coupling of solute and solvent flows within membrane pores, and changes in the diffusional flux of solute across a membrane which are referable to flow-dependent changes in solute concentration profiles within unstirred layers (e.g., Fig. 3). The results of these calculations, summarized in Table X, indicate clearly that unstirred layer effects are adequate to explain the observed flux asymmetry for urea in toad urinary bladder.

In short, it appears that, when appropriate corrections (Hays and Franki, 1970) for unstirred layer effects in toad bladder are made, the ADH-dependent

TABLE X
Analysis of Apparent Solvent Drag for Urea in Antidiuretic Hormone (ADH)-Treated Toad Urinary Bladder[a]

J_v (cm^3 sec^{-1} cm^2 × 10^5)	Urea flux ratio		
	Observed	Predicted	
		Unstirred-layer effect	Solvent-drag effect
6.94	1.73	1.78	2.65

[a]The J_v value and the observed urea flux ratio are from Leaf and Hays (1962). The predicted flux ratios for either an unstirred-layer effect or a solvent-drag effect were computed as described previously (Andreoli and Schafer, 1976; Andreoli et al., 1971) using the unstirred layer thickness of 1055 × 10^{-4} cm (Table I) estimated by Hays and Franki (1970) for toad urinary bladder. Adapted from Andreoli and Schafer (1976).

values of P_f, P_{D_w}, the P_f/P_{D_w} ratio, and the activation energy for water diffusion (either with or without hormone), are virtually the same in cortical collecting tubules and in the epithelial cell layer of toad urinary bladder. We have argued that, in cortical collecting tubules, the discrepancy between P_f and P_{D_w} is referable to cellular diffusion constraints (cf. Section IV). Likewise, in toad urinary bladder, Parisi and Piccini (1973) have provided direct evidence for significant hindrances to water diffusion within the epithelial cell layer. Thus it may be that, both in cortical collecting tubules and toad urinary bladder, osmotic water flow across apical membranes is diffusional in nature; and in both epithelia, comparable molecular events may account for ADH-dependent increments in water transport.

VII. A Parallel, Solubility-Diffusion Model for ADH Action

We now consider a model that attempts to rationalize water and solute permeation through luminal membranes of cortical collecting tubules, and the effects of ADH on these processes. For convenience, the following is a summary of the major issues for analysis.

1. In cortical collecting tubules, luminal plasma membranes are the rate-limiting site for water and solute permeation (Figs. 4 and 5; Table V). The same consideration applies to amphibian epithelia (Maffly et al., 1960).

2. In cortical collecting tubules, the ADH-dependent P_f/P_{D_w} ratio (Table II) is referable to cellular constraints to diffusion (Table VI); P_f is not significantly affected by the osmotic transient phenomenon (Fig. 6) and provides an accurate index to the rate of water diffusion across luminal membranes, i.e., P_f equals P_w^m in Eq. (8); lumen to bath osmosis involves a transcellular pathway (Section IV, F); and cellular constraints to diffusion may be accounted for in terms of a reduction in the fractional area of the cell layer available for water transport (Fig. 7). In the toad urinary bladder, P_f, P_{D_w}, and the P_f/P_{D_w} ratio across the epithelial cell layer are very nearly the same, after correction for unstirred layer effects (Hays and Franki, 1970), as the values in Table II; the epithelial cell layer of toad urinary bladder hinders significantly the diffusion of THO (Parisi and Piccini, 1973); and as indicated in Section VI, the water permeation pathway is dissociable from, and in parallel with, the urea and Na$^+$ transport routes in apical plasma membranes.

3. In cortical collecting tubules, ADH increases the diffusion rate of moderately lipophilic species through luminal membranes (Table VII), but the P_{D_i}/P_f ratio remains quite small (Tables II and VII); i.e., the rate of water diffusion through luminal membranes is appreciably greater than the diffusion rate for moderately lipophilic solutes, either with or without ADH. Entirely comparable

results obtain in toad urinary bladder (Levine et al., 1976; Pietras and Wright, 1975).

4. In cortical collecting tubules, the activation energies for water and moderately lipophilic solute diffusion are, respectively, 9–10 and 16–20 kcal mole^{-1}, either with or without ADH (Tables VIII and IX). In toad urinary bladder the activation energy for water diffusion is approximately 10 kcal mole^{-1}, with or without ADH (Hays et al., 1971); to our knowledge, activation energy data for moderately lipophilic solute diffusion are not available.

In order to arrive at some tentative conclusions about the molecular nature of the water and solute permeation pathways, we shall consider that luminal plasma membranes of cortical collecting tubules share the lipid bilayer arrangement common to the molecular architecture of a wide variety of biological plasma membranes and synthetic, entirely-lipid bilayer membranes. Thus by analyzing water and solute permeation across cortical collecting tubules in terms of comparable processes in synthetic lipid bilayer membranes, it might be possible to inquire about the extent to which transport processes solely through the hydrophobic regions of luminal membranes can account for ADH-mediated water and solute permeation in collecting tubules, or whether more complex models—involving, for example, specialized, protein-containing regions—are required. Two characteristics of transport events in lipid bilayer membranes and in cortical collecting tubules will be compared: the ratio of water to nonelectrolyte permeability, i.e., the P_f/P_{D_i} ratio; and the activation energies for water and solute permeation.

A. ANALYSIS OF THE P_f/P_{D_i} RATIO

It seems clear, based on current evidence (Andreoli and Troutman, 1971; Cass and Finkelstein, 1967; Hanai and Haydon, 1966; Price and Thompson, 1969; Redwood and Haydon, 1969; Reeves and Dowben, 1970; Träuble, 1971), that osmotic water flow across synthetic bilayer membranes occurs by a solubility-diffusion process; in other words, P_f represents the true permeability coefficient for water diffusion across synthetic bilayer membranes. Given the assumption that osmotic water flow across apical membranes of the collecting tubule, and probably of the toad urinary bladder, is diffusional, it is relevant to evaluate water and nonelectrolyte discrimination in these membranes with respect to synthetic bilayer membranes.

Such a comparison is presented in Table XI. Three representative bilayer membranes are presented, having P_f and P_{D_i} values comparable to those observed in other bilayer membrane systems (Cass and Finkelstein, Finkelstein and Cass, 1967; Graziani and Livne, 1972; Poznansky et al., 1976; Price and Thompson, 1969; Redwood and Haydon, 1969; Reeves and Dowben, 1970). Table XI

TABLE XI

A Comparison of P_f and P_{D_i} in Cortical Collecting Tubules and Synthetic Lipid Bilayer Membranes[a]

System	Solute	Solute β_{oil}	P_{D_i} (cm sec^{-1} × 10^4)	P_f (cm sec^{-1} × 10^4)	P_f/P_{D_i}	References[b]
Cortical collecting tubules						
−ADH	Urea	0.00014	0.03	20	666	1–5
−ADH	Acetamide	0.0008	0.034	20	588	1–5
−ADH	Thiourea	0.0012	0.03	20	666	1–5
−ADH	Butyramide	0.01	0.14	20	142	6; Table VIII
+ADH	Urea	0.00014	0.02	197	9850	1–5
+ADH	Acetamide	0.0008	0.036	197	5472	1–5
+ADH	Thiourea	0.0012	0.04	197	4925	1–5
+ADH	Butyramide	0.01	0.32	197	615	6; Table VII
Lipid Bilayer membranes						
Sheep red cell lipid/decane (planar)	Acetamide	0.0008	0.83	22.9	22.7	11
Lecithin/decane (planar)	Urea	0.00014	0.036	19	527	7–9
	Acetamide	0.0008	1.43	19	13	7–9
	Thiourea	0.0012	0.046	19	413	7–9
Lecithin/chloroform-decane (spherical)	Acetamide	0.0008	0.24	13	54	10
	Valeramide	0.023	1.83	13	7.1	10

[a] The values of P_f and P_{D_i} are for 22–25°C. Adapted from Al-Zahid et al. (1977).
[b] Key to references: (1) Burg et al., 1970. (2) Grantham and Burg, 1966. (3) Schafer and Andreoli, 1972a. (4) Schafer and Andreoli, 1972b. (5) Schafer et al., 1974b. (6) Al-Zahid et al., 1977. (7) Gallucci et al., 1971. (8) Lippe, 1969. (9) Vreeman, 1966. (10) Poznansky et al., 1976. (11) Andreoli et al., 1969.

indicates clearly that the ADH-independent P_f in collecting tubules is the same as in bilayer membranes, while the ADH-dependent P_f is approximately 10-fold greater than in bilayer membranes. However, bilayer membranes are appreciably more permeable to solutes having β_{oil} values $\geqslant 0.0008$, e.g., acetamide, than collecting tubules with or without ADH. Consequently the P_f/P_{D_i} ratio is appreciably greater in collecting tubules than in bilayer membranes in three circumstances: (a) When the solute is acetamide, primarily because this solute permeates bilayer membranes better than collecting tubules. (b) Sufficient data are not available in bilayer membranes for the solutes listed in Table XI, but the P_f/P_{D_i} ratio for valeramide (a solute having a β_{oil} and molecular weight similar to butyramide) in bilayer membranes is 20–36 times less than the P_f/P_{D_i} ratio for butyramide in collecting tubules. (c) The P_f/P_{D_i} ratios in cortical collecting tubules with ADH are uniformly at least 10-fold greater than in collecting ducts.

In other words, using synthetic bilayer membranes as a frame of reference, it is difficult to envision ADH-mediated effects on the hydrophobic regions of luminal membranes which permit, simultaneously, at least 10-fold higher rates of water diffusion but appreciably lower permeation rates for nonelectrolytes having β_{oil} values $\geqslant 0.0008$, than in synthetic bilayers. Rather, we propose (Al-Zahid et al., 1977) that ADH-mediated water and nonelectrolyte diffusion across luminal membranes may involve parallel pathways: one for water and the other for moderately lipophilic solutes. The water-diffusion pathway, although appreciably more permeable to water than synthetic bilayer membranes, virtually excludes small, relatively hydrophilic solutes, such as urea, thiourea, and acetamide. The regions for nonelectrolyte diffusion are less permeable than synthetic bilayer membranes to solutes having β_{oil} values $\geqslant 0.0008$.

B. ANALYSIS OF ACTIVATION ENERGIES

Additional support for such a view may be derived by comparing the activation energies for water and nonelectrolyte diffusion through synthetic lipid bilayer membranes with those in cortical collecting tubules. In bilayer membranes, the activation energy for water diffusion (termed E_A^w) may be expressed as the sum of two terms: the energy required for a water molecule to break the four hydrogen bonds formed with neighboring water molecules (termed $E_A^{H}w$); and a nonhydrogen bond term (termed $E_A^{D}w$), i.e., the energy for water diffusion within the membrane (Cohen, 1975; Price and Thompson, 1969; Redwood and Haydon, 1969). It is also probable that the activation energy for nonelectrolyte diffusion (termed E_A^s) across a bilayer membrane may be rationalized in terms of the same process (de Gier et al., 1971; Gallucci et al., 1971), with $E_A^{H}s$ and $E_A^{D}s$ representing the components of E_A^s due, respectively, to rupture of hydrogen bonds in solution and diffusion within the membrane. Cohen (1975) computed 1.8 kcal mole^{-1} to be the activation energy per hydrogen bond in aqueous solution, either for water or for solutes. And in liposomes, he found that $E_A^D w$ and $E_A^D s$ varied depending either on liposome composition or, for liposomes of a particular composition, whether the liposomes were above or below their transition temperature; but it is important that, for liposomes of a given composition at the same temperature, $E_A^D s$ and $E_A^D w$ were the same.

In cortical collecting tubules, the ADH-dependent activation energies for water and solute permeation were, respectively, 8.9 and 16.6–19.6 kcal mole^{-1} (Tables VIII and IX). Taking four, three, and two as the number of aqueous hydrogen bonds for, respectively, water, isobutyramide or butyramide, and antipyrine, these E_A^W and E_A^s data yield, following the analysis applied to synthetic bilayer membranes, apparent values of 0.9 and 10.6–15.6 kcal mole^{-1} for, respectively, $E_A^D w$ and $E_A^D s$ for luminal membranes of cortical collecting tubules. Two alternative explanations may account for this disparity: either $E_A^D w$ and $E_A^D s$ differ within the hydrophobic regions of luminal membranes or water traverses a different pathway than moderately lipophilic species in these membranes. Thus, to the extent that activation energy measurements in liposomes [where $E_A^D w$ and $E_A^D w$ are the same for a given membrane (Cohen, 1975)] provide a frame of reference for assessing transport processes in the hydrophobic regions of luminal membranes, we may infer that, in the latter, water and moderately lipophilic solutes traverse parallel permeation pathways.

C. PROPERTIES OF THE MODEL: THE EFFECT OF ADH

Finally, we speculate on the nature of the transport pathways in luminal membranes of cortical collecting tubules and on the effects of ADH on these pathways. Since ADH increased the permeation rates for moderately lipophilic solutes without affecting the activation energy for solute permeation, it seems improbable that ADH produced a *generalized* increase in the fluidity of the hydrocarbon matrix in luminal membranes (for example, by shortening, branching, or unsaturation of the fatty acid chains of phospholipids), since, in comparison with synthetic bilayers, a generalized fluidity increase should have been accompanied by a fall in E_A^s due to a reduction in $E_A^D s$ (Cohen, 1975). We suggest that nonelectrolyte diffusion across cortical collecting tubules may proceed via discrete hydrophobic pathways in luminal membranes, and that ADH increases the number of such pathways. A corollary of this hypothesis is that the entire hydrophobic region of luminal membranes may be a mosaic structure containing (a) relatively fluid (or disorganized) hydrophobic pathways for butyramide (and presumably isobutyramide and antipyrine) permeation, whose number may be increased by ADH; and (b) more ordered hydrophobic regions unaffected by ADH, which have lower permeation rates and appreciably higher activation energies for transport of moderately lipophilic solutes. Indeed, given the fact that luminal membranes are less permeable than bilayer membranes to moderately lipophilic solutes, it is probable that the hydrophobic regions of luminal membranes which are unaffected by ADH are rather tightly organized, i.e., highly condensed.

We consider next the pathway for water diffusion in luminal membranes. This process differs in at least two major respects from water diffusion in synthetic bilayers: the ADH-dependent P_f in luminal membranes exceeds reported P_f values in synthetic bilayers (Table XI); and in luminal membranes with or without ADH, the apparent E_A^Ds for nonelectrolytes exceeds the apparent E_A^Dw (cf. above), while in liposomes of a given composition, E_A^Dw equals E_A^Ds (Cohen, 1975). Using synthetic bilayers as a frame of reference, two types of water-conducting pathways might be envisioned to account for these results.

First, luminal membranes might contain regions where membrane phospholipids are in a highly disorganized state. Thus it is conceivable that, for such pathways, E_A^Hw is 8 kcal mole^{-1}, i.e., diffusing water molecules are fully dehydrated. But because of regional disorganization of phospholipid molecules and attendant increases in membrane fluidity, the rate of water diffusion is quite high and E_A^Dw quite low in these pathways with respect to the remaining hydrophobic regions of luminal membranes. Clearly, steric constraints within these regions would need to be sufficiently great to exclude solutes, such as urea, to which luminal membranes are virtually impermeable, even with ADH (Tables III and IV).

Alternatively, water might diffuse through luminal membranes via small aqueous channels. In that case, water molecules might not need to be fully dehydrated in order to diffuse through these channels, and E_A^Hw might then be appreciably less than 8 kcal mole^{-1}. For example, in bilayer membranes gramicidin A dimers form aqueous channels having radii of approximately 2 Å (Haydon and Hladky, 1972; Urry et al., 1971) and water/urea selectivity ratios in excess of 10^3 (Finkelstein, 1974). Thus a 2-Å radius channel in luminal membranes, similar to the gramicidin A dimer in bilayer membranes, might account for both the high ADH-dependent P_f/P_{D_i} ratios for urea, acetamide, and thiourea (Table XI) and the lower activation energy for water with respect to nonelectrolyte permeation (Tables VIII and IX).

Such a channel, if present in luminal membranes, would need to have an appreciably lower Na$^+$ conductance than a gramicidin A channel. Finkelstein (1974) found that the water permeability of a gramicidin A channel was 1.2×10^{-14} cm^3 sec^{-1}. Hence, for an ADH-dependent P_f of 186×10^{-4} cm sec^{-1} (Table II), one requires 1.6×10^{12} gramicidin A channels per cm^2 of luminal membrane; since the Na$^+$ conductance of a gramicidin A channel is 6×10^{-12} ohm^{-1} (Finkelstein, 1974), this number of channels would give a luminal membrane resistance of 0.1 ohm cm^2. But the transepithelial resistance of cortical collecting tubules, with or without ADH, is in excess of 800 ohm cm^2 (Helman et al., 1971), and Helman (1973) has found that luminal plasma membranes have appreciably higher electrical resistances than either junctional complexes or basolateral membranes. Accordingly, if ~2-Å radius aqueous channels having the same water permeability as gramicidin A channels constituted the route for ADH-mediated water diffusion in luminal plasma membranes, the Na^+ conductance of such a putative channel would need to be approximately 10^{-4} of the Na^+ conductance of a gramicidin A channel.

It is not possible at present to rule out the possibility that the water permeation pathways in luminal membranes involve highly disorganized, phospholipid regions which exclude urea, thiourea, and acetamide by steric hindrance. But based on observations in synthetic lipid bilayer membrane systems, where water and nonelectrolyte permeation ordinarily vary pari passu (Table XI) (Cass and Finkelstein, 1967; Finkelstein and Cass, 1967; Graziani and Livne, 1972; Poznansky et al., 1976; Price and Thompson, 1969; Redwood and Haydon, 1969; Reeves and Dowben, 1970), it seems more probable that aqueous channels, sufficiently narrow to exclude urea, thiourea, and acetamide having unit Na^+ conductances at least 10^{-4} less than a gramicidin A channel account for water diffusion through luminal surfaces. And since ADH did not affect the activation energy for water diffusion (Table VIII), we conclude that the effect of ADH on water permeation in these tubules depended on increasing the number of water channels in luminal membranes.

Thus in summary, it may be argued that, in cortical collecting tubules and possibly other ADH-sensitive tissues, water and moderately lipophilic solutes traverse luminal plasma membranes by parallel diffusion pathways (Fig. 8): narrow aqueous channels for water and hydrophobic regions for moderately lipophilic solutes. Both regions, but particularly the hydrophobic pathway for moderately lipophilic solute transport, may be in a higher entropy state than the

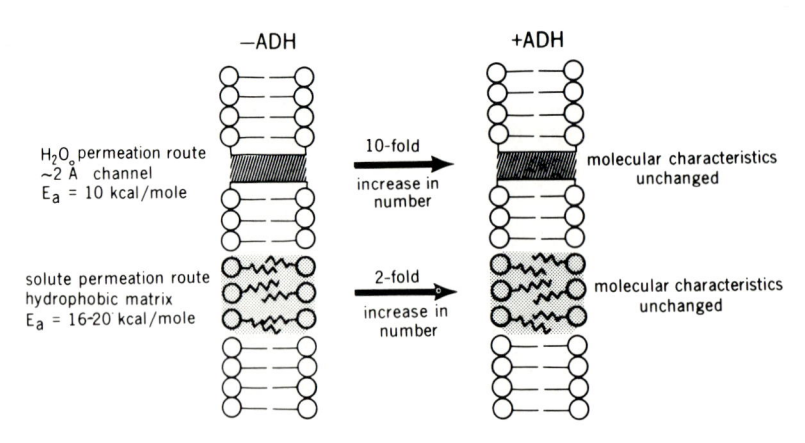

FIG. 8. A schematic model for the action of antidiuretic hormone (ADH) on luminal plasma membranes of hormone-sensitive epithelia.

remaining, impermeant regions of luminal membranes. And ADH increases the number of these specialized pathways without necessarily affecting the ordering of the regions of luminal membranes that are impermeant to water and solutes. Thus the total ADH-dependent change in the entropy of luminal membranes will depend on magnitude of the hormone-mediated increase in the fractional area occupied by the specialized water and solute permeation pathways.

ACKNOWLEDGMENTS

We acknowledge contributions of Drs. C. S. Patlak and G. Al-Zahid, who collaborated with us in some of the work to which we refer, and of Miss S. L. Troutman, who conducted many of the experiments. During the course of the investigations conducted in this laboratory, T. E. Andreoli was a recipient of a Research Career Development Award from the National Institutes of Health (5-K04-GM18161) and J. A. Schafer was an Established Investigator of the American Heart Association (71-177). These investigations were supported by research grants from the American Heart Association (75-805), the National Science Foundation (BMS 74-13645), and the National Institutes of Health (5-R01-AM14873, 5-T01-HL05951, and 1-T32-GM 07195). Some of the studies mentioned were conducted while the authors were members of the Department of Physiology and Pharmacology and the Department of Medicine at Duke University Medical School.

REFERENCES

Al-Zahid, G., Schafer, J. A., Troutman, S. L., and Andreoli, T. E. (1977). *J. Membr. Biol.* **31**, 103.
Andersen, B., and Ussing, H. H. (1957). *Acta Physiol. Scand.* **39**, 228.
Andreoli, T. E. (1973). *Kidney Int.* **4**, 337.
Andreoli, T. E., and Schafer, J. A. (1976). *Annu. Rev. Physiol.* **39**, 451.
Andreoli, T. E., and Troutman, S. L. (1971). *J. Gen. Physiol.* **57**, 464.
Andreoli, T. E., Dennis, V. W., and Weigl, A. M. (1969). *J. Gen. Physiol.* **53**, 133.
Andreoli, T. E., Schafer, J. A., and Troutman, S. L. (1971). *J. Gen. Physiol.* **57**, 479.
Bentley, P. J. (1968). *J. Physiol. (London)* **195**, 317.
Bentley, P. J. (1974). *J. Pharmacol. Exp. Ther.* **189**, 563.
Berl, T., Harbottle, J. A., and Schrier, R. W. (1974). *Kidney Int.* **6**, 247.
Berliner, R. W. (1976). *Kidney Int.* **9**, 214.
Berliner, R. W., and Bennett, C. M. (1967). *Am. J. Med.* **42**, 777.
Biber, T. U. L., and Curran, P. F. (1968). *J. Gen. Physiol.* **51**, 606.
Burg, M., Grantham, J., Abramow, M., and Orloff, J. (1966). *Am. J. Physiol.* **210**, 1293.
Burg, M., Helman, S. L., Grantham, J., and Orloff, J. (1970). *In* "Urea and the Kidney" (B. Schmidt-Nielson and D. W. S. Kerr, eds.), pp. 193–199. Excerpta Med. Found. Amsterdam.
Capraro, V., and Bernini, G. (1952). *Nature (London)* **169**, 454.
Carasso, N., Favard, P., and Bourguet, J. (1973). *J. Microsc. (Paris)* **18**, 383.
Cass, A., and Finkelstein, A. (1967). *J. Gen. Physiol.* **50**, 1765.
Civan, M. M. (1970). *J. Theor. Biol.* **27**, 387.
Cohen, B. E. (1975). *J. Membr. Biol.* **20**, 205.
Collander, R., and Bärlund, H. (1933). *Acta Bot. Fenn.* **11**, 1.
Colton, C. K. (1967). "Artificial Kidney–Chronic Uremia Program," National Institute of

Arthritis and Metabolic Disease, National Institutes of Health, U.S.P.H.S. Bethesda, Maryland. (Federal Clearinghouse Accession No. PB 182–281.)
Cort, J. H., Schück, O., Stříbrná, J., Škopková, J., Jošt, K., and Mulder, J. L. (1975). *Kidney Int.* **8**, 292.
Dainty, J. (1963). *Adv. Bot. Res.* **1**, 279.
Dainty, J., and House, C. R. (1966a). *J. Physiol. (London)* **182**, 66.
Dainty, J., and House, C. R. (1966b). *J. Physiol. (London)* **185**, 172.
de Gier, J., Mandersloot, J. G., Hupkes, J. V., McElhaney, R. N., and van Beer, W. P. (1971). *Biochim. Biophys. Acta* **233**, 610.
Diamond, J. M. (1966). *J. Physiol. (London)* **183**, 83.
DiBona, D. R., and Civan, M. M. (1973). *J. Membr. Biol.* **12**, 101.
Douglas, W. W., and Poisner, A. M. (1964). *J. Physiol. (London)* **172**, 1.
Dousa, T. P., and Barnes, L. D. (1974). *J. Clin. Invest.* **54**, 252.
Dreifuss, J. J. (1975). *Ann. N. Y. Acad. Sci.* **248**, 184.
Dreifuss, J. J., Kalnins, I., Kelly, J. S., and Ruf, K. B. (1971). *J. Physiol. (London)* **215**, 805.
Dunn, F. L., Brennan, T. J., Nelson, A. E., and Robertson, G. L. (1973). *J. Clin. Invest.* **52**, 3212.
du Vigneaud, V. (1969). *Johns Hopkins Med. J.* **124**, 53.
Edwards, R. W., Kitau, M. J., Chard, T., and Besser, G. M. (1973). *Br. Med. J.* **3**, 375.
Erlij, D., and Martinez-Palomo, A. (1972). *J. Membr. Biol.* **9**, 229.
Farquhar, M. G., and Palade, G. E. (1963). *J. Cell Biol.* **17**, 375.
Finkelstein, A. (1974). In "Drugs and Transport Processes" (B. A. Callingham, ed.), pp. 241–250. Univ. Park Press, Baltimore, Maryland.
Finkelstein, A., and Cass, A. (1967). *Nature (London)* **216**, 717.
Finn, A. L. (1971). *J. Gen. Physiol.* **57**, 349.
Finn, A. L. (1975). *J. Gen. Physiol.* **56**, 503.
Finn, A. L., and Rockoff, M. L. (1971). *J. Gen. Physiol.* **57**, 326.
Forte, L. R., Chao, W. T. H., Walkenbach, R. J., and Byington, K. H. (1972). *Biochem. Biophys. Res. Commun.* **49**, 1510.
Franz, T. J., and van Bruggen, J. T. (1967). *J. Gen. Physiol.* **50**, 933.
Franz, T. J., Galey, W. R., and van Bruggen, J. T. (1968). *J. Gen. Physiol.* **51**, 1.
Frindt, G., and Burg, M. B. (1972). *Kidney Int.* **1**, 224.
Frömter, E., and Diamond, J. M. (1972). *Nature (London), New Biol.* **235**, 9.
Galey, W. R., and van Bruggen, J. T. (1970). *J. Gen. Physiol.* **55**, 220.
Gallucci, E., Micelli, S., and Lippe, C. (1971). *Arch. Int. Physiol. Biochim.* **79**, 881.
Ganote, C. E., Grantham, J. J., Moses, H. L., Burg, M. B., and Orloff, J. (1968). *J. Cell Biol.* **36**, 355.
Ginzburg, B. F., and Katchalsky, A. (1963). *J. Gen. Physiol.* **47**, 403.
Goetz, K. L., Bond, G. C., and Smith, W. E. (1974). *Proc. Soc. Exp. Biol. Med.* **145**, 277.
Gottschalk, C. W. (1961). *Physiologist* **4**, 35.
Gottschalk, C. W. (1964). *Am. J. Med.* **36**, 670.
Grantham, J. J. (1970). *Science* **168**, 1903.
Grantham, J. J., and Burg, M. B. (1966). *Am. J. Physiol.* **211**, 255.
Grantham, J. J., Ganote, C. E., Burg, M. B., and Orloff, J. (1969). *J. Cell Biol.* **41**, 562.
Graziani, Y., and Livne, A. (1972). *J. Membr. Biol.* **7**, 275.
Green, K., and Otori, T. (1970). *J. Physiol. (London)* **207**, 93.
Hanai, T., and Haydon, D. A. (1966). *J. Theor. Biol.* **11**, 370.
Handler, J. S., and Orloff, J. (1963). *Am. J. Physiol.* **205**, 298.
Handler, J. S., and Orloff, J. (1971). *Ann. N. Y. Acad. Sci.* **185**, 345.

Haydon, D. A., and Hladky, S. B. (1972). *Q. Rev. Biophys.* **5**, 187.
Hays, R. M. (1972a). *J. Membr. Biol.* **10**, 367.
Hays, R. M. (1972b). *Curr. Top. Membr. Transp.* **3**, 330-366.
Hays, R. M., and Franki, N. (1970). *J. Membr. Biol.* **2**, 263.
Hays, R. M., and Leaf, A. (1962a). *J. Gen. Physiol.* **45**, 905.
Hays, R. M., and Leaf, A. (1962b). *J. Gen. Physiol.* **45**, 933.
Hays, R. M., Franki, N., and Soberman, R. (1971). *J. Clin. Invest.* **50**, 1016.
Helman, S. I. (1973). *Abstr. 6th Annu. Meet. Am. Soc. Nephrol.* p. 49.
Helman, S. I., and Miller, D. A. (1974). *Am. J. Physiol.* **226**, 1198.
Helman, S. I., Grantham, J. J., and Burg, M. B. (1971). *Am. J. Physiol.* **220**, 1825.
Holz, R., and Finkelstein, A. (1970). *J. Gen. Physiol.* **56**, 125.
Janáček, K., and Rybová, R. (1970). *Pfluegers Arch.* **318**, 294.
Janáček, K., Rybová, R., and Slaviková, M. (1972). *Biochim. Biophys. Acta* **288**, 221.
Jard, S., Bourguet, J., Favard, P., and Carasso, N. (1971). *J. Membr. Biol.* **4**, 124.
Kachadorian, W. A., Wade, J. B., and DiScala, V. A. (1975). *Science* **190**, 67.
Kedem, O., and Katchalsky, A. (1958). *Biochim. Biophys. Acta* **27**, 229.
Kinne, R., and Schwartz, I. L. (1977). *In* "Disturbances in Body Fluid Osmolality" (T. E. Andreoli, J. J. Grantham, and F. C. Rector, Jr., eds.). Am. Physiol. Soc., Bethesda, Maryland (in press).
Koefoed-Johnsen, V., and Ussing, H. H. (1953). *Acta Physiol. Scand.* **28**, 60.
Kokko, J. P., and Rector, F. C., Jr. (1972). *Kidney Int.* **2**, 214.
Läuger, P., Richter, J., and Lesslauer, W. (1967). *Ber. Bunsenges. Phys. Chem.* **71**, 906.
Leaf, A., and Dempsey, E. (1958). *J. Biol. Chem.* **235**, 2160.
Leaf, A., and Frazier, H. S. (1961). *Prog. Cardiovasc. Dis.* **4**, 47.
Leaf, A., and Hays, R. M. (1962). *J. Gen. Physiol.* **45**, 921.
Leaf, A., Anderson, J., and Page, L. B. (1958). *J. Gen. Physiol.* **41**, 657.
Levine, S. D., Franki, N., and Hays, R. M. (1973a). *J. Clin. Invest.* **52**, 1435.
Levine, S. D., Franki, N., and Hays, R. M. (1973b). *J. Clin. Invest.* **52**, 2083.
Levine, S. D., Levine, R. D., Worthington, R. E., and Hays, R. M. (1975a). *Clin. Res.* **23**, 432A.
Levine, S. D., Worthington, R. E., and Hays, R. M. (1975b). *Clin. Res.* **23**, 368A.
Levine, S. D., Franki, N., Einhorn, R., and Hays, R. M. (1976). *Kidney Int.* **9**, 30.
Lindley, B. D., Hoshiko, T., and Leb, D. E. (1964). *J. Gen. Physiol.* **47**, 773.
Lippe, C. (1969). *J. Mol. Biol.* **39**, 669.
Loeschke, K., Bentzel, C. J., and Csáky, T. S. (1970). *Am. J. Physiol.* **218**, 1723.
Maffly, R. H., Hays, R. M., Lamdin, E., and Leaf, A. (1960). *J. Clin. Invest.* **39**, 630.
Marsh, D. J., and Azen, S. P. (1975). *Am. J. Physiol.* **228**, 71.
Michelakis, A. M. (1970). *Proc. Soc. Exp. Biol. Med.* **135**, 13.
Morel, F., and Bastide, F. (1965). *Biochim. Biophys. Acta* **94**, 609.
Morgan, T., Sakai, F., and Berliner, R. W. (1968). *Am. J. Physiol.* **214**, 574.
Nernst, W. (1904). *Z. Phys. Chem. (Leipzig)* **47**, 52.
Pappenheimer, J. R. (1953). *Physiol. Rev.* **33**, 387.
Parisi, M., and Piccini, Z. F. (1973). *J. Membr. Biol.* **12**, 227.
Pennell, J. P., Lacy, F. B., and Jamison, R. L. (1974). *Kidney Int.* **5**, 337.
Pietras, R. J., and Wright, E. M. (1975). *J. Membr. Biol.* **22**, 107.
Poznansky, M., Tong, S., White, P. C., Milgram, J. M., and Solomon, A. K. (1976). *J. Gen. Physiol.* **67**, 45.
Price, H. D., and Thompson, T. E. (1969). *J. Mol. Biol.* **41**, 443.
Rector, F. C., Jr. (1977). *In* "Disturbances in Body Fluid Osmolality" (T. E. Andreoli, J. J.

Grantham, and F. C. Rector, Jr., eds.). Am. Physiol. Soc., Bethesda, Maryland (in press).
Redwood, W. R., and Haydon, D. A. (1969). *J. Theor. Biol.* **22**, 1.
Reeves, J. P., and Dowben, R. M. (1970). *J. Membr. Biol.* **3**, 123.
Robbins, E., and Mauro, A. (1960). *J. Gen. Physiol.* **43**, 523.
Robertson, G. L. (1974). *Annu. Rev. Med.* **25**, 315.
Robertson, G. L., and Athar, S. (1976). *J. Clin. Endocrinol. Metab.* **42**, 613.
Robertson, G. L., Athar, S., and Shelton, R. L. (1977). *In* "Disturbances in Body Fluid Osmolality" (T. E. Andreoli, J. J. Grantham, and F. C. Rector, Jr., eds.). Am. Physiol. Soc., Bethesda, Maryland (in press).
Rocha, A. S., and Kokko, J. P. (1974). *Kidney Int.* **6**, 379.
Sachs, H. (1967). *Am. J. Med.* **42**, 687.
Sawyer, W. H. (1967). *Am. J. Med.* **42**, 678.
Sawyer, W. H., Acosta, M., Balaspiri, L., Judd, J., and Manning, M. (1974a). *Endocrinology* **94**, 1106.
Sawyer, W. H., Acosta, M., and Manning, M. (1974b). *Endocrinology* **95**, 140.
Schafer, J. A., and Andreoli, T. E. (1972a). *J. Clin. Invest.* **51**, 1264.
Schafer, J. A., and Andreoli, T. E. (1972b). *J. Clin. Invest.* **51**, 1279.
Schafer, J. A., Patlak, C. S., and Andreoli, T. E. (1974a). *J. Gen. Physiol.* **64**, 201.
Schafer, J. A., Troutman, S. L., and Andreoli, T. E. (1974b). *J. Gen. Physiol.* **64**, 228.
Scharrer, E., and Scharrer, B. (1954). *Recent Prog. Horm. Res.* **10**, 183.
Schrier, R. W., Berl, T., Anderson, R. J., and McDonald, K. M. (1977). *In* "Disturbances in Body Fluid Osmolality" (T. E. Andreoli, J. J. Grantham, and F. C. Rector, Jr., eds.). Am. Physiol. Soc., Bethesda, Maryland (in press).
Schwartz, I. L., Schlatz, L. J., Kinne-Saffran, E., and Kinne, R. (1974). *Proc. Natl. Acad. Sci. U.S.A.* **71**, 2595.
Shuchter, S. H., Franki, N., and Hays, R. M. (1973). *J. Membr. Biol.* **14**, 177.
Smulders, A. P., Tormey, J. M., and Wright, E. M. (1972). *J. Membr. Biol.* **7**, 164.
Solomon, A. K., and Gary-Bobo, C. M. (1972). *Biochim. Biophys. Acta* **255**, 1019.
Stephenson, J. L. (1972). *Kidney Int.* **2**, 85.
Sutherland, E. W. (1961-1962). *Harvey Lect.* **57**, 17.
Taylor, A. (1977). *In* "Disturbances in Body Fluid Osmolality" (T. E. Andreoli, J. J. Grantham, and F. C. Rector, Jr., eds.). Am. Physiol. Soc., Bethesda, Maryland (in press).
Taylor, A., Mamelak, M., Reaven, E., and Maffly, R. (1973). *Science* **181**, 347.
Teorell, T. (1936). *J. Biol. Chem.* **113**, 735.
Träuble, H. (1971). *J. Membr. Biol.* **4**, 193.
Urry, D. W., Goodall, M. C., Glickson, J. D., and Mayers, D. F. (1971). *Proc. Natl. Acad. Sci. U.S.A.* **68**, 1907.
Ussing, H. H. (1966). *Ann. N. Y. Acad. Sci.* **137**, 543.
Ussing, H. H. (1969). *Q. Rev. Biophys.* **1**, 365.
Ussing, H. H., and Windhager, E. E. (1964). *Acta Physiol. Scand.* **61**, 484.
Ussing, H. H., and Zerahn, K. (1951). *Acta Physiol. Scand.* **23**, 110.
Verney, E. B. (1947). *Proc. R. Soc. London, Ser. B* **135**, 25.
Vreeman, H. J. (1966). *Proc. K. Ned. Akad. Wet. Ser. B.* **69**, 542.
Walter, R., Rudinger, J., and Schwartz, I. L. (1967). *Am. J. Med.* **42**, 653.
Walter, R., Smith, C. W., Mehta, P. K., Boonjarern, S., Arruda, J. A. L., and Kurtzman, N. A. (1977). *In* "Disturbances in Body Fluid Osmolality" (T. E. Andreoli, J. J. Grantham, and F. C. Rector, Jr., eds.). Am. Physiol. Soc., Bethesda, Maryland (in press).

Wang, J. H. (1954). *J. Am. Chem. Soc.* 76, 4755.
Wang, J. H., Robinson, C. V., and Edelman, I. S. (1953). *J. Am. Chem. Soc.* 75, 466.
Wang, J. H., Anfinsen, C. B., and Polestra, F. M. (1954). *J. Am. Chem. Soc.* 76, 4763.
Wright, E. M., Smulders, A. P., and Tormey, J. M. (1972). *J. Membr. Biol.* 7, 198.

DISCUSSION

J. C. Orr: In your determination of activation energies, did you measure rates only at 25° and 37°C, or did you examine a range of temperature? I am concerned that the tubule membrane might undergo a phase transition in the range of 25–37°C or close to that range.

T. E. Andreoli: There might well be a phase transition. The problem is the following: above 37°C the tubule becomes exceedingly unstable; below 25°C the measurements are so low that one cannot make them; and in the range 25–37°C one cannot measure reliable differences for ADH-independent P_f values, even in paired observations on a large number of tubules. So we call our measurements apparent activation energies and note that there might well be a break in the Arrhenius plots. The operative fact, though, is that most phase transitions in membranes occur well below 20°C; in anuran epithelia where these issues have been evaluated, there is no bend in the Arrhenius curve until one reaches 15°C.

J. C. Orr: I gather that your assumption is that the water molecules must be stripped of all their hydrogen bonds on going into the membrane, or for instance could dimeric water or trimeric water get through and therefore alter your conclusions?

T. E. Andreoli: In a synthetic bilayer, the evidence is quite strong that water molecules must be stripped, or dehydrated, prior to diffusion through the membrane. The E_A of 10 kcal mole^{-1} in renal tubules could, in the limit, represent two sets of processes. In one, the water molecules break all hydrogen bonds prior to entering the channel and the activation energy for movement within the channel is only 1–2 kcal mole^{-1}. A more likely possibility is that water does not have to break as many hydrogen bonds, and that the channel might well contain regions where the water molecule can form and reform hydrogen bonds in much the same way as it does in bulk solution. In that instance, 8 kcal per mole would not have to be expended in letting the water molecule make the transition from bulk solution to the membrane channel. So, a channel could in principle account for either type of water diffusion.

J. C. Orr: By tunneling, did you mean quantum mechanical tunneling? Water is too big for that, and the distances too great.

T. E. Andreoli: No. Picture a channel lined with either carbonyl oxygens or OH groups, that is, similar to gramicidin A channels or amphotericin B channels. The water molecule might simply then move along the channel, forming partial hydrogen bonds along the channel length. Is that acceptable?

J. C. Orr: Yes, thank you.

I. C. P. Smith: In your model for the solute route in the absence of the hormone, could the route be via some region of phospholipid that has shorter chains or more double bonds, i.e., greater fluidity than the rest of the membrane?

T. E. Andreoli: We envision the hydrophobic region as being a mosaic: it has a parallel, maybe highly condensed, region; and pathways, which are more permeant to solutes, that for the sake of simplicity we infer are relatively fluid regions. We have no other information.

I. C. P. Smith: I think that is consistent with current thinking on membrane structure. Could it be possible that the increase in permeability induced by the hormone is due to an increase in the degree of phase separation within the membrane—an increase in the degree of heterogeneity of the membrane. In other words, is there some very specific interaction between the hormone and the membrane such as to cluster phospholipids containing fatty acids with double bonds or short chains?

T. E. Andreoli: But that is a bias. Although I tend to believe that this may be the case, all one can say from the data is that the solute permeations sites are more fluid. The activation energy argument says that the hormone must increase the number of water and solute permeation sites. Beyond that, it becomes a very speculative argument.

I. C. P. Smith: What I am building up to is that this can be tested. It can be tested either by nuclear magnetic resonance or, much easier and just as good in this case, by electron spin resonance spin labels. I am thinking of, say, a labeled phosphatidylcholine with unsaturated fatty aryl chains. Thus one knows exactly what type of lipid one is looking at, and, if there is phase separation of this type of lipid, one can measure it directly. Any increase in the degree of phase separation induced by the hormone would also be directly measurable. The idea of permeability and phase discontinuities is pretty acceptable now. If one looks at the plots of permeability against temperature, one of ten seems an enormous burst right at the region where the phases are separating during a gel-to-liquid crystal transition. What about lysine vasopressin acting on the permeability of phosphatidylserine? Would you say that this is a good model system for the behavior that you have been studying?

T. E. Andreoli: Graziani and Livne have done it, but I think the signal-to-noise ratio in those experiments is quite low. We have tried vasopressin in bilayers but have had trouble reproducing those data.

S. Korenman: To resonate further with this hypothesis, since this is a cAMP-mediated event and since Dousa *et al.* have demonstrated that there is membrane protein kinase which is activated in relation to these events, I think we must assume that phosphorylation of membrane protein plays a central role in whatever changes in the structure of the membrane take place, but I wonder whether you could comment on your view of those elements of the mechanism of the action of the hormone.

T. E. Andreoli: In the tubule, one can reproduce with the dibutyryl or 8-bromo analogs of cAMP virtually all the effects that one sees with ADH. The major question that you ask is what is the biochemical link between the second messenger and the final transport event. There is good evidence that altering the kinetics of phosphorylation, dephosphorylation reactions in luminal membranes has a central role. A second major line of argument has to do with microtubule and microfilament assemblies as mediators of the ADH effect. How these will all sort out in the future is imponderable at the moment. I would like to know what the answer is, but I do not think anyone as yet does know with certainty.

S. Korenman: Somewhere in the fog of my scientific background, I recall that Pauling showed a long time ago that water is more tightly structured close to surfaces, like the surface of a cell membrane, than it is in open solution, and that therefore the hydrogen-bonded structure of water is very close to ice in the vicinity of the membrane. Is it possible that a surface phenomenon making the water more liquid rather than more solid could play a role in the ADH effect?

T. E. Andreoli: Pauling has suggested this, and Sheraga has shown that hydrophobic proteins that are coiled do result in an ordering of water. It is why I pointed out earlier that unstirred layer phenomenology is precisely that; one could have, for example, a 10-Å layer of structured water adjacent to the surface that could act in exactly the same fashion. There is experimentally one problem with that interpretation: if one looks at water permeation through a synthetic bilayer membrane, which is highly structured in terms of its phospholipid organization but is devoid of protein, one finds that the rate of water permeation or diffusion through the bilayer can be accounted for entirely in terms of the water solubility-diffusion properties of bulk lipid phases. To put it another way, the bilayer acts as a

60-Å-thick layer of olive oil, precisely as Overton, Hober, Collander, and Barlund predicted. From there on, it becomes intuitive. Are there proteins situated on the surface of the membrane whose ordering could be changed? Mueller has argued that that is precisely how excitation occurs: channel-forming peptides sit on the surface; you apply a gating voltage and drive these into the membrane, where they form an ion-conducting channel. The problem though is that, even in such an instance, the rate of water solubility diffusion in an ADH-treated tubule is on the order of 300 μm per second, and I know of no bilayer model that could account for that magnitude of water solubility diffusion. The closest you can get is on the order of 50 μm per second by making bilayers out of highly unsaturated fat. Something else therefore might account for the ADH effect, but your original point remains plausible.

D. Rodbard: Several years ago, workers studied the possibility that an enzymelike testicular hyaluronidase might be involved in depolymerizing polysacchardies, presumably increasing permeability through the intercellular junctions. A number of your observations mitigate against this possibility, and I wonder whether that hypothesis is dead at this time, or whether there is a possibility that this kind of mechanism might account for the changing in the number of apparent water-permeability sites?

T. E. Andreoli: I cannot comment, except to say that our data would indicate that the mode of water transport is through luminal membranes, not junctional complexes. I have no quantitative estimate of how much hyaluronidase is present in the cortical collecting duct.

D. Rodbard: Is there any tissue surrounding the basement membrane?

T. E. Andreoli: No. There is only a single basement membrane surrounding the epithelium.

D. Rodbard: Conceivably, that could modulate the number of sites available to the cell membrane.

T. E. Andreoli: No, I do not think so, because the basement membrane is remarkably leaky: it even sieves proteins quite well. It must be the luminal plasma membrane that is rate-limiting to water permeation, and I think the data in virtually all hormone-sensitive epithelial coincide on that point.

A. D. Rogol: You have spoken a lot about activation energy, but what about looking at it from the point of view of how much energy the cell itself must generate to allow these transport processes to occur. What then is the cost in terms of energy expenditure at the basal rate and at the maximum rate of both water and solute flow? Can the cell make enough energy to move all these molecules, and, if so, what percentage of the cell's metabolic fuels must be expended to perform these transport processes?

T. E. Andreoli: There are no data in the collecting duct per se. To move water, one does not require energy, provided a chemical gradient for net water flux exists. I do not think that anyone has any idea at the moment of the approximate stoichiometry between moles of cAMP made vis-à-vis moles of water transported. In terms of the average amount of energy that a cell spends in ion-transport processes, Edelman's observations with thyroid hormone and its role in thermogenesis indicate that about one-third of metabolic energy, in cells like the liver, is spent driving sodium–potassium pumps.

K. Sterling: I liked your conclusion, namely, that the number of water permeation sites go up with vasopressin and also the physical chemistry on which this depends. Several members of this conference have complained that I have not asked facetious enough questions, so I will do so now. You are, I believe, the head of the Renal Hemodialysis Program at the Medical College of Alabama. Does not this whole presentation lead you to conclude that the most efficient extracorporeal hemodialysis would employ, let us say, some form of collecting ducts treated with vasopressin? Dr. Wilbur Sawyer, our chairman, points out that *Necturus* has a collecting duct 30 μm in diameter and a few millimeters long.

Of potentially greater potential practical utility would be anuran bladders, such as that of the African goliath frog or giant marine toad. Their bladders could be sutured end to end in cylindrical form to provide better exchange than Visking casing, especially after being treated with vasopressin.

T. E. Andreoli: You would get a whopping rejection response.

R. O. Greep: In the experiments where you showed the swelling of the tubule cells and the intercellular spaces, this was at the histological level, and that seems rather prosaic in comparison with your other highly sophisticated techniques. Have you not looked at these cells with electron microscopy or perhaps the electron microprobe?

T. E. Andreoli: We have not. Grantham, Burg, and Ganote have looked at the cells under EM, and the results under EM are virtually the same as the results with simple histology. To my knowledge, and I have spoken with Claude Lechene directly about this issue, I do not think electron probe microanalysis is yet at the point where one can begin looking at tubular cellular contents. A very central issue, one that many people are anxious about, is to try to get electronprobe data on solute concentrations within intercellular spaces. Clearly, that is a very desirable long-term goal.

J. S. Roberts: Your argument that intercellular spaces seem not to play a role in collecting duct physiology is compelling. Why do you think this role is absent here but present, for example, in proximal tubular epithelium? Is it fortuitous, or is it necessary because of the different function of the collecting duct?

T. E. Andreoli: That is a teleological question, which I can only answer teleologically. The collecting duct, like many other electrically tight epithelia, including the mammalian and the amphibian urinary bladder, is concerned with transporting rather small volumes of fluid or solute, distinguishing or discriminating between water and solutes, and maintaining very steep transepithelial gradients. For this, an epithelium is needed that has a really tight shunt pathway, which is precisely our line of argument. Some years ago, Frömter and Diamond surveyed many epithelia, and classed them as electrically tight or electrically leaky. The latter, such as gallbladder, the proximal tubule, and the small intestine, are coarse adjusters of homeostasis: they transport very large volumes of fluid and solute isotonically and do not establish very appreciable transepithelial concentration gradients. In fact, one finds that, in general, the degree of leakiness parallels the absolute magnitude of the transport function; so it is that the proximal tubule, which probably transports more salt and water per unit luminal area than most other mammalian epithelia, has the lowest transepithelial resistance, on the order of 5 ohm-cm^2, which says that the shunt pathway is wide open. It is a teleological argument, but I think a persuasive one.

Clinical Significance of Circulating Proinsulin and C-Peptide

ARTHUR H. RUBENSTEIN, DONALD F. STEINER,
DAVID L. HORWITZ, MARY E. MAKO,
MARSHALL B. BLOCK, JEROME I. STARR,
HIDESHI KUZUYA, AND FRANCO MELANI

Departments of Medicine and Biochemistry, University of Chicago, Chicago, Illinois

I. Introduction

Until 1967 it was believed that the two chains of insulin were synthesized separately and then joined by means of two disulfide bonds at a postribosomal site. However, studies by Steiner and his colleagues (Steiner and Oyer, 1967; Steiner *et al.*, 1969) using slices of a pancreatic islet cell tumor demonstrated the presence of a large molecular weight, single chain precursor molecule, which was named proinsulin. This protein consists of the insulin A and B chains linked by an additional polypeptide segment of approximately 30 to 35 amino acids, depending on the species (Fig. 1). It seems probable that the proinsulin connecting segment facilitates the formation of the native structure of the insulin molecule by ensuring the correct pairing of the cysteine residues during forma-

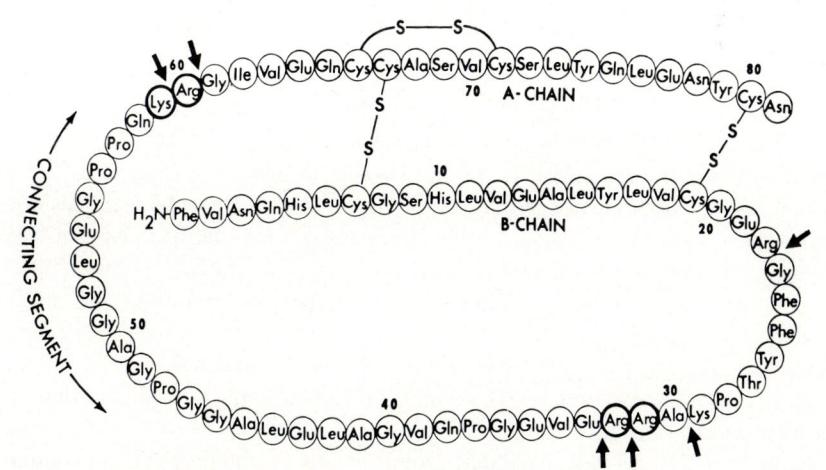

FIG. 1. Covalent structure of bovine proinsulin. Arrows indicate sites of tryptic cleavage. Reproduced in modified form from Nolan *et al.* (1971).

tion of the interchain disulfide bonds. After its transport to the Golgi apparatus of the beta cell, proinsulin is condensed into membrane-enclosed granules. The enzymes involved in the transformation of proinsulin to insulin are probably localized in the granules or their membranes, and the conversion process takes place as the granules mature and move toward the plasma membrane of the cell (Fig. 2). These proteolytic enzymes cleave proinsulin at specific sites where two pairs of basic amino acids join the C-peptide to the insulin chains (Kemmler and Steiner, 1970; Kemmler et al., 1971, 1973). The major products of this reaction are insulin and the connecting peptide (C-peptide). These two proteins are retained together within the secretion granules and subsequently liberated in equimolar amounts during exocytosis of beta-cell granules (Fig. 2).

II. Circulating Proinsulin-Like Components

A. MEASUREMENT OF PROINSULIN IN SERUM

A number of studies of normal human plasma or urine samples have indicated the presence of small amounts of immunoreactive material similar in molecular weight to proinsulin (Rubenstein et al., 1968; Roth et al., 1968). Although it would be advantageous to measure human proinsulin and its intermediate fractions (proinsulin-like components, PLC) by direct immunoassay in unextracted serum, this has not been possible. The reasons lie in the cross-reactivity of proinsulin with insulin (Fig. 3), on the one hand, and the C-peptide on the other. As all three of these peptides have been identified in the circulation, a preliminary step is required to separate them from each other. The most commonly used approach has involved gel filtration of serum followed by measurement of the column fractions in the insulin immunoassay. In our initial studies we extracted serum insulin and proinsulin into acid ethanol and separated the two peptides by gel filtration on a Bio-Gel P-30 column equilibrated in 3 M acetic acid (Melani et al., 1970a). The initial reason for choosing this technique was the reluctance to gel filter serum in neutral or alkaline buffers in which polymerization or aggregation of insulin might occur. In fact, this does not appear to be a problem. Another advantage is the ability to extract large volumes of serum and yet separate the hormones on relatively small columns. Furthermore, it is easier to characterize the separated proinsulin and insulin under these conditions when most of the other serum proteins have been removed. The most obvious disadvantage of the method is the length of time required for the procedure and the limitation on the number of samples that can be analyzed by one laboratory.

Roth et al. (1968) have separated proinsulin and insulin on 1×50 cm columns of Sephadex G-50 fine, equilibrated in a Veronal buffer containing human serum

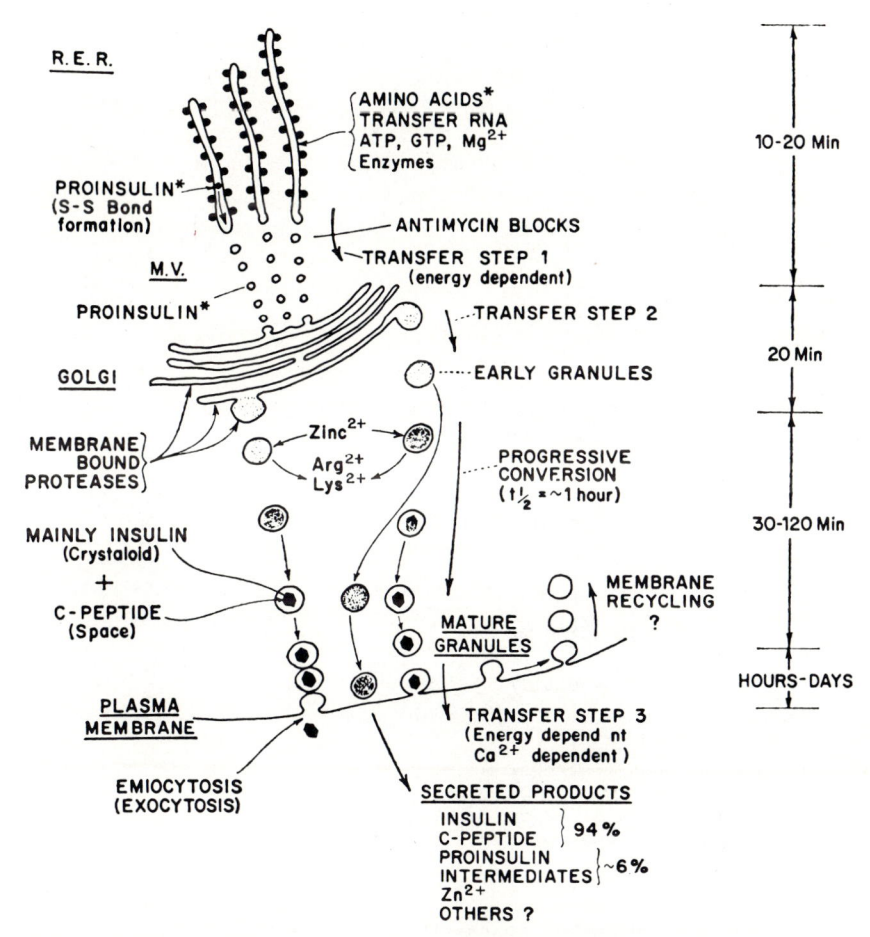

FIG. 2. Schematic summary of the insulin biosynthetic machinery of the pancreatic beta cells. R.E.R., rough endoplasmic reticulum; M.V., microvesicles. Reproduced from Steiner and Rubenstein (1973).

albumin, rabbit fraction II, and toluene. One or two milliliters of serum are applied directly to the column, fraction sizes of 1.0–1.5 ml are collected, and 0.4–0.8 ml aliquots are taken for immunoassay. We have modified this method to use a column of Bio-Gel P-30, equilibrated in the borate buffer that we use in the immunoassay (Fig. 4). Fractions can be collected directly into the immunoassay tubes, thus obviating the need for further pipetting at this stage. The void volume is determined by the elution position of [^{125}I] albumin or blue dextran 2000, while the salt peak is marked by Na ^{125}I. The column is calibrated with

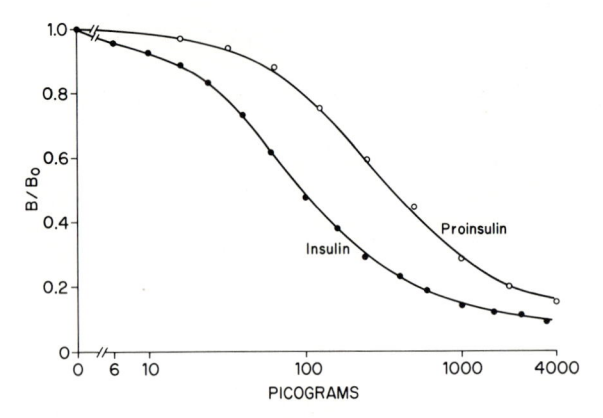

FIG. 3. Cross-reaction of human proinsulin and insulin in the insulin radioimmunoassay system (porcine insulin antiserum; porcine [^{125}I] insulin tracer).

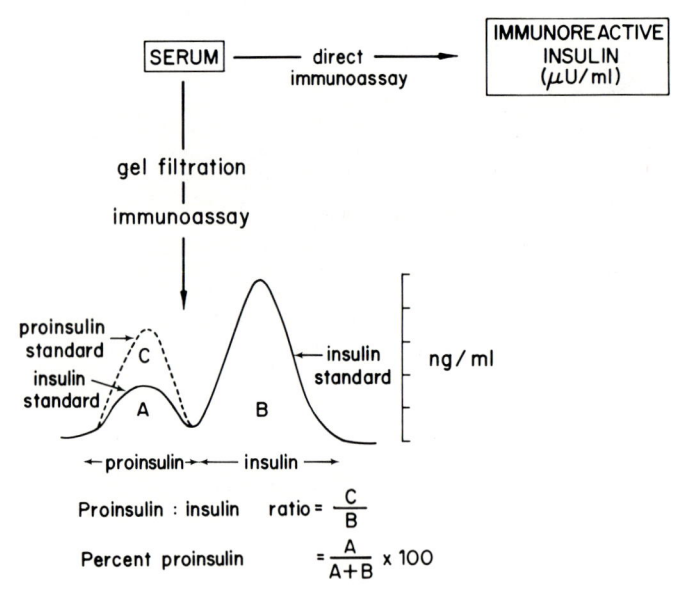

FIG. 4. A schematic diagram illustrating the measurement of proinsulin and insulin in serum. Serum is measured directly in the immunoassay (immunoreactive insulin, IRI). It is then gel-filtered to separate the proinsulin and insulin, and each fraction is assayed in the insulin radioimmunoassay. The early eluting peak (proinsulin-like-component) is read from the human proinsulin standard curve (curve C), or alternatively from the human insulin standard (curve A). The sum of the individual fractions in each peak (corrected for volume) is the proinsulin and insulin concentration, respectively. The relationship between proinsulin and insulin has been expressed in two ways: the proinsulin to insulin ratio, with each read from its appropriate standard; and the percentage proinsulin, where both peptides have been read from the insulin standard. Reproduced from Starr and Rubenstein (1974).

tracers of [^{125}I]proinsulin and [^{125}I]insulin. Because certain preparations of these labeled hormones may not elute identically with the native proteins, it may be preferable to determine the characteristics of the column by assaying the elution positions of unlabeled insulin (2 ng) and proinsulin (2 ng).

When serum is directly applied to these columns, recoveries have been essentially complete. In order to calculate the absolute level of proinsulin and insulin, fractions of the earlier eluting peak are read from a human proinsulin standard, while those comprising the second peak are measured against a human insulin standard. As the supply of human proinsulin is limited at present, many investigators have expressed the values of proinsulin in terms of the insulin standard (Fig. 4).

Another method for separating insulin from proinsulin has been described by Kitabchi and his co-workers (Kitabchi *et al.*, 1971; Duckworth *et al.*, 1972). These investigators have used an enzyme that is relatively specific for the proteolytic degradation of insulin, but not proinsulin. Measuring samples in an insulin assay before and after incubation with this enzyme (insulin-specific protease, ISP) should enable one to determine the relative concentrations of the two peptides. The accuracy of this method is limited, however, especially at low serum immunoreactive insulin (IRI) concentrations, because the degradation of insulin is generally incomplete (Fig. 5). This may be due, in part, to the presence

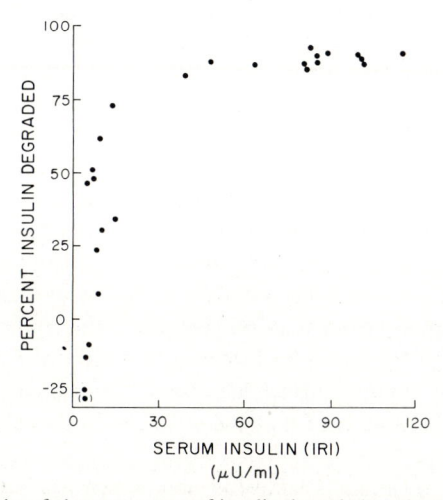

FIG. 5. Relationship of the percentage of insulin degraded to the concentration of insulin initially present in each serum sample. In sera containing more than 50 µU/ml, more than 85% of the insulin was destroyed by the insulin-specific protease. However, a smaller percentage of insulin was degraded in sera containing less than 20 µU/ml, and the results were highly variable. From Starr *et al.* (1975).

of noncompetitive inhibitors of the enzyme in plasma (Cresto *et al.*, 1974; Starr *et al.*, 1975).

B. CHARACTERIZATION OF CIRCULATING PROINSULIN-LIKE COMPONENTS

In most studies PLC has been measured after gel filtration of sera on Sephadex or Bio-Gel columns equilibrated in acetic acid, borate, or Veronal buffers. Because of its higher molecular weight, PLC elutes before insulin and may be identified in either the insulin or human C-peptide immunoassays. However, these methods do not differentiate the two-chain proinsulin intermediates (Steiner *et al.*, 1968) from the single-chain precursor, and additional techniques are required to demonstrate their presence in the circulation.

Lazarus *et al.* (1972) have described the presence of a proinsulin intermediate, in addition to proinsulin, in the serum of a patient with a surgically documented carcinoma of the pancreatic islets. The fasting total IRI concentration was approximately 1000 μU/ml, and the percentage PLC was 85%. An acid–alcohol extract of serum was electrophoresed on polyacrylamide gel, and three peaks were identified. Two of these had mobilities corresponding to insulin and proinsulin, while the third peak ran in an intermediate position. The authors suggested that the migration behavior of this peak was compatible with desdipeptide proinsulin (Chance, 1972). Although the biological activity of the PLC was almost 50% that of insulin, it is difficult to be certain of this result because of problems in standardizing the proinsulin components.

In addition to this intermediate form, other components that react in the insulin assay have been identified in sera of patients with islet cell tumors. Thus Gorden *et al.* (1971a) have described a proinsulin-like component that eluted ahead of the proinsulin marker on a 1.5×90 cm column of Sephadex G-50 in a subject with an islet cell carcinoma. The PLC isolated from this patient's serum did not cross-react identically with a porcine insulin standard, whereas dilutions of plasma PLC from healthy subjects and other patients with tumors were indistinguishable from the porcine standard in this assay system. Its biological activity was three times greater than porcine proinsulin, and it was converted to insulin by exposure to trypsin. It is of interest that the serum insulin component also did not exhibit immunological identity with a porcine insulin standard. Because of the small amounts of material available, further characterization was not achieved.

Yet another form of immunoreactive insulin, which has been named "big, big insulin," has been described in the plasma of an insulinoma suspect by Yalow and Berson (1973). This component had a molecular weight greater than 100,000, was immunochemically identical to human insulin, was more basic than porcine or human insulin, and was rapidly transformed by trypsin to an insulin-like component. Sramkova *et al.* (1975) have subsequently reported that this component yielded ordinary 6000 molecular weight insulin when extracted with acid–ethanol. The high-molecular-weight immunoreactive insulin-like material in this patient was apparently due to the production of an abnormal insulin-binding globulin by a plasmacytoma.

The presence of a high-molcular-weight immunoreactive component in an acid–ethanol extract of an insulinoma was described by Melani et al. (1970b). This material, which eluted in the void volume of the Bio-Gel P-30 column, comprised 10% of the total immunoreactive insulin-like material. It reacted in the C-peptide assay but was not converted to insulin or proinsulin by trypsin. In additional studies on insulinoma patients, we have noted variable amounts of material eluting in the void volume of Bio-Gel P-30 columns that react with insulin antibodies. Yalow and Berson (1973) also have found small amounts (0.7–1.0%) of such components in acid–ethanol extracts of two islet cell adenomas and a normal pancreas. The origin and significance of these components is still uncertain, and further work is necessary in order to characterize them fully. Nunes-Corra et al. (1974) studied a patient who suffered from severe hypoglycemic attacks, presumably on the basis of an islet cell adenoma. In addition to insulin and proinsulin, they found a third peak in the patient's serum with a molecular weight of 24,000. The material in this peak reacted with insulin antibodies and had approximately 50–100% of the biological activity of insulin. It seems unlikely that this component could be closely related to either preproinsulin or proinsulin since neither form could be expected to have such a high level of biological activity.

C. CIRCULATING PROINSULIN LEVELS IN NORMAL SUBJECTS AND DISEASE STATES

The mean fasting proinsulin levels in normal subjects is 0.16±0.2 ng/ml (Mako et al., 1973). In studies using a human insulin standard for measurement of both proinsulin and insulin, PLC comprises 15% of the total immunoreactive insulin concentration (range 0–22%). After oral glucose, the levels of proinsulin rise slowly and peak later than insulin (Fig. 6). When expressed as a percentage of the insulin concentration, a decline from the fasting value is observed during the first 15–60 minutes (Gorden and Roth, 1969). Thereafter proinsulin contributes an increasing amount to the immunoreactive insulin level. Obese patients with hyperinsulinemia have raised fasting concentrations of proinsulin and a greater absolute increase after glucose than subjects of normal weight (Melani et al., 1970a). However, the high levels of proinsulin coexist with raised insulin concentrations, so that the relative proportions of the two polypeptides are generally in the same range observed in healthy subjects (Gordon and Roth, 1969).

Great interest has been expressed in the possibility that an altered proinsulin: insulin ratio might occur in patients with diabetes mellitus. Initial results in a limited number of both normal weight and obese patients with mild diabetes

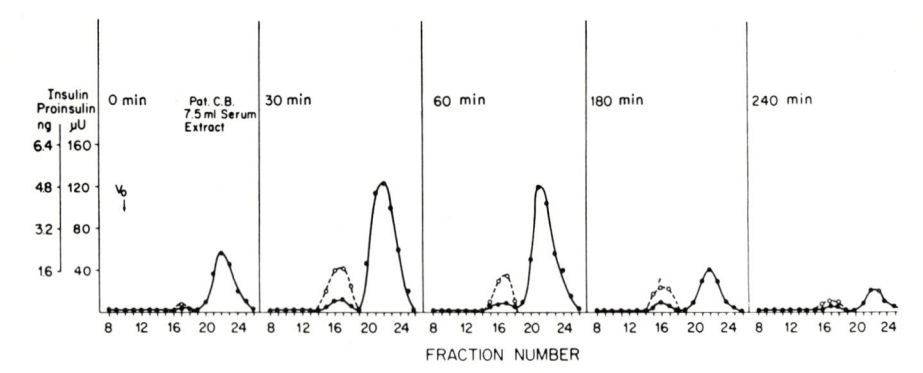

FIG. 6. Gel filtration patterns of sera from a healthy subject separated on Sephadex G-50 1×50 cm) columns equilibrated in 1 *M* acetic acid. The samples were taken in the fasting state and at various times following the administration of 100 gm of glucose. From Melani *et al.* (1970a).

characterized only by glucose intolerance, demonstrated basal and postglucose responses indistinguishable from control subjects (Melani *et al.*, 1970a; Gorden and Roth, 1969; Gordon *et al.*, 1971b). More recently, Duckworth *et al.* (1972), using the insulin-specific protease to measure insulin, reported that both obesity and carbohydrate intolerance were associated with slightly increased PLC levels, but that the coexistence of the two conditions, especially in older diabetics, was marked by significantly elevated PLC concentrations and a rise in the PLC:IRI ratio. In contrast, children with mild diabetes and elevated immunoreactive insulin levels have proinsulin values within the normal range (Fig. 7) in both the fasting and stimulated state (Rosenbloom *et al.*, 1975; Burghen *et al.*, 1976). Similar results have been reported in late-gestational diabetic pregnancies (Phelps *et al.*, 1975; Kuhl, 1976). PLC levels were slightly elevated during gestation but did not constitute an abnormal proportion of the total insulin immunoreactivity in nondiabetic or mildly diabetic women, since the insulin concentration increased in parallel (Fig. 8).

There are a number of conditions that are characterized by an elevated proinsulin:insulin ratio. An increase in PLC with age was reported by Duckworth and Kitabchi (1976). Gorden *et al.* (1972) showed that in six patients with severe hypokalemia of diverse etiologies the PLC formed a greater proportion of the total IRI in the basal and poststimulated state. These subjects were glucose intolerant and exhibited delayed and low insulin responses, and the high PLC percentage was at least partly a result of their insulinopenic state. Correction of the hypokalemia reversed this abnormality. In severe diabetics (Fig. 8) and other patients with low fasting insulin levels, a similar situation may be found (Gordon *et al.*, 1974; Phelps *et al.*, 1975). Mako *et al.* (1973) have pointed out that the absolute concentration and percentage PLC in the basal state in patients with

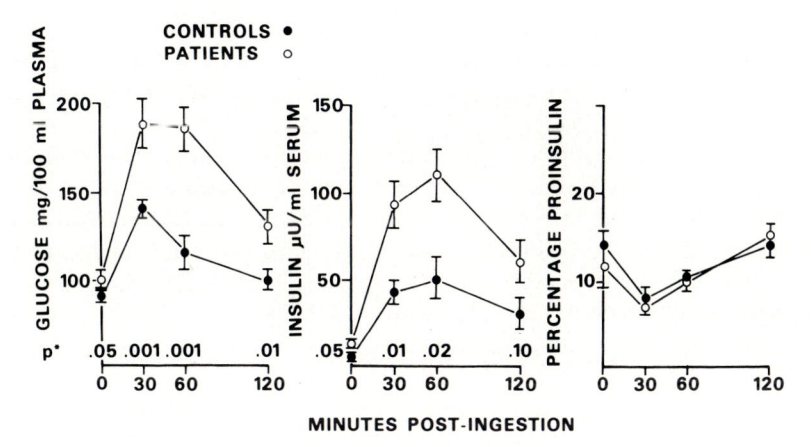

FIG. 7. Plasma glucose and serum insulin concentrations and percentage of immunoassayable insulin activity represented by proinsulin in seven controls and seven chemical-diabetic children and adolescents during oral glucose tolerance testing. P^* = difference between control and patient means. From Rosenbloom et al. (1975).

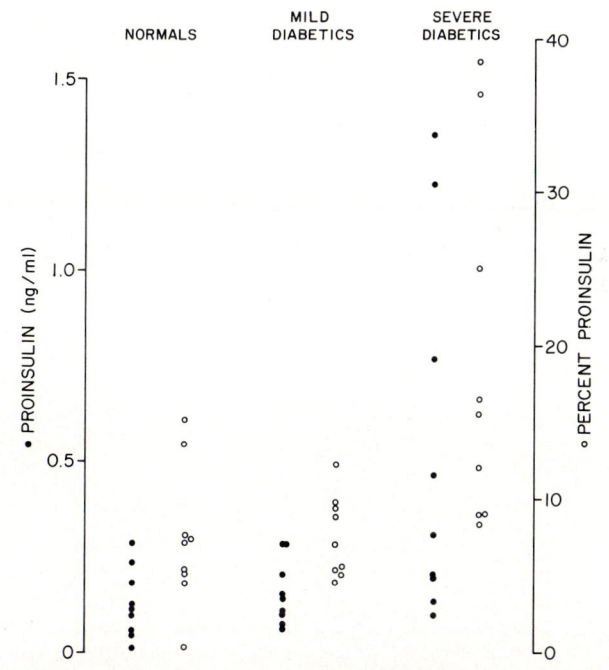

FIG. 8. Plasma proinsulin after an overnight fast in late pregnancy. Individual values have been depicted in absolute terms (ng/ml) by the filled circles, and as a percentage of the total immunoreactive insulin recovered during column fractionation of plasma (percentage proinsulin) by the open circles. From Phelps et al. (1975).

chronic renal failure are markedly elevated and that the values may overlap those found in subjects with islet cell tumors. The reason for this finding lies in the critical role of the kidney as the major organ involved in proinsulin degradation (Katz and Rubenstein, 1973).

The major clinical significance of a raised serum proinsulin concentration has been in the diagnosis of pancreatic islet cell tumors (Fig. 9). We have studied 17 patients with beta-cell adenomas and two with carcinomas (Rubenstein *et al.*, 1974). The absolute basal PLC concentrations varied between 0.23 and 17.48 ng/ml. Only three patients overlapped the normal range (0.038–0.45 ng/ml). The percentage of PLC ranged from 2.9 to 71 (normal values 3.0 to 22.0%), and only four of the insulinoma group fell within the range of the control subjects (Fig.

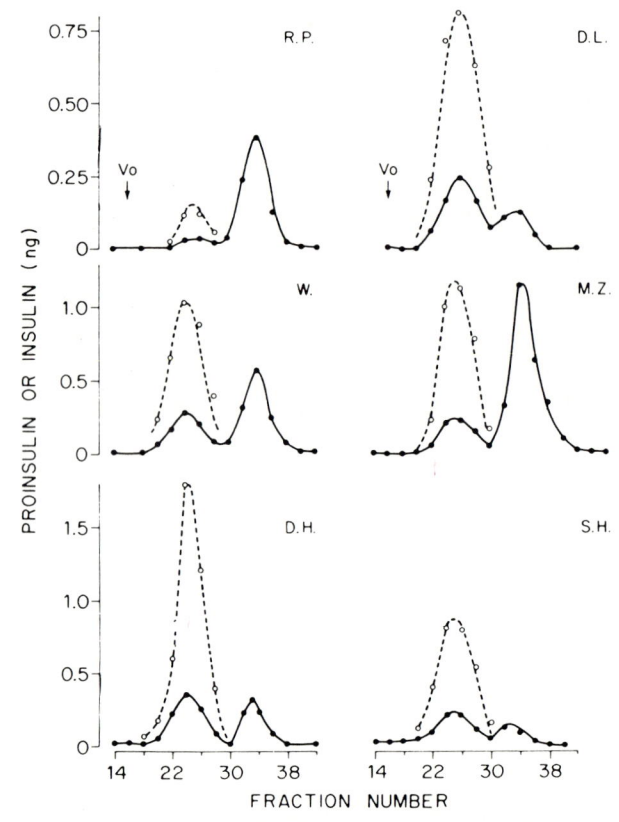

FIG. 9. Gel filtration patterns of proinsulin-like components (PLC) and insulin in a control subject (top left) and five patients with islet cell tumors. The sera were gel filtered on Bio-Gel P-30 columns equilibrated in a borate albumin buffer and measured in the insulin immunoassay. The dotted line represents the PLC value read from the human proinsulin standard. Note the different scales on the ordinate. From Rubenstein *et al.* (1974).

10). These results are similar to those of Sherman *et al.* (1972), who described 3 of 21 islet cell tumor patients with basal IRI concentrations within their normal range, but only 1 with a normal proinsulin concentration. Four subjects had percentage PLC which overlapped their controls. The findings in these two studies are representative of the conclusions in a number of other reports (Blackard *et al.*, 1970; Goldsmith *et al.*, 1969; Pearson *et al.*, 1972; Lazarus *et al.*, 1970; Gorden *et al.*, 1971b; Gutman *et al.*, 1971; Creutzfeldt *et al.*, 1973).

We have considered the possibility that serum PLC estimations may be useful in differentiating benign beta-cell tumors from carcinomas. The basal PLC percentage was 58 and 70 in two of our patients with malignant tumors, but values in this range (above 50%) were also noted in five subjects with adenomas. Similar results were found by Sherman *et al.* (1972) (two patients with malignant tumors had percentage PLC of 62 and 76, while five adenoma cases also had values higher than 50%). The patient with a malignant tumor had the highest percentage PLC (89%) in the study of Gutman *et al.* (1971), and Gorden *et al.* (1971b) reported values of 38, 46, 53, and 78% in four such patients. Although Blackard *et al.* (1970) did not measure basal PLC, the percentage 2 hours after

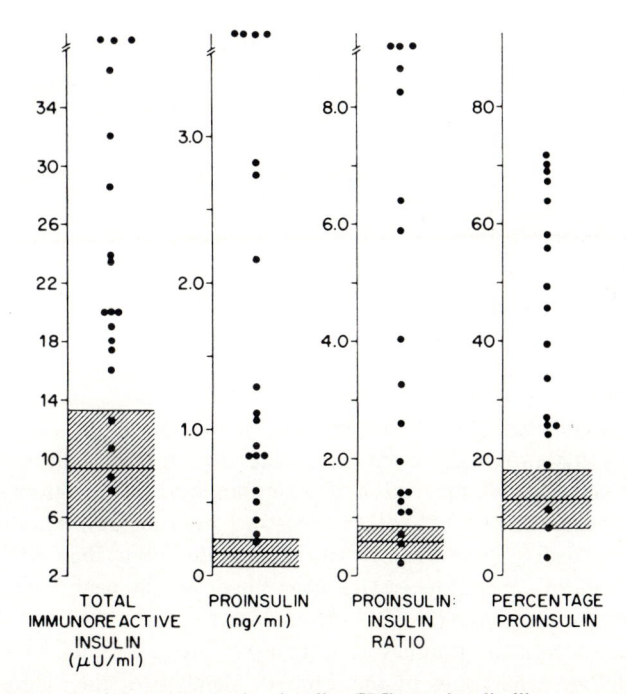

FIG. 10. Basal total immunoreactive insulin (IRI), proinsulin-like components, proinsulin:insulin ratio, and percentage of proinsulin in 19 patients with islet cell tumors. The mean ± 1 SD of 46 control subjects is shown in the hatched area. From Rubenstein *et al.* (1974).

oral glucose was 53%. The finding of Pearson *et al.* (1972) is also of interest in this regard, because the percentage PLC in their patient with an islet cell carcinoma rose to 80% over an 8-week period. The authors concluded that the percentage PLC may rise as loss of tumor differentiation occurs. Nevertheless, as many drugs and continuous glucose infusions were administered during this time, the results should be interpreted with caution. Very high PLC values in a further two patients with carcinomas have also been reported by Taylor *et al.* (1970) and Lazarus *et al.* (1970).

These results suggest that most patients with malignant islet cell tumors do have a markedly elevated percentage PLC. However, a significant number of subjects with adenomas also fall into this high range. On the other hand, it seems that the finding of a low percentage PLC in a patient with an islet cell tumor does favor the diagnosis of a benign lesion.

D. FAMILIAL HYPERPROINSULINEMIA

For some years we had analyzed plasma samples from many diabetic patients in the hope of finding a defect in the conversion of proinsulin to insulin. We were not successful until I visited Dr. Kenneth Gabbay in Boston and had the opportunity to investigate a young man who had presented 5 years previously with a blood glucose of 41 mg/100 ml. The circumstances of this episode indicated that fasting and alcohol intake may have precipitated the attack. However, a number of tests were carried out to exclude other causes for the hypoglycemia. The glucose values during tolbutamide and intravenous glucose tests were normal, but the levels of serum immunoreactive insulin were elevated.

Fasting serum specimens obtained from the propositus and a normal volunteer were subjected to Bio-Gel P-30 column chromatography (Fig. 11). In contrast to the normal serum, the serum of the propositus showed a markedly increased proinsulin peak, with a smaller insulin peak eluting in later fractions. The proinsulin peak in the propositus constituted 85% of total "insulin" immunoreactivity eluted from the column, in contrast to 12% in the normal control serum (when the proinsulin levels were read from a human insulin standard curve). Our range for the proportion of proinsulin in 62 fasting serum specimens from healthy subjects is 3.5 to 22.0%. The true proinsulin levels in the propositus are four to five times higher and constitute more than 95% of total circulating "insulin" immunoreactivity when the proinsulin peak is read against a human proinsulin standard (Fig. 11, bottom panel).

Figure 12 presents the findings of a survey of blood samples obtained from members of four generations of the kindred. Members of the kindred demonstrating hyperproinsulinemia were found in all four generations and ranged from 2 to 69 years in age. The proportion of male and female affected and unaffected progeny was the same, and the abnormality was transmitted by both sexes (Fig.

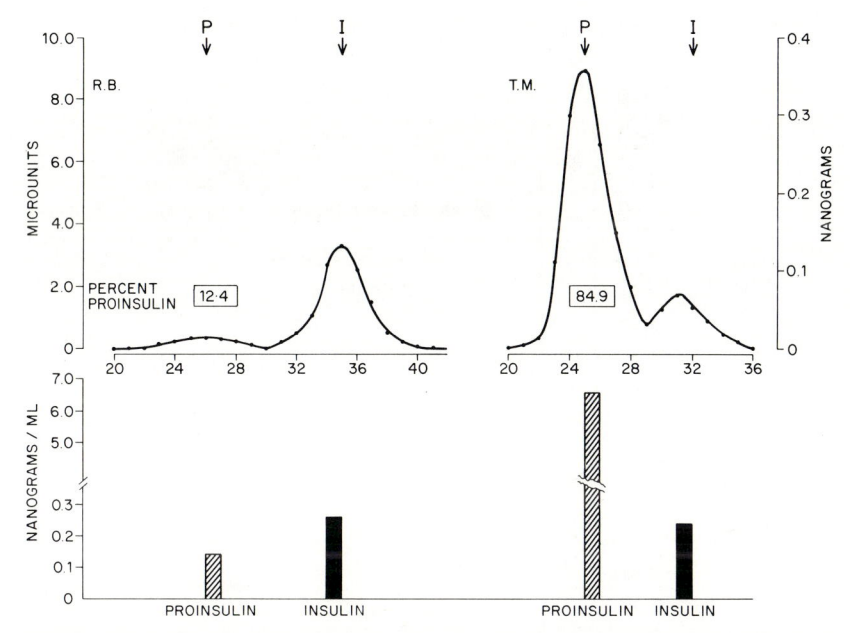

FIG. 11. Proinsulin and insulin peaks, resolved by chromatography and read from a human insulin standard, are shown in the top panels. The results are expressed as microunits (or nanograms) per tube. The serum on the left was taken from a normal subject, and that of the propositus is shown on the right. The proinsulin fraction is expressed as a percent of total immunoreactivity (insulin standard). The bottom panels show absolute amounts of proinsulin (read from a human proinsulin standard) and insulin (read from a human insulin standard) in 1.0 ml of fasting serum. From Gabbay et al. (1976).

12). The proinsulin fraction in the affected members ranged from 74 to 92% (mean of 85%±0.3 SEM) despite wide variation in the levels of direct serum immunoreactive insulin. All the unaffected family members had normal levels of proinsulin, and the percentage contribution of proinsulin to total immunoreactive insulin in these subjects was below 30%. The pattern of inheritance indicates that familial hyperproinsulinemia is transmitted as an autosomal dominant defect.

It seems probable that this kindred represents a genetic defect in proinsulin synthesis or its cleavage (or both) within the beta cell. Pending biochemical definition of the proinsulin in this kindred, the possible existence of partially cleaved proinsulin must also be considered (e.g., proinsulin intermediates cleaved at either B or A chain links with the C-peptide). Such proinsulin intermediates could not be distinguished from proinsulin by the gel-filtration techniques used.

The fortuitous clinical circumstances leading to this discovery in our patient suggest that familial hyperproinsulinemia is an asymptomatic defect. In addition,

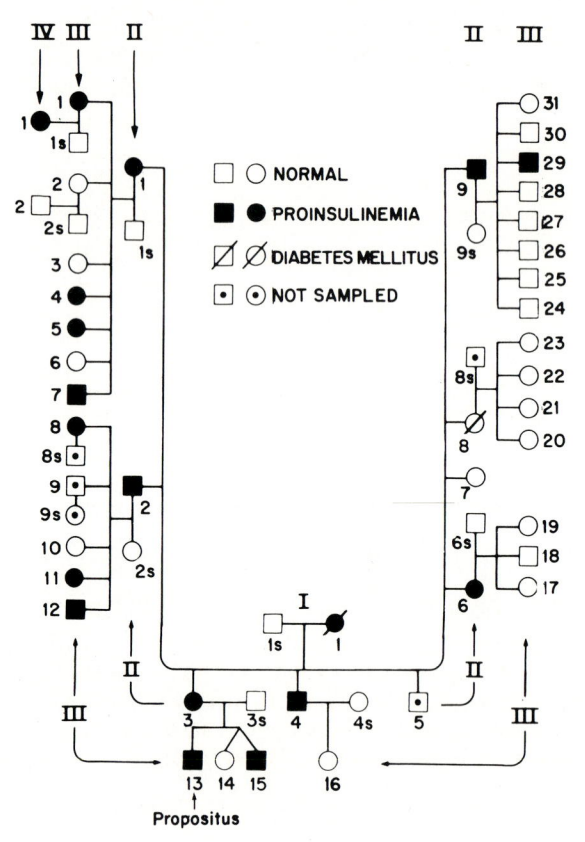

FIG. 12. Members of four generations (I–IV) with hyperproinsulinemia. The propositus (III-13) is indicated by the arrow. An autosomal dominant pattern of inheritance of the defect is indicated by the findings. From Gabbay *et al.* (1976).

this disorder of the beta-cell function does not seem to be related to the development of diabetes mellitus. Chemical diabetes mellitus is known to be present in one affected member of the family, but was also present in one unaffected member. However, definitive conclusions regarding the lack of association of hyperproinsulinemia and diabetes mellitus await more detailed investigation of the chemical nature of the proinsulin and carbohydrate tolerance in the affected family members.

E. PERIPHERAL METABOLISM OF PROINSULIN

Proinsulin comprises 2–9% of the immunoreactive insulin-like material in normal pancreas (Rastogi *et al.*, 1970; Sando *et al.*, 1972; Sando and Grodsky,

1973). This value is similar to that found in the portal vein of man (Horwitz et al., 1975b), but much lower than in peripheral serum. This discrepancy can be explained by the slower metabolism of proinsulin compared to insulin. Thus the immunological half-life of intravenously injected porcine proinsulin was 18 to 20 minutes, and that of insulin 8 and 6 minutes, in baboons and swine, respectively (Stoll et al., 1971). Similar values were found after injection of bovine proinsulin and insulin in dogs (Rubenstein et al., 1970a). Sonksen et al. (1973) infused porcine insulin and proinsulin into healthy subjects and showed a mean metabolic clearance rate of 13.3 ml/kg per minute for insulin and 3.1 ml/kg per minute for proinsulin with mean half-lives of 4.4 and 25.6 minutes, respectively. The half-disappearance time of endogenous proinsulin (18–25 minutes) was markedly slower than that of insulin (3–4 minutes) in three patients after removal of their islet cell tumors (Starr and Rubenstein, 1974) (Fig. 13). It should also be noted that there is no evidence for conversion of proinsulin to insulin in the circulation (Rubenstein et al., 1969b).

In studies in rats, the metabolic clearance rate (MCR) of bovine insulin (16.4±0.4 ml/min) was significantly greater than that of proinsulin (6.7±0.3 ml/min) (Fig. 14). The MCR of both polypeptides was independent of plasma levels over a wide range of steady-state plasma concentrations varying from 1 to 15 ng/ml. In contrast to the differences in their MCR, the renal disposition of the two polypeptides was similar, being characterized by high extraction and very low urinary clearance (Katz and Rubenstein, 1973). The renal arteriovenous differences of proinsulin and insulin averaged 36 and 40%, respectively, and was linearly related to their arterial concentration between 2 and 25 ng/ml. The fractional urinary clearance never exceeded 0.6%, indicating that more than 99% of the amount filtered was sequestered in the kidney.

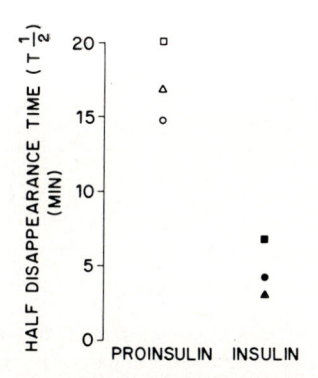

FIG. 13. The half-disappearance times of proinsulin and insulin following removal of islet cell tumors in three patients. The mean value for proinsulin was 17.2 minutes, and for insulin 4.8 minutes. From Starr and Rubenstein (1974).

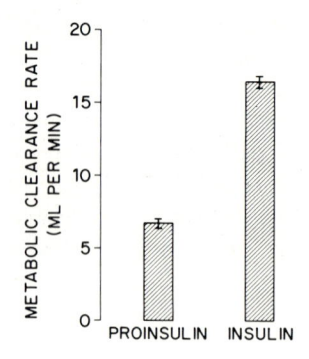

FIG. 14. Comparisons of the metabolic clearance of bovine insulin and proinsulin in rats.

On the other hand, studies on the removal of bovine proinsulin and insulin by the isolated perfused rat liver have shown that the hepatic extraction of proinsulin is considerably slower than that of insulin, at both high and low concentrations (Fig. 15). That this finding was probably not related to the heterologous system used was demonstrated by comparing the degradation of labeled rat and bovine proinsulin by both rat and bovine liver homogenates (Fig. 16) (Rubenstein et al., 1972). Stoll et al. (1970) have also shown, using an isolated perfused liver, that porcine monocomponent insulin was cleared with a $T_{1/2}$ of 17 minutes, while there was no significant clearance of porcine proinsulin.

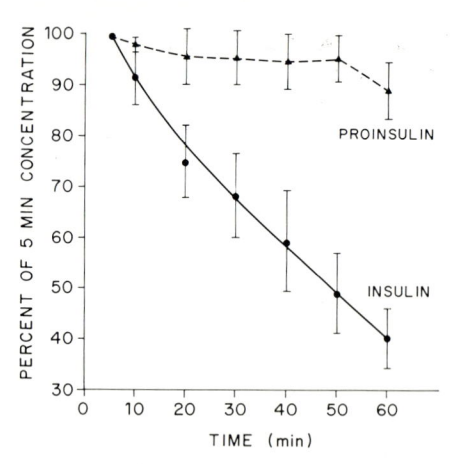

FIG. 15. Disappearance curves of high concentrations of insulin and proinsulin in the isolated perfused rat liver. The concentrations 5 minutes after the addition of the hormones were 168–215 ng/ml, and the results are expressed as a percentage of these levels. From Rubenstein et al. (1972).

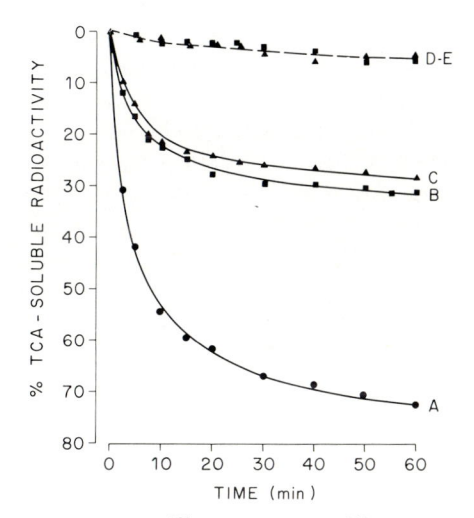

FIG. 16. Degradation of bovine [^{131}I]insulin (A), [^{131}I]proinsulin (B), and rat [^{131}I] proinsulin (C) by rat liver homogenates. The inhibition of rat (D) and bovine (E) [^{131}I]proinsulin degradation by 50 μg of bovine proinsulin is shown. From Rubenstein *et al*. (1972).

A protease that is relatively specific for insulin, and does not significantly degrade proinsulin, has been described by Brush (1971), who isolated the enzyme from muscle, and by Burghen *et al*. (1972), who found a similar enzyme in liver. Studies by Kitabchi and Stentz (1972) on the degradation of insulin and proinsulin by rat organ homogenates showed variation in the relative ability of different tissues to degrade insulin and proinsulin. Only pancreas and kidney homogenates degraded immunoreactive proinsulin at a rate greater than 10% that of insulin. Whether or not this particular enzyme (ISP) is important in the degradation of insulin under physiological conditions is still uncertain. The problem of defining the specific cellular site where insulin degradation takes place under physiological circumstances is important and will undoubtedly require innovative experiments to solve.

III. Circulating C-Peptide

Although the insulin immunoassay has been widely applied to the study of circulating insulin levels in healthy subjects and in diabetics managed with diet or oral hypoglycemic agents, the method has proved less useful in insulin-treated diabetic patients. The reasons for this include the development of circulating insulin antibodies in response to repeated injections of bovine and/or porcine insulin and the immunological similarity of these species of insulin compared to the human hormone. Despite the fact that a number of investigators have devised ingenious methods to circumvent these problems, there is little information about the natural history of beta-cell function in insulin-treated diabetics. The development of an immunoassay for human C-peptide has provided an alternative means of studying beta cell secretory capacity in these patients (Fig. 17). Proteolytic conversion of proinsulin results in the formation of equimolar concentrations of C-peptide and insulin (Kemmler and Steiner, 1970; Kemmler *et al.*, 1971; Steiner *et al.*, 1974), which are stored within the beta-cell granules and subsequently liberated together during exocytosis (Rubenstein *et al.*, 1969a). Circulating levels of C-peptide thus reflect beta-cell secretory activity and, because neither insulin, nor insulin antibodies interfere with the C-peptide immunoassay, its measurement has opened up new approaches for the study of beta-cell function in a variety of physiological and pathological circumstances.

A. METHODOLOGICAL CONSIDERATIONS

Widespread availability of the human C-peptide immunoassay has been hindered by several factors. First, because of the poor cross-reactivity between human C-peptide and all the other C-peptides which have been studied (Rubenstein *et al.*, 1970b), it has been appreciated that human C-peptide must be used for standard, radioactive label, and as antigen for the production of antibodies. Second, human C-peptide has a relatively low molecular weight (3021 daltons)

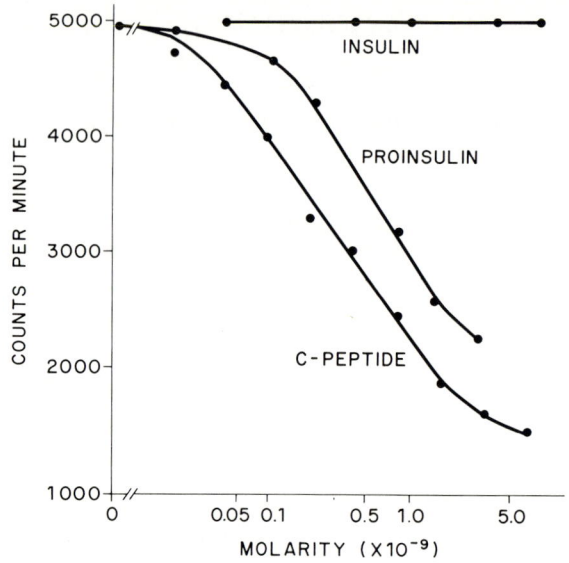

FIG. 17. Standard curve of human C-peptide and proinsulin with an antiserum raised to human C-peptide. Insulin does not cross-react in this system.

and thus tends to be a poor antigen. Even when coupled to large proteins, such as bovine serum albumin, immunization with C-peptide does not consistently result in a high-titer antiserum in either rabbits or guinea pigs. It is possible that a lack of rigid secondary or tertiary structure (Frank and Veros, 1968; Markussen, 1971) renders this peptide poorly immunogenic.

A number of investigators have succeeded in developing assays that are suitable for measuring the concentration of human C-peptide in serum (Melani *et al.*, 1970c; Block *et al.*, 1972a; Faber *et al.*, 1976; Heding *et al.*, 1974; Heding, 1975; Heding and Rasmussen, 1975). However, there are still many problems with these methods that need to be appreciated in order to interpret and compare the results from different laboratories. Because of the difficulty in purifying sufficient natural human C-peptide, most assays utilize standards of synthetic C-peptide. Under these circumstances, it is important to confirm the immunological identity of the serum or urine C-peptide and the synthetic C-peptide standard. Labeling is complicated by the absence of tyrosine residues in the C-peptide, necessitating the addition of this amino acid to the natural or synthetic molecule before iodination with ^{125}I. Heding (1975) has drawn attention to additional potential sources of error in the determination of serum C-peptide. She has shown that there is considerable variation in the displacement of the tracer from different C-peptide antisera by sera from long-term insulin-treated juvenile diabetics. As certain antisera give zero values with these serum samples, it seems probable that some antisera to human C-peptide are capable of reacting to varying degrees with substances in serum that are different from C-peptide and proinsulin (Heding, 1975). Similar results were obtained by Faber *et al.* (1976). Recently, both Heding *et al.* (1974) and Heding (1975) and Kuzuya *et al.* (1977a) have obtained evidence that immunoreactive serum C-peptide may exist in more than form. Gel filtration of sera revealed a smaller molecular weight peptide in addition to the major component which corresponded to the pancreatic C-peptide standard. The immunoreactivity of this smaller peptide differed markedly when assayed with various antisera, and its presence may thus partially explain the wide range in serum C-peptide concentrations reported by different investigators.

Because the serum proinsulin concentration is much lower than that of C-peptide and many C-peptide antisera react less well with proinsulin than with C-peptide, the contribution of proinsulin to serum C-peptide immunoreactivity is very small under most circumstances. However, in patients with islet cell tumors, hypokalemia or chronic renal failure, the proinsulin level may rise appreciably (see Section I). Removal of proinsulin (and insulin) can be accomplished by incubating these sera with Sepharose-coupled insulin antibodies, and the C-peptide can then be determined directly in the supernatant (Heding, 1975). In insulin-treated diabetic patients, proinsulin binds to circulating insulin antibodies (Block *et al.*, 1972a) (Fig. 18). Because the half-life of the proinsulin–

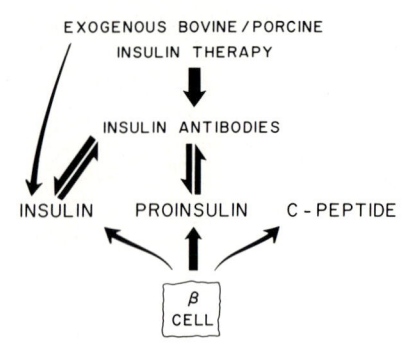

FIG. 18. Schematic representation of the binding of endogenously secreted proinsulin (and insulin) to circulating antibodies formed in response to bovine and/or porcine insulin therapy. The secreted C-peptide does not bind to these antibodies. From Block *et al.* (1972a).

antibody complex is markedly prolonged compared to that of proinsulin itself, antibody-bound proinsulin may accumulate to the point where the proinsulin concentration is substantially higher than that of C-peptide (Fig. 19) (Fink *et al.*, 1974). In samples from these diabetics, direct measurement of C-peptide immunoreactivity may indicate that beta-cell secretion is taking place, but the values cannot be compared to those in control subjects or be used to quantitate insulin secretion accurately. In this regard, Kuzuya *et al.* (1977b) have evaluated the use of polyethelene glycol for precipitating insulin antibodies from sera, followed by measurement of C-peptide in the supernatant. This simple procedure can also be used for the determination of free insulin levels and has considerable potential for the evaluation of insulin-requiring diabetic patients (Fig. 20).

B. CIRCULATING C-PEPTIDE

1. Healthy Subjects

C-peptide, together with insulin and proinsulin, is secreted by the pancreatic beta cells into the portal circulation and must pass through the liver before entering the peripheral blood. In order to relate peripheral concentrations to beta-cell secretion, simultaneously determined portal and peripheral blood levels (Fig. 21) were compared in subjects whose portal blood was obtained by umbilical vein catherization prior to operation (Horwitz *et al.*, 1975b). Following stimulation by intravenous glucose or arginine, both insulin and C-peptide reached peak concentrations at 90–120 seconds after the onset of the stimulus, but the peak did not occur until 2–5 minutes in the peripheral circulation. In portal blood, the relative increase over basal values for insulin was greater than

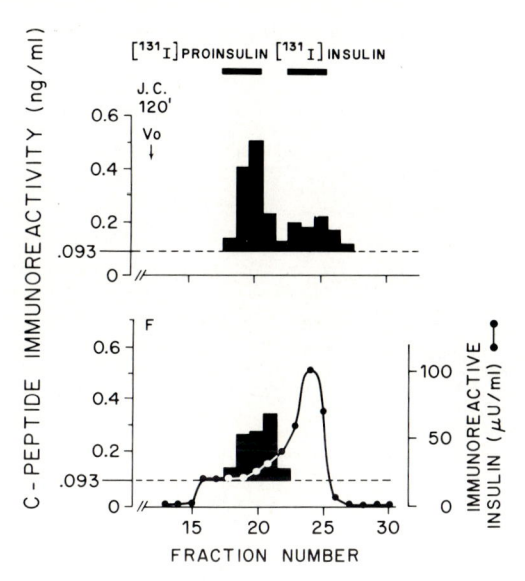

FIG. 19. Gel filtration of acid–ethanol extracts of serum samples from an adult-onset insulin-requiring diabetic taken in the fasting state and 2 hours after glucose. The block bars indicate the results with the C-peptide assay (C-peptide standard) and the line (●————●) shows the results with the insulin assay. The column was calibrated with [^{131}I]proinsulin and [^{131}I]insulin (top of figure). From Block *et al.* (1972a).

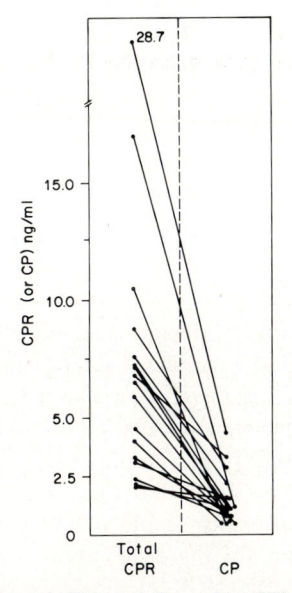

FIG. 20. Comparison of total CPR and C-peptide (CP) concentration in sera containing insulin antibodies. The values determined by direct assay (total CPR) were compared with those measured after polyethylene glycol precipitation (C-peptide). The sera were obtained from insulin-treated diabetics, a patient with insulin autoimmune syndrome, a patient with factitious hypoglycemia, and newborn babies from insulin-treated diabetic mothers. From Kuzuya *et al.* (1977b).

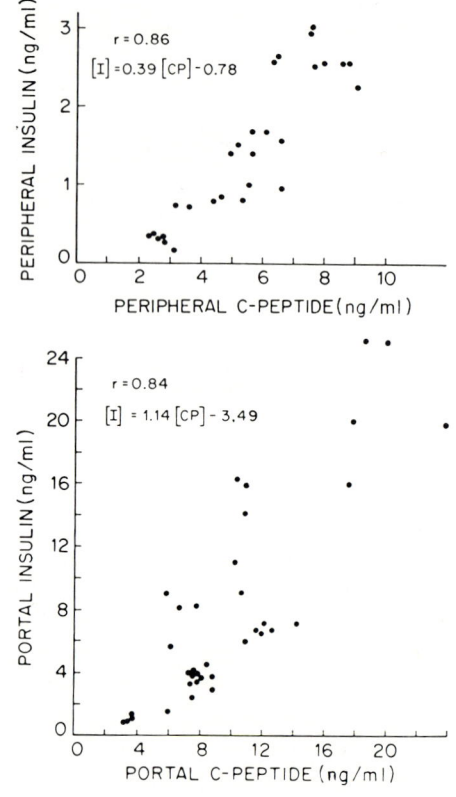

FIG. 21. Correlation between insulin and C-peptide (CP) concentration in both portal and peripheral blood during and after glucose infusions in three subjects. Product moment correlation coefficients (r) and least-squares regression equations are shown. From Horwitz et al. (1975b).

that for C-peptide, while in the peripheral circulation this difference was not as great. In general, both portal and peripheral serum concentrations of C-peptide were greater than those of insulin on a molar basis, but at times of peak secretion insulin and C-peptide were present in portal blood in nearly equimolar quantities. These findings confirm that insulin and C-peptide are released by the beta cell in equimolar concentrations, but this relationship is not preserved in the peripheral circulation. Differences in the metabolic clearance rates (MCR) of insulin and C-peptide account for the different ratio of these two peptides in peripheral blood (Fig. 22). The MCR of insulin is much faster than that of C-peptide in rats (16.4±0.4 ml vs 4.6±0.2 ml per minute) (Katz and Rubenstein, 1973). In man, insulin disappears from the circulation with a half-life of approximately 4–5 minutes, while that of C-peptide is considerably longer.

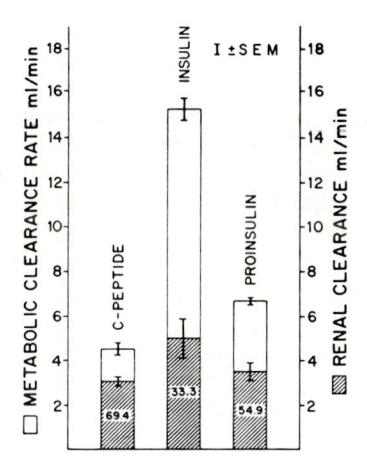

FIG. 22. Metabolic clearance rates of insulin, proinsulin, and C-peptide. The contribution of the kidney to the total metabolic clearance rate of the peptides is shown by the hatched area. From Katz and Rubenstein (1973).

Furthermore, substantial amounts of insulin are removed in its initial transit through the liver, while the hepatic extraction of C-peptide is very low or negligible (Stoll et al., 1970). These metabolic differences alter the equimolar ratio of C-peptide to insulin in the portal vein to values greater than five in the peripheral circulation (Heding and Rasmussen, 1975; Horwitz et al., 1975b). The ratio decreases under non-steady-state conditions when beta cell secretion is stimulated by glucose or arginine. Nevertheless, the correlation of the two polypeptides in the systemic circulation is reproducible (Fig. 21), so that C-peptide concentrations provide an accurate picture of insulin secretion under most circumstances.

Fasting C-peptide concentrations in healthy subjects range between 0.9 and 3.5 ng/ml (Block et al., 1972a; Heding and Rasmussen, 1975; Kaneko et al., 1974). After the administration of oral glucose, the levels rise approximately 5- to 6-fold to 4.4–7.6 ng/ml (Fig. 23). Because of its slower metabolism, C-peptide levels tend to remain elevated after those of insulin have returned to basal values. Serum C-peptide levels have also been measured in six normal pregnant women over a 24-hour period. The mean value of 18 samples taken over a 24-hour period in these subjects ranged from 2.53 to 3.81 ng/ml (Lewis et al., 1976).

2. Diabetic Patients

Because the insulin immunoassay cannot readily distinguish endogenously secreted insulin from exogenously administered bovine or porcine insulin in insulin-requiring diabetic patients, measurement of C-peptide has proved to be extremely helpful in assessing their beta-cell secretory function. The poor

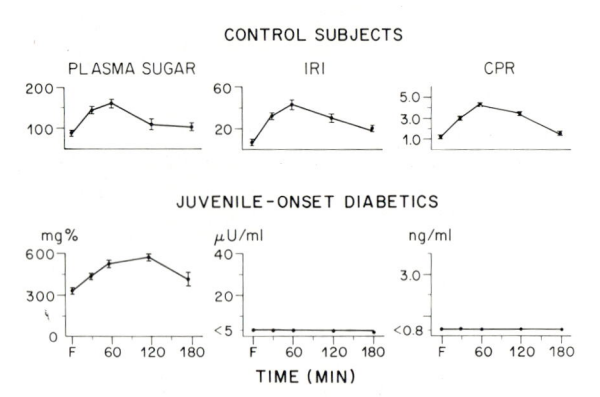

FIG. 23. Plasma sugar (mg/100 ml), immunoreactive insulin (IRI), and C-peptide (CPR) levels during an oral glucose tolerance test in control subjects (above) and five juvenile-onset diabetics at the time of diagnosis of their disease (below). From Block et al. (1972a).

cross-reactivity between human, bovine, and porcine proinsulins and C-peptides (Rubenstein et al., 1970b) has indicated that the assay for human C-peptide will not be affected by small amounts of these species of proinsulin, which may be present as impurities in commercial insulin preparations (Steiner et al., 1968).

Both insulin and C-peptide levels are very low or unmeasurable during episodes of ketoacidosis in adult-onset diabetics (Fig. 24) (Block et al., 1972b). Similarly, C-peptide was undetectable in five newly diagnosed untreated juvenile-onset diabetics with fasting blood sugars above 200 mg/100 ml (Fig. 23) (Block et al., 1972a). Low C-peptide values were also found in 10 untreated diabetics with circulating insulin levels below 30 μU/ml during an oral glucose tolerance test (Heding and Rasmussen, 1975). Following correction of the ketoacidosis or severe hyperglycemic by insulin therapy, partial recovery of beta-cell function occurred as assessed by increases in serum C-peptide (Block et al., 1972b). Sequential studies in three insulin-treated diabetic patients showed significant increases in serum C-peptide during the phase of clinical remission (Block et al., 1973b), while clinical relapse was again associated with decreasing serum C-peptide concentrations (Fig. 25). Heding and Rasmussen (1975) have described three patients who had been treated with insulin for varying periods of time (3 months to 9 years), but whose therapy had been recently discontinued. A small increase in C-peptide in response to glucose was noted. The test was repeated after 9 months in one of the patients, and a considerable improvement in the C-peptide response was observed. These studies indicate that beta-cell secretory ability may be temporarily impaired, and that partial recovery in their function may occur. At present there is little information about the factors that may be

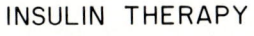

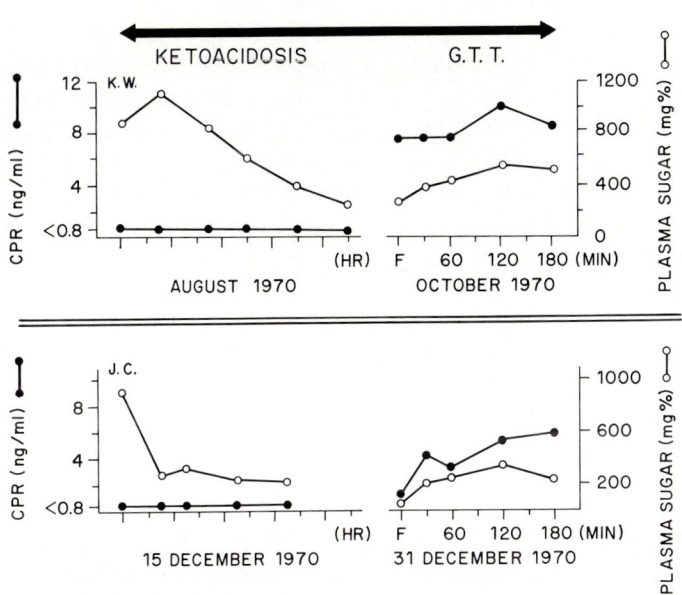

FIG. 24. Serum C-peptide and plasma sugars (mg/100 ml) during the initial 10 hours of therapy for ketoacidosis and 12 weeks (KW) and 2 weeks (JC) later during oral glucose tolerance test (GTT). From Block *et al.* (1972b).

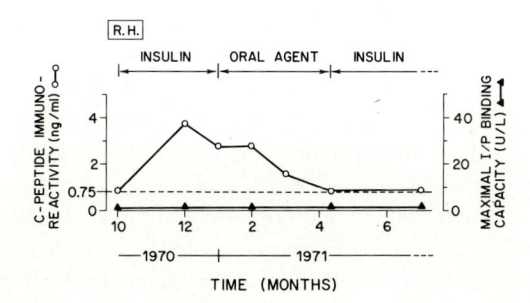

FIG. 25. Serial values of serum C-peptide immunoreactivity and insulin-binding capacity at the time of diagnosis, during remission, and at the time of exacerbation of the disease. The dashed line represents the lower limit of sensitivity of the C-peptide assay. From Block *et al.* (1973b).

important in affecting this process, but long-term studies can now be undertaken with the aim of identifying them. Complete recovery from insulin-dependent diabetes is a rare but fascinating clinical event, and the C-peptide assay has proved very useful in monitoring the course of such patients. An example of such a patient with well documented mumps virus infection and diabetic ketoacidosis was described by Block et al. (1973a). After clinical recovery from mumps, her need for insulin gradually diminished, with eventual return of normal carbohydrate tolerance. Her serum insulin could not be measured because of the development of circulating insulin antibodies, but glucose and C-peptide responses were normal in both standard and cortisone-primed glucose tolerance tests.

Measurement of serum C-peptide in patients with established diabetes has shown varying degrees of beta-cell secretory impairment. Five ketosis-prone juvenile diabetics who had been treated with insulin for longer than 5 years had no measurable C-peptide in either the basal or poststimulatory state (Block et al., 1972a). However, subsequent studies have indicated that C-peptide may be detected in many of these patients during the first years of their disease. In a series of 35 juvenile diabetics, Grajwer et al. (1977) found that 19 of 21 patients with a duration of diabetes of less than 5 years had detectable levels of circulating C-peptide, while only 6 of 14 patients whose disease had been present for longer than 5 years had similar values. This finding suggests that the loss of beta-cell secretory capacity in juvenile-onset diabetes is not abrupt, but continues for several years after the diabetes becomes clinically manifest.

We have been interested in the question of whether retention of beta-cell secretory ability in diabetic patients, albeit at a lower than normal level, has any clinical significance. Preliminary studies have been performed in which C-peptide levels in stable and unstable adult diabetics have been compared. The patients were classified as stable or unstable on the basis of diurnal plasma and urine glucose variability. The stable group had greater C-peptide levels in both the basal state and following stimulation with either oral glucose or arginine (Reynolds et al., 1974). In the unstable group, basal C-peptide levels were at the lower limit of sensitivity of the assay, and no response to either glucose or arginine could be detected. In addition, these unstable diabetics also showed a severe alpha-cell defect as measured by minimal glucagon responses to hypoglycemia (Reynolds et al., 1974). In a second study, Grajwer et al. (1977) measured C-peptide levels in 35 patients with juvenile onset diabetes. Those patients with the highest C-peptide concentrations all fell into the adequately controlled group, while individuals with very low values were found in both the adequately and poorly controlled categories. These authors concluded that residual insulin secretion in diabetic patients may facilitate good control, but absent beta-cell reserve is not always associated with poor control. Lewis et al. (1976) came to similar conclusions in their studies of pregnant diabetic women.

In those patients without significant beta-cell reserve, meticulous attention to diet and exogenous insulin therapy succeeded in achieving excellent control of their blood sugar concentrations.

The C-peptide assay has also been applied to further our understanding of beta-cell function in neonates of insulin-treated diabetic mothers (Block et al., 1974; Phelps et al., 1977). These infants represent a group in whom neonatal hypoglycemia is frequent and profound, and in whom accelerated glucose utilization has been demonstrated. However, precise documentation of hyperinsulinism has not been possible because of interference in the immunoassay for insulin caused by the transplacental passage of insulin antibodies generated in the mother in response to the administration of commercial insulin preparations. Basal C-peptide and its response to intravenous glucose were examined 2–4 hours after birth in eight such infants and nine controls (Phelps et al., 1977). Following glucose, serum C-peptide and immunoreactive insulin increased in parallel in the normal infants, peak values for both peptides being observed in the late plasma specimens. On the contrary, in five infants of diabetic mothers (IDM), maximal increments in C-peptide occurred in the earliest plasma specimen secured, i.e., at 2 or 10 minutes after glucose. Values for glucose disposal in these five IDM exceeded the highest rates found in other IDM and in the normal infants. These relationships suggest that the greater acute islet responsiveness of the offspring of some insulin-treated mothers may contribute, at least in part, to their increased capacity for glucose disposition.

3. Hypoglycemic Disorders

It is of interest that one of the most useful applications of the C-peptide assay has been to facilitate the diagnosis of various hypoglycemic disorders, including islet cell tumors and factitious hypoglycemia occurring in diabetic and nondiabetic subjects. In keeping with methods used for the diagnosis of most hyperfunctioning endocrine tumors, it would seem useful to be able to show failure of insulin suppression in the presence of hypoglycemia in patients with insulinomas. However, this approach has not been widely used because controlled hypoglycemia in man is best produced by an infusion of exogenous insulin. However, because of structural similarities between bovine/porcine and human insulins, it is difficult to measure endogenously secreted insulin under these conditions. This problem has been circumvented by Turner and Johnson (1973), who produced hypoglycemia in man by administering fish insulin. A fall in the endogenous serum insulin concentration was demonstrated using an antibody that reacted with human, but not fish, insulin. Measurement of C-peptide during hypoglycemia induced by the infusion of porcine insulin provides an alternative method for evaluating inhibition of beta-cell secretion (Horwitz et al., 1975a). We have shown impaired suppression of endogenous insulin (C-peptide) secretion in 11 of 12 patients with surgically proved insulinomas, and similar results have been reported by Heding *et al.* (1975). Although insulinomas and nesidioblastosis are rare in diabetics, they must be distinguished from other causes of spontaneous, fasting hypoglycemia, such as mesenchymal tumors or hepatic or renal failure, as well as from non-insulin-mediated disorders, such as alcohol-induced hypoglycemia. The C-peptide assay offers great potential in this regard, because the major diagnostic difficulties occur in diabetic patients who have been treated with insulin. In such patients, estimation of C-peptide may be used instead of insulin to document the presence of endogenous hyperinsulinism (Sandler *et al.*, 1975).

Another difficult diagnostic problem in both diabetic and nondiabetic patients is distinguishing hyperinsulinemia due to endogenous insulin secretion from that due to the surreptitious injection of exogenous insulin. This distinction may be made by measurement of both immunoreactive insulin and C-peptide in serum taken when the patient is hypoglycemic. Because endogenously secreted insulin is accompanied by the liberation of C-peptide, while exogenous insulin will suppress endogenous secretion and hence lower the serum C-peptide concentration (Horwitz *et al.*, 1975a), serum insulin and C-peptide will show similar elevations in pancreatic hyperinsulinism, while C-peptide will be low in relationship to the serum insulin level following injection of exogenous insulin. This observation has been of great help in diagnosing several cases in which patients were surreptitiously administering insulin to themselves (Couropmitree *et al.*, 1975; Service *et al.*, 1975). It is also worth mentioning that although the finding of insulin antibodies in the serum of a nondiabetic individual was previously believed to be incontrovertible evidence of exogenous insulin administration, this interpretation may be incorrect. Reports from Japan and Norway have suggested that insulin autoantibodies occur in certain individuals, some of whom may present with spontaneous hypoglycemia (Folling and Norman, 1972; Hirata and Ishizu, 1972).

C-peptide measurements are useful in the postoperative evaluation of patients who have undergone "total" pancreatectomy for nesidioblastosis or carcinoma. Significant C-peptide levels indicate the presence of residual pancreatic tissue. If the pancreatectomy was performed for removal of an insulinoma, increasing C-peptide levels suggest either a recurrence of the tumor or the presence of functioning metastases. Serum insulin measurement cannot always be used in this situation, because these patients often require exogenous insulin.

C. URINARY C-PEPTIDE

Although peptide hormone concentrations in blood reflect most accurately the minute-to-minute changes in their secretion, measurement of urine levels may be advantageous in reflecting average serum values over a period of time. Furthermore, this approach may also be useful in situations where repeated blood sampling is difficult, such as in small children. While assay of insulin in urine has been shown to be feasible (Rubenstein, 1969), its value is diminished because only a small fraction of the total amount of insulin secreted appears in the urine. Because the urinary clearance of C-peptide is much higher than that of insulin in animals (Katz and Rubenstein, 1973), a number of investigators have considered the utility of measuring urine C-peptide in man.

In 28 nondiabetic subjects with normal kidney function or various renal diseases, C-peptide clearance was independent of creatinine clearance over a range of 6 to 190 ml/min (Horwitz et al., 1977). Urine C-peptide clearance (5.1±0.6 ml/min) was greater than that of insulin (1.1±0.2 ml/min), and the total quantity of C-peptide excreted in the urine per day represented 5% of pancreatic secretion, as compared to only 0.1% in the case of insulin. Healthy subjects excreted 36±4 µg of C-peptide per 24 hours, while this value in juvenile-onset diabetics was only 1.1±0.5 µg. Adult-onset diabetics excreted 24±7 µg/24 hours, the range overlapping the excretory rates of both normal subjects and juvenile-onset diabetics. Two insulin-requiring, adult-onset diabetics showed significant beta cell reserve during the course of acute infections (Horwitz et al., 1977).

Kaneko et al. (1975) studied the urinary excretion of C-peptide in healthy subjects and diabetic patients before and after the administration of oral glucose or intravenous tolbutamide. The mean basal C-peptide level in the controls was 26.2 ng per milligram of creatinine, and the postglucose value was 66.7 ng per milligram of creatinine. There was a marked decrease in both basal and poststimulatory urinary C-peptide values in newly diagnosed diabetics with elevated fasting blood sugar concentrations over 200 mg/dl and in insulin-treated patients. Furthermore, a close correlation between the increments in plasma and urinary C-peptide levels was observed.

D. FUTURE CONSIDERATIONS

The widespread availability of methods for the measurement of insulin and C-peptide in the plasma of healthy subjects and patients with disorders of glucose homeostasis will undoubtedly continue to provide new and useful information concerning their physiology, diagnoses, and natural history. The most common clinical indications for measuring C-peptide levels are summarized in Table I (Horwitz et al., 1976). However, in order to obtain the maximal information from these new techniques, one should be aware of the potential problems in interpreting the results in insulin-treated diabetic patients who have retained some residual beta-cell secretory capacity. The state and nature of the beta-cell peptides found in these patients' circulation are shown in Fig. 26. C-peptide remains free in the plasma and provides an accurate measure of beta-cell secretion. Endogenously secreted proinsulin and insulin exist both free

TABLE I
Clinical Indications for C-Peptide Measurement

Hypoglycemic states
  1. Diagnosis of insulinoma (or ? beta-cell hyperplasia) in insulin-requiring diabetics
  2. Diagnosis of insulinoma (suppression test)
  3. Diagnosis of surreptitious injection of insulin
Euglycemic state
  1. Demonstration of remission phase or "recovery" from diabetes
Hyperglycemic states
  1. Follow-up evaluation after pancreatectomy
  2. Evaluation of the "brittle" diabetic

---

and bound to insulin antibodies. The injected insulin usually consists of a mixture of bovine and porcine species with a low contaminating amount of bovine and porcine proinsulin. There is, however, no C-peptide in commercial insulin preparations, because this peptide is lost in the crystallization process. The exogenously administered insulins also exist in the free state and bound to insulin antibodies. In addition, bovine and porcine proinsulins also stimulate the production of specific antibodies directed toward their connecting segments in many patients. These species of proinsulin will thus exist both in the free state and bound to insulin antibodies, as well as to specific antibodies directed to their

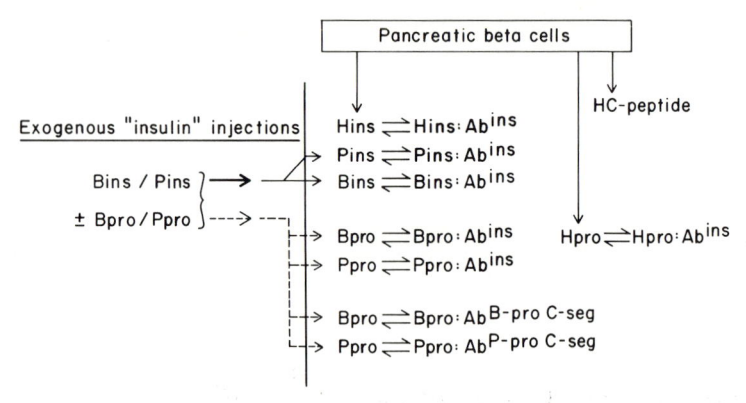

FIG. 26. Circulating antibodies in insulin-treated diabetics. Both endogenous insulin and proinsulin are bound to insulin antibodies (AB^{ins}). Antibodies directed to bovine or porcine proinsulin connecting segment (AB$^{B\text{-}proC\text{-}seg}$, AB$^{P\text{-}proC\text{-}seg}$) bind the homologous species of proinsulin specifically. Endogenous human C-peptide is not bound to any of these antibodies. It should be noted that insulin antibodies may be more specific for bovine or porcine insulin, but this distinction is not made in the figure. H, human; B, bovine; P, porcine; ins, insulin; pro, proinsulin; pro C-seg, proinsulin connecting segment.

connecting peptide regions. The development of methods to identify and quantify these various peptides and their antibodies with precision has provided powerful tools for diagnostic and investigative purposes.

ACKNOWLEDGMENTS

The authors are indebted to many students and colleagues who have contributed to various aspects of this work, including A. Cruz, P. Blix, R. Bergenstal, J. Scarlett, R. Rosenfield, D. Juhn, and J. Karlin. Various aspects of this research have been supported by grants from the U.S. Public Health Service (AM 13941, AM 13914, and AM 19206), the Juvenile Diabetes Foundation, the Bertha and Henry Brownstein Foundation, Eli Lilly and Company, and the University of Chicago Diabetes–Endocrinology Center (AM 17046).

Arthur H. Rubenstein is a recipient of an Established Investigator Award from the American Diabetes Association, and David L. Horwitz holds a Research Career Development Award.

We wish to thank Ms. R. Roberts for her assistance in preparing this manuscript.

REFERENCES

Blackard, W. G., Garcia, A. R., and Brown, C. L. (1970). *J. Clin. Endocrinol. Metab.* **31**, 215.

Block, M. B., Mako, M. E., Steiner, D. F., and Rubenstein, A. H. (1972a). *Diabetes* **21**, 1013.

Block, M. B., Mako, M. E., Steiner, D. F., and Rubenstein, A. H. (1972b). *J. Clin. Endocrinol. Metab.* **35**, 402.

Block, M. B., Berk, J. E., Fridhandler, L. S., Steiner, D. F., and Rubenstein, A. H. (1973a). *Ann. Intern. Med.* **78**, 663.

Block, M. B., Rosenfield, R. L., Mako, M. E., Steiner, D. F., and Rubenstein, A. H. (1973b). *N. Engl. J. Med.* **288**, 1148.

Block, M. B., Pildes, R. S., Mossabboy, N. A., Steiner, D. F., and Rubenstein, A. H. (1974). *Pediatrics* **53**, 923.

Brush, J. S. (1971). *Diabetes* **20**, 151.

Burghen, G. A., Kitabchi, A. E., and Brush, J. S. (1972). *Endocrinology* **91**, 633.

Burghen, G. A., Eheldorf, J. N., Truoy, R. L., and Kitabchi, A. E. (1976). *J. Pediatr.* **89**, 48.

Chance, R. E. (1972). *Proc. Congr. Int. Diabetes Fed., 7th, 1971* p. 292.

Couropmitree, C., Freinkel, N., Hagel, T. C., Horwitz, D. L., Metzger, B., Rubenstein, A. H., and Hahnel, R. (1975). *Ann. Intern. Med.* **82**, 201.

Cresto, J. C., Lavine, R. L., Fink, G., and Recant, L. (1974). *Diabetes* **23**, 505.

Creutzfeldt, W. R., Arnold, C., Creutzfeldt, C., Deuticke, U., Frerichs, H., and Track, N. S. (1973). *Diabetologia* **9**, 217.

Duckworth, W. C., and Kitabchi, A. E. (1976). *J. Lab. Clin. Med.* **88**, 259.

Duckworth, W. C., Kitabchi, A. E., and Heinemann, M. (1972). *Am. J. Med.* **53**, 418.

Faber, O. K., Markussen, J., Naithani J. K., and Binder, C. (1976). *Hoppe-Seyler's Z. Physiol. Chem.* **357**, 751.

Fink, G., Cresto, J. C., Gutman, R. A., Lavine, R. L., Rubenstein, A. H., and Recant, L. (1974). *Horm. Metab. Res.* **6**, 439.

Folling, I., and Norman, N. (1972). *Diabetes* **21**, 814.

Frank, B. H., and Veros, A. J. (1968). *Biochem. Biophys. Res. Commun.* **32**, 155.

Gabbay, K. H., DeLuca, K., Fisher, J. N., Mako, M. E., and Rubenstein, A. H. (1976). *N. Engl. J. Med.* **294**, 911.
Goldsmith, S. J., Yalow, R. S., and Berson, S. A. 969). *Diabetes* **18**, 340.
Gorden, P., and Roth, J. (1969). *J. Clin. Invest.* **48**, 2225.
Gorden, P., Freychet, P., and Nanken, H. (1971a). *J. Clin. Endocrinol. Metab.* **33**, 983.
Gorden, P., Sherman, B., and Roth, J. (1971b). *J. Clin. Invest.* **50**, 2113.
Gorden, P., Sherman, B. M., and Simopoulos, A. P. (1972). *J. Clin. Endocrinol. Metab.* **34**, 235.
Gorden, P., Hendricks, C. M., and Roth, J. (1974). *Diabetologia* **10**, 469.
Grajwer, L. A., Pildes, R. S., Horwitz, D. L., and Rubenstein, A. H. (1977). *J. Pediatr.* **90**, 42.
Gutman, R. A., Lazarus, N. R., Penhos, J. C., Fajans, S. S., and Recant, L. (1971). *N. Engl. J. Med.* **284**, 1003.
Heding, L. G. (1975). *Diabetologia* **11**, 541.
Heding, L. G., and Rasmussen, S. M. (1975). *Diabetologia* **11**, 201.
Heding, L. G., Larsen, V. D., Markussen, J., Jorgensen, K. H., and Hallund, O. (1974). *Horm. Metab. Res., Suppl.* **5**, 40.
Heding, L., Turner, R. C., and Harris, E. (1975). *Diabetes* **24**, Suppl. 2, 412.
Hirata, Y., Ishizu, H. (1972). *Tohoku J. Exp. Med.* **107**, 277.
Horwitz, D. L., Rubenstein, A. H., Reynolds, C., Molnar, G. D., and Yanaihara, N. (1975a). *Horm. Metab. Res.* **71**, 449.
Horwitz, D. L., Starr, J. I., Mako, M. E., Blackard, W. G., and Rubenstein, A. H. (1975b). *J. Clin. Invest.* **55**, 1278.
Horwitz, D. L., Rubenstein, A. H., and Katz, A. I. (1977). *Diabetes* **26**, 30.
Horwitz, D. L., Kuzuya, H., and Rubenstein, A. H. (1976). *N. Engl. J. Med.* **295**, 207.
Kaneko, T., Oka, H., Munemura, M., Oda, T., Yamashita, K., Suzuki, S., Yanaihara, N., Hashimoto, T., and Yanaihara, C. (1974). *Endocrinol. Jpn.* **21**, 141.
Kaneko, T., Munemura, M., Oka, H., Oda, T., Suzuki, S., Yasuda, H., Yanaihara, N., Nakagawa, S., and Makabe, K. (1975). *Endocrinol. Jpn.* **22**, 207.
Katz, A. I., and Rubenstein, A. H. (1973). *J. Clin. Invest.* **52**, 1113.
Kemmler, W., and Steiner, D. F. (1970). *Biochem. Biophys. Res. Commun.* **41**, 1223.
Kemmler, W., Peterson, J. D., and Steiner, D. F. (1971). *J. Biol. Chem.* **246**, 6786.
Kemmler, W., Steiner, D. F., and Borg, J. (1973). *J. Biol. Chem.* **248**, 4544.
Kitabchi, A. E., and Stentz, F. B. (1972). *Diabetes* **21**, 1091.
Kitabchi, A. E., Duckworth, W. C., Brush, J. S., and Heinemenn, M. (1971). *J. Clin. Invest.* **50**, 1792.
Kuhl, C. (1976). *Diabetologia* **12**, 295.
Kuzuya, H., Blix, P. M., Horwitz, D. L., Steiner, D. F., Rubenstein, A. H., Binder, C., and Faber, O. K. (1977a). *J. Clin. Endocrinol. Metab.* (in press).
Kuzuya, H., Blix, P. M., Horwitz, D. L., and Rubenstein, A. H. (1977b). *Diabetes* **26**, 22.
Lazarus, N. R., Tanese, T., Gutman, R., and Recant, L. (1970). *J. Clin. Endocrinol. Metab.* **30**, 273.
Lazarus, N. R., Gutman, R. A., Panhos, J. C., and Recant, L. (1972). *Diabetologia* **8**, 131.
Lewis, S. B., Wallin, J. D., Kuzuya, H., Murray, W. K., Coustan, D. R., Daane, T. A., and Rubenstein, A. H. (1976). *Diabetologia* **12**, 343.
Mako, M. E., Block, M., Starr, J., Nielson, E., Friedman, E., and Rubenstein, A. (1973). *Clin. Res.* **21**, 631.
Markussen, J. (1971). *Int. J. Protein. Res.* **3**, 201.
Melani, F., Rubenstein, A. H., and Steiner, D. F. (1970a). *J. Clin. Invest.* **49**, 497.

Melani, F., Ryan, W. G., Rubenstein, A. H., and Steiner, D. F. (1970b). *N. Engl. J. Med.* **283**, 713.
Melani, F., Rubenstein, A. H., Oyer, P. E., and Steiner, D. F. (1970c). *Proc. Natl. Acad. Sci. U.S.A.* **67**, 148.
Nolan, C., Margoliasch, E., Peterson, J. D., and Steiner, D. F. (1971). *J. Biol. Chem.* **246**, 2780.
Nunes-Correa, J., Lowy, C., and Sonksen, P. H. (1974). *Lancet* **1**, 837.
Pearson, M. J., Larkins, B. G., and Martin, F. I. R. (1972). *Metab., Clin. Exp.* **21**, 551.
Phelps, R. L., Bergenstal, R., Freinkel, N., Rubenstein, A. H., Metzger, B. E., and Mako, M. (1975). *J. Clin. Endocrinol. Metab.* **41**, 1085.
Phelps, R. L., Freinkel, N., Rubenstein, A. H., Metzger, B. E., Kuzuya, H., Boehm, J. J., and Molsted-Pedersen, L. (1977). *J. Clin. Endocrinol. Metab* (in press).
Rastogi, G. K., Letarte, J., and Fraser, T. R. (1970). *Lancet* **1**, 7.
Reynolds, C., Horwitz, D. L., Molnar, G. D., Rubenstein, A. H., and Taylor, W. F. (1974). *Diabetes* **23**, Suppl. 1, 343.
Rosenbloom, A. L., Starr, J. I., Juhn, D., and Rubenstein, A. H. (1975). *Diabetes* **24**, 753.
Roth, J., Gorden, P., and Pastan, I. (1968). *Proc. Natl. Acad. Sci. U.S.A.* **61**, 138.
Rubenstein, A. H. (1969). *J. Am. Med. Assoc.* **209**, 254.
Rubenstein, A. H., Cho, S., and Steiner, D. F. (1968). *Lancet* **1**, 697.
Rubenstein, A. H., Clark, J. L., Melani, F., and Steiner, D. F. (1969a). *Nature (London)* **224**, 697.
Rubenstein, A. H., Melani, F., Pilkis, S., and Steiner, D. F. (1969b). *Postgrad. Med. J.* **45**, Suppl., 476.
Rubenstein, A. H., Mako, M., Steiner, D. F., Brown, D., and Pullman, T. N. (1970a). *J. Lab. Clin. Med.* **76**, 868.
Rubenstein, A. H., Mako, M., Welbourne, W. P., Melani, F., and Steiner, D. F. (1970b). *Diabetes* **19**, 546.
Rubenstein, A. H., Pottenger, L. A., Mako, M., Getz, G. S., and Steiner, D. F. (1972). *J. Clin. Invest.* **51**, 912.
Rubenstein, A. H., Mako, M. E., Starr, J. I., Juhn, D. J., and Horwitz, D. L. (1974). *Proc. Congr. Int. Diabetes Fed., 8th, 1973* p. 736.
Sandler, R., Horwitz, D. L., Rubenstein, A. H., and Kuzuya, H. (1975). *Am. J. Med.* **59**, 730.
Sando, H., and Grodsky, G. M. (1973). *Diabetes* **22**, 354.
Sando, H., Borg, J., and Steiner, D. F. (1972). *J. Clin. Invest.* **51**, 1476.
Service, F. J., Rubenstein, A. H., and Horwitz, D. L. (1975). *Mayo Clin. Proc.* **50**, 697.
Sherman, B. M., Pek, S., Fajans, S. S., Floyd, J. C., and Conn, J. W. (1972). *J. Clin. Endocrinol. Metab.* **35**, 271.
Sonksen, P. H., Tompkins, C. V., Srivastava, M. C., and Nabarro, J. D. (1973). *Clin. Sci. Mol. Med.* **45**, 633.
Sramkova, J., Pav, J., and Engelberth, O. (1975). *Diabetes* **24**, 214.
Starr, J. I., and Rubenstein, A. H. (1974). *J. Clin. Endocrinol. Metab.* **38**, 305.
Starr, J. I., Juhn, D. J., Rubenstein, A. H., and Kitabchi, A. E. (1975). *J. Lab. Clin. Med.* **86**, 631.
Steiner, D. F., and Oyer, P. E. (1967). *Proc. Natl. Acad. Sci. U.S.A.* **57**, 473.
Steiner, D. F., and Rubenstein, A. H. (1973). *Proc. Midwest Conf. Endocrinol. Metab., 8th,* p. 43.
Steiner, D. F., Hallund, O., Rubenstein, A. H., Cho, S., and Bayliss, C. (1968). *Diabetes* **17**, 725.
Steiner, D. F., Clark, J. L., Nolan, C., Rubenstein, A. H., Margoliash, E., Aten, B., and Oyer, P. E. (1969). *Recent Prog. Horm. Res.* **25**, 207.

Steiner, D. F., Kemmler, W., Tager, H. S., and Peterson, J. D. (1974). *Fed. Proc., Fed. Am. Soc. Exp. Biol.* **33**, 2105.
Stoll, R. W., Touber, J. L., Menahan, L. H., and Williams, R. H. (1970). *Proc. Soc. Exp. Biol. Med.* **3**, 894.
Stoll, R. W., Touber, J. L., Winterscheid, L. C., Ensinck, J. W., and Williams, R. H. (1971). *Endocrinology* **88**, 714.
Taylor, S. G., Schwartz, T. B., Zannini, J. J., and Ryan, W. W. (1970). *Arch. Intern. Med.* **126**, 654.
Turner, R. C., and Johnson, P. C. (1973). *Lancet* **1**, 1483.
Yalow, R. S., and Berson, S. A. (1973). *Metab., Clin. Exp.* **22**, 703.

DISCUSSION

R. H. Unger: In the experiments in which C-peptide was used in an attempt to determine whether insulin would feed back negatively on its secretion, do you have evidence that in the absence of hypoglycemia this occurs? Another question is whether you think that C peptide may have any effect on insulin secretion, as I believe has been recently suggested by a Japanese group.

A. Rubenstein: These are important questions. We have examined the possibility that C peptide or insulin may inhibit its own secretion independently of the prevailing blood sugar level. Preliminary studies that Dr. Horwitz and I have carried out with Dr. Lillienquist have indicated that insulin may inhibit its own secretion. The blood sugar of normal subjects was kept constant while the insulin level was raised to various levels by exogenous infusion. Plasma levels of C peptide declined under these circumstances. It is a difficult experiment to do, and I really do not want to sound too confident of the result. As Dr. Unger pointed out, this has been a controversial question for a number of years and you can find an equal number of studies that support and refute the hypothesis. As he mentioned, there is a recent paper by Yasuda *et al.* which indicates that one of the rat C peptides, but not the other, inhibits insulin release in isolated islets. I think the question is still open. In any event the effect seems to be rather small compared to substrate inhibition, and I think that is one of the reasons why the results have been equivocal.

M. B. Dratman: You mentioned, Dr. Rubinstein, that normal individuals and those with mild diabetes seem to have similar ratios of proinsulin to insulin, whereas those with severe and decompensated diabetes exhibit a higher proinsulin to insulin ratio. In addition to the possibility that the beta cell itself might be sick, under circumstances of diabetic decompensation, it may be that your observations indicate a diminished rate of proinsulin disposal, secondary to latent chronic renal defects, that may be present although not necessarily obvious in patients with severe decompensated diabetes.

A. Rubenstein: We excluded patients who had overt renal failure. All these patients had normal blood urea nitrogen (BUN) levels. Plasma levels of proinsulin rise only when there is a significant reduction in glomerular filtration rate (GFR) and renal plasma flow. Thus, although there may be a reduction in GFR in the presence of a normal BUN, I do not think there would be a significant rise in proinsulin under these circumstances. Therefore, although theoretically that is a very important possibility, I think it probably was not a factor in these subjects.

R. Yalow: In the severe diabetic one can expect, because of continuous hyperglycemia, continuous secretion of the maximal amount of proinsulin and insulin, that these patients are able to put out. Therefore there exists a steady-state condition where the absolute levels of the two are governed by their secretion rates and the relative turnover times. In those patients who are not maximally stimulated, there can well be spurts of insulin secretion at irregular intervals. For several hours after every secretory spurt there is a higher ratio of insulin to proinsulin. Our interpretation as to the higher fractional amount of proinsulin in the severe diabetic is that it can be accounted for in part at least by continuous stimulation of both components.

A. Rubenstein: I agree.

J. A. Parsons: Do you have any observations on the possibility that rising blood levels of insulin may increase the hepatic extraction of insulin? Dr. Stevenson, in my laboratory, has been doing some infusions of insulin into the portal circulation of unanesthetized dogs for comparison with infusions in peripheral veins. In the intermediate range of rates of infusion, he frequently sees that infusions into the portal bed cause a fall in the peripheral insulin concentration. Apart from the possibility of feedback on endogenous secretion that you are talking about, a change in extraction might, of course, explain such a finding.

A. Rubenstein: That is a good question which is difficult to answer definitively, because I do not think there is conclusive evidence in the literature. The extraction of insulin in the perfused rat liver system is a saturable process. However, in the intact animal, the question arises whether hepatic insulin extraction may be dependent upon the physiologic situation which gives rise to the enhanced insulin release. In fact, Dr. James Field has pointed out that the extraction of insulin may rise as the insulin concentration increases relative to the fasting state in dogs given intraduodenal glucose. As far as I remember, this was not the same when the portal vein insulin level was raised by the infusion of exogenous hormone.

R. Yalow: We tried to ask and answer that question about 20 years ago with studies on the turnover of labeled insulin administered along with different loads of unlabeled insulin. The rate of turnover of the labeled insulin did not change until it was given with very large doses of stable insulin. For instance in 4-kg rabbits there was a slowing of the disappearance rate of radioactive insulin, when administered with 500 units of unlabeled insulin. It would appear that the liver, together with other degrading organs, is capable of removing enormous amounts of insulin in the intact animal.

J. Carter: You spoke about the honeymoon period in juvenile diabetics. I wonder if you have any comparable studies in adults with maturity-onset diabetes. As you know, if such patients come in very symptomatic, they will often go through a similar course where, once they come under control, the insulin requirements may decline or actually disappear. Do you have any studies with the C-peptide assay on pancreatic function during the time they first come in and when they are going through this recovery period? If there is a change in insulin secretion, do you wish to speculate on what is happening to their beta cells during that time?

A. Rubenstein: We do have a considerable amount of information about adult-onset diabetics, which I have not had time to show. These patients are markedly hyperglycemic but without ketosis and are brought under control by a variety of therapeutic interventions, such as diet, oral agents, or insulin. When the blood sugar returns to normal, most of these patients exhibit partial recovery of beta-cell secretory function. We have followed these patients in a sequential fashion, and many of them retain this ability to secrete insulin for prolonged periods of time. This contrasts with juvenile-onset diabetics, most of whom secrete insulin only during the honeymoon period, which tends to be short-lived, for 1 month, 3 months, or 6 months. The difference between the two types of diabetes is very fascinating. We believe that the diabetics who are easy to control and possibly will develop less severe long-term complications are those who retain this secretory capacity. We would like to learn the nature of the process which continues in juvenile-onset diabetics and leads to continued destruction of beta cells.

K. Sterling: In text Fig. 23, showing some ketotic subjects with a zero C-peptide level, were those all juvenile onset cases with ketoacidosis?

A. Rubenstein: They were all subjects less than 20 years of age.

K. Sterling: Do you have similar data on adults with an initial diagnosis of diabetes and ketoacidosis?

A. Rubenstein: We do, and the finding is extremely similar. I think it is clear that patients who present in ketoacidosis have almost no detectable circulating insulin, whether juvenile-onset or adult-onset. The point, though, is that with older people, the recovery seems to be much more significant and long lasting; but at the time of presentation the insulin and C-peptide levels are rather similar in the two groups.

K. Sterling:, This is a very interesting validation of what Brush had apparently just guessed at, as you mentioned—I think in the 1940s—long before radioimmunoassay methods had been developed. When I was a house officer at your institution, the University of Chicago, a contemporary, a young resident, abruptly developed diabetes mellitus; I guess he may have been aged 25 or so; they gave him rather high doses of insulin, keeping him slightly hypoglycemic at all times, on the theory (following Brush) that his ultimate insulin requirement would be much less than if he got conventional therapy. Now I wonder if, in the light of your present-day thinking, this would still be something you would advise, let us say, in a young physician who developed diabetes abruptly, with or without ketoacidosis.

A. Rubenstein: I have not heard of that individual. The question is whether we could maintain beta-cell function for a longer period of time, and maybe to a greater degree, if we did control the blood sugar very tightly early on in the course of the disease? We do not often accomplish tight control in these diabetics. Many newly diagnosed diabetics are taught how to administer insulin and are then sent home. I am sure that those of you who have looked after these patients will agree that their blood sugars are seldom within the normal range. Dr. Jackson and others have suggested that it would be advantageous to maintain their blood sugars as close to the normal range as possible, or even mildly hypoglycemic. For a long time people have overlooked this idea, but if it is clinically reasonable to accomplish it in terms of the patient's life style, etc., I, for one, think it would be a worthwhile clinical experiment to undertake under carefully controlled conditions. The results may be interesting; I do not know whether the complete loss of beta-cell function that occurs in these patients could be prevented to some extent by that kind of clinical intervention. It is possible that it would not be affected, but I do not think that we really know the answer as yet.

M. M. Martin: Is there enough evidence now to settle the argument, which has been going on for some time, as to whether children with abnormal glucose tolerance tests who are entirely asymptomatic should be treated with insulin. Most of us have not treated them until such time as they become symptomatic. Would treatment with insulin maintain endogenous secretion for a much longer period of time and possibly delay on the onset of acute decompensated diabetes?

R. Yalow: May I suggest that these questions that relate to the therapy of juvenile diabetes are not very relevant to the discussion. I think we have exhausted the time to be spent on that topic.

R. Guillemin: One would expect that in normal individuals administration of somatostatin would likely affect the secretion of the proinsulins, the C-peptides, very much as it does on insulin. This is not necessarily so in the several types of patients that you have described. What sort of effect on secretion of proinsulin, C-peptides, and so on has been seen either by your group or others after administration of somatostatin?

A. Rubenstein: This is a study that we would be most eager to undertake. Unfortunately, we have not yet been able to do so. I think that it would be worthwhile to compare beta-cell suppression with this peptide with our other results.

D. Steiner: I wish to congratulate Dr. Rubenstein on his excellent presentation, and to add that it has been both a privilege and a great pleasure to work with him and his associates over the years.

You alluded to some work of Susan Terris which showed that insulin degradation and removal by the liver seems to be dependent largely upon prior receptor binding [S. Terris and D. Steiner, *J. Biol. Chem.* 250, 8389 (1975); *J. Clin. Invest.* 57, 885 (1976)]. This is an important concept and one that seems to have greater generality for peptide hormones that are not degraded in the blood. If receptor trapping of insulin is the rate-determining step in its degradation by tissues such as the liver, then one can easily understand the large mass of data that has accumulated with insulin and various derivatives of insulin. This indicates that the less active derivatives are extracted and degraded at a slower rate by the liver and have prolonged half-times in the plasma. The less active derivatives of insulin always have a higher apparent activity *in vivo* than *in vitro* when compared to insulin. In this regard, I would like to ask whether, just as you and Gabbay have shown, an inherited structural abnormality, presumably in the proinsulin molecule, may lead to accumulation of insulin in the blood, and might it equally be possible that some nonobese diabetics who have hyperinsulinemia without hyperproinsulinemia, might also have a defect in their insulin molecules? Insofar as this structural change might reduce the receptor binding ability of the insulin to, say, 10 or 20% of normal, would this not only lead to an increased secretion of the hormone, but tend to self-compensate by means of a parallel reduction in its extraction and removal by the liver, and perhaps by other tissues as well? What do you think about this possibility?

A. Rubenstein: The idea that the insulin molecule itself may be abnormal is intriguing. It has been extremely difficult to show this convincingly experimentally. The amounts of insulin that one can extract from plasma are very small, even with scaled-up methods and affinity chromatography. One can make immunological comparisons and measure biological responses in terms of immunological activity, as Dr. Jesse Roth and others have done. But small differences, which are difficult to interpret, may be detected. Nevertheless, I think there will be some patients who are shown to have this problem. In patients with mild hyperinsulinism, one might postulate that an amino acid change in the insulin may have very little effect on the molecule, and this would be very difficult to detect unless one had the opportunity to study the pancreas. Under these circumstances one could extract the insulin and determine its amino acid sequence. I think that this is an extremely important area that should not be neglected when we try to elucidate the causes of diabetes.

R. S. Yalow: I would like to answer Dr. Steiner's question a little bit differently. In fact it was shown that the liver enzyme degrading system does distinguish between proinsulin and insulin. This by itself can account for the difference in degradation rates of proinsulin and insulin without requiring differences in receptor sites. Receptor sites for degradation of biologically active hormones are not consistent with our observation that the biologically inactive gastrin peptide, gastrin 1-13, is turned over at the same rate as the biologically active peptide, gastrin 3-17. Probably the degradation of a peptide hormone is enzymically determined and has relatively little to do with the receptor sites involved in biologic activity.

D. Steiner: That is indeed the classical view of endocrinologists, i.e., that "much is secreted and little is chosen," so to speak, to actually bind with receptors, the rest being degraded elsewhere by a presumably irrelevant mechanism. Of course, this is the case where there are enzymes either in the plasma or elsewhere in the circulation that degrade the hormones. But in numerous other cases where there seems to be no degradation in the circulation one might anticipate that hormone degradation could be governed primarily by receptor interactions, as we now believe is the case with insulin. Something akin to this process may occur with parathyroid hormone where the biologically active N-terminal fragment disappears extremely rapidly from the circulation while the inactive carboxyl terminal part of the molecule lingers in the circulation for a much longer period of time. However, while it is true that proinsulin is a poor substrate for a liver insulin degrading enzyme, this is clearly not the case with the almost inactive derivative, desalanyl desasparaginyl insulin. This analog is degraded equally as rapidly as insulin by the liver enzyme system but it is not degraded appreciably by the intact liver, nor is it capable of binding to insulin receptors. Thus in examining larger numbers of analogs one finds that degradation rates correlate well with binding affinity but not necessarily with enzyme susceptibility. It seems likely that these peptides must somehow traverse the barrier between the plasma and the degrading enzymes, i.e., the plasma membrane. Dr. Terris' results suggest that it is the receptor that bridges this gap in some way that is not yet clearly understood mechanistically.

A. Rubenstein: I think that the last point you mentioned is most important. A great deal depends on the accessibility of the hormone to the enzyme. At the moment there is no convincing evidence that this enzyme, insulin-specific protease (ISP), is on the plasma membrane. It has been isolated from the cytosol, and unless one imagines that insulin and proinsulin traverse the membrane independently of binding to it, the membrane receptor would control its delivery to the cytosol and degrading enzymes. This question has not been answered.

R. S. Yalow: Your own data on the survival of insulin and proinsulin in the uremic where the kidney degrading system has been lost results in the different turnover times because the liver enzyme degrading systems are different for the two hormonal forms. This observation demonstrates the significance of the enzyme degrading systems. I would disagree with Dr. Steiner, who seems to suggest that these degrading systems are due to nonspecific proteases. As Dr. Arthur Mirsky showed many years ago, insulinase is insulinase and other peptides do not competitively inhibit its action. There is a different enzyme for ACTH and probably equally specific ones for each of the peptide hormones. There are not nonspecific proteases.

J. F. Habener: You mentioned some recent observations indicating the existence of heterogeneity in the proinsulin-like peak obtained from gel filtration of sera. Would you care to elaborate further on the nature of that heterogeneity?

A. Rubenstein: Experiments designed to elucidate the nature of the components in the 9000 molecular weight peak have shown considerable heterogeneity; in addition, we have shown that the C-peptide is heterogeneous as well. In terms of the 9000 molecular weight peak, the groups of Dr. L. Recant and Dr. R. Williams identified single-chain proinsulin, as well as some intermediates that are two-chain 9000 MW substances, by polyacrylamide gel electrophoresis. In these latter molecules, the C-peptide is still attached to either the A or B chain, but there is a single cleavage at the junction of the C-peptide and one insulin chain. Another possibility is the presence of a component with a chymotryptic split in the C-peptide region itself. In terms of the C-peptide itself, we have now measured serum C-peptide concentrations with different antisera, similarly to Dr. Yalow and Berson's studies with parathyroid hormone. The values obtained with some antisera differ considerably compared to others. It thus appears that there are fragments of the C-peptide in plasma, which may arise in the circulation after secretion. However we do not know whether that is correct, or whether the cleavage may occur within the beta cell.

R. S. Yalow: Since insulinoma patients and your subject with familial hyperproinsulinemia often present with hypoglycemia, is it not likely that the proinsulin-like peak contains a hormonal form other than the relatively biologically inactive proinsulin?

A. Rubenstein: Dr. Yalow's point is that the intermediate components have much higher biological activity than the single-chain proinsulin itself. In this particular family, the insulin levels tend to be within the normal range and most of the individuals were not hyperglycemic or hypoglycemic. It is thus probable that their proinsulin component had little biological activity. The one individual did present with hypoglycemia, but it was probably on another basis.

L. Bullock: You mentioned the heterogeneity of the C-peptide between different species in contrast to the conservative nature of the insulin portion of the molecule. In view of this heterogeneity, is it necessary to raise antibodies against the C-peptide of each species or is there some cross reactivity? I am particularly interested in the guinea pig. The uniqueness of guinea pig insulin and the difficulty of isolating it for antibody production makes study of beta-cell function in this species difficult. I wonder what could be done with the C-peptide.

A. Rubenstein: That is an excellent question. Perhaps I will answer part of it, and then ask Don Steiner to help me in terms of the guinea pig C-peptide. Approximately 10 species of C-peptide have been sequenced, and differences in the sequence of almost 50% of the amino acid residues have been identified. In our initial studies we were disappointed because we could not use bovine or porcine C-peptide antibodies to measure human C-peptide. We have investigated the cross reactivity of various species of C-peptide and have found them to be very low. We have concluded that one needs specific antisera to each particular C-peptide for their precise measurement. I know nothing about the guinea pig C-peptide, but Don Steiner could say a word about it.

D. Steiner: The guinea pig C-peptide is no more different from the other known mammalian C-peptides than they are from each other. Thus the structural anomaly in the guinea pig lies solely in the insulin and not in the C-peptide. To state this another way, the apparent mutation acceptance rate in the C-peptide region of the guinea pig proinsulin is the same as that in other proinsulins, but the mutation acceptance rate in the insulin part of the molecule is much higher than in other mammalian insulins, suggesting some relaxation of selective pressures toward this part of the proinsulin molecule in this and several related species. Why this has occurred remains a mystery!

G. Cahill: I would like to make a comment on the elevation of insulin. Going back and reading the papers of Donough O'Brien and Joe Kimmel that certain diabetics did have a slightly different insulin as reflected by both its biological activity and its inactivation by a degrading system. We know now that insulin will undergo fairly readily the Maillard reaction whereby glucose reacts with free amino groups. The demonstration of A_{1c} hemoglobin being simply a glycosylated hemoglobin and the fact that probably many other proteins in the body react similarly make one wonder whether insulin itself inside the beta cell on exposure to the high glucose levels can itself become glycosylated. We also know that glycosylated proteins tend to have longer half-lives and also may be less biologically active. Have you or Don Steiner looked to see whether you are isolating insulin or proinsulin that may have become glycosylated somewhere on the molecule?

A. Rubenstein: I know that Don Steiner investigated whether there was any carbohydrate in normal proinsulin, but did not find any.

D. Steiner: So far no one has found any glycosylated components to my knowledge. Dr. Lernmark and I did some biosynthetic studies with isolated rat islets but found no incorporation of glucosamine or fucose into either the proinsulin or the insulin components. However, in a similar study with slices from a human insulinoma, we did find some glucosamine counts in the insulin fraction, but we failed to verify that the radioactive components were indeed located within carbohydrate side chains. It could be that glycosylation does occur in certain tumors or perhaps even normally in certain species, depending on the availability of accepting sites for carbohydrate side chains in the proinsulin molecules. It is an intriguing idea and one that would bear further looking into.

D. A. Fisher: In the family with the postulated defect in C-peptide, have you measured C-peptide in serum of subjects with elevated proinsulin concentrations? Second, do these family members have any tendency to hypoglycemia or a late peak in effective insulin as their proinsulin slowly degrades?

A. Rubenstein: Preliminary studies have indicated that the C-peptide concentration is proportional to that of insulin. We think the abnormality is in the junctional region and that the conversion is slower than under normal conditions. This would result in granules containing increased amounts of this precursorlike material. I would stress that all the family members are quite healthy and would not have come to medical attention if it had not been for the initial presentation of the propositus.

R. S. Yalow: My comments relate to how some studies similar to those reported by Dr. Rubenstein can be performed without C-peptide assay values to measure endogenous insulin in the presence of endogenous antibodies. Insulin antisera with sensitivites of the order of 1 pg/ml are readily available. This would permit insulin determinations at 2.5 units per milliliter of plasma to be performed at a dilution of 1:100 with labeled human insulin. It is only in the rare case of insulin resistance that antibodies are detectable with labeled human insulin in plasma diluted 1:100. It is most unusual therefore not to be able to perform direct insulin determination in the vast majority of insulin-treated subjects because of the competition of endogenous antibody. The second comment deals with what you might call hypoglycemia factitious. Many of us have antisera that recognize the difference between human and the beef–pork insulins used for therapy. With these antisera, one can readily distinguish between insulinoma-induced hypoglycemia and surreptitious insulin administration. We have detected two such cases in our own hospital. This is a method readily available in many laboratories.

A. Rubenstein: I have no argument with that. I wish to point out, however, that you need to withdraw insulin therapy in order to use the first approach that you mentioned. If you measure C-peptide, this is not necessary. With regard to your second point, I agree that it is an alternative approach. Measurement of C-peptide complements these techniques.

E. M. Bogdanove: I would like to extend the discussion that was going on between Dr. Steiner, Dr. Yalow, and yourself with regard to the mechanism of the metabolic clearance differences between insulin and C-peptide. Your remark that the hormone has to get to the enzymes before it can be degraded by them reminded that Ira Goldfine has been reporting [I. D. Goldfine and G. J. Smith, *Proc. Natl. Acad. Sci. U.S.A.* **73**, 1427 (1976)] that insulin probably does get into liver cells (and indeed into their nuclei). More recently, I understand Gordon Niswender has a paper (T. Chen, J. Able, and G. Niswender, *Cytobiologia*, submitted for publication) showing that LH gets into corpus luteum cells, and Janet Nolin has reported [J. Nolin and R. Witorsch, *Endocrinology* **99**, 949 (1976)] that prolactin gets into mammary gland cells. Do you have, or think it might be worthwhile to get, data regarding the relative abilities of the C-peptide and insulin to get into hepatic cells and into intracellular loci in renal tissue? Might not such information help to explain the differential in metabolic clearance?

A. Rubenstein: Thank you for your comment. As far as I am aware Dr. I. Goldfine has shown binding of insulin to hepatic nuclei, but I am not sure that he has convincingly shown that a significant amount of insulin traverses the cell and binds to the nuclear membrane in a whole-cell preparation. I may be wrong, but that is my interpretation of his work to date.

E. M. Bogdanove: I cannot argue with your conservative evaluation of Goldfine's findings, but I think the histologic evidence from the other two studies is more convincing on this point.

A. Rubenstein: I think that it is an important point. The question still remains whether a peptide hormone penetrates a cell by traversing the plasma membrane by way of binding to a receptor, or whether it can enter independently of that receptor. I just do not think that question is answered yet, even though both mechanisms are possible. None of the discussions we have had would preclude either possibility. The experimental data at present are convincing that insulin does not enter the cell except through binding to the receptor. It is a very interesting and important point that really should be studied in depth.

Glucagon and the A Cells

R. H. UNGER, P. RASKIN,
C. B. SRIKANT, AND L. ORCI

Veterans Administration Hospital and the University of Texas Southwestern Medical School, Department of Internal Medicine, Dallas, Texas, and the Institute of Histology and Embryology, Geneva, Switzerland

I. Introduction

Only once before in the 54 years since its discovery by Kimball and Murlin (1923) has glucagon been a subject for review in these Proceedings. In 1957 Dr. Piero Foá summarized the evidence, then rather nonspecific and indirect, that glucagon was the second hormone of the pancreas (Foá et al., 1957). However, not until the present decade has the possibility that glucagon might be a hormone of importance in health or disease been considered seriously. In this period the introduction by Berson and Yalow (Berson et al., 1956) of the technology required for the development of a radioimmunoassay for glucagon (Unger et al., 1959) and, more recently, the availability of a glucagon suppressant (Koerker et al., 1974), somatostatin, discovered by Dr. Guillemin and his associates (see Brazeau et al., 1973), have made possible the accumulation of compelling evidence in support of this view. It is now widely accepted that the glucagon-secreting A cells are important components of the organ that regulates nutrient flux, the islets of Langerhans (Unger and Orci, 1977a). Moreover, malfunction of A cells has been demonstrated to be present in all forms of spontaneous and experimental diabetes thus far studied (Aguilar-Parada et al., 1969; Müller et al., 1971), which has led to the proposal that abnormal glucagon secretion may be an important contributing pathogenetic factor to the metabolic derangements of diabetes. This review will consider the present status of such evidence.

II. The Function and Structure of the Normal and Diabetic Islets of Langerhans

A. PHYSIOLOGIC ROLES

In mammals and other highly developed organisms the maintenance of health and optimal functional capacities requires that the delivery of the key nutrients to the individual organ systems of the body be at all times closely matched to their needs. The abundance of well-adapted organisms that survive and thrive on this planet, despite unpredictable and often hostile conditions, indicates that problems of nutrient delivery have been solved successfully by most species. In advanced forms of life this requires orchestration of nutrient flux by a regulatory system capable, on the one hand, of sensing fuel requirements of the various tissues and of initiating the metabolic processes necessary to meet them, and, on the other hand, of sensing the availability and effecting the rapid disposal of exogenous nutrients into tissues in which they can be incorporated into macromolecules—glycogen, fat, and protein. Glucagon and insulin are clearly capable of achieving these metabolic functions. Because they exert biologically opposing actions upon the target tissues upon which they both act, principally the liver, their relative concentrations can determine the direction and magnitude of the disposition of key nutrients under a given set of circumstances. The studies of Park and Exton (1972) and of Mackrell and Sokol (1969) first provided the basis for the idea that a high insulin–low glucagon mixture (Unger, 1971), such as normally occurs in response to carbohydrate-containing meals, will promote the anabolic disposition of the exogenous nutrients through increased glycogenesis, lipogenesis, and protein biosynthesis, while a low insulin–high glucagon mixture favors the catabolic processes, glycogenolysis, lipogenesis, ketogenesis, and gluconeogenesis (Fig. 1). In starvation and/or during

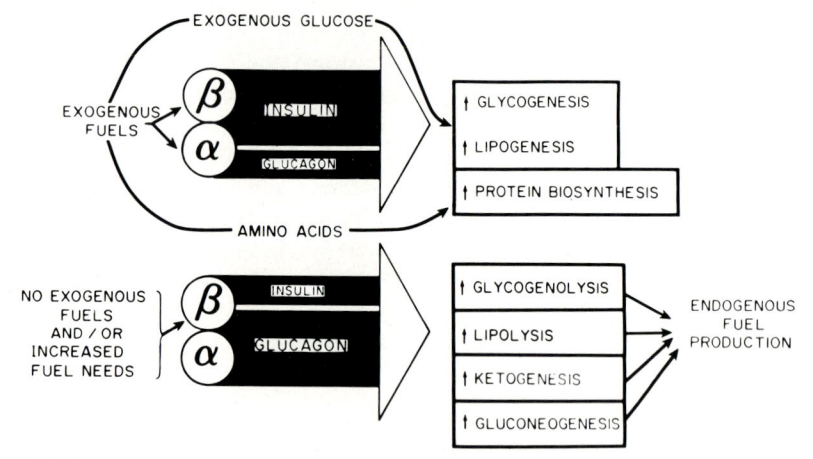

FIG. 1. Schematization of the normal response of the alpha-beta cell couple to an abundance of exogenous fuels (upper portion) and to a deficiency of exogenous fuels and/or increased fuel needs. The high insulin–low glucagon mixture secreted during fuel abundance promotes glycogenesis and lipogenesis from glucose and protein biosynthesis from amino acids. The low insulin–high glucagon secretory mixture in the second instance promotes increased glycogenolysis, gluconeogenesis, lipolysis, and ketogenesis. Reprinted, with permission of *Diabetes,* from Unger (1971).

increased need for fuels a persistently low insulin–high glucagon mixture can maintain the required rate of glucose production and prevent hypoglycemia through increased glycogenolysis and gluconeogenesis (Fig. 1). However, the use of amino acids to produce glucose is soon reduced as ketone production increases and replaces glucose (Cahill *et al.,* 1966; Marliss *et al.,* 1972) as a fuel for the central nervous system and other tissues, thereby conserving body proteins (Cahill, 1971).

As Cahill (1971) has emphasized, the homeostasis of glucose, uniquely among the key nutrients, is characterized by relative lack of change in its extracellular fluid concentration even though wide changes in glucose flux rates may be taking place. Thus, the plasma glucose levels of healthy young individuals range between 60 and 150 mg/100 ml irrespective of the magnitude of carbohydrate intake or of glucose utilization. On *a priori* grounds it can be reasoned that a homeostatic system capable of altering glucose flux rates without permitting major perturbations in glucose concentration would require a push–pull system of biologic antagonists that can independently vary glucose influx to and efflux from the extracellular space so as to assure their overall equality, thereby keeping glucose concentration within narrow limits irrespective of changes in flux rates (Unger, 1976a). These functions are ascribed to insulin and glucagon in the schema depicted in Fig. 2, which is basic to the concept of a bihormonal system of glucoregulation by the islets of Langerhans. The system must, of course, have a glucose sensor that can, in response to small changes in glycemia,

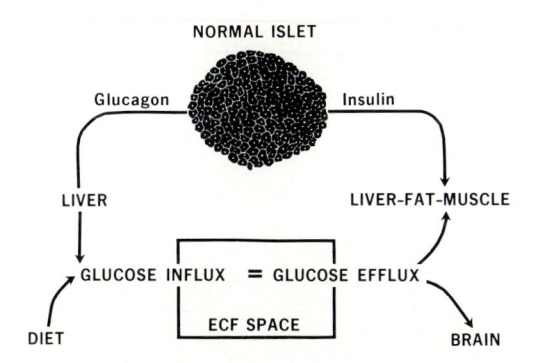

FIG. 2. Schematization of the glucoregulatory function in the normal islet. Sensing changes in glucose concentration in the extracellular (ECF) space, the normal islet responds by secreting a mixture of insulin and glucagon which maintains glycemia within the so-called normal range irrespective of flux rates, so that in the overall sense glucose influx and efflux are ultimately equal. Any increase in glucose efflux, whether induced by exercise or aminogenic insulin secretion, is matched by a glucagon-mediated increase in glucose influx to prevent hypoglycemia; any dietary increase in glucose influx is matched by an insulin-mediated increase in glucose efflux to prevent hyperglycemia. Reprinted, with permission of *Archives of Internal Medicine,* from Unger and Orci (1977a).

alter the composition of the insulin–glucagon secretion mixture so as to provide the required change in glucose flux without allowing either hypo- or hyperglycemia to ensue. In the resting state, total glucose production (glucagon-mediated plus glucagon-independent glucose production) obviously is equal to total glucose utilization (insulin-mediated plus insulin-independent glucose utilization), so that glucose concentration is constant.

In normal subjects any increase in glucose utilization, whether the result of exercise or of protein-induced insulin secretion, will be matched by an increase in glucagon-mediated hepatic glucose production sufficient to prevent hypoglycemia. Conversely, any sudden increase in glucose influx as the result of carbohydrate ingestion will elicit a prompt increase in insulin-mediated glucose utilization sufficient to limit the magnitude and duration of hyperglycemia (Unger, 1976a). The narrow range of glycemic change observed in normal subjects (Fig. 3) speaks for the precision with which this system operates in the majority of human and subhuman species (Raskin and Unger, 1977).

B. PATHOPHYSIOLOGIC ROLE IN DIABETES

Certain disorders of fuel delivery are characterized by impairment in the capacity of the islets of Langerhans to sense and/or to respond to perturbations in glycemia. In one such disorder, diabetes mellitus, an increase in glucose entry

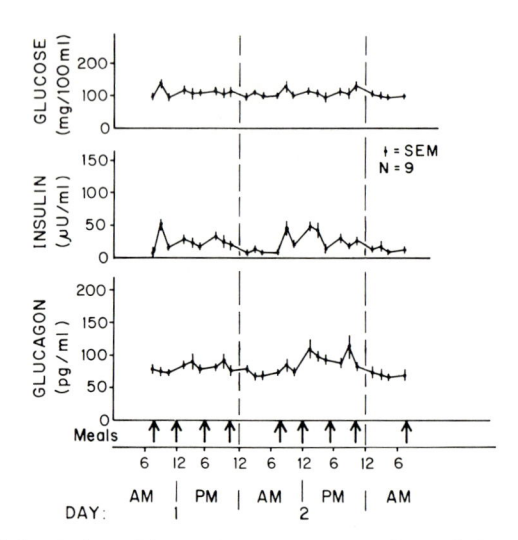

FIG. 3. Around-the-clock profiles of plasma glucose, insulin, and glucagon (mean ± SEM) in nine normal subjects. Blood samples were obtained every 2 hours. Reprinted, with permission of *Diabetes,* from Unger (1976a).

into the extracellular space, whether derived from ingested glucose ("exogenous hyperglycemia") or from increased hepatic glucose production ("endogenous hyperglycemia") fails to elicit the instant and quantitatively appropriate insulin release that in nondiabetics effectively limits the magnitude and duration of any rise in glycemia. Instead, in diabetes the hyperglycemia persists until the glucose is cleared at a subnormal rate by a relatively fixed rate of glucose uptake and by urinary excretion.

A defect either in glucose-sensing by the islets, in insulin secretion, or in insulin sensitivity of the target tissues is obviously, then, the *sine qua non* of the diabetic derangement. The hyperglycemia resulting from delayed disposal of ingested or infused glucose (*exogenous* hyperglycemia) is entirely the consequence of the insulin-related defect and is probably not dependent upon the presence of glucagon. The proposed pathogenetic role of glucagon in diabetic glucoregulation is limited to *endogenous* hyperglycemia, defined as a rise in plasma glucose occurring in the absence of carbohydrate ingestion. Severe endogenous hyperglycemia seems to require the presence of glucagon (or other glycogenolytic substances) in quantities that are high relative to the level of insulin (Unger and Orci, 1975; Dobbs *et al.*, 1975; Sakurai *et al.*, 1975). Studies in man (Gerich *et al.*, 1975a) and in totally depancreatized dogs (Sakurai *et al.*, 1975) reveal that, when plasma immunoreactive glucagon (IRG) levels are suppressed to approximately 40 pg/ml by somatostatin infusion, endogenous hyperglycemia of more than 160 mg/100 ml does not occur, even in the total absence of insulin—at least not during 18 hours of observation (Gerich *et al.*, 1975a). This suggests that while insulin deficiency may by itself augment hepatic glucose production above the rate of utilization, the augmentation is insufficient to cause the massive hyperglycemia that occurs in the presence of glucagon. (Perhaps in the absence of glucagon renal excretion of glucose is sufficient to prevent endogenous hyperglycemia from rising much above 160 mg/100 ml.)

In addition to the contribution of glucagon to the diabetic defect in glucoregulation, it now seems that glucagon is essential for the development of ketoacidosis. In insulin-deprived juvenile-type diabetics, suppression of IRG levels by somatostatin to approximately 40 pg/ml markedly attenuates hyperketonemia (Gerich *et al.*, 1975a), supporting the demonstration by McGarry *et al.* (1975) that glucagon enhances the ketogenic capacity of the liver. It would thus appear that glucagon is essential for the marked hepatic overproduction of both glucose and ketones that characterizes severe insulin deficiency. The metabolic defects attributable entirely to insulin deficiency by itself and independent of glucagon include *exogenous* hyperglycemia (i.e., impaired carbohydrate tolerance) and increased levels of free fatty acid (Table I); the diabetic defects that appear to require the presence of glucagon include severe endogenous hyperglycemia and ketoacidosis (Table I). For these reasons it has been hypothesized (Unger and Orci, 1975) that glucagon is "essential" for the complete metabolic syndrome of

TABLE I
Contribution of Islet Cell Abnormalities to the Metabolic Abnormalities of Diabetes Mellitus

Abnormalities	Insulin deficiency	Relative IRG[a] excess
↓ Glucose utilization	++++	0
↑ Hepatic glucose production	+	++++
↑ Lipolysis	++++	+ (?)
↑ Ketogenic capacity of liver	+ (?)	++++

[a]IRG, immunoreactive glucagon.

severe diabetes previously attributed entirely to the direct consequences of the insulin deficiency alone.

C. MORPHOFUNCTIONAL RELATIONSHIPS OF THE NORMAL AND DIABETIC ISLETS

Recent advances in our understanding of the morphology of the Langerhans microorgans now permit consideration of possible relationships between certain of their anatomical features and their functions. The morphologic features which strike us as having potential functional significance are listed in Table II.

1. Deployment of Islets

These include, first, the deployment of islets as individual microorgans embedded within a firm retroperitoneal organ, rather than as a single large endocrine organ. This certainly reduces the possibility of their mass destruction by a

TABLE II
Postulated Morphofunctional Relationships

Morphologic feature	Functional implication
1. Deployment into separated microorgans	? Reduced vulnerability to attack
2. Specific heterocellular arrangements	Permits direct cell-to-cell feedbacks (see Fig. 14)
3. Entry of vessels and nerves in tricellular area	May be the control center
4. Tight junctions	May guide hormones to appropriate destinations
5. Gap junctions	May provide for intercellular coordination

single pathologic process, while affording intimate intercellular relationships considered below.

2. Topographic Relationships of Cell Types

The endocrine cells are organized into topographically specific patterns (Fig. 4). In man and in the rat they form an outer islet mantle composed of A cells [in the paraduodenal islets the A-cell mantle is replaced by a mantle of pancreatic polypeptide-containing cells (Orci, 1977)] one to three layers in thickness and comprising approximately 25% of the endocrine population, a central mass of B cells that makes up more than 60% of the cell population, and intervening somatostatin-containing D cells forming approximately 10% of the endocrine

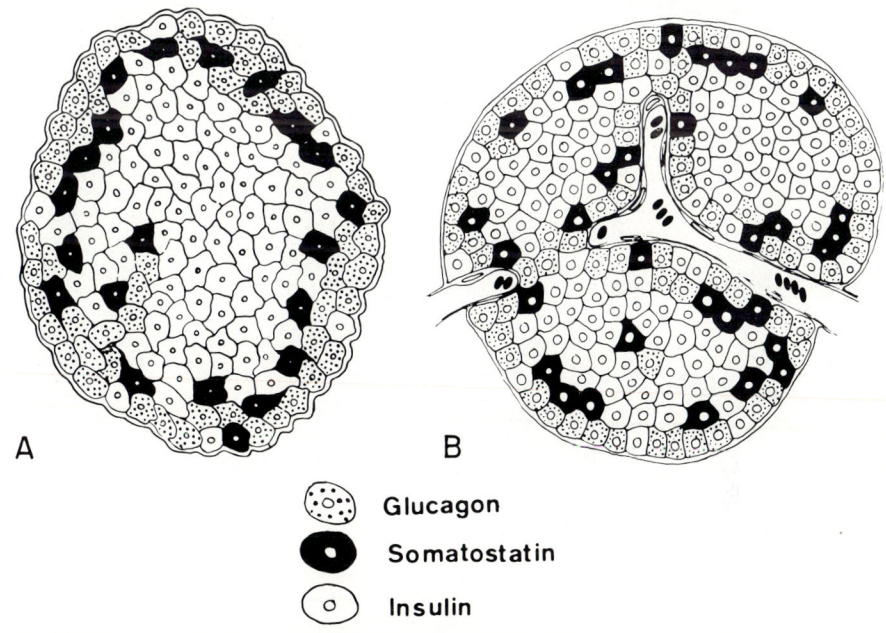

FIG. 4. (A) Schematic representation of the number and distribution of insulin-, glucagon-, and somatostatin-containing cells in the normal rat islet. Note the characteristic position of most glucagon- and somatostatin-containing cells at the periphery of the islet, surrounding the centrally located insulin-containing cells. Cell types in the islet for which a characteristic function and/or morphology is not defined are intentionally omitted. Reprinted, with permission of *The Lancet,* from Orci and Unger (1975).

(B) Schematic representation of the number and distribution of insulin-, glucagon-, and somatostatin-containing cells in the normal human islet. Large vascular channels penetrate the islet and are followed by glucagon- and somatostatin-containing cells. This pattern divides the total islet mass into smaller subunits, each of which contains a center formed mainly of insulin cells and surrounded by glucagon and somatostatin cells. Cell types for which a definite function and/or morphology has not yet been determined are intentionally omitted.

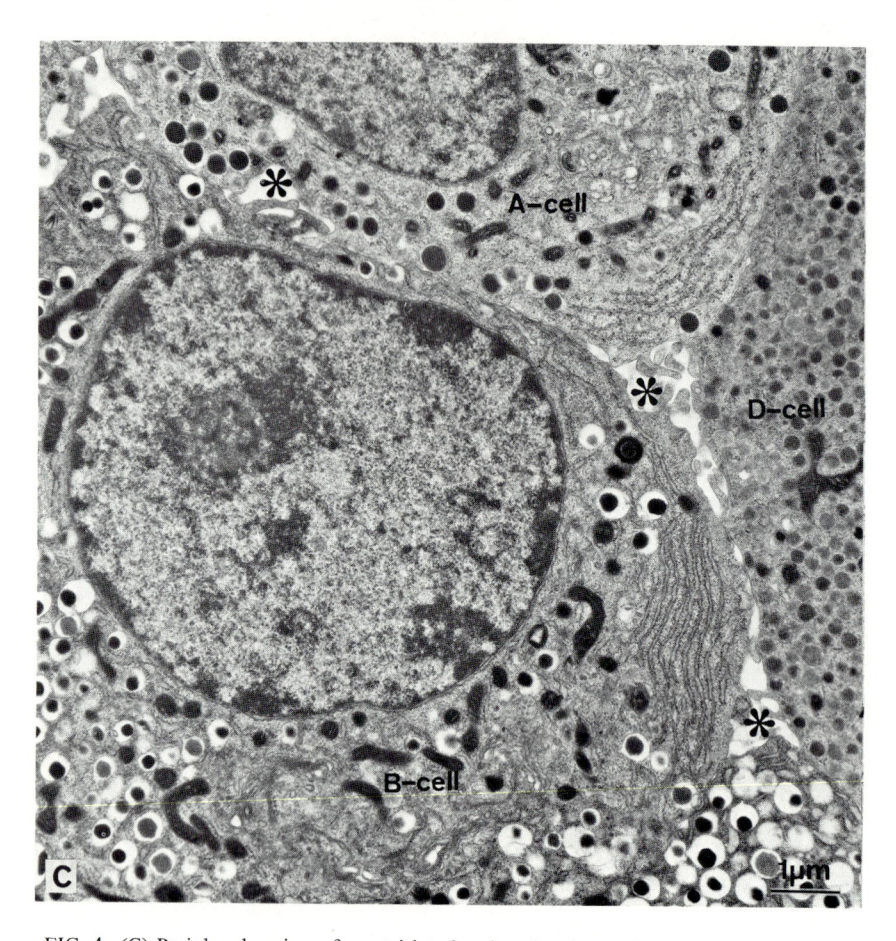

FIG. 4. (C) Peripheral region of a rat islet showing the close relationship between glucagon-containing cell (A cell), somatostatin-like material-containing cell (D cell), and insulin-containing cell (B cell). The asterisks indicate intracellular spaces.

community (Orci *et al.,* 1976). A cells and D cells are almost always in close juxtaposition to one another and to the peripherally situated layer of B cells so that probably the majority of both A and D cells have contacts with one or two other cell types; by contrast, except for the peripherally located cells, most B cells are in contact only with other B cells (Orci and Unger, 1975). Thus, in man and in the rat the islet has an outer heterocellular "cortex" and a central B cell "medulla." If direct "paracrine" hormone-cell interactions take place at the local level, they would obviously be maximal in this heterocellular area. Indeed, as indicated in Table III, the secretory products of these three cell types, insulin,

TABLE III
Actions of Islet Cell Secretory Products upon Secretion by
Other Islet Cells

Secretory product	B cell	A cell	D cell
Insulin	—	↓	?
Glucagon	↑	—	↑
Somatostatin	↓	↓	—

glucagon, and somatostatin, can directly influence the secretion of one or more of the neighboring cells. Samols, who first demonstrated that glucagon stimulates insulin secretion (Samols *et al.*, 1965), has long championed an insulin–glucagon feedback (Samols *et al.*, 1972) and has lately presented convincing *in vitro* evidence of a negative effect of endogenous insulin upon glucagon secretion (Samols and Harrison, 1976a). [Evidence that insulin inhibits glucagon secretion in the absence of an increase in glycemia has also been reported in man (Raskin *et al.*, 1975).] Somatostatin has been clearly shown to block both insulin (Alberti *et al.*, 1973) and glucagon (Koerker *et al.*, 1974; Mortimer *et al.*, 1974) secretion, and, according to Samols, its inhibition of glucagon occurs at far lower doses than its effect on insulin secretion (Samols and Harrison, 1976b). Since studies with the perfused dog pancreas (Fig. 5) suggest that exogenous (Patton *et al.*, 1976a) and perhaps endogenous (Patton *et al.*, 1976b) glucagon stimulate the release of somatostatin, an A cell–D cell feedback circuit (Orci and Unger, 1975) could be a physiologic reality. The effect of insulin on somatostatin release remains unclear, but thus far no effect has been demonstrated in the perfused pancreas of nondiabetic animals (Patton *et al.*, 1976a). However, glucagon-mediated somatostatin release could well prevent glucagon-mediated insulin secretion during endogenous glucagon secretion intended to prevent hypoglycemia.

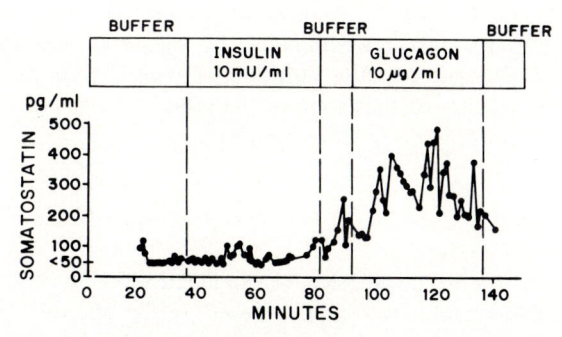

FIG. 5. The effect of perfusion of the isolated dog pancreas with high doses of insulin and glucagon upon the concentration of immunoreactive somatostatin in the effluent vein.

3. Neurovascular Elements

A third anatomic feature with possible functional implications (Table II) is the entrance of afferent blood vessels and nerves in the heterocellular area of the islets (Fujita, 1976). This could signify that this region is designed to receive the neural and blood-borne signals that influence the responses of the remaining regions of the microorgan, and that perhaps it constitutes the so-called "pacemaker" area (Meissner, 1976). The fact that glucose-sensing by both A and B cells can be attenuated by sympathetic nerve stimulation (Marliss *et al.*, 1973) could be construed as further support for special functions in that region—perhaps glucose sensing and a "pacemaker" function. Conceivably the stress-induced changes in the A-cell and B-cell responses, relative hyperglucagonemia and hypoinsulinemia, are mediated by adrenergic signals to this region of the islet. The loss of glucose-sensing during stress or, at least, loss of glucose-responsive secretory behavior by the A and B cells, is believed to play a major role in the maintenance of stress hyperglycemia, which may be a vital insular adjustment designed to enhance glucose availability for the glucose-requiring central nervous system during periods of cerebral hypoperfusion.

4. Tight Junctions

A fourth anatomical feature with functional implications (Table II) is the network of tight junctions (Fig. 6A) that compartmentalizes the extracellular space of the islets (Orci *et al.*, 1973a); these partitions are demonstrably impermeable to secretory products of the islets (Orci *et al.*, 1973b) and can, therefore, create zones or channels between islet cells. Such structures might influence the destination of the various secretory products of the islet cells, i.e., determine whether a hormone reaches receptors on neighboring cells or whether it is barred from extensive local interactions and directed rather into effluent capillaries for export to remote targets. If, for example, glucagon had direct access by diffusion to all the B cells in the islet in which it was secreted, the resulting outpouring of insulin would induce hypoglycemia and would defeat the very purpose of glucagon secretion, which is to prevent hypoglycemia. Perhaps by constant remodeling of the extensive labyrinth of tight junctions the islet influences the disposition of its various products.

5. Gap Junctions

Finally, demonstration in the islets of gap junctions (Orci *et al.*, 1973a, 1975) (Table II), low-resistance pathways believed to be the site of electrotonic coupling, provides anatomical support for the notion that the islet is a functional syncytium and that signals may be transmitted between homologous and heterologous islet cells without entering the intercellular space. Such contacts could provide a means of coordinating the total hormonal output of asynchronously secreting B cells and A cells and of producing the required hormonal mixture. The reciprocal oscillations of insulin and glucagon reported by Goodner *et al.* (1976) and the electrophysiologic studies of Meissner (1976) are compatible with such a function (Fig. 6B).

C. THE GLUCOSE-SENSING FUNCTION OF THE A CELL

Despite the evidence that in intact animals the A cell responds to changes in glycemia (Ohneda *et al.*, 1969) and, like the B cell, is influenced by the α-anomer of D-glucose more than by the β-anomer (Matschinsky and Pagliara, 1976), there is a circumstantial basis for doubt that the A cell has intrinsic glucose-sensing capacity. Whenever A cells are situated apart from B cells, as in glucagon-secreting tumors (Tiengo *et al.*, 1976), in the gastric fundus of the dog (Blazquez *et al.*, 1976) and in the pancreatic islets of juvenile-type diabetics (Unger *et al.*, 1970), they seem incapable of responding to changes in glucose (Table IV). Yet, in each of these three circumstances the A cells are suppressed by insulin and are stimulated by amino acids (Tiengo *et al.*, 1976; Blazquez *et al.*, 1976; Unger *et al.*, 1970). This could be interpreted as evidence that A cells lack a glucose sensor of their own, or that if they do possess a glucose sensor, it is functionally dependent upon insulin. Alternatively, the A cell response to glycemic change could be passive, mediated by the B cell with which the A cells of the islets normally enjoy proximity. Thus, the B cell, perhaps through insulin release or by some other signal, seems to play an essential role in glucose sensing by the A cell; whether it is a permissive role involving a glucose-mediated signal or whether the B cell transmits the signal itself in response to glycemic change is unknown. But the signal or the permissive factor could well be insulin released into the intercellular space that separates the A cell from the B cell (Fig. 6).

III. Extrapancreatic A Cells and Immunoreactive Glucagon (IRG)

It has now been well established that A cells are present in the canine gastric fundus. These cells contain secretory granules indistinguishable from those of the A cells of the pancreas (Larsson *et al.*, 1975; Baetens *et al.*, 1976a), and they stain positively with immunofluorescent techniques specific for glucagon (Larsson *et al.*, 1975; Baetens *et al.*, 1976a). At the ultrastructural level their secretory granules are found to contain immunoperoxidase coupled to specific glucagon antibodies (Fig. 7). A cells have been identified in one adult human gastric fundus by means of immunoperoxidase staining (Muñoz *et al.*, 1977), but

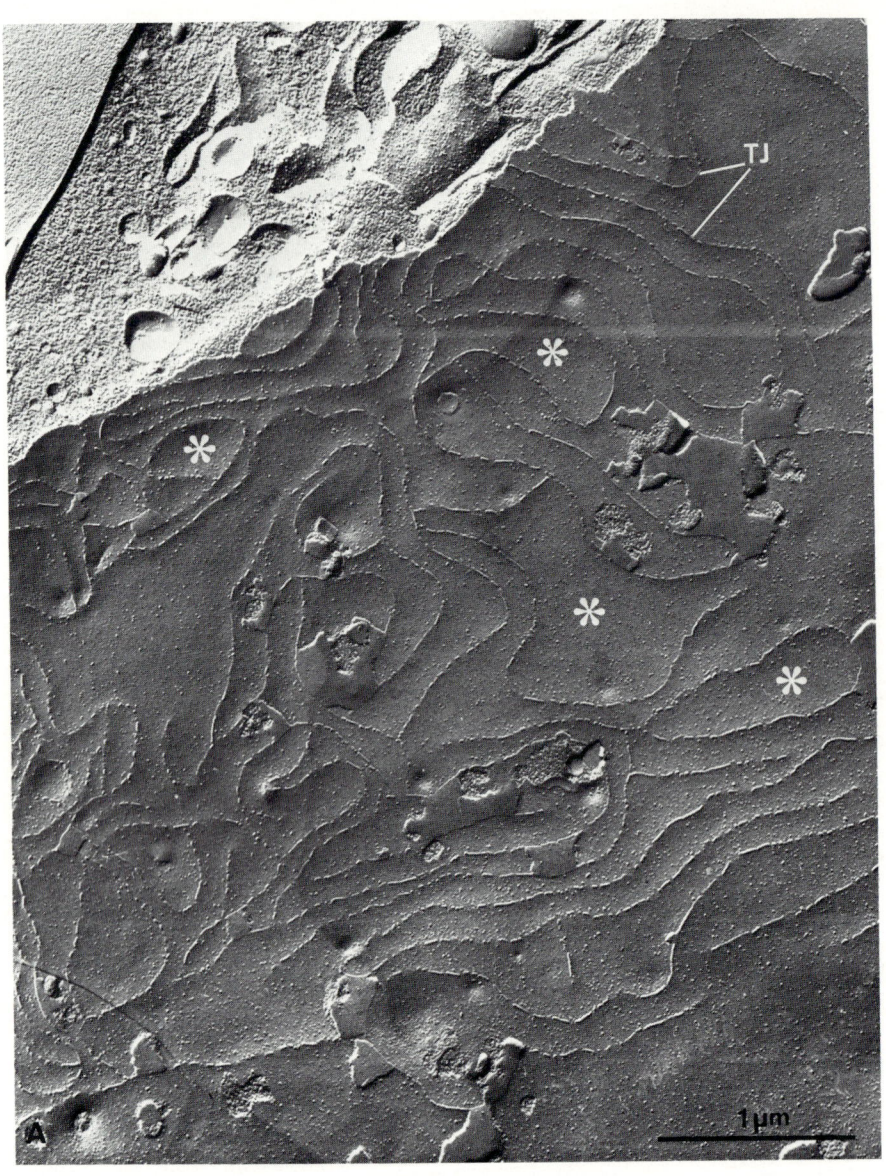

FIG. 6. (A) Freeze-fracture replica of an adult human pancreatic islet. The exposed face of an islet cell plasma membrane contains a network of variagated tight junctional (TJ) fibrils. The fibrils have been correlated with zones of fusion between the outer leaflets of the two adjacent cell membranes enclosing the intercellular space. The fibrils thus delimit various domains in the membrane (asterisk). Such domains could represent specific areas of the membrane preferentially exposed to or protected from external influences. The fact that such junctions can be modified experimentally clearly indicates that they are labile differentiations, and probably reflects their ability to modulate constantly the intercellular relationships.

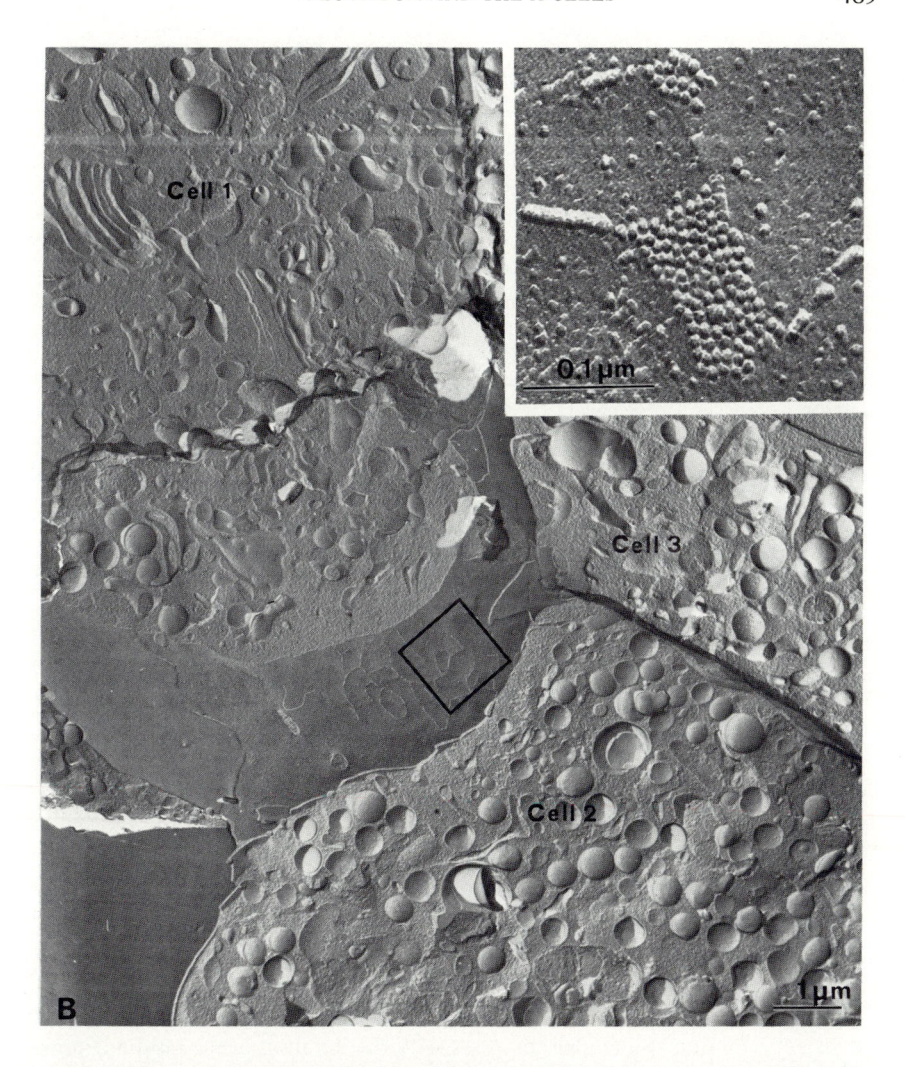

FIG. 6. (B) Freeze-fracture replica of an islet isolated from sulfonyl urea-treated rat. The fracture process has split the plasma membranes of two adjacent cells, one of which (Cell 1) appears poorly granulated and tentatively identified as a B cell, the other (Cell 2) is well granulated and tentatively identified as an A cell. The short fibrils (TJ) represent elements of a poorly developed tight junction. In the area outlined by the rectangle and seen at higher magnification in the inset, one sees aggregates of particles characteristic of gap junctions.

TABLE IV
Response of A Cells Not in Contact with B Cells

Challenge	Human glucagonoma	Canine gastric fundus	Juvenile diabetic islets
Change in glycemia	0	0	0
↑ In insulin	↓	↓	↓
Arginine, i.v.	↑	↑	↑

other attempts to demonstrate them in the adult human stomach have thus far been unsuccessful. However, if dog stomach, which immediately after death contains an abundance of A cells and glucagon, is permitted to remain at room temperature for 1 hour, they too become difficult to identify. A cells have been observed by electron microscopy in the stomach of human fetuses (Fig. 7B) (L. Orci, unpublished observations) and in the duodenum of human adults (Sasagawa et al., 1973), and Knudsen et al. (1975) have identified pancreas-type glucagon immunoreactivity in the human colon by means of immunofluorescent staining methods. Reports of glucagon-containing cells in the salivary glands of several species have recently emerged from Lawrence's laboratory (Lawrence et al., 1976). It seems clear that in animals, and in all probability in man, glucagon-containing cells are present outside of the pancreas.

Under normal circumstances it appears that a contribution of glucagon from extrapancreatic A cells to the circulating immunoreactive glucagon (IRG) pool is meager. Even by direct catheterization of the effluent vein draining the canine gastric fundus of normal dogs, a net contribution of IRG was observed only during the unphysiologic experimental maneuver of arginine stimulation (Muñoz-Barragan et al., 1976). However, in the insulin-deprived state, as first noted by Vranić et al. (1974), Matsuyama and Foá (1975), and Mashiter et al. (1975), substantial glucagonemia is present in insulin-deprived, totally depancreatized dog. Mashiter et al. (1975) showed that the response of extrapancreatic glucagon secretion to arginine increases with the duration of the insulin deprivation, a "functional hypertrophy." In confirming their studies we have observed that insulin in small doses causes a dramatic suppression of extrapancreatic glucagon secretion to undetectable levels (Dobbs et al., 1975). We conclude, therefore, that extrapancreatic A cells, in contrast to those residing within the islets of Langerhans, produce little or no IRG in the presence of adequate quantities of insulin, but that in the insulin-deprived state, whether induced by total pancreatectomy (Dobbs et al., 1975; Blazquez et al., 1976) or by alloxan-induced destruction of B cells (Blazquez et al., 1977), the gastric fundus may be an important source of plasma IRG in the dog.

Whether the same holds true in man remains to be determined. Difficulties in identifying extractable IRG or identifying A cells in postmortem specimens of human fundus do not necessarily exclude the existence of gastric glucagon in totally depancreatized humans. As was pointed out above, the canine fundus also contains little extractable IRG after several hours at room temperature. It is, therefore, possible that, as in the dog, the IRG reported to circulate in totally depancreatized subjects who are relatively insulin deficient (Palmer *et al.*, 1976) may be of fundic origin. Conceivably, in totally depancreatized patients, particularly if hypoinsulinemia is frequent, hyperplasia of fundic A cells may take place; such hyperplasia has been shown to occur in dogs (L. Orci and M. Vranić, unpublished observations).

There is other evidence that extrapancreatic glucagon is present in man, as in other species, whatever its source may be. Müller and co-workers (1974) found low IRG levels in two totally depancreatized patients even though they exhibited normal fasting blood glucose levels—evidence that they were well insulinized. Palmer *et al.* (1976) reported three totally depancreatized patients with basal IRG levels in the normal or high-normal range; in two of the three patients IRG rose during arginine stimulation. The arginine-stimulated rise in plasma IRG was associated with an increase in hyperglycemia, suggesting that extrapancreatic IRG is biologically active and that in the absence of adequate insulin it may contribute to endogenous hyperglycemia. But conflicting evidence has emerged from the laboratory of Barnes and Bloom (1976). They report no IRG in the circulation of several totally depancreatized patients. Most of their patients had received insulin on the afternoon before the specimens were drawn and, in view of the exquisite sensitivity of extrapancreatic A cells to the administration of insulin it is possible that this suppressed circulating IRG levels. Moreover, the radioimmunoassay which they used was one in which the patient's own plasma, "stripped" by immunoabsorbance, was present in the glucagon standards; such a system could not measure biologically active IRG moieties not removed by immunoabsorbance. In view of the heterogeneity of pancreatic and extrapancreatic IRGs and GLIs, this technologic difference could explain the difference between their findings and those of Palmer *et al.* (1976).

IV. The Immunoreactive Glucagons

A. THE IRGs OF THE PANCREATIC A CELLS

The true A cells of the islets of Langerhans are known to release their IRG-containing granules through a process of exocytosis (Orci *et al.*, 1970), but the precise events relating to the biosynthesis and control of the secretory process have not as yet been elucidated. Acid-alcohol extracts of whole canine

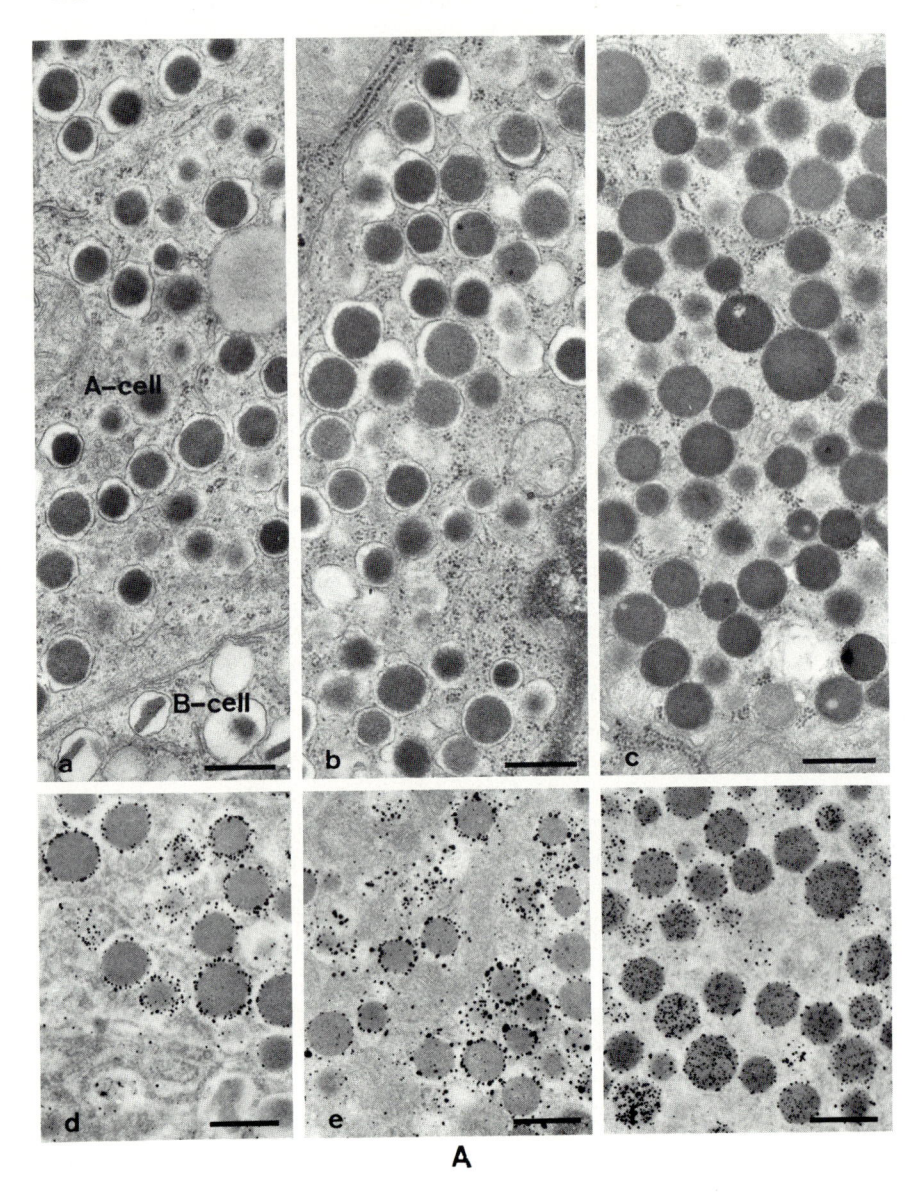

FIG. 7. (A) Comparison between an A cell from the endocrine pancreas (a), a gastric A cell (b), and a gastric A like cell (c) showing the similarities and the differences among respective secretory granules. The bar represents 0.5 μm. Reprinted, with permission of *Journal of Cell Biology*, from Baetens *et al.* (1976a).

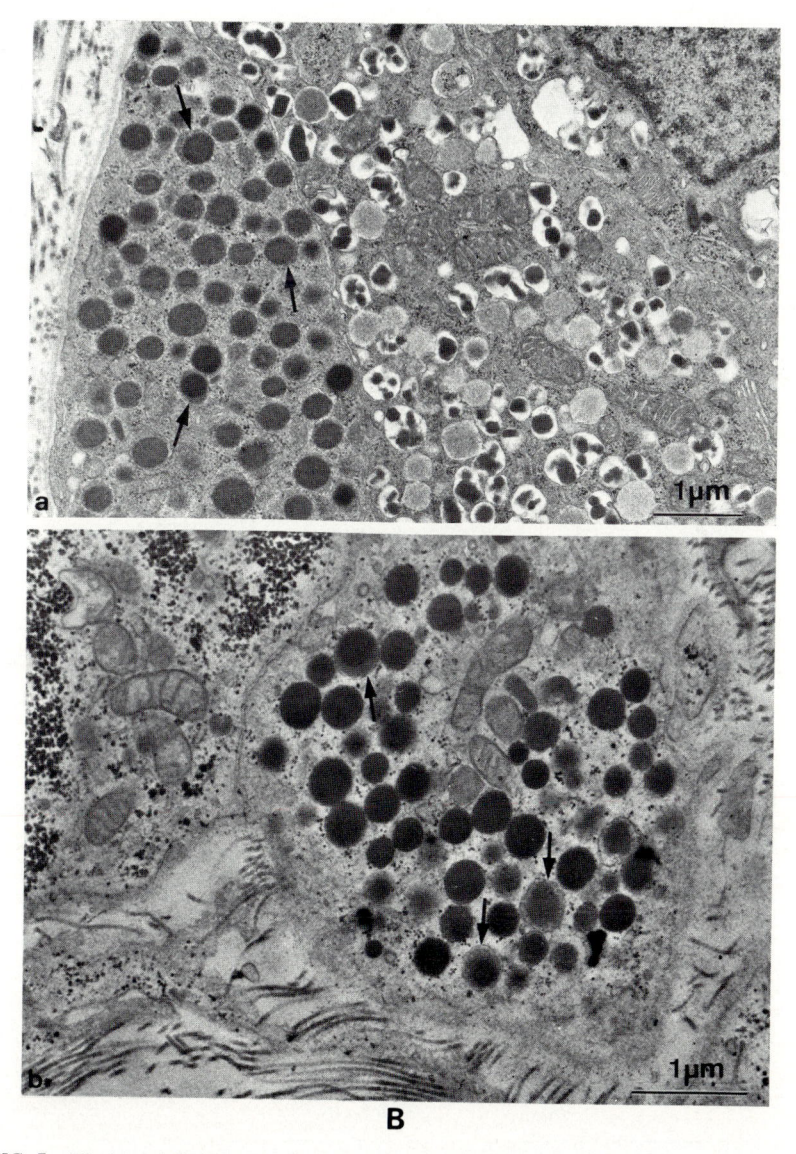

FIG. 7. (B) (a) Adult human pancreatic A_2-cell (glucagon). (b) Endocrine cell from the fundic mucosa of a 13 weeks human embryo. Notice the similarity between the secretory granules (arrows) of the two cells.

pancreas and of isolated rat islets reveal the presence of several IRG moieties (Table V). The most abundant immunoreactive peak corresponds on gel filtration to the [^{125}I] glucagon marker. It cannot be differentiated from crystalline beef–pork glucagon either by its glycogenolytic activity in the isolated perfused rat liver system (Rigopoulou et al., 1970), by its ability to activate adenylate cyclase in and displace [^{125}I] glucagon from isolated rat liver membranes (Srikant et al., 1977), or by its isoelectric point (Srikant et al., 1977). A smaller immunoreactive fragment estimated to have a molecular weight of about 2000, is frequently recovered in such preparations, but its properties have not been tested (Srikant et al., 1977). An IRG fraction estimated to have a molecular weight of approximately 9000 (Rigopoulou et al., 1970) had, on the basis of studies of Noe and Bauer (1971), been considered as a possible glucagon precursor and is sometimes referred to as proglucagon. Tager and Steiner (1973) have identified the primary structure of a strongly basic 37 amino acid fragment of so-called proglucagon that contains the 29-amino acid primary sequence of bovine glucagon, the remaining 8 residues being at the C terminus. Trakatellis et al. (1975) have reported that angler fish "proglucagon" is a 78-amino acid single-chain polypeptide from which glucagon is liberated by tryptic cleavage.

We have observed in sonicated preparations of rat islets, but not in acid-alcohol extracts of pancreas, a still larger moiety estimated to have a molecular weight of more than 45,000 (C. B. Srikant, R. E. Dobbs, and R. H. Unger, unpublished observations). Tager has also reported an IRG of about this size (Tager and Markese, 1976) and, in addition, several workers (Hellerström et al., 1974; Tung et al., 1977) have found IRGs intermediate in size between this and the IRG^{9000}. It is not clear whether these larger IRGs are preprohormones or artifacts. Suffice it to say that, in all likelihood, glucagon-secreting tissues contain, in addition to the biologically active product, a spectrum of prohormones ranging in size to at least 45,000 daltons. It is noteworthy that IRG^{9,000} is devoid of glycogenolytic activity (Rigopoulou et al., 1970) and fails to activate adenylate cyclase or to bind to glucagon receptors (Srikant et al., 1977), although it binds to antibodies directed against the C-terminal portion of the

TABLE V
Biologic Activities (as % of Glucagon Activity) of Canine Pancreas Immunoreactive Glucagons (IRG)

Pancreatic	Glycogenolytic	Adenylate cyclase	[^{125}I] Glucagon displacing
IRG^{2000}	?	?	?
Glucagon	100	100	100
IRG$^{\sim 9000}$	0	0	~10

glucagon molecule. Apparently, then, presence of the C-terminal residue interferes with the interaction of the molecule with its receptor far more than with antibodies directed toward the C-terminal 24–29 segment of the glucagon molecule.

B. THE IRGs OF THE GASTRIC FUNDUS

Fundic IRGs in the dog have been similarly studied (Table VI) (Srikant et al., 1977). Most of the fundic IRG appears in the eluates corresponding to the [^{125}I]glucagon marker. It has a pI of 6.4 as compared with 6.5 for pancreatic glucagon. Interestingly, it may be even more powerful in terms of its biologic activity than immunoequivalent quantities of pancreatic glucagon. Its glycogenolytic activity is somewhat greater than that of immunoequivalent quantities of pancreatic glucagon, and its ability to activate adenylate cyclase in isolated rat liver membranes far exceeds that of immunoequivalent quantities of pancreatic glucagon. However, its capacity to displace [^{125}I]glucagon does not differ significantly from pancreatic glucagon. The data suggest that some very subtle difference in fundic glucagon renders it biologically more effective than pancreatic glucagon relative to its affinity for both its receptor and its antibody— perhaps a "super-glucagon." These findings may have implications with respect to the biologic potency of circulating IRG in the totally depancreatized individual, particularly if, in addition to a circulating "super-glucagon," "up-regulation" (Gavin et al., 1974) of hepatic glucagon receptors is induced by the concomitant chronic hypoglucagonemia.

A yet larger IRG with an estimated molecular weight of 65,000 was also identified in the canine fundus. Its pI was 6.4, and it had the same glycogenolytic activity as immunoequivalent amounts of crystalline glucagon (Srikant et al., 1977). In addition, its capacity to activate adenylate cyclase and to displace [^{125}I]glucagon did not differ significantly from immunoequivalent quantities of glucagon. Neither 8 M urea nor incubation in acetic acid altered its molecular

TABLE VI
Biologic Activities (as % of Glucagon Activity) of Canine Stomach Immunoreactive Glucagons (IRG)

Fundus	Glycogenolytic	Adenylate cyclase	[^{125}I]Glucagon displacing
IRG^{2000}	?	?	?
IRG^{3500}	120	180	107
IRG^{9000}	0	0	~10
IRG^{65000}	110	83	98

size as determined by gel filtration. The dilution slopes for all of the foregoing IRG moieties were superimposable. Clearly, the relationship of the IRGs to one another remains to be determined. It is noteworthy that the largest IRG yet recovered from tissue, IRG^{65000} from the fundus, is biologically active whereas the 9000 molecular piece is devoid of glycogenolytic activity, at least in the liver.

C. PLASMA IRGs

The heterogeneity of plasma IRG is even more perplexing (Valverde et al., 1974, 1975). There is an IRG^{3500}, which rises briskly during hypoglycemia and stimulation with arginine and becomes unmeasurable during suppression with glucose or somatostatin. This component is the one most prominently increased during insulin deficiency. In addition, there is an IRG ^{9,000}, which is somewhat elevated in hypersecretory states such as phlorizin-hypoglycemia and insulin deficiency. There is also a small moiety of approximately 2000 daltons. Finally, there is "big plasma glucagon" (BPG), first described by Valverde and co-workers (1974), in human and dog plasma. It is estimated to have a molecular weight of approximately 180,000 and resembles an IgG (Valverde and Villanueva, 1976). Incubation with 8 M urea does not alter its molecular size, but tryptic hydrolysis results in the appearance of the smaller IRG components of plasma (Valverde et al., 1974). In most normal and diabetic subjects, BPG contributes about 40 pg equivalents per milliliter to total plasma IRG, but it is markedly elevated in some patients and may result in a high whole-plasma IRG value. Ensinck and Palmer (1976) first described a normal healthy subject with a negative family history of diabetes mellitus in whom BPG levels of $\sim$3000 pg/ml were observed. Elevations in BPG were found in other members of his kindred and the trait was believed to be transmitted as an autosomonal dominant. In preliminary studies, C. B. Srikant, E. Ipp, P. Raskin, and R. H. Unger (unpublished observations) have observed BPG elevations in several diabetic patients. Eluates from the plasma of patients with high levels of BPG were found to have glycogenolytic activity, while eluates from patients without increased BPG were devoid of such activity. In two experiments the BPG values in such patients increased during arginine stimulation. The chemical nature of BPG, its physiologic and pathophysiologic significance and its tissue or tissues of origin remain to be elucidated.

It should be kept in mind that measurements of whole-plasma IRG, while usually reflecting the level of IRG^{3500} (true glucagon), may in certain instances be the result of elevations of IRG moieties other than true glucagon. For example, it is now well established that the high IRG levels observed in patients with renal failure is, in part, the result of a disproportionately high IRG ^{9000} (Kuku et al., 1976), in addition to a high IRG^{3500} level. In patients with glucagonoma IRG^{9000} as well as IRG^{3500} is increased (Valverde et al., 1976).

V. The Role of Glucagon in Diabetic Hyperglycemia

While the importance of glucagon as a glucoregulatory hormone in normal man and other mammals is now widely accepted, its contribution to the hyperglycemia of diabetes is at present a matter of ongoing controversy. Since conventional treatment of diabetes only rarely converts the hectic glycemic profile of the diabetic patient into the narrow glycemic profile of nondiabetics, depicted in Fig. 3, the contribution of the hectic pattern of plasma glucagon, typified by the patient in Fig. 8, to this hyperglycemia is of potential practical importance. Therapeutic measures that would correct the hyperglucagonemia might provide more normal glucoregulation; reduction in hyperglycemia, in turn, could help prevent or delay complications of the diabetic disorder.

A. EFFECTS OF INCREASED GLUCAGON IN INSULIN-TREATED DIABETICS

The evidence against an important role for glucagon in the hyperglycemia of diabetes is based largely on the studies of Sherwin *et al.* (1976a). They were unable to demonstrate deterioration in the glycemic control of two juvenile diabetics during a constant infusion of crystalline glucagon. While acknowledging that glucagon is biologically potent in the absence of insulin, they argue that in the presence of insulin hyperglucagonemia is without demonstrable influence on the blood sugar. They attribute this lack of effect not only to overriding biologic

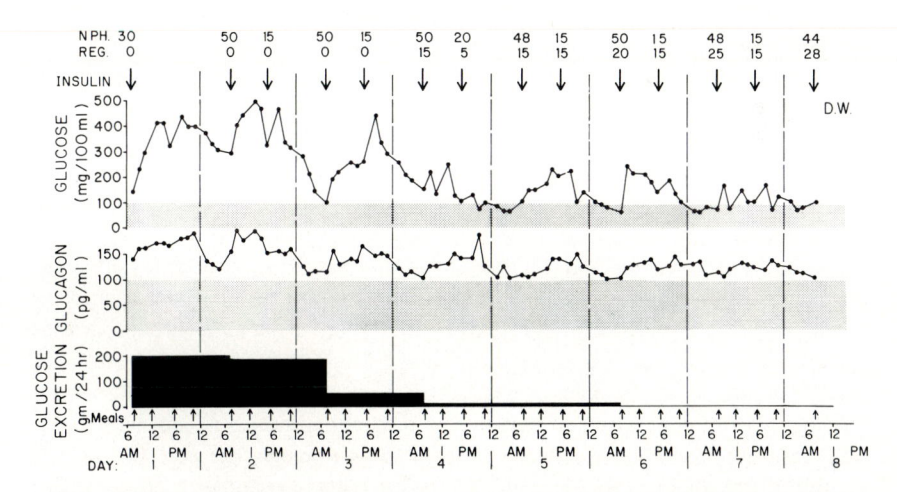

FIG. 8. Around-the-clock profiles of plasma glucagon, insulin, and glucose measured in blood specimens obtained at 2-hour intervals and 24-hour glucose excretion. At a total insulin dose of 103 units on day 7 of hospitalization, postprandial hyperglycemia and hyperglucagonemia were still present.

actions of insulin upon liver, but to the now well-established (Cherrington and Vranić, 1974; Sherwin et al., 1976a) evanescence of glucagon's effect of hepatic glucose production independent of insulin—perhaps the result of "down-regulation" (Gavin et al., 1974) of glucagon's hepatic receptors. Finally they ascribe the dramatic improvement in the postprandial hyperglycemia of diabetic patients observed during somatostatin infusion (Gerich et al., 1975b) not to suppression of postprandial hyperglucagonemia, but to direct somatostatin-induced delay or reduction of glucose absorption from the gastrointestinal tract (Wahren and Felig, 1976). If their appraisal of glucagon's contribution to diabetic hyperglycemia is correct, efforts to develop a pharmacologic means of suppressing glucagon secretion would seem pointless.

However, in our opinion the weight of evidence continues to favor the view that glucagon plays an important role in the abnormal glucoregulation of diabetes—particularly in the insulin-requiring forms of diabetes. We attribute the failure of Sherwin et al. to note a deleterious effect of constant infusions of glucagon in the two juvenile diabetics to certain experimental conditions: (1) the glucagon levels were maintained at levels too low to overcome the effect of high levels of insulin which occurred periodically in their patients, causing precipitous declines in plasma glucose into the hypoglycemic range; (2) the infused glucagon contained 4 ml of whole blood per 100 ml of infusate, causing a substantial loss of biologic activity in excess of immunologic activity (Unger, 1976b); and (3) the administration of glucagon as a constant infusion rather than in the phasic bursts that characterize the secretion of endogenous glucagon. We have studied six juvenile-type diabetics (Raskin and Unger, 1976), in which glucagon was administered subcutaneously so as to produce phasic patterns of hyperglucagonemia ranging from 100 to more than 1000 pg/ml; a significant increase in hyperglycemia was observed but, more strikingly, a massive increase in glycosuria occurred in every patient examined (Fig. 9). In addition, urea and ketone excretion increased significantly. We, therefore, attribute the findings of Sherwin et al. (1976a) to the particular experimental conditions employed, rather than to impotence of glucagon in the presence of insulin.

Just as hyperglucagonemia induced by administration of exogenous glucagon can increase hyperglycemia when insulin levels are normal but fixed because of B-cell insufficiency (Raskin and Unger, 1976), so an increase in endogenous glucagon secretion can aggravate the hyperglycemia in such patients (Aguilar-Parada et al., 1969; Palmer et al., 1976; Barnes and Bloom, 1976; Unger, 1976b); the administration of arginine, a powerful stimulus of glucagon secretion, results in a rise in plasma glucose of approximately 60 mg/100 ml in a period of 80 minutes (Unger et al., 1970). Since arginine itself is not a gluconeogenic substrate, the increase in glucose is quite probably the consequence of the glycogenolytic action of the endogenous hyperglucagonemia. Indeed, as shown in Table VII, whenever endogenous hyperglucagonemia is

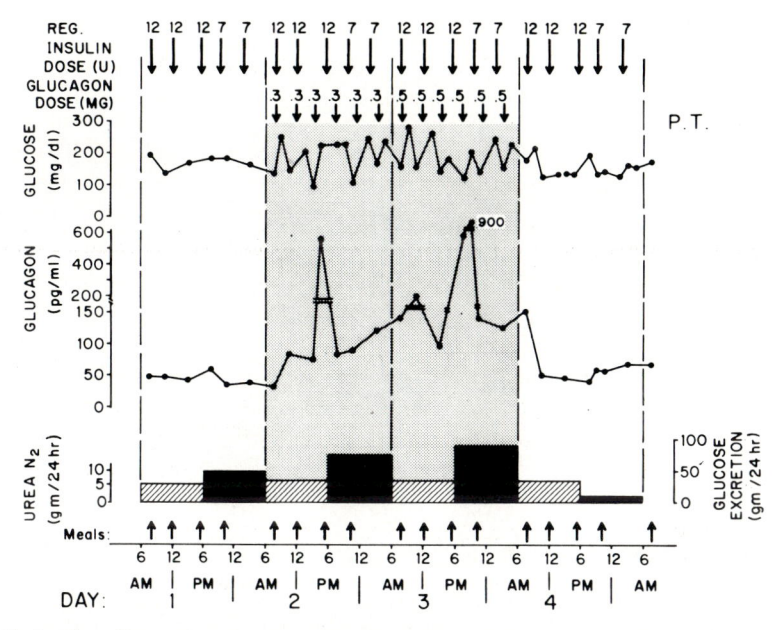

FIG. 9. The effects of exogenous hyperglucagonemia induced by subcutaneous injections of glucagon upon the profile of plasma glucose and glucagon and upon 24-hour glucose excretion (solid bars) and 24-hour urea excretion (hatched bars) with five injections of regular insulin per day. The two days of hyperglucagonemia were characterized by increased lability of hyperglycemia and progressive increase in glycosuria, both of which receded when glucagon was discontinued.

induced, whether by arginine infusion in juvenile-type diabetics or totally depancreatized human subjects, or in juvenile diabetics by the ingestion of a mixed meal, glucose levels rise; whenever under these same circumstances the rise in glucagon is blocked, whether by somatostatin or by insulin, no change in plasma glucose occurs (Unger, 1976c). These facts argue strongly that a rise in endogenous glucagon is associated with a rise in plasma glucose, unless an increase in insulin can compensate for the effects of hyperglucagonemia, or the fixed level of insulin is sufficiently high to override the hepatic actions of glucagon. It has been argued that the amelioration in postprandial hyperglycemia that is associated with the administration of somatostatin, first reported by Gerich *et al.* (1975b), is the consequence not of glucagon suppression, but rather of a somatostatin-induced delay in gastric emptying and/or glucose absorption (Wahren and Felig, 1976). However, in insulin-deprived dogs (Sakurai *et al.*, 1975) the infusion of somatostatin prevents both the hyperglucagonemia and hyperglycemia otherwise induced by alanine, a somatostatin effect clearly inde-

TABLE VII
Relationship of Changes in Glucagon and Glucose in Insulinized Human Diabetics

	Δ Glucagon (pg/ml)	Δ Glucose (mg/100 ml)
Juvenile diabetics		
Arginine infusion (Unger et al., 1970)	↑325	↑80
Arginine infusion (Barnes and Bloom, 1976)	↑300	↑59
Mixed meal plus insulin (Gerich et al., 1975b)	↑ 58	↑84
Mixed meal plus insulin plus somatostatin (Gerich et al., 1975b)	↓ 55	↓50
Totally depancreatized patients		
Arginine infusion (Barnes and Bloom, 1976)	0	0
Arginine infusion (Palmer et al., 1976)	↑ 61	↑48

pendent of gastrointestinal actions. While this does not exclude a gastrointestinal component in the dramatic amelioration of glycemia often observed in human juvenile diabetics during the continuous administration of somatostatin, it does indicate that at least part of this amelioration may be mediated by prevention of stimulated glucagon secretion.

Nor does the evidence that glucagon's effect on the liver is transient exclude its importance in diabetic hyperglycemia. Endogenous glucagon released in phasic bursts and causing transient increases in hepatic glucose production can in the absence of a compensatory increase in insulin levels add to the extracellular space a quantity of glucose which, as in the case of ingested glucose, is cleared at a subnormal rate, thus contributing to the hyperglycemia—abnormal "endogenous glucose tolerance." Any addition of glucose, whether of dietary or of hepatic origin, to this slowly turning over glucose pool will be reflected by disproportionate hyperglycemia.

B. EFFECTS OF DECREASED GLUCAGON IN INSULIN-DEPRIVED DIABETICS

Perhaps the most convincing evidence in support of glucagon's contribution to both endogenous hyperglycemia and ketoacidosis comes from the studies of Gerich et al. (Fig. 10) showing that blockade of hyperglucagonemia greatly attenuates both the endogenous hyperglycemia and ketonemia which otherwise occur in juvenile diabetics when the administration of insulin is abruptly terminated (Gerich et al., 1975a). Concomitant replacement infusion of glucagon together with the somatostatin completely reverses its action.

And in insulin-deprived dogs with established steady-state hyperglycemia, the suppression of glucagon with somatostatin (Sakurai et al., 1975) results in a

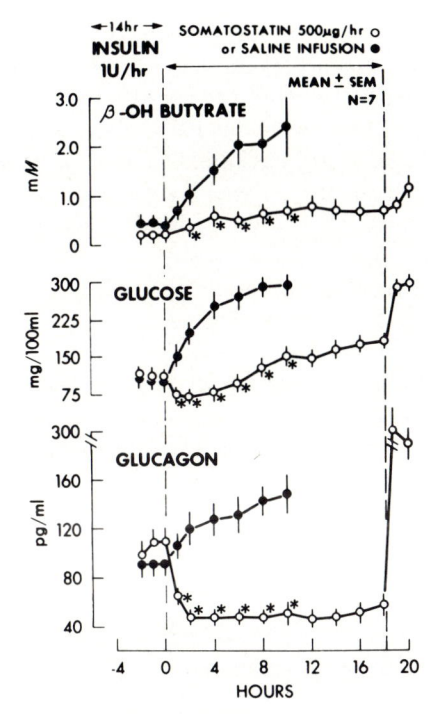

FIG. 10. The effect of sudden discontinuation of a 14-hour insulin infusion in a group of seven juvenile diabetics upon plasma β-hydroxybutyrate, glucose, and glucagon on two different occasions. On one occasion a control infusion of normal saline was maintained after insulin had been discontinued; a rise in β-hydroxybutyrate, glucose, and glucagon was observed. On the other occasion, somatostatin was infused, lowering plasma glucagon levels to below 50 pg/ml; under these circumstances the rise in β-hydroxybutyrate and in plasma glucose was markedly attenuated for a period of 18 hours. Upon discontinuation of somatostatin glucagon rebounded whereupon β-hydroxybutyrate and glucose levels also increased. Asterisk (*) indicates $p < 0.01$. Reprinted, with permission of *New England Journal of Medicine,* from Gerich *et al.* (1975a).

progressive decline in plasma glucose levels, presumably because the reduction in glucagon secretion has lowered hepatic glucose production below the rate of glucose removal. Glucagon replacement infusion completely reverses this decline in hyperglycemia.

VI. Therapy with Somatostatin

TREATMENT OF DIABETES WITH SOMATOSTATIN

Somatostatin has been employed as an adjunct to insulin treatment in juvenile-type diabetics both by Gerich (1976) and by our own group (Raskin, 1976).

Whereas conventional treatment with insulin alone, even in large doses administered subcutaneously two or three times daily, failed to control postprandial hyperglycemia (Fig. 8), somatostatin infused continuously in a dose of 2 mg/day for 5 days was accompanied by a reduction in the glycemic profile to within the range of normal (Fig. 11). Upon discontinuing the somatostatin, a rebound in both plasma glucagon levels and postprandial hyperglycemia was observed on the same insulin dose. Gerich (1976) has reported equally spectacular results in such patients. Measuring plasma glucose at hourly intervals around the clock, they succeeded in eliminating hyperglycemia using doses of insulin less than half of that which without somatostatin had failed to prevent postprandial hyperglycemia. Although these remarkable results may in part be the result of somatostatin-induced changes in gastric emptying and splanchnic blood flow, as suggested by Wahren and Felig (1976), it seems probable that, at least in part, the improvement is a consequence of glucagon suppression and reduction in hepatic glucose production during mixed meals. While the efficacy of somatostatin in the therapy of diabetes remains difficult to evaluate, it is clear that the results are promising, leading to the hope that more potent analogs of somatostatin with more selective effects on glucagon secretion and a greater resistance to degradation may be proved of value.

The rationale of therapeutic glucagon suppression in diabetes mellitus is based on the twin assumptions that glucagon-mediated hyperglycemia is a contributing

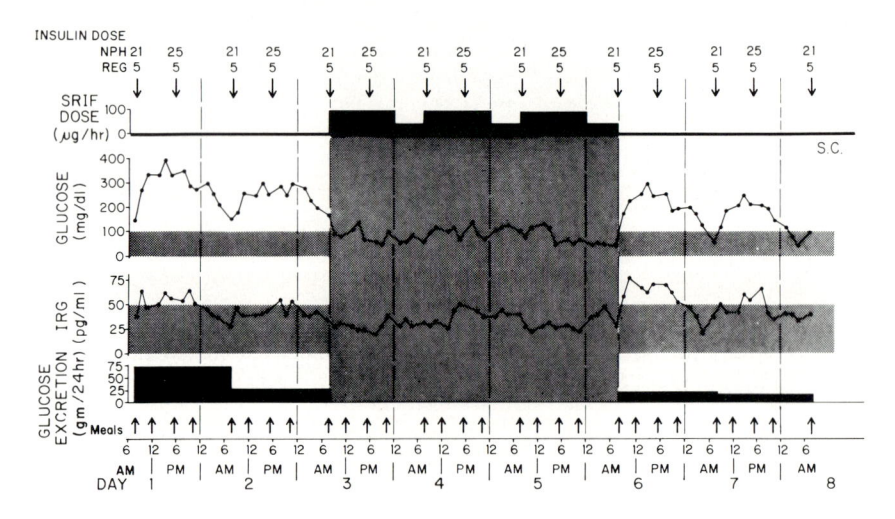

FIG. 11. The effect of continuous somatostatin (SRIF) infusion for 3 days upon the profile of plasma glucose and immunoreactive glucagon (IRG) and 24-hour excretion in a 29-year-old female juvenile-type diabetic. During the period of somatostatin infusion represented by the vertically stippled area, blood sugars remained within the normal range and plasma IRG was significantly lowered.

factor to diabetic hyperglycemia and that it cannot effectively be eliminated even by aggressive treatment with insulin. Indeed, while glucagon levels can be significantly reduced by conventionally administered insulin in high doses, postprandial hyperglycemia appears to persist in most juvenile and adult-type diabetics even at insulin doses in excess of 200 units per day given in three divided doses. Even if a further increase in insulin dose were to correct the abnormal glucagon secretion, which in the juvenile diabetic appears to be related to the absence of B cells and/or their secretory product insulin in the islets, such large quantities of insulin would result in an abnormally high rate of glucose utilization, while at the same time preventing hepatic glucose production, a combination that would inevitably result in hypoglycemia. It is, therefore, theoretically preferable to give a dose of insulin which raises glucose utilization only to normal and to control inappropriate glucagon-mediated bursts of hepatic glucose production by means of a somatostatin-like agent capable of suppressing glucagon secretion without influencing glucose utilization.

VII. The Etiology of A-Cell Dysfunction in Diabetes Mellitus

It is widely accepted that in virtually every known form of diabetes mellitus, spontaneous or acquired, hyperglucagonemia is present at least relative to the concomitant hyperglycemia (Aguilar-Parada *et al.*, 1969; Müller *et al.*, 1971; Dobbs *et al.*, 1975; Vranić *et al.*, 1974; Matsuyama and Foá, 1975; Mashiter *et al.*, 1975). This means that the normally negative glucagon response to hyperglycemia is absent or at least diminished, either as a result of impaired sensing of glycemic change or because of inadequate response of A-cell secretion to the sensing signal. Not only does hyperglycemia fail to lower plasma IRG levels in diabetics, but hypoglycemia fails to stimulate it (Gerich *et al.*, 1973), suggesting a profound disturbance in the glucoregulatory function of the diabetic A cell. A second characteristic abnormality in A-cell function observed in diabetics is an exaggerated response to aminogenic stimulation with arginine (Unger *et al.*, 1970) or alanine (Wise *et al.*, 1973). Whether this constitutes a separate abnormality or is merely a manifestation of a generally augmented A-cell functional capacity resulting from autonomy from glycemic control is not known at the present time. It seems highly likely that aminogenic stimulation of the A cell involves a direct action on the A cell itself rather than mediation from outside the A cell; this view is based on the fact that A cells secreting tumors (Tiengo *et al.*, 1976), the A cells of the gastric fundus (Blazquez *et al.*, 1976a), and the A cells of the juvenile diabetic islets (Unger *et al.*, 1970) all respond in qualitatively similar fashion to the infusion of arginine (Table IV). However, it is clear that the magnitude of this response is exaggerated by the absence of insulin; the glucagon response to arginine infusion or to a protein meal is significantly greater in the juvenile diabetic than in the nondiabetic patient, but this abnormality can be corrected to normal by the concomitant infusion of insulin (Raskin et al., 1976); interestingly, however, arginine- and protein-induced hyperglycemia persists despite normalization of protein-induced glucagon secretion, evidence that unless additional insulin is available even a normal glucagon increment will induce hyperglycemia.

In the adult-type hyperinsulinemic patient in whom within-islet insulin may be normal, the glucagon response to arginine and to protein feeding are also exaggerated, but in such patients insulin—even in pharmacologic doses—does not reduce the response (Raskin et al., 1976). If such patients have islets resembling those of the ob/ob mouse, in which insulin-containing cells are abundant but somatostatin-containing cells very sparse (Baetens et al., 1976b), one might ascribe the exaggerated response to arginine to a lack of inhibition by aminogenic somatostatin release. Perhaps, as depicted in Fig. 12, the secretion of the

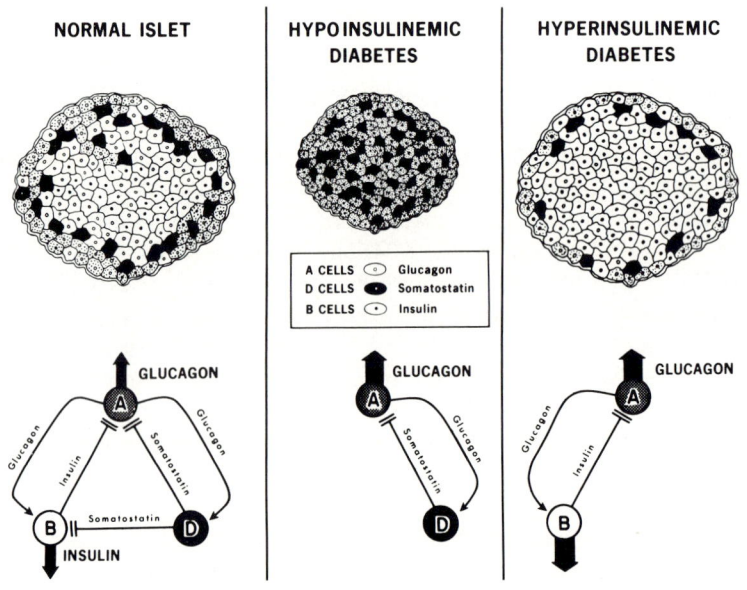

FIG. 12. Schematization of local paracrine actions of the secretory products of A, B, and D cells (lower panels) in the normal islet, the islet of hypoinsulinemic diabetes in relationship to the morphologic abnormalities (upper panels). In the hypoinsulinemic form of diabetes, the restraining influence of insulin on A-cell function is lacking (Orci et al., 1976) and only somatostatin is present to inhibit glucagon secretion; exogenous insulin in sufficient concentrations seems to correct the excessive glucagon production (Raskin et al., 1976). In the hyperinsulinemic form of diabetes, at least in mice (Baetens et al., 1976b) (humans have not as yet been studied), insulin-containing B cells are abundant but somatostatin-containing D-cells sparse, suggesting lack of the restraining influence of somatostatin on A-cell function, only insulin being present to inhibit glucagon secretion. This may explain why exogenous insulin fails to correct the abnormal A-cell responses of hyperinsulinemic diabetes of man. Reprinted, with permission of *Diabetes,* from Unger and Orci (1977b).

normal A cell is restrained both by local insulin (Samols and Harrison, 1976a) and somatostatin (Koerker *et al.*, 1974) release, particularly during hyperaminoacidemia, and, unless both peptides are present, an exaggerated A-cell response to stimuli will occur. In the hypoinsulinemic diabetic, insulin will restore the response to normal (Raskin *et al.*, 1976), but in the hyperinsulinemic diabetic insulin has little effect and perhaps only somatostatin plus insulin would correct it.

Insulin, in supraphysiologic quantities does, indeed, render the exaggerated response of plasma IRG to arginine and to a protein meal indistinguishable from that of the nondiabetics (Raskin *et al.*, 1976). The IRG response to a glucose meal is also greatly improved by insulin in normal concentrations and is reduced to normal by supraphysiologic bolus injections of insulin (Aydin *et al.*, 1977a). The need for supraphysiologic insulin to correct the A-cell response of juvenile-type diabetics could be attributed to insensitivity of the A-cell to insulin and/or to glucose, the result of some intrinsic loss of glucose-sensing capacity which it may share with the B cell; however, more attractive to us at this time is the concept, previously espoused by Weir *et al.* (1976), of a simple within-islet insulin deficiency. The appeal of this view is based on several considerations: (1) the evidence that insulin, either directly or by permissively augmenting glucose entry into the A cell, can reduce plasma IRG levels in such patients; (2) the fact that in nondiabetics the vast majority of A cells appear to have direct contact with B cells but such contacts are lacking in the juvenile diabetic state; (3) the fact that supraphysiologic bolus injections of insulin, which might qualitatively simulate the pattern of rapid insulin release into the extracellular space around the A cells, restore the IRG pattern to normal, at least in the absolute sense. The so-called "insulin-independent abnormality" of A-cell function that we had referred to in earlier studies (Raskin *et al.*, 1975) may in retrospect be secondary to insulin lack, not in the plasma, but at the level of the extracellular space of the A cells. Figure 13 compares the juvenile diabetic islet with that of the nondiabetic, and Table VIII reveals a total absence of insulin-containing cells and relative within-islet hyperplasia of A cells and D cells in such patients (Orci *et al.*, 1976). Conceivably, the increased number of D cells fails to inhibit the augmented A-cell activity when within-islet insulin is absent (Fig. 12). In other words, the anatomical disorganization of the juvenile diabetic islet of Langerhans with loss of direct A-cell contacts with contiguous B cells and D cells may constitute the underlying lesion in juvenile diabetes that does not respond to insulin replacement in physiologic amounts but requires supraphysiologic doses.

In hyperinsulinemic adult-onset diabetics in whom an abundance of insulin-containing B cells is presumed to be present (Fig. 12), glucose-sensing function is probably present, though obtunded. Unlike the case in the juvenile-type diabetic, even supraphysiologic doses of exogenous insulin fail to improve the exaggerated glucagon response to amino acids (Fig. 14) (Raskin *et al.*, 1976) and protein (Aydin *et al.*, 1977b). It would seem, then, that the normal restraint of

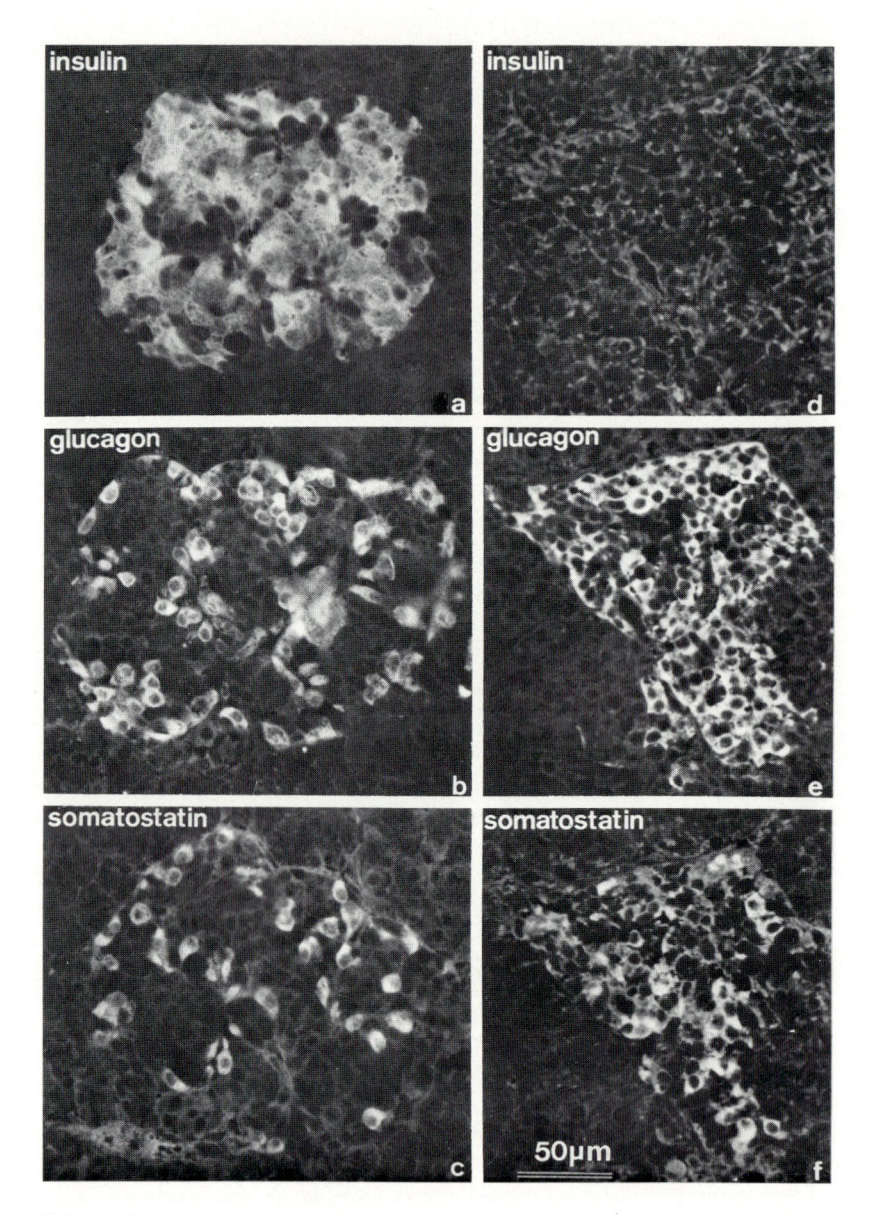

FIG. 13. Consecutive serial sections of pancreases of human normal and of chronic juvenile-type diabetic subjects processed for indirect immunofluorescence. (a–c) Distribution of insulin, glucagon, and somatostatin-containing cells in a normal human islet. (d–f) Demonstrates the striking topographical and numerical changes of these three cell types in an islet of a chronic juvenile-type diabetic subject.

TABLE VIII
Volume Density of Insulin-, Glucagon-, and Somatostatin-
Immunofluorescent Cells in the Islets of Chronic Juvenile-Type
Diabetic Subjects and of Nondiabetic Subjects

	Insulin	Glucagon	Somatostatin
Controls (n = 4)	0.624 ± 0.022	0.240 ± 0.024	0.092 ± 0.012
Diabetics (n = 2)	0	0.688 ± 0.029[a]	0.238 ± 0.024[a]

[a] $P < 0.001$ that value is the same as for controls.

aminogenic glucagon release may require both insulin release by neighboring B cells, which are plentiful in hyperinsulinemic diabetes, and somatostatin release by neighboring D cells, which, although not yet studied in man, are very sparse in hyperinsulinemic ob/ob diabetic mice (Baetens *et al.*, 1976b).

To summarize the foregoing, in the juvenile diabetic in whom the normal A-cell contacts with B cells are absent and D cells increased, exogenous insulin in

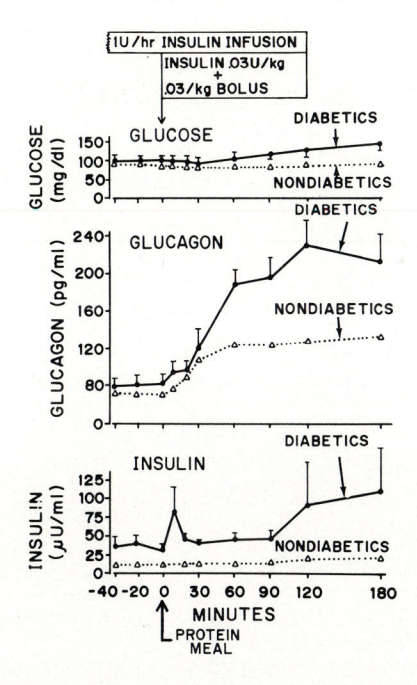

FIG. 14. Ineffectiveness of insulin infused at 1 U per hour for 14 hours before the experiment plus a supplementary insulin dose given during the ingestion of a protein meal in restoring to normal the exaggerated protein-induced rise in plasma immunoreactive glucagon in adult-type diabetics. (Unpublished observations of I. Aydin, P. Raskin, and R. H. Unger.)

suprapbysiologic quantities corrects exaggerated aminogenic secretory response and lowers glucagon levels to normal, although it fails to recreate the lost capacity of the A cell to sense glycemic change. By contrast, in the adult type hyperinsulinemic diabetic, in whom it is expected that insulin-containing B cells are plentiful and D cells sparse, the exaggerated aminogenic response of glucagon secretion is unaffected by even very high levels of exogenous insulin while the glucose sensing capacity appears to be at least partially preserved. Perhaps disturbances in A-cell secretory activity in diabetes are the consequence of a paucity of A-cell contacts with both somatostatin- and insulin-producing cells, and the particular characteristics of the disturbances are determined by the type of A-cell-inhibiting cell that is lacking.

ACKNOWLEDGMENTS

This work was supported by VA Institutional Research Support Grant 549-8000-01; National Institutes of Health Grants AM 02700-16, I-ROI-AM 18179, and I-MOI-RR 00633; Pfizer Laboratories, New York; Bristol Myers Company, New York; Mead Johnson Center, Evansville, Indiana; Dr. Karl Thomae GmbH, Germany; Merck, Sharpe and Dohme, Rahway, New Jersey; CIBA-GEIGY Corporation, Summit, New Jersey; The Upjohn Company, Kalamazoo, Michigan; Eli Lilly, Indianapolis, Indiana; 30K Rabbit Fund; and Fonds National Suisse de la Recherche Scientifique Grant No. 3.553.75.

The authors wish to thank the following persons for their expert technical assistance: Virginia Harris, Kay McCorkle, Margaret Bickham, Cathy Mitchell, Danny Sandlin, Loretta Clendenen, Sara Innis, Helen Gibson, and John Diffie.

For secretarial assistance, the authors express their thanks to Susan Freeman and Billie Godfrey.

REFERENCES

Aguilar-Parada, E., Eisentraut, A. M., and Unger, R. H. (1969). *Am. J. Med. Sci.* **257**, 415.
Alberti, K. G. M. M., Christensen, N. J., Christensen, S. E., Hansen, A. P., Iversen, J., Lundbaek, K., Seyer-Hansen, K., and Orskov, H. (1973). *Lancet* **2**, 1299.
Aydin, I., Raskin, P., and Unger, R. H. (1977a). (In preparation.)
Aydin, I., Raskin, P., and Unger, R. H. (1977b). *Clin. Res.* **25**, 31A.
Baetens, D., Rufener, C., Srikant, C. B., Dobbs, R. E., Unger, R. H., and Orci, L. (1976a). *J. Cell Biol.* **69**, 455.
Baetens, D., Coleman, D. L., and Orci, L. (1976b). *Diabetes* **25**, 344.
Barnes, A. J., and Bloom, S. R. (1976). *Lancet* **1**, 219.
Berson, S. A., Yalow, R. S., Bauman, A., Rothschild, M. A., and Newerly, K. (1956). *J. Clin. Invest.* **35**, 1970.
Blazquez, E., Muñoz-Barragan, L., Patton, G. S., Orci, L., Dobbs, R. E., and Unger, R. H. (1976). *Endocrinology* **99**, 1182.
Blazquez, E., Muñoz-Barragan, L., Patton, G. S., Dobbs, R. E., and Unger, R. H. (1977). *J. Lab. Clin. Med.* (in press).
Brazeau, P., Vale, W., Burgus, R., Ling, N., Butcher, M., Rivier, J., and Guillemin, R. (1973). *Science* **179**, 77.
Cahill, G. F., Jr. (1971). *Diabetes* **20**, 785.

Cahill, G. F., Jr., Herrera, M. G., Morgan, A. P., Soeldner, J. S., Steinke, J., Levy, P. L., Reichard, G. A., Jr., and Kipnis, D. M. (1966). *J. Clin. Invest.* **45**, 1751.
Cherrington, A., and Vranić, M. (1974). *Metab., Clin. Exp.* **23**, 729.
Dobbs, R. E., Sakurai, H., Faloona, G. R., Valverde, I., Baetens, D., Orci, L., and Unger, R. H. (1975). *Science* **187**, 544.
Ensinck, J. W., and Palmer, J. P. (1976). *Metab., Clin. Exp.* **25** (Suppl.), 1409.
Foá, P. P., Galansino, G., and Pozza, G. (1957). *Recent Prog. Horm. Res.* **13**, 473.
Fujita, T. (1976). *Int. Symp. GEP Endocr. Syst., 1977* (in press).
Gavin, J. R., III, Roth, J., Neville, D. M., Jr., DeMeyts, P., and Buell, D. N. (1974). *Proc. Natl. Acad. Sci. U.S.A.* **71**, 84.
Gerich, J. E. (1976). *Metab., Clin. Exp.* **25** (in press).
Gerich, J. E., Langlois, M., Noacco, C., Karam, J. H., and Forsham, P. H. (1973). *Science* **182**, 171.
Gerich, J. E., Lorenzi, M., Bier, D. M., Schneider, V., Tsalikian, E., Karam, J. H., and Forsham, P. H. (1975a). *N. Engl. J. Med.* **292**, 985.
Gerich, J. E., Lorenzi, M., Karam, J. H., Schneider, V., and Forsham, P. H. (1975b). *J. Am. Med. Assoc.* **234**, 159.
Goodner, C. J., Walike, B. C., Koerker, D. J., Brown, A. C., Ensinck, J. W., Chideckel, E. W., Palmer, J., and Kalnasy, L. W. (1976). *Diabetes* **25**, 340.
Hellerström, C., Howell, S. L., Edwards, J. C., Andersson, A., and Östenson, G. C. (1974). *Biochem. J.* **140**, 13.
Kimball, C. P., and Murlin, J. R. (1923). *J. Biol. Chem.* **58**, 337.
Knudsen, J. B., Holst, J. J., Asnoes, S., and Johansen, A. (1975). *Acta Pathol. Microbiol. Scand., Sect. A* **83**, 741.
Koerker, D. J., Ruch, W., Chideckel, E., Palmer, J., Goodner, C. J., Ensinck, J., and Gale, C. C. (1974). *Science* **184**, 482.
Kuku, S. F., Zeidler, A., Emmanouel, D. S., Katz, A. I., Rubenstein, A. H., Levin, N. W., and Tello, A. (1976). *J. Clin. Endocrinol. Metab.* **42**, 173.
Larsson, L. I., Holst, J., Hakanson, R., and Sundler, F. (1975). *Histochemistry* **44**, 281.
Lawrence, A. M., Tan, S., Hojvat, S., Kirsteins, L., and Mitton, J. (1976). *Metab., Clin. Exp.* **25** (in press).
McGarry, J. D., Wright, P., and Foster, D. (1975). *J. Clin. Invest.* **55**, 1202.
Mackrell, D. J., and Sokal, J. E. (1969). *Diabetes* **18**, 724.
Marliss, E. B., Aoki, T. T., and Cahill, G. F., Jr. (1972). *In* "Glucagon. Molecular Physiology, Clinical and Therapeutic Implications" (P. J. Lefebvre and R. H. Unger, eds.), p. 123. Pergamon, Oxford.
Marliss, E. B., Girardier, L., Seydoux, J., Wollheim, C. B., Kanazawa, Y., Orci, L., Renold, A. E., and Porte, D., Jr. (1973). *J. Clin. Invest.* **52**, 1246.
Mashiter, K., Harding, P. E., Chou, M., Mashiter, G. D., Stout, J., Diamond, D., and Field, J. B. (1975). *Endocrinology* **96**, 678.
Matschinsky, F. M., and Pagliara, A. S. (1976). *Metab., Clin. Exp.* **25**, 1203.
Matsuyama, T., and Foá, P. P. (1975). *Proc. Soc. Exp. Biol. Med.* **147**, 97.
Meissner, H. P. (1976). *Nature (London)* **262**, 502.
Mortimer, C. H., Turnbridge, W. M. G., Carr, D., Yeomans, L., Lind, T., Coy, D. H., Bloom, S. R., Kastin, A., Mallinson, C. N., Besser, G. M., Schally, A. V., and Hall, R. (1974). *Lancet* **1**, 697.
Müller, W. A., Faloona, G. R., and Unger, R. H. (1971). *J. Clin. Invest.* **50**, 1992.
Müller, W. A., Brennan, M. F., Tan, M. H., and Aoki, T. T. (1974). *Diabetes* **23**, 512.
Muñoz, L., Rufener, C., Srikant, C. B., Dobbs, R. E., Shannon, W. A., Jr., Baetens, D., and Unger, R. H. (1977). *Horm. Metab. Res.* **9**, 37.

Muñoz-Barragan, L., Blazquez, E., Patton, G. S., Dobbs, R. E., and Unger, R. H. (1976). *Am. J. Physiol.* **231**, 1057.
Noe, B. D., and Bauer, G. E. (1971). *Endocrinology* **89**, 642.
Ohneda, A., Aguilar-Parada, E., Eisentraut, A. M., and Unger, R. H. (1969). *Diabetes* **18**, 1.
Orci, L. (1977). Cf. Discussion following paper by Floyd *et al.*, this volume.
Orci, L., and Unger, R. H. (1975). *Lancet* **2**, 1243.
Orci, L., Stauffacher, W., Renold, A. E., and Rouiller, C. (1970). *Acta Isot.* **10**, 171.
Orci, L., Unger, R. H., and Renold, A. E. (1973a). *Experientia* **29**, 1015.
Orci, L., Amherdt, M., Henquin, J. C., Lambert, A. E., Unger, R. H., and Renold, A. E. (1973b). *Science* **180**, 647.
Orci, L., Malaisse-Lagae, F., Ravazzola, M., Rouiller, C., Renold, A. E., Perrelet, A., and Unger, R. H. (1975). *J. Clin. Invest.* **56**, 1066.
Orci, L., Baetens, D., Rufener, C., Amherdt, M., Ravazzola, M., Studer, P., Malaisse-Lagae, F., and Unger, R. H. (1976). *Proc. Natl. Acad. Sci. U.S.A.* **73**, 1338.
Palmer, J. P., Werner, P. L., Benson, J. W., and Ensinck, J. W. (1976). *Metab., Clin. Exp.* **25** (Suppl.), 1483.
Park, C. R., and Exton, J. H. (1972). *In* "Glucagon. Molecular Physiology, Clinical and Therapeutic Implications" (P. J. Lefebvre and R. H. Unger, eds.), p. 77. Pergamon, Oxford.
Patton, G. S., Dobbs, R., Orci, L., Vale, W., and Unger, R. H. (1976a). *Metab., Clin Exp.* **25** (Suppl), 1269.
Patton, G., Ipp, E., Dobbs, R. E., Orci, L., Vale, W., and Unger, R. H. (1976b). *Life Sci.* **19**, 1957.
Raskin, P. (1976). *Metab., Clin. Exp.* **25** (Suppl.), 124.
Raskin, P., and Unger, R. H. (1976). *Diabetes* **25**, 341.
Raskin, P., and Unger, R. H. (1977). Submitted for publication.
Raskin, P., Fujita, Y., and Unger, R. H. (1975). *J. Clin. Invest.* **56**, 1132.
Raskin, P., Aydin, I., and Unger, R. H. (1976). *Diabetes* **25**, 227.
Rigopoulou, D., Valverde, I., Marco, J., Faloona, G., and Unger, R. H. (1970). *J. Biol. Chem.* **245**, 496.
Sakurai, H., Dobbs, R. E., and Unger, R. H. (1975). *Metab., Clin. Exp.* **24**, 1287.
Samols, E., and Harrison, J. (1976a). *Metab., Clin. Exp.* **25** (Suppl.), 1211.
Samols, E., and Harrison, J. (1976b). *Metab., Clin. Exp.* **25** (Suppl.), 1265.
Samols, E., Marri, G., and Marks, V. (1965). *Lancet* **2**, 415.
Samols, E., Tyler, J., and Marks, V. (1972). *In* "Glucagon. Molecular Physiology, Clinical and Therapeutic Implications" (P. J. Lefebvre, and R. H. Unger, eds.), p. 151. Pergamon, Oxford.
Sasagawa, T., Kobayashi, S., and Fujita, T. (1973). *In* "Gastro-Entero-Pancreatic Endocrine System" (T. Fujita, ed.), p. 31. Igaku Shoin, Ltd., Tokyo.
Sherwin, R. S., Fisher, M., Hendler, R., and Felig, P. (1976a). *N. Engl. J. Med.* **294**, 455.
Sherwin, R. S., Wahren, J., and Felig, P. (1976b). *Metab., Clin. Exp.* **25** (in press).
Srikant, C. B., McCorkle, K., and Unger, R. H. (1977). *J. Biol. Chem.* **251**, 1847.
Tager, H. S., and Markese, J. (1976). *Metab., Clin. Exp.* **25** (Suppl.), 1343.
Tager, H. S., and Steiner, D. F. (1973). *Proc. Natl. Acad. Sci. U.S.A.* **70**, 2321.
Tiengo, A., Fedele, D., Marchiori, E., Nosadini, R., and Muggeo, M. (1976). *Diabetes* **25**, 408.
Trakatellis, A. C., Tada, K., Yamaji, K., and Gardiki-Kouidou, P. (1975). **14**, 1508.
Tung, A. K., Rosenzweig, S. A., and Foá, P. P. (1976). *Proc. Soc. Exp. Biol. Med.* **153**, 344.
Unger, R. H. (1971). *Diabetes* **20**, 834.
Unger, R. H. (1976a). *Diabetes* **25**, 136.
Unger, R. H. (1976b). *Metab., Clin. Exp.* **25** (Suppl.), 1523.

Unger, R. H. (1976c). *N. Engl. J. Med.* **294,** 1239.
Unger, R. H., and Orci, L. (1975). *Lancet* **1,** 14.
Unger, R. H., and Orci, L. (1977a). *Arch. Intern. Med.* (in press).
Unger, R. H., and Orci, L. (1977b). *Diabetes* **26,** 241.
Unger, R. H., Eisentraut, A. M., McCall, M. S., Keller, S., Lanz, H. C., and Madison, L. L. (1959). *Proc. Soc. Exp. Biol. Med.* **102,** 621.
Unger, R. H., Aguilar-Parada, E., Müller, W. A., and Eisentraut, A. M. (1970). *J. Clin. Invest.* **49,** 837.
Valverde, I., and Villanueva, M. L. (1976). *Metab., Clin. Exp.* **25** (Suppl.), 1393.
Valverde, I., Villanueva, M. L., Lozano, I., and Marco, J. (1974). *J. Clin. Endocrinol. Metab.* **39,** 1020.
Valverde, I., Dobbs, R. E., and Unger, R. H. (1975). *Metab., Clin. Exp.* **24,** 1021.
Valverde, I., Lemon, H. M., Kessinger, A., and Unger, R. H. (1976). *J. Clin. Endocrinol. Metab.* **42,** 804.
Vranič, M. Pek, S., and Kawamori, R. (1974). *Diabetes* **23,** 905.
Wahren, J., and Felig, P. (1976). *Clin. Res.* **24,** 461A.
Weir, G. C., Knowlton, S. S., Atkins, R. F., McKennan, K. X., and Martin, D. B. (1976). *Diabetes* **25,** 275.
Wise, J. K., Hendler, R., and Felig, P. (1973). *N. Engl. J. Med.* **288,** 487.

DISCUSSION

E. M. Bogdanove: I have two rather simple questions. One has to do with the increase in delta cells in the juvenile diabetic. Do you attribute any particular significance to that other than that it is a passive, relative increase in the alpha and delta cells resulting from the disappearance of the beta cells? The other question has to do with the somatostatin infusions, which clearly lower glucagon levels and at the same time ameliorate the hypoglycemia in the diabetic. However, somatostatin was originally named for its effect on growth hormone. Do you have any data on whether these somatostatin infusions also reduce somatotropin secretion, and, if so, whether growth hormone plays any role in this sort of hypoglycemia?

R. Unger: These are both very important questions that Dr. Orci may prefer to answer. The first portion of the question relates to the increase in delta cells in the pancreas, which he and his colleagues have observed: would you like to answer that Dr. Orci?

L. Orci: In human subjects, the evaluation of islet cell population has been done solely in relation to the islet: indeed, in all cases, only a limited sample of the pancreas was available for examination. The analysis of the whole pancreas was, however, possible in a group of rats 16 months after the induction of diabetes by a single injection of streptozotocin. In these animals, we were able to show that the diminution of insulin-containing cells was accompanied by a significant increase in somatostatin-containing cells, not only relative to the islet, but for the whole pancreas as well [L. Orci, B. Baetens, C. Rufener, M. Amherdt, M. Ravazzola, P. Studer, F. Malaisse-Lagae, and R. H. Unger, *Proc. Natl. Acad. Sci. U.S.A.* **73,** 1338–1342 (1976)].

We believe that any alteration of islet architecture and, hence, of cellular interrelationships may also affect per se the function of each islet taken separately as a microorgan.

R. Unger: I think, if I understood the question, that it concerned our interpretation of the presence of increased numbers of delta cells in the light of increased glucagon secretion. One interpretation might be that D cell hyperplasia constitutes a valiant, but futile, compensatory effort to restrain the hyperfunction of A cells that are separated from contact with insulin. On the other hand, we do not know how many A cells come in contact with the D cells. Are they in clusters? Are they well dispersed? The other question: Could some of these effects been mediated by growth hormone? What is the effect of continuous somatostatin infusion on human growth hormone levels? Dr. Gerich has studied this and, to the best of my knowledge, is convinced that changes in growth hormone induced by the administration of somatostatin cannot explain any of the results observed.

J. E. Rall: I wonder if you would say something about the antibodies you have used and whether you think you are measuring largely glucagon derived from pancreatic secretion or whether there is a component of nonpancreatic glucagon involved. I refer to steady-state as well as stimulated levels of glucagon.

R. Unger: That is a very obviously vital question. If I can answer in a very superficial way, the antibody that we use is known as 30K antibody. It is a so-called non-cross-reacting antibody, and it is quite specific for pancreatic-type glucagon. We are trying to break ourselves of the habit of using the term "glucagon," which we think should be restricted to the 29-amino acid polypeptide characterized by Dr. Bromer, and use instead, for want of a better name, immunoreactive glucagon, or IRG, to refer to the material measured by so-called specific antiglucagon antibodies. Such antibodies, according to Dr. Lise Heding, recognized the 24 to 29 portion of the glucagon molecule. This piece is presumably present in a whole family of heterogeneous IRGs that range in molecular weight from 2000 to as high as 180,000 in the plasma. In tissues we have found IRGs of as much as 65,000 MW. There are A cells in tissues outside the pancreas that are in all respects indistinguishable from pancreatic A cells. These cells must contain the entire spectrum of the IRG family, indistinguishable either chemically or biologically by any test we have available to us from what we get out of the pancreas, except that it appears that fundic glucagon could be a "superglucagon." Its ability to activate adenylate cyclase at immunoequivalent concentrations is approximately twice that of similarly extracted pancreatic IRG^{3500}.

Dr. Isabel Valverde, using 30K antiglucagon serum has characterized the various IRG fractions that circulate in the plasma. The IRG that elutes with glucagon ^{125}I marker is probably true glucagon. It is the most active moiety; it rises when one stimulates with arginine; when one suppresses with glucose or somatostatin, this is what disappears. And in the vast majority of diabetics, this is the fraction that is high. Then there is a fraction that we now call IRG^{9000}. This is about the size of what was first described in pancreatic extracts by Dr. Rigopoulu in our laboratory as large glucagon immunoreactivity, and subsequently Noe and Bauer believed it to be proglucagon. It does not bind to isolated hepatocyte membranes and does not activate adenylate cyclase. It has no glycogenolytic activity in the isolated perfused liver. As Dr. Rubenstein's group has shown, this fraction is high in renal insufficiency, and Valverde has also demonstrated it to be increased in a glucagonoma patient. Finally, in the void volume there is material that Valverde refers to as "big plasma glucagon" or BPG, and we'll use that designation. This material has been considered an interference factor—something of methodologic nuisance. However, in some diabetics who always exhibit plasma IRG levels of 300 or 400 pg/ml a void volume peak of IRG with a molecular weight of about 180,000 on rechromatography may be the cause for the high IRG levels in two such patients. Moreover, when stimulation with arginine cause a rise in plasma IRG from 300 pg/ml to over 500 pg/ml; the bulk of the increase in whole plasma IRG was the result of an increase in BPG. The glucagon levels were not high. To find out whether this material is biologically active, Dr. C. B. Srikant has perfused BPG in the Mortimore isolated perfused liver preparation. He has found, in a group of six diabetic patients with the mean BPG immunoreactivity of about 1 ng, a glycogenolytic response equivalent to 5 ng of crystalline glucagon. Control samples of plasma with normal BPG levels were devoid of glycogenolytic activity.

Ensinck and Palmer in Seattle have reported the case of a medical student, now a physician, whose plasma contained large amounts of BPG. We found no glycogenolytic activity in his BPG. Many members of his family have this trait, which Palmer and Ensinck believe to be an autosomal dominant. There is no history of diabetes in his family. The other patients with high BPGs that we have tested have all been diabetics, and in their BPG glycogenolytic activity was present. We think that there may be more high BPG levels among the Pima Indian diabetic population than in the controls that we have studied so far. This is a strange protein, and its significance remains to be determined. It is, in a way, analogous to a long-acting thyroid stimulator that cross-reacted with an anti-TSH serum.

G. Cahill: Were these treated diabetics?

R. Unger: The ones we have studied were treated with oral agents only.

J. Dupré: We have been interested in the paradoxical elevated glucagon levels you have illustrated because of our obsession with gut hormones, and also because we have shown that gastric inhibitory polypeptide (GIP) is potentially glucagonotropic in animals. As you know, there has been difficulty in showing this effect in man, but we have recently demonstrated this action. We also have some evidence regarding the biological effects of glucagon in the presence of GIP.

The demonstration of the glucagonotropic action of GIP in physiological dosage is shown in Fig. A, which illustrates a study in human cirrhotics with portacaval anastomoses, selecting patients with high basal glucagon levels whom we regard as hypersecretors of glucagon. In these subjects the intravenous infusion of GIP consistently produced an increment in immunoreactive glucagon (IRG) that is of the order of 200 pg/ml and, on chromatography, was found to be 3500 MW IRG. The two further subjects, both shunted cirrhotics who did not show basal hyperglucagonemia, did not respond to GIP. This action of GIP in man fits with the notion that this hormone might stimulate the paradoxical rises of IRG in diabetes. A striking thing about these experiments and others in man has been that there has been no rise in glucose when endogenous glucagon is stimulated, and Fig. B shows something bearing on that.

We had earlier found that GIP is not lipolytic, despite its structural similarity with secretin VIP and glucagon, so we examined the interactions of glucagon and GIP in fat cells in case

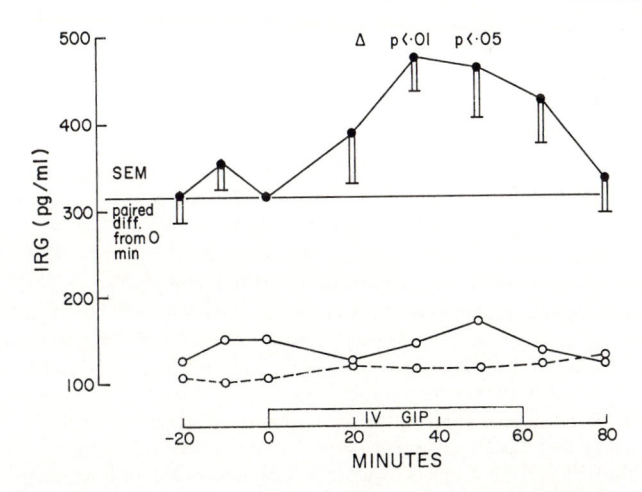

FIG. A. Glucagonotropic action of gastric inhibitory polypeptide (GIP; 2 μg/minute i.v.) infused in subjects with liver cirrhosis and portacaval anastomoses having hyperglucagonemia (> 200 pg/ml; $n = 6$) and normal plasma glucagon ($n = 2$). IRG, immunoreactive glucagon.

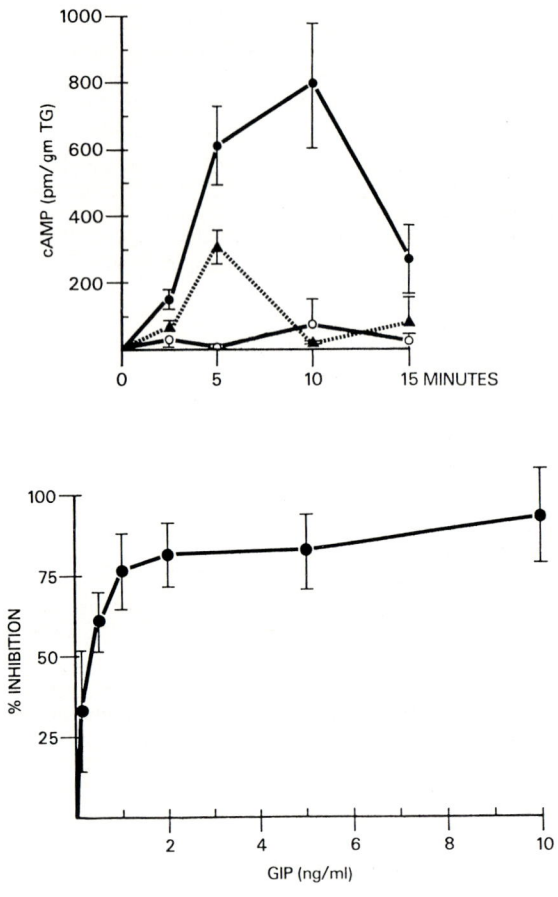

FIG. B. *Top:* Effect of glucagon (100 pg/ml, ●———●) or gastric inhibitory polypeptide (GIP; 50 ng/ml, ○———○) or both (▲– –▲) on generation of cyclic AMP (cAMP) in adipocytes. *Bottom:* Inhibition by GIP of fat cell cAMP response to glucagon (100 pg/ml). There was a 10-minute incubation in the presence of theophylline (1 mM).

we might be able to demonstrate competitive inhibition of glucagon's effects. We were able to inhibit lipolysis and to displace glucagon from its receptors with GIP. These experiments, however, were done with supraphysiologic concentrations of the hormones. In the study illustrated in Fig. B, you will see at the top the effect of glucagon or GIP, or the two combined, on cyclic AMP generation in adipocytes, with 100 pg of glucagon and with 50 ng of GIP per milliliter. The top line shows the stimulation with glucagon; the bottom flat line shows the nil effect of GIP alone, and with the combination you can see that there was significant inhibition of glucagon's reactions. The lower panel shows that if the lowest concentration of GIP that can modulate the actions of glucagon, here again using 100 pg of glucagon and 50 pg of GIP per milliliter, inhibited approximately 50% of the action of the glucagon. This concentration of GIP is right within the physiologic range. We therefore wonder whether GIP is a physiologic modulator of glucagon's actions. Obviously the liver is the organ to look at. We have reproduced the membrane-binding effect in liver cell membranes, but have not yet tested the biologic effects in liver cells.

Have you in your work noticed situations in which, with stimulation of the gut, the expected effects of endogenous glucagon are not seen? That is, have you observed a situation in which immunoreactive glucagon stimulated by meals is apparently not having its expected effect on glucose metabolism?

R. Unger: Yes, there is one such circumstance. When one puts a pure fat meal into the duodenum of the dogs, one observes a rise in plasma IRG, no change in insulin, and yet no change in glucose. This has not been explained adequately. With respect to your comment on the paradoxical effect of hyperglycemia upon plasma IRG, you are not suggesting that GIP would be involved in the paradoxical effect observed with intravenous glucose?

J. Dupré: No, but we have found the paradoxical effect more readily elicited with oral glucose than with intravenous glucose, and there are paradoxical responses to ingestion of galactose, and triglyceride in human diabetics, which are compatible with GIP being the responsible agent.

G. Cahill: Does GIP go up with a fat meal?

J. Dupré: Yes, this is the most potent stimulus.

J. Geller: From your discussion it would seem that the local hormone concentrations within the beta cell were very important in ultimately regulating glucose homeostasis. At least I got that impression, since you mentioned that isolated alpha cells were not responsive to glycemic stimuli. Is the glycemic stimulus, that is, the message to the alpha cell, via somatostatin or mediated by some other mechanism within the islet cells? Do you envisage somatostatin as playing an important physiologic role in the ultimate regulation of glucose homeostasis?

R. Unger: I cannot answer that. It is clearly an important question because there are always D cells next to A cells wherever one looks, even in the glucagon-producing tumors that Dr. Orci has examined. He has been astounded by the high number of somatostatin-containing cells. I really would not be surprised if somatostatin were involved in the mediation of the glucose effect.

R. S. Bernstein: I am concerned about the discrepancy between your results and those of Drs. Sherwin and Felig on the response to exogenous glucagon in diabetics. In Dr. Sherwin's work they infused glucagon continuously and achieved levels, I believe, around 400 pg/ml, which certainly is within the range where one would expect to see effects of glucagon. They showed no deterioration of diabetic control. There are differences in the experimental design between their work and yours. The main difference is that they gave the hormone by continuous intravenous infusion, rather than by subcutaneous route, and that you achieved much higher levels in pulses. Might part of the difference between your results be that you achieved these very high pharmacological levels in pulses? I think the group at Vanderbilt has shown that the effects of glucagon are transient in terms of hepatic gluconeogenesis and glycogenolysis, and maybe this is why there is the difference between your work and theirs.

R. Unger: With respect to the second question, it is well known now and has been established by Vranić and Cherrington and a number of groups, and most recently by Phil Felig and John Wahren and the Liljenquist group that the actions of glucagon administered as a continuous infusion are evanescent. On the other hand, phasic bursts of glucagon appear to have more sustained effects is due, at least in the nondiabetic, in part to insulin. But above and beyond that there is an apparent loss of sensitivity to glucagon in the liver, which could involve the process of "down-regulation" as suggested by the work of Dr. Jesse Roth and his colleagues. However, with respect to glucagon's role in diabetic hyperglycemia, a transient dumping of glucose into the extracellular fluid of a diabetic with its diminished efflux capacity will contribute to endogenous hyperglycemia. This can be viewed, if you will, as impaired "endogenous glucose tolerance." With respect to the discrepancy between Phil Felig's findings and ours, I think there are many differences in experimental design that would explain the differences. It is true that our IRG levels were much higher than Phil's, at least transiently, but in looking at the available data to see whether ours are really too high, we find in stimulated alloxan-diabetic dogs portal vein levels ranging from 700 to 1500 pg/ml. Blackard has reported in normal man during stimulation with arginine values over 1000 pg/ml. So we do not think that the concentration necessarily must be 250 or 500 or 750 to stimulate the situation in the portal vein of poorly controlled diabetics. Finally, we have repeated the Felig study using the same dose of glucagon as he used and we find marked deterioration in diabetic control.

Y. C. Patel: First of all, a comment on one of the questions raised earlier about pancreatic somatostatin in streptozotocin diabetes. As A. Orci has found by morphometry, we have also found by direct radioimmunoassay measurement that the islet content of somatostatin in rats made diabetic with streptozotocin is strikingly increased, and this applies not only to individual islets but also to the whole pancreas. The concentration of somatostatin per whole pancreas at 4 weeks and 6 weeks after streptozotocin is strikingly increased. I was very interested in your somatostatin secretion data, and would like to show some of our results on somatostatin secretion from isolated islets (Table A); These studies were done recently, and also in Orci's laboratory. When groups of 20 rat islets are incubated for 1 hour, there is a spontaneous basal release of somatostatin that is readily measurable by radioimmunoassay. In a preliminary study using this system, we screened a large number of agents for possible effect on somatostatin release. Under basal conditions shown in Table A in controls, there is release of somatostatin of the order of 23.6 and 29 pg per islet per hour, representing approximately 4–5% of the total islet content. We have found that dopamine at a concentration of 10^{-6} M significantly stimulates the release of somatostatin, but, contrary to your findings in the isolated perfused dog pancreas, we would not find any reproducible effect of arginine or glucagon in this system. Dibutyryl cAMP has been the most potent secretogogue that we have found so far, a concentration of 5 mM producing a doubling of somatostatin release. We have not found any reproducible effect of glucose, insulin, BPP, isoproterenol, epinephrine or, carbamylcholine.

P. G. Condliffe: Do you have any idea as to the nature of the signals exchanged between the A and B cells? If the cells communicate directly, as Dr. Orci suggests, via the gap junctions, could insulin or glucagon or some derivative of the prohormones be the messen-

TABLE A
Somatostatin (SRIF) Release from Isolated Islets

Conditions	SRIF release[a] (pg/islet/hr)
Control	23.6 ± 4.4
Dopamine, 10^{-6} M	39.5 ± 4.8[b]
Arginine, 5 mM	31.9 ± 7.0
Control	29 ± 6.2
Glucagon 0.3 μM	43.6 ± 4.5
dbcAMP 5 mM	59.1 ± 4.1[c]

[a] mean ± SE ($n = 8$).
[b] $p < 0.05$.
[c] $p < 0.001$.

ger? Or, since these are rather large molecules to move quickly between the cells, would it be more likely that there are smaller compounds involved.

R. Unger: No, there is no information whatsoever on that, other than that, in certain tissues, molecules of 500 MW or less can move from one cell to a neighboring cell without entering the medium. But we cannot say what their role in the islets may be. It is just that the anatomical equipment has been identified. It might serve some coordinating function and make the islet a sort of functional syncitium, but all you can do is speculate.

G. Cahill: In some subjects, the better you control the diabetes the more the glucagon levels come down to normal. In others, and this is a very heterogeneous population, glucagon levels tend toward normal, but it must be emphasized that "normal control is not achievable."

A. Rubenstein: I wish to raise a point about peripheral and portal vein glucagon levels to which you have alluded and which has been discussed at some length by Felig and others. They have concluded that glucagon is not extracted to any appreciable extent by the liver. However, I wonder if you have any data to bear on whether there is active hepatic extraction of glucagon.

R. Unger: Yes, we and other people have data. I think Phil Felig's data in man gave a ratio of portal to peripheral of about 1.3. Other people have gotten higher values. If we chromatograph the fractions in the portal and the peripheral vein, we find that although glucagon extraction is still considerably lower than we had anticipated, it is above 1.3, closer to 1.8. In certain nonbasal circumstances, you get big gradients, and on stimulation there is, even after the period of equal equilibration, a high ratio. When you measure whole plasma you are measuring a mixture of IRGs, each of which has a different extraction ratio across the liver.

A Newly Recognized Pancreatic Polypeptide; Plasma Levels in Health and Disease[1]

JOHN C. FLOYD, JR., STEFAN S. FAJANS,
SUMER PEK, AND RONALD E. CHANCE

The Department of Internal Medicine (Division of Endocrinology and Metabolism and the Metabolism Research Unit), The University of Michigan, Ann Arbor, Michigan; and The Lilly Research Laboratories, Indianapolis, Indiana

I. Introduction

A pancreatic polypeptide of 36 amino acid residues with hormonal properties was recently discovered independently by Kimmel and associates at the University of Kansas (Kimmel *et al.*, 1968, 1973, 1974, 1975; Kimmel and Pollack, 1975; Langslow *et al.*, 1973; Hazelwood, 1973; Hazelwood *et al.*, 1973) and by Chance and associates at the Lilly Research Laboratories (Chance, 1972; Chance and Jones, 1974; Chance *et al.*, 1975, 1977a; Lin and Chance, 1972, 1974a,b; Lin *et al.*, 1973, 1974). The chicken polypeptide isolated and characterized by Kimmel's group has been tentatively designated avian pancreatic polypeptide, or aPP. Similarly, the mammalian polypeptides studied by Chance *et al.*, particularly the bovine and human homologs, have been designated bPP and hPP, respectively.

II. Chemistry and Species Homology of Pancreatic Polypeptides

The avian and mammalian polypeptides were isolated from acid-alcohol extracts of pancreas and were purified by ion-exchange and gel-filtration chromatographic procedures commonly used in purifying glucagon and insulin (Chance and Jones, 1974; Kimmel *et al.*, 1968, 1975; Chance *et al.*, 1977a). The amount of pancreatic polypeptide extractable from the pancreas relative to the amounts of extractable glucagon and insulin varies among species; in chicken pancreas the

---

[1] Work supported in part by U.S. Public Health Service grants AM-02244 and AM-00888, National Institute of Arthritis, Metabolism and Digestive Diseases; RR-42, General Clinical Research Program; 5P11-GM 15559, National Institutes of General Medical Sciences; by The American Diabetes Association–Michigan Affiliate; The Upjohn Company, Kalamazoo, Michigan; Pfizer, Inc., New York, New York; and the Eli Lilly Company, Indianapolis, Indiana.

polypeptide is more abundant than insulin (Kimmel *et al.*, 1968) whereas in pig, sheep, and cow pancreases it is present in amounts equal to or greater than glucagon (R. E. Chance, unpublished observations, 1976). The average molecular weight of the pancreatic polypeptides is 4200.

The amino acid sequences of aPP and bPP are compared in Fig. 1. Of the 36 positions, there are 16 at which the amino acid residues are identical and there are other positions at which substitutions are conservative. In addition, certain structural features have been preserved, such as the location of proline residues (positions 2, 5, 8) and acidic residues (positions 10, 11, 15) in the amino-terminal segment and the basic carboxyl terminus (Kimmel *et al.*, 1975). The four mammalian polypeptides isolated from bovine, ovine, porcine, and human pancreases (Fig. 2) have nearly identical amino acid sequences (Chance and Jones, 1974; Chance *et al.*, 1977a). The ovine pancreatic polypeptide, oPP, differs from bPP only at position 2, at which serine replaces proline; and the porcine pancreatic polypeptide, pPP, differs from bPP at position 6, at which valine replaces glutamine. The human pancreatic polypeptide, hPP, has valine at position 6, an asparagine at either position 10 or 11, and an aspartic acid at position 23. A structural feature common to the 5 species of PP studied so far is the blocked carboxyl-terminal residue at position 36, which is occupied by tyrosine amide. This is a physicochemical property suggestive of a polypeptide hormone, since amidated carboxyl-terminal residues appear to occur only in polypeptide hormones.

Computer analysis of the primary structures of the pancreatic polypeptides, which employed alignment statistics for identifying related protein sequences, was performed by J. F. Mornex (personal communication, 1976), who found evidence that avian, bovine, ovine, porcine, and human pancreatic polypeptides are homologous. There was, however, no conclusive evidence that any of these pancreatic polypeptides are homologs of secretin, glucagon, gastric inhibitory polypeptide (GIP), gastrin, cholecystokinin, or vasoactive intestinal polypeptide (VIP), or of peptides that are hypothetical ancestors of the former three hormones.

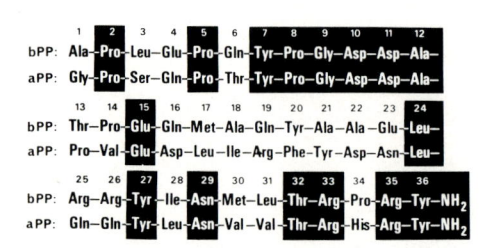

```
        1    2    3    4    5    6    7    8    9   10   11   12
bPP:   Ala- Pro- Leu- Glu- Pro- Gln- Tyr- Pro- Gly- Asp- Asp- Ala-
aPP:   Gly- Pro- Ser- Gln- Pro- Thr- Tyr- Pro- Gly- Asp- Asp- Ala-

       13   14   15   16   17   18   19   20   21   22   23   24
bPP:   Thr- Pro- Glu- Gln- Met- Ala- Gln- Tyr- Ala- Ala- Glu- Leu-
aPP:   Pro- Val- Glu- Asp- Leu- Ile- Arg- Phe- Tyr- Asp- Asn- Leu-

       25   26   27   28   29   30   31   32   33   34   35   36
bPP:   Arg- Arg- Tyr- Ile- Asn- Met- Leu- Thr- Arg- Pro- Arg- Tyr-NH₂
aPP:   Gln- Gln- Tyr- Leu- Asn- Val- Val- Thr- Arg- His- Arg- Tyr-NH₂
```

FIG. 1. Amino acid sequences of avian pancreatic polypeptide (aPP) and bovine pancreatic polypeptide (bPP).

```
 1   2   3   4   5   6   7   8   9   10  11  12
Ala-Pro-Leu-Glu-Pro-Gln-Tyr-Pro-Gly-Asp-Asp-Ala-
 13  14  15  16  17  18  19  20  21  22  23  24
Thr-Pro-Glu-Gln-Met-Ala-Gln-Tyr-Ala-Ala-Glu-Leu-
 25  26  27  28  29  30  31  32  33  34  35  36
Arg-Arg-Tyr-Ile-Asn-Met-Leu-Thr-Arg-Pro-Arg-Tyr-NH$_2$
```

	2	6	10	11	23
Bovine (bPP):	Pro	Gln	Asp	Asp	Glu
Ovine (oPP):	Ser				
Porcine (pPP):	Val				
Human (hPP):	Val		Asx	Asx	Asp

FIG. 2. Amino acid sequence of bovine pancreatic polypeptide (bPP); comparison with sequences of ovine, porcine, and human pancreatic polypeptides.

III. Biological Effects of aPP and bPP

The physiologic functions of these pancreatic polypeptides are unknown. Studies with aPP in chicken (Hazelwood, 1973; Hazelwood et al., 1973; Langslow and Hazelwood, 1975; McCumbee and Hazelwood, 1976) and bPP in dogs (Lin et al., 1973, 1974; Lin and Chance, 1972, 1974a,b) suggest possible gastrointestinal functions. aPP was found to be a potent gastric secretagogue in chickens (Hazelwood et al., 1973; Langslow and Hazelwood, 1975) whereas in dogs bPP was found to be a potent inhibitor of pancreatic enzyme secretion induced by cholecystokinin plus secretin (Lin et al., 1973, 1974; Lin and Chance, 1974b; Chance et al., 1975). In the chicken, injected aPP induces hepatic glycogenolysis and a decrease in plasma glycerol, but does not induce a change in plasma glucose (Hazelwood et al., 1973). Similarly in the dog there is no change in blood glucose when bPP is administered (Lin and Chance, 1972). The physiologic relevance of these findings in the dog and the chicken remains conjectural. The search for the physiologic meaning of these polypeptides has been facilitated by antisera raised against aPP (Langslow et al., 1973; Larsson et al., 1974), bPP (Chance et al., 1977b; Larsson et al., 1976b), and hPP (Chance et al., 1977b; Larsson et al., 1975b, 1976b). These antisera have been used extensively in both immunohistochemical studies (Larsson et al., 1974, 1975a,b, 1976b; L.-I. Larsson, G. Boder, and W. N. Shaw, personal communication, 1976; W. Gepts, personal communication, 1976; Polak et al., 1976; Baetens et al., 1976; Gersell et al., 1976; P. Heitz, J. M. Polak, S. R. Bloom, and A. G. E. Pearse, personal communication, 1976; Rufener et al., 1976) and radioimmunoassay development, particularly the hPP radioimmunoassay (Floyd and Fajans, 1976a,b; Floyd et al., 1975, 1976a,b, 1977; Adrian et al., 1976; T. E. Adrian, S. R. Bloom, M. G. Bryant, J. M. Polak, P. Heitz, and A. J. Barnes, personal communication, 1976; Chance et al., 1975; Polak et al., 1976; Schwartz et al., 1976).

IV. Distribution of Pancreatic Polypeptide in Tissues

A. IMMUNOHISTOCHEMICAL LOCALIZATION AND ULTRASTRUCTURE OF PP CELLS

Immunohistochemical studies show that aPP is stored in an endocrine-type cell disseminated in the exocrine parenchyma of the chicken pancreas (aPP cells are seldom found in islets of chicken pancreas) and different from islet cell types A, B, and D (Larsson *et al.*, 1974). hPP (Larsson *et al.*, 1975b, 1976b) and several other mammalian pancreatic polypeptides (Larsson *et al.*, 1976b) have also been localized to an endocrine-type cell different from islet A, B, and D cells. The PP cells are found both in the peripheral part of the islets and scattered throughout the exocrine parenchyma. The relative frequency of PP cells in these two locations may vary from one species to another (Larsson *et al.*, 1976b). Figure 3 shows immunofluorescent PP cells in the periphery of an islet in the duodenal lobe of hamster pancreas (Larsson *et al.*, 1976b). Extrainsular PP cells situated in ductal epithelium of mouse pancreas are evident in Fig. 4 (Larsson *et al.*, 1976b). In man hPP cells were found in the periphery of islets and also scattered in the exocrine pancreatic parenchyma and in the epithelium of small and medium-sized ducts (Larsson *et al.*, 1975b, 1976b). That hPP cells are located

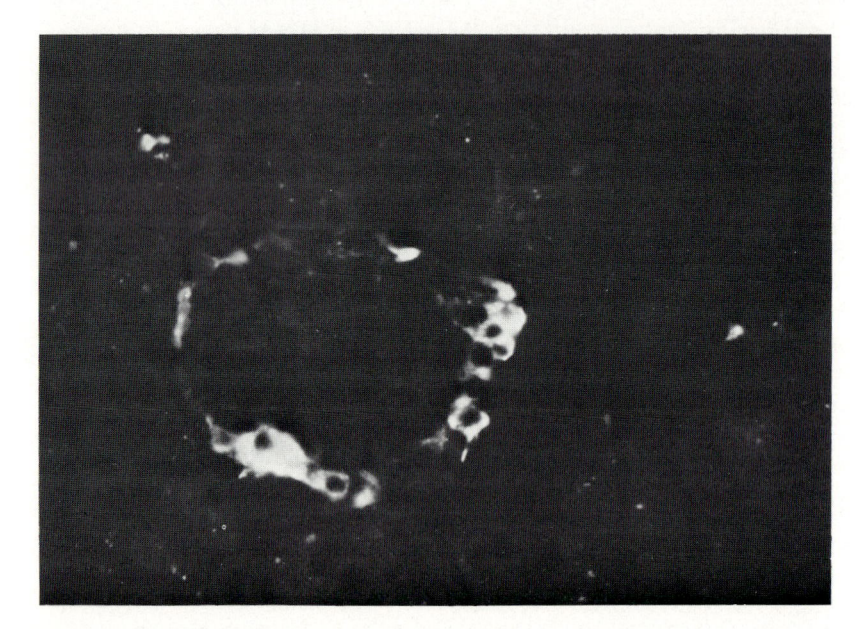

FIG. 3. Immunofluorescent pancreatic polypeptide (PP) cells in the peripheral portion of an islet in the duodenal lobe of hamster pancreas. Rabbit anti-hPP serum and fluorescein-labeled sheep antirabbit IgG were used. From Larsson *et al.* (1976b), with permission.

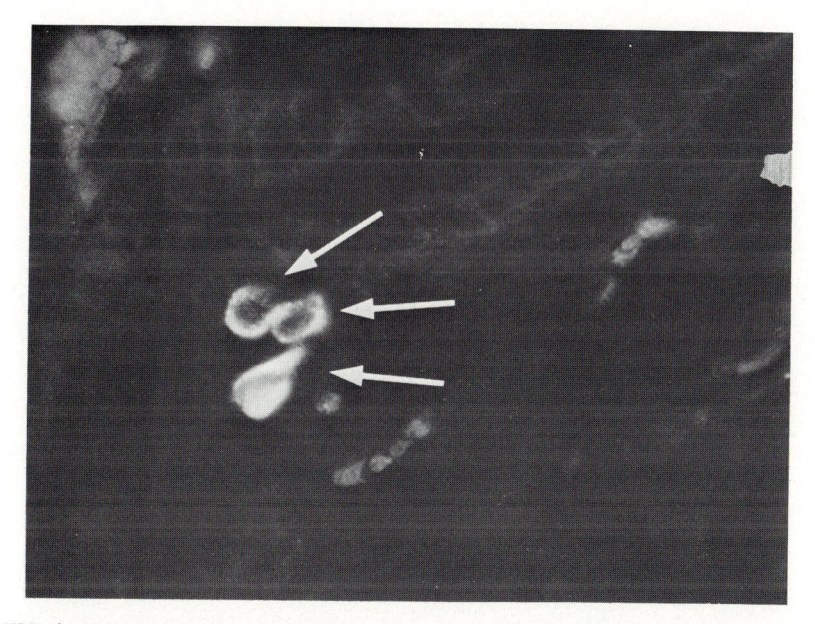

FIG. 4. Immunofuorescent pancreatic polypeptide (PP) cells in ductal epithelium of mouse pancreas. Rabbit anti-hPP serum and fluorescein-labeled sheep antirabbit IgG were used. From Larsson et al. (1976b), with permission.

both within and outside of the islets has been confirmed recently by Gersell and associates (1976).

In some species, e.g., dog, PP cells are more abundant in the portion of the pancreas adjacent to the duodenum (Larsson et al., 1976b; Baetens et al., 1976; Gersell et al., 1976), a location at which in the dog extractable PP also is particularly high (Chance et al., 1977b; Gersell et al., 1976). The PP cells of dog (and cat) are probably identical with the previously described F (or X) cells (Larsson et al., 1976b; Baetens et al., 1976) found mainly in the uncinate process of the dog (Munger et al., 1965). In human fetal pancreas, Larsson et al. (1975b) found more hPP cells in the head than in the tail of the pancreas. A study designed to determine hPP cell number and/or hPP concentration in various parts of adult human pancreas has not been reported.

Ultrastructural examination of PP cells of the cat performed by Larsson and associates (1976b) reveals cytoplasmic granules of moderate to high electron density with the granule membrane closely applied to the dense core (Fig. 5). In osmicated specimens they found the mean diameter of hPP cell granules to be 125 nm whereas the diameters of D-, A-, and B-cell granules were 230, 235, and 300 nm, respectively. By application of immunohistochemistry at the ultrastructural level, hPP was shown to be contained in the hPP cell granules (Fig. 5).

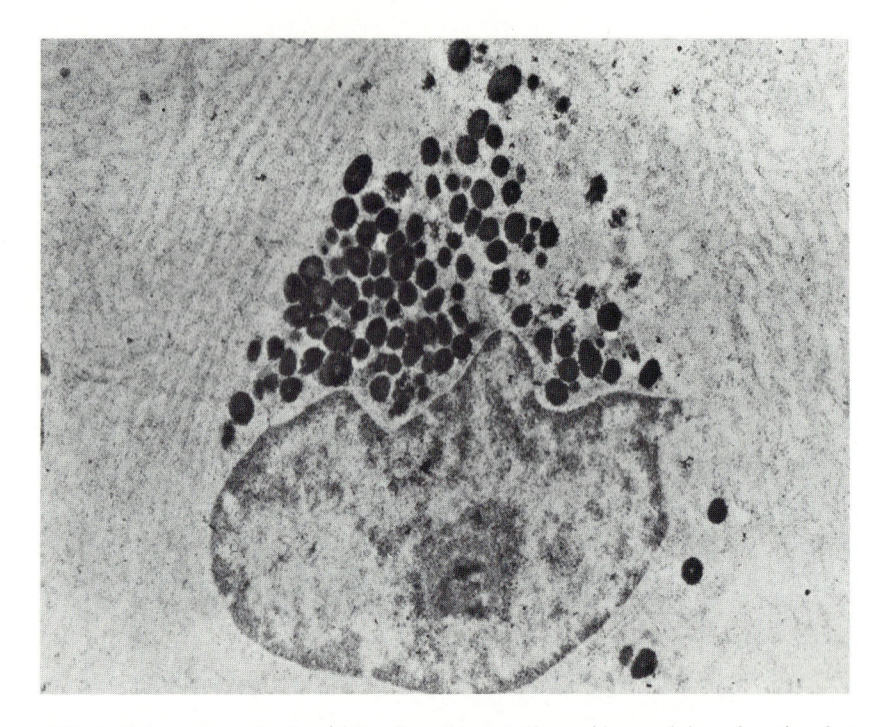

FIG. 5. Cat pancreas, duodenal lobe. Peroxidase–antiperoxidase staining of section from glutaraldehyde-fixed, nonosmicated specimen (human pancreatic polypeptide antiserum, diluted 1:200). The cytoplasmic granules are heavily loaded with electron-dense material ($\times 11,000$), indicating the presence of pancreatic polypeptide. From Larsson et al. (1976b), with permission.

B. hPP IN EXTRACTS OF GASTROINTESTINAL TISSUES AND PANCREAS

Pancreatic polypeptide has been found in parts of the gastrointestinal tract outside the pancreas (Larsson et al., 1976b; Baetens et al., 1976; Adrian et al., 1976). Larsson et al. (1976b) have identified PP cells in the gastric mucosa of the opossum and dog. In the dog, Baetens et al. (1976) identified PP cells in gastric and ileal mucosa. In subhuman primates low concentrations of PP cells and small amounts of extractable PP immunoreactivity were detected in stomach, duodenum, and ileum (Adrian et al., 1976; Bloom et al., 1976); however, 93% of gastrointestinal tract PP immunoreactivity was found in the pancreas. It seems probable that in man some hPP cells will be found in parts of the gastrointestinal tract outside the pancreas.

In three patients who had had total pancreatectomy we have found basal plasma levels of hPP to be 11, 11, and 9 pg/ml, levels that are about one-fifth the mean of younger healthy subjects and at the limit of sensitivity of our assay. Adrian, Bloom, and associates reported that hPP is undetectable in plasma of pancreatectomized patients and further that in such patients plasma levels of hPP do not rise after the ingestion of mixed meals (Adrian *et al.*, 1976; Bloom *et al.*, 1976), a potent stimulus to hPP secretion. Thus, the pancreas appears to be the major source of hPP which circulates in the basal state and of that which increases acutely in plasma when food is eaten. Whether the hPP cells in the islets and those among the acinar cells of the pancreas both participate in the response to food ingestion is unknown.

V. Radioimmunoassay of hPP

A double-antibody radioimmunoassay was used to measure concentrations of hPP. Purified hPP was used to prepare standards and tracer and also as immunizing antigen for the rabbit antiserum used in the assay (Chance *et al.*, 1977b). Typically, the least amount of hPP detectable is 1.9 pg. The assay is specific in that the following hormones do not cross-react in the assay: aPP, 0.1 μg/ml; GIP 10 μg/ml; VIP, 100 μg/ml; gastrin 4 μg/ml; cholecystokinin, 38 μg/ml; secretin, 19 μg/ml; glucagon, 0.1 μg/ml; insulin, 0.1 μg/ml; porcine proinsulin, 100 μg/ml; somatostatin, 100 μg/ml.

VI. Regulation of Secretion of hPP in Healthy Subjects

A. BASAL LEVELS IN HEALTHY SUBJECTS

Basal plasma levels of hPP were measured in 120 young control subjects (Floyd *et al.*, 1976a,b, 1977; Floyd and Fajans, 1976a,b). Their mean age was 25 years ± 3.4 (SD) and mean percent ideal body weight was 98 ± 10 (SD). hPP was detectable in plasma of each subject. The mean plasma concentration of hPP was 51 pg/ml; the standard deviation was 27 pg/ml. The levels of these subjects will be compared to the levels of some groups of patients. More recently basal plasma hPP has been measured in a group of older University faculty members who were ostensibly well but had responded to invitations for physical and laboratory health examinations. Among these older patients whose ages encompassed a wider range of years than did those of the young control subjects there was a significant correlation of hPP with age. A correlation was also demonstrable when the data of the younger and older subjects were combined ($r = 0.59, p < 0.01$). The combined data are displayed by decade of age in Fig. 6. The increase in plasma hPP with age is evident. Because of this, in the presentation of findings of patient groups, age-adjustments of the data will have been made.

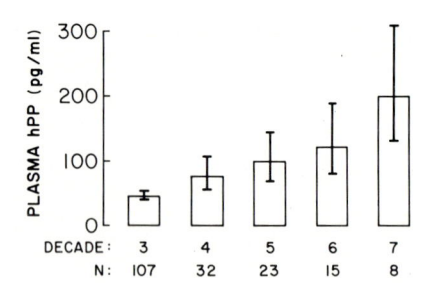

FIG. 6. Plasma levels of human pancreatic polypeptide (hPP) in 185 healthy subjects grouped according to decades of age (mean plasma hPP and 95% confidence limits). Number of subjects per decade is indicated below bar. Distribution of basal hPP levels is best characterized as log normal. Heights of bars are geometric means and the 95% confidence limits are asymmetric.

B. INGESTION OF PROTEIN MEALS

Plasma levels of human pancreatic polypeptide before and after beginning the ingestion of 500 gm of cooked ground beef by healthy subjects are shown in Fig. 7. Mean basal hPP was 57 pg/ml. At 5 minutes after beginning the ingestion of beef (Floyd et al., 1975, 1976b, 1977) plasma hPP was 229 pg/ml (in some tests it had increased at 3 minutes). At 10 minutes hPP was 400 pg/ml, and at 240 minutes mean plasma hPP was 580 pg/ml. This level was sustained for two additional hours in 2 subjects in whom more prolonged observations were made.

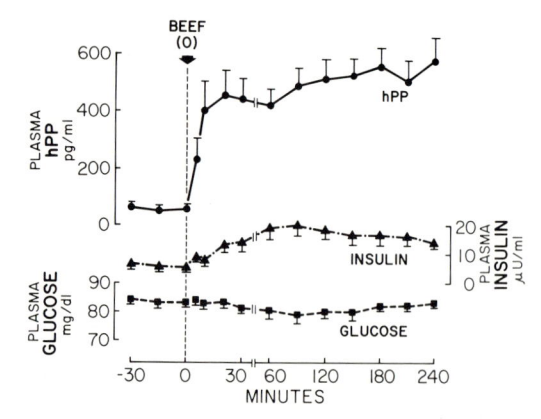

FIG. 7. Effect of the ingestion of cooked ground beef (500 gm) upon plasma levels of human pancreatic polypeptide (hPP), insulin, and glucose in 10 healthy subjects (mean ± SE).

Increase in plasma immunoreactive insulin was also evident at 5 minutes and thereafter. Plasma glucose transiently declined by 5 mg/dl. These findings (Floyd *et al.*, 1975, 1976b, 1977) have been confirmed recently by Schwartz *et al.* (1976), who found a rapid 8-fold increase in plasma hPP in healthy subjects who ingested mixed meals that were protein-rich. Plasma levels of immunoreactive glucagon measured during the experiments shown in Fig. 7 are shown in Fig. 8 (Floyd *et al.*, 1976b, 1977). Mean basal levels of hPP and glucagon were 57 and 62 pg/ml. The first discernible increase in mean plasma glucagon was at 30 minutes whereas for hPP it was at 5 minutes. The maximal increase in plasma hPP above the basal levels was about 3 times that of plasma glucagon.

C. STUDIES ON MECHANISM OF PROTEIN-INDUCED INCREASE IN PLASMA LEVEL OF hPP

1. Ingestion of Fat

In an effort to determine whether fat, present to some extent in the protein meals, would stimulate increase in plasma hPP, 100 ml of the dietary supplement Lipomul (Upjohn Co.) was fed to healthy subjects (Fig. 9). The basal level of hPP was 40 pg/ml. At 5 minutes after beginning the ingestion, plasma hPP had increased significantly to 68 pg/ml and at 20 minutes was 122 pg/ml, after which it declined slightly to remain at about 100 pg/ml through 240 minutes. This amount of Lipomul contains about 70 gm of fat, whereas the protein meals contained about 50 gm. Thus, these results suggest that fat has a modest stimulatory effect but per se could account for only a minor part of the stimulus for the release of hPP which results from the ingestion of meat. In contrast to the modest stimulatory effect on plasma hPP noted with ingestion of lipid, we noted no significant increase in 3 subjects who received lipid administered

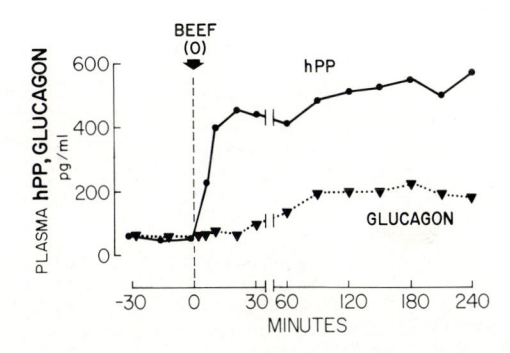

FIG. 8. Effect of the ingestion of cooked ground beef (500 gm) upon plasma levels of human pancreatic polypeptide (hPP) and glucagon in 10 healthy subjects.

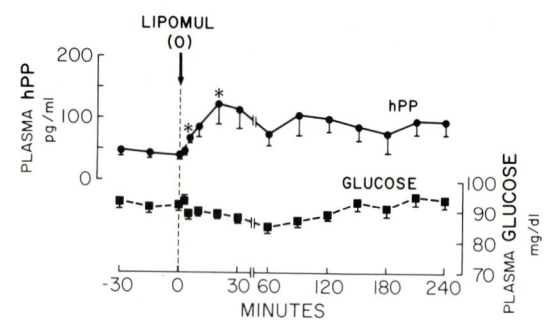

FIG. 9. Effect of the ingestion of Lipomul (100 ml) upon plasma levels of human pancreatic polypeptide (hPP) in 8 healthy subjects (mean ± SE). Asterisk signifies statistically significant difference from 0 minute value.

intravenously [405 ml Intralipid (Cutter Co.) administered over 90 minutes]. Kimmel and associates found in the chicken also that the ingestion of fat was much less effective than was protein in causing increase in plasma levels of PP (Kimmel and Pollock, 1975).

2. Intravenous Administration of Amino Acids

Amino acids were infused to determine whether rising plasma levels of certain amino acids induce increases in plasma levels of hPP (Floyd et al., 1976b, 1977). Individual L-amino acids (arginine, leucine, alanine) were infused intravenously at the dosage of 2.4 mmoles/kg body weight (maximal dose 172 mmoles), and a mixture of 10 essential amino acids (Floyd et al., 1966) was infused at the dosage of 2.9 mmoles/kg (maximal dosage 212 mmoles or 30 gm). All infusions lasted 30 minutes. After the start of the 10-amino acid infusion (Fig. 10) mean

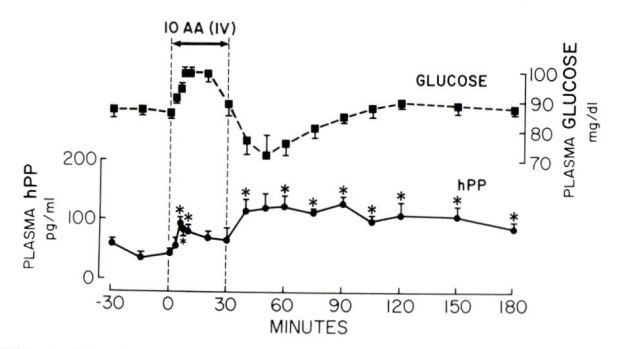

FIG. 10. Effect of the intravenous administration of a mixture of 10 essential amino acids (10 AA) upon plasma levels of human pancreatic polypeptide (hPP) and glucose in 8 healthy subjects (mean ± SE). Dosage of 10 AA 2.9 mmoles per kilogram body weight (maximal dosage 212 mmoles or 30 gm). Asterisk signifies statistically significant difference from basal hPP (average of −15 and 0 minute values).

plasma hPP increased from a basal level of 45 pg/ml to a modest but significant peak of 95 pg/ml at 7 minutes. hPP then declined toward the basal level, but after 30 minutes began to rise again to reach 117 pg/ml at 40 minutes, at which level it remained through 150 minutes. The area under the hPP curve from 0 to 180 minutes was 14 times the area observed during infusions of 300 ml of normal saline (9 subjects). Plasma glucose rose above then fell below basal levels at 40 minutes concomitantly with an increase of hPP. During the infusion of arginine (Fig. 11) plasma glucagon rose promptly and at 30 minutes peaked at 365 pg/ml, but there was no concomitant increase in plasma hPP (Floyd *et al.*, 1976b, 1977). Thereafter plasma levels of hPP rose modestly to about 175 pg/ml. The cause of these modest increases in plasma hPP is unclear. In experiments in which normal saline was infused, there were increases in plasma hPP at 150 and 180 minutes similar to those shown here but without changes in plasma glucose. After the infusion of alanine and the infusion of leucine there were only late increases in plasma hPP and they were even more modest than those of the arginine experiments (Floyd *et al.*, 1976b, 1977). Although the infusion of the amino acid mixture leads to significant increases in plasma levels of hPP, it seems unlikely that the large increases in plasma hPP that begin 3–5 minutes after starting the ingestion of meat are, in the main, the result of direct stimulation by amino acids upon hPP cells.

3. *Other Possible Mediations through Which the Ingestion of Protein Effects Elevation of Plasma hPP*

The ingestion of meat might initiate release of hPP by means of certain stimuli initiated in the gastrointestinal tract. Thus, protein meals were fed to 5 patients

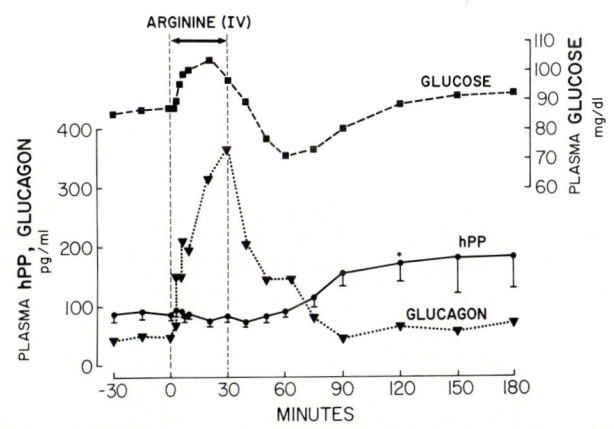

FIG. 11. Effect of intravenous L-arginine (0.41 g/kg body weight, maximal dosage 30 gm) upon plasma human pancreatic polypeptide (hPP), glucagon, and glucose in 5 healthy subjects (mean ± SE). Asterisk signifies statistically significant difference from basal hPP (average of −15 and 0 minute values).

with pernicious anemia, patients who have achylia gastrica (Fig. 12). These subjects were elderly and on the average were able to eat only 290 gm of cooked ground beef as compared to 500 gm eaten by the healthy subjects. Nevertheless, mean plasma hPP rose by 5 minutes and reached a peak of 400 pg/ml at 30 minutes. This prompt increase at 5 minutes was noted in 4 of the 5 patients, and by 10 minutes it was noted in all. It appears, therefore, that the production of hydrochloric acid (and pepsinogens) is not requisite for ingested protein to stimulate release of hPP.

To determine whether secreted hydrochloric acid might contribute to protein-induced secretion of hPP, 200 ml of 0.1 N hydrochloric acid was instilled into the duodenum of each of 2 healthy subjects over periods of 30 minutes (Fig. 13). In one subject, levels remained at about 100 pg/ml during the 2 hours of observation. In the second subject plasma hPP increased about 50 pg/ml at 75 minutes, some 45 minutes after the infusion of acid had been discontinued. These subjects had shown prompt, large, and sustained increases of hPP when protein had been ingested. Therefore a duodenal effect of secreted hydrochloric acid does not appear to be a mediator of or be necessary for protein-induced increase in plasma hPP.

An additional gastrointestinal effect induced by protein ingestion is the secretion of gastrin. Plasma levels of hPP were determined in subjects to whom pentagastrin was administered intravenously (Fig. 14). Mean basal plasma hPP was 68 pg/ml. At 2 minutes after the pentagastrin bolus, plasma hPP increased significantly to 203 pg/ml after which it declined toward base line. At this dose of pentagastrin it can be expected that plasma levels reached are equivalent to plasma levels of gastrin that are much higher than occur in physiological circumstances. It seems likely that lower, more physiological doses would induce even more modest, if any, increases in plasma hPP. It will require further study to establish whether or not gastrin when secreted acutely in physiological

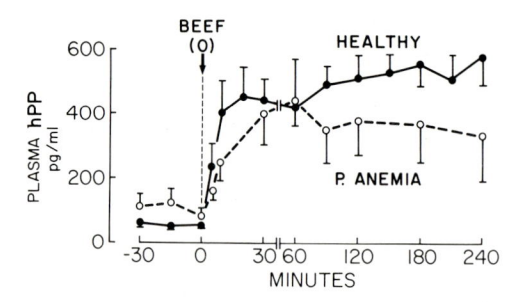

FIG. 12. Effect of the ingestion of protein meal upon plasma levels of human pancreatic polypeptide (hPP) in 5 patients with pernicious anemia (average meal 290 gm of cooked ground beef), and 10 healthy subjects (500 gm of cooked ground beef). Mean ± SE.

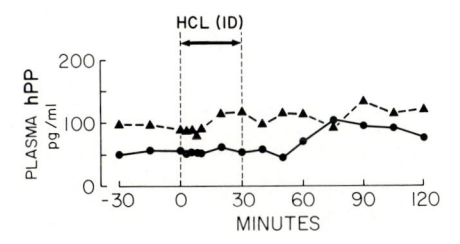

FIG. 13. Effect of the instillation of 200 ml of 0.1 N hydrochloric acid by tube into the duodenum; effect upon plasma levels of human pancreatic polypeptide (hPP) in 2 healthy subjects.

circumstances stimulates release of hPP. Chronic elevation of plasma levels of endogenous gastrin evidently does not stimulate increased secretion of hPP. In patients with pernicious anemia, basal plasma levels of gastrin characteristically are elevated 15 to 20 times normal. The mean basal level of hPP of the pernicious anemia patients (Fig. 12) was only 85 pg/ml.

4. Neural Effects

Protein-enriched meals may stimulate release of hPP in part by the generation of a neural stimulus to the PP-secreting cells. Recently, Schwartz and associates (1976) have presented evidence to suggest that in man an intact vagus nerve is requisite for the full expression of protein-induced increase in plasma hPP. They observed that after truncal vagotomy in duodenal ulcer patients there was a markedly reduced serum hPP response (no significant increase) during the first 15 minutes after beginning the ingestion of a protein-rich mixed meal. Their

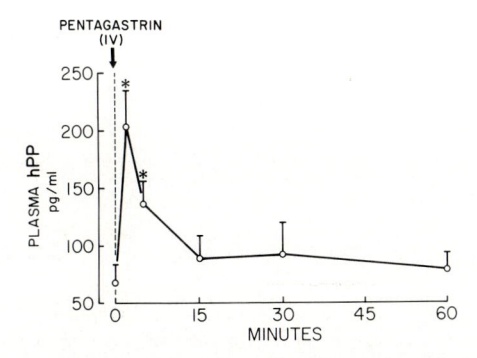

FIG. 14. Effect of the intravenous injection of pentagastrin (0.5 μg per kilogram of body weight) in 5 healthy subjects. Mean ± SE. Asterisk signifies statistically significant difference from 0 minute level.

suggestion that the vagus nerve is important in the early rapid increase of hPP after food ingestion was strengthened by their finding in the pig that the electrical stimulation of the vagus nerve was associated with acute increase in levels of PP in blood (Schwartz et al., 1976). Thus at least part of the stimulus for protein-induced release of hPP appears to be neural and to be mediated by the vagus nerve. They have also (Fig. 15) found that the perfused porcine pancreas secretes PP in response to acetylcholine ($D_{50} \sim 10^{-7}$ M) and that atropine blocks this response (T. W. Schwartz, personal communication, 1976).

5. *Effect of Somatostatin*

The effects of somatostatin upon nutrient-induced increase in plasma levels of hPP were determined because the infusion of somatostatin causes inhibition of the release of insulin and glucagon, as well as because hPP cells and somatostatin

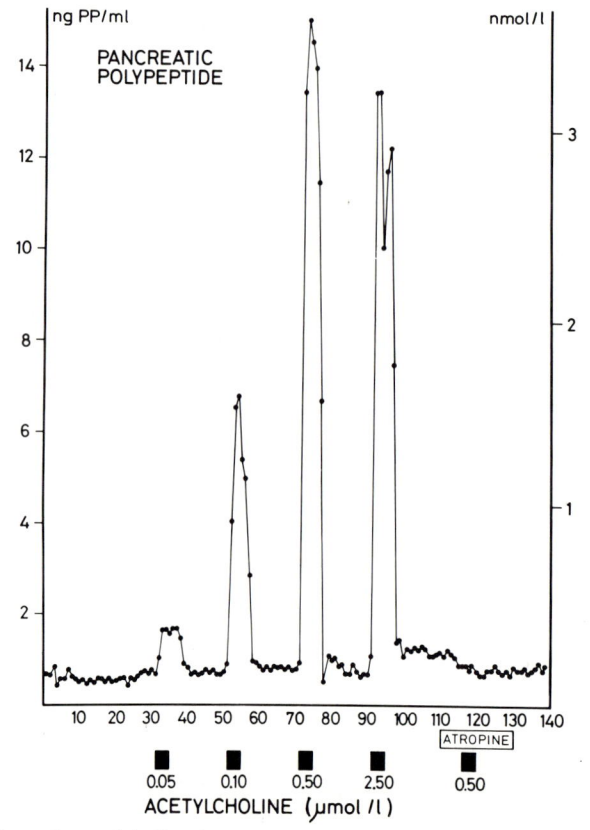

FIG. 15. Effect of acetylcholine and atropine on PP secretion by the perfused porcine pancreas. From T. W. Schwartz (personal communication, 1976).

cells are in close proximity in the periphery of the islets. In a collaborative study Dr. John Gerich infused somatostatin into a 69-year-old patient with maturity-onset type of diabetes mellitus, and we measured hPP in samples of the patient's plasma (Fig. 16). On the control day before the somatostatin infusion, basal plasma hPP was 225 pg/ml. After breakfast hPP increased to a peak of 1000 pg/ml at 9:30 A.M. and to peaks of 1200 pg/ml after lunch and 1300 pg/ml after dinner. On day 3 of a 3-day infusion of somatostatin (lower panel) at 8:30 A.M. basal plasma hPP was lowered to 85 pg/ml, and there was a complete absence of increases in plasma hPP following ingestion of breakfast, lunch, and dinner. On the day after the cessation of somatostatin, postprandial levels of hPP were 3000, 2000, and 1400 pg/ml (Fig. 17). These levels were higher than those on the control day before somatostatin (Fig. 16). Thus the mechanisms by which nutrient ingestion, including protein, stimulate increase in plasma hPP are blocked by somatostatin. The similarity of postprandial increases in plasma glucose on the control days and on the somatostatin day suggests that the

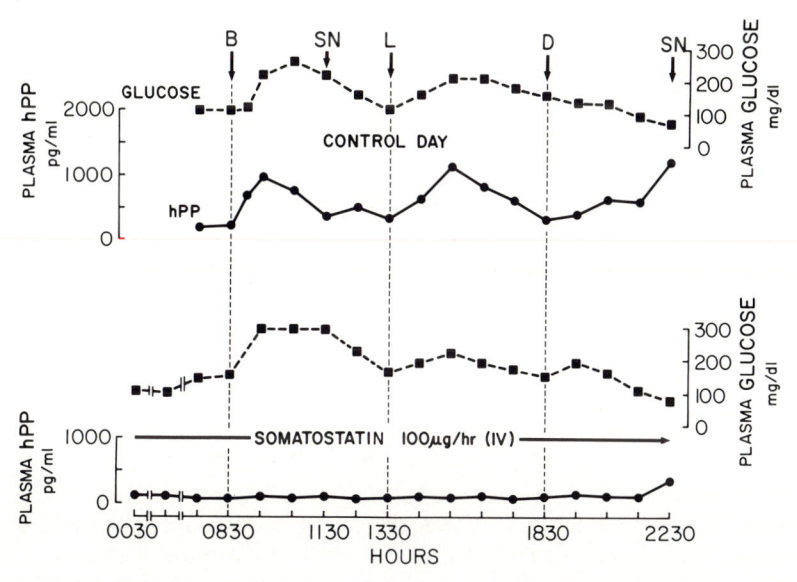

FIG. 16. Infusion of somatostatin in a 69-year-old male with maturity-onset type of diabetes. Insulin and a prescribed diet (mixed meals B, L, D and carbohydrate snacks SN) were begun several days before these studies and were continued during control and somatostatin days. The control day preceding the 3-day somatostatin infusion is shown in the upper panel and the last day of somatostatin in the lower panel. The insulin doses on the presomatostatin control day were 10 U Regular plus 22 U NPH at 8 A.M. and 10 U Regular plus 14 U NPH at 6 P.M. On the first somatostatin day the insulin dose was changed to 5 U Regular plus 10 U NPH at 8 A.M. and 6 U Regular plus 5 U NPH at 6 P.M. (In collaboration with Dr. John Gerich, who performed the patient studies.)

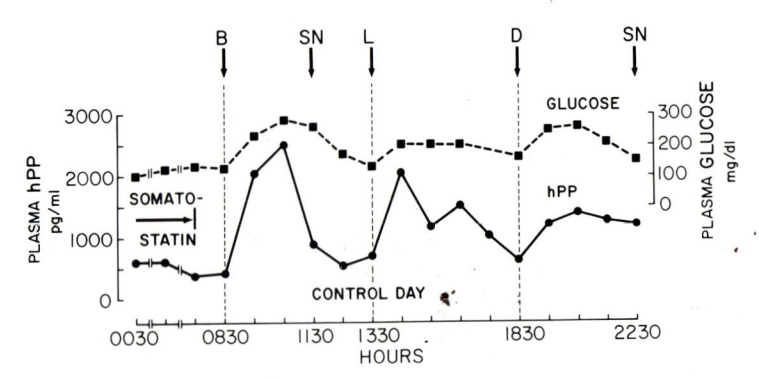

FIG. 17. Control day after cessation of somatostatin. Same patient as in Fig. 16; see legend of Fig. 16 concerning diet and insulin dosage.

somatostatin effect upon hPP is not dependent to an important extent upon any slowing of gastric emptying.

D. INGESTION AND INTRAVENOUS ADMINISTRATION OF GLUCOSE

After the ingestion of glucose (Floyd *et al.*, 1976b, 1977), hPP rose significantly from a basal level of 68 pg/ml to 127 pg/ml at 15 minutes (Fig. 18), and then declined to remain near 100 pg/ml through 150 minutes. Thereafter hPP rose to a peak level of 445 pg/ml at 240 minutes. After first rising, mean plasma glucose fell to 52 mg/dl at 210 minutes. The increases in hPP at 210 and at 240 minutes are similar in timing to those of growth hormone at 210 and 250 minutes (Fig. 19). Mean plasma glucagon did not change. Before glucose was infused basal plasma hPP was 54 pg/ml (Fig. 20). During the glucose infusion hPP fell to 44 pg/ml at 30 minutes and persisted at about this level through 75 minutes, when hPP was 41 pg/ml (Floyd *et al.*, 1976b).* During the 60 minutes of saline infusion (9 healthy subjects), there was no decrease in the mean level of hPP. The modestly reduced levels during hyperglycemia induced by the infusion of intravenous glucose (Fig. 20) contrast with the modestly increased plasma levels of hPP induced by the ingestion of glucose (Fig. 18). After the infusion of glucose was stopped (Fig. 20) plasma glucose fell to basal and then to below basal levels. Concomitantly plasma hPP rose to levels above basal. Kimmel and

---

*The results of six additional infusions of glucose in healthy subjects have not confirmed this finding. Additional study of the effect of glucose infusion upon plasma hPP is needed.

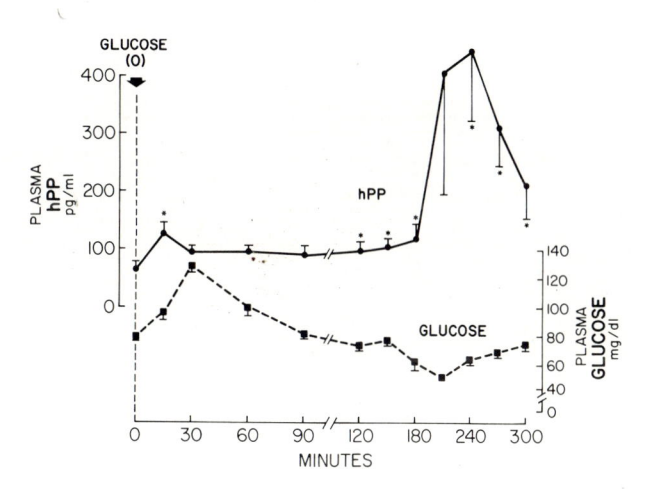

FIG. 18. Effect of the ingestion of glucose (1.75 gm per kilogram ideal body weight) upon plasma levels of human pancreatic polypeptide (hPP) and glucose in 9 healthy subjects (mean ± SE). Asterisk signifies statistically significant difference from the 0 minute hPP value.

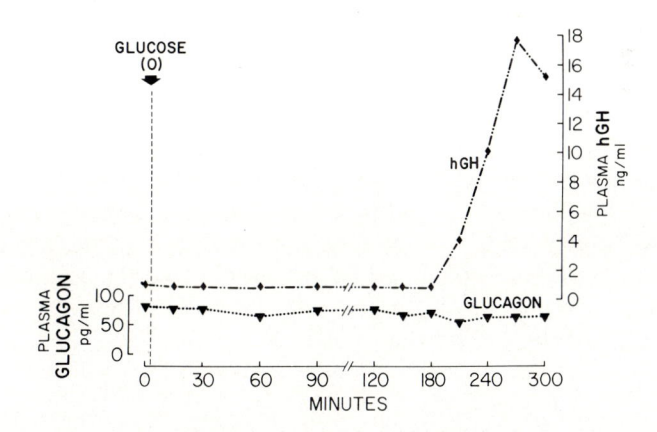

FIG. 19. Effect of the ingestion of glucose (1.75 g/kg ideal body weight) upon plasma levels of glucagon and growth hormone (hGH) in 9 healthy subjects.

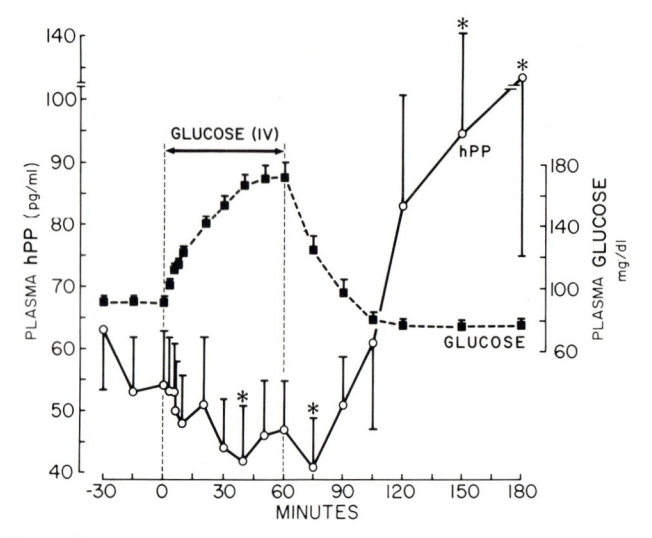

FIG. 20. Effect of intravenous glucose (30 gm) upon plasma levels of glucose and human pancreatic polypeptide (hPP) in 13 healthy subjects (mean ± SE). Asterisk signifies statistically significant difference from basal hPP (average of −15 and 0 minute values).

associates observed in the chicken, as in man (Floyd *et al.*, 1976b), that there was a transient increase in PP after oral glucose (Kimmel *et al.*, 1974; Kimmel and Pollock, 1976) and a decrease during intravenous glucose (J. R. Kimmel, personal communication, 1976).

E. HYPOGLYCEMIA INDUCED BY INSULIN INJECTION

The association of decreases in plasma glucose with subsequent increases in plasma hPP, as seen after oral and intravenous glucose, is demonstrated also after the administration of insulin (Fig. 21). Plasma glucose decreased to a nadir of 44 mg/dl at 25 minutes and was followed by an increase in plasma hPP to 219 pg/ml at 40 minutes. These patterns of plasma glucose and hPP were accentuated when a higher dosage of insulin, 0.1 U/kg, was injected (Fig. 22); the nadir of plasma glucose was 21 mg/dl whereas with the lower dosage it had been 44 mg/dl. Concomitantly the increase of hPP was much greater, to 1300 pg/ml, compared to 219 pg/ml seen with the lower dosage. With both dosages the nadir of plasma glucose preceded the maximal increase in hPP by 10–15 minutes. In six additional experiments in which insulin, 0.1 U/kg, was injected decrease in blood glucose was prevented by infusion of glucose; in these experiments there was no change in plasma levels of hPP. The greatly increased levels of hPP

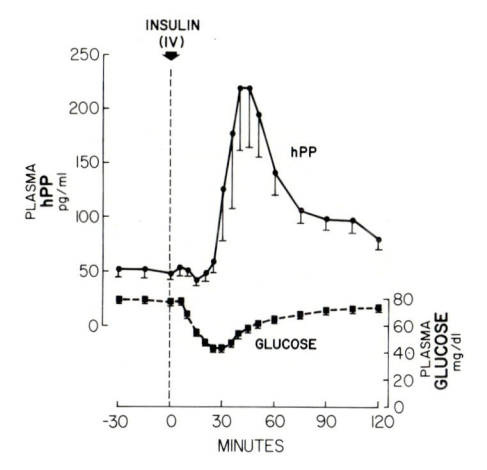

FIG. 21. Effect of intravenous insulin (0.03 U/kg body weight) upon plasma levels of glucose and human pancreatic polypeptide (hPP) in 12 healthy subjects (mean ± SE).

associated with acute hypoglycemia (Floyd *et al.*, 1976b, 1977) contrast with the levels which may be observed during chronic hypoglycemia. In a patient who had a non-insulin-secreting hemangiopericytoma and associated fasting hypoglycemia the mean fasting plasma glucose of nine mornings was 33 mg/dl; the mean plasma hPP was 47 pg/ml.

After the administration of insulin (Fig. 23), increases in levels of glucagon were of lesser magnitude but began earlier than those of hPP (Floyd *et al.*,

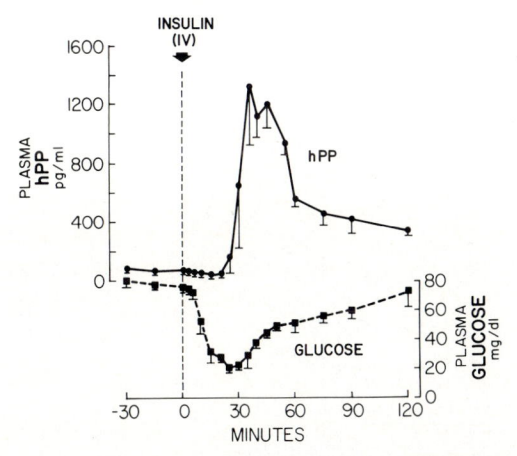

FIG. 22. Effect of intravenous insulin (0.1 U/kg body weight) upon plasma levels of glucose and human pancreatic polypeptide (hPP) in 3 healthy subjects (mean ± SE).

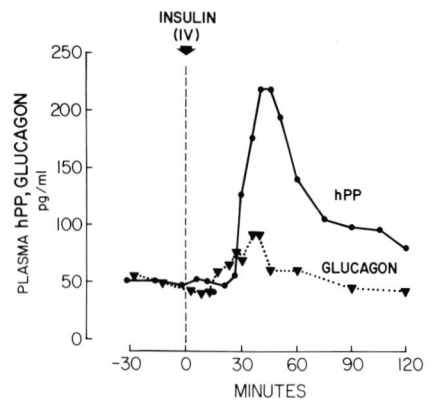

FIG. 23. Effect of intravenous insulin (0.03 U/kg body weight) upon plasma levels of glucagon and human pancreatic polypeptide (hPP) in 12 healthy subjects.

1976b, 1977). (Following the ingestion of protein meals, glucagon increases also had been of lesser magnitude but had begun later than those of hPP.) After insulin administration (Floyd et al., 1977) the patterns of changes in growth hormone and hPP are similar (Fig. 24), as had been the case late after the ingestion of glucose (Figs. 18 and 19).

F. EFFECT OF FASTING

During a 72-hour fast (Floyd et al., 1976b, 1977) mean plasma glucose fell progressively to the 62nd hour and rose modestly by the 69th hour (Fig. 25). As

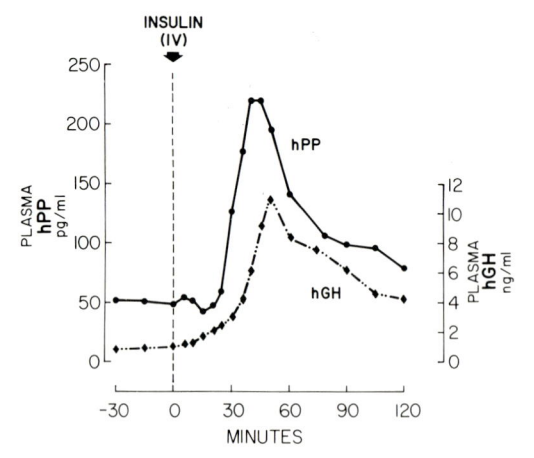

FIG. 24. Effect of intravenous insulin (0.03 U/kg body weight) upon plasma hPP and growth hormone (hGH) in 12 healthy subjects.

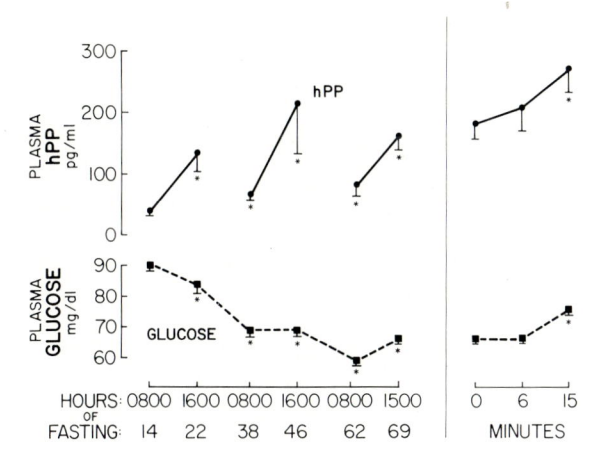

FIG. 25. Effect of prolonged fasting (left-hand panel) and of exercise at hour 7 of fasting (right-hand panel) upon plasma human pancreatic polypeptide (hPP) and glucose in 13 healthy females (mean ± SE). Asterisks in left-hand panel signify statistically significant difference from the 8 A.M. value after 14 hours of overnight fasting. Asterisks in right-hand panel signify statistically significant difference from the 0 minute value.

hours of fasting increased, so did plasma levels of hPP. At the 62nd hour mean plasma hPP at 8 AM was 84 pg/ml, which was twice the initial level. A diurnal variation in plasma hPP also is evident in that plasma levels are higher each afternoon than on the preceding and succeeding mornings. The effects of exercise upon plasma levels of hPP is seen in the right-hand panel. At about the 70th hour of fasting, the subjects exercised to their maximal capabilities for about 15 minutes. At zero minutes, plasma hPP was 189 pg/ml. After exercise hPP rose to 275 pg/ml; plasma glucose also rose significantly.

The summary and conclusions from these studies in healthy subjects are as follows:

1. hPP can be detected consistently in plasma of healthy subjects.

2. The ingestion of a protein meal stimulates the release of hPP as well as the release of other islet hormones. The rapidity of the increase in plasma hPP after protein ingestion and the occurrence of only modest increases during intravenous infusions of the mixture of 10 essential amino acids suggest that the gastrointestinal tract is important in the mediation of the effect of protein to stimulate release of hPP; the nervous system may be involved in this effect.

3. There are large differences in the ability of ingested nutrients to stimulate release of hPP. The effects of ingested glucose and fat are small and somewhat transient. The effect of ingested protein is large and prolonged.

4. Acute hypoglycemia, fasting, and exercise are associated with increases of hPP in a manner somewhat analogous to the concomitant increases of plasma

glucagon and growth hormone. Acute hyperglycemia induced by intravenous glucose is associated with decreases in plasma hPP.

5. The changes in plasma hPP demonstrated in these studies suggest that this newly recognized islet hormone may have physiological regulatory functions.

VII. hPP in Disease States

A. INSULIN-SECRETING AND "NONFUNCTIONING" ISLET CELL TUMORS

HPP was extracted from control pancreases and insulinomas with acid–ethanol, and the concentration of hPP was expressed as micrograms per gram of wet tissue (Floyd *et al.*, 1975, 1976a; Floyd and Fajans, 1976a). The mean concentration in the 8 control pancreases was 0.86 µg/gm (Fig. 26). The mean concentration of hPP in the 26 insulinomas was 0.36 µg/gm, which was significantly less than that of control pancreases. The concentration of glucagon also was lower than in the control pancreases. The lower levels of hPP and glucagon might be expected, since typical insulinomas are composed mainly of beta cells. An additional insulinoma, however, contained 10.5 µg of hPP per gram. This concentration was 6 times greater than the mean plus 2 SD of the control pancreases. When these data were categorized as to islet pathology (solitary adenoma, adenoma plus microadenomatosis, carcinoma), there were no differences in tumor concentrations of hPP among the categories including one additional patient who had nesidioblastosis (Fig. 27). Five patients had multiple endocrine adenomatosis type I (MEA-I) including the patient whose tumor had the highest concentration of hPP, 10.5 µg/gm. The results of the 4 other patients with MEA-I were similar to those of the non MEA-I patients.

Basal plasma levels of hPP in patients with benign insulinomas (Floyd *et al.*, 1975, 1976a; Floyd and Fajans, 1976a) are shown in Fig. 28. In contrast to the

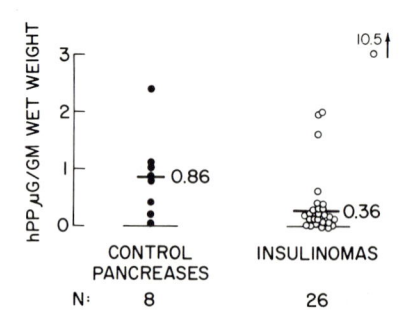

FIG. 26. Concentrations of hPP in control (postmortem) pancreases and in insulinomas. $P < 0.05$.

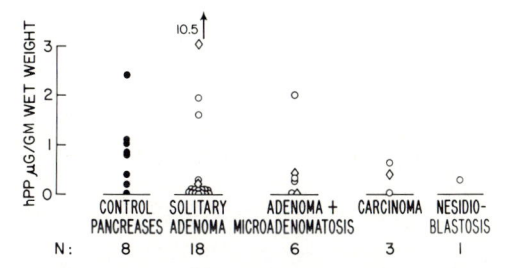

FIG. 27. Concentrations of human pancreatic polypeptides (hPP) in control (postmortem) pancreases and in insulinomas, categorized as to islet pathology, and in nesidioblastosis. Diamond symbol denotes multiple endocrine adenomatosis type I.

mean of healthy subjects, 51 pg/ml, the mean of the insulinoma patients was 249 pg/ml. The range of their concentrations was from 40 to 880 pg/ml. Seven of these 19 patients were not hypoglycemic at the time these measurements were made. In 15 patients, plasma hPP exceeded the range of the young control subjects. When the levels of the insulinoma patients were age-adjusted 10 were supernormal (≥95 percentile value of 185 healthy subjects). The patient whose adenoma contained a supernormal concentration of hPP had a normal plasma level.

Figure 29 shows basal plasma levels of hPP before and after removal of insulinomas. The mean preoperative level of 260 pg/ml was not lowered by operation, but on the contrary at a mean of 15 days postoperatively, mean plasma hPP was 316 pg/ml. The acute effect of the operative removal of an insulinoma is shown in Fig. 30. After removal of the tumor, plasma insulin fell in both portal and peripheral venous plasma; in contrast, plasma levels of hPP rose. We have concluded therefore, that the major source of elevated plasma hPP in a

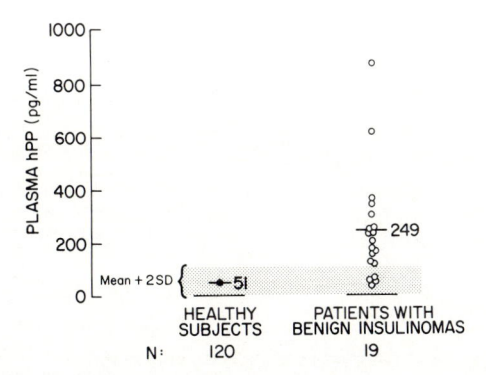

FIG. 28. Basal plasma concentrations of human pancreatic polypeptides (hPP) in patients with benign insulinomas. Stippled area encompasses mean + 2 SD of basal levels of 120 healthy control subjects of mean age 25 years.

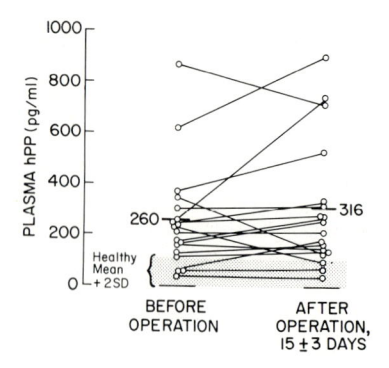

FIG. 29. Basal plasma concentrations of human pancreatic polypeptide (hPP) in 18 patients before and after removal of insulinomas. See also legend to Fig. 28.

patient with a benign insulinoma is not the tumor, nor is the elevation due to any hypoglycemia or hyperinsulinemia (Floyd et al., 1975, 1976a; Floyd and Fajans, 1976a). This conclusion is corroborated by the findings of Larsson and associates (Larsson, 1976; Larsson et al., 1976a), who reported that in many patients with different types of endocrine pancreatic tumors there was hyperplasia of hPP cells in the pancreatic parenchyma outside the tumors.

Table I shows findings in patients with sporadic islet carcinoma (Floyd et al., 1976a; Floyd and Fajans, 1976a). Plasma hPP ranged from 377 to 1580 pg/ml, and in each was elevated as compared to the young control subjects; it was elevated also when age-adjusted. In the 57- and 60-year old patients there was neither hyperinsulinemia nor hyperglucagonemia. These two islet carcinomas had been considered to be nonfunctioning.

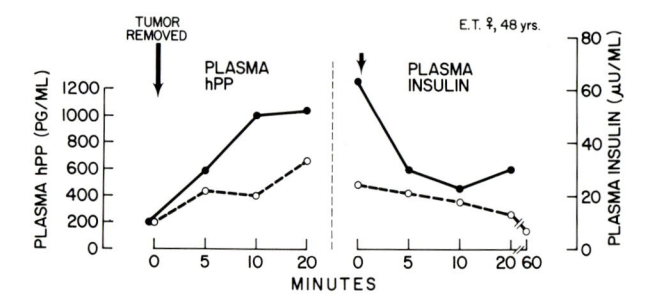

FIG. 30. Acute effect upon portal (●——●) and peripheral (○– – –○) vein plasma concentrations of human pancreatic polypeptide (hPP) and insulin of the operative removal of an insulinoma.

TABLE I
Hormone Levels in Sporadic Islet Carcinoma and Healthy Control Subjects[a]

Sex	Age (years)	hPP[b] (pg/ml)	Glucagon (pg/ml)	Insulin (μU/ml)
M	27	377	188	131
M	72	1580	96	38
M	57	476	51	16
F	60	1216	49	6
Healthy mean:		51	61	9
(N = 120) Mean + 2 SD:		110	145	19

[a]Mean age 25 years.
[b]Human pancreatic polypeptide.

B. MULTIPLE ENDOCRINE ADENOMATOSIS TYPE I

In each of 3 patients with islet carcinoma who are affected by familial multiple endocrine adenomatosis type I (Floyd *et al.*, 1976a; Floyd and Fajans, 1976a), plasma hPP was elevated (Table II). The 64-year-old had an hPP level of 844 pg/ml, but neither hyperinsulinemia nor hypoglycemia had been present. She also had been considered to have no hormonal abnormality. Her brother, age 50, had elevated preoperative plasma hPP of 533 pg/ml and glucagon of 844 pg/ml. Although his level of plasma insulin was normal, he had autonomous insulin production and mild, intermittent hypoglycemia. Subtotal pancreatectomy elim-

TABLE II
Islet Carcinoma and Familial Multiple Endocrine Adenomatosis Type I

Family, sex	Age (years)	hPP[a] (pg/ml)	Glucagon (pg/ml)	Insulin (μU/ml)
"W", F	49	511	164	19
"L", F	64	884	101	15
"L", M	50	Preop: 533	844	11
		Postop: 592	92	8
Healthy mean:		51	61	9
(N = 120)[b] Mean + 2 SD:		110	145	19

[a]Human pancreatic polypeptide.
[b]Mean age, 25 years.

inated the hypoglycemia and hyperglucagonemia and ostensibly the islet carcinoma as well. Nevertheless plasma hPP has remained persistently elevated in the range of 592 pg/ml for the past 3 years since his operation without his having evidence of hyperinsulinemia, hypoglycemia, or hyperglucagonemia.

The findings in patients who had islet carcinoma and familial multiple endocrine adenomatosis type I led to the measurement of plasma levels of hPP in other members of their families. All three members of the "W" family had had hyperparathyroidism (Table III). The hPP level of the 49-year-old patient with islet carcinoma was 511 pg/ml (see also Table II). The hPP level of her brother, age 48, was 1366 pg/ml, and on other occasions as high as 2200 pg/ml. A third sibling had had two microscopic islet adenomas removed. She remains hypergastrinemic although her levels of hPP, glucagon, and insulin are normal.

Elevated hPP levels were found in 2 of 6 members of generation I of the "L" family (Table II). Twelve members of generation II were studied (Table IV). An older sibling who had no other manifestation of MEA-I had a level of hPP of 160 pg/ml, which on age adjustment is probably supernormal (>90th percentile value of 185 healthy subjects). In 2 siblings who had been affected with insulinoma and hyperparathyroidism hPP was normal as well as in each of 8 apparently unaffected family members, whose mean level was 46 pg/ml. In generation III, an 18-year-old family member who has no other manifestation of MEA-I had a basal level of hPP of 1770 pg/ml (Table V). A 23-year-old family member

TABLE III
"W" Family: Multiple Endocrine Adenomatosis Type I

Condition	Age (years)	hPP[a] (pg/ml)	Glucagon (pg/ml)	Insulin (µU/ml)
Hyperparathyroidism[b] Islet carcinoma Lipoma[b]	49	511	164	19
Hyperparathyroidism[b]	48	1366	334	25
Hyperparathyroidism[b] Islet adenomas[b]	43	103	90	5
Healthy mean: ($N = 120$)[c] Mean + 2 SD:		51 110	61 145	9 19

[a]Human pancreatic polypeptide
[b]Postoperative
[c]Mean age of healthy controls, 25 years.

TABLE IV
Generation II, "L" Family: Multiple Endocrine Adenomatosis Type I

		Plasma levels		
Condition	Age (years)	hPP[a] (pg/ml)	Glucagon (pg/ml)	Insulin (μU/ml)
OU[b]	44	160	67	20
OU	40	131	55	5
Islet adenoma[c] +	30	84	39	11
hyperparathyroidism[c]	27	48	100	10
"Unaffected"	(N = 8)			
Mean:	22	46	69	11
Range:	13–38	2–79	37–106	5–22

[a]Human pancreatic polypeptide.
[b]OU, otherwise unaffected.
[c]Postoperative.

affected by hyperparathyroidism and hyperprolactinemia had an elevated plasma hPP of 139. The 10 evidently unaffected members of the third generation had a mean level of 39 pg/ml.

In this general connection we have made an additional observation of interest. Two of three members of a family evidently affected only with subcutaneous lipomas had elevated basal plasma hPP levels (438, 341, 99 pg/ml). When these siblings ingested protein meals of 75, 500, and 500 gm, respectively, plasma hPP

TABLE V
Generation III, "L" Family: Multiple Endocrine Adenomatosis Type I

		Plasma levels		
Condition	Age (years)	hPP[a] (pg/ml)	Glucagon (pg/ml)	Insulin (μU/ml)
OU[b]	18	1770	129	11
Hyperparathyroidism + hyperprolactinemia	23	139	–	14
Probably 'unaffected'	17	110	46	11
"Unaffected"	(N = 10)			
Mean	14	39	72	10
Range	9–20	3–89	38–157	1–21

[a]Human pancreatic polypeptide.
[b]OU, otherwise unaffected.

increased to 2260, 9570, and 974 pg/ml, respectively, the former two responses being grossly supernormal.

Elevation in plasma concentration of hPP is an endocrine manifestation in some patients with benign insulinomas, in some with islet carcinomas previously considered to be nonfunctioning as well as a manifestation in some patients with functioning islet carcinomas (Floyd et al., 1975; 1976a; Floyd and Fajans, 1976a). In this connection Polak et al. have reported that hepatic and lymph node metastases of a patient with a pancreatic endocrine tumor contained high concentrations of hPP immunoreactivity (Polak et al., 1976); they considered this as indicating that hPP was a primary product of the tumor and that the finding of elevated plasma levels of hPP in this patient and in other patients with islet tumors was further evidence that the tumors abnormally produced hPP. Nevertheless the source of the elevated plasma levels of hPP seen in some patients with islet tumors remains uncertain. To date no patient has been reported to have had a fall of plasma hPP to normal following removal of an islet tumor.

The summary and conclusions of these studies are as follows:

1. In patients with benign insulinomas, the major source of elevated plasma hPP is not the insulinoma, nor is the elevation due to any hypoglycemia or hyperinsulinemia.

2. Elevation of plasma hPP is an endocrine manifestation in some patients (with or without MEA type I) with islet carcinomas previously considered as nonfunctioning or as functioning.

3. Elevated plasma levels of hPP occur in patients with MEA type I and may indicate involvement by family members who otherwise are not known to be affected.

4. These findings widen the scope of recognized endocrine disturbance associated with benign sporadic insulinomas, islet carcinoma, and MEA type I.

C. DIABETES MELLITUS

Plasma levels of hPP were determined in 282 patients with diabetes mellitus (Floyd and Fajans, 1976b). The levels of 70% of the patients were below 400 pg/ml, and the levels of 90% were below 1000 pg/ml, and the levels of 90% were below 1000 pg/ml; the levels of the upper 10th percentile ranged from 1000 to greater than 10,000 pg/ml (Fig. 31). Thus the range of this upper 10% of the patients was 10 times the range of the other 90% of the patients. Further study of the plasma of these patients revealed that about 80% had the capacity to bind hPP. This binding was the mechanism responsible for most of these apparently very high levels of hPP. Presumably, it indicates the presence of circulating antibody due to the antigenic effect of pancreatic polypeptide present in the therapeutic insulin preparations with which these patients were being treated.

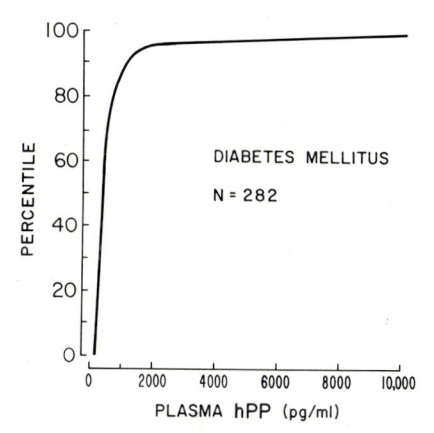

FIG. 31. Cumulative sample distribution of basal plasma concentrations of human pancreatic polypeptides (hPP) in 282 patients with diabetes mellitus (see text).

We have estimated that such preparations contain less than 0.01% PP. Evidence of antibody to PP may appear as early as 2 months after the onset of insulin therapy, but none may be detectable even after 25 years of insulin treatment. No obvious clinical correlate with plasma binding of hPP has been noted.

In the light of the finding of plasma binding of hPP among this special group of patients with hPP levels above 1000 pg/ml, the plasmas of all diabetic patients being measured for hPP were screened for antibody to PP. Evidence of anti-PP antibody was found only among insulin-treated patients. It was present in 64% of juvenile-onset type diabetic patients and in 14% of insulin-treated maturity-onset type diabetic patients. The data to follow were obtained from study of the patients who were shown not to have evidence of circulating anti-PP antibody. In them, as had been seen for healthy subjects, there is a positive correlation of age and basal plasma level of hPP. Therefore in the presentation to follow the data will have been normalized to age 40 years using the respective regression equations of the healthy subjects and of the diabetic patients.

Mean plasma hPP in 204 diabetic patients was 120 pg/ml (Fig. 32). This contrasts with the 85 pg/ml of 190 healthy subjects. The level of the diabetic patients is significantly higher than that of the healthy subjects. The mean basal plasma level of glucose in the diabetic patients was 158 mg/dl, and in the healthy subjects, 82 mg/dl. The hPP levels of the healthy subjects are also contrasted with the levels of the diabetic patients grouped as to maturity-onset and juvenile-onset types of diabetes (Fig. 33). The mean basal hPP level of 33 patients with juvenile-onset type of diabetes (JOD) is 170 pg/ml, and of 157 patients with maturity-onset type of diabetes (MOD) is 113 pg/ml as compared to the level of the control subjects, 85 pg/ml. The mean level of the JOD is

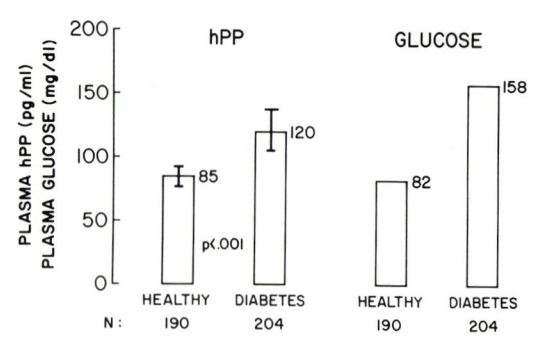

FIG. 32. Mean basal plasma levels of human pancreatic polypeptide (hPP) and glucose (mean and 95% confidence limits) in diabetic patients who had no evidence of circulating anti-PP antibody and healthy control subjects. Results for both groups adjusted to age 40 years.

significantly higher than that of the MOD and that of the healthy subjects. The basal levels of hPP of these MOD grouped according to the modes of their treatment are compared to the level of the healthy subjects in Fig. 34. The level of the MOD treated with diet plus insulin, 130 pg/ml, was significantly greater than the level of the control subjects. The data are also shown with the patients grouped according to clinical type and mode of treatment (Fig. 35). In the right-hand panel, as viewed from left to right the metabolic state of these groups becomes less compensated as evidenced by clinical type and therapy: MOD and diet, MOD and diet plus sulfonylurea, MOD and diet plus insulin, and JOD and diet plus insulin. Also there are progressively higher mean basal levels of plasma glucose 135 through 222 mg/dl. Correspondingly, mean plasma hPP becomes

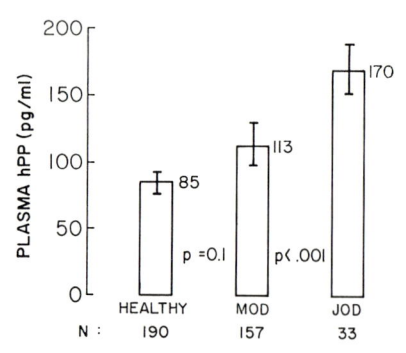

FIG. 33. Mean basal plasma levels of hPP (mean and 95% confidence limits) in healthy subjects, in patients with maturity-onset type (MOD), and juvenile-onset type (JOD) diabetes.

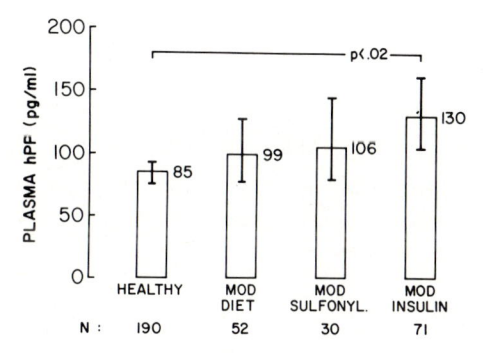

FIG. 34. Mean plasma levels of hPP (mean and 95% confidence limits) in healthy subjects and in treatment groups of maturity-onset type diabetes (MOD).

progressively higher, 99 through 170 pg/ml. As clinical severity of diabetes increases so do mean basal plasma levels of glucagon, 105 through 179 pg/ml and so do plasma levels of growth hormone 2.2 through 3.4 pg/ml (Fig. 36).

The findings of elevated plasma levels of hPP in patients with diabetes (Floyd and Fajans, 1976b) is corroborated by the findings of W. Gepts (personal communication, 1976) in the pancreases of such patients. Gepts used either rabbit anti-hPP or anti-bPP serum and the immunocytochemical peroxidase–antiperoxidase technique of Sternberger. In the pancreas of a patient who had had maturity-onset type of diabetes, hPP cells are increased in number (Figs. 37–39). This patient was a 74-year-old male who had had known diabetes for 21 years and had been treated with chlorpropamide. In the pancreas of a patient who had

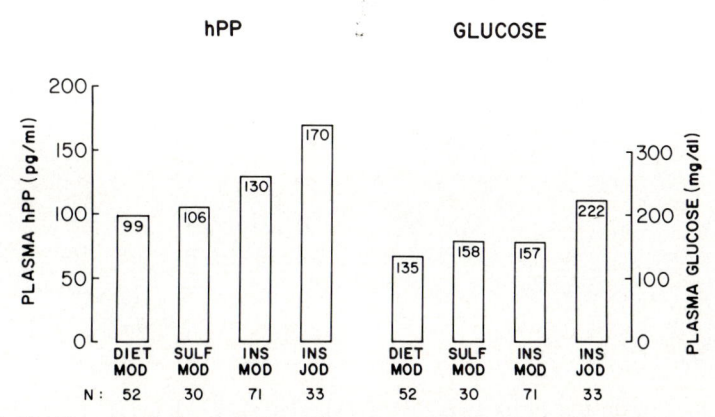

FIG. 35. Mean basal plasma levels of human pancreatic polypeptide (hPP) and glucose in diabetic patients grouped according to their clinical type and therapy (see text).

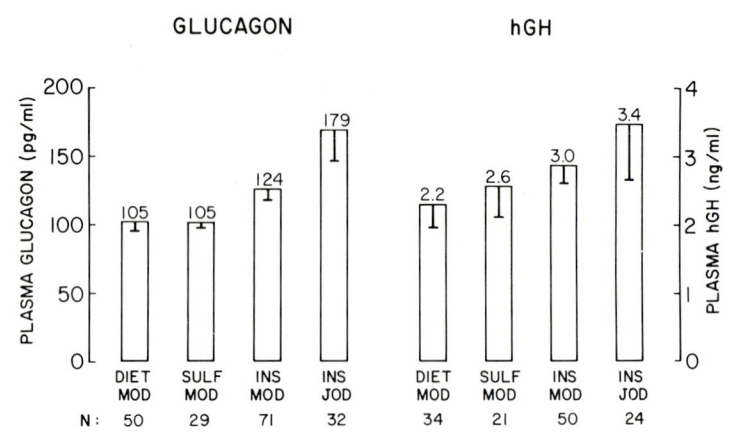

FIG. 36. Mean basal plasma levels of glucagon and growth hormone (hGH) in diabetic patients grouped according to their clinical type and therapy (see text).

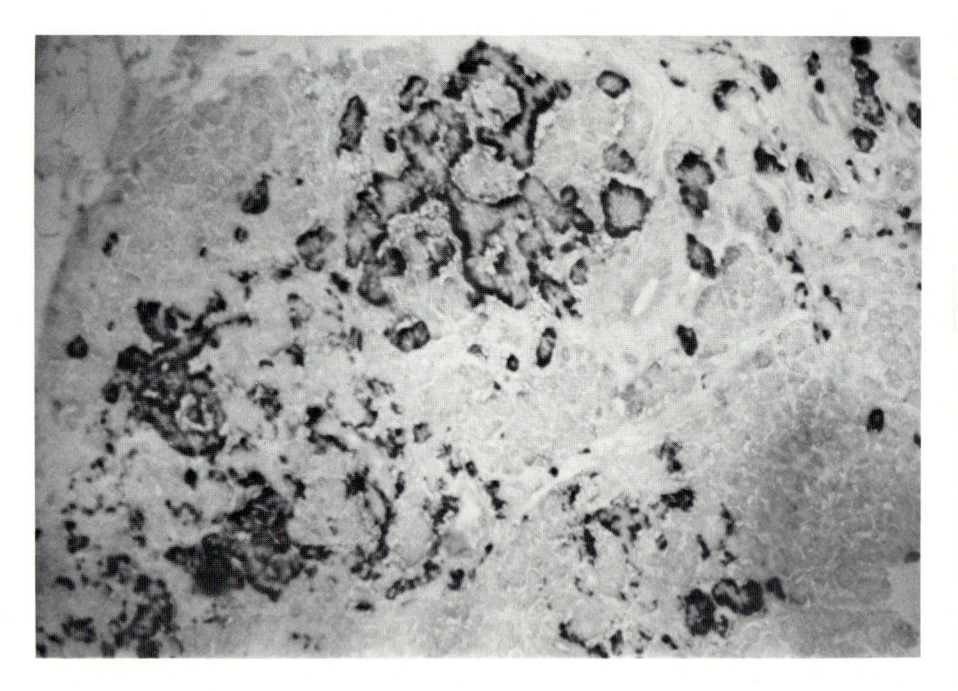

FIG. 37. Local area of human pancreatic polypeptide (hPP)-cell hyperplasia in the pancreas of a maturity-onset diabetic. Peroxidase-antiperoxidase-soluble complex method of Sternberger using specific antibodies to hPP (gift of R. Chance, Lilly Research Laboratories, Indianapolis, Indiana). × 140. From W. Gepts (personal communication, 1976).

FIG. 38. Islet of Langerhans in a maturity-onset diabetic (same case as Fig. 37). Peroxidase-antiperoxidase-soluble complex method of Sternberger using specific antibodies to human pancreatic polypeptide (gift of R. Chance, Lilly Research Laboratories, Indianapolis, Indiana). × 350. From W. Gepts (personal communication, 1976).

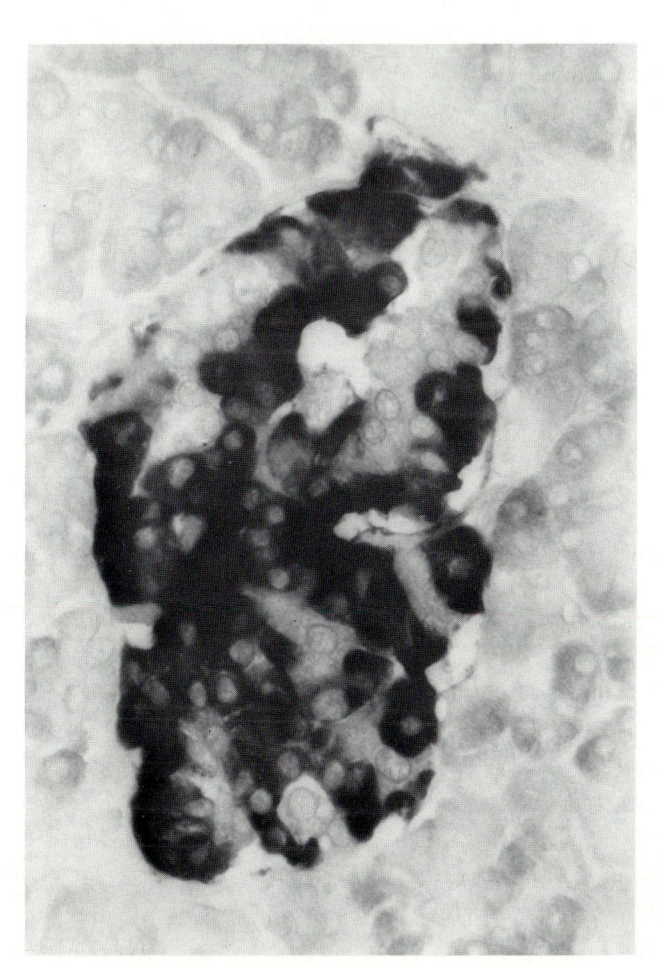

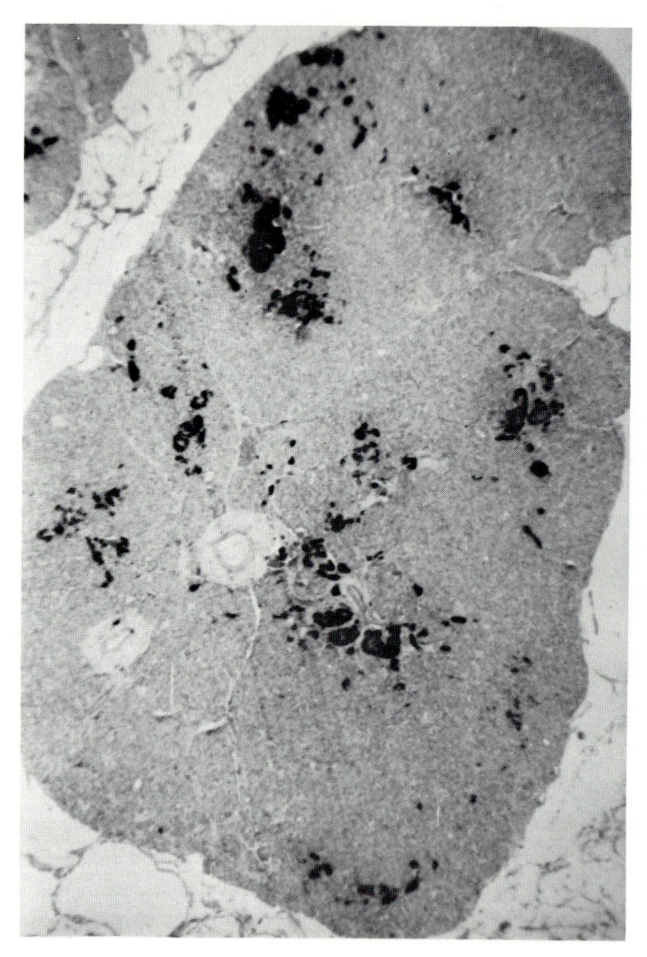

FIG. 39. Hyperplasia of human pancreatic polypeptide (hPP) cells in a lobule of the pancreas of a maturity-onset diabetic (same case as Fig. 37). Peroxidase-antiperoxidase-soluble complex method of Sternberger using specific antibodies to hPP (gift of R. Chance, Lilly Research Laboratories, Indianapolis, Indiana). × 35. From W. Gepts (personal communication, 1976).

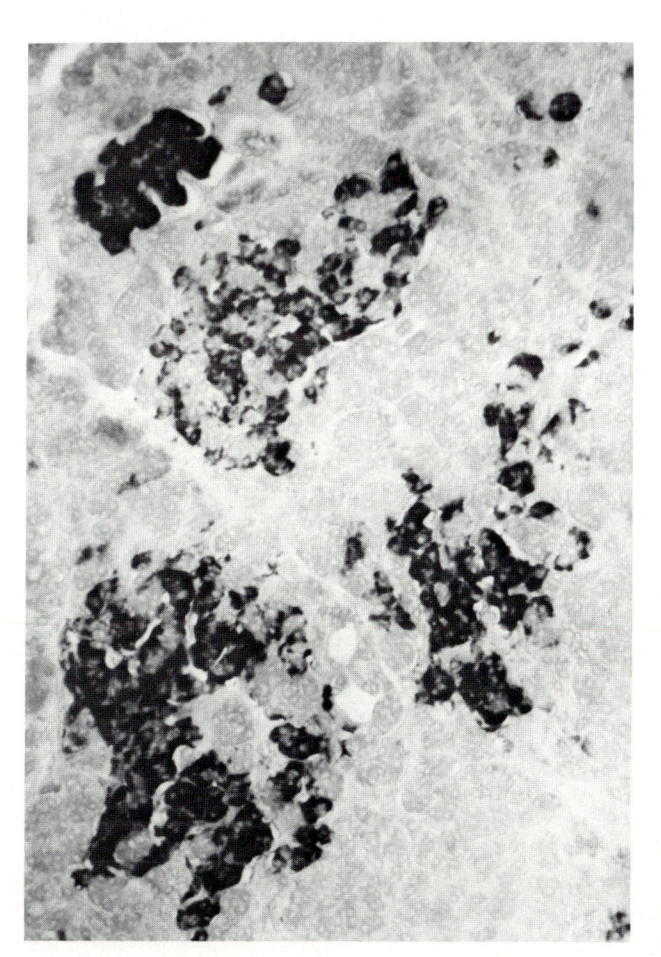

FIG. 40. Hyperplasia of human pancreatic polypeptide (hPP) cells in the pancreas of a chronic juvenile (insulin-dependent) diabetic. Known duration of diabetes: 17 years. Peroxidase-antiperoxidase-soluble complex method of Sternberger using specific antibodies to hPP (gift of R. Chance, Lilly Research Laboratories, Indianapolis, Indiana). × 70. From W. Gepts (personal communication, 1976).

had juvenile-onset type diabetes for 17 years, an increased number of hPP cells is also visible (Fig. 40). This patient was 37 years old at his death. For further data on the pathology of this case, see Gepts (1965). W. Gepts (personal communication, 1976) interprets his findings provisionally as showing hyperplasia of the hPP cells, which is more prominent in pancreas of patients with juvenile-onset than with maturity-onset type diabetes. However, an increased number of hPP cells in pancreas is not specific for genetic diabetes; increased numbers of hPP cells are found in pancreas of animals with experimental diabetes, in pancreas of patients with pancreatitis (W. Gepts, personal communication, 1976), and in uninvolved pancreas of patients with islet tumors (Larsson, 1976; Larsson *et al.*, 1976a).

The summary of and conclusions from these studies in diabetic patients are as follows:

1. Mean basal plasma level of human pancreatic polypeptide is elevated in patients with diabetes mellitus, in particular in the more severe of the maturity-onset type requiring treatment with insulin and in the juvenile-onset type of diabetes mellitus.

2. Levels of hPP tend to increase with the clinical severity of diabetes as ascertained by degree of fasting hyperglycemia.

3. The plasmas of some insulin-treated diabetic patients bind hPP, probably the effect of antibody produced in consequence of treatment with insulin preparations containing pancreatic polypeptide.

4. Basal plasma levels of hPP increase with age and this must be considered when interpreting levels in patients with diabetes mellitus, in patients suspected of having pancreatic islet tumors, or familial multiple endocrine adenomatosis type I.

5. These findings in diabetic patients may lead to an increased understanding of the pathogenesis or sequelae of the diabetes syndrome(s).

The delineation of aspects of the regulation of the secretion of human pancreatic polypeptide in healthy subjects (Floyd *et al.*, 1975, 1976b, 1977; Floyd and Fajans, 1976b), together with the identification of disease states in which plasma concentrations are aberrant (Floyd *et al.*, 1975, 1976a; Floyd and Fajans, 1976a,b), should stimulate the performance of further studies aimed at delineating physiological functions of hPP. This should be possible when there is enough pure hPP for metabolic studies. Information obtained from such studies might provide new insight into human physiology, possibly into the pathogenesis of some disease states and help place human pancreatic polypeptide with complete credentials into the ranks of the recognized hormones.

REFERENCES

Adrian, T. E., Bloom, S. R., Bryant, M. G., Polak, J. M., and Heitz, P. (1976). *Gut* **17**, 393.
Baetens, D., Rufener, C., and Orci, L. (1976). *Experientia* **32**, 785.

Bloom, S. R., Adrian, T. G., Bryant, M. G., Polak, J. M., and Heitz, P. (1976). *Prog. Int. Congr. Endocrinol., 5th, 1976* Abstract No. 833, p. 344.
Chance, R. E. (1972). *Diabetes* **21**, Suppl. 2, 536.
Chance, R. E., and Jones, W. E. (1974). U.S. Patent 3,842,063.
Chance, R. E., Lin, T. M., Johnson, M. G., Moon, N. E., Evans, D. C., Jones, W. E., and Koffenberger, J. E. (1975). *57th Annu. Meet. Endocr. Soc.* Abstract No. 265, p. 183.
Chance, R. E., Johnson, M. G., Koffenberger, J. E., Jr., and Jones, W. E. (1977a). In preparation.
Chance, R. E., Moon, N. E., and Johnson, M. G. (1977b). In preparation.
Floyd, J. C., Jr., and Fajans, S. S. (1976a). *Prog. Int. Congr. Endocrinol., 5th, 1976* Abstr. No. 834, p. 344.
Floyd, J. C., Jr., and Fajans, S. S. (1976b). *Diabetes* **25**, Suppl. 1, 330.
Floyd, J. C., Jr., Fajans, S. S., Conn, J. W., Knopf, R. F., and Rull, J. (1966). *J. Clin. Invest.* **45**, 1487.
Floyd, J. C., Jr., Chance, R. E., Hayashi, M., Moon, N. E., and Fajans, S. S. (1975). *Clin. Res.* **23**, 535 (abstr.).
Floyd, J. C., Jr., Fajans, S. S., Hayashi, M., Chance, R. E., and Moon, N. E. (1976a). *58th Annu. Meet. Endocr. Soc.* Abstract No. 20, p. 66.
Floyd, J. C., Jr., Fajans, S. S., and Pek, S. (1976b). *Clin. Res.* **24**, 485 (abstr.).
Floyd, J. C., Jr., Fajans, S. S., and Pek, S. (1977). *Trans. Assoc. Am. Physicians* **89**, 146.
Gepts, W. (1965). *Diabetes* **14**, 619.
Gersell, D. J., Greider, M. H., and Gingerich, R. L. (1976). *Diabetes* **25**, Suppl. 1, 364.
Hazelwood, R. L. (1973). *Am. Zool.* **13**, 699.
Hazelwood, R. L., Turner, S. D., Kimmel, J. R., and Pollock, H. G. (1973). *Gen. Comp. Endocrinol.* **21**, 485.
Kimmel, J. R., and Pollock, H. G. (1975). *Fed. Proc., Fed. Am. Soc. Exp. Biol.* **34**, 454 (abstr.).
Kimmel, J. R., and Pollock, H. G. (1976). *Proc. Int. Congr. Endocrinol., 5th, 1976* Abstract No. 828, p. 342.
Kimmel, J. R., Pollock, H. G., and Hazelwood, R. L. (1968). *Endocrinology* **83**, 1323.
Kimmel, J. R., Pollock, H. G., and Hazelwood, R. L. (1973). In "Atlas of Protein Sequence and Structure" (M. O. Dayhoff, ed.), Vol. 5, Suppl. 1, p. 5. Natl. Biomed. Res. Found., Washington, D.C.
Kimmel, J. R., Pollock, H. G., and Hazelwood, R. L. (1974). *56th Annu. Meet. Endocr. Soc.* Abstract No. 403, p.257.
Kimmel, J. R., Hayden, L. J., and Pollock, H. G. (1975). *J. Biol. Chem.* **250**, 9369.
Langslow, D. R., and Hazelwood, R. L. (1975). *Diabetologia* **11**, 357 (abstr.).
Langslow, D. R., Kimmel, J. R., and Pollock, H. G. (1973). *Endocrinology* **93**, 558.
Larsson, L.-I. (1976). *Lancet* **2**, 149.
Larsson, L.-I., Sundler, F., Håkanson, R., Pollock, H. G., and Kimmel, J. R. (1974). *Histochemistry* **42**, 377.
Larsson, L.-I., Sundler, F., and Håkanson, R. (1975a). *Diabetologia* **11**, 357 (abstr.).
Larsson, L.-I., Sundler, F., and Håkanson, R. (1975b). *Cell Tissue Res.* **156**, 167.
Larsson, L.-I., Schwartz, T., Lundqvist, G., Chance, R. E., Sundler, F., Rehfeld, J. F., Grimelius, L., Fahrenkrug, J., Schaffalitzky de Muckadell, O., and Moon, N. (1976a). *Am. J. Pathol.* **85**, 675.
Larsson, L.-I., Sundler, F., and Håkanson, R. (1976b). *Diabetologia* **12**, 211.
Lin, T. M., and Chance, R. E. (1972). *Gastroenterology* **62**, 852 (abstr.).
Lin, T. M., and Chance, R. E. (1974a). In "Endocrinology of the Gut" (W. Y. Chey and F. P. Brooks, eds.), p. 143. Chas. B. Slack, Inc., Thorofare, New Jersey.
Lin, T. M., and Chance, R. E. (1974b). *Gastroenterology* **67**, 737.

Lin, T. M., Chance, R. E., and Evans, D. (1973). *Gastroenterology* **64**, 865 (abstr.).
Lin, T. M., Evans, D. R., and Chance, R. E. (1974). *Gastroenterology* **66**, 852 (abstr.).
McCumbee, W. D., and Hazelwood, R. L. (1976). *58th Annu. Meet. Endocr. Soc.* Abstract No. 478, p. 296.
Munger, B. L., Caramia, F., and Lacy, P. E. (1965). *Z. Zellforsch. Mikrosk. Anat.* **67**, 776.
Polak, J. M., Bloom, S. R., Adrian, T. E., Heitz, P., Bryant, M. G., and Pearse, A. G. E. (1976). *Lancet* **1**, 328.
Rufener, C., Baetens, D., and Orci, L. (1976). *Experientia* **32**, 919.
Schwartz, T. W., Rehfeld, J. F., Stadil, F., Larsson, L.-I., Chance, R. E., and Moon, N. (1976). *Lancet* **1**, 1102.

DISCUSSION

M. Lippman: Did you look for hPP in urine? Have you looked for it in pancreatic duct fluid? Do you know whether it is present in pancreatic adenocarcinoma or in any other malignancy in terms of aberrant production or secretion rates?

J. C. Floyd: No.

A. M. Spiegel: Have you investigated the role of catecholamines on the release of this polypeptide specifically in relation to your studies with insulin hypoglycemia? Have you excluded an indirect mechanism through catecholamines in this regard?

J. C. Floyd: We have not excluded that mechanism in insulin hypoglycemia. Our preliminary studies suggest adrenergic modulation of hPP release in that there is a fall of the basal level of hPP when one blockades β receptors and simultaneously stimulates α receptors (infusion of epinephrine plus propranolol).

A. M. Spiegel: The second question was stimulated by this association in one patient with MEA and hyperparathyroidism and elevated levels of this polypeptide. Have you investigated any kind of relationship between the polypeptide and calcium metabolism? For example, several other GI hormones and polypeptides have effects, such as the release of calcitonin by gastrin. Apparently, you do not have sufficient material available to test this type of effect. Have you investigated any other aspects of this?

J. C. Floyd: We have infused calcium in patients with insulinomas and found no change in the level of hPP in plasma. In some patients who appear to have familial hyperparathyroidism, hPP is elevated (observations in collaboration with Dr. S. J. Marx).

L. S. Jacobs: These are elegant studies. To pursue the catechol, the possibility is probably worth pointing out that, specifically in the incidence of insulin-induced hypoglycemia, the rise in catecholamines is the earliest identifiable hormonal event—occurring before the rise in growth hormone, before the rise in other counterregulatory hormones—and therefore would sit very nicely in the time scale that you showed. Furthermore, the rise following insulin in plasma catecholamines is of the order of about 50-fold above base line. So it remains, at least in my mind, as a possibility; therefore, I would like to know whether you have tested epinephrine or acetycholine in the *in vitro* study in which you showed acetylcholine atropine blockade of the acetylcholine stimulation.

Have you had enough of the material to test plasma disappearance, since all the levels you showed, especially in diabetics, relate to presumed secretion? I think that is very reasonable, nonetheless; we have been hearing more and more about hyperglycemia modifying the structure of proteins, and we know that glycosylation may alter plasma half-lives, as Dr. Vaitukaitis has reminded us. Have you had enough material to test it *in vivo* in terms of plasma disappearance?

J. C. Floyd: The infusion of epinephrine (5 μg/minute for 60 minutes) in three healthy subjects was not associated with increase in plasma hPP. The infusion of epinephrine plus phentolamine (stimulation of β receptors, blockage of α receptors) in three healthy subjects gave a similar result. It seems unlikely that the increase in plasma hPP induced by the injection of insulin is, in the main, mediated by an adrenergic mechanism. S. R. Bloom has said that the increase in plasma hPP induced by insulin injection is blocked by the prior administration of atropine.

We have not studied half-life by injecting hPP. Dr. J. Kimmel tells me that the half-time of disappearance from plasma of injected aPP in the chicken is of the order of 4–6 minutes.

L. S. Jacobs: Within your suppressibility, it must be very short.

J. C. Floyd: It would appear from the rate of decrease in plasma hPP after increases induced by insulin injection producing hypoglycemia, that the *in vivo* plasma half-life is quite short. From such observations I would guess that it would be less than 10 minutes, on the order of 5 minutes.

P. Licht: I am not familiar with the nature of your patient population, but the variations in normal levels as well as the relation to food prompts me to wonder whether this might be related in any way to cultural or racial considerations. Cultural differences might be genetic differences, but the possibility also exists that consistent dietary differences could have prolonged effects on sensitivity or resting levels of the hPP.

J. C. Floyd: You are asking specifically whether there might be some explainable difference among the diabetic patients who had the elevated levels as opposed to those who do not?

P. Licht: Yes, but also variations among normals. Your data on the islet tumors suggest that this is a familial genetic marker for one of the MEA syndromes, supporting the possibility of genetic population.

J. C. Floyd: We have no precise information as to the composition of the diets of the healthy subjects. It seems possible that antecedent diet could influence basal levels of hPP in plasma. We do not believe elevated plasma hPP among MEA-I family members is any more, or less, of a genetic marker than is, for example, elevated plasma insulin or elevated plasma gastrin.

J. R. Kimmel: We work with chickens, and we ran into a peptide (polypeptide hormone III, PH III) completely by accident while trying to purify chicken insulin. We could not get the peptide away from insulin, but when we got it out it was pure. We subsequently followed up by determining the amino acid sequence and by developing a radioimmunoassay. With regard to the things that stimulate secretion of the hormone we have done many of the same experiments that Dr. Floyd has reported here for humans. In general, what stimulates secretion in humans also stimulates secretion in the bird. Figure A demonstrates gel filtration patterns of chicken plasma and an acid–alcohol extract of chicken plasma. In each fraction we determined the content of the pancreatic polypeptide by radioimmunoassay. The arrow labeled "insulin" shows about where insulin would emerge in this gel filtration pattern. In the plasma, the label "125-I PH III" shows where the hormone emerges when ^{125}I-labeled hormone is filtered. The material we immunoassay in plasma emerges essentially as one peak in gel filtration patterns, although there may be a larger component at the void volume of the column.

Figure B shows what happens to the plasma levels of pancreatic polypeptide if one allows chickens that have been fasted for 26 hours to eat ad libitum. Blood samples were collected at the times indicated. There is a 10–20-fold increase in plasma level that peaks at about 1 hour. During the first hour these chickens feed very vigorously, and after this the rate at which they feed slows down. We then follow the levels through the day, and it is apparent that they level off at around 8 ng/ml. Note also that plasma levels are considerably higher in these chickens than in humans. The most important thing I would like to point out is the striking difference in these two age groups.

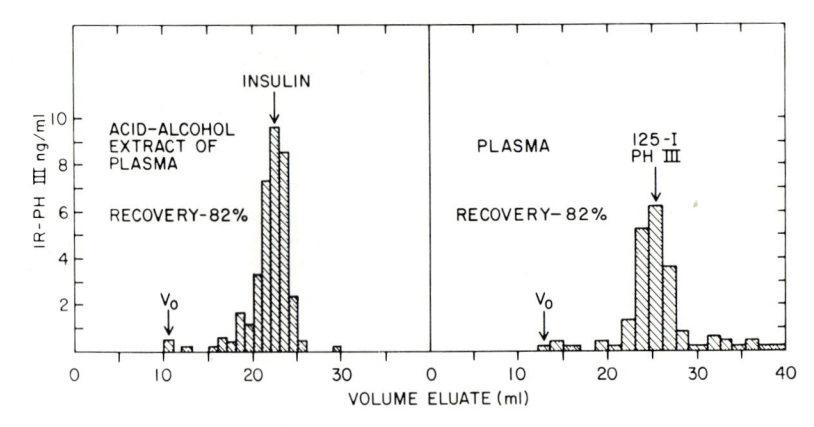

FIG. A. Gel filtration patterns of chicken plasma on Sephadex G-50 SF, 1 × 45 cm column, at 25°C, 0.04 M phosphate, pH 7.4, and 0.1% albumin.

From this, we went to studies in which we put measured amounts of food into the crop of chickens. We studied protein, fat, and carbohydrate and found that protein stimulated secretion the best. In these studies, however, the rise in plasma level was never as great as when the birds fed ad libitum, so we thought there might be some cephalic effect. The results of a study done in connection with this are shown in Fig. C. We fasted for 24 hours two groups of chickens and then collected plasma samples at time 0. The control group is represented by the open bars; the experimental group by hatched bars. It can seem that in both groups plasma levels are very low after completion of the fast. At that time we made food available to both groups of chickens in identical boxes, but for one group we had a transparent cover on the feed box. The food was packed up close to the cover. These chickens tried as vigorously to get at the food as did the ones that could actually obtain food. After 25 minutes the chicken gave up in despair and quit trying to eat; We took a sample of blood at that point. In the birds that could get food it is apparent that plasma

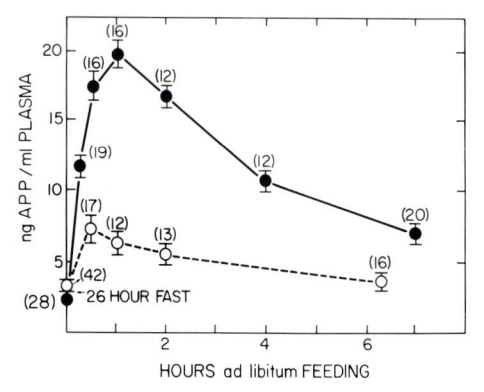

FIG. B. Immunoreactive avian pancreatic polypeptides (aPP) in plasma of fasted-refed chickens. ●, Females 8–10 weeks old; ○, females 16–22 weeks old.

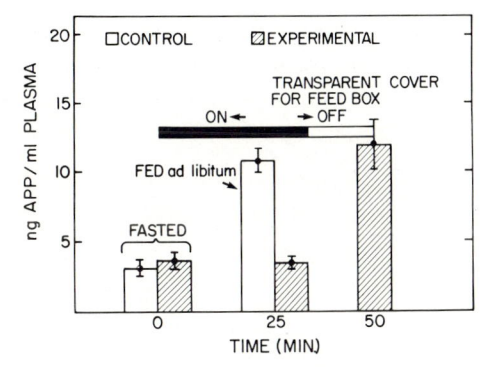

FIG. C. Immunoreactive avian pancreatic polypeptides (aPP) in plasma of fasted sham-fed chickens.

levels increased, while in the others plasma levels did not change. We then took the lid off the feed box and let the experimental group eat, and it is apparent that they were capable of responding. However, there does not appear to be an effect associated with the anticipation of eating, the sight of food, or the sight of other birds eating.

After having ascertained that protein, fat, and carbohydrate cause a rise in plasma level when infused into the crop, we then elected to try to find from which segment of the gut the signal came to trigger this response. In order to do so, we anticipated having to use anesthetized birds. Figure D shows the results of our control experiments, which is about as far as we got. We infused into the group represented by open circles a 20% suspension of Startena; this is a standard 30 ml dose that we put into the crop. We observed the expected rise in plasma level, but in our anesthetized birds nothing happened at all. This indicates that pentobarbital anesthesia completely blocks the secretory response. We now know that atropine will do the same thing. Our avian physiologist friends told us that the food never got out of the crop in these birds; it never got down to the portion of the gut from which the stimulus could arise, so we got at that problem by the method shown in Fig. E.

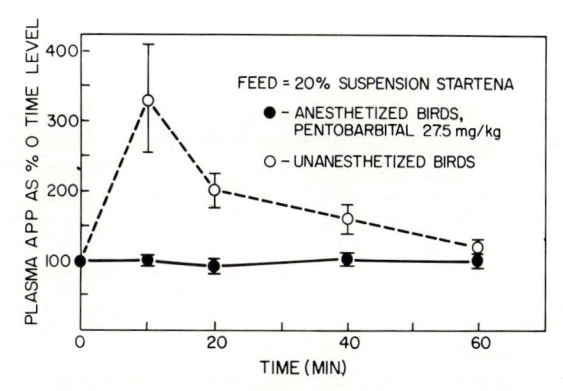

FIG. D. Effect of anesthesia on avian pancreatic polypeptide (aPP)-secretory response to feeding in female chickens 8–12 weeks old.

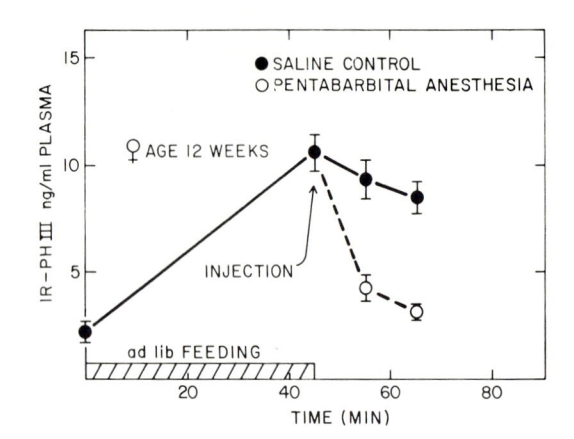

FIG. E. Effect of pentobarbital on plasma level of peptide PH III in fed chickens.

In this case we fasted birds for 24 hours, collected plasma, then allowed them to eat for 40 minutes. At the end of that time we collected plasma samples, and then divided the birds into two groups. One group we injected with saline and then just took food away, and the plasma levels trickled downward, so to speak; the other group we injected with pentobarbital, and the plasma levels dropped precipitously (Fig. E). The important thing about this is that, since we have now determined the half-life of this peptide in chickens to be about 4.5 minutes, we know that the fall in plasma level is at a rate just about equivalent to that half-life. Therefore, it appears that pentobarbital anesthesia has completely shut off, or virtually shut off, secretion of this hormone. We have not yet done this experiment with atropine, but we need to try this next. We autopsied these birds and food was distributed throughout the gut.

We made some effort to try to find out what happens to metabolites after injection of this hormone; although we now know that the injections we used were at pharmacological levels, we did observe hepatic glycogenolysis without blood sugar change. In fasted birds there is about a 30% drop in circulating free fatty acids, and there is a decrease in plasma glycerol. The most striking change is a drop in plasma amino acids, in particular alanine. This is quite significant, and we need to go back and check this with physiological doses.

S. Korenman: Several things about this are particularly fascinating, and one of the particular features is that there is so much homology between this protein over a wide variety of species, suggesting that to keep this homology must be very important. One aspect that you did not report was whether you have studied the possibility that this is a prohormone. I do not recall whether this is possible; the structure went by so fast. Have you studied trypsin digests of this material to see whether it has a more potent biological effect.

J. C. Floyd: I am going to ask Dr. Chance to answer this question. He has shown that removal of the carboxyl-terminal amino acid residue markedly reduces some biological effects of bPP.

G. S. Cahill: Dr. Chance, there appears to be more species differences in PP than in glucagon. Am I correct?

R. E. Chance: In answer to Dr. Cahill's question first, glucagon does appear to be more protected among species than other polypeptide hormones. The six mammalian glucagons studied appear to have identical amino acid sequences (bovine, human, and porcine actually determined, whereas camel, rabbit, and rat are assumed to be identical owing to identical amino acid compositions). There are minor variations within the avian species studied (chicken, turkey, and duck). With regard to the primary structures of the five pancreatic polypeptides studied so far (see text Figs. 1 and 2), there are a few differences among the mammalian species, but many differences between the avian and mammalian species.

In answer to Dr. Korenman's question: we found during our studies [R. E. Chance, M. G. Johnson, J. E. Koffenberger, and W. E. Jones, in preparation] that the Arg^{35} -Tyr^{36} -NH_2 bond was the most rapidly hydrolyzed followed by the Arg^{26} -Tyr^{27}, Arg^{25} -Arg^{26}, and Arg^{33} -Pro^{34} bonds in order of sensitivity. We found that removal of the C-terminal Tyr^{36} -NH_2 residue resulted in loss of biological activity in several gastrointestinal tests [T. M. Lin, D. R. Evans, and R. E. Chance, *Gastroenterology* 66, 852 (abstr.) (1974); T. M. Lin and R. E. Chance, *ibid.* 67, 737], which seemingly rules out the idea of a trypsin-activated prohormone. We cannot rule out the possibility that bPP, etc., is a precursor molecule acted on by another enzyme system.

G. S. Cahill: I would like to ask Dr. Hazelwood to give some comparative comments.

R. L. Hazelwood: Dr. Kimmel, Dr. Chance, and I are pleased to see that this hormone with which we have been working now has a disease state or states associated with it. One of the earliest things we did with aPP was to establish physiological effects by taking birds and cannulating them appropriately to collect gastric fluid over a period of time after aPP injection. We see in Fig. F a sampling of three different doses (12.5, 25, and 50 μg/kg). The response was immediate as the hormone was injected into a peripheral wing vein. It was very dramatic; by the time the syringe was emptied of aPP one could actually see an increase in

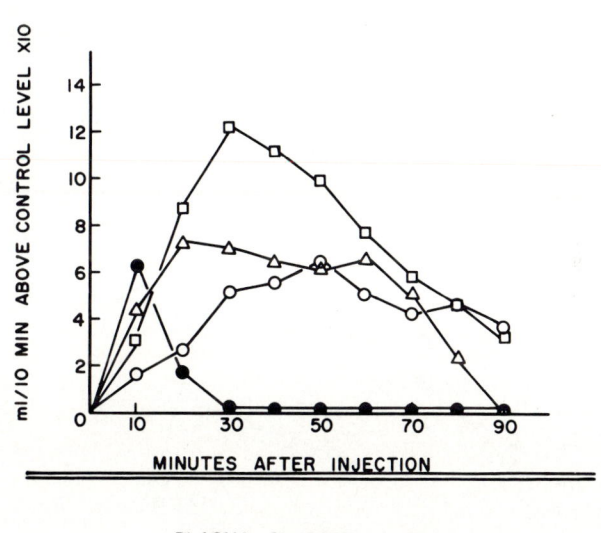

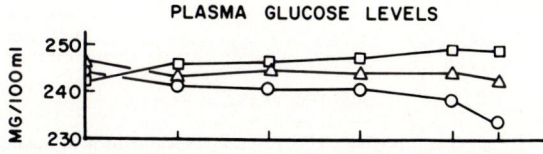

FIG. F. Proventricular (secretory) response to avian pancreatic polypeptide (aPP): ○, 12.5 μg/kg (n = 7); △, 25.0 μg/kg (n = 5); □, 50.0 μg/kg (n = 5); ●, pentagestrin 1 μg/kg (n = 6).

number of drops of gastric fluid (secretion) coming out of the cannula placed between the proventriculus and the gizzard. As shown on the bottom part of this graph, at no time, regardless of the doses of aPP, was there any perturbation of plasma glucose levels.

To see whether this was merely a fluid volume response, we also checked the contents of this volume, and in Fig. G one sees that not only free acid, but also pepsin and total protein concentration, respond immediately. These increases in concentration per sample collected over a 90-minute period after the injection were in this case in response to 25 μg of aPP administered per kilogram of body weight. Again, the fluid volume was increased (see above the abscissa of Fig. G). All data here are expressed as above control or basal levels.

These data were presented several years ago, and we are now looking at three different levels of tissue organization in an effort to elucidate more precisely the physiological response to APP. We are interested in possible intrapancreatic control of aPP secretion as well as the effect of aPP on glucagon and insulin secretion. Figure H is an example of recent work done with Dr. Langslow of Edinburgh in which we sampled many peripheral parameters as well as pancreatic glucagonlike activity and insulin levels in plasma after aPP injection. One can see once again, in response to a 50 μg per kilogram body weight dose of aPP, that plasma glucose is not disturbed (filled circles). Note that free fatty acids are invariably depressed within 2.5 minutes after the aPP injection and that uric acid levels are increased slightly but significantly. Skipping to the bottom of Fig. H, we can see that aPP apparently (at this dose at least) has no effect whatsoever on plasma insulin levels. The response of glucagon is seen in diagram d of Fig. H, where we have plotted the C-terminal GLI as measured by Dr. Keith Buchanan in his laboratory on a "double-blind" basis. "Pancreatic glucagon," so called, is the uppermost line (filled circle). aPP immediately induces a sharp increase in glucagonlike activity. The control values are the open circles at the bottom.

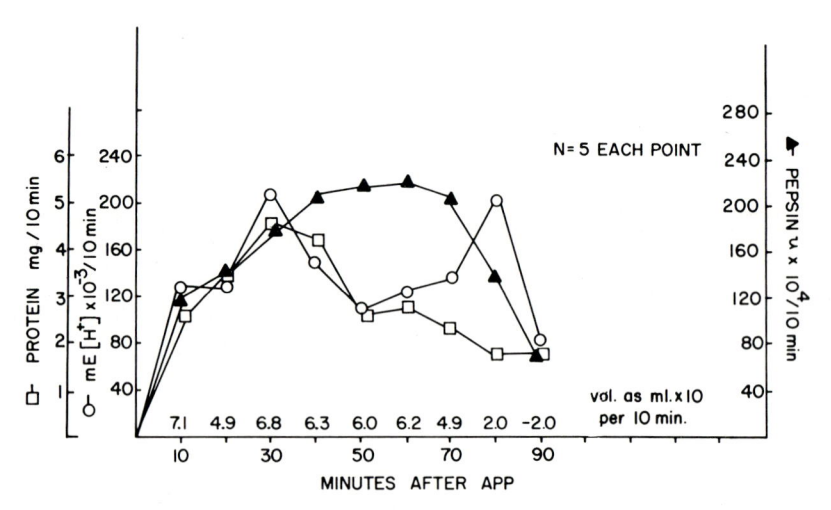

FIG. G. Gastric [H$^+$], pepsin, protein, and volume response to avian pancreatic polypeptide (aPP), 25 μg/kg, expressed as above control levels.

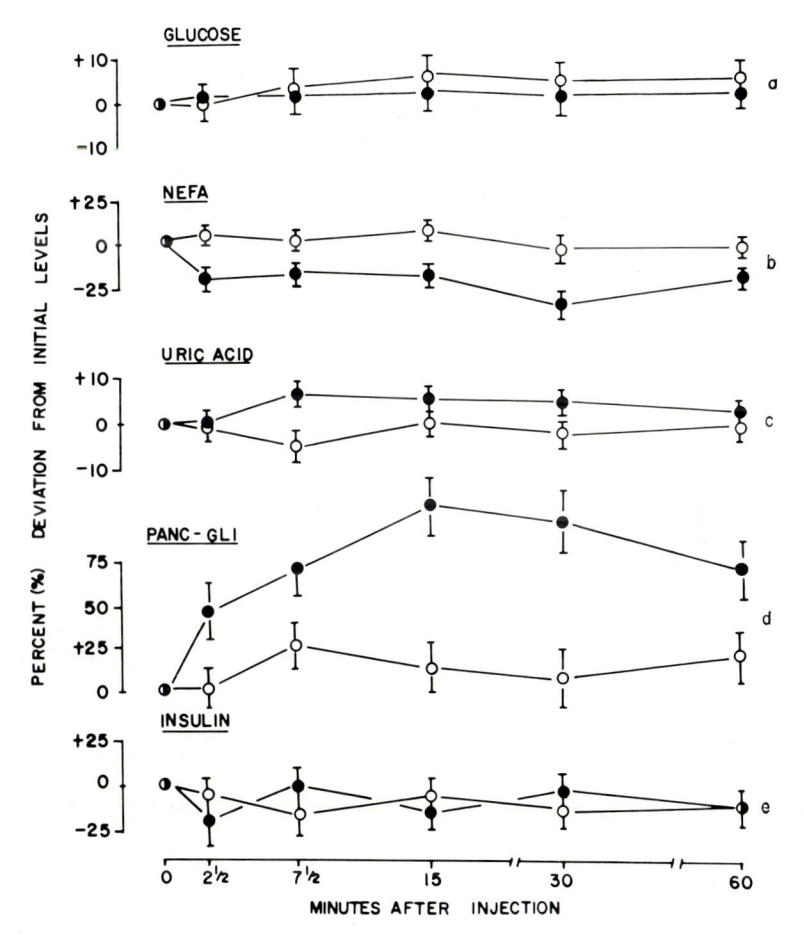

FIG. H. Effect of avian pancreatic polypeptide (aPP), 50 μg per kilogram body weight, on various plasma parameters in adult chickens. ○, Controls (n = 11); ●, aPP (n = 15). (a) Glucose, 227 mg/100 ml; (b) nonesterified fatty acids, 513 μM; (c) uric acid, 5.23 mg/100 ml; (d) pancreatic glucagonlike activity, 0.63 ng/ml; (e) insulin, 1.18 ng/ml.

A second approach that we have been using recently is to look at the isolated avian hepatocyte and adipocyte for binding properties and possible mechanisms of action of aPP. Mr. McCumbee in our laboratory has had very little success in binding studies with the hepatocyte; we have had reasonably good luck—there is good binding with the isolated adipocyte of chick source.

In Table A is an example of the antilipolytic action of aPP on glucagon-stimulated lipolysis. The control values are to the far left, where there is no aPP present. Basal levels are 0.48. aPP at 25 ng/ml or at 100 ng/ml causes remarkable depression of glucagon-induced lipolysis. Neither pentagastrin nor insulin, at two different levels, alter the lipolytic trend

TABLE A
Inhibition of Glucagon-Stimulated Lipolysis in Chicken Adipocytes[a]

Control	aPP[b] (ng/ml)		Pentagastrin (ng/ml)		Insulin (ng/ml)	
0 aPP	25	100	17.5	175	4	40
0.48 ±0.02	0.36[c] ±0.01	0.32[d] ±0.04	0.50 ±0.02	0.52 ±0.04	0.49 ±0.05	0.50 ±0.02

[a] Glycerol release = μmoles per 100 mg dry weight per hour. Each flask of cells contained 5 ng of glucagon per milliliter. $N = 4$.
[b] aPP, avian pancreatic polypeptides.
[c] = $P < 0.001$.
[d] = $P < 0.01$.

from the basal level seen at the far left. We interpret this to mean, therefore, that aPP at the adipocyte probably has an antilipolytic effect.

Further studies with aPP are those undertaken by Laurentz in our laboratory on the possible trophic action of APP on the avian gut. Following the guideline shown by Grossman's laboratory in California and Johnson's in Houston, we wondered if the depancreatized animal would have alterations in gastric secretion or gastric mucosal composition. Definitely, we found partial proventricular atrophy, we found a decrease in DNA and RNA concentrations, etc. Granted, we took out many peptides when the pancreas came out, and our supply of aPP was/is not luxurious enough that one could initiate a replacement regimen. Therefore, we turned to a developing chick embryo system and injected aPP, following various parameters of protein anabolism in the gut. Laurentz observed a definite increase in the incorporation of [^{14}C] leucine, increases in DNA and RNA levels, and so forth, in the proventriculus and the upper duodenum of the developing chick in response to graded doses of aPP. We suggest, then, the possibility of a trophic action of aPP on the developing chick gut.

G. S. Cahill: Is aPP lipogenic in the chicken, Dr. Hazelwood? It seems to do everything that insulin does in the chicken except on glycogen.

R. L. Hazelwood: Yes, the data that we have on hand, as well as those gathered partially in Dr. Kimmel's laboratory with Dr. Joe Hayden, and in my laboratory with Mr. McCumbee, indicate aPP to be hypoglycerolemic, it decreases nonesterified fatty acids and it has antilipolytic activity at the adipocyte. There is also an increase in lipid formation in the liver and possibly an increase in triglyceride release to the plasma.

G. J. Macdonald: Your interesting data in regard to chicken aPP causes me to ask if you have considered age and sex differences beyond the pullet and hen to include the cockerel and rooster? My second question deals with the increases of aPP and hPP found upon feeding digestible substances. Have you fed nondigestible substances, such as cellulose, for the purpose of loading the stomach to determine whether aPP and hPP changes are in response to your filling the stomach?

J. C. Floyd: We have not had our subjects ingest nondigestible material.

J. R. Kimmel: When I was talking about what stimulated secretion of the hormone, I meant to mention that we did two types of control experiments. In one set of chickens, we put a balloon in the crop, blew it up with 30 ml of water, and observed that nothing happened to the plasma level. We also used a suspension polyacrylamide gel in 30 ml of saline, and that did not affect the plasma level.

R. L. Hazelwood: We have an answer to the other question about aPP levels in various types of birds, I believe. Using the immunoassay developed by Langslow and Kimmel, we find (see Table B) in the laying chicken that the levels of aPP are almost 6 mg per 100 gm of fresh tissue and that the insulin levels are much less in the same pancreas. The 1-day-old chicken is the champion of the group in respect to aPP levels; in the duck and turtle, there is merely a trace when anti-aPP antibody is used. These are all old data, but still they indicate varying levels depending on the type of animal assayed. Note that in the 1-day chick hatchout the chick is virtually on a 90% fat diet. The yolk sac is not completely absorbed, so for another 5 or 6 days, the chick will be on a heavy lipid diet. In further answer to your question, we find that the younger bird is more sensitive to injections of APP than the older birds regardless of what parameter we test. We find that the female is probably more sensitive than the male, and that the fasted animal is more sensitive to aPP than is the fed animal.

J. S. Roberts: Is there any evidence? The experiment you would have to do would involve the injection of the polypeptide to indicate whether or not the polypeptide may affect satiety in any of these species, would it not?

J. C. Floyd: We have not administered any PP to man. We can say that some patients who have had very high levels (over 1000 pg/ml) are mildly obese, so obviously they are not getting satiated by hPP.

J. S. Roberts: Have you given enough in the chickens to see whether there is an anorexigenic effect in the hormone?

R. L. Hazelwood: I think we can say from Dr. Kimmel's graphs (Figs. A–E) that as the animal became satiated the levels fell off. However, I think we need an immunofluorescent study here, possibly checking the hypothalamic areas.

H. Papkoff: Could any of the people here, working with the pancreatic polypeptide, tell me if the pituitary has any influence over its regulation; does hypophysectomy, for instance, affect its release?

J. C. Floyd: We do not have information on the effects of hypophysectomy in man.

L. Orci: In November 1975, we received from Dr. R. Chance anti-bPP (bovine pancreatic polypeptide) serum, to perform immunofluorescent studies. Some data collected in our laboratory refer to the geographical distribution of PP-containing cells in pancreas. We first

TABLE B
Species Comparison of Pancreatic Insulin and Avian Pancreatic Polypeptide (aPP) Content [a]

Species	aPP (mg/100 gm)	Insulin (mg/100 gm)
Chicken, Laying	5.9	1.5 (equiv. pork)
Capon	4.2	2.4
1 Day old	8.2	–
Duck, adult ♀	6.0	5.6
Turtle, adult ♀	1.22	–
Bovine, adult ♀	Trace	20

[a] aPP was not detected in crude pancreatic extracts of pig, cat, guinea pig, rattlesnake, toad, bullfrog. All extracts had positive insulin values.

applied anti-bPP serum to the dog pancreas. To our surprise, the highest proportion in PP-containing cells was found in the uncinate process, a part of dog pancreas that is in close relationship with the duodenum.

We then turned our attention to human, mouse (Fig. I), and rat pancreas, and we were struck by the following observations. In some islets, PP-cells were scarce, and the incubation of a consecutive serial section with antiglucagon revealed numerous glucagon-containing cells. On the other hand, there were islets presenting with a reverse pattern, that is, numerous PP-containing cells and very few glucagon-containing cells. This finding prompted us to study systematically the entire pancreas of the rat; from the study we surmised that islets rich in PP cells were more frequent in the duodenal part of the pancreas. In order to assess this supposition, we isolated islets separately from the duodenal and from the splenic part of the pancreas, and the pelleted islets were processed for immunofluorescence. Figure J compares pelleted islets from the head of the pancreas (a, b) to islets from the rest of the pancreas (c, d). There is clearly a mirror image between the distribution of PP- and glucagon-cells depending whether the islets come from the head or the tail of the pancreas. In this context, it is worth recalling that the uncinate process of the dog, mentioned above as containing many PP cells, is virtually devoid of glucagon cells.

Another finding of interest is that the topographical distribution of PP- and glucagon-rich islets seems to be underlined by a precise pattern of vascularization. Indeed, after injection

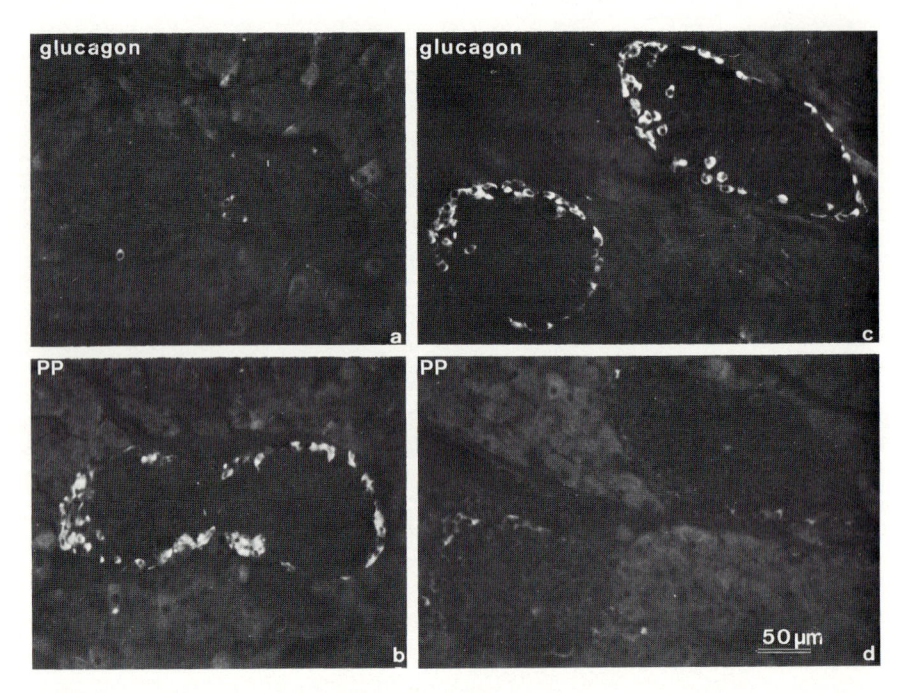

FIG. I. Normal mouse (C57BL/6J): serial sections of an islet from the head (a,b) and an islet from the tail (c,d), after application of antiglucagon, and antibovine pancreatic polypeptide (PP) sera. Note the inverse relationship between PP and glucagon in the two islets.

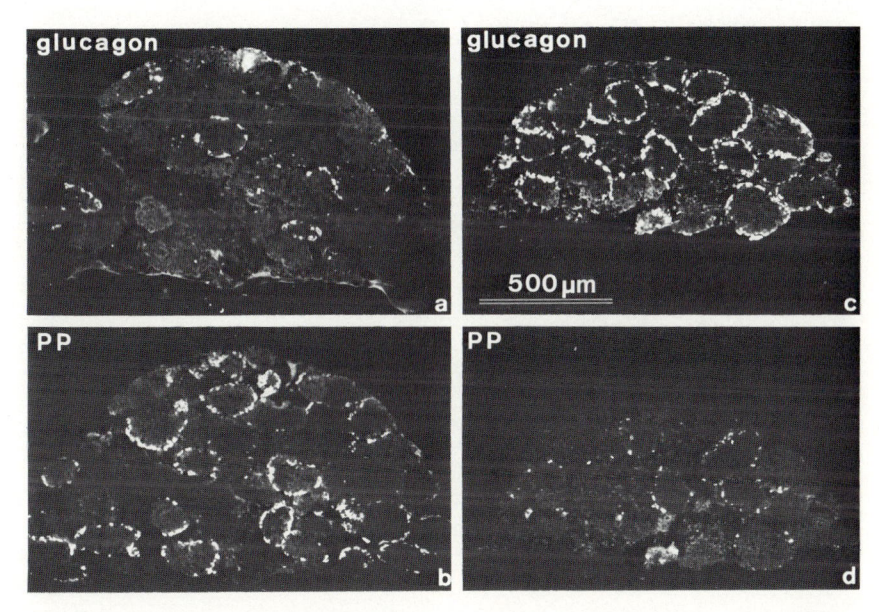

FIG. J. Rat pancreas. (a,b) Pelleted islets from the head of the pancreas. (c,d) Pelleted islets from the rest of the pancreas.

of India ink in the superior mesenteric artery, the two lower thirds of the head of the pancreas are stained (Fig. K). This region corresponds to the irrigation territory of the inferior pancreaticoduodenal artery; the other, unstained, regions of the pancreas are supplied by the gastroduodenal and splenic arteries, both branches of the celiac trunk. PP-rich and glucagon-poor islets are in the territory of the pancreaticoduodenal artery, and PP-poor, glucagon-rich islets belong to areas irrigated by the gastroduodenal and splenic arteries.

A last comment about PP cells is that, in a recent study of ob/ob (C57BL/6J) mice, characterized by islet hyperplasia, hyperinsulinemia, and hyperglycemia, we found a reduction of almost 65% of the PP cells in the islets.

G. Cahill: Dr. Orci, are you suggesting that it could be the same type cell that in one area is making glucagon and in another area making PP? Or are they two discretely different cells that are mutually exclusive in the same islets?

L. Orci: There are undoubtedly two distinct cell types: the glucagon-containing cell is the A_2 cell, whereas the PP-containing cell probably corresponds to the so-called F cell [C. Rufener, D. Baetens, and L. Orci, *Experientia* **32**, 919–920 (1976)].

R. E. Chance: Two years ago my co-workers and I obtained results that support Dr. Orci's observations. We were aware of the studies by B. L. Munger, F. Caramia, and P. E. Lacy [*Z. Zellforsch. Mikrosk. Anat.* **67**, 776 (1965)] which indicated that islets in the uncinate process of the canine pancreas (in the duodenal lobe) contained few, if any, glucagon-producing A cells. They also noted another cell type that was particularly abundant in the islets from this region of the pancreas, and they called it the F cell. In October, 1974, Dr. Larsson [see L.-I. Larsson *et al.* (1976)] indicated in a personal communication that he had observed

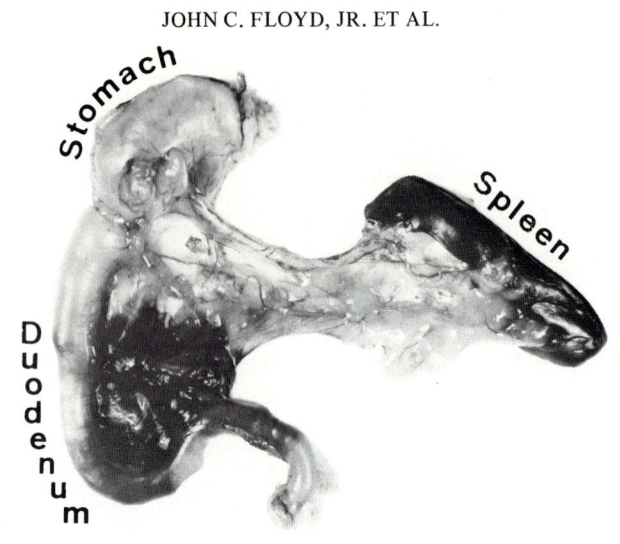

FIG. K. After the injection of India ink into the superior mesenteric artery, the black staining is limited to the inferior two-thirds of the head of the pancreas and part of duodenum. Most of the islets present in this region were of the PP type (see Figs. I and J).

significant immunofluorescence in the canine uncinate process using our bPP and hPP antisera. We promptly confirmed Dr. Larsson's immunocytochemical observations using the hPP radioimmunoassay on acid–alcohol extracts of two general sections of the canine pancreas–the uncinate-containing duodenal lobe and the tail-containing splenic lobe. The duodenal lobe contained about 4-fold more PP than the splenic lobe (55 versus 13 µg per gram of tissue) and, as expected, less glucagon (0.4 versus 4.3 µg/gm). The insulin content was similar for each section (51 versus 61 µg/gm). More recent studies in our laboratory in collaboration with Drs. Gingerich and Greider in St. Louis [see *Diabetes* 25, Suppl. 1; 329 (1976)] indicate that the uncinate process per se may have as much as 10 times the content of cPP as the tail region, further supporting the hypothesis that cPP is the F-cell product.

Although the physiologic function of pancreatic polypeptide is unknown, numerous studies in our laboratory during the past few years by Dr. Lin suggest that pancreatic polypeptide may have gastrointestinal function. A potent and consistent action of bPP when administered intravenously to dogs is its inhibition of pancreatic enzyme secretion induced by cholecystokinin plus secretin with infusion doses ranging from 1 to 5 µg/kg per hour. Pancreatic volume and bicarbonate are also inhibited under these conditions.

Finally, I would like to illustrate that pancreatic polypeptide is not just a figment of our imagination. For example, a convenient procedure for obtaining bPP is illustrated in Fig. L, which is a typical elution profile resulting from the gel filtration of an acid–alcohol extract fraction (pH 6.5 ether–alcohol precipitate) from approximately 500 gm of fresh bovine pancreas. As shown by polyacrylamide disc-gel electrophoresis (20% to 0.2% gels, pH 8.5), insulin elutes mainly in fractions H and I. BPP elutes immediately after the main insulin peak and is found in the more basic band in fraction J. Both glucagon and trypsin inhibitor are also present in fraction J and are indistinguishable from bPP by electrophoresis. bPP can be purified further by additional chromatographic procedure. Concerning the question of a bPP precursor, I would like to point out that we have recently examined the various

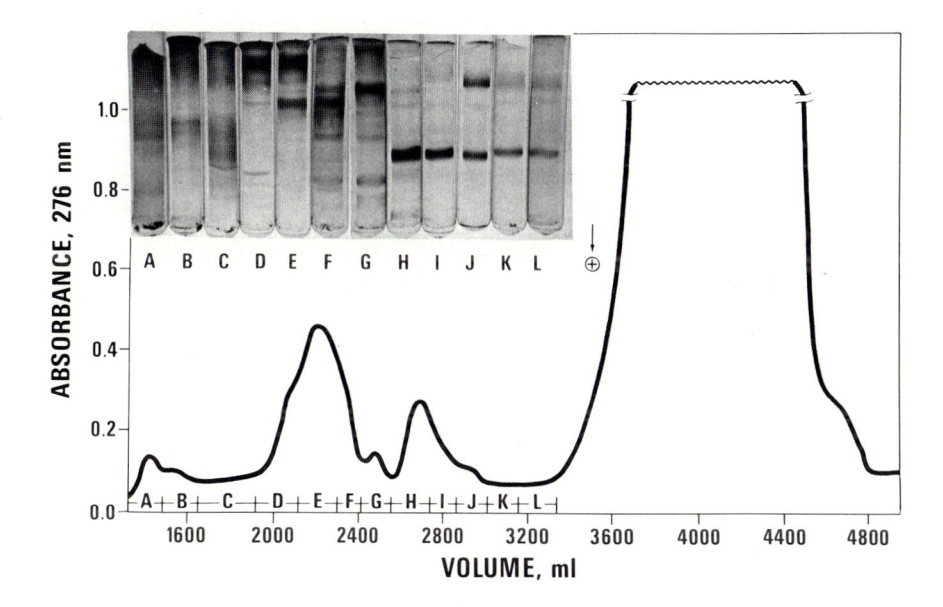

FIG. L. Sephadex G-50 (fine) elution profile of an ether–alcohol precipitate from an acid–ethanol extract of approximately 500 gm of bovine pancreas. The gel filtration was conducted on a 5 × 200 cm column in 1 M acetic acid at 4°C. The fractions were pooled as shown, lyophilized, and examined by polyacrylamide disc-gel electrophoresis (20% to 0.2% gels, pH 8.5).

fractions from the elution profile in Fig. L and have preliminary evidence that peak G contains a small, but significant amount of a bPP-like substance as detected by the bPP RIA, suggesting that a bPP precursor may exist, as would be expected from discussions earlier at this conference (see Habener *et al.*, this volume).

R. E. Frisch: I have a question about the chicken. Are there differences in effects of fasting in the aPP levels in the immature and mature female?

J. R. Kimmel: I do not think so. I think the fasting plasma levels are approximately the same. I would have to go back and look at the data carefully, but I do not believe there are any differences.

S. J. Marx: I would like to sound a brief cautionary note before we conclude that hPP may invariably be a marker for the MEN type 1 syndrome. We have been looking at a group of 26 patients from four kindreds with clear-cut MEN type 1. We determined, with the collaboration of several co-workers, the basal peripheral concentrations of a number of peptide hormones including glucagon, gastrin, parathyroid hormone, prolactin, calcitonin, and the α and β subunits of hCG. Dr. Floyd analyzed a group of these samples; in about 70% of the patients with previously established MEN type 1, there were elevated peripheral levels of hPP; this means that about 30% of the patients did not have elevated levels. This would lead us to suspect that, if this peptide is applied as a marker of the syndrome, it will have a significant false negative rate, probably above 30% in undiagnosed carriers of the gene.

J. C. Floyd: I thank Dr. Marx for emphasizing this point.

P. K. Donahoe: Is the variation that one sees in the peripheral cell islet production of pancreatic polypeptide related to the early embryonic division of the pancreas into the dorsal and ventral anlagen? It might be fruitful to study the fetal pancreas to see whether the adult distribution of pancreatic polypeptide cells correlates with their appearance in the embryonic ventral anlagen.

J. C. Floyd: Dr. Larson finds in human fetal pancreas many more PP cells than in the adult. I am not aware that a systematic study has been made looking for possible different concentrations of hPP in different portions of human pancreas. So, whether or not the findings in the dog with respect to the increased concentration of PP in the duodenal lobe apply in man, I do not know.

L. Orci: In the embryo, the pancreas is formed by the fusion of two distinct primordia, the so-called ventral and dorsal buds. The major part of the pancreas derives from the dorsal bud, whereas the ventral bud will give rise to the two lower thirds of the head. It is in the part derived from the ventral primordium that PP cells seem to be especially numerous. We are conducting at the present time experimental studies aiming specifically at clarifying this point.

J. C. Floyd: My point was that it has not been demonstrated in man that the distribution of the PP cells is as you have described it in the dog.

Steroid Hormone Actions in Tissue Culture Cells and Cell Hybrids—Their Relation to Human Malignancies

E. Brad Thompson,* Michael R. Norman[†]
and Marc E. Lippman[‡]

*Laboratory of Biochemistry, and [‡]Medicine Branch, National Cancer Institute
National Institutes of Health, Bethesda, Maryland

The search for understanding of the cell-inhibitory actions of steroids with glucocorticoid action has been a long and frustrating one, one that is still continuing. At these conferences, for example, reports have been made by many distinguished investigators on various aspects of the problem (Pearson and Eliel, 1951; Kaplan et al., 1954; Blecher and White, 1959; Glenn et al., 1963; Makman et al., 1967; Yamamoto et al., 1976). Despite this long period of work and the development of a number of theories attempting to explain how glucocorticoids act to inhibit vital cell processes, no generally accepted model has yet emerged (Makman et al., 1967, 1970, 1971; Munck, 1971; Mosher et al., 1971; Turnell and Burton, 1974, 1975; Behrens et al., 1974; Borthwick and Bell, 1975; Stevens and Stevens, 1975).

Two facts about steroids in general and glucocorticoids in particular are agreed upon, however. The first is that these hormones frequently act as inducers of new protein synthesis. Since the landmark observations of Lin and Knox (1957, 1958) showing increased hepatic tryptophan oxygenase and tyrosine aminotransferase activity in the rat liver after administration of corticosteroid, a wealth of studies have been carried out showing increased amounts of tissue-specific protein synthesis in many tissues (Pitot and Yatvin, 1973). Indirect experiments suggested that accumulation of specific mRNA accounted for the increased tyrosine aminotransferase and tryptophan oxygenase synthesis (Kenney, 1962; Granner et al., 1968; Greengard and Acs, 1962; Beck et al., 1972). In a few cases detailed studies have gone further by formally showing steroid-provoked increases in specific proteins to be due to accumulation of specific mRNA, measured by increased translatable mRNA (Chan et al., 1973; Feigelson et al.,

---

[†]Permanent address: Department of Chemical Pathology, King's College Hospital, Denmark Hill, London, England.

1975, 1976; Schimke *et al.*, 1975; Palmiter, 1975), or by annealing with message-complementary DNA (O'Malley and Means, 1974; Cox *et al.*, 1974; Schimke *et al.*, 1975; Martial *et al.*, 1976). The way in which these mRNAs accumulate in induced cells remains an area of active research.

The second general fact concerning steroids is that many, indeed nearly all, of their actions appear to require, as a first step, high-affinity binding to intracellular receptor proteins (Thompson and Lippman, 1974; King and Mainwaring, 1974; Yamamoto and Alberts, 1976). The quantitative as well as qualitative specificity of steroids for their receptors in target cells helps greatly in removing the confusion that existed hardly more than a decade ago over the apparently multiple sites and types of action for steroid hormones (for example, see review by Glenn *et al.*, 1963).

One strong line of evidence for the essential role of these receptors comes from studies showing that when lymphoid or fibroblastic cells, normally inhibited by glucocorticoids, are selected for resistance to such steroids, the resistant cells nearly always have quantitatively or qualitatively altered receptors (Rosenau *et al.*, 1972; Kaiser *et al.*, 1974; Lippman and Thompson, 1974; Yamamoto *et al.*, 1976; Thompson and Norman, 1976). In addition, several careful studies in rodent cell lines or fresh thymocytes have shown good correlation in sensitive cells between the specificity and capacity of various steroids' binding to their receptors, and the ability of those steroids to inhibit various cell processes (Makman *et al.*, 1968; Schaumburg and Bojesen, 1968; Hackney *et al.*, 1970; Baxter *et al.*, 1971; Munck and Wira, 1971; Munck *et al.*, 1972; Schaumberg, 1972).

We decided that because of the central role of the glucocorticoid receptor, measurements of such receptors in *human* leukemias might provide information of value. First, no data were available at that time that clearly demonstrated a role for glucocorticoid receptors in any action of glucocorticoids in humans. Second, the ability to predict which leukemic patients would respond to glucocorticoid treatment would be of obvious clinical value. In the bygone era of steroid-only treatment of various leukemias, it was found that some leukemias, such as acute lymphoblastic leukemia, frequently responded favorably—at least temporarily—to corticosteroid treatment, whereas others, such as acute myeloblastic leukemia, responded much less often (Pearson and Eliel, 1951; Fessas *et al.*, 1954; Ranney and Gellhorn, 1957; Granville *et al.*, 1958; Shanbrom and Miller, 1962; Roath *et al.*, 1964). Nowadays, with multidrug treatment, usually including steroids, a test that would allow one to avoid including an unnecessary and potentially dangerous hormone in the therapeutic combination would obviously be valuable. Besides the fact that high-dose steroids are capable of producing peptic ulcers, psychoses, poor wound healing, negative nitrogen balance, diabetes, susceptibility to infection, and so forth, it simply is not good pharmacology to treat with an ineffective drug. Third, the leukemias offer the scientific advantage of a relatively homogeneous preparation of the actual cells upon which the steroid acts. Thus comparison of interactions of hormone receptor with effects on cell processes can be made directly, yielding information about dosage effects and correlations between filling receptor sites and cellular responses. As we began these experiments, the measurement of estrogen receptors in breast cancer had already begun to be established as a very useful aid in distinguishing estrogen-sensitive from estrogen-insensitive tumors and since then, this fact has been widely confirmed (Jensen et al., 1971a; McGuire et al., 1975). The encouraging results in breast cancer further spurred us to examine glucocorticoids and the leukemias.

Consequently, we obtained from the National Institutes of Health leukemia service, leukophoresis specimens of cells from patients with several types of leukemia and examined these cells for (1) the presence of glucocorticoid receptors, (2) the response of the cells *in vitro* to glucocorticoids, and (3) the clinical response of the patients to the combined drug therapy then being used at NIH, therapy that included high-dose glucocorticoids (Lippman et al., 1973a,b).

The first group of patients were those with juvenile onset acute lymphoblastic leukemia, diagnosed by standard morphologic criteria. No patient had received glucocorticoid therapy within 10 days of obtaining cells for study. Our first task was to verify an assay method in this species and tissue. We chose to use the competitive binding assay as employed by Baxter and Tomkins (1971) and Rousseau *et al.* (1972). In this method, cell extracts are incubated with radioactive steroid in the presence or absence of excess "competing" nonradioactive steroid. After incubating a suitable length of time, unbound labeled steroid is removed by a brief exposure to charcoal, and specifically bound steroid is determined by the difference between "competed" and "uncompeted" samples. Figure 1 shows that, over the concentration range of protein in the extracts examined, there was a linear relation between specifically bound steroid and protein concentration.

Since the accepted minimal criteria for steroid receptors are saturability, limited capacity, and specificity for steroids with appropriate activity, we next sought to define these properties in extracts from these human leukemic lymphoblasts. Table I shows a study of steroid specificity. To extracts containing $3.8 \times 10^{-7} M$ [^{3}H] dexamethasone (formula: 1,4-pregnadiene-9-fluoro-16α-methyl-11β,17α,21-triol-3,20-dione), a 100-fold higher concentration of a number of nonradiolabeled steroids was added. As Table I shows, some of these steroids successfully competed for the binding of the labeled glucocorticoid; others did not. For comparison, data from the literature on the ability of these steroids to function as inhibitory hormones in rodent lymphoid systems or as inducers of tyrosine aminotransferase in HTC (rat hepatoma) cells are shown as well. As can be seen, those steroids with biologic activity as glucocorticoids or

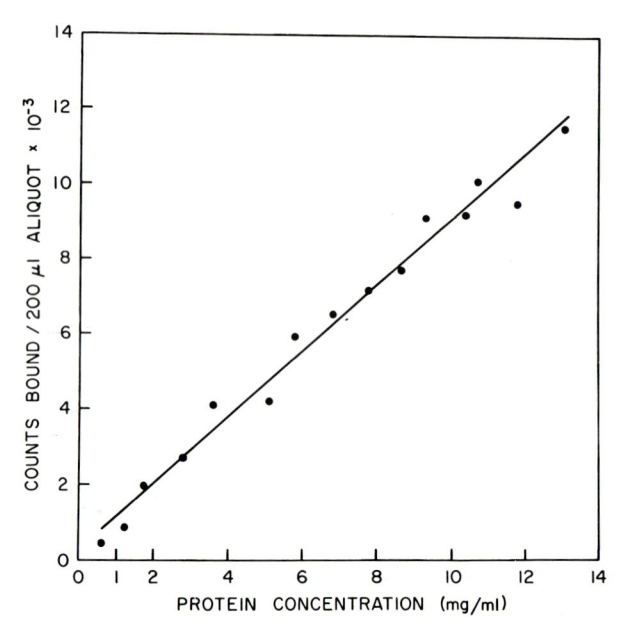

FIG. 1. Relation of specific binding at 4°C to cytoplasmic extract concentration. Specifically bound [^{3}H]dexamethasone was determined using a steroid concentration of 3.4 × 10^{-7} M and dilutions of an acute lymphoblastic leukemia blast cell cytoplasmic extract. From Lippman *et al.* (1973b). Reprinted with permission.

antiglucocorticoids were those that best prevented the binding of dexamethasone in the human cell extracts. The criterion of specificity seemed to be met.

As for capacity and saturability, Fig. 2 shows a typical study on a cell extract from one patient's cells in which increasing concentrations of [^{3}H] cortisol were added in the competitive binding assay. The data show that the "competable" counts, representing specifically bound steroid, approached saturation at about 2 × 10^{-6} M cortisol. The data shown by the open circles and dashed line indicate the nonspecifically bound, "uncompetable" cortisol, the background in the assay. In the inset is the Scatchard plot (Scatchard, 1949) of the specific binding data. Its slope allows computation of an apparent dissociation constant, K_d, of 1.3 × 10^{-8} M; the straight line suggests a single class of binding sites; and from the X intercept the approximate number of sites per cell, 1.5 × 10^4, can be derived. [Note: Owing to a misprint not corrected in proof, the number of sites per cell is given as 1.5 × 10^5 in Lippman *et al.* (1973b).]

From these data it seems that human acute leukemic lymphoblasts possess receptors for native and synthetic glucocorticoids. Few of these receptors, however, would be occupied by physiologic concentrations of the hormone. In

TABLE I
Comparison of Binding Affinity for Lymphoblast Binding Protein with Biologic Activity of the Steroids in Inhibitory (Lymphoid) and Stimulatory (Hepatoma) Systems[a]

Competing steroid	% [^{3}H]Dexamethasone[b] bound in presence of 100-fold excess of competing steroid	Biologic activity of competing steroid in lymphoid systems[c]	Biologic activity in hepatoma tissue culture cells[c]
Dexamethasone	0	Optimal inducer	Optimal inducer Suboptimal inducer[d]
Progesterone	0	Anti-inducer[d]	Anti-inducer
5β-Dihydrocortisol	0	Unknown	Unknown
Hydrocortisone	0	Optimal inducer	Optimal inducer
Triamicinolone	6	Optimal inducer	Optimal inducer[e]
Aldosterone	12	Unknown	Optimal inducer
Prednisolone	22	Optimal inducer[e]	Optimal inducer[e]
Deoxycorticosterone	24	Suboptimal inducer	Optimal inducer
Corticosterone	24	Optimal inducer	Optimal inducer
Cortexolone	31	Anti-inducer	Unknown
2α-Hydroxycortisol	40	Inactive	Unknown
17α-Methyltestosterone	40	Unknown	Anti-inducer
Spironolactone	57	Unknown	Unknown
Prednisone	69	Inactive[f] Inactive[d]	Unknown
Testosterone	74	Anti-inducer	Anti-inducer
Cortisone	84	Inactive[f]	Anti-inducer
19-Nortestosterone	84	Unknown	Unknown
17β-Estradiol	94	Inactive[f]	Anti-inducer
Etiocholanolone	100	Unknown	Unknown
Tetrahydrocortisol	100	Inactive	Inactive
Androstenedione	100	Inactive	Inactive

[a]Table reproduced, with permission, from Lippman *et al.* (1973b). Detailed references are given therein.
[b]Dexamethasone concentration was 3.8 × 10^{-7} M.
[c]The classification of biologic activity is that of Samuels and Tomkins, 1970.
[d]Conflicting results published.
[e]Unpublished results.
[f]Because of the method used in assaying biologic effect, the possibility that these steroids are anti-inducers cannot be excluded.

the body it seems very probable that the free, not the total (free plus protein-bound) steroid, is available to the cellular receptors (Yates *et al.*, 1974). Free circulating cortisol has been estimated to range between 7.6 × 10^{-9} M and 5 × 10^{-8} M. Stimulation with ACTH could raise the level to about 1.5 × 10^{-7} M (Baumann *et al.*, 1975). Since certain synthetic glucocorticoids do not bind to transcortin, at pharmacologic concentrations they might be expected to reach

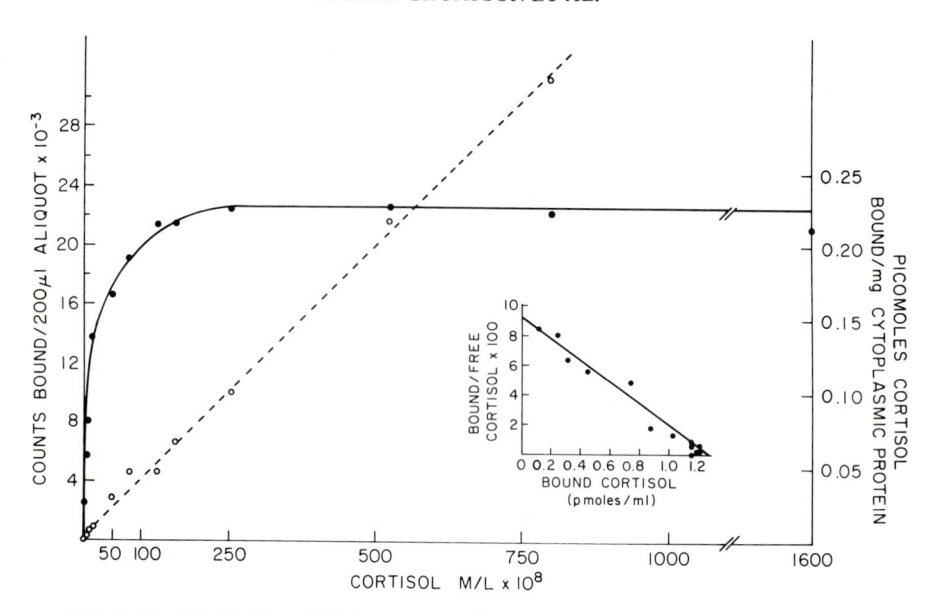

FIG. 2. Specific binding of [^{3}H] cortisol at 4°C to cytoplasmic extract from lymphoblasts from a patient with acute lymphoblastic leukemia (●———●). The free steroid concentration is determined by subtracting the bound steroid from the initial amount added to the incubation. Nonspecifically bound noncompetible cortisol (○– – –○), is the background. The inset shows a Scatchard plot of the specific binding curve. From Lippman *et al.* (1973b). Reprinted with permission.

circulating concentrations adequate to saturate the receptors. Besides, in animal cells it has been demonstrated that several synthetic steroids have higher affinity for glucocorticoid receptors than do the natural hormones. This we found to be the case here also. A saturation study such as that just shown using [^{3}H] dexamethasone, not hydrocortisone, yielded a dissociation constant indicating 10-fold higher affinity. Saturation of receptors occurred at about $2.5 \times 10^{-7} M$ (for example, see Fig. 3A). Parenterally administered, 5 mg of dexamethasone could produce plasma concentrations of $5 \times 10^{-7} M$ (Chard *et al.*, 1966; Claman, 1972), twice that needed to saturate the receptors.

It remained for us to see whether the direct cellular effects of the hormone would correlate with receptor occupancy. This indeed seemed to be the case. As the study proceeded, we found that the cells from most patients contained receptors, but cells of a few patients did not. Figure 3 shows a pair of such patients. In Fig. 3A, the line drawn through the filled circles shows the saturation curve for [^{3}H] dexamethasone specifically bound in the cytoplasmic fraction of one patient's lymphoblasts. By way of contrast, Fig. 3B shows similar data for a patient whose cells yielded an extract with little or no specific binding. The data represented by the triangles in this figure show one measure of

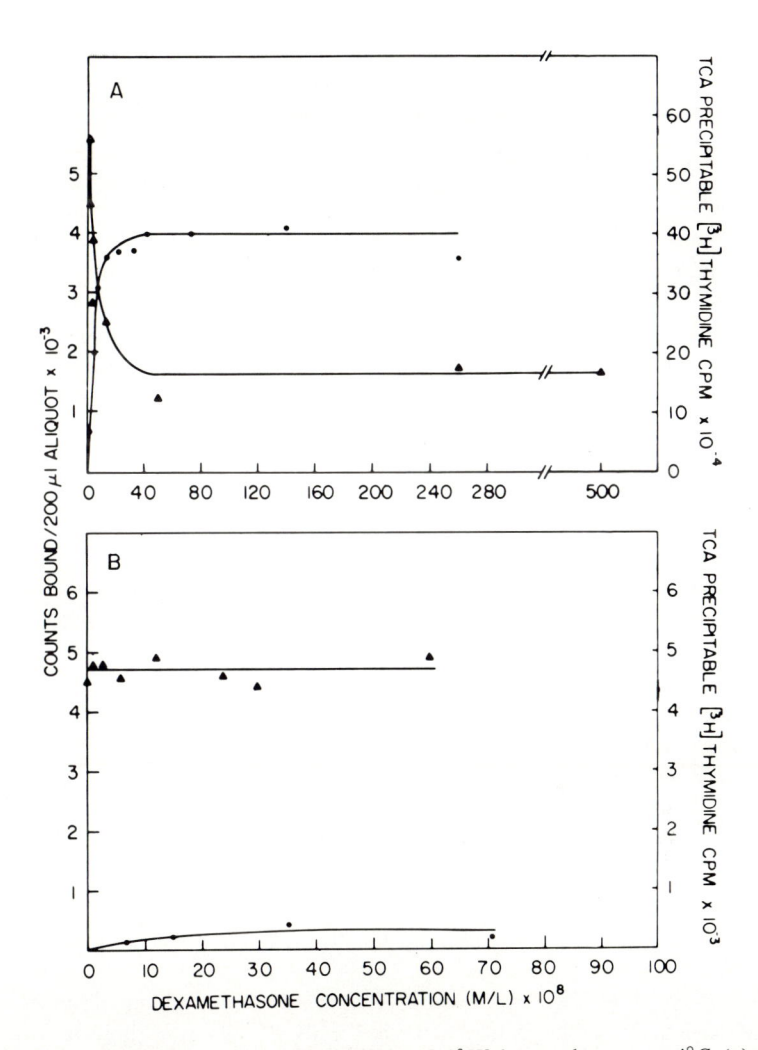

FIG. 3. (A) Comparison of specific binding of [^{3}H]dexamethasone at 4°C (●) with inhibition of [^{3}H]thymidine incorporation at 37°C (▲) as a function of dexamethasone concentration. These data are for lymphoblasts from a patient with untreated acute lymphoblastic leukemia. (B) Similar comparison to that of (A) except that in this case the lymphoblasts were from a patient with acute lymphoblastic leukemia, who clinically was resistant to glucocorticoids. (●) Specific binding to steroid-binding protein; (▲) [^{3}H]thymidine incorporation. From Lippman et al. (1973b). Reprinted with permission.

biological effect on these cells in the same two patients. Whole cells were incubated at 37°C in tissue culture medium and exposed for 18 hours to varying concentrations of the steroid. Then [^{3}H] thymidine was added to the medium; 2 hours later the cells were collected and washed, and the radioactivity precipitable by cold 5% trichloroacetic acid was determined. As the figure shows, the cells that contained receptor demonstrated striking inhibition of nucleoside uptake into precipitable material, presumably DNA, whereas the receptorless cells did not. In addition, there was good agreement between the concentration of hormone which saturated the receptor sites in the sensitive cells and that which inhibited thymidine uptake. The agreement seems especially good, considering that the receptor assay was done on cytosol extracts at 0°C whereas the thymidine uptake was done on whole cells at 37°C. All four patients studied with little or no demonstrable receptors in their lymphoblasts showed resistance to dexamethasone inhibitory effects *in vitro*, similar to that shown in Fig. 3B.

Retrospective evaluations of clinical responses were carried out on a total of 36 patients with acute lymphoblastic leukemia, and these were correlated with quantitative studies on their lymphoblasts' glucocorticoid receptor content. Figure 4 shows a summary of these data. The patients fell into three classes. In all 22 untreated patients, readily demonstrable receptors were found, and remissions were obtained in all with the combined glucocorticoid-containing chemotherapy. Six patients in relapse were found still to possess receptors, and in all these a further remission was induced. In 6 additional patients in relapse after repeated drug-induced remissions, virtually no receptor (specific binding) was demonstrable, and these were not responsive to the steroid-containing drug combination. One patient's cells were obtained for study at several times in the course of the disease. We found that on first admission this patient had responded well to the glucocorticoid-containing therapy and that there had been a second remission to such therapy when the patient reappeared in relapse. Cells obtained at both occasions had receptors. Later, however, the patient appeared in a relapse to which no response to the glucocorticoid-containing therapy was obtained. Cells obtained just prior to the therapy that failed no longer showed receptors.

Konior *et al.* (1976) have further characterized the lymphoblasts of acute lymphoblastic leukemia as T cells or null cells. T cells were those that formed EAC rosettes, while null cells formed neither EAC nor E rosettes and lacked receptors for the Fc portion of the immunoglobulin molecule. Of 46 patients so studied, about 70% possessed null cells, and the other 30% T cells, as their circulating blasts. The null cells were found, on the average, to contain about 7500 steroid receptor sites per cell, whereas the T cells contained an average of about 2500 sites per cell. These values were significantly different ($P < 0.001$) from one another. Although the reason is not known, it is an interesting clinical fact that null-cell leukemias respond much more favorably to therapy than do

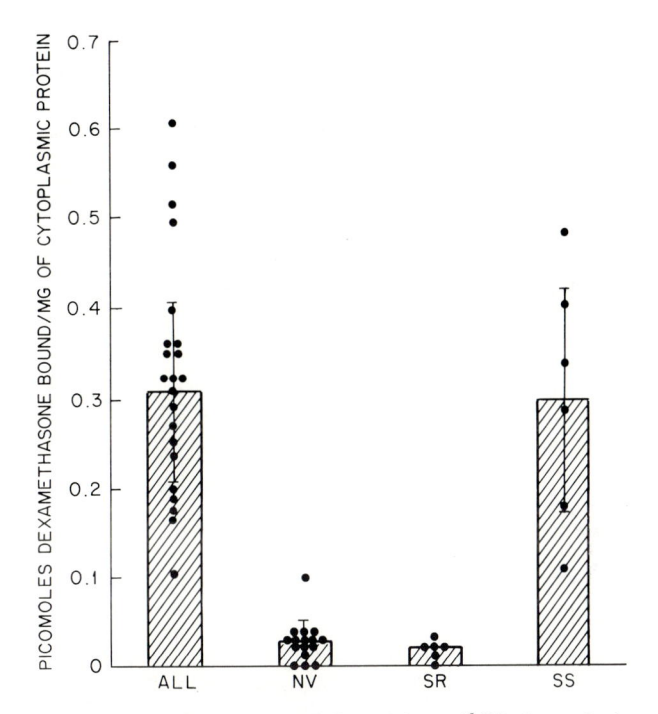

FIG. 4. The total steroid-binding protein (SBP) activity at 4°C is shown for lymphoblasts from 22 untreated patients with acute lymphoblastic leukemia (ALL) ± 1 SD; lymphocytes from 16 normal volunteers (NV) ± 1 SD; and lymphoblasts from 6 steroid-unresponsive patients (SR); and from 6 patients who had received some steroid therapy but were still steroid responsive (SS) ± 1 SD. Steroid concentrations used were greatly in excess of those needed to saturate the SBP. Some of these patients have been previously presented. From Lippman *et al.* (1973b). Reprinted with permission.

T-cell leukemias. Whether or how these differences observed in receptor number are related to the differences in clinical course is not at all clear.

We also had the opportunity to study cells from 16 patients with acute myeloblastic leukemia (Lippman *et al.*, 1975). Only one of the 16 displayed high-affinity, low-capacity sites, which saturated at about 2×10^{-7} M dexamethasone. In this patient's cells, close correlation was found between the saturation of receptors and inhibition of thymidine uptake and incorporation (Fig. 5), as had been the case with the sensitive cells from lymphoblastic leukemia. Here also, there was specificity in competition studies for glucocorticoids (Table II). Two other patients with myeloblastic leukemia supplied cells with some "competable" binding, but only at concentrations of steroid high above those likely to be achieved even by high-dose steroid therapy. The cells of the remaining 13 cases lacked receptors.

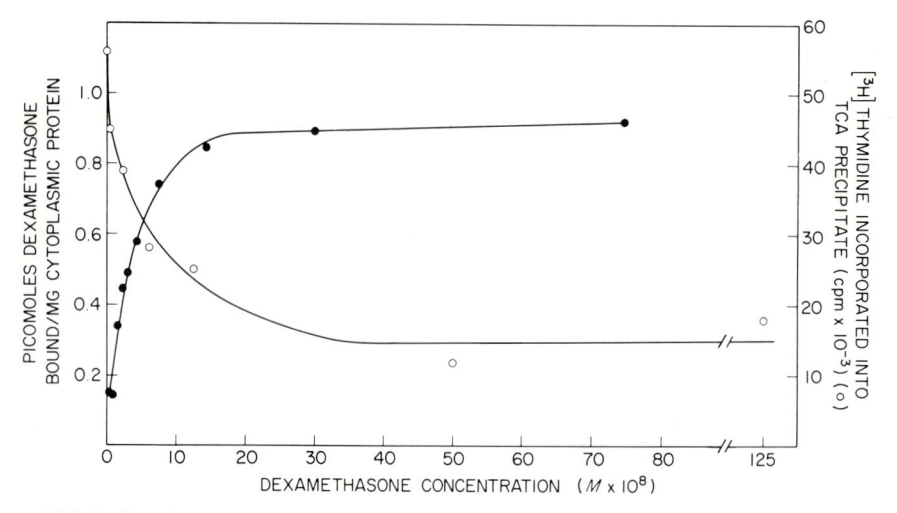

FIG. 5. Correlation between saturation of glucocorticoid receptors and inhibition of DNA synthesis in receptor-rich leukemic blasts from a patient with acute myelogenous leukemia. Binding [^{3}H]dexamethasone to specific cytosol sites (●), and [^{3}H]thymidine incorporation into trichloroacetic acid (TCA)-precipitable material (○). From Lippman *et al.* (1975). Reprinted with permission.

TABLE II
Specificity of the Steroid-Binding Protein from Myeloblasts of a Patient with Acute Myeloblastic Leukemia[a]

Steroid	Inhibition of [^{3}H] dexamethasone binding (%)
Dexamethasone	100[b]
Prednisolone	100
Cortisol	98
Aldosterone	95
Triamcinolone	95
Deoxycorticosterone	72
Corticosterone	65
19-Nortestosterone	54
Prednisone	48
17 β-Estradiol	39
17 α-Methyltestosterone	28
Etiocholanolone	13
Tetrahydrocortisol	3

[a] Reproduced from Lippman *et al.* (1975). Reprinted with permission.
[b] A 100-fold excess of various unlabeled steroids was added in the standard assay system and their ability to compete with radiolabeled dexamethasone was quantified. Dexamethasone inhibition of its own binding is arbitrarily set at 100%.

Although we were unable to carry out retrospective clinical analysis on these patients as we had done on the acute lymphoblastic leukemic patients, it is instructive to note that in the era of treatment in which steroids alone were tested for therapeutic effect in the leukemias, patients with myeloblastic leukemia only occasionally showed favorable responses (Ranney and Gellhorn, 1957; Roath et al., 1964). Our receptor and *in vitro* studies are certainly consistent with the view that steroid receptors in leukemic cells are necessary for (adequate) therapeutic response to glucocorticoids. We should note that Gailani et al. (1973) have similar data. In their studies, cells from 3 of 3 patients with acute lymphoblastic leukemia were receptor-positive while in only 2 of 6 patients with acute myeloblastic leukemia could cells with glucocorticoid receptors be demonstrated.

We conclude from the evidence in these studies that glucocorticoid-responsive human leukemic lymphoblasts possess specific receptor sites for glucocorticoids. Glucocorticoid-resistant blasts often show greatly reduced steroid-binding activity. The occupation of these steroid-binding sites by active glucocorticoids correlates with the pharmacologic actions of the hormones, and these actions are exerted directly on the leukemic blast itself. There is agreement between the affinity of the binding protein for various steroids and their biological potency (Thompson et al., 1976). Although these studies were retrospective, in acute lymphoblastic leukemia there appeared to be good correlation also with the presence of the cellular receptors and the clinical response of the patients to therapy that included glucocorticoids. Therefore the determination of steroid-specific binding capacity may provide a rapid *in vitro* test for predicting steroid responsiveness in human leukemic blasts. Nevertheless, we anticipate that occasionally, there will occur receptor-positive glucocorticoid-resistant cases, for a variety of reasons, both theoretical and empirical.

Theoretically, the glucocorticoid receptor is only one step in a chain of events between the entry of steroid into the cell and the ultimate expression of specific genes. Thus its presence should be necessary, but not always sufficient for the steroid to act. A later step might be blocked. One empirical reason is that we have identified in tissue culture, cells with these properties. Two human leukemic cell lines, obtained from the American Type Culture Collection and the ARK-A mouse leukemic cell line were found to possess steroid receptor sites that were normal by a number of criteria; yet the cells were entirely resistant to inhibition by dexamethasone (Lippman et al., 1974). Table III summarizes some of the properties of these cells. Although some test not yet applied may yet show the receptors of these cells to be abnormal, the point of these studies is that by several standard, accepted criteria they appear normal. In addition, we have isolated a few lines of HTC cells that appear to have entirely normal receptor activity but are induced by glucocorticoids only slightly or not at all (Thompson et al., 1976). These cells possess the corticosteroid-inducible en-

TABLE III

Properties of Three Lines of Glucocorticoid-Resistant Leukemic Cells That Apparently Contain Glucocorticoid Receptors[a]

Cell responses to glucocorticoid	
Inhibition of glucose uptake	No
Inhibition of uridine incorporation	No
Inhibition of thymidine incorporation	No
Steroid-Binding Proteins	
Present	Yes
Concentration/saturation	Normal range
Nuclear binding	Yes
Steroid specificity	Yes

[a]Conclusions based on data of Lippman *et al.* (1974).

zyme, tyrosine aminotransferase. The "noninducible" clones still contain the enzyme at basal levels; i.e., the gene is there and is functioning, but the induction response is lacking. In these cases, we speculate that the difficulty lies in some late step of the response to the steroids, a step subsequent to the binding of steroid to receptors in the cytoplasm and movement to, followed by binding in, the nucleus. These instances seem to be the exception, rather than the rule, but they may prove to be valuable for exploring the later steps of steroid action. In the majority of cases, however, it would seem that the presence of glucocorticoid receptor implies sensitivity to the steroid, and certainly absence of or altered receptors (i.e., lowered affinity for steroid) strongly imply lack of steroid sensitivity.

Perhaps more odd is the frequency with which lowered or altered receptors are found, *in vivo* and *in vitro,* as the apparent mechanism for steroid resistance. We would further like to know the origin of the resistant, receptorless cell found in the repeatedly treated patient with acute lymphoblastic leukemia or in many patients with acute myeloblastic (and other) leukemias even prior to treatment. Feeling that questions such as this might be approached in a tissue culture model system, we sought a human lymphoid cell line that was sensitive to the lympholytic effect of glucocorticoids. We have found such a line and have begun its characterization (Thompson and Norman, 1976). This we believe to be the first demonstration of glucocorticoid sensitivity in a human lymphoblastic cell line, and we are excited about its potential usefulness. Known as CEM cells, these cells were originally isolated in 1964 from a patient with acute lymphoblastic leukemia by Foley and co-workers (1965). Because they possessed human thymic lymphocyte-specific antigens, they were classified by Smith *et al.* (1974) as thymus derived. We obtained our original stock of sensitive uncloned CEM cells from the laboratory of Dr. Dean Mann, NC1. Supplies of CEM cells from

American Type Culture Collection, as we have reported previously, were found to be steroid resistant (Lippman *et al.*, 1974). We presume that this apparent contradiction comes merely from diverging cell populations during long periods in culture.

Figure 6 shows the growth curve of the sensitive CEM cells in the presence or absence of 10^{-6} M dexamethasone. As the data show, there is a marked, but not complete, inhibition of growth of the uncloned cells. We suspected that this might be due to a mixed population of sensitive and resistant cells, accumulated in this long-term culture cell line. Accordingly, the original line was cloned in the absence of added steroids, with a cloning efficiency of 40%. Seventeen clones were isolated and tested for sensitivity to steroid. As predicted, we found that the majority of the clones were sensitive to the inhibitory action of dexamethasone, while a few were not.

Figure 7 shows the growth curves of two clones. The upper two curves show the growth, with or without dexamethasone, of the resistant clone CEM C1, and

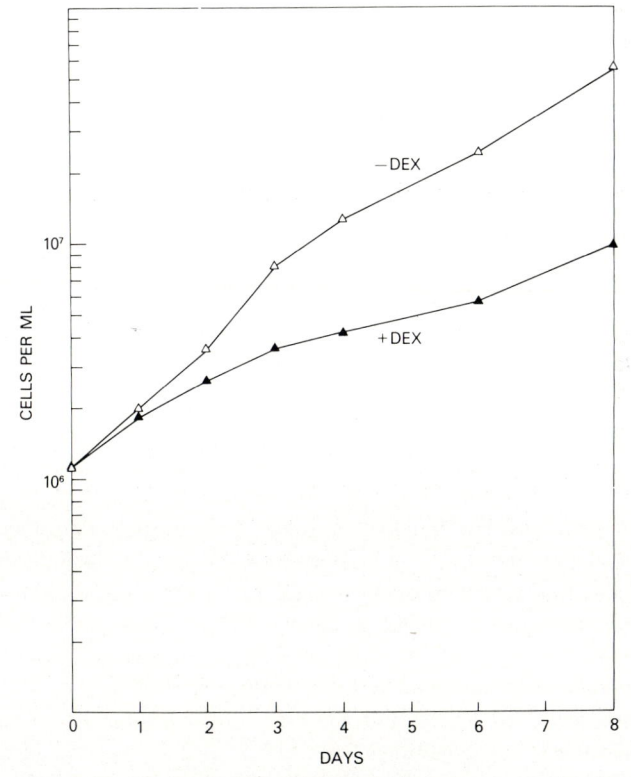

FIG. 6. Growth of human leukemic CEM cells in the presence or the absence of 10^{-6} M dexamethasone (DEX). From Norman and Thompson (1976).

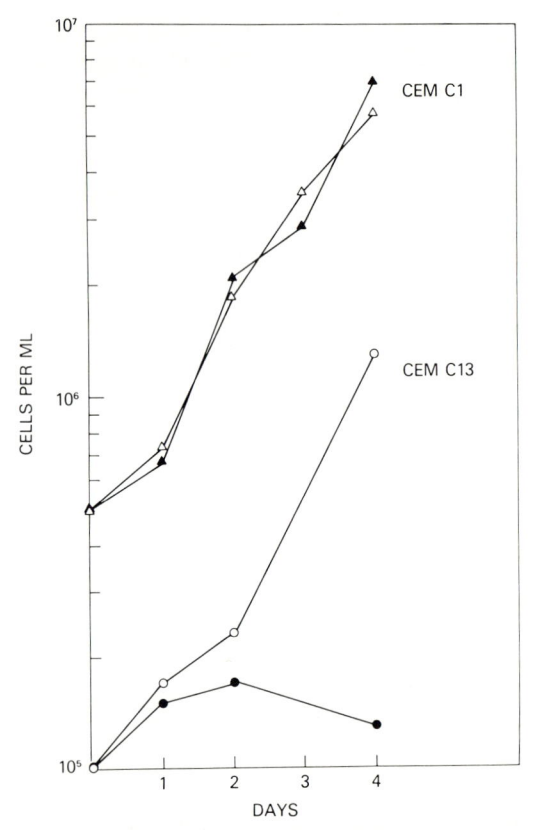

FIG. 7. Growth of a resistant (CEM C1) and a sensitive (CEM C13) clone of human leukemic cells in 10^{-6} M dexamethasone. ○, △: Without steroid; ●, ▲: with steroid. From Norman and Thompson (1976).

the lower two curves show the sensitive clone CEM C13, whose growth is completely inhibited by the steroid. The dramatic effect of the hormone on C13 cells is best depicted by Fig. 8, which shows a phase-contrast photomicrograph of control cells compared with cells treated for 4 days with dexamethasone. This photo illustrates the late phase of the steroidal destruction of these cells. Note that one cell in the treated population appears to have been spared. Presumably from cells such as this, resistant populations arise. Sequential Coulter counter scans of a dexamethasone-treated sensitive culture showed a progressive reduction in cell size at 6, 12, 24, 36, and 48 hours after addition of steroid. By 48 hours, this had reached a reduction of nearly 50% in cell volume, relative to the controls. At the later times, a subfraction corresponding to very small particles

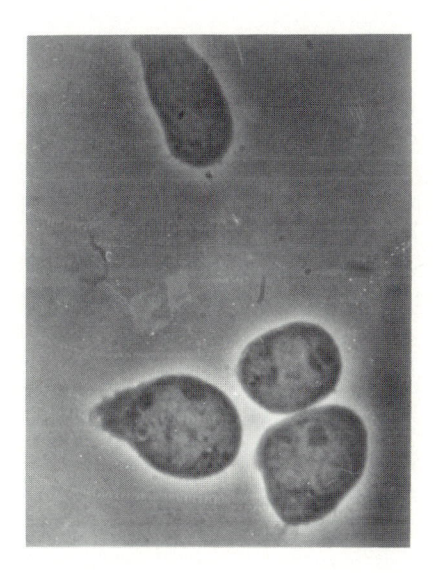

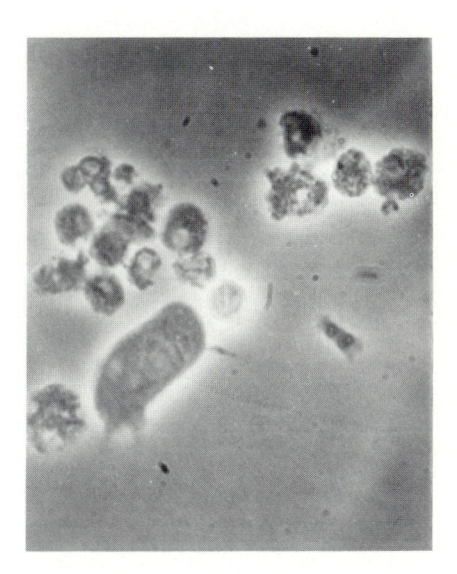

FIG. 8. Appearance of steroid-sensitive CEM cells after 4 days' exposure to 10^{-6} M dexamethasone. Phase-contrast photomicrograph. Left, control cells; right, steroid-treated cells. From Norman and Thompson (1976).

appeared, presumably reflecting the accumulation of microcells and cell debris such as that shown in the photomicrograph.

We next sought to define the kinetics and dose-relationships of these effects. The dose-response of the overall growth-inhibitory effect is shown in Fig. 9. Measurement of DNA showed an inhibition of growth with as little as 5×10^{-8}

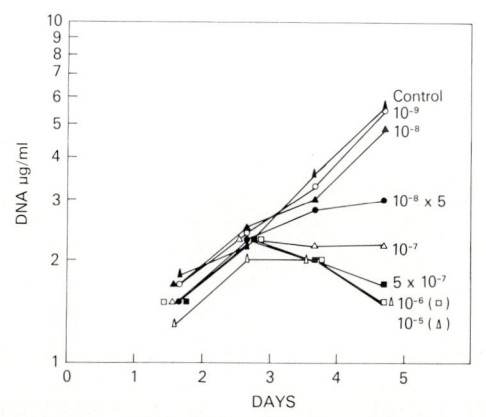

FIG. 9. Growth of sensitive CEM cells in various concentrations of dexamethasone. From Norman and Thompson (1976).

M dexamethasone, although maximal inhibition required 10 times higher concentration.

Flow microfluorometry was used to determine whether cell growth was blocked in a specific phase of the cell cycle. In this technique, mithramycin, which binds to DNA, is added to randomly growing cells. The fluorescence of bound mithramycin, which is proportional to DNA content, is measured in single cells as they flow past a laser beam. Thus cells in G_1, which have a diploid DNA content, can easily be distinguished from those which have replicated their DNA but not yet divided. Figure 10 shows cells of sensitive clone C13 accumulated in G_1 after exposure to micromolar dexamethasone. Control cells growing logarithmically have constant proportions of their populations in each phase of the cell cycle. After 12 hours of exposure to steroid, cells progressively accumulate in G_1 and there is a corresponding depletion of cells in S. These data indicate that the process by which steroids inhibit growth and ultimately cause the dramatic cellular effects already shown involves a block in the growth cycle, preventing cells from replicating DNA. Dose-response studies of this effect on the cell cycle show a marked similarity to those on total growth. Figure 11 shows such an experiment. The data are shown as the percentage of cells in G_1 vs S after 24 hours of treatment with increasing concentrations of dexamethasone. As was the case with overall growth, an effect

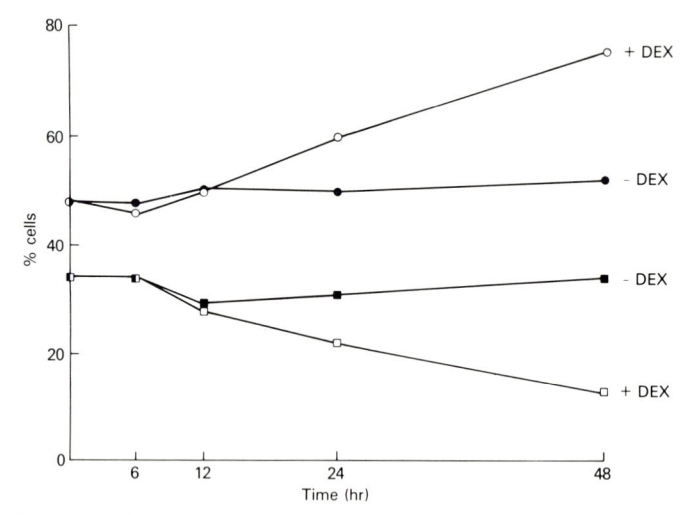

FIG. 10. Flow microfluorometry of CEM cells. A culture of steroid-sensitive cells was divided in half. One half was exposed to $10^{-6} M$ dexamethasone, and the other half served as control. At various times, samples were withdrawn, exposed to mithramycin and examined for stage of growth cycle by flow microfluorometry. Upper two curves (circles), cells in G1. Lower two curves (squares), cells in S. ○, □: Cells treated with dexamethasone (+ DEX); ●, ■, control cells (−DEX). From Norman and Thompson (1976).

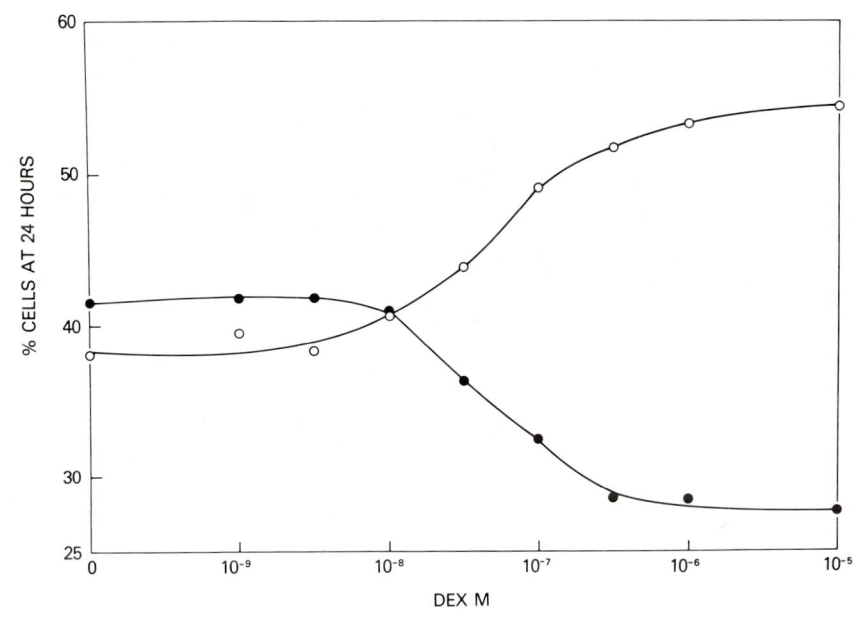

FIG. 11. Dose-response of CEM cells exposed to increasing doses of dexamethasone and examined for position in growth cycle 24 hours later by flow microfluorometry, as in Fig. 10. Abscissa, molarity of dexamethasone (DEX M). ○, Cells in G_1 phase; ●, cells in S phase. From Norman and Thompson (1976).

is first seen with something over $10^{-8}\,M$ steroid, and the maximum effect occurs at a little above $10^{-7}\,M$.

The processes of RNA, DNA, and protein synthesis were next investigated. Figure 12 shows the polysome patterns on sucrose gradients of CEM cells at 0, 24, and 48 hours after addition of $10^{-6}\,M$ dexamethasone. Immediately apparent is the progressive loss of polysomal material, with a corresponding increase in monomers. These changes would be compatible with either defective initiation or decreased availability of mRNA. Inhibition of thymidine incorporation was equally dramatic. The kinetics of [^{3}H] thymidine incorporation into trichloroacetic acid-precipitable material by a steroid-sensitive clone of CEM cells at two cell concentrations are shown in Fig. 13. It is clear from the data there that the inhibition of thymidine uptake by steroids is highly dependent upon culture conditions. In the more dilute cells, there was a rapid inhibition of uptake, which progressed to total inhibition by 5–6 hours. Control thymidine uptake was much lower and ceased sooner in the heavier culture, and there was no inhibition of this lowered amount of thymidine uptake. We are presently investigating the reason for this difference, since it could well be pertinent to the *in vivo* therapeutic response,

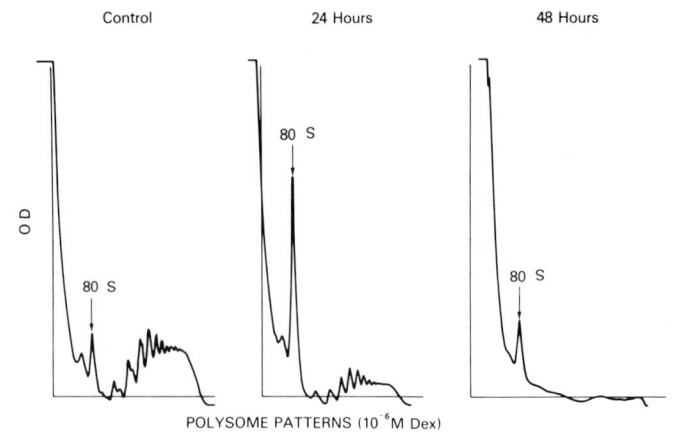

FIG. 12. Polysome patterns of CEM cells on 10 to 50% linear sucrose gradients. Left-hand panel shows polysomes from control cells compared with those from cells 24 hours (middle panel) and 48 hours (right-hand panel) after addition of 10^{-6} M dexamethasone to the cultures. Ordinate (OD), absorbance at 280 nm. From Norman and Thompson (1976).

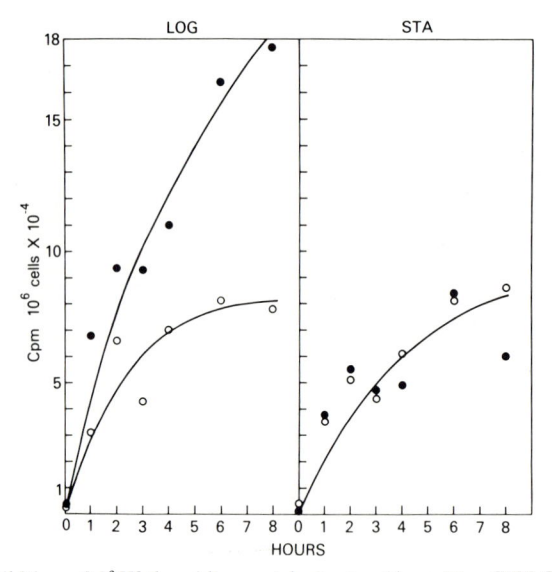

FIG. 13. Inhibition of [^{3}H] thymidine uptake in steroid sensitive CEM C13 cells. A dense culture of cells was divided and fed fresh growth medium at a ratio of 3 ml of cells to 1 ml of medium (STA) or 1 ml of cells to 2 ml of medium (LOG). One day later, each culture was divided, and to one half of each 10^{-6} M dexamethasone was added. Cell counts were 4.6 × 10^6 cells/ml for STA controls (●, right-hand panel), 4.3 × 10^6 cells/ml for STA, steroid-treated (○, right-hand panel), 1.8 × 10^6 cells/ml for LOG control (●, left-hand panel), and 2.8 × 10^6 cells/ml for LOG, steroid-treated (○, left-hand panel). Simultaneously added to each flask was 1 μCi of [^{3}H] thymidine (New England Nuclear, 20 Ci/mmole) per milliliter. Thereafter at the intervals shown, 0.1-ml aliquots in triplicate were removed from each flask, pipetted onto filter papers, and soaked in 10% trichloroacetic acid. After washing extensively in cold 5% TCA, the papers were dried in ethanol, ether, and air, and then were assayed for radioactivity by liquid scintillation counting.

or lack of it, to glucocorticoids. Possible factors which might be relevant include cell density, nutritional state, growth phase, and metabolism of the steroid.

The relation of these pheonomena to steroid receptors of these cells was next investigated. Whole-cell steroid binding at 37°C was determined by a slight modification of the competitive binding assay of Sibley and Tomkins (1974). The data in Fig. 14 show an increase in specifically bound steroid with dexamethasone concentration for cells of two sensitive clones, but there was no consistent specific binding in the steroid-resistant clone. The inset shows the Scatchard plots of the data for the sensitive clones. Each set of data falls on a single straight line, indicating a single class of binding sites, with about 20,000 sites per cell. Note that the apparent dissociation constants (calculated by least squares fit), seem to correspond well with the dose-response curves for steroid effect, shown previously. Receptor saturation and pharmacologic effect therefore seem to go together.

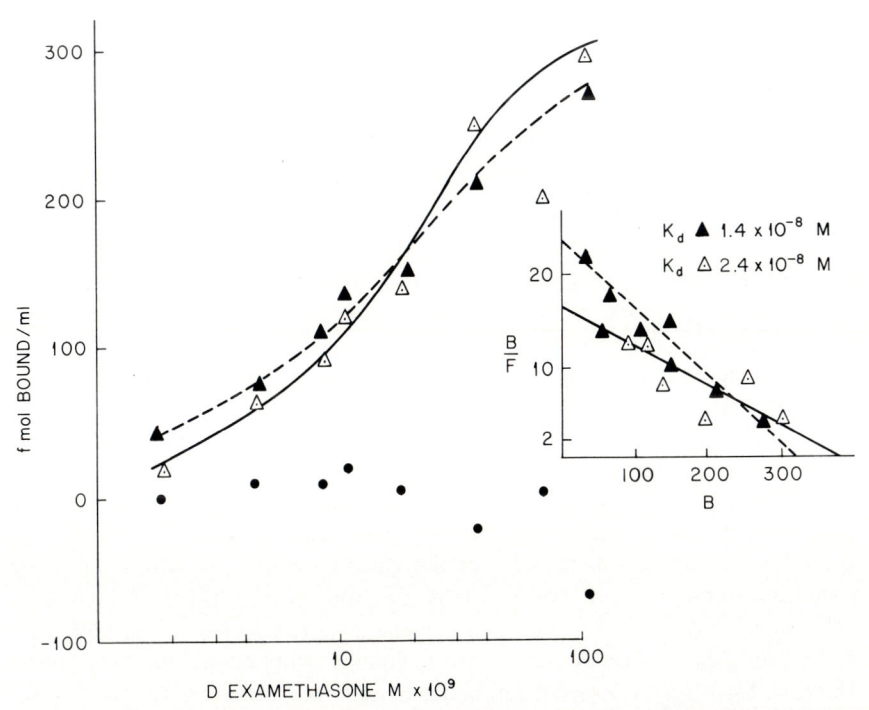

FIG. 14. Specific glucocorticoid binding in CEM cells. Sensitive clones CEM C7 (▲) and CEM C13 (△) and resistant clone CEM C1 (●) were assayed for specific glucocorticoid binding sites for [^{3}H]dexamethasone by the whole-cell binding method of Sibley and Tomkins (1974). Plotted on the ordinate is specifically bound dexamethasone; and on the abscissa, concentration of [^{3}H]dexamethasone. Inset: Scatchard plots and calculated apparent dissociation constants (K_d) for the two receptor-containing sensitive clones. From Norman and Thompson (1976).

In sum, we have described the first characterization of a human leukemic cell culture line which is killed by steroids and have defined some of the basic parameters of this system. The human cell line has been shown to display many of the classic responses to corticosteroids observed in rat thymocytes, rodent lymphoid tumors, and rodent cell lines. In it, as with rodent cells, there seems to be a correlation between saturation of steroid receptors and steroid effect.

One question about standard multidrug therapy for leukemia is raised by our preliminary data. In such therapy, remissions are induced by the administration, virtually simultaneously, of cycle-active drugs, such as vincristine, methotrexate, daunorubicin, and 6-mercaptopurine, and high-dose corticosteroids. Is it appropriate to treat simultaneously with glucocorticoids and cycle-specific drugs, when the population of cells upon which the latter act is diminished by treatment with glucocorticoids? We have suggested previously that it might make more rational therapy to separate by an appropriate, brief, period the administration of glucocorticoids from that of chemotherapeutic agents (Lippman and Thompson, 1973). These data in tissue culture seem to lead toward the same conclusion.

Studies of Steroid Action by the Use of Somatic Cell Hybrids

Thus far we have been discussing the question of steroid sensitivity and resistance within the limits of those cells that appear to have natural sensitivity. Another, broader, question is: What is the reason for the varied tissue and cellular sensitivity to steroids? In the early days of work on sex steroid receptors, it was frequently pointed out that responsive tissues possessed such receptors whereas nonresponsive tissues did not (Jensen *et al.,* 1971b; Jensen and DeSombre, 1972). At first blush, one might conclude that all that was necessary to explain the specific cellular responses to a given steroid hormone was possession of the appropriate receptor. On second thought it is obvious that this cannot be true. The different responses by various receptor-containing cells to a given steroid require additional levels of control. This is true of all steroids, but is particularly striking in the case of the glucocorticoids, for which receptors have been found in many tissues. The many tissue-specific effects of glucocorticoids range from induction of various tissue-specific enzymes to inhibition of DNA synthesis, and as we have been discussing, lymphocytolysis. Thus, while the receptor theory of steroid action has helped to visualize the first step in the path by which steroids exert their multiple effects, it has not solved entirely the problem of specificity of response.

It is widely believed that the necessary additional controls may be exerted at the level of the nucleus, by molecules that define the genes to be affected by the entering steroid–receptor complex. Such intranuclear controls could be imagined to be negative or positive (Fig. 15). In fact, none as yet have been convincingly defined in biochemical terms. An example of negative control might be a histone or other protein, covering genes that otherwise would have high affinity for the steroid–receptor complex. This concept supposes that the steroid-receptor complex inherently possesses high affinity for appropriate gene sites. What specifically differentiated cells would do is block the genes that are not to respond to the hormone from interacting with the complex, leaving available the genes "correct" for that cell type.

Positive control supposes that some molecule intervenes between the hormone–receptor complex and the genes to be affected. This molecule would bind to the complex and thereby confer upon it cistron specificity. An array of such control molecules would be required to provide specific linkage to the variety of specific genes that we know steroids can affect. (The question of how differentiated cells determine which control molecules to produce is not addressed by these models.) A number of experiments have been carried out trying to identify such molecules, and candidates have been found in both the histone and nonhistone fractions of the nuclear proteins (O'Malley et al., 1972; Spelsburg et al., 1972; Puca et al., 1974). Of course, small molecules could be part of the controls operative in either model (Cake et al., 1976).

Note that this positive control model predicts that cells differentiate by losing or failing to make the intermediary adapter molecules. In a cell with functional steroid receptor, the choice of specifically induced products, whether few or

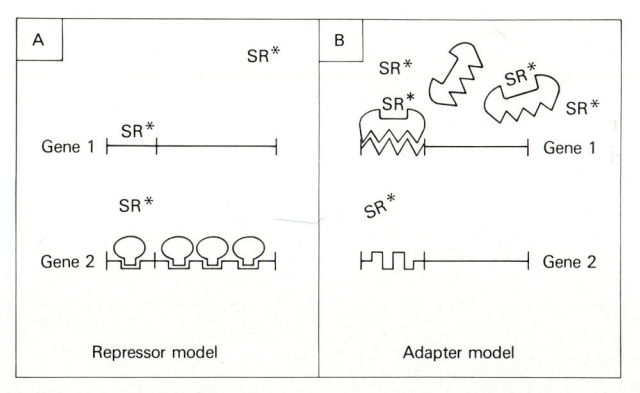

FIG. 15. Models for two types of possible intranuclear control over steroid–receptor complex interactions with DNA. (A) A pure negative-control system, with specifically designed proteins blocking access of activated complex (SR*) by binding to one gene, while leaving open another for interaction with SR*. (B) A pure positive-control model, in which activated steroid receptor complex (SR*) in the nucleus binds to a specific site on an intermediary acceptor molecule, which in turn specifies, by virtue of its second, gene-specific site, the gene to which the ternary complex will bind. Overlapping models are possible.

many, would depend entirely on the presence of the positively acting adapters there.

The possibility also exists that what appears to us as a single receptor for each type of steroid, for example, "the" glucocorticoid receptor, may in fact be heterogeneous in subtle ways beyond our means of present detection. This would provide yet another route to cellular specificity of steroid actions.

One approach to understanding the ways in which cells control their responses to steroids is somatic cell hybridization. This technique allows one to ask the basic genetic question of whether a given process is dominant or recessive in the hybrid. Furthermore, as the hybrid replicates, chromosomes are lost, and sometimes it is possible to correlate the loss or return of a function with chromosome loss (Fig. 16).

The models above make different predictions about the steroid responses of cell hybrids. The adapter protein model predicts that combining cells with different patterns of response will result in expression of both sets of responses, since the positively acting adapters defining both sets would be present. Similarly, if specific subsets of receptors are responsible for various responses, hybrids of dissimilar cells would be additive. On the other hand, in hybrids between negatively controlled cells, the repressing molecules might act cis or trans; that is they might act only on their parental DNA or on the genes of the other cell as well. In the latter case, the steroid-inducible functions would be shut off.

We decided several several years ago to study glucocorticoid action by the use of such hybrids. Our data, and that of others, suggests that there may be specific negative control elements over steroid-evoked cellular responses. Most of the work on steroid effects in cell hybrids has been on the inducible enzyme

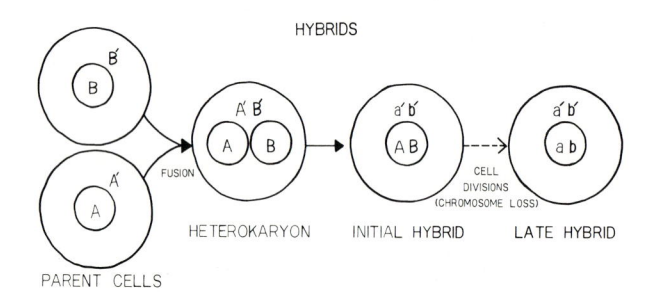

FIG. 16. Somatic cell hybridization scheme. Large circles represent the plasma membrane; inner circles, nuclei. Cytoplasmic content is shown by A' and B' and chromosomal content by A and B. Immediately after cell fusion and before nuclear fusion, virtually all the components of both parents are present. As DNA synthesis proceeds, a small fraction of heterokaryons survive to become hybrids as the nuclei fuse. Cytoplasmic elements from both parents are lost or altered (A'; B'; a', b'), while at the earliest stage, chromosomal content may remain intact (or nearly so). As the hybrid continues to divide, chromosomes from one or both parents are lost (A, B; a, b).

tyrosine aminotransferase (TAT). We elected, initially, to see whether this enzyme was expressed in cell heterokaryons (Thompson and Gelehrter, 1971). For the hormone responsive parent we chose HTC cells, the rat hepatoma cell line which I established while in the laboratory of the late Gordon Tomkins (Thompson et al., 1966). These were the second established line of cells showing steroidal induction of a specific hepatic enzyme. The first was the H4-11-E line, begun by Pitot et al. (1964). In them, TAT induction has been studied extensively. Among other things, in HTC cells it has been shown that induction is linked with receptor occupancy by an active glucocorticoid, that enzyme induction is due to increased enzyme synthesis, and that upon removal of the inducing glucocorticoid, enzyme synthesis returns to basal levels (Granner et al., 1968, 1970; Samuels and Tomkins, 1970; Rousseau et al., 1972). The hepatoma from which this line derived was originally produced by prolonged feeding of a chemical carcinogen to rats of the inbred buffalo rat strain (Morris et al., 1963). As the other parent for the cross, therefore, we chose a clone of BRL cells, epithelial, diploid cells grown from the liver of the same strain of rat (Coon, 1969) in the belief that this would provide at least a close genetic background. Fusion was enhanced by adding UV-inactivated Sendai virus to the mixture of the two parental cells.

Since we were to examine the cells at the heterokaryon stage, a means of determining the presence of TAT in each cell was needed. Consequently, we developed a histochemical stain for the enzyme, based on reduction of a tetrazolium dye by p-hydroxyphenyl pyruvate, the product of the TAT reaction (Thompson and Tomkins, 1971). In order to locate true heterokaryons, one cell line was chronically exposed to [^{3}H] thymidine prior to fusion; thus by radioautography we could identify heterokaryons as those cells that contained a mixture of labeled and unlabeled nuclei. By carrying out the histochemical reaction on the same cells, enzyme activity could be discerned over each cell. The TAT-positive HTC cells had been preinduced by exposure to 10^{-6} M dexamethasone, and the cells were kept in the steroid after fusion, to maximize any enzyme activity. The slides, when examined 24 hours after fusion, showed a mixture of cell types. The smaller HTC cells, carrying the thymidine-labeled nuclei, had reacted strongly. The BRL cells showed little or no tyrosine aminotransferase activity, a fact that was confirmed separately by biochemical and immunological studies. All identifiable heterokaryons were examined, and all showed minimal reaction for the aminotransferase, resembling that in the enzyme-negative, noninducible BRL parent.

To better quantitate this effect, instead of carrying out radioautography and histochemistry on the same slide, duplicate slides were run, one for each process. Independent assessment was then made for enzyme reactivity and for multinucleated cells. In this case, the nuclei of the BRL cells, the TAT noninducible parent, were prelabeled with thymidine; therefore we predicted that any cell

TABLE IV
Comparison of Tyrosine Aminotransferase (TAT) Activity with Source of Nuclei in Multinucleate Cells after Fusion to Inducible (HTC) and Uninducible (BRL-62) Cells[a]

Input cell ratio[b]	Histochemical assay for TAT					Source of nuclei			
	Total multinucleate cells examined	TAT −	TAT +	TAT ?	Total multinucleate cells examined	All nuclei labeled	Some labeled	No. of nuclei labeled
A								
HTC:BRL-62 2:1	500 (100%)	289 (58%)	201 (40%)	10 (2%)	400 (100%)	162 (40.5%)	102 (25.5%)	136 (34%)
							(66%)	
B								
HTC:BRL-62 10:1	500 (100%)	208 (41.6%)	282 (56.4%)	10 (2%)	500 (100%)	85	100	315
							(37%)	(63%)

C				
HTC:BRL-62	125		147	
2:1	(100%)		(100%)	
	83	38	4	
	(66%)	(30%)		
		39	54	51
			(65%)	(35%)

[a] From Thompson and Gelehrter (1971).

[b] Sparsely seeded cultures of BRL-62 cells were grown for 3 days in the presence of [^{3}H] thymidine (5 μCi per 75-cm^2 flask of New England Nuclear NET 27, 1 mCi/ml, 3.6 mg/ml). HTC cells were similarly grown, without label. Both cultures were then treated with trypsin and counted. To portions of medium containing 5 × 10^5 HTC cells, BRL-62 cells were added to give 10:1 or 2:1 ratios of HTC cells to BRL-62 cells. The mixtures were pelleted by centrifugation and then suspended in 0.3 ml of phosphate-buffered saline (pH 7.6) containing either 600 (A and B above) or 1200 (C above) hemagglutinating units of UV-inactivated Sendai virus. After incubation at 0°C with occasional shaking for 60 min, the cells were collected by centrifugation and seeded onto glass cover slips in petri dishes. Growth medium at 37°C, containing 10 μM dexamethasone phosphate, was added, and the cultures were incubated at 37°C in an air–CO$_2$ incubator for 24 hours, after which the cover slips were removed, and the cells on several slips (for each variable) were examined radioautographically and histochemically. Control studies were done at the same time, in which cells were incubated without virus or each cell line was fused with itself. The data from these, not shown in the table, showed that the fusion rate was about 10% with either cell line, and that neither self-fusion nor simply carrying the cells through the procedure without fusion altered the TAT content of either cell line, as determined histochemically.

The data in the table were obtained by scoring only the multinucleate cells in the virus-treated cultures containing HTC and BRL-62 cells, in the ratios indicated. First, the multinucleate cells in the histochemical reaction were scored for the presence or the absence of TAT; then the parallel culture, which had been radioautographed, was examined. Multinucleate cells containing all, one or more, or no labeled nuclei were counted consecutively to avoid inadvertent bias. Then the fraction of cells containing one or more BRL-62 (labeled) nuclei was compared with the fraction that failed to show TAT activity.

with one or more labeled nuclei should give a negative histochemical reaction. Some of these would be true HTC × BRL heterokaryons, while others would be BRL × BRL homokaryons. Table IV shows that, within experimental variability, the prediction was met, as varying ratios of HTC and BRL cells were fused to produce differing proportions of heterokaryons. In an effort to rule out simple inactivation of TAT by something in the BRL cell, extracts of HTC and BRL cells were mixed and biochemically assayed for enzyme activity. Table V shows that there was no inactivation of the enzyme under these conditions. If an inactivator of TAT is contained within BRL cells, it did not survive the extraction conditions.

From these experiments we would postulate that the noninducible, tyrosine aminotransferase-negative parent possesses the trans-dominant ability to shut off

TABLE V

Tyrosine Aminotransferase Activity in Extracts from HTC Cells and BRL-62 Cells[a]

	Volume of cell Extract (μl)[b]		Net A_{331} /10 min
	HTC	BRL-62	
Expt. 1: Assayed immediately after mixing	5	–	0.231
	5	5	0.228
	5	10	0.255
	5	25	0.254
	5	50	0.232
	–	50	0.010
	10	–	0.521
	–	25	0.000
Expt. 2: Mixtures held at 37°C for 50 min, then assayed	5	–	0.241
	5	5	0.250
	5	25	0.250
	5	50	0.238
	–	50	0.000

[a] From Thompson and Gelehrter (1971).

[b] Extracts of HTC and BRL-62 cells in 0.1 M potassium phosphate buffer (pH 7.6) were obtained by sonically disrupting cells that had previously been exposed for 18 hours to 10 μM dexamethasone phosphate in growth medium. The extracts contained, in addition to disrupted cells, 2 mg/ml of crude bovine plasma albumin, 0.5 mM 2-oxoglutarate, and 0.2 mM pyridoxal phosphate. Aliquot portions of each were assayed singly, or after mixing. The amount of enzyme activity is expressed by the A_{331} /10 min, a measure of the product, P-hydroxyphenylpyruvate, formed. The HTC cell extract contained 1.2 mg/ml and the BRL-62 extract 0.3 mg/ml of protein.

induction of a specific enzyme in the presence of the inducing steroid. The effect seems rapid and specific. If the enzyme in the hybrid were decaying at its normal half-life of a few hours, the 24-hour interval between hybridization and our assays would have been sufficient for the level of preinduced enzyme to have fallen below detectable levels. Finally, the regulation we observed had to have occurred when the gene for tyrosine aminotransferase was still present, since both nuclei were still intact and had not yet undergone the first round of replication after fusion.

At nearly the same time that we were doing these experiments, Schneider and Weiss (1971) were studying in classic cell hybrids the other inducible hepatoma then available. They found, as early as fused clones could be grown up to sufficient quantities for assay, that such hybrids lacked TAT inducibility. At this point, the hybrids retained nearly all the chromosomes of each parent. Although it was impossible to be sure that the chromosome containing the TAT gene had not been lost preferentially, this seemed an unlikely explanation. In subsequent work, Weiss and her colleagues formally showed that in some hybrids at least, the gene had been present but unexpressed by observing the return of inducibility after a period of growth during which some of the non-inducible parent's chromosomes had been lost (Weiss and Chaplain, 1971; Weiss et al., 1975; Weiss, 1975). Perhaps a negative control function had been encoded on one of those chromosomes, or perhaps some more general balance of cell chromosome constituents had been altered, permitting reexpression of TAT induction.

Several studies have now confirmed the basic finding that TAT is not expressed in fresh hybrids. We have repeated this type of experiment in crosses of HTC with 3T3 cells (Benedict et al., 1972) and HTC with L cells (Levisohn and Thompson, 1973). In every single hybrid clone examined in our laboratory we have found total loss of TAT activity. In our crosses, and in Weiss' later work (1975), tests with anti-TAT antiserum showed no immunologically identifiable enzyme to be present in hybrids wherein the enzyme was rendered noninducible. Weiss and her colleagues have also shown the extinction and later reexpression of corticosteroid-inducible alanine aminotransferase (Sparkes and Weiss, 1973; Weiss, 1975), and Rintoul and Morrow (1975) have demonstrated loss of inducibility of tryptophan oxygenase in hybrids between noninducible LMTK⁻ and inducible fetal liver cells. Croce et al. (1973, 1975) have presented data from crosses between a hepatoma line containing inducible TAT and human diploid fibroblasts, suggesting that the human X chromosome contains the information responsible for shutting off the TAT inducibility. Table VI summarizes some of these results.

The only exception to the total extinction of the TAT gene so far reported occurred in crosses between the inducible hepatoma Faza and mouse lymphoid (Lc) cells (Table VI). In this cross, all the clones showed a marked reduction in enzyme content, but continued inducibility. Electrophoretic evidence suggested

TABLE VI
Expression of Glucocorticoid-Inducible Hepatic Enzymes in Somatic Cell Hybrids

Enzyme	Parental cells Inducible	Parental cells Noninducible	Hybrids	Reexpression reported	References
Tyrosine aminotransferase	HTC (Buffalo rat) hepatoma	BRL-62 (Buffalo rat) diploid liver epithelium	Noninducible[a]	No	Thompson and Geleherter (1971)
	FU5 (rat)	3T3 (mouse)	Noninducible	No	Schneider and Weiss (1971)
	FU5-5 (rat)	BRL-1 (rat)	Noninducible	Yes	Weiss and Chaplain (1971)
	HTC (rat)	3T3 (mouse)	Noninducible	No	Benedict et al (1972)
	HTC (rat)	L (mouse)	Noninducible	No	Levisohn and Thompson (1973)
	FU5AH (rat)	KOP (human)	Noninducible	Yes	Croce et al. (1973)
	Faza 967 (rat)	DON (Chinese hamster)	Noninducible	Yes	Weiss et al. (1975)
	FU5AH (rat)	CI-ID (mouse)	Noninducible	No	Croce et al. (1975)
	Faza (rat)	Lc (mouse)	Inducible at reduced level	No	Brown and Weiss (1975)
Alanine aminotransferase	Faza 967 (rat)	DON (Chinese hamster)	Noninducible	Yes	Weiss et al. (1975)
Tryptophan oxygenase	FLC (mouse)	LMTK⁻ (mouse)	Noninducible	No	Rintoul et al. (1973a)

[a]Heterokaryons.

that both rat and mouse enzyme were being coexpressed (Brown and Weiss, 1975). Since lymphoid cells normally do not synthesize this enzyme, their TAT gene must have been activated in the hybrids. This is the only cross reported between lymphoid and liver-derived cells, and it may be that in such cells a different set of control elements is found from those in the fibroblastic and epithelioid parents used for the other crosses.

These experiments taken together suggest that steroid-specific responses may be tightly regulated by dominant negative control elements in cell hybrids. Nuclear proteins in the acidic fraction have recently been shown to mask progesterone receptor binding sites (Webster et al., 1976). Perhaps such proteins are involved in the dominant control seen over glucocorticoid induced functions. We are engaged in a number of studies seeking answers to the following questions. Will all glucocorticoid-induced products prove to be negatively controlled in this way? What about the inhibitory responses evoked by glucocorticoids in certain cells? Will effects induced by other steroids also be lost in appropriate hybrids? Are there specific chromosomes involved in this regulation?

Pertinent to these problems are our experiments using HTC × L cell hybrids. Table VII lists some of the pertinent properties of these parental lines. Both contain glucocorticoid-specific receptors (Hackney et al., 1970; Baxter and Tomkins, 1971; Rousseau et al., 1972) and steroid-inducible glutamine synthetase activity. L cells are inhibited by glucocorticoids; they show reduced growth, clone less efficiently, and biochemically, show marked reductions in DNA and RNA synthesis and lowered uptake of glucose. Later, protein synthesis also is diminished (Pratt and Aranow, 1966). Steroid receptors are essential for these effects, and steroid-resistant receptorless L cells have been isolated (Hackney et al., 1970; Lippman and Thompson, 1974). For fusion, the thymidine kinase-minus L cell line LB82 was chosen. L cells lack TAT, and the enzyme is not induced in them by glucocorticoids (Levisohn and Thompson,

TABLE VII
Some Characteristics of HTC and L Cells

L cells		HTC cells
−	Thymidine kinase (LB82)	+
−	Inducible hepatic tyrosine aminotransferase	+
−	Production of C'2	+
−	Isozyme 1 lactate dehydrogenase	+
+	Isozyme 2	−
+	Chromosome centromeres stained by Hoechst 33258 or Giemsa	−
+	Hypoxanthine-guanine phosphoribosyltransferase (HTC-H1)	−
+	Inhibition of thymidine uptake by glucocorticoids (LA9, L929)	−
+	Cytoplasmic glucocorticoid receptor	+
+	Inducible glutamine synthetase	+

1973). HTC cells do contain inducible TAT. An HTC subclone, HTC-Hlc, which lacks the enzyme hypoxanthine–guanine phosphoribosyltransferase, was used in these fusions. Hybrids were created by mixing the cells either in the presence or in the absence of killed Sendai virus. Selection was carried out using medium containing hypoxanthine, methotrexate, and thymidine (Szybalska and Szybalski, 1962). As soon as colonies were large enough they were picked and grown in separate dishes. Several colonies, each originating in a separate dish, were chosen on the basis of morphology for further study. [Striking differences in the growth and morphology of H × L clones we have noted before (Levisohn and Thompson, 1973), and detailed reports on the appearance of growth inhibition, delayed tumor formation, and the membrane proteins of such hybrids have been published (Lyons and Thompson, 1977), but are beyond the purview of the present report.]

Chromosomal analysis of these new hybrid clones showed that at our earliest examination, two contained enough chromosomes to account for nearly all the chromosomes of both parental lines, while two others were considerably reduced in number. Thus we conclude that already the process of chromosomes loss had begun in clones V4a and V5 (Table VIII and Fig. 17). Note that the HTC parent used in these hybrids was 2S, that is it contained an average of 110 chromosomes, twice the usual number for this cell line. By use of special staining techniques, mouse chromosomes can be distinguished from rat chromosomes by virtue of the intense staining "C-banding" in the mouse centromeric region (Caspersson and Zech, 1973). When these methods were applied to the hybrids, both mouse and rat chromosomes could be distinguished (Fig. 18). As would be expected from a rat × mouse cross, these hybrids seemed to lose rat chromosomes in preference to mouse. All the hybrids, regardless of total chromosomes

TABLE VIII
Chromosome Content of HTC Cells, L Cells, and H × L Hybrids

Parent or clone	Average chromosome number
Parents	
HTC-H1	110
LB82	50
Hybrid clones	
02	135
07	133
07SF2	69
07Ag3	97
V4a	95
V5	98

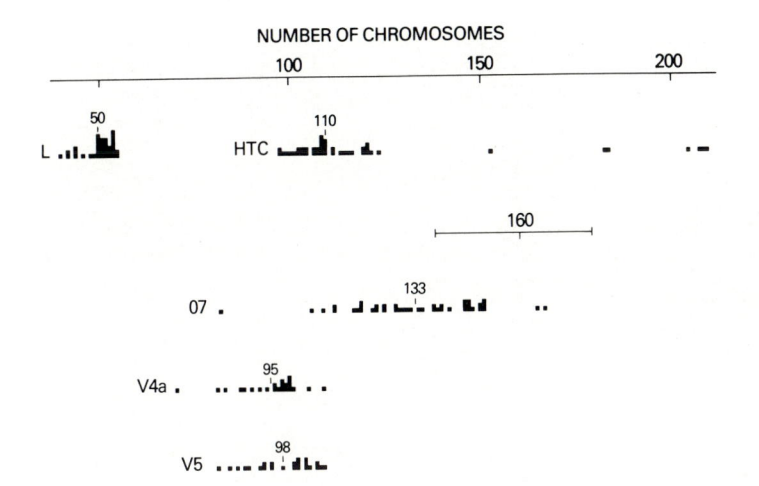

FIG. 17. Histograms of HTC-H1, LB82, and 3 hybrid clones of HTC × LB82. From Lyons and Thompson (1977). Reprinted with permission.

number, retained virtually all their mouse chromosomes. The hybrid nature of the clones was also verified by examination of LDH isozymes. HTC and L cells possess distinctive and different LDH isozymes, and the hybrids showed both, plus the expected array of subunit combinations (Lyons and Thompson, 1977). As in other crosses, TAT was not expressed in these hybrids, and could not be induced by treatment with dexamethasone (Table IX).

One obvious possible explanation for these results we seem to have ruled out. From all the studies on steroid resistance developed in tumor cells, *in vivo* and *in vitro*, one might expect that loss or inactivation of steroid receptors would be a likely mechanism through which steroid nonresponsiveness in hybrids might be regulated. This, however, does not seem to be the case. The H × L hybrids seem to have a full complement of receptor, as is shown in Fig. 19. In fact, we have presented evidence that H × L hybrids contain glucocorticoid receptors specific for each parental cell type (Lippman and Thompson, 1974). Croce *et al.* (1975) also presented evidence for continued receptor presence in their rat hepatoma hybrids. Furthermore, we have examined a series of Dr. Weiss' hybrid clones, which she kindly supplied, and found glucocorticoid binding sites in every one (E. B. Thompson, unpublished). Thus in no case has the mechanism of dominant loss of glucocorticoidal TAT induction been shown to be due to loss of receptor function in the hybrid. The control seems to be distal to the receptor. Evidence not only that are the receptors there but that they are functional comes from the fact that glutamine synthetase, an inducible enzyme present in both L cells

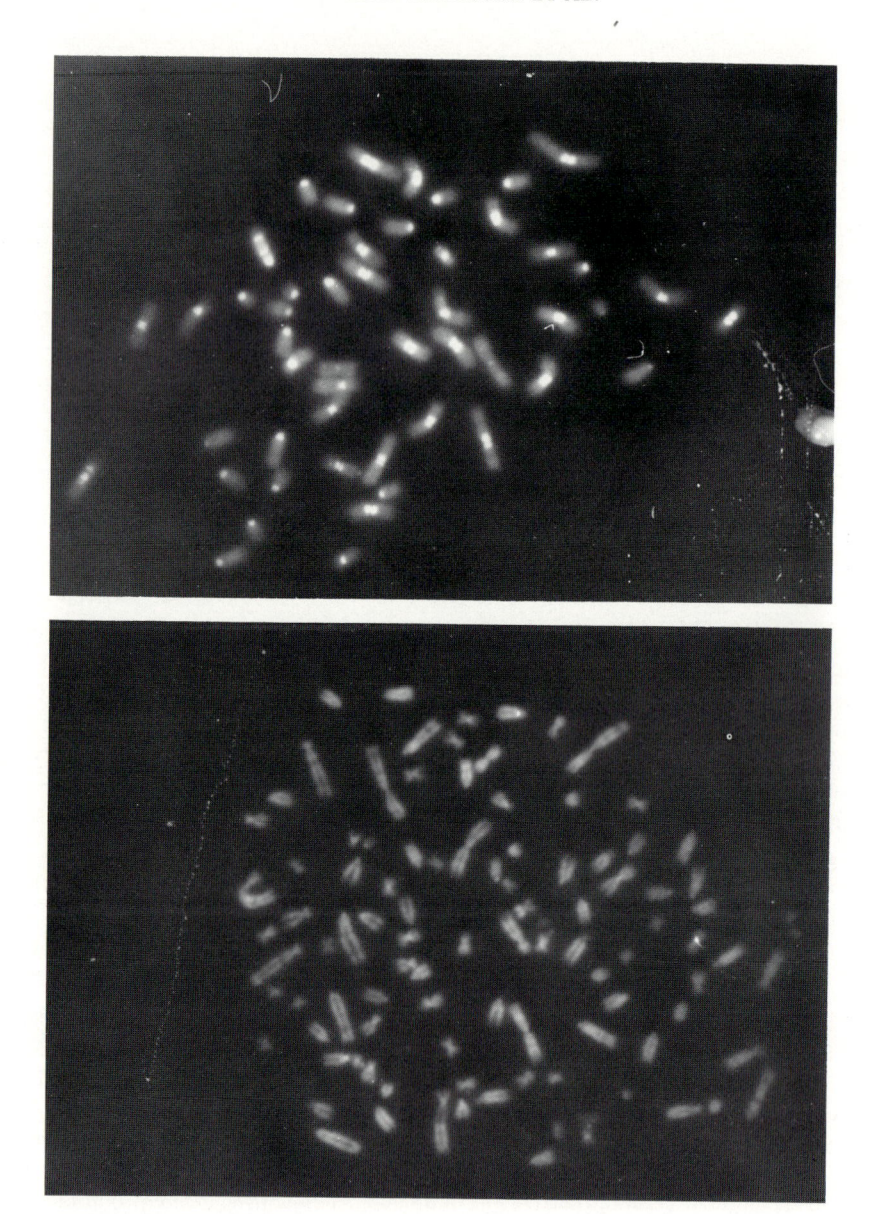

FIG. 18. Hoechst 33258 fluorescent staining of LB82 (top) and HTC-Hl (bottom) chromosomes, demonstrating C-banding in the mouse L-cell chromosomes (bright centromeric fluorescence in chromosomes top panel). Staining by modification of method of Gropp et al. in Caspersson and Zech (1973). From E. B. Thompson and P. Garner, unpublished results.

TABLE IX
Tyrosine Aminotransferase Activity of HTC and L Cells and Their Hybrids[a]

	Parent cells			Hybrids					
		Expt. 1[b]	Expt. 2[b]		Expt. 1[b]	Expt. 2[b]		Expt. 1[b]	Expt. 2[b]
HTC-H1	C	7.2	4.0	07 C	0.3	0	02 C	–	0
	D	41.8	51.1	D	0.7	0	D	–	0
LB82	C	1.0	–	07SF2 C	–	0.8	V4a C	0.4	0
	D	1.1	–	D	–	1.4	D	0.4	0
				07Ag3 C	–	0	V5 C	0	0
				D	–	0	D	0	0

[a] Cells were plated in duplicate 60-mm dishes in growth medium with (D) or without (C) 10^{-6} M dexamethasone overnight, after which they were collected and assayed as described by Levisohn and Thompson (1973).
[b] Activity is expressed as milliunits per minute per milligram of protein.

and HTC cells (Table VII) continues to be induced in the hybrids (Table X). The induction of this enzyme has not been carefully correlated with receptor occupancy by steroid, as has TAT, but one would assume that active steroid–receptor complexes are necessary as with other inducible systems. Thus the regulation which seems to be demonstrable by the use of hybrids appears to be quite specific.

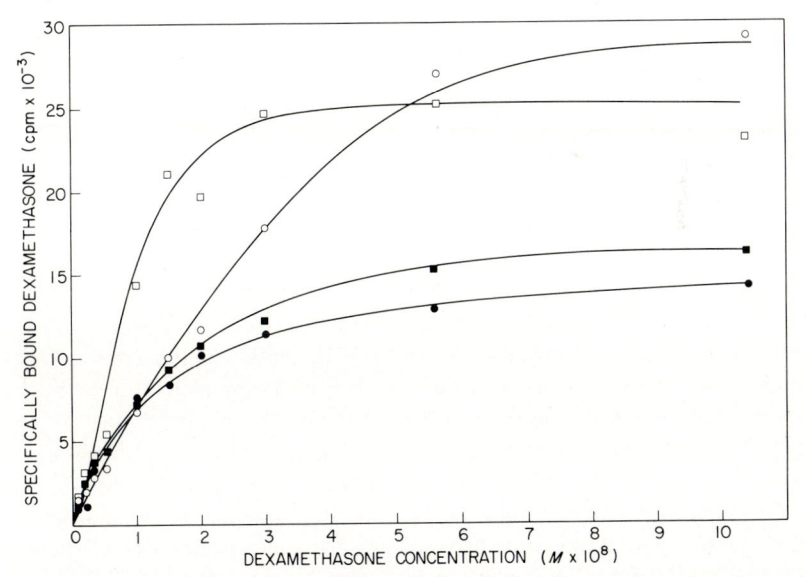

FIG. 19. Glucocorticoid receptor binding of [^{3}H]dexamethasone in cytosols from HTC cells (●), L cells (■), HL5 hybrid cells (□), and a mixture of HTC and L cells (○). From Lippman and Thompson (1974). Reprinted with permission.

TABLE X
Fold Induction of Glutamine Synthetase in HTC Cells, L Cells, and H × L Hybrids

Cells[a]	Fold induction (induced/control)
Parents	
HTC-H1	6
LB82	3.3
Hybrids	
02	7.5
07	5.0
07Ag3	6.3
07SF2	4.2
V4a	4.8
V5	N.D.

[a] Cells were grown and treated with dexamethasone as for Table IX. Glutamine synthetase was assayed by a slight modification of the method of Barnes *et al.* (1974). Fold induction represents glutamine synthetase specific activity, induced ÷ basal.

In preliminary experiments, we have begun to examine the inhibitory responses typical of L cells in the HTC × L cell hybrids. Using as a marker the inhibition of thymidine uptake by glucocorticoids, our initial results suggest that the inhibitory responses, as well as the inducible enzyme TAT, may be shut off in early hybrids containing essentially all the chromosomes of both parents. Whether the inhibition of L cells and lymphoid cells by glucocorticoids will prove to be entirely lost in hybrids with noninhibited cells will be the focus of research in this area in the next few years. The properties of the regulation observed may be summarized as follows: There seems to be dominant negative regulation of glucocorticoid-induced responses in somatic cell hybrids. If a specific response is missing from one of a pair of parents, it will be shut off in hybrids between those parents. The dominant control is trans-effective. Glucocorticoid-induced functions that are shared by both parents continue to be expressed in the hybrids. As chromosomes are lost during growth of the hybrids, missing functions may return, provided that the chromosomes lost are those from the dominant, negative-controlling parent. It may be, as has been suggested, that a specific chromosome contains a regulatory gene for each function, or the explanation may be more complex. For instance, a subtle balance of chromosomes, cytoplasm, and chromatin proteins may be involved. The control seems to be exerted distal to the function of active steroid receptor. Finally, the control that seems to be occurring is not explained by the positive-acting acceptor protein mechanisms popularly under consideration in models of steroid hormone action. As noted above, the specific acceptor molecules proposed in such models would act positively and one would therefore predict that in hybrids, a function missing from one of two parents would be expressed, not extinguished, in the hybrid. As we have seen, this is not the case for the examples studied so far.

That the examples studied so far are extremely limited is obvious. The use of somatic cell hybrids as a way of examining cellular regulatory mechanisms is a recent technique, and most of the work with this method is as yet merely observational. Generalizations are often easy when data is limited. Certainly control of negative and positive nature has been observed in hybrids, and as well, dramatic effects of cytoplasm on nuclei have been seen (Harris *et al.*, 1966; Bolund *et al.*, 1969; Harris, 1970; Gurdon, 1974). Whether the regulation of glucocorticoid effects observed represents specific effects by specific molecules and will lead to information about the means by which differentiated cells govern themselves only additional experiments will tell.

REFERENCES

Barnes, P. R., Hersh, R. T., and Kitos, P. A. (1974). *In Vitro* 9, 230.
Baumann, G., Rappaport, G., Lemarchand-Beraud, T., and Felber, J. (1975). *J. Clin. Endocrinol. Metab.* 40, 462.
Baxter, J. D., and Tomkins, G. M. (1971). *Proc. Natl. Acad. Sci. U.S.A.* 68, 932.
Baxter, J. D., Harris, A. W., Tomkins, G. M., and Cohn, M. (1971). *Science* 171, 189.
Beck, J., Beck, G., Wong, K. Y., and Tomkins, G. M. (1972). *Proc. Natl. Acad. Sci. U.S.A.* 69, 3615.
Behrens, U. J., Mashburn, L. T., Stevens, J., Hollander, V. P., and Lampen, N. (1974). *Cancer Res.* 34, 2926.
Benedict, W. F., Nebert, D. W., and Thompson, E. B. (1972). *Proc. Natl. Acad. Sci. U.S.A.* 69, 2179.
Blecher, M., and White, A. (1959). *Recent Prog. Horm. Res.* 15, 391.
Bolund, L., Ringertz, N. R., and Harris, H. (1969). *J. Cell Sci.* 4, 71.
Borthwick, N. M., and Bell, P. A. (1975). *FEBS Lett.* 60, 396.
Brown, J. E., and Weiss, M. C. (1975). *Cell* 6, 481.
Cake, M. H., Goidl, J. A., Parchman, L. G., and Litwack, G. (1976). *Biochem. Biophys. Res. Commun.* 71, 45.
Caspersson, T., and Zech, L., eds. (1973). *In* "Chromosome Identification." Academic Press, New York.
Chan, L., Means, A. R., and O'Malley, B. W. (1973). *Proc. Natl. Acad. Sci. U.S.A.* 70, 1870.
Chard, R. J., Smith, E. K., and Hartmann, J. R. (1966). *Proc. Am. Assoc. Cancer Res.* 7, 13. (abstr.).
Claman, H. N. (1972). *N. Engl. J. Med.* 287, 388.
Coon, H. G. (1969). *Carnegie Inst. Washington, Yearb.* 67, 419.
Cox, R. F., Haines, M. E., and Emtage, S. (1974). *Eur. J. Biochem.* 49, 225.
Croce, C. M., Litwack, G., and Koprowski, H. (1973). *Proc. Natl. Acad. Sci. U.S.A.* 70, 1268.

Croce, C. M., Litwack, G., and Koprowski, H. (1975). *In* Gene Expression and Carcinogenesis in Cultured Liver" (L. E. Gerschenson and E. B. Thompson, eds.), p. 325. Academic Press, New York.

Feigelson, P., Beato, M., Colman, P., Kalimi, M., Killiwick, L. A., and Schutz, G. (1975). *Recent Prog. Horm. Res.* 31, 213.

Feigelson, P., Kurtz, D., Sippel, R., and Ansah-iadom, X. (1976). *Abstr. Short Commun. Poster Presentations, Int. Congr. Endocrinol., 5th, 1976* Abstract No. 34, p. 14.

Fessas, P., Wintrobe, M., Thompson, R., and Cartwright, G. (1954). *Arch. Intern. Med.* 94, 384.

Foley, G. E., Lazarus, H., Farber, S., Uzman, B. G., Boone, B. A., and McCarthy, R. E. (1965). *Cancer* 18, 522.

Gailani, S., Minowada, J., Silvernail, P., Nussbaum, A., Kaiser, N., Rosen, F., and Shimoaka, K. (1973). *Cancer Res.* 33, 2653.

Glenn, E. M., Miller, W. L., and Schlagel, C. A. (1963). *Recent Prog. Horm. Res.* 19, 107.

Granner, D. K., Hayashi, S., Thompson, E. B., and Tomkins, G. M. (1968). *J. Mol. Biol.* 35, 291.

Granner, D. K., Thompson, E. B., and Tomkins, G. M. (1970). *J. Biol. Chem.* 245, 1472.

Granville, N., Rubio, F., Jr., Unugur, A., Schulman, E., and Dameshek, W. (1958). *N. Engl. J. Med.* 259, 207.

Greengard, O., and Acs, G. (1962). *Biochim. Biophys. Acta* 61, 652.

Gurdon, J. B. (1974). *In* "The Control of Gene Expression in Animal Development," p. 160. Harvard Univ. Press, Cambridge, Massachusetts.

Hackney, J. F., Gross, S. R., Aronow, L., and Pratt, W. B. (1970). *Mol. Pharmacol.* 6, 500.

Harris, H. (1970). *In* "Cell Fusion," p. 108. Harvard Univ. Press, Cambridge, Massachusetts.

Harris, H., Watkins, J. F., Ford, C. E., and Schoefl, G. I. (1966). *J. Cell Sci.* 1, 1.

Jensen, E. V., and DeSombre, E. R. (1972). *Annu. Rev. Biochem.* 41, 203.

Jensen, E. V., Block, G. E., Smith, S., Kyser, K., and DeSombre, E. R. (1971a). *Natl. Cancer Inst., Monogr.* 34, 55.

Jensen, E. V., Numata, M., Brecher, P. I., and DeSombre, E. R. (1971b). *Biochem. Soc. Symp.* 32, 133.

Kaiser, N., Milholland, R. J., and Rosen, F. (1974). *Cancer Res.* 34, 621.

Kaplan, H. S., Nagareda, C. S., and Brown, M. B. (1954). *Recent Prog. Horm. Res.* 10, 293.

Kenney, F. T. (1962). *J. Biol. Chem.* 237, 3495.

King, R. J. B., and Mainwaring, W. I. P. (1974). *In* "Steroid-cell Interactions," p. 430. Univ. Park Press, Baltimore, Maryland.

Konior, G. S., Lippman, M. E., Johnson, G. E., and Leventhal, B. G. (1977). *Cancer Res.* (submitted for publication).

Levisohn, S. R., and Thompson, E. B. (1973). *J. Cell. Physiol.* 81, 225.

Lin, E. C. C., and Knox, W. E. (1957). *Biochim. Biophys. Acta* 26, 85.

Lin, E. C. C., and Knox, W. E. (1958). *J. Biol. Chem.* 233, 1186.

Lippman, M. E., and Thompson, E. B. (1973). *Lancet* 1, 1198.

Lippman, M. E., and Thompson, E. B. (1974). *J. Biol. Chem.* 249, 2483.

Lippman, M. E., Halterman, R. H., Perry, S., Leventhal, B. G., and Thompson, E. B. (1973a). *Nature (London), New Biol.* 242, 157.

Lippman, M. E., Halterman, R. H., Leventhal, B. G., Perry, S., and Thompson, E. B. (1973b). *J. Clin. Invest.* 52, 1715.

Lippman, M. E., Perry, S., and Thompson, E. B. (1974). *Cancer Res.* 34, 1572.

Lippman, M. E., Perry, S., and Thompson, E. B. (1975). *Am. J. Med.* 59, 224.

Lyons, L. B., and Thompson, E. B. (1977). *J. Cell. Physiol.* 90, 179.

McGuire, W. L., Carbone, P. P., and Vollmer, E. P., eds. (1975). "Estrogen Receptors in Human Breast Cancer." Raven, New York.
Makman, M. H., Nakagawa, S., and White, A. (1967). *Recent Prog. Horm. Res.* **23**, 195.
Makman, M. H., Dvorkin, B., and White, A. (1968). *J. Biol. Chem.* **243**, 1485.
Makman, M. H., Nakagawa, S., Dvorkin, B., and White, A. (1970). *J. Biol. Chem.* **245**, 2556.
Makman, M. H., Dvorkin, B., and White, A. (1971). *Proc. Natl. Acad. Sci. U.S.A.* **68**, 1269.
Martial, J. A., Baxter, J. D., Goodman, H. M., and Seeburg, P. H. (1977). *Proc. Natl. Acad. Sci. U.S.A.*, **74** (in press).
Morris, H. P., Wagner, B. P., Ray, F. E., Stewart, H. L., and Snell, K. C. (1963). *J. Natl. Cancer Inst.* **30**, 143.
Mosher, K. M., Young, D. A., and Munck, A. (1971). *J. Biol. Chem.* **246**, 654.
Munck, A. (1971). *Perspect. Biol. Med.* **14**, 265.
Munck, A., and Wira, C. (1971). *Adv. Biosci.* **7**, 301.
Munck, A., Wira, C., Young, D. A., Mosher, K. M., Hallahan, C., and Bell, P. A. (1972). *J. Steroid Biochem.* **3**, 567.
Norman, M. R., and Thompson, E. B. (1976). Submitted for publication.
O'Malley, B. W., and Means, A. R. (1974). *Science* **183**, 610.
O'Malley, B. W., Spelsberg, T. C., Schrader, W. T., Chytil, F., and Steggles, A. W. (1972). *Nature (London)* **235**, 141.
Palmiter, R. D. (1975). *Cell* **4**, 189.
Pearson, O. H., and Eliel, L. P. (1951). *Recent Prog. Horm. Res.* **6**, 373.
Pitot, H. C., and Yatvin, M. B. (1973). *Physiol. Rev.* **53**, 228.
Pitot, H. C., Paraino, C., Morse, P. A., Jr., and Potter, V. R. (1964). *Natl. Cancer Inst., Monogr.* **13**, 229.
Pratt, W. B., and Aronow, L. (1966). *J. Biol. Chem.* **241**, 5244.
Puca, G. A., Sica, V., and Nola, E. (1974). *Proc. Natl. Acad. Sci. U.S.A.* **71**, 979.
Ranney, H. M., and Gellhorn, A. (1957). *Am. J. Med.* **22**, 405.
Rintoul, D., and Morrow, J. (1975). *In* "Gene Expression and Carcinogenesis in Cultured Liver" (L. E. Gerschenson and E. B. Thompson, eds.), p. 311. Academic Press, New York.
Rintoul, D., Colofiore, J., and Morrow, J. (1973a). *Exp. Cell Res.* **78**, 414.
Rintoul, D., Lewis, R. F., and Morrow, J. (1973b). *Biochem. Genet.* **9**, 375.
Roath, S., Israels, M., and Wilkinson, J. (1964). *Q. J. Med.* **33**, 257.
Rosenau, W., Baxter, J. D., Rousseau, G. G., and Tomkins, G. M. (1972). *Nature (London), New Biol.* **237**, 20.
Rousseau, G. G., Baxter, J. D., and Tomkins, G. M. (1972). *J. Mol. Biol.* **67**, 99.
Samuels, H. H., and Tomkins, G. M. (1970). *J. Mol. Biol.* **52**, 57.
Scatchard, G. (1949). *Ann. N. Y. Acad. Sci.* **51**, 660.
Schaumburg, B. P. (1972). *Biochim. Biophys. Acta* **261**, 219.
Schaumburg, B. P., and Bojesen, E. (1968). *Biochim. Biophys. Acta* **170**, 172.
Schimke, R. T., McKnight, G. S., Schapiro, D. J., Sullivan, D., and Palacios, R. (1975). *Recent Prog. Horm. Res.* **31**, 175.
Schneider, J. A., and Weiss, M. C. (1971). *Proc. Natl. Acad. Sci. U.S.A.* **68**, 127.
Shanbrom, E., and Miller, S. (1962). *N. Engl. J. Med.* **266**, 1354.
Sibley, C. H., and Tomkins, G. M. (1974). *Cell* **2**, 221.
Smith, R. W., Blaese, R. M., Hathcock, K. S., Buell, D. N., Edelson, R. L., and Lutzner, M. A. (1974). *In* "Lymphocyte Recognition and Effector Mechanisms" (K. Lindahl-Kiessling and D. Osoba, eds.), p. 127. Academic Press, New York.
Sparkes, R. S., and Weiss, M. C. (1973). *Proc. Natl. Acad. Sci. U.S.A.* **70**, 377.

Spelsberg, T. C., Steggles, A. W., Chytil, F., and O'Malley, B. W. (1972). *J. Biol. Chem.* **247**, 1368.
Stevens, J., and Stevens, Y. (1975). *J. Natl. Cancer. Inst.* **54**, 1493.
Szybalska, E. H., and Szybalski, W. (1962). *Proc. Natl. Acad. Sci. U.S.A.* **48**, 2026.
Thompson, E. B., and Gelehrter, T. D. (1971). *Proc. Natl. Acad. Sci. U.S.A.* **68**, 2589.
Thompson, E. B., and Lippman, M. E. (1974). *Metab. Clin. Exp.* **23**, 159.
Thompson, E. B., and Norman, M. R. (1976). *Abstr., Short Commun. Poster Presentations, Int. Congr. Endocrinol., 5th, 1976* Abstract No. 139, p. 56.
Thompson, E. B., and Tomkins, G. M. (1971). *J. Cell Biol.* **49**, 921.
Thompson, E. B., Tomkins, G. M., and Curran, J. F. (1966). *Proc. Natl. Acad. Sci. U.S.A.* **56**, 296.
Thompson, E. B., Aviv, D., and Lippman, M. E. (1977). *Endocrinology* **100**, 406.
Turnell, R. W., and Burton, A. F. (1974). *Cancer Res.* **34**, 39.
Turnell, R. W., and Burton, A. F. (1975). *Mol. Cell. Biochem.* **9**, 175.
Webster, R. A., Pikler, G. M., and Spelsberg, T. C. (1976). *Biochem. J.* **156**, 409.
Weiss, M. C. (1975). *In* "Gene Expression and Carcinogenesis in Cultured Liver" (L. E. Gerschenson and E. B. Thompson, eds.), p. 346. Academic Press, New York.
Weiss, M. C., and Chaplain, M. (1971). *Proc. Natl. Acad. Sci. U.S.A.* **68**, 3026.
Weiss, M. C., Sparkes, R. S., and Bertoloti, R. (1975). *Somatic Cell Genet.* **1**, 27.
Yamamoto, K. R., and Alberts, B. M. (1976). *Annu. Rev. Biochem.* **45**, 721.
Yamamoto, K. R., Gehring, U., Stampfer, M. R., and Sibley, C. H. (1976). *Recent Prog. Horm. Res.* **32**, 3.
Yates, F. E., Marsh, D. J., and Maran, J. W. (1974). *In* "Medical Physiology" (D. D. Mountcastle, ed.), Vol. 2, p. 1696. Mosby, St. Louis, Missouri.

DISCUSSION

S. J. Marx: You were very precise in maintaining the possible distinction between the dexamethasone-binding proteins and the glucocorticoid receptors. Is there, in fact, evidence to suggest that these dexamethasone-binding proteins are related to or different from the cytoplasmic receptors thought to be involved in the action of glucocorticoids at physiologic concentrations?

E. B. Thompson: The only evidence is that they are saturated by pharmacologic rather than physiologic levels of steroid.

J. Furth: I have two questions: The crux of leukemia therapy with cortisone and radiation is that, no matter how sensitive the original leukemias are to these agents, after long-continued therapy a new cell type emerges that is resistant to them. Your Jack Robbins, while a student at Cornell, tackled this problem. He concluded that this is due to the emergence (mutation) of a new cell type rather than to natural selection of some preexisting cell. What progress has been made in settling this problem?

E. B. Thompson: We hope to use the CEM cell line that I described to get at the question of the origin of resistant cells. At this point, all I can say is that we have done some preliminary cloning studies and find a very high rate of development of resistant cells in this population, a rate of about 1 in 10^6, higher than a "normal" mutation rate. We plan to extend this study, using known mutagens, to see whether we can develop a variety of resistant populations and then do hybridization studies to look at their properties: Work of this sort has been begun in rodent cell lines by Gehring *et al.* [Yamamoto, K. R., Gehring, U., Stampfer, M. R., and Sibley, C. H., *Recent Prog. Horm. Res.* **32**, 3 (1976)].

J. Furth: My second question relates to the difference between lymphoid leukemia and lymphoma. I though that the lymphoma cells were autoantigenic. This was a rather heretic view because in those times we thought that the natural tumors were never antigenic. But we now know that many of them are. It is conceivable that lymphoma cells are localized, grow by continguity, antibodies in blood keeping them out of circulation.
Experimental lymphoma cells grafted on preirradiated (immune-depressed) animals behave like leukemia cells. What is the current concept as to the basic difference between leukemia and lymphoma? The lymphoma cells are accepted as neoplastic, the leukemias were thought to be mere hyperplasias.

E. B. Thompson: The solution of the problem of metastases, which I believe is the problem you are addressing, is not known, nor is the specific difference between the lymphosarcoma and the leukemic cell.

M. Lippman: I would like to show and briefly discuss four graphs. We have been interested in studying normal human lymphocytes for glucocorticoid receptor activity. Figure A shows a typical binding curve of tritiated dexamethasone to intact human peripheral blood lymphocytes. The whole cell assay technique which we employ does not distinguish between cytoplasmic binding and binding at other subcellular sites. Nonetheless, specific, saturable high affinity binding is demonstrable. Normal human circulating lymphocytes are relatively unresponsive to glucocorticoids. Lymphocytes treated with phytohemag-

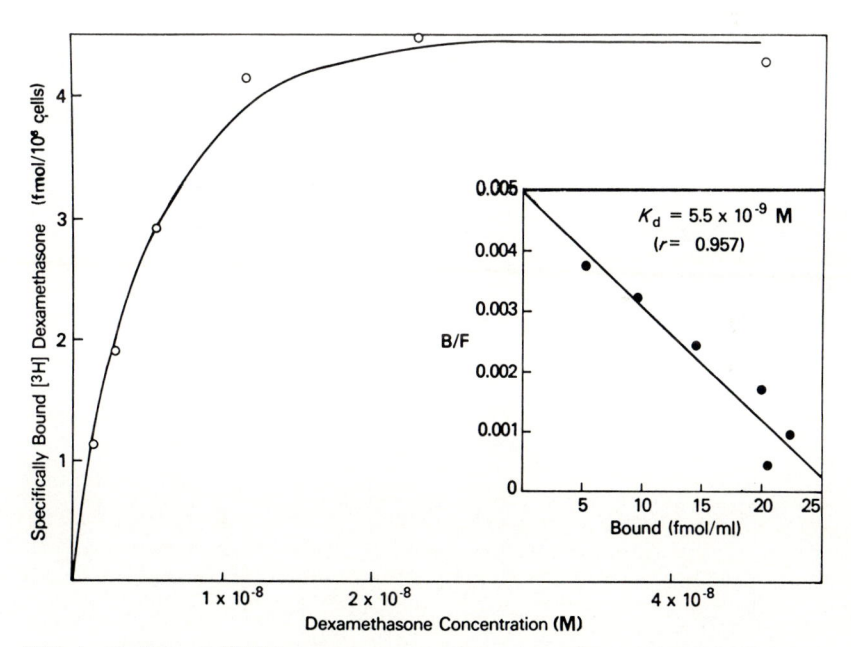

FIG. A. Binding of [^{3}H] dexamethasone to human peripheral blood lymphocytes. A Scatchard plot of the same binding data is shown in the inset. A whole-cell binding technique is employed in which the cells are incubated with tritiated steroid with or without an excess of unlabeled competitor for 3 hours at 22°C. Unbound steroid is removed by washing the cells with ice cold phosphate-buffered saline three times.

glutinin (PHA) are strongly inhibited by glucocorticoids. We wondered whether PHA might be an inducer of glucocorticoid receptor in these cells. Figure B shows the effect of PHA treatment on purified human peripheral blood lymphocytes as a function of time. Typically, between 10 and 16 hours we see a 2- to 4-fold increase in specific glucocorticoid receptor sites per cell. Nine separate experiments are summarized in Fig. C. As shown, there is about a 2.5-fold induction of specific glucocorticoid receptor sites by PHA. In studies not reviewed here, we have been able to show that this induced receptor activity is identical in binding affinity and steroid-binding specificity to that found in unstimulated lymphocytes. Furthermore, this induction by PHA can be blocked by simultaneous treatment with either actinomycin D or cycloheximide.

Finally, there is excellent agreement between concentrations of glucocorticoid that bind to receptor and those that induce inhibition of nucleoside incorporation. In Fig. D concentrations of dexamethasone that half-maximally inhibit thymidine incorporation are closely equivalent to those that saturate about half of the receptor sites. Thus, glucocorticoid receptor sites are demonstrable in human peripheral blood lymphocytes. Binding activity is inducible by PHA stimulation and, finally, there is good agreement between concentrations of glucocorticoid that bind to receptor and those that induce biologic effects.

E. B. **Thompson**: With respect to Dr. Marx' question about the nature of the receptors in the leukemic blasts: In Fig. D it appears that one is getting the biological effect in these cells at less than 10^{-9} M dexamethasone concentration, whereas it took something like 2×10^{-8} M dexamethasone to produce the same effect and to saturate the receptor in the leukemic blasts.

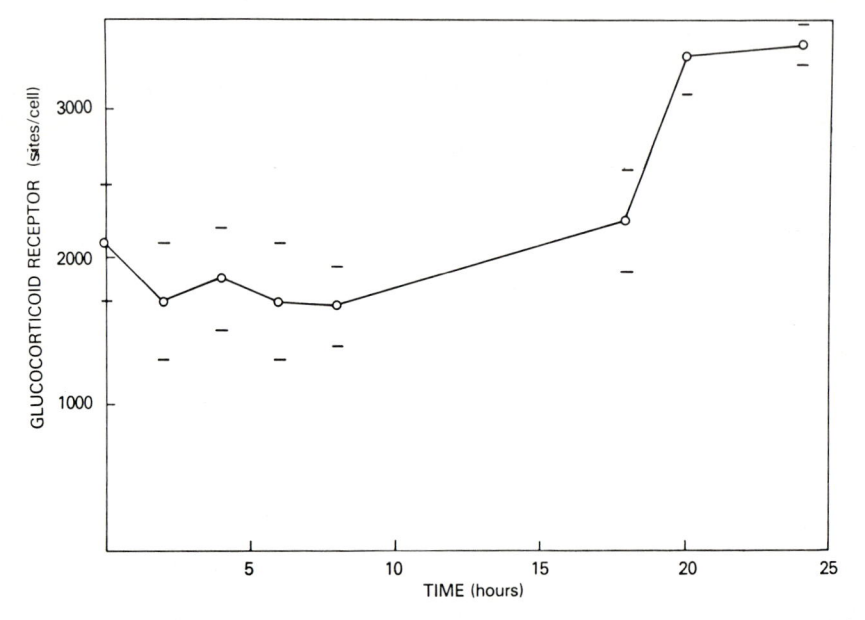

FIG. B. Time course of glucocorticoid receptor induction by phytohemagglutinin (PHA) in human peripheral blood lymphocytes. Specific binding was assessed as outlined for Fig. A. Results are means of triplicate determinations plus or minus one standard deviation.

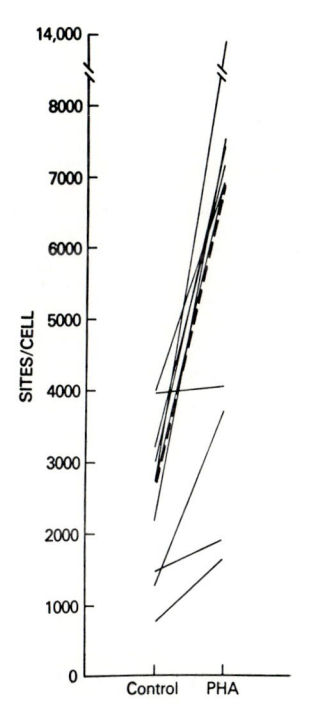

FIG. C. Effects of phytohemagglutinin (PHA) on glucocorticoid receptor activity in human peripheral blood lymphocytes. Cells were stimulated with PHA for 18 hours. Specific binding was assessed as described for Fig. A. The median value for nonstimulated and PHA-stimulated lymphocytes is shown by the dashed line.

V. P. Hollander: There is evidence that glucocorticoids destroy lymphoid tissue by some action on the cell surface. Transport processes are inhibited at an early event in glucocorticoid action. Dr. Behrens, in our laboratory, demonstrated a difference in alcian blue staining of the cell surface between glucocorticoid sensitive and resistant cells. Lymphocytolysis would be expected to inhibit thymidine incorporation, would it not? I wonder whether thymidine incorporation studies do not represent such a late step in cell destruction that the information is not relevant to glucocorticoid-mediated lymphocytolysis.

E. B. Thompson: I think those points are well taken. I would dispute with you only as to the use of the term "relevant" or "meaningful." Lymphocytolysis is the culmination of the inhibitory process. The cell breaks up and is gone. As far as I know there are six or eight notions now of how that process is brought about. But it is my opinion that none of us know as yet the initial, key, or necessary first step in the process. Therefore, one can measure a given end point for steroid action which is convenient. For these studies we chose thymidine incorporation as a convenient end point against which to titrate steroid action. As you well know, the suggestion is that in cells killed by steroids something is induced that is lethal for the cell. It may be that the changes you and others observe in the cell membrane are the result of such induction.

A. Segaloff: I am disturbed by some of this, particularly since you mentioned the problem of the ethics of multiple agent chemotherapy in acute lymphatic leukemia. I am

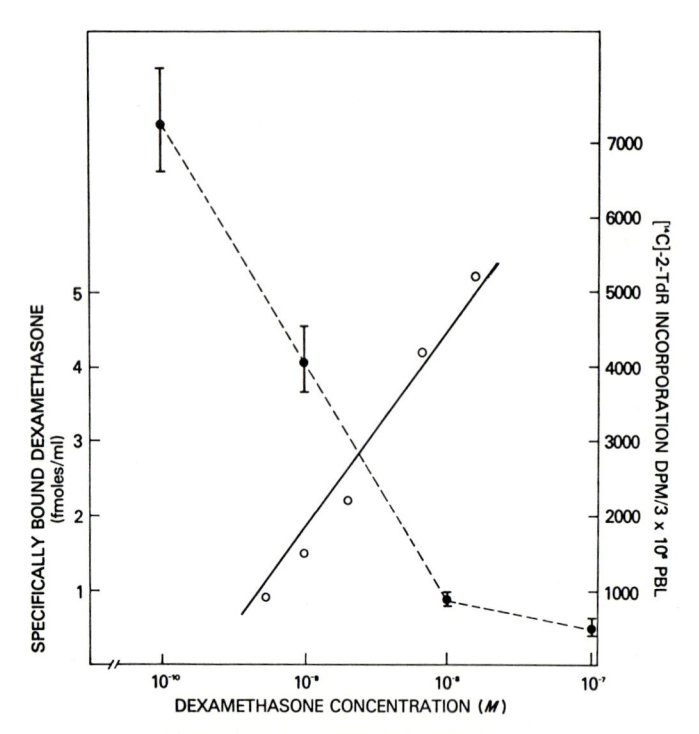

FIG. D. Comparison of dexamethasone binding to human peripheral blood lymphocytes (PBL) and inhibition of thymidine incorporation. Binding (o——o) is assessed as described for Fig. A. •– – –•, Inhibition of thymidine incorporation. Results are means of quadruplicate determinations plus or minus one standard deviation.

troubled by what I consider to be a very good example of the ostrich phenomenon, which we have just seen. For practical purposes people treating acute leukemia use prednisone as the only corticoid, something I have deplored for many years. Yet many of these phenomena that you have reported were shown in studies using dexamethasone which has different properties with respect to receptors on the data you just reported. Prednisone is not one of the really good competitors and is not toally converted to prednisolone. I would like to know what happens if you use prednisone, since I think that getting hematologists to change is impossible.

E. B. Thompson: It has long been established that prednisone per se is not a very effective glucocorticoid and that it does not bind well to steroid receptors. However, in the body, it is rapidly converted to prednisolone, which is an active glucocorticoid and does bind to receptors with virtually the same affinity as dexamethasone. In the figure regarding specificity you would have noted no doubt that prednisolone is a very effective competitor. Therefore, I think it is entirely appropriate to use any steroid that shows such high binding properties correlated with physiological or pharmacological actions.

A. White: Would you comment on Table I, which showed data indicating the inhibitory effects of various steroids on the binding of dexamethasone? The data as I recall indicated a remarkable difference between cortisone and corticosterone. You also had effective inhibition of dexamethasone binding by progesterone. It was shown a number of years ago in our laboratory that progesterone inhibits the actions of hydrocortisone on a number of parameters of rat lymphoid cells. The question therefore really relates to structure–activity relationships with regard to steroids that inhibit binding of dexamethasone. The second question relates to whether, after such binding studies, you have recovered the steroid and demonstrated that it has not been metabolically altered. I make this point because a number of years ago Dougherty and Berliner demonstrated that, in contrast to normal mouse lymphoid cells, malignant cells possessed a very active 20-hydroxysteroid dehydrogenase, which rapidly converted the active corticoid into an inactive corticoid, they postulated that this might be a basis for some of the resistance of the malignant cell to the lymphocytolytic action of the corticoid. Third, it is a well known phenomenon that in the line of T-cell maturation the early T cell is remarkably more hydrocortisone-sensitive than is the mature T cell. Have you had an opportunity to explore whether this is related to a loss of receptor during the T-cell maturation process?

E. B. Thompson: The first point was with regard to progesterone and its ability to inhibit the binding of steroid in the lymphoblasts: I agree, the point you made is exactly correct. It *is* a question of structure and function. It is known that progesterone binds to the glucocorticoid receptor in a very large number of systems and that in those systems it almost universally acts as a competitive inhibitor of glucocorticoid action. Tomkins, Baxter, Rousseau, and their colleagues, in fact, have proposed a model suggesting that it is progesterone binding to receptor that prevents allosteric transition of the receptor and therefore prevents its movement to the cell nucleus—a model for which there are some supporting data. Therefore, in this study I would say that progesterone is doing the same thing and that its ability to compete for the receptor with effective glucocorticoids fits its action in other glucocorticoid-sensitive systems.

Your second point concerned metabolic alterations of the steroids. Could this account for resistance in some of the resistance cell lines? That is a very important point and in any study on steroid action should be considered. (For that matter it has long been suggested that steroids were perhaps altered to more active forms in certain tissues or cells.) We had shown previously in HTC cells, for example, that the steroid is found unaltered after it has carried out is physiological effect. One of the reasons why we use dexamethasone in the studies presented here is its relative insensitivity to metabolizing enzymes. We have not in every patient, however, carried out studies to show that there was no metabolism. We have examined the steroid-containing medium from treated cells in several cases by placing it on a sensitive cell line and showing that it still contained inducing steroid.

Will you please repeat your third question?

A. White: The question was whether, inasmuch as the early T cell is hydrocortisone sensitive, there is, during the maturation of the early T cell to the immunologically competent T cell, an alteration in the number of dexamethasone receptors.

E. B. Thompson: Could one do a time study of maturing T cells showing a change in the steroid receptors? I would refer back to the studies that Dr. Lippman alluded to, in which they examined normal T cells derived from peripheral blood and found that the whole cells did have steroid binding; so did B cells for that matter. However, it was difficult to see cytoplasmic receptors in those cells. Whether the receptors were localized entirely in the nucleus or not has not been formally shown, but one suspects that that is the case. It is also possible that in preparing cytoplasmic extracts the receptor is destroyed. In blasts the receptor content per cell was higher, and it was easy to show cytoplasmic receptor.

M. V. Nekola: How long do the CEM cells remain responsive *in vitro*? Do they, like a number of other cell lines, have to be grown *in vivo* to maintain their responsiveness *in vitro*? If you would grow the resistance cells *in vivo*, would they become responsive *in vitro*?

E. B. Thompson: These cells have been *in vitro* for years. They are human cells, mind you. This is the first study of glucocorticoid effects on an established line of human leukemic cells, and because of their species of origin, reincubating them *in vivo* requires some very special methods. It might be possible to cycle them in the nude mouse, for example, but, in fact, we have not done so. The sensitive cells had been cultured several years before we obtained them and still contained a majority of sensitive cells. The resistant clones we have just begun to develop in the last year, so I cannot answer the question of how long they will remain resistant, or whether in the absence of steroid they will again begin to make receptors. I think that is one of the important questions.

C. W. Bardin: In cells with low receptor activity, have you had an opportunity to determine whether there is decreased normal receptor, or altered steroid binding? In addition, do you have a cell line with normal receptor which does not respond to steroid?

E. B. Thompson: In searching for a cell line to work on *in vitro*, we looked at a number of cell lines that were steroid resistant. As I mentioned in passing, several of these do contain receptors that appear normal by the following criteria: Quantity, the steroid concentration at which they are saturated, their affinity for steroid, and their ability to bind to nuclei. In fact, in earlier days they would have been accepted as perfectly all right, and we have as yet found no lesion in these receptors. One of these lines was the CEM line we obtained from the American Type Culture Collection. The sensitive CEM line, we were given by Dr. D. Mann, NCI. In addition, we have developed sublines of tyrosine aminotransferase noninducible HTC cells which contain perfectly normal receptors [Thompson, E. B., Aviv, D., and Lippman, M. E., *Endocrinology* **100**, 406 (1977)]. They translocate to the nucleus *in vivo* and so forth; yet the cells were uninducible. So I think there are clear examples of cells that, at our current level of sophistication, contain normal receptors and are not responsive to the effects of steroids, lympholytic or enzyme-inducing. We predict that as people look at cells obtained directly from more cases in the blood dyscrasias, they will find occasionally some that are receptor positive, but steroid resistant as well. Why these are the minority I do not know, because of all the cases published so far, it seems that the vast majority *in vivo* that are steroid resistant are receptorless.

As to your other question whether these receptor positive cells are resistant to all the actions of glucocorticoids or just to their lympholytic effects; all I can say is that they are resistant in the sense that they grow perfectly well in the presence of steroids; that their macromolecular synthesis is unaffected by steroids. Those are all the parameters we have looked at. In the case of the HTC cells with normal receptor but noninducible for tyrosine aminotransferase, we are now engaged in a collaboration with Drs. T. D. Gelehrter and D. K. Granner to see whether the half dozen or so other inducible properties in the cells are also blocked. The other question about altered receptor: In their studies, Yamamoto, Baxter, Harris, Tomkins, and colleagues showed in a number of papers that mouse lymphoid cells selected for resistance to glucocorticoids often had apparently normal receptors. But these receptors upon further examination were altered; i.e., they had lesser affinity for steroid or would not move to the nucleus and so forth.

In our cases of resistant cells where there was enough receptor to examine, albeit in a lower than normal amount, it appeared that it has essentially the same affinity for steroid as did the sensitive ones. One of the enigmas about development of resistance to steroids is why a lowered quantity of steroid receptor sometimes correlates with resistance.

V. P. Hollander: When tumors are resistant *in vivo* to the action of glucocorticoids, the tumor cells *in vitro* are usually resistant to glucocorticoid effects on glucose, uridine, and other transport processes. Yet, receptor differences are often not as clear-cut as these effects on transport processes. I believe that the discrepancy results from technical difficulty in measuring corticosteroid receptor, and that is the hardest steroid receptor to measure because of inherent instability and differences in optimal conditions for different tumors.

E. B. Thompson: I think the point is well taken.

H. L. Bradlow: If you take nuclei from cells that are resistant to dexamethasone, and incubate them with the cytosol receptor steroid complex from cells that are responsive, do you get nuclear uptake? Also in the further case where the cytoplasmic receptor is present, but there is no response, can you get nuclear uptake?

E. B. Thompson: What is "this case"?

H. L. Bradlow: You mentioned that you sometimes have cells where there is a cytoplasmic receptor, but the cells are not responsive to the hormone. Is this a failure of nuclear uptake?

E. B. Thompson: It is not a failure of nuclear uptake measured by *in vitro* nuclear uptake. We examined three cell lines, two human and an AKRa mouse leukemia. In each case there was *in vitro* nuclear uptake of the cytoplasmic receptor. The other question you asked was whether there was ever an instance in which the nucleus seemed to be unable to take up steroid: we have not found one like that.

J. R. Pasqualini: In your system, when you look for the binding of dexamethasone or cortisone, do you find some binding in the nuclei; if so, do you have any idea in what nuclear fraction is the binding?

E. B. Thompson: When we found nuclear binding, where is it in the nucleus? We did not attempt to subfractionate the nucleus, and therefore we cannot say whether it was on the periphery or on the inside. Only by analogy with other systems would we imagine it to be with the chromatin fraction, but we have no direct data bearing upon that.

A. Martin: Have you done any binding pattern studies on the chromosomes in the cell lines to see whether there is a difference between the resistant and nonresistant forms?

E. B. Thompson: We are just setting up to do chromosome ideograms, and that is one of the advantages of turning to the human system. Clones can be obtained that are quasi-diploid, and with that we hope to do exactly what you suggest. The further advantage is that in hybrid studies one is able to follow the loss of specific chromosomes rather than groups of chromosomes as in the mouse/rat crosses.

The Role of Hormones on Digestive and Urinary Tract Carcinogenesis

RICHARD S. YAMAMOTO AND ELIZABETH K. WEISBURGER

Carcinogen Metabolism and Toxicology Branch, Division of Cancer Cause and Prevention, National Cancer Institute, National Institutes of Health, Bethesda, Maryland

I. Introduction

Cancer is a group of diseases found in man and in all other animal species. Cancer research involves many specialties, including physiology, biochemistry, genetics, immunology, virology, nutrition, and pathology. Work with experimental animals has led to an understanding of some of the causes of cancer and the growth and development of this disease. Application of the knowledge gained from experimental studies may eventually lead to control of human neoplasms.

Differences in the distribution of tumors between the sexes were noted early. Obviously, because of the dissimilarities in the growth and development of the primary and secondary accessory sex organs and the production of the hormones to stimulate these organs, there will be differences in the distribution of cancer. However, it is of interest that females tend to have a higher incidence of accessory sex organ tumors, such as breast, uterine, and ovarian cancer, as shown in Fig. 1. The only cancer in the male which has comparable incidence rates is prostatic cancer. This phenomenon has been attributed to the cyclic crossover changes in the hormonal milieu. The suggestion has been made that the bicyclic phase of the female pituitary and the sex hormones provided greater opportunity to lose control of the delicate balance maintained by these organs, particularly involving the hypothalamus. In the male, the secretion of hormones is not in such rigid balance and does not require such a feedback type of mechanism.

The perplexing situation concerns neoplasia of non-sex organs, males having a higher incidence than females. Particularly involved are neoplasms of the digestive tract, including liver and nonendocrine pancreas, and the urinary tract. The sex differences in incidence show that the hormonal makeup must have some influence on carcinogenesis. This presentation attempts to explain some of the causes of the differences.

The differences seen in man occur also in the animal kingdom, in particular in rodents, in which most experimental work has been done. The differences in the

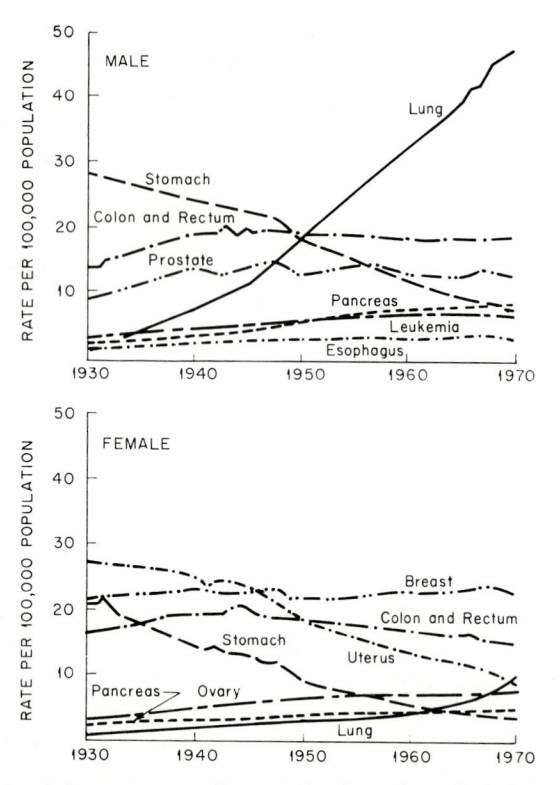

FIG. 1. Time trends in cancer mortality rates by site and sex: United States, 1930–1970 (age-adjusted to U.S. 1940 population). From "Cancer Facts and Figures" (American Cancer Society, New York, 1974).

incidence of spontaneous tumors, particularly of the sex organs, have been found in females. Mammary tumors, being comparable to breast cancer in women, have been well studied. Because of the site and size, mammary tumors can be readily studied experimentally. The mammary tumor virus (MTV), or milk factor of Bittner, produces mammary tumor only in female mice, even though the males do have a rudimentary mammary gland. Instead of mammary tumors, the males of these susceptible strains tend to develop spontaneous virus-free hepatoma. The female MTV-free mice also developed hepatoma at a very late stage and also later mammary tumors free of virus (Heston and Vlahakis, 1968).

Even in the case of induced carcinogenesis, sex differences in tumor incidence are apparent. Female rats in particular tend to be less susceptible to the toxicity and carcinogenicity of chemicals. Therefore, the sex hormones will be used as the starting point of differences in this discussion of the role of hormones in chemical carcinogenesis.

II. General Considerations

In general, the format of this discussion is as follows: (1) steps in chemical carcinogenesis starting from the administration of carcinogen through the eventual growth and development of carcinoma; (2) hormonal characteristics in chemical carcinogenesis; (3) chemical carcinogenesis, particularly in and of the liver, starting with metabolism of the carcinogens and tumor development; (4) the role of hormones in chemical carcinogenesis of the digestive and urinary tracts; and (5) a summary.

A general scheme of the steps in cancer formation starting with the administration of a chemical carcinogen is shown in Fig. 2. Most of the chemical carcinogens will require metabolic activation to an "ultimate" carcinogen unless they are direct-acting carcinogens. This is one area where the hormones will play a role, as will be shown later. The ultimate carcinogen will attach itself to specific receptors, probably DNA, RNA, or protein, thereby altering the receptor. The first covalent interactions of chemical carcinogens with proteins of the target tissues were noted by Miller and Miller (1947). Studies were then made on the reaction between alkylating agents and nucleic acids (Wheeler and Skipper, 1957). Covalent binding of activated chemical carcinogens with the macromolecules (DNA, RNA, and/or protein) *in vivo* has been established (Farber, 1968; Miller and Miller, 1966). Which of the macromolecules is the tumor-causing receptor still awaits confirmation. After these initial steps latent tumor cells start developing, and after some time growth of the tissue occurs very rapidly and a well-differentiated tumor develops. The growth and development of the tumor cells is the second stage where a hormonal effect can occur. Eventually the tumors progress into rapidly growing undifferentiated cancer, which continues to grow progressively larger.

```
Chemical carcinogen
      ↓    Metabolic activation
Ultimate carcinogen
      ↓    Attachment to
           specific receptor
Altered receptor
      ↓    Expression
Latent tumor cell
      ↓    Growth, development
Differentiated tumor
      ↓    Progression
Undifferentiated cancer
```

FIG. 2. Steps in chemical carcinogenesis. From Weisburger (1973).

III. Hormones in Carcinogenesis

The role of hormones in carcinogenesis is quite general. Hormones participate as possible "carcinogens," yet not as direct carcinogens. Hormones seem to have an intermediary role in the promotion of activating or detoxifying mechanisms, as cocarcinogens participating together with a carcinogen, as stimulating or maintaining a cancerous growth, and also in promoting the growth and development of tumors. Hormones will be discussed here as "carcinogens," as controllers for the metabolism of a carcinogen, as modifiers of carcinogenesis, as maintainers of tumors, and as involved in metabolism itself.

A. HORMONES AS "CARCINOGENS"

As carcinogens, hormones act quite differently from other chemical carcinogens. Most of the hormones, natural or synthetic, cause tumors to develop not by direct contact with a target organ, but by affecting a second endocrine gland, causing an atrophy or lack of function in the second gland. The uniqueness of natural hormones as "carcinogens" is that they are also natural constituents of the body that undergo metabolism. Carcinogenesis may be caused by the imbalance produced by the presence of excess hormones, which may disrupt the feedback mechanism. Consequently, there is a lack of inhibition of the target organ, which continues to produce substances until the organ loses control (Bielschowsky and Horning, 1958; Gardner, 1953).

Chronic administration of estrogen resulted in the development of large tumors of the anterior pituitary gland in rats as well as in many strains of mice (Gardner et al., 1957). The consequence of lack of inhibitory feedback will cause the pituitary gland to increase secretion of hormones until they lose control of production and the gland becomes tumorous. Depending on the particular cells involved, the pituitary secretes mammotropic, growth, and/or adrenocorticotropic hormones (Furth, 1961; Furth et al., 1956). The pituitary tumor MtT/F4 will be discussed later.

The synthetic hormone diethylstilbestrol (DES) is not only estrogenic, but also carcinogenic. Unexpectedly, DES produced tumors in mouse testes (Andervont et al., 1961). Chronic administration of DES induced renal tumors in male hamsters and in the ovariectomized female hamsters. This phenomenon has not been described in other animals and is of great interest, as the tumor induced is in an organ that is not normally regarded as belonging to the endocrine system (Kirkman, 1959).

The prolonged administration of purified growth hormone in female rats has resulted in tumors of the lymphatic tissues, reproductive organs, and adrenals (Moon et al., 1950a,b,c). However, no tumors were evident when hypophysectomized rats were similarly administered growth hormones (Moon et al., 1951).

Growth hormone thus caused carcinogenesis by hormonal imbalance rather than as a self-sufficient carcinogen.

B. EFFECT OF HORMONES ON METABOLISM OF CARCINOGENS

Hormones play a large role in the metabolism of drugs (Kato, 1974). Therefore it is not difficult to see that the metabolism of chemical carcinogens, a group of xenobiotics, will also be modified by the lack or the excess of hormones. There may be great differences in the metabolism of carcinogens in male and female animals. For example, excretion of a dose of N-hydroxy-2-fluorenylacetamide (N-OH-FAA) was greater in the female than in the male (Lotlikar *et al.*, 1964; E. K. Weisburger *et al.*, 1964). Therefore in the male more of the carcinogen remained bound to cellular macromolecules, a step in the progression toward neoplasia.

C. HORMONES AS MODIFYING FACTORS IN CARCINOGENESIS

One of the most important roles of hormones is their modifying action in chemical carcinogenesis. Indirect evidence of hormonal influences on carcinogenesis is available from earlier knowledge of sex differences in spontaneous tumors. For instance, C3H mice are prone to mammary tumor (caused by mammary tumor virus, MTV). The males, not affected by the presence of mammary tumor, show a high rate of spontaneous hepatoma. Not until MTV-free female mice were developed, were spontaneous hepatomas found in the very old females of this strain. As chemical carcinogens were artificially used to induce carcinomas, various sex differences were noticed. Therefore, to examine the difference between the sexes, the endocrine organs were ablated to lower certain hormone concentrations, and administration of hormones permitted a study of the effect of excesses. Thus hypophysectomy, adrenalectomy, thyroidectomy, castration, etc., were performed to test the modifying ability of the missing hormones.

The role of hormones as modifiers in chemical carcinogensis is 2-fold. (1) The hormone can modify the metabolism of the chemical carcinogens analogous with various xenobiotics, and (2) the hormone can modify the cause of tumor formation after the precancerous lesion has been induced.

D. HORMONE-DEPENDENT TUMORS

Hormone-dependent tumors are those, such as breast and prostate tumors, that can be made to regress by the removal of the appropriate endocrine gland, by castration, by adrenalectomy, or by hypophysectomy (Boyland, 1961). It seems that, although descendants of normal cells have some of the properties of their nonmalignant ancestors, malignant cells resemble each other more closely than the organs from which they arose. One of the properties, however, which a few tumors exhibit, is dependence on the presence of hormone for their growth and development (Huggins, 1963).

E. THE METABOLISM OF HORMONES

The metabolism of the hormones themselves may play an important role in their ability to modify carcinogenesis. The steroid hormones particularly may vary in their ability to exert influence, depending on this factor (Toh, 1972). Hormones implanted subcutaneously are effective, but when placed in the spleen, the hormone is metabolized and thus deactivated. Similarly, a metabolizable hormone in the presence of a carcinogen may compete for the metabolizing enzymes and thereby reduce the level of activation of the carcinogen.

IV. Chemical Carcinogenesis

Since 1775, when Sir Percivall Pott, a prominent English surgeon, attributed cancer of the scrotum in chimney sweeps to long exposure and intimate contact with soot (Potter, 1963), the field of chemical carcinogenesis has come a long way. However, until early this century all attempts to reproduce cancer in animals failed. In 1915, two patient Japanese investigators, Yamagiwa and Ishikawa (1915) conclusively developed skin cancer in rabbits at the site of application of coal tar. By 1930, British scientists under the leadership of Sir Ernest Kennaway had isolated a pure compound, benzo[a]pyrene, from tar (Kennaway and Hieger, 1930). Now we have a long list of individual chemicals that will produce cancer of various types and in different organs in animals.

A. EXPERIMENTAL CHEMICAL CARCINOGENESIS

There are many steps involved in cancer formation with chemicals. To induce tumors experimentally, an effective chemical has to be chosen. This chemical may be active in its own right or may have to be activated metabolically to be effective. The chemical may also have to find its way to the proper target organ. At the target tissue, a receptor is then altered for expression of the change that occurs. In carcinogenesis, be it physical or chemical, there is usually a latent period when nothing seems to be happening. Then, through the growth and development of the cells and tissues, abnormal growth occurs eventually. A rather well differentiated tumor may become an undifferentiated malignant tumor.

B. CHEMICAL CARCINOGENS

Chemical carcinogens can be divided into two groups, the direct-acting carcinogens and the procarcinogens. Examples of the latter are the polycyclic hydrocarbons, aromatic amines, nitrosamines, and many natural products. An interesting group of carcinogenic organic compounds is comprised of the amines, which include aminoazo dyes and aromatic amines. Other effective carcinogens are nitrosamines, nitrosamides, and certain hydrazines. To complicate matters, some aromatic amines may or may not be active, depending on the attachment of the amine or the acetamide group. In the case of 2- and 4-fluorenylacetamide, 2-fluorenylacetamide is very active whereas the 4-compound is inactive.

C. ACTIVATING METABOLISM OF CARCINOGENS

The metabolism of the carcinogens is important in the formation of an active carcinogen. Here the terms procarcinogen, proximate carcinogen, and ultimate carcinogen are considered. An example is in the metabolism of N-2-fluorenylacetamide (FAA; procarcinogen) to its N-hydroxyl derivative (proximate carcinogen), then to its sulfate ester (ultimate carcinogen) (Miller and Miller, 1969b). We were able to show that N-hydroxylation of FAA can be largely inhibited with acetanilide, both by prevention of hepatocellular carcinoma formation and by the study of urinary metabolites. However, it was also interesting to find that esterification of N-OH-FAA to the active sulfate was also prevented by acetanilide. In the latter case, it was the deficiency of the sulfate ions, as shown by the restitution of carcinogenicity by addition of excess sulfate to satisfy the p-hydroxyacetanilide, a sulfate trap formed from acetanilide. Since the restoration of a carcinogenic effect was not complete, there must be other factors playing a role (Weisburger et al., 1972). The hormones play a role in the metabolism of the carcinogens, as will be shown later.

As mentioned earlier regarding "procarcinogen to ultimate carcinogen," most carcinogens have to be activated. Of the two general groups of carcinogens, the direct acting do not require any activating metabolism and they may affect tissues at the site of injection. Those requiring biochemical transformation, indirect-acting carcinogens, may affect tissues distant from the point of application. As long as they can be absorbed and are stable enough, dietary feeding is the simplest and is a relevant method of administration. In the former group are the alkylating agents and epoxides, whereas in the second group are many other carcinogens, amines and many natural products, including aflatoxins and cycasin, that require metabolic activation, particularly by the liver.

The term "activating" metabolism is necessary, since the usual purpose of metabolism of an organic compound is to detoxify the compound and keep it soluble until it can be removed from the body through the bile, urine, or perspiration or in the expired air. Many toxic materials may be detoxified by biochemical transformation into hydroxylated compounds. These inactivated hydroxylated compounds are solubilized for removal by combining with glucosiduronic, sulfuric, or phosphoric acid. This procedure is also effective in keeping a carcinogen inert and soluble for transfer to tissue sites.

Among the carcinogens, the aromatic amines have been studied intensively, particularly the azo dyes and 2-fluorenylacetamide (FAA; or acetylaminofluorene, AAF). The metabolism of FAA has been reviewed many times (Miller, 1970; Miller and Miller, 1969a; Weisburger and Weisburger, 1963) and will be discussed briefly (see Fig. 3).

FAA is carcinogenic to liver and other tissues in the rat, mouse, rabbit, hamster, dog, cat, and fowl, but not in the guinea pig. FAA is rapidly metabolized to its hydroxylated derivatives in the liver. Most of the hydroxylation is on the ring positions, 1-, 3-, 5-, 6-, 7-, and 8. None of these hydroxylated derivatives were found to be carcinogenic (Miller *et al.*, 1960; Morris *et al.*, 1960), and they represent a means of detoxifying and removing the compounds from the body. However, the N-hydroxylated compound *N*-hydroxy-*N*-2-fluorenylacetamide (*N*-OH-FAA) was found to be a stronger carcinogen than FAA (Miller *et al.*, 1964), especially at the site of application in rat, mouse, hamster, rabbit, and guinea pig (Miller *et al.*, 1961, 1964). Although FAA was not carcinogenic in guinea pig, *N*-OH-FAA was carcinogenic in this species. It was later found that guinea pig liver had little or no ability to N-hydroxylate FAA (Miller

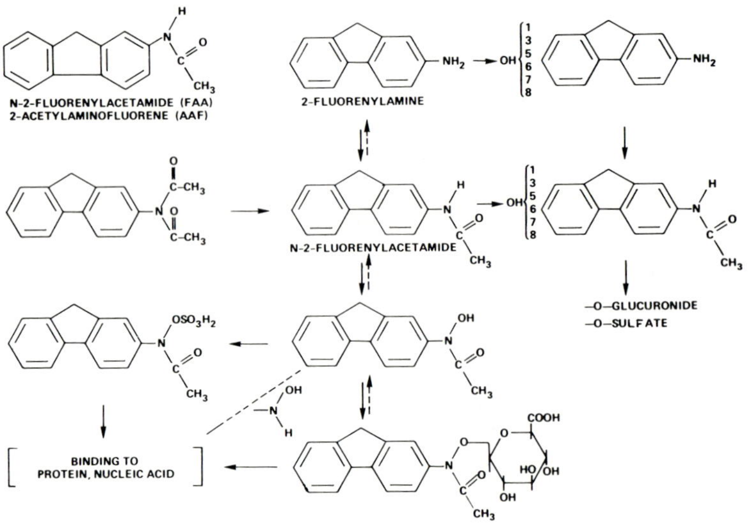

FIG. 3. Metabolites of *N*-2-fluorenylacetatamide.

et al., 1964). In the same manner, it was shown that, by synthetic N-hydroxylation, 7-hydroxy-FAA, a noncarcinogenic metabolite of FAA (Hoch-Ligeti, 1947; Morris *et al.*, 1960), was converted into a potent carcinogen for the rat.

As a sequel to studies on the urinary excretion of FAA metabolites, including *N*-OH-FAA and ring-hydroxylated derivatives, we investigated the effect of a competitive inhibitor on FAA metabolism and carcinogencity. We first used acetanilide (phenylacetamide), a compound with fewer rings than FAA. In a typical experiment with rats, we found that acetanilide prevented FAA hepatocarcinogenesis; urinary analysis for *N*-OH-FAA showed that much less FAA was converted to *N*-OH-FAA in the presence of acetanilide. Thus acetanilide prevented formation of sufficient *N*-OH-FAA to induce tumors. Not being satisfied with the explanation of lack of N-hydroxylation, we fed acetanilide with *N*-OH-FAA to rats. The result was that the acetanilide also inhibited *N*-OH-FAA carcinogenesis. It was suggested that acetanilide is metabolized to *p*-hydroxyacetanilide, a good sulfate trap, thereby causing a deficiency of sulfate needed for activation of *N*-OH-FAA to the "ultimate" carcinogen. The importance of sulfotransferase had been established (DeBaun *et al.*, 1968, 1970a,b; King and Phillips, 1968). By supplementation of our acetanilide–*N*-OH-FAA diet with sulfate, we were able to reinstate part of the carcinogenicity of *N*-OH-FAA. The sulfate supplementation did not augment FAA carcinogenicity (Table I).

At present, the *N*-hydroxysulfate ester is thought to be an "ultimate" carcinogen. A possible mechanism by which the sulfate derivative is attached to the macromolecules, RNA, DNA, and proteins, has been postulated (King and Phillips, 1968). There are others who proposed that the *N*-hydroxyglucuronide

TABLE I
Hepatocarcinogenesis: Inhibition by Acetanilide and Augmentation by Sodium Sulfate

Diet supplements[a]			No. of rats	Liver pathology		Liver cancer (%)
				Hyperplasia	Cancer	
FAA	AA	SO$_4$				
+	−	−	2	0	2	100
+	+	−	10	9	1	10
+	+	+	10	10	0	0
N-OH-FAA	AA	SO$_4$				
+	−	−	15	0	15	100
+	+	−	22	19	3	17
+	+	+	37	20	17	46
+	+	PO$_4$	12	11	1	8

[a]*N*-2-Fluorenylacetamide (FAA), 0.03%; *N*-hydroxy-FAA (*N*-OH-FAA), 0.032%; SO$_4$, 0.84 or 2.52; PO$_4$, 2.25%.

of FAA may be the ultimate carcinogen (Irving and Wiseman, 1969; Irving et al., 1967a,b) since some of the organs susceptible to FAA, such as the mammary gland and Zymbal's gland, did not have sulfotransferase activity (Irving et al., 1971).

D. RESPONSE OF ANIMAL SPECIES OR STRAINS TO CARCINOGENS

Rats and mice, the most convenient species for most laboratories, can differ in responses within their own species as much as do different species. This is true particularly with mice, where many different strains have been developed. Some of the strains, such as DBA or C57BL are quite resistant to lung adenoma development, however, the Swiss or A strains are very susceptible. With respect to colon carcinogenesis, the C57BL strain is resistant to dimethylhydrazine (DMH) carcinogenesis but susceptible to azoxymethane (AOM) carcinogenesis. Other strains, such as the DBA/2, are susceptible to both DMH and AOM.

In some cases, metabolic studies offer an explanation for the differing response of various species to a carcinogen. The inactivity of FAA in guinea pigs, referred to earlier, was readily explained by metabolic studies showing the absence of N-hydroxylation.

Organotropism of Carcinogens. Organotropism pertains to action at a specific organ site. The organotropism of nitrosamines and nitrosamides particularly, was reported by Druckrey et al. (1967). A carcinogen can be nonspecific, and tumors may be found in liver, mammary gland, or ear duct, on skin, or as leukemia. An animal given such a compound may have one or several types of tumor in early stages, but the animal may die from the more advanced tumor before other types can arise. On the other hand, some of the nitrosamines or nitrosamides have been very specific, such as diethylnitrosamine for the liver, N-methyl-N'-nitro-N-nitrosoguanidine for stomach, and diisopropanolnitrosamine for pancreatic duct carcinogenesis. N-Butyl-N-4-hydroxybutylnitrosamine, a metabolite of dibutylnitrosamine, produced urinary bladder tumors consistently.

E. ENVIRONMENTAL FACTORS

The most important environmental factors involved in carcinogenesis are diet and food intake, particularly so when the carcinogen is administered in the diet. Carcinogens usually tend to lower food intake, and thereby decrease body weight gains. However, in the case of growth hormone administration, where there is an increase in food intake and growth, more carcinogen is ingested. Lowering of food intake has not interfered with hepatocarcinogenesis. As an example, rats fed N-OH-FAA were also given growth hormone. The rats on the higher level of N-OH-FAA, 160 ppm, ate less and, although the response was slightly different, the results were as predicted, with many more tumors and after shorter time than at the lower rate of 80 ppm. The toxic effect and regeneration of liver tissues was greater. More about these effects will be mentioned later.

F. CELLULAR GROWTH AND DEVELOPMENT AS A RESPONSE TO CARCINOGEN

As mentioned in the preceding section, constant increase in weight does not mean that the tumor is developing. In reality, growth of the liver is a response to the toxicity of the carcinogen, true for most of the hepatic carcinogens. Therefore, one does not have to administer the carcinogen continuously to necropsy, but can stop entirely so that the liver cells can recover. One reason is to overcome the toxic reaction; another is to be able to compare the tumors histologically with the surrounding normal organ tissues, which are absent when the carcinogens are administered to the time of necropsy.

A thorough histological study of the growth and development of hepatocellular carcinoma with the use of the aromatic amine derivative, N-2-fluorenyldiacetamide (FDAA) has been accomplished (Reuber, 1965b; Reuber and Firminger, 1963; Stewart and Snell, 1959). Although the work has been mainly on the inbred A × C strain, the observations can be carried over to other strains, such as the F344, BUF, or even noninbred strains. Tumor formation seems to follow a similar pattern, starting with (a) a few cells as a focus of hyperplasia, (b) an area of hyperplasia, at which time some areas may turn hydropic and may possibly decrease in size, whereas most of the hyperplastic areas continue to expand, compress, and eventually constrict the adjacent liver cord to form (c) hyperplastic nodules. The cells in the nodules are identical to those in the hyperplastic area. These well differentiated cells developed into atypical cells and become (d) small hepatocellular carcinomas. It takes the small hepatocellular carcinoma (hyperplastic nodules with atypia) too long to develop into (e) a well developed hepatocellular carcinoma. Thus, it is believed that the latter developed on its own. Hepatocellular carcinomas are classified histologically into highly differentiated, well differentiated, and poorly differentiated, the well differentiated predominating. A small number of cholangiomas and cholangiocarcinomas appear to be present in hepatomas induced by aromatic amines, but cholangiocarcinomas are more commonly induced by certain aminoazo dyes.

V. Hormones in Chemical Carcinogenesis

A. HORMONES IN LIVER CARCINOGENESIS

Aside from the hormone-dependent and hormone-producing organs, the liver has been most intensely studied in regard to the effect of hormones. Hepatocarcinogens used for such studies have been 4-dimethylaminoazobenzene (DAB), its more potent derivative, 3'-methyl-4-dimethylaminoazobenzene (3'-Me-DAB), N-2-fluorenylacetamide (FAA), N-2-fluorenyldiacetamide (FDAA), and more recently, diethylnitrosamine.

In the discussion to follow, the sex hormones and sex organs will be first considered, then the pituitary and its hormones, and finally other hormones. In the discussion, all the hormones involved in liver carcinogenesis will be mentioned.

1. Sex Hormones in Liver Carcinogenesis

There is a striking difference in the incidence of cancer between males and females. Apart from the reproductive organs and thyroid cancers, men always have a higher incidence of cancer than women (Ashley, 1969a,b; Shimkin, 1965). Sex differences in the incidence of both spontaneous and induced tumors were also found in animals. The study of these differences may give us clues as to the role of some intrinsic factors involved in carcinogenesis (Butler, 1964).

Primary liver cancer is more predominant in males than in females. This holds not only in higher-incidence areas such as Africa and Asia, but also in low-incidence areas such as Europe and America. Of particular interest is that the etiologies of liver tumors are quite different, possibly reflecting varied exposure to environmental carcinogens. In Africa, hepatoma develops rather early in life, whereas in Asia liver tumors are found at older ages.

The sex difference in incidence of liver tumors in laboratory animals is quite marked. Male rats are much more prone to have chemically induced tumors with azo dyes (Baba and Takayama, 1961; Rumsfeld et al., 1951) or FAA (Bielschowsky, 1961; Kirby, 1947; Morris and Firminger, 1956; Sidransky et al., 1961) than are female rats. Male mice have spontaneous hepatoma sooner and at a higher incidence than do female mice (Andervont, 1950; Hancock and Dickie, 1969; Heston and Vlahakis, 1968). Chemically induced hepatomas develop in male mice earlier and in larger number than in female mice (Biancifiori, 1970; Kirschbaum, 1957; Leathem, 1951).

Together with the sex differences, the strain susceptibility to hepatocarcinogenesis is very marked. Fortunately, early in our studies with rats we started to use the inbred F344 strain rats. The purpose was chiefly to maintain the transplantable mammotropic tumor (MtT). However, we found that the time period for the development of liver tumors was less, only 5-6 months and with smaller doses of carcinogens in this strain. For example, where 300 ppm of FAA was needed with Sprague-Dawley, Buffalo, or ACI rats, 200 ppm sufficed to produce hepatocellular carcinoma in 100% of the F344 strain rats. However, disadvantages of using the F344 rats were their susceptibility to respiratory diseases and the development of interstitial cell tumors of the testes in males over 15 months of age.

The sex difference in incidence was maintained regardless of the route of administration. In male animals, liver tumors developed earlier, more rapidly, and at lower dosages whether the route of administration was per os, by gavage, in drinking water, or parenterally. Female rats were able to tolerate higher doses of the liver carcinogens but eventually did develop hepatoma, for the resistance to chemical carcinogenesis was not complete. In most of these experiments, the dosages and the time periods were controlled to bring out the differences. By controlling the conditions and the health of the animals as closely as possible, generally using the least number of animals permissible statistically and watching the groups carefully, we have been very successful in reproducing our results.

Castration of male rats decreased liver tumor incidence from FAA, which was restored by administration of testosterone (Bielschowsky, 1961; Firminger and Reuber, 1961; Morris and Firminger, 1956). The effect on tumorigenesis by N-FDAA in male rats was dependent on age. Castration of older rats (52 weeks old) completely prevented the development of hepatocellular carcinoma, but administration of testosterone to these castrates did not restore the susceptibility to carcinoma as in young rats and produced only mild cirrhosis (Reuber, 1976). Administration of testosterone to intact male rats also intensified tumor development (Firminger and Reuber, 1961; Stasney *et al.*, 1947).

In order to test whether the increase in tumorigenesis from testosterone was due to its anabolic property, norethandrolone, an anabolic hormone, was tested. Norethandrolone is a synthetic hormone that has been reported to possess equal anabolic potency with only 16% of the androgenic property of testosterone (Saunders and Drill, 1956, 1957). Since either of the procarcinogens FAA or FDAA, administered by others (Firminger and Reuber, 1961; Stasney *et al.*, 1947), necessitated metabolic activation, N-OH-FAA (80 ppm) was used, and norethandrolone (70 ppm) was fed after a preliminary 10-week carcinogen feeding. In this scheme norethandrolone was to act as a promoter, or as an enhancer in tumorigenesis rather than as a stimulator of activating metabolism of the carcinogen.

Neoplastic changes were seen only in the liver, owing to the relatively short 21-week experimental period and the low dose of carcinogen, only 80 ppm. All the rats on a noncarcinogenic diet were free of lesions. Therefore, Tables II and III list only those rats on the carcinogenic diet during the first 10 weeks on experiment. The results show that there is a slight promoting action of norenthandrolone on liver carcinogenesis after the initial damage was done by the carcinogen alone. Similar experiments with testosterone showed that testosterone did have some promoting action (Firminger and Reuber, 1961).

Male rats receiving estrogens or diethylstilbestrol showed some resistance to carcinogenesis (Reuber and Firminger, 1962). Castration of male rats and treatment with diethylstilbestrol developed an animal more resistant to hepatocarcinogensis (Morris and Firminger, 1956). In contrast, progesterone made intact or

TABLE II
Body and Organ Weights of F344 Rats Fed Norethandrolone

Sex	Diet[a] N-OH-FAA	Nor-E	No. of rats	Body (gm)	Liver (gm/100 gm)	Testes (gm)	Adrenal (mg)	Pituitary (mg)
Male	−	−	8	268	4.0	2.59	29.5	7.5
	−	+	12	232	3.9	1.48	27.6	6.3
	+	−	8	254	6.8	2.68	31.7	7.3
	+	+	12	220	6.7	1.60	27.8	5.7
						Ovaries (mg)		
Female	−	−	8	180	3.3	53.9	36.6	12.2
	−	+	12	190	3.7	25.3	25.2	6.2
	+	−	8	168	4.7	50.3	34.7	10.3
	+	+	12	178	5.8	20.4	24.3	6.4

[a]Diet: 0–10th week N-hydroxy-N-2-fluorenylacetamide (N-OH-FAA), 80 ppm; 10–11th week, diet A; 11th–21st week, N-OH-FAA, 80 ppm: nor-E—norethandrolone, 70 ppm.

TABLE III
Histologic Picture of Livers of Rats Prefed N-OH-FAA for 10 Weeks

Sex	Number of rats	Treatment[a] N-OH-FAA	Nor-E	No hyperplasia	Area of hyperplasia	Hyperplastic nodules	Small hepatoma	Hepatoma
Male	8	−	−	−	5	2	1	−
	12	−	+	−	3	8	1	−
	8	+	−	−	1	6	1	−
	12	+	+	−	−	4	5	3
Female	8	−	−	6	2	−	−	−
	12	−	+	9	3	−	−	−
	8	+	−	−	8	−	−	−
	12	+	+	−	9	3	−	−

[a]Treatment during weeks 11 to 21 of experiment; during 10th week all rats were on the control diet A. N-OH-FAA is carcinogen N-hydroxy-N-2-fluorenylacetamide; Nor-E is anabolic agent norethandrolone.

castrated male rats more susceptible to liver carcinogenesis (Reuber and Firminger, 1962).

As mentioned earlier, females were quite resistant to hepatocarcinogenesis under conditions that were effective for male rats. This difference seems to be essentially quantitative, not qualitative, as higher dosages, longer administration, or a longer waiting period will also produce tumors in most females with these carcinogens. Thus administration of testosterone increased the tumorigenic action of FDAA in both intact and spayed female rats (Firminger and Reuber, 1961). Spayed female rats were as resistant to FDAA carcinogenesis as intact females (Morris and Firminger, 1956). Upon administration of diethylstilbestrol to female rats, there was no decrease in tumorigenesis (Reuber and Firminger, 1962). Subcutaneous implantation of testosterone pellets in neonatally castrated male and female rats produced liver tumors when FAA was administered by stomach tube. However, intrasplenic implantation of testosterone pellets failed to exert any effect, probably owing to the inactivation of testosterone by the liver (Toh, 1972).

We became interested in the effect of neonatal injection of androgen and estrogen on liver tumorigenesis by FAA (E. K. Weisburger et al., 1968; J. H. Weisburger et al., 1966). Instead of a simple case of excess testosterone or estradiol injection, the neonatal injection of a single dose of these hormones changed the response of these rats during their adult stage (Levine, 1966).

There is evidence that the brain, particularly the hypothalamus, is involved in sex differentiation. Once that stage is passed, approximately 5 days after birth, the testosterone secretion starts and the animal becomes a male (Barraclough, 1961). If testis is removed before that, the pup tends to become a female and the mammalian behavior patterns are basically female. The male pattern must be induced by the action of testosterone on the brain of newborns (Whalen, 1965).

In order to test the role of hormones on hepatocarcinogenesis, we have studied neonatal estrogen administration and the subsequent response of these animals to hepatocarcinogenesis. Since the males are more susceptible to carcinogensis than the females, we thought we had a good tool. In order to bypass the effect on the activating metabolism of the carcinogen, we used N-OH-FAA, then known as the more proximal carcinogen.

F344 strain rats were injected with 100 μg of estradiol benzoate per pup within 24 hours of birth (up to 5 days is sufficient). At weaning age (about 4 weeks in F344), we placed the rats on three different diets, 0, 80, and 160 ppm N-OH-FAA in the diet for 16 weeks, then held them for another 10 weeks on control diet.

The results in Table IV showed that both males and females treated with estrogen at birth had atrophic gonads, which did not seem to be affected by the carcinogen. The "male" treated rats did not gain as well as the control males, but

TABLE IV
Effect of Neonatal Injection of Estradiol Benzoate on N-Hydroxy-N-2-fluorenylacetamide (N-OH-FAA) Liver Carcinogenesis

Sex	Conc. N-OH-FAA (ppm)	Estradiol[a]	Number of rats	Body (gm)	Liver (gm/100 gm)	Gonads	Hyperplasia	Cancer
Male						Testes (gm)		
	0	+	10	284	2.54 ± 0.10	1.54 ± 0.25	0	0
	80	+	10	249	5.82 ± 0.40	1.47 ± 0.19	2	8
	80	−	10	254	4.76 ± 0.26	2.54 ± 0.07	5	5
	160	+	15	236	7.81 ± 0.44	1.51 ± 0.19	2	13
	160	−	11	248	7.46 ± 0.41	2.76 ± 0.03	1	10
Female						Ovaries (mg)		
	0	+	12	211	2.51 ± 0.06	26.2 ± 3.20	0	0
	80	+	13	206	3.87 ± 0.14	23.1 ± 1.80	13	0
	80	−	8	178	2.84 ± 0.07	51.6 ± 2.50	8	0
	160	+	22	185	5.30 ± 0.20	24.9 ± 1.60	13	9
	160	−	16	173	3.58 ± 0.10	54.0 ± 1.60	16	0

[a] Estradiol benzoate at 100 µg within 24 hours after birth.

the "females" gained more weight than their control counterparts. The susceptibility to hepatocarcinogenesis was greater in the neonatally treated groups. The "males" on 160 ppm did not respond additionally, for they were maximally stimulated. However, at 80 ppm, there was some enhancement of tumorigenesis. In the "females," 80 ppm even among the treated animals seemed too low to be effective, but at 160 ppm, the level was sufficient to cause enhancement in tumorigenesis. The change in the estradiol-treated "females" is difficult to understand. The results were unlike those from earlier work, where administration of estradiol to weanling rats failed to enhance tumorigenesis (Firminger and Reuber, 1961; Reuber and Firminger, 1962).

In the second experiment, testosterone propionate (TP) and estradiol benzoate (EB) were used. The newborn pups were sexed and divided equally for the injections of 500 µg of TP and 250 µg of EB. The protocol was similar to that of the first experiment except that only one level of carcinogen (160 ppm of N-OH-FAA) was used. The results after 16 weeks of carcinogenic diet and 10 more weeks of control diet are shown in Table V. The pituitary, ovaries, uterus, and testes, except adrenal glands, all atrophied in both treated groups. The carcinogenic diet itself had no effect on the organ weights, except for the liver. The liver histopathology showed that neonatal estradiol treatment decreased the

TABLE V
Body and Endocrine Weights[a]

			Male				Female			
Group	Diet[b]	No.	Body wt. (gm)	Testes (gm)	Pit. (mg)	No.	Body wt. (gm)	Ovaries (mg)	Uterus (mg)	Pit. (mg)
Testosterone	A	10	303	1.8	7.3	10	222	24	208	8.5
propionate	K	24	230	1.6	6.6	18	194	24	286	8.6
Estradiol	A	26	253	0.61	7.9	10	193	18	99	6.2
benzoate	K	26	208	0.65	7.0	18	176	18	89	5.8
Vehicle	A	8	281	2.5	7.9	10	172	44	475	14.5
control	K	12	221	2.4	6.9	16	160	50	483	12.0

[a]Pit. = pituitary. Adrenal weights were not included because of no difference. For details see Weisburger et al. (1968).
[b]Diet A = control, Diet K = 160 ppm N-OH-FAA.

number of rats with carcinomas in the "male" but gave a variable result in the "female." TP did increase the incidence of carcinoma in the "female" group, with material numbers of cancer in the "males" (Table VI).

Metabolism studies of N-OH-FAA in the neonatally treated rats were done on rats which had been on the carcinogen for 8 weeks. The amount of radioactivity (expressed in microcuries per mole of metabolites per milliliter) circulating in the plasmas matched those reported earlier. Although the blood values were quite variable, the plasma levels were constant, with lower values in the rats on carcinogenic diet. The treated males on diet A showed consistently higher levels

TABLE VI
Liver Histopathology

			Liver pathology		
			Hyperplasia		
		Liver wt.			Cancer
Sex	Group	(gm)	0	+	(%)
Male	Testosterone propionate	13.2	–	5	19 (79)
	Estradiol benzoate	9.0	–	19	7 (36)
	Control	13.3	–	1	11 (91)
Female	Testosterone propionate	9.5	–	14	4 (22)
	Estradiol benzoate	5.9	7[a]	9	2 (11)
	Control	5.9	–	16	– (0)

[a]No hyperplasia, but toxic changes.

than the control males and all the females on diet A. In the urinary excretion studies (Fig. 4), the rats on diet K excreted more radioactivity than those on diet A, the control females on the carcinogenic diet excreting the most. However, the excess in the female was mostly in the glucuronide fraction, which turned out to be the *N*-OH-FAA conjugate, as shown in Fig. 5. The male rats synthesized and excreted more of the sulfate fraction. Most of the nonexcreted carcinogen was retained and bound to the tissues.

In order to further study the metabolism of the carcinogen, we have undertaken a study of some enzyme systems involved in the metabolism of the carcinogen and the binding to the sites of carcinogen action. Thus far, we have studied only β-glucuronidase and acetyltransferase activities. For β-glucuronidase assay we have used Fishman's method (1965) involving the hydrolysis of phenolphthalein glucuronide. The acetyltransferase activity was measured by a modified method of Lotlikar and Luha (1971) using 2-fluorenamine and acetyl-CoA. The first analysis was made when the rats were 7 weeks old, just before they were placed on their respective diets (Fig. 6). The male rats had higher acetyltransferase and kidney β-glucuronidase activities. (Figs. 7 and 8) On the other hand, in female rats liver β-glucuronidase was more active. The two liver enzymes increased in both females and males maintained on the dietary carcinogen. The kidney β-glucuronidase increased in animals on the *N*-OH-FAA diet.

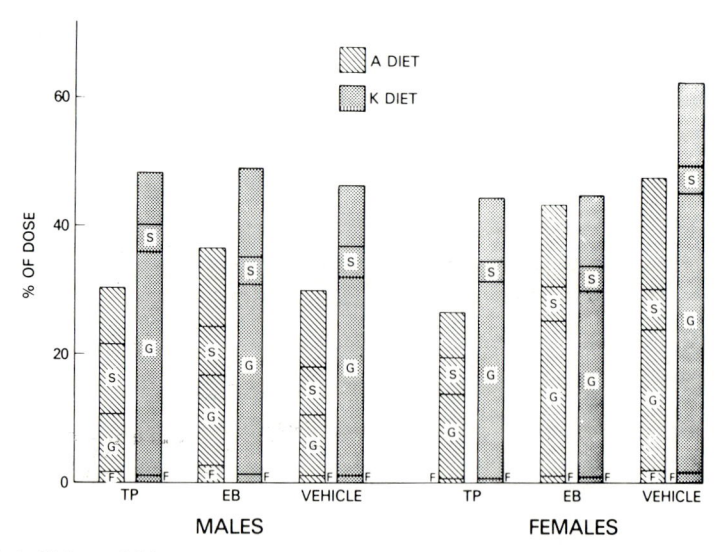

FIG. 4. Urinary 24-hour excretion of metabolites in rats treated neonatally with *N*-hydroxy-*N*-2-fluorenylacetamide. G, glucuronide fraction; S, sulfate fraction; F, free compound.

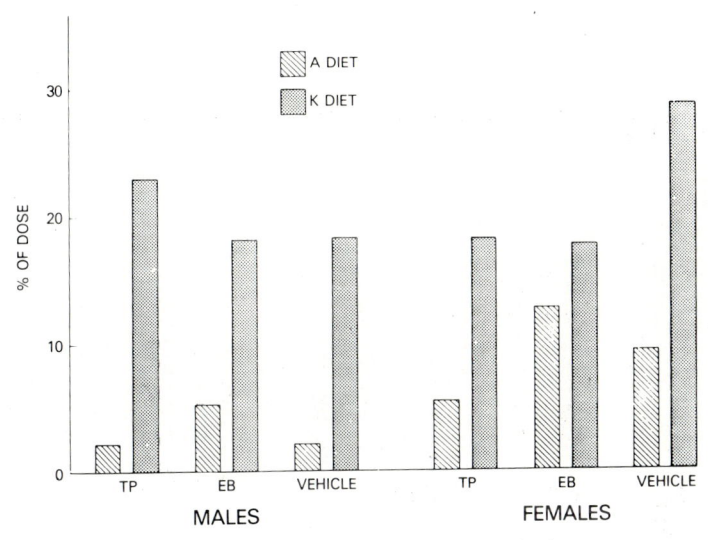

FIG. 5. Glucuronide fraction from Fig. 4.

The TP-treated rats responded with slightly lower enzyme activities when on carcinogenic diet. There was no trend in the enzymes to correlate with tumorigenesis in the newborn steroid-treated rats. Further work will be done with more direct-acting enzymes, such as sulfotransferase or glucuronyltransferase.

2. Pituitary Hormones in Liver Carcinogenesis

The pituitary gland seems to exert its effect on carcinogenesis in a dual fashion by regulating the activating metabolism and influencing the growth and development of the latent tumor cells. Hypophysectomized mice had a considerably lower incidence of carcinoma than intact mice when treated with 3,4-benzo[a]pyrene (Andervont et al., 1961). The lowered incidence made the authors think that the difference was quantitative rather than qualitative. Moon and his co-workers (1950a,b,c) observed that prolonged administration of growth hormone in female rats resulted in tumors of the lymphatic tissues, reproductive organs, and adrenals. In contrast, there were no tumors evident when hypophysectomized rats were used (Moon et al., 1951). Moon et al. (1952) failed to induce tumors in hypophysectomized rats with implantation of 3-methylcholanthrene. The carcinogenic action of 3'-MeDAB was completely inhibited by hypophysectomy (Griffin et al., 1953a,b; 1955a,b; Robertson et al., 1953, 1954).

Hypophysectomy seems to block the carcinogenic process in the liver, particularly as shown with 3'-Me-DAB (Griffin et al., 1953a,b; Robertson et al., 1953,

FIG. 6

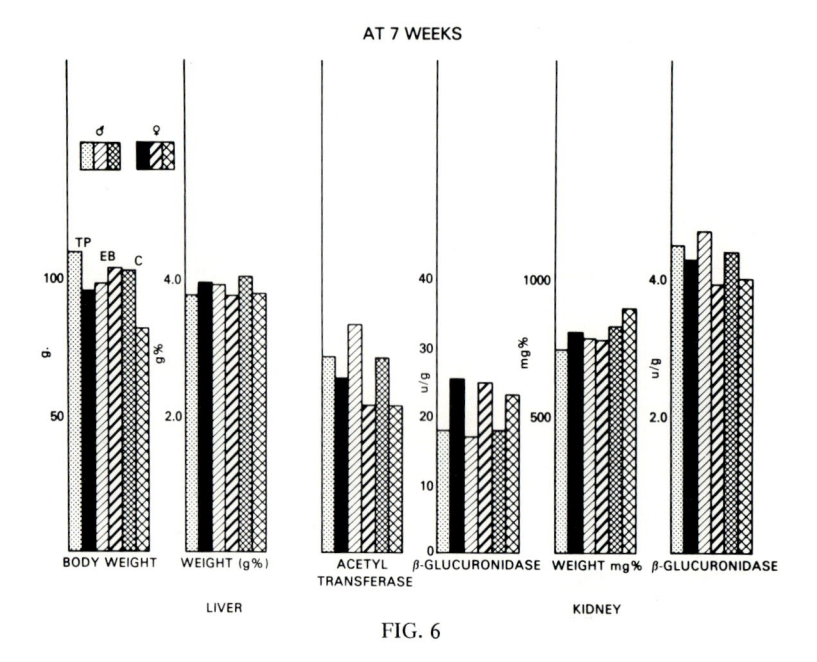

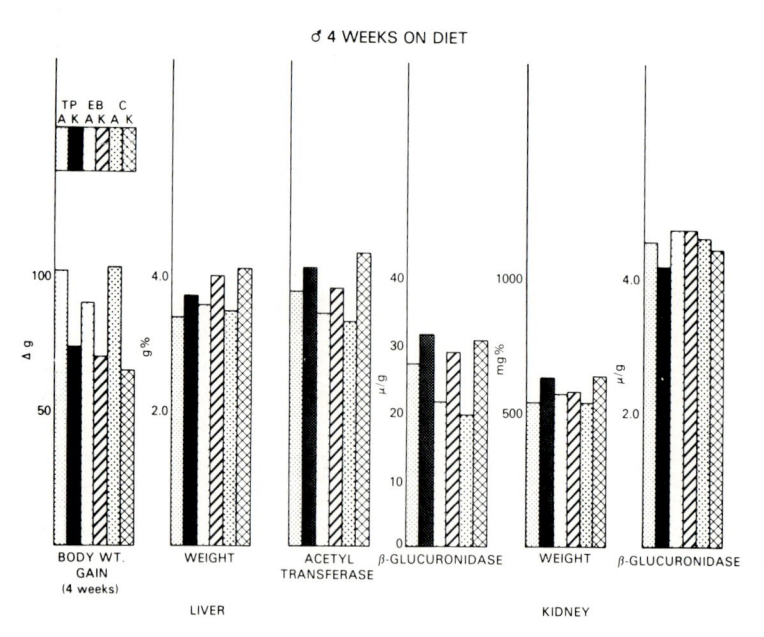

FIG. 7

FIG. 8

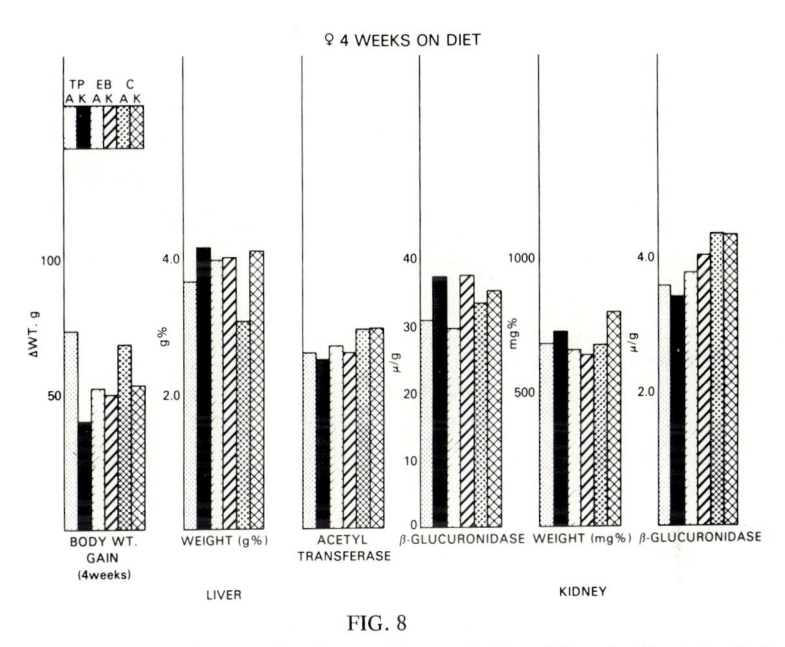

FIGS. 6–8. Acetyltransferase and β-glucuronidase activities of 7-week-old rats just before they were placed on their respective diets (Fig. 6) and of male rats (Fig. 7) and female rats (Fig. 8) after 4 weeks on diet.

1954), with aflatoxin (Goodall and Butler, 1969), and with FAA (Reuber, 1969; Skoryna, 1955), but not with dimethylnitrosamine (Lee and Goodall, 1968). However, evidence has appeared showing conflicting results on the effect of the pituitary on the growth of established tumors or on the development of transplanted tumor. Hypophysectomy inhibits the growth of some tumors, but other tumors continue to grow. Gardner (1942) found that spontaneous mammary tumors continued to develop in hypophysectomized mice and fibrosarcomas could be successfully transplanted in hypophysectomized rats (Loeffler, 1952).

The time of hypophysectomy is important in liver carcinogenesis. Hypophysectomy performed 14 weeks after the start of carcinogen feeding (Skoryna, 1955), prevented tumor development, but not at 23 weeks. Neither did it prevent the conversion of FDAA-induced hyperplastic hepatic cells into malignant tumors (Reuber, 1969).

The extension of the duration of feeding the carcinogen-containing diet to over twice the usual 13 weeks, i.e., 28 weeks (Griffin et al., 1953a,b), did not induce any hepatoma. Lower food intake with consequent decrease in the intake of 3'-Me-DAB was not the cause, for pair-fed intact animals had the usual liver damage at 10 weeks and tumors at 13 weeks (Robertson et al., 1953).

Administration of growth hormones to hypophysectomized rats restored some of the hepatocarcinogenic effects of 3′-Me-DAB (Griffin et al., 1955a,b; Robertson et al., 1954). However, tumors were diagnosed as cholangiomas, whereas intact animals had both cholangiomas and hepatocellular carcinomas (Robertson et al., 1954). Adrenocorticotropin, gonadotropin, and thyrotropin were also able to produce some effect (Robertson et al., 1954). On the other hand, adrenocorticotropin did not affect FDAA carcinogenicity (O'Neal et al., 1958). Thyrotropin was ineffective in restoring tumors in hypophysectomized rats fed 3′-Me-DAB (Griffin et al., 1955a,b).

As a source of pituitary hormones we used pituitary tumor MtT/F4 to test its effect on FAA hepatocarcinogensis. The work described below was done in our laboratory, which was then headed by Dr. John Weisburger (Shirasu et al., 1966, 1967a,b; Weisburger, 1968; J. H. Weisburger et al., 1964; J. H. Weisburger, unpublished data).

The choice of a pituitary tumor came about when there was no supply of rat pituitary hormones. The use of bovine hormones was then prevalent, but we felt that, aside from the immunological problem, the specificity was not sufficient. The other benefit of a pituitary tumor is that the source of hormones and the secretion of hormones were constant. Injection of hormones usually produces a pulsating condition that reaches a high level at the time of administration and the lowest level just before the next injection. For carcinogenesis studies, which require months of administration, a constant level is desirable. Other techniques which have been used, but with greater effort, are constant injections, Rose's technique using a constant hormone secreter, or implantation of pellets. This latter works very well with steroids but is questionable for protein hormones because of stability problems.

We have been very successful in maintaining and transplanting (for which we are very thankful to Drs. Jacob Furth and Robert Bates) the pituitary tumor MtT/F4 in F344 strain rats (Bates et al., 1962; Furth et al., 1956). All our transplantations have been successful.

We chose N-OH-FAA as the carcinogen for several reasons. (a) It was an effective hepatocarcinogen, less toxic in the diet and thus gave very reproducible results. (b) It was a more proximal carcinogen than FAA or FDAA, thus bypassing the influences on the activation necessary for carcinogenesis in the mammary gland, Zymbal's gland, and kidney.

We asked the following questions: (i) Will pituitary hormones secreted constantly enhance carcinogenesis? (ii) Will the hormones promote carcinogenesis after carcinogenesis is initiated? (iii) Do the hormones play a role in the activating mechanism of procarcinogens? (iv) Do the hormones play a role in the growth and development of carcinoma? (v) What is the MtT effect on castrates?

Experimentally, (a) MtT was implanted in rats fed a carcinogen-containing diet at the same time. (b) Carcinogen was first fed for 8 weeks; then the rats were implanted with MtT and continued on carcinogen diet for 10 more weeks. (c) Carcinogen was first fed for 12 weeks followed by MtT implantation and then diet A. (d) Metabolism of the carcinogen was examined under the first regimen. (e) Comparison was made with rats administered purified bovine hormone. In all these experiments, we used N-OH-FAA at the 0.016% (or 160 ppm) level in a semipurified diet (diet K) and, as a control, the semipurified diet (diet A). Both male and female F344 rats were used in all the experiments, started on experiment at 4 weeks of age.

The results of the experiments are shown in Table VI, and Table VII shows the body and organ weights of the various groups of rats. MtT depressed weight gain. As previously reported, diet K was quite toxic and decreased the weight gain. The MtT starts to grow 1 week after implantation and grows quite rapidly, so that at about 10–15 weeks the rats begin to die from exhaustion from the ACTH effect. The responses of females are similar, except that the carcinogenic diet does not affect their growth as it does that of males. The organ weights of the diet A and diet K groups did not differ much either with or without MtT.

The liver weights of both males and females in the MtT-implanted groups increased tremendously. The kidneys of the MtT-implanted rats were also enlarged. However, the carcinogen-diet group with implanted MtT had slightly smaller kidneys, but there was much more damage than with only MtT implantation. Both the females and males in this group were polydipsic and polyuric. The MtT-implanted rats also had extreme proteinuria due to albumin. Conversely, the plasma was completely depleted of albumin but had increased levels of γ-globulin. The adrenals weighed seven to eight times as much as the controls. There is a possibility that the high content of ACTH caused the increased

TABLE VII
Body and Organ Weights—13 Weeks Treatment

Sex	N-OH-FAA[a]	MtT[b]	No. of rats	Body (gm)	Liver (gm)	Kidney (gm)	Testes (gm)	Adrenals (mg)	Pituitary (mg)
Male	−	−	10	249	8.1	1.58	2.54	37	7.2
	+	−	7	182	15.0	1.20	2.27	32	6.3
	−	+	13	147	13.2	5.19	0.66	449	6.0
	+	+	12	110	10.9	2.26	0.42	295	5.0
							Ovaries (mg)		
Female	−	−	10	163	5.0	1.05	58.4	48	11.1
	+	−	9	119	5.9	0.87	35.9	31	6.3
	−	+	15	140	11.2	3.86	46.6	381	7.0
	+	+	16	106	10.4	1.95	30.8	353	4.6

[a] N-Hydroxy-N-2-fluorenylacetamide.
[b] Pituitary tumor MtT/F4.

metabolic rate and early death of the MtT-implanted rats. Very little or no abdominal fat was seen in the MtT-implanted rats. The testes of the MtT-implanted rats atrophied and did not grow as was expected. The ovaries showed no differences even though there seemed to be an excess of lactogenic hormones. The pituitaries of the females atrophied, being comparable in size to those of the males. There was a tremendous increase in plasma volume due to both an increase in blood volume and a decrease in hematocrit.

As seen in Table VIII the results from liver histopathology showed that 13 weeks on N-OH-FAA at 160 ppm did not suffice to induce hepatoma in all the rats. Usually the rats were kept about 10 more weeks on control diet for growth and development of tumors. Livers of the MtT group then showed hyperplastic nodules with small hepatoma and carcinoma. The females had very few nodular hyperplasia, which rapidly turned into small hepatoma in the presence of MtT. Thus MtT enhanced the carcinogenic effect of N-OH-FAA by shortening the time for induction of hepatoma.

During the course of the studies, we noticed that the rats implanted with MtT did not gain weight in spite of the higher food and water intake. The animals did feel slightly warmer. We thus determined the body weights, body temperature, food and water intake, and urinary output, as given in Fig. 9.

To clarify the enhancement of N-OH-FAA carcinogenicity with MtT, studies on the metabolism of N-OH-FAA were made in rats fed N-OH-FAA and given MtT. The results show that in the male rats 50% of the N-OH-FAA excreted in the urine was either the free compound or the glucuronide of N-OH-FAA. Of the glucosiduronic acid metabolites, about 40% combined directly with N-OH-FAA, very little being detoxified through ring hydroxylation. In the female, 52.2% of

TABLE VIII
Liver Histopathology—13 Weeks Treatment

Sex	N-OH-FAA[a]	MtT[b]	No. of rats	Normal	Hypertrophy	Hyperplastic nodules	Carcinoma
Male	−	−	10	10	−	−	−
	+	−	7	−	−	5	2
	−	+	13	−	13	−	−
	+	+	12	−	−	−	12
Female	−	−	10	10	−	−	−
	+	−	9	−	−	9	−
	−	+	15	−	15	−	−
	+	+	16	−	−	2	14

[a] N-Hydroxy-N-2-fluorenylacetamide.
[b] Pituitary tumor MtT/F4.

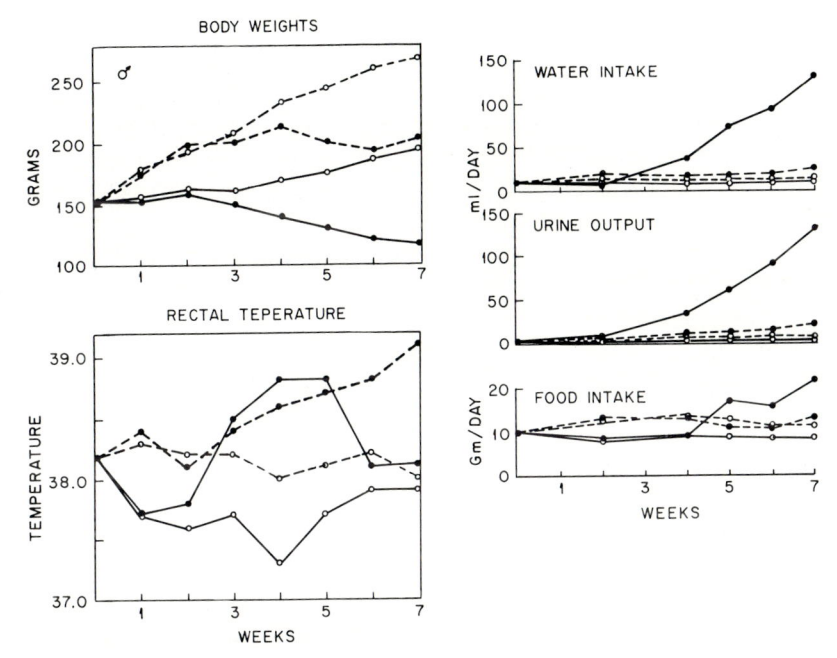

FIG. 9. Determinations in rats implanted with pituitary tumor MtT/F4. ○– – –○, diet A; ○———○, diet K; ●– – –●, diet A + testosterone; ●———●, diet K + testosterone.

the injected dose was found in the glucosiduronic acid moiety, of which 34.0% of the dose was as the *N*-OH-FAA glucuronide. Thus the maintenance of elevated amounts of the proximate carcinogen *N*-OH-FAA and lower conversion to other metabolites may contribute to the enhancement of hepatoma formation in both the female and the male rats.

In vitro studies of dehydroxylation and deacetylation of *N*-OH-FAA in liver homogenates from rats on the 4 experimental regimens, showed that the activity of both dehydroxylase and deacylase was lower in the MtT-implanted group, another explanation for the maintenance of a high concentration of the proximate carcinogen.

In order to test whether MtT could enhance and further develop carcinoma in rats, a study was set up to feed only the carcinogen-containing diet and then implant MtT. The first study resulted in the rapid death of the young rats. These results prompted another study of feeding the carcinogen for 8 weeks and then implanting MtT to test whether carcinogenesis could be enhanced. The results are shown in Tables IX and X. The organ weights were similar to those found in relation to the different regimens in the 13-week study. However, the body weights were greater in females of the MtT-implanted groups as compared to the

TABLE IX
Body and Organ Weights—18 Weeks on Experiment

Sex	N-OH-FAA[a]	MtT[b]	No. of rats	Body (gm)	Liver (gm)	Kidney (gm)	Testes (gm)	Adrenals (mg)	Pituitary (mg)
Male	−	−	5	257	7.8	1.53	2.63	32.0	7.4
	+	−	6	210	8.0	1.28	2.46	31.3	6.7
	−	+	5	189	13.1	4.26	1.08	318.0	6.9
	+	+	11	165	16.4	5.11	0.48	393.0	6.2
							Ovaries (mg)		
Female	−	−	4	170	5.3	1.02	56.0	42.0	12.4
	+	−	5	153	5.4	0.99	49.7	35.4	10.4
	−	+	8	231	16.6	5.28	53.2	407.0	8.4
	+	+	8	188	16.1	5.08	44.8	434.0	6.9

[a] N-Hydroxy-N-2-fluorenylacetamide.
[b] Pituitary tumor MtT/F4.

control groups. With respect to histopathology, all the male carcinogen-fed rats had well developed carcinomas. In the females, all the livers of the carcinogen-fed controls were normal, showing that the level of carcinogen produced hyperplastic lesions that disappeared instead of continuing to develop into carcinoma even if kept for a long time (Reuber, 1965b). However, the female rats im-

TABLE X
Liver Histopathology—18 Weeks on Experiment

Sex	N-OH-FAA[a]	MtT[b]	No. of rats	Normal	Hypertrophy	Hyperplastic nodules	Carcinoma
Male	−	−	5	5	−	−	−
	+	−	6	−	−	−	6
	−	+	5	−	5	−	−
	+	+	11	−	−	−	11
Female	−	−	4	4	−	−	−
	+	−	5	5	−	−	−
	−	+	8	−	8	−	−
	+	+	8	−	−	−	8

[a] N-Hydroxy-N-2-fluorenylacetamide.
[b] Pituitary tumor MtT/F4.

planted with MtT after carcinogen feeding did develop carcinoma. From this study we deduced that MtT implantation did have some promoting effect.

The demonstration that certain pituitary factors accelerated liver formation in rats fed N-OH-FAA and implanted with MtT led to an investigation of the participation of the gonadal system. Weanling F344 strain rats were castrated at 4 weeks of age. At 12 weeks of age, the intact and castrated rats were divided into two groups, one being fed 160 ppm of N-OH-FAA (diet K) and the other the control diet (diet A) for 12 weeks, after which all rats received the control diet. After a week on control diet, MtT was implanted in the designated groups. The effect of the regimen was observed 15 weeks later. The body and organ weights (Tables XI–XIV) were similar to those in the previous groups. However, inhibition of carcinogenesis in castrates contributed to keeping the liver at normal or smaller size. The spleen and heart were enlarged in the presence of MtT, owing to enhancement of the hematopoietic system. The endocrine organ weights were similar to those in previous studies, however, the adrenal response to ACTH in the male castrates was greater than in the intact male rats. The pituitary gland response to castration was, as expected, larger in the male but smaller in the female castrates. The growth of the MtT was the same in both male and female castrates, whereas the MtT usually grows larger and faster in the female.

The liver histopathology (Table XV) showed similar results as the previous study, where MtT enhanced carcinogenicity of N-OH-FAA in the intact female whereas the males were maximally affected even without the MtT implant. Thus MtT did enhance the hepatocarcinogenicity of the proximal carcinogen N-OH-FAA in both castrated male and female rats. However, castration, as shown by many previous workers (Firminger and Reuber, 1961; Morris and Firminger,

TABLE XI
Body and Organ Weights—Castrate Male F344—MtT[a]

Subject	MtT 0–12	MtT on 13	No. of rats	Body, 28 wk	Liver (gm/100 gm)	Kidney (gm/100 gm)	Spleen (mg/100 gm)	Heart (mg/100 gm)
Intact	A	–	8	331	2.9	0.53	160	247
	K	–	9	299	6.0	0.60	184	268
	A	+	7	318	6.0	1.72	289	398
	K	+	9	268	9.0	2.16	298	490
Castrate[a]	A	–	6	270	2.4	0.50	172	258
	K	–	5	216	3.0	0.57	174	260
	A	+	5	252	7.8	2.02	329	411
	K	+	7	208	9.2	2.2	264	418

[a]Castrated at 4 weeks; MtT = pituitary tumor MtT/F4.

TABLE XII
Endocrine Organ Weights—Castrate Male F344—MtT[a]

Subject	0–12	MtT on 13	No. of rats	Testes (gm)	Adrenals (mg)	Pituitary (mg)	MtT (gm/100 gm)
Intact	A	–	8	2.78	35.3	8.2	–
	K	–	9	2.81	32.7	8.3	–
	A	+	7	0.77	278.8	8.2	2.4
	K	+	9	0.74	285.7	7.9	2.4
Castrate	A	–	6	–	30.7	13.8	–
	K	–	5	–	38.2	11.6	–
	A	+	5	–	413.4	11.4	3.8
	K	+	7	–	385.2	10.6	3.6

[a]MtT = pituitary tumor MtT/F4.

TABLE XIII
Body and Organ Weights of Intact and Castrated Female F344 Rats—MtT[a]

Subject	0–12	MtT on 13	No. of rats	Body (gm)	Liver (gm/100 gm)	Kidney (gm/100 gm)	Spleen (mg/100 gm)	Heart (mg/100 gm)
Intact	A	–	10	198	3.2	0.62	179	294
	K	–	12	192	3.2	0.60	199	290
	A	+	9	271	7.1	1.49	379	419
	K	+	12	244	7.4	1.68	404	455
Castrate	A	–	6	208	2.2	0.53	180	263
	K	–	5	216	2.8	0.51	174	264
	A	+	5	274	5.8	1.10	308	353
	K	+	6	219	8.0	2.01	361	473

[a]Castrated at 4 weeks; started on experiment at 12 weeks. MtT = pituitary tumor MtT/F4.

TABLE XIV
Endocrine Organ Weights—Castrates ± MtT[a] Female F344

Subject	0–12	MtT on 13	No. of rats	Ovaries (mg)	Adrenals (mg)	Pituitary (mg)	MtT (gm/100 gm)
Intact	A	–	10	54.8	41.5	15.4	–
	K	–	12	54.5	41.0	14.3	–
	A	+	9	51.9	266.1	12.2	5.4
	K	+	12	47.9	290.4	11.5	4.3
Castrates	A	–	6	–	34.5	11.1	–
	K	–	5	–	33.3	10.7	–
	A	+	5	–	242.7	11.1	3.4
	K	+	6	–	292.8	9.9	3.9

[a]MtT = pituitary tumor MtT/F4.

TABLE XV
Sequential Action of MtT[a]

| | | | Liver histopathology | | | | |
| | | | Hyperplasia | | | | |
Subject	MtT	No. of rats	No.	Area	Nodular	Small hepatoma	Carcinoma
Male, intact	−	9	−	−	1	3	5
	+	7	−	−	1	4	3
Male, castrate	−	10	−	7	2	1	−
	+	8	−	2	1	4	2
Female, intact	−	12	8	2	2	−	−
	+	12	−	2	5	2	3
Female, castrate	−	5	3	2	−	−	−
	+	6	−	1	2	1	2

[a] 12 Weeks on diet K, 18 weeks on diet A. Pituitary tumor MtT/F4 (MtT) was implanted during 13th week.

1956; Reuber and Firminger, 1962; Rumsfeld et al., 1951), usually decreased the response of hepatocarcinogens.

Further work in our laboratory was done in an attempt to identify the particular hormone or hormones that were involved in the above findings (Shirasu et al., 1967a). Previous studies (Bates et al., 1962; Furth, 1961; Takemoto et al., 1962) established that Furth MtT/F4 produced large amounts of ACTH, growth hormone, and prolactin. Thus both ACTH and growth hormone were tested, assuming that prolactin stimulated the mammary gland and not the liver. ACTH was shown to elevate the rectal temperature and decrease the weight of the rats. However, growth hormone maintained a higher level of glucosiduronic acid conjugates and concomitantly the level of N-OH-FAA glucosiduronic acid.

These studies show that the increase in tumorigenesis through secretion of hormones by MtT may be like the promoting effect of croton oil in skin carcinogenesis. The MtT effect may be that of maintaining a higher level of the carcinogen, especially in the studies where MtT was implanted at the same time the carcinogen was administered. However, in the latter studies, where the MtT was implanted its hormonal action came after the removal of most of the carcinogen, demonstrating that the MtT effect enhanced the growth and development of the tumor. Many attempts to find an initiation-promotion system similar to that of skin carcinogenesis (Berenblum, 1941) for epithelial tissue have not been too successful. With an initiation-promotion system, we may be able to differentiate carcinogens from promoters.

The promoting power of growth hormone on the liver shows that the liver can hypertrophy and also have hyperplasia (histologically seen in MtT-treated animals). In the presence of altered cells, there could be tumor formation as reported in the liver regeneration studies by Glinos *et al.* (1951).

3. Other Hormones

Adrenalectomy has been reported to decrease the carcinogenicity of FDAA in female rats, but only slightly in the males (Firminger and Reuber, 1961; Reuber, 1965a). The administration of cortisone to either adrenalectomized or intact male and female rats did not alter this depression in hepatocarcinogenesis (Fishman, 1965; Reuber, 1969), but deoxycorticosterone acetate had a striking protective effect in both adrenalectomized and intact rats (Firminger and Reuber, 1961; Reuber, 1965a; Symeonidis *et al.*, 1954).

Thyroidectomy completely inhibited the hepatocarcinogenicity of FAA (Bielschowsky, 1958; Bielschowsky and Hall, 1953). However, thyroidectomy has to be done prior to or up to 14 weeks for FAA (Bielschowsky and Hall, 1953) or 23 weeks for FDAA (Reuber, 1969) after the start of carcinogen feeding to be effective. Feeding 0.1% thyroid powder restored the carcinogenicity of FDAA in thyroidectomized rats (Reuber, 1965c), as did iodide supplementation (Goodall, 1965). None of the steroids, including progesterone, androsterone, norethandrolone, or cortisone, or the pituitary hormones examined were completely effective in overcoming the protective effect of thyroidectomy (Reuber, 1965c). The anabolic hormones, such as growth hormone and norethandrolone, did advance the lesions to nodular hyperplasia (Reuber, 1965c).

B. HORMONES IN URINARY TRACT CARCINOGENESIS

1. Experimental Renal Carcinogenesis

Human renal tumors are quite uncommon, however; the incidence in men is 4.7 and in women 2.8 per 100,000. Spontaneous kidney tumors in mice and rats are also low. Experimentally prolonged treatment with estrogen led to the induction of kidney tumors in male hamsters but rarely in female hamsters (Kirkman and Bacon, 1952; Kirkman and Wurster, 1957). However, there are no fundamental genetic differences between the kidneys of male and female hamsters. Renal tumors can also be induced by estrogen treatment in immature or ovariectomized adult female hamsters. Tumors were also produced in female hamsters if administration of estrogen was started during metestrus, indicating that the high level of progesterone normally prevented the induction of kidney tumors in estrogen-treated female hamsters (Kirkman, 1959, 1974).

2. Urinary Tract Carcinogenesis

Spontaneous urinary tract cancer is comparatively rare in domesticated animals (Cotchin, 1956) owing to the fact that they are slaughtered at a very young age. Urinary tract cancer seems to be found in older animals. Bladder cancer is also uncommon in laboratory animals (Bullock and Curtis, 1930). However, the paucity of spontaneous cancer is a great advantage when testing chemicals for carcinogenicity, for even low incidences of induced tumors become significant.

A sex difference was found in the survey by Ashley (1969a,b)—namely, that women were more resistant to bladder cancer, the female to male ratio of bladder tumors being as low as 0.28.

The difference in response between male and female animals to induction of urinary bladder tumors has been observed, but found to be quite variable. The presence of foreign body in the mouse stimulated the formation of tumors, though simultaneous action of the two is not necessary (Clayson *et al.*, 1967). Thus the interpretations of bladder carcinogenesis will be difficult as long as certain carcinogens are used.

N-Butyl-*N*-hydroxybutylnitrosamine (BBN), a hydroxylated derivative of dibutylnitrosamine (DBN) was found to be a good urinary bladder carcinogen (Druckrey *et al.*, 1967). When rats were administered BBN in drinking water and injected with testosterone propionate 3 times per week for 20 weeks, there was an enhancement of urinary bladder carcinogenesis (Kono *et al.*, 1973). However, estriol had no effect on enhancement of BBN tumorigenesis but seemed to protect the animals.

C. ROLE OF HORMONES ON CARCINOGENESIS IN OTHER ORGANS

There are many other hormones that were not discussed here. One group of particular interest at this time are the gastroenteropancreatic hormones, such as gastrin, serotonin, histamine, which have been studied in glandular stomach carcinogenesis. The administration of gastrin but not serotonin, histamine, or insulin during the process of *N*-methyl-*N'*-nitro-*N*-nitrosoguanidine gastric carcinogenesis induced poorly differentiated adenocarcinoma and scirrhous carcinoma in the glandular stomach (Tahara and Haizuka, 1975). This biological model seems to be very important for the study of comparable human gastric cancer. In Japan, where stomach cancer is 5 to 6 times as prevalent as in the United States, many of the gastric cancer patients are young and tend to have the scirrhous type (a very hard cartilaginous carcinoma). Thus this model enables the study of the morphogenesis of this type of cancer in comparison with the predominant undifferentiated adenocarcinoma found mostly in older people.

Cancer of the pancreas as shown in the American Cancer Society's chart shows a greater increase in men than in women. Similar observations were made by

Muir (1961). Spontaneous pancreatic tumors were also found predominantly in male rats (Rowlatt and Roe, 1967) and in male hamsters (Fortner, 1961). The newly synthesized pancreatic duct carcinogen, diisopropanolnitrosamine, induced tumors in Syrian golden hamsters (Kruger et al., 1974; Pour et al., 1974). The female hamsters withstood the toxicity better and had a longer latency period, demonstrating again the difference in response between the sexes.

VI. Summary

In summary, it seems that, within the steps in cancer formation as presented, the hormones play a role at two stages: (a) the activation of the chemical carcinogen to the "ultimate" carcinogen, and (b) possibly during the growth and development of the latent tumor cells into a differentiated tumor. The interplay of hormones is so delicate and all-encompassing that it is difficult to conclude specifically which stage is affected more.

We have shown that the anabolic hormones are possibly working in the area of growth and development, while others affect the activating steps. More effort is needed to delineate the specific roles of hormones in chemical carcinogenesis.

REFERENCES

Andervont, H. B. (1950). *J. Natl. Cancer Inst.* **11**, 581.
Andervont, H. B., Shimkin, M. B., and Canter, H. Y. (1961). *Acta Unio Int. Cancrum* **17**, 105.
Ashley, D. J. B. (1969a). *Br. J. Cancer* **23**, 21.
Ashley, D. J. B. (1969b). *Br. J. Cancer* **23**, 26.
Baba, T., and Takayama, S. (1961). *Gann* **52**, 73.
Barraclough, C. A. (1961). *Endocrinology* **68**, 62.
Bates, R. W., Milkovic, S., and Garrison, M. M. (1962). *Endocrinology* **71**, 943.
Berenblum, I. (1941). *Cancer Res.* **1**, 807.
Biancifiori, C. (1970). *J. Natl. Cancer Inst.* **44**, 943.
Bielschowsky, F. (1944). *Br. J. Exp. Pathol.* **25**, 1.
Bielschowsky, F. (1958). *Br. J. Cancer* **12**, 231.
Bielschowsky, F. (1961). *Acta Unio Int. Cancrum* **17**, 121.
Bielschowsky, F., and Hall, W. H. (1953). *Br. J. Cancer* **7**, 358.
Bielschowsky, F., and Horning, E. S. (1958). *Br. Med. Bull.* **14**, 106.
Boyland, E. (1961). *Prog. Exp. Tumor Res.* **2**, 145.
Bullock, F. D., and Curtis, M. R. (1930). *J. Cancer Res.* **14**, 1.
Butler, W. H. (1964). *Br. J. Cancer* **18**, 756.
Clayson, D. B., Lawson, T. A., and Pringle, J. A. S. (1967). *Br. J. Cancer* **21**, 755.
Clayton, C. C., and Baumann, C. A. (1949). *Cancer Res.* **9**, 575.
Cotchin, E. (1956). *Commonw. Bur. Anim. Health, Rev. Ser.* **4**.
Cramer, J. W., Miller, J. A., and Miller, E. C. (1960). *J. Biol. Chem.* **235**, 885.

DeBaun, J. R., Rowley, E. C., Miller, E. C., and Miller, J. A. (1968). *Proc. Soc. Exp. Biol. Med.* **129**, 268.
DeBaun, J. R., Miller, E. C., and Miller, J. A. (1970a). *Cancer Res.* **30**, 577.
DeBaun, J. R., Smith, J. Y. R., Miller, E. C., and Miller, J. A. (1970b). *Science* **167**, 184.
Druckrey, H., Preussmann, R., Ivankovic, S., and Schmähl, D. (1967). *Z. Krebsforsch.* **69**, 103.
Evans, E. S., Taurog, A., Koneff, A. A., Potter, G. D., Chaikoff, I. L., and Simpson, M. E. (1960). *Endocrinology* **67**, 619.
Farber, E. (1968). *Cancer Res.* **28**, 1859.
Firminger, H. I., and Reuber, M. D. (1961). *J. Natl. Cancer Inst.* **27**, 559.
Fishman, W. H. (1965). *Methods Horm. Res.* **4**, 273.
Fortner, J. G. (1961). *Cancer Res.* **21**, 1491.
Furth, J. (1961). *Fed. Proc., Fed. Am. Soc. Exp. Biol.* **20**, 865.
Furth, J., Clifton, K. H., Gadsden, E. L., and Buffet, R. F. (1956). *Cancer Res.* **16**, 608.
Gardner, W. U. (1942). *Cancer Res.* **2**, 476.
Gardner, W. U. (1953). *Adv. Cancer Res.* **1**, 173.
Gardner, W. U., Pfeiffer, C. A., and Trentin, J. J. (1957). *In* "Physiopathology of Cancer" (F. Homburger, ed.), p. 152. Harper (Hoeber), New York.
Glinos, A. D., Bucher, N. L. R., and Aub, J. C. (1951). *J. Exp. Med.* **93**, 313.
Goodall, C. M. (1965). *Endocrinology* **76**, 1027.
Goodall, C. M., and Butler, W. H. (1969). *Int. J. Cancer* **4**, 422.
Griffin, A. C., Rinfret, A. P., and Corsiglia, V. F. (1953a). *Cancer Res.* **13**, 77.
Griffin, A. C., Rinfret, A. P., O'Neal, M., and Robertson, C. H. (1953b). *Proc. Am. Assoc. Cancer Res.* **1**, 21.
Griffin, A. C., O'Neal, M. A., Robertson, C. H., Hoffman, H. E., and Richardson, H. L. (1955a). *Acta Unio Int. Cancrum* **11**, 671.
Griffin, A. C., Richardson, H. L., Robertson, C. H., O'Neal, M. A., and Spain, J. D. (1955b). *J. Natl. Cancer Inst.* **15**, 1623.
Gustafsson, J. A., and Stenberg, A. (1976). *Science* **191**, 203.
Hancock, R. L., and Dickie, M. M. (1969). *J. Natl. Cancer Inst.* **43**, 407.
Heston, W. E., and Vlahakis, G. (1968). *J. Natl. Cancer Inst.* **40**, 1161.
Hoch-Ligeti, C. (1947). *Br. J. Cancer* **1**, 391.
Huggins, C. (1963). *J. Am. Med. Assoc.* **186**, 481.
Irving, C. C., and Wiseman, R., Jr. (1969). *Cancer Res.* **29**, 812.
Irving, C. C., Wiseman, R., Jr., and Young, J. M. (1967a). *Cancer Res.* **27**, 838.
Irving, C. C., Wiseman, R., Jr., and Hill, J. T. (1967b). *Cancer Res.* **27**, 2309.
Irving, C. C., Janss, D. H., and Russell, L. T. (1971). *Cancer Res.* **31**, 387.
Kato, R. (1974). *Drug. Metab. Rev.* **3**, 1.
Kennaway, E. L., and Hieger, I. (1930). *Br. Med. J.* **1**, 1044.
King, C. M., and Phillips, B. (1968). *Science* **159**, 1351.
Kirby, A. H. M. (1947). *Br. J. Cancer* **1**, 68.
Kirkman, H. (1959). *Natl. Cancer Inst., Monogr.* **1**, 1.
Kirkman, H. (1974). *Cancer Res.* **34**, 2728.
Kirkman, H., and Bacon, R. L. (1952). *J. Natl. Cancer Inst.* **13**, 757.
Kirkman, H., and Wurster, D. H. (1957). *Proc. Am. Assoc. Cancer Res.* **2**, 221.
Kirschbaum, A. (1957). *Cancer Res.* **17**, 432.
Kono, N., Sasaki, N., Umetsu, R., and Otani, T. (1973). *Kobe J. Med. Sci.* **19**, 151.
Korteweg, R., and Thomas, F. (1939). *Am. J. Cancer* **37**, 36.

Kruger, F. W., Pour, P., and Althoff, J. (1974). *Naturwissenschaften* **61**, 328.
Leathem, J. H. (1951). *Cancer Res.* **11**, 266.
Lee, K. Y., and Goodall, C. M. (1968). *Biochem. J.* **106**, 767.
Levine, S. (1966). *Sci. Am.* **214**, 84.
Loeffler, J. B. (1952). *Cancer* **5**, 161.
Lotlikar, P. D., and Luha, H. (1971). *Biochem. J.* **123**, 287.
Lotlikar, P. D., Enomoto, M., Miller, E. C., and Miller, J. A. (1964). *Cancer Res.* **24**, 1835.
Lower, G. M., Jr., and Bryan, G. T. (1973). *Biochem. Pharmacol* **22**, 1581.
Miller, E. C., and Miller J. A. (1947). *Cancer Res.* **7**, 468.
Miller, E. C., and Miller J. A. (1966). *Pharmacol. Rev.* **18**, 805.
Miller, E. C., Miller, J. A., and Hartmann, H. A. (1961). *Cancer Res.* **21**, 815.
Miller, E. C., Miller, J. A., and Enomoto, M. (1964). *Cancer Res.* **24**, 2018.
Miller, J. A. (1970). *Cancer Res.* **30**, 559.
Miller, J. A., and Miller E. C. (1969a). *Jerusalem Symp. Quantum Chem. Biochem.* **1**, 237.
Miller, J. A., and Miller, E. C. (1969b). *Prog. Exp. Tumor Res.* **11**, 273.
Miller, J. A., Cramer, J. W., and Miller, E. C. (1960). *Cancer Res.* **20**, 950.
Miller, W. L., and Baumann, C. A. (1951). *Cancer Res.* **11**, 634.
Moon, H. D., Simpson, M. E., Li, C. H., and Evans, H. M. (1950a). *Cancer Res.* **10**, 297.
Moon, H. D., Simpson, M. E., Li, C. H., and Evans, H. M. (1950b). *Cancer Res.* **10**, 364.
Moon, H. D., Simpson, M. E., Li, C. H., and Evans, H. M. (1950c). *Cancer Res.* **10**, 549.
Moon, H. D., Simpson, M. E., Li, C. H., and Evans, H. M. (1951). *Cancer Res.* **11**, 535.
Moon, H. D., Simpson, M. E., and Evans, H. M. (1952). *Science* **116**, 331.
Morris, H. P., and Firminger, H. I. (1956). *J. Natl. Cancer Inst.* **16**, 927.
Morris, H. P., Velat, C. A., Wagner, B. P., Dahlgard, M., and Ray, F. E. (1960). *J. Natl. Cancer Inst.* **24**, 149.
Muir, C. S. (1961). *Br. J. Cancer* **15**, 30.
O'Neal, M. A., Hoffman, H. E., Dodge, B. G., and Griffin, A. C. (1958). *J. Natl. Cancer Inst.* **21**, 1161.
Potter, M. (1963). *Natl. Cancer Inst., Monogr.* **10**, 1.
Pour, P., Kruger, F. W., Althoff, J., Cardesa, A., and Mohr, U. (1974). *Am. J. Pathol.* **76**, 349.
Reuber, M. D. (1965a). *J. Natl. Cancer Inst.* **34**, 587.
Reuber, M. D. (1965b). *J. Natl. Cancer Inst.* **34** 697.
Reuber, M. D. (1965c). *J. Natl. Cancer Inst.* **35**, 959.
Reuber, M. D. (1966). *J. Natl. Cancer Inst.* **36** 775.
Reuber, M. D. (1968). *Cancer Res.* **28**, 2177.
Reuber, M. D. (1969). *J. Natl. Cancer Inst.* **43**, 445.
Reuber, M. D. (1975). *Gann Monogr. Cancer Res.* **17**, 301.
Reuber, M. D. (1976). *Eur. J. Cancer* **12**, 137.
Reuber, M. D., and Firminger, H. I. (1962). *J. Natl. Cancer Inst.* **29**, 933.
Reuber, M. D., and Firminger, H. I. (1963). *J. Natl. Cancer Inst.* **31**, 1407.
Richardson, H. L., Griffin, A. C., and Rinfret, A. P. (1953). *Cancer* **6**, 1025.
Richardson, H. L., O'Neal, M. A., Robertson, C. H., and Griffin, A. C. (1954). *Cancer* **7**, 1044.
Robertson, C. H., O'Neal, M. A., Griffin, A. C., and Richardson, H. L. (1953). *Cancer Res.* **13**, 776.
Robertson, C. H., O'Neal, M. A., Richardson, H. L., and Griffin, A. C. (1954). *Cancer Res.* **14**, 549.
Rowlatt, U., and Roe, F. J. C. (1967). *J. Natl. Cancer Inst.* **39**, 18.
Rumsfeld, H. W., Jr., Miller, W. L., Jr., and Baumann, C. A. (1951). *Cancer Res.* **11**, 814.

Saunders, F. J., and Drill, V. A. (1956). *Endocrinology* **58**, 567.
Saunders, F. J., and Drill, V. A. (1957). *Proc. Soc. Exp. Biol. Med.* **94**, 646.
Shimkin, M. B. (1965). *Cancer Res.* **25**, 1363.
Shirasu, Y., Grantham, P. H., Yamamoto, R. S., and Weisburger, J. H. (1966). *Cancer Res.* **26**, 600.
Shirasu, Y., Grantham, P. H., and Weisburger, J. H. (1967a). *Int. J. Cancer* **2**, 59.
Shirasu, Y., Grantham, P. H., Weisburger, E. K., and Weisburger, J. H. (1967b). *Cancer Res.* **27**, 81.
Sidransky, H., Wagner, B. P., and Morris, H. P. (1961). *J. Natl. Cancer Inst.* **26**, 151.
Skoryna, S. C. (1955). *Proc. Can. Cancer Res. Conf.* **1**, 107.
Stasney, J., Paschkis, K. E., Cantarow, A., and Rothenberg, M. S. (1947). *Cancer Res.* **7**, 356.
Stewart, H. L., and Snell, K. C. (1959). *In* "The Physiopathology of Cancer" (F. Homburger, ed.), 2nd ed., p. 85. Harper (Hoeber), New York.
Symeonidis, A., Mulay, A. S., and Burgoyne, F. H. (1954). *J. Natl. Cancer Inst.* **14**, 805.
Tahara, E., and Haizuka, S. (1975). *Gann* **66**, 421.
Takemoto, H., Yokoro, K., Furth, J., and Cohen, A. I. (1962). *Cancer Res.* **22**, 917.
Toh, Y.-C. (1972). *J. Natl. Cancer Inst.* **48**, 113.
Whalen, R. E. (1965). *J. Comp. Physiol. Psychol.* **57**, 175.
Weisburger, E. K., Grantham, P. H., and Weisburger, J. H. (1964). *Biochemistry* **3**, 808.
Weisburger, E. K., Yamamoto, R. S., Glass, R. M., Grantham, P. H., and Weisburger, J. H. (1968). *Endocrinology* **82**, 685.
Weisburger, J. H. (1968). *Mol. Pathol. Carcinogenesis N.Z. Med J.* **67**, 44.
Weisburger, J. H. (1973). *In* "Cancer Medicine" (J. F. Holland and E. Frei, III, eds.), p. 45. Lea & Febiger, Philadelphia, Pennsylvania.
Weisburger, J. H., and Weisburger, E. K. (1963). *Clin. Pharmacol. Ther.* **4**, 110.
Weisburger, J. H., Pai, S. R., and Yamamoto, R. S. (1964). *J. Natl. Cancer Inst.* **32**, 881.
Weisburger, J. H., Yamamoto, R. S., Korsis, J., and Weisburger, E. K. (1966). *Science* **154**, 673.
Weisburger, J. H., Yamamoto, R. S., Williams, G. M., Grantham, P. H., Matsushima, T., and Weisburger, E. K. (1972). *Cancer Res.* **32**, 387.
Wheeler, G. P., and Skipper, H. E. (1957). *Arch. Biochem. Biophys.* **72**, 465.
Yamagiwa, K., and Ichikawa, K. (1915). *Mitt. Med. Fak. Univ. Tokyo* **15**, 295.

DISCUSSION

L. Bullock: I would like to mention some work that we have been doing which is related to what you presented today. We have been working in collaboration with Dr. David Brusick on the hormonal control of mouse kidney microsomal activation of dimethylnitrosamine to a mutagen as a screening test for carcinogenicity. We employed the Bruce Ames test using a histidine-requiring mutant of *Salmonella typhimurium* to evaluate the mutagenicity of the microsomally activated drug. Initially, two interesting observations had been made: (1) Microsomes from different mouse strains varied in their ability to activate dimethylnitrosamine, and this correlated with the incidence of kidney tumors. This suggested that this *in vitro* test for mutagenicity did indeed correlate with *in vivo* carcinogenicity of dimethylnitrosamine. (2) Kidney microsomes from males were several orders of magnitude more active than those from females, suggesting that androgens may influence the mutagenic or carcinogenic potential of a substance in some tissues, in this case the kidney. To elucidate further this latter observation, we used the androgen-insensitive male mouse, which lacks androgen receptors, and its normal littermates. Kidney microsomes from normal males had higher activating potentials than those from females, and the activity of microsomes from androgen-insensitive mice was similar to those from females. Castration of the normal male brought the activity of its microsomes down to female levels. Treatment with androgens, increased the activating capability of microsomes from both males and females to levels greater than that seen in the normal male mouse. By contrast, the activity of microsomes from androgen insensitive mice was not affected by androgen treatment. This suggests that the role of androgens in influencing microsomal activation of dimethylnitrosamine is a specific one and is mediated via the androgen receptor. We extended this study to include a progestin, medroxyprogesterone acetate (MPA), which is androgenic in the mouse kidney. MPA induced the activating capabilities of microsomes from normal females but was inactive in the androgen-insensitive mouse. So this is another example of an androgenic effect of MPA which is probably mediated via the androgen receptor. More important, however, I think this is an example of how two different steroids, androgens and progestins, may influence the carcinogenecity of some drugs in at least one tissue, the mouse kidney.

R. S. Yamamoto: In that respect you will probably find that the estrogen will inhibit some of these reactions. The estrogens have some inhibitive activity on testosterone. On the other hand, progesterone has been found to be like the androgens, so it is on the anabolic side of the female cycle.

L. Bullock: We have not looked at castrated females or the effect of estrogen per se.

N. B. Schwartz: I think it is dangerous to use the particular two experimental models that you used, with neonatal estrogen or testosterone treatment, and assume that one has converted a male into a female or a female into a male. In fact, the work that has been done using either of those two experimental treatments shows that there are some shifts that are relatively permanent but there are some that are different. In any case, one has not done either of those "conversions." The female with that dose of testosterone may be on the borderline of cycling or not cycling, and the male with that dose of estrogen may be capable of normal spermatogenesis. Measurements of circulating steroids later on in life when you are testing your carcinogenic agents may reveal some abnormal steroids but does not necessarily reveal reductions in testosterone in the male, or reductions of estradiol in the females.

R. S. Yamamoto: As you know it is quite difficult to eliminate all the possibilities. The hormones play a role in three different areas and the hormonal milieu is responsive to changes so that by using the newborn treatment you get a sort of intersex that might give us a better clue than the removal of the organs, which then eliminate the stimulus to the other organs that may play an important functional role.

N. B. Schwartz: Fortunately there are some very easy ways of maintaining steady levels of steroids in either the female or the male. This is by implanting Silastic capsules containing the steroid; one can set the level of these steroids virtually at anything from zero to very high doses and know that they are steady. This might be an interesting steroid background on which you might test your carcinogens.

R. S. Yamamoto: Yes, I think the steroids are good for such implantation, but, if you are planning to use growth hormone or some other protein hormones, the stability of these proteins will be in question.

J. Furth: What I have to say will harmonize the speaker with the discussors. There is a fundamental difference between carcinogens and hormones. Carcinogens alter the DNA or DNA-chromatin and the altered DNA reproduces itself. As your experiments demonstrate, the hormonal action is reversible.

Utilizing estrogens (female and a carcinogenic hydrocarbon with an affinity for the mammary gland), you produced mammary tumors. A small dose of carcinogen produces latent tumor cells. Adding the hormonal stimulus, you produced mammary tumors. With removal of the ovary, the source of the stimulating hormone, mammary tumors of the dependent type will regress. In contrast, you selected AAF, a hepatocarcinogen favored by male androgens, and you produced hepatocarcinomas. Latent liver cancer cells induced by AAF can be reactivated by androgens. These experiments illustrate the fundamental nature of carcinogens and the role of hormones in it.

R. S. Yamamoto: Because of your excellent work on mammary tumorigenesis, Dr. Furth, I tried to keep away from tumorigenesis of organs in which the hormones are directly involved. However, at this meeting I realized that there are few organs that are not affected by hormones.

J. Furth: May I add to what I said before: as a source of large quantities of isologous hormones you used the transplantable tumor F4, which is trihormonal, having mammotropic, somatotropic, and adrenotropic activities. It might have been better to utilize monomorphous tumors. Analyzing your data, it seems that the prolactin secreted accounts for the mammary gland stimulation; the growth hormone may have been an overall stimulus of tumor growth: the influence of adrenotropins is uncertain.

SUBJECT INDEX

A

A cells
  extrapancreatic, 487–491
  glucagon and, 477–517
    immunoreactive, 487–491
  glucose-sensing function of, 487
Acromegaly, somatostatin role in, 4
ACTH
  biosynthetic precursors of, 284
  fragments of, in neurons, 18
  for water and solute permeation, 413–415
    analysis of, 423–424
    antidiuretic hormone effects on, 417–418
Adenomatosis, multiple endocrine type, pancreatic polypeptides in, 543–546
Adolescents, osmoregulation in, 384
Adrenergic nervous system, fetal susceptibility in, 116
Alanine aminotransferase, glucocorticoid induction of, 598
Amniotic fluid, thyroid hormones in, 115
Amphibia
  epithelia of, antidiuretic hormone effects on, 415–420
  gonadotropin effects on
    reproduction, 196–198, 218–219
    steroid production, 212–214, 221–222
  pituitaries from, 172
Anamnesis, in Graves' disease, 46
Angiotensin
  carbon-13 NMR studies on, 318, 323–324
  in neurons, 18
Antibodies, hormonelike activity of, 53
Antidiuresis
  vasopressin in, 367–369
  syndrome associated with, 369–372
Antidiuretic hormone (ADH), 387–434
  activity of, parallel, solubility-diffusion model for, 420–427
  effects on
    collecting tubules and amphibian epithelia, 415–420
    hydrophilic solute transport coefficients, 399–403
    permeation by moderately lipophilic solutes, 412–413
    transcellular vs. paracellular water flow, 411–412
    water and solute permeation rates, 398–413
  in water homeostasis, 387–434
Antimüllerian hormone, 117–167
  bioassay of, 123–125
  biochemical studies of, 148–159
  from calf testis, 136
  evidence for, 126–128
  gel filtration of, 148–151
  initiation of production of, 132–133
  ion-exchange chromatography of, 157–159
  loss of activity of, 133–134
  macromolecular nature of, 148–151
  membrane permeability to, 148–151
  in müllerian regression, 120–123
  physiological studies on, 120–147
  in postnatal testicular tissue, 134
  protein nature of, 145–146
  regulatory mechanisms for, 145–147
  site of production of, 135–145
  species specificity of, 128–132
  testicular, chronological evolution of, 132–134
  ultracentrifugation studies on, 154–156
Apical membranes, 389
  in water permeation pathway, 403–406
APUD concept relation to neurons and endocrine cells, 15
Arginine vasopressin, function of, in health and disease, 333–434
Ascites cell-free system parathyroid mRNA translation in, 276–279
Avian pancreatic polypeptide (aPP)
  amino acid sequence of, 520
  biological effects of, 521, 563–565
Axon
  dendrite connections with, 11
  hypothalamic peptide effects on, 9–10

B

Basolateral membranes, 389
Behavior, endorphin effects on, 9
Big plasma glucagon, occurrence of, 512–513

SUBJECT INDEX

Birds
  antimüllerian hormone in, 130–132
  gonadotropin effects on
    reproduction, 195–196, 217–218
    steroid production, 209–212
  pituitaries from, 171–172
Bladder, carcinogenesis in, hormone role in, 647
Bombesin, 14
Bovine pancreatic polypeptide (bPP)
  amino acid sequence of, 521
  biological effects of, 521
Breast cancer, luteinizing hormone releasing factor effects on, 20
Brittle diabetes
  C-peptide in, 464
  somatostatin in, 6
N-Butyl-N-hydroxybutylnitrosamine (BBN), as urinary tract carcinogen, 647

C

C-peptide
  circulating levels of, 451–462
    methods for, 452–454
  clinical significance of, 435–475
  in diabetes, 457–461
  in hypoglycemia, 461–462, 464
  in urine, 462–463
Caerulein, 14
cAMP, neuronal, 14
cAMP assay, of thyroid-stimulating immunoglobulins, 32–33, 57
Cancer
  human, steroid hormone action and, 571–653
  inappropriate antidiuresis in, 369–372
Carbamazepine, vasopressin in antidiuretic effect of, 367–369
Carbon-13 nuclear magnetic resonance, of hormone structure and function, 309–332
Carbon-13 nuclei, properties of, 309
Carcinogenesis
  chemical, 619
  of digestive and urinary tract, role of hormones in, 617–653
Carcinogens, activating metabolism of, 623–627
Chemical carcinogenesis
  in animal strains, 626
  cellular growth and development in, 627
  environmental factors in, 626–627
  experimental, 622
  hormones in, 627–648
  steps in, 619
Chlorpropamide, vasopressin in antidiuretic effect of, 367–369
CNS thyrotropin-releasing factor effects on, 13
Collecting tubules, antidiuretic hormone effects on, 406–410, 415–420
Crowding Graves' disease and, 45, 55
Cybernin, hypothalamic peptide as, 17

D

D cells, 14
Dendrites axon connections with, 11
Diabetes
  A-cell dysfunction in, 503–508
  C-peptide in, 457–461
  insulin therapy of, glucagon effects on, 497–501
  islets of Langerhans in, 480–482
  juvenile-type, somatostatin role in, 4–5
  pancreatic polypeptides in, 546–554
  proinsulin levels in, 441–444
  somatostatin therapy of, 501–503
  vasopressin effects in, 375
Digestive tract, hormone effects on carcinogenesis in, 617–653
Dual-barrier hypothesis, of antidiuretic action on collecting tubules, 415–417

E

Emesis, vasopressin and, 357–359
Emotional stress, Graves' disease etiology and, 46–48, 55
Endocrine cells, APUD concept and, 15
α-Endorphin, 1
β-Endorphin, 1
  morphinelike activity of, 20
Endorphins
  biological effects of, 6–9
  distribution of, in body, 14, 24
  as morphine substitute, 27
Enkephalins, distribution of, 14
Estradiol, in hepatic carcinogenesis, 632–633
Estrogens, carcinogenesis and, 652–653

F

Fasting effect on pancreating polypeptides, 538–540
Fat effects on pancreatic polypeptides, 527–528
Female gonadotropin effects on reproduction in, 214–222
N-2-Fluorenylacetamide, metabolites of, 624
Follicle-stimulating hormone (FSH)
 amino acid composition of, 185
 binding of, 223–230
 carbohydrate composition of, 184
 fractionation of, 177
 function of, 169

G

Gastric fundus, immunoreactive glucagons and, 495–496
Gastrin
 in cerebral cortex cells, 18
 distribution of, in body, 14
 prohormonal forms of, 284
Gel filtration, of antimüllerian hormone, 151–154
Glucagon(s)
 A cells and, 477–517
 fractionation of, 177
 immunoreactive, 487–496, 512
  A cells and, 487–491
  of gastric fundus, 495–496
  of plasma, 496–497
 prohormonal forms of, 284
 activity and susceptibility to, in fetus, 116
 enzymes induced by, 598
Glucose, effects on pancreatic polypeptides, 534–536
Glutamine synthetase, induction of, in somatic cell hybrids, 604
Gonadotropin(s), 169–248
 amino acid composition of, 183–185
 amino-terminal analysis of, 183–185
 assessment of purity of, 182–183
 binding of, 222–234
 bioassay of, 172–176
 biochemical characterization of, 182–190
 carbohydrate analysis of, 183
 effects on
  female reproduction, 214–222
  male reproduction, 191–214
 identification of, 179–181
 immunochemistry of, 189–190
 ion-exchange chromatography of, 178
 physiological actions of, 190–222
 subunit structure of, 186–189
Graves' disease
 circulating immune complexes in, 44–46
 etiology of, 46–48
 LATS in, 29–57
 sex differences in, 54

H

Heat, extreme, Graves' disease and, 55
Hepatocarcinogenesis, acetanilide inhibition of, 625
Heroin, endorphin substitute for, 27
Histidine, tautomeric forms of, in peptide hormones, 319
Hormones
 in carcinogenesis, 620–622
 in liver, 627–646
 as "carcinogens," 620–621
 -dependent tumors, 621–622
 effect on, carcinogen metabolism, 621
 metabolism of, 622
 redefinition of, 16–17
HTC cells characteristics of, 599
Human pancreatic polypeptides
 radioimmunoassay of, 525
 regulation of, 525–540
 in tissue extracts, 524–528
Human thyroid adenyl cyclase stimulator (H-TACS), assay of, 32
Human thyroid-stimulating immunoglobulin G
 hyperthyroidism and, 42–44
 isoelectric focusing of, 37–41
 zoological specificity of, 34–37
Human thyroid stimulator (HTS), assay of, 72
Hydrophilic solute transport coefficients, antidiuretic hormone effects on, 399–403
Hyperglycemia C-peptide in, 464
Hyperproinsulinemia, familial, proinsulin in, 446–448
Hyperthryoidism, thyroid-stimulating immunoglobulin G and, 42–44
Hypoglycemia, pancreatic polypeptides in, 536–538

Hypothalamic peptides, importance of, 1–28
Hypothalamic-pituitary-thyroid system
  ontogenesis of, 59–116
  steroid effects on, 107
Hypothalamus, maturation of, 70, 82, 94, 102

I

Immunoglobulin G, human thyroid-stimulating, see Human thyroid-stimulating immunoglobulin G
Insulin
  biosynthesis of, scheme for, 437
  in serum, measurement of, 438
Iodothyronines, placental transfer of, 60–68
Ion-exchange chromatography
  of antimullerian hormone, 157–159
  of gonadotropins, 178
Islets of Langerhans
  cell types in, 483
  in diabetes, 480–482
  function and structure of, 477–487
  gap junctions in, 486–487
  morphological relationships of, 482–487
  neurovascular elements of, 486
  physiologic roles of, 477–480
  tight junctions in, 486
  tumors of, pancreatic polypeptides in, 540–542
Isoelectric focusing, of human thyroid-stimulating immunoglobulin G

J

Juvenile diabetics
  islet cell dysfunction in, 507–508
  somatostatin in therapy of, 501–502

K

Kidney, carcinogenesis in, hormone role in, 646

L

L cells characteristics of, 599
LATS (long-acting thyroid stimulator), 29–57
  cytochemical bioassay of, 41–42
  in Graves disease, 29–57
LATS-P, assay of, 32
Leukemia, steroid-binding proteins from myeloblasts of, 580
β-Lipoproteins (β-LPH)
  endorphins and, 7–8
  fragments of in neurons, 18
γ-Lipoproteins (γ-LPH), as hormone precursors, 284
Lithium, effect on thyroid, 55–56
Liver, carcinogenesis of, hormones in, 627–646
Luteinizing hormone (LH)
  amino acid composition of, 184–187
  binding of, 230–234
  biological activities of, 18
  carbohydrate composition of, 183
  comparative studies on, 175
  fractionation of, 177
  function of, 169
  sedimentation coefficients of, 186
Luteinizing hormone releasing factor (LRF), 1–3
  analogs of, 2
  carbon-13 NMR studies on, 324–325
Lymphocytes, in Graves' disease, 47–48, 50–51

M

Male, gonadotropin effects on reproduction in, 191–214
Mammary tumor virus, 618
Mediastinoma, hypothalamic peptides in, 15
α-Melanocyte-stimulating hormone, 1
  prohormonal forms of, 284
  sequences of, 15
β-Melanocyte-stimulating hormone, prohormonal forms of, 284
Membranes, measurement of water and solute transport across, 389–390
Molecular mobility, carbon-13 spin-lattice relaxation times as indicators of, 319–325
Morphine, endorphin substitute for, 27
mRNA
  from parathyroid
    identification of pre-parathyroid hormone using, 266–276

SUBJECT INDEX

translation of, in ascites cell-free system, 276–279
Müllerian ducts
　embryology of, 118–120
　indifferent or ambisexual stage of, 118–119
　regression of, testis role in, 120–123
　sex differentiation of, 119–120

N

Nausea, vasopressin and, 357–359
Neural ectoderm, peptide hormone-producing cells from, 16
Neuroblastomas, vasointestinal peptide in, 15
Neuroendocrine control, maturation of, 96–98, 102
Neurons
　APUD concept and, 15
　hypothalamic peptide effects on, 9–10, 12–13
Neurotensin, 1
　distribution of, in body, 14
　sequence studies on, 27
Nuclear magnetic resonance, of hormone structure and function, 309–332

O

Ovaries, gonadotropin effects on, 214–219
Ovulation, gonadotropin effects on, 214–219
Oxytocin, carbon-13 NMR studies on, 314–315, 323–324

P

Pancreas
　carcinogenesis in
　　hormone role in, 648
　　somatostatin and, 5
Pancreatic polypeptide(s), 519–570
　amino acid effects on, 528–529
　biological effects of, 521
　chemistry and species homology of, 519–520
　in diabetes, 546–554
　in disease states, 540–554
　distribution of, in tissues, 522–525
　fat effects on, 527–528
　in humans, see Human pancreatic polypeptide
　hypoglycemia and, 536–538
　neural effects on, 531–532
　after protein ingestion, 526–527
　radioimmunoassay of, 525
　regulation of, 525–540
　somatostatin effects on, 532–534
Paracrine secretion
　of hormones, 17
　of hypothalamic peptides, 17
Parathyroid gland, secretory productions of, 292–294
Parathyroid hormone (PTH), 249–308
　amino acid sequences of, 251
　biological and immunological properties of, 261–263
　biosynthesis of, 249–308
　　regulation, 288–292
　　scheme for, 287
　in parathyroid gland slices, 253–256
　precursors of, 252–253, 266–276
　　in endocrine and nonendocrine secretory tissues, 283–285
　　enzymic conversion, 259–261
　　identification, 279–283
　　primary structure, 256–259
　primary structures of, 250–252
Pars intermedia, cells of, APUD series and, 16
Peptide hormones, NMR studies on conformation of, 313–326
Permeation, by water and solutes, activation energies for, 413–415
Phylogeny, of animal orders, 237
Physalemin, 14
Pituitaries
　collection of, 171–172
　embryogenesis of
　　in humans, 70
　　in rat, 94
　　in sheep, 82
Pituitary hormones
　comparative studies on, 170–191
　effect on an antimüllerian hormone regulation, 145
　fractionation of, 176–181
　in hepatic carcinogenesis, 637–646

Pituitary portal system, maturation of, 70–71, 95–96
Placenta, thyroid hormone transfer of, 60–68
Poly-L-lysine, carbon-13 NMR studies on, 321–322
Portal system
  maturation of
    in rat, 95–96
    in sheep, 82
Pregnancy, steroid effects on thyroid function in, 107
Prehormones, functions of, 300
Pre-proparathyroid hormone, 266–276
  function of, in parathyroid cellular transport, 285–288
  identification of, 279–283
Prohormones, functions of, 300
Proinsulin
  characterization of, 440–441
  circulating levels of, 436–451
  clinical significance of, 435–475
  covalent structure of, 435
  in diabetes, 468–469
  in familial hyperproinsulinemia, 446–448
  in normal and disease states, 444–446
  peripheral metabolism of, 448–451
  in serum, 436–440
Proline, cis and trans isomers of, in peptide hormones, 315–318
Proparathyroid hormone
  enzymic conversion of, 259–261
  function of, in cellular transport, 285–288
  precursor of, 266–276
  primary structure of, 256–259
  radioimmunoassays of, 263–266
Pseudohermaphroditism, müllerian regression and, 126

R

Rat, thyroid function maturation in, 93–100, 102
Reproduction
  gonadotropin effects on, 191–214
Reptiles
  gonadotropin effects on
    reproduction, 191–195, 214–217
    steroid production, 198–209, 219–221
  pituitaries from, 171
Retinopathy, diabetic, somatostatin in, 6

S

Sedimentation constant, of antimullerian hormone, 156
Sertoli cells, tissue culture of, 140–145
Sex hormones, in hepatic carcinogenesis, 627–646
Sheep
  hypothalamus maturation in, 82, 102
  thyroid function development in, 82–93
Somatic cell hybrids, steroid hormone studies using, 590–605
Somatostatin, 1
  biological effects of, 3–6
  diabetes therapy with, 501–503
  distribution of, in body, 3–4
  effects on pancreatic polypeptides, 532–534
  in neurons, 18
  release of, from isolated islets, 516
Spermatogenesis, gonadotropin effects on, 191–214
Steroid hormones
  effect on thyroid function in pregnancy, 107
  human cancer and, 571–615
  studies on, using somatic cell hybrids, 571–615
Stomach, cancer of, hormones and, 647
Stress, Graves' disease etiology and, 46–48, 55
Substance P, 1
  distribution of, in body, 14
  in neurons, 18

T

Testis
  of calf, microdissection of, 135–138
  gonadotropin effects on, 191–198
    steroid production, 198–214
  of humans, irradiation studies on, 138–140
  role in müllerian regression, 120–123
Testosterone, in hepatic carcinogenesis, 632–633, 652
Thermogenesis, thyroid hormones in, 59–60
Thyroid hormones
  placental transfer of, 60–68
  in thermogenesis, 59–60

SUBJECT INDEX

Thyroid function
 development of
  in humans, 69–81
  in sheep, 82–93
Thyroid gland
 embryogenesis of
  in humans, 69–70
  in rats, 93–94
  in sheep, 82
Thyroid-stimulating immunoglobulins (TSI),
  assay of, 32
Thyrotropin (TSH)
 in development of neuroendocrine
  control, 71–77, 82–85
 in fetal pituitary and serum, 71–72
 gonadotropic effects of, 244–245
 placental transfer of, 60–68
Thyrotropin-releasing factor (TRF), 1–3
 analog of, 2
 effects on CNS, 13
Thyrotropin-releasing hormone (TRH)
 placental transfer of, 60–68
Tryptophan oxygenase, glucocorticoid
  induction of, 598, 602
Tumors
 hormone-dependent, 621
Tumors, pancreatic, somatostatin effects on, 5
Tyrosine aminotransferase
 induction of, 598
 in multinucleate somatic cell hybrids, 594–596

U

Urinary tract
 carcinogenesis in, hormone role in, 646–647
 hormone effects on carcinogenesis in, 617–653

Unstirred layer problem in water transport, 392–398
Urine
 arginine vasopressin in, 362–364
 C-peptide in, 462–463

V

Vasointestinal peptide (VIP)
 distribution of in body, 14
 in neuroblastomas, 15
 in neurons, 18
Vasopressin, 1, 333–385
 in antidiuresis, 367–369
  inappropriate, syndrome of, 369–372
 in diabetes insipidus, 364–367
 distribution and clearance of, 359–362
 emesis and, 357–359
 in health and disease, 333–434
 in interaction of osmotic and hemodynamic control systems, 351–354
 osmotic control of, 342–347
 pathophysiology of, 364–372
 physiologic control of, 342–364
 pressor control effects on, 350–351
 in urine, 362–364
 volume control, effects on, 348–350

W

Water flow, transcellular versus, paracellular,
 antidiuretic hormone effects on,
 411–412
Water homeostasis
 antidiuretic hormone in, 387–434
 biophysical aspects of, 389–398
 nomenclature in, 388–389
Water permeation pathway, antidiuretic
  hormone effects on, 403–406
Wheat germ, use in pre-parathyroid
  hormone studies, 266–271

CUMULATIVE SUBJECT INDEX

A

Acetylcholine, role of, in mechanism of nerve activity, 1, 1
Acetylcholine system, in neural function, 5, 37
Acid mucopolysaccharides, effects of hormones on, 14, 427
ACTH
  mechanism of action, 27, 433
  metabolism of adrenal cell cultures and, 26, 623
ACTH and cortisone
  experimental studies with, in neoplastic disease, 6, 373
  physiologic effects of, in man, 6, 315
ACTH peptides, biological properties of, 18, 1
ACTH secretion
  mechanism of stimulation of, 11, 83
  regulation of, 7, 75
Adaptation, diseases of, 8, 117
Adenyl cyclase system, role of, in hormone action 21, 623
Adipose tissue, influence of hormones on, as center of fat metabolism, 16, 467
Adrenal
  in cancer, 8, 273
  pituitary hormone administration and, metabolic changes following, 4, 229
  reproductive glands and, embryogenesis of, 6, 1
  response to ACTH in flowing system, 27, 607
Adrenal cortical metabolism, fluid and, clinical aspects of, 7, 469
Adrenal cortical thyroid, growth hormones and, effects of in fasting metabolism, 4, 153
Adrenal cortical tumors, pathologic physiology of, 2, 345
Adrenal cortex
  essential hypertension and, 32, 377
  reference to, with experimental endocrine tumors, 5, 383
  role of, in stress of human subjects, 1, 123
  in salt and water metabolism, 6, 247
  steroid hydroxylation by, 25, 523
  stress and, roles of, in protein metabolism, 6, 277
Adrenal cortex hormones, biosynthesis of, studies on, 7, 255
Adrenal cortex steroids
  chemistry of, 6, 195
  partial synthesis of, some advances in, 1, 83
Adrenal disease, urinary steroids in, 3, 103
Adrenal function in mental disease, 4, 291
Adrenal glands, problems of fetal endocrinology, 22, 541
Adrenal hormones, metabolism of, 3, 103
Adrenal medulla, 12, 27
Adrenal secretory product, nature and biogenesis of, 6, 215
Adrenal steroids, mode of action on lymphocytes, 23, 195
Adrenal steroid biogenesis, disorders of, 23, 375
Adrenal steroid hormones, some aspects of biogenesis of, 12, 79
Adrenocortical hormones, bioassay of, 8, 87
Adrenocortical physiology, studies of brain metabolism in relation to, 10, 29
Adrenocortical steroids
  metabolic effects of, *in vivo* and *in vitro,* 19, 207
  micromethods for analysis of, 8, 51
  miscible pool and turnover rate of, 15, 231
β-Adrenergic receptors, adenylate cyclase coupled, 32, 597
Aging men and women, steroid metabolism in, 11, 307

CUMULATIVE SUBJECT INDEX

Alcoholic steroids, estimation of, 9, 251
Aldosterone
  adrenocortical secretion of, 22, 381
  biochemical mechanism of action, 24, 1
  control and physiologic action of, 15, 311
  factors which influence secretion of, 15, 275
  isolation, chemistry, and physiology of, recent progress in, 11, 183
  mode of action of, 22, 431
Aldosterone secretion
  control of, 19, 311
  and metabolism of, in secondary hyperaldosteronism, 17, 293
  studies of, in relation to electrolyte balance in man, 12, 175
Amphenone and related compounds, studies on, 11, 119
Androgen biosynthesis, 19, 251
Androgens
  metabolism of, 2, 179
  by tissues, 4, 65
  secretion and interconversion of, 19, 275
Androgen metabolism, in male pseudohermaphroditism, 29, 65
Anesthesia and unanesthetized trauma, endocrine changes after, 13, 511
Angiotensin, a renal hormone, 18, 167
Animals, chemical communication among, 19, 673
Anterior hypophysis, neurovascular regulation, 29, 161
Anterior pituitary function
  hypothalamic neurohormones and, 24, 497
  influence of central nervous system on, 13, 67
  neural control of, 24, 439
Anterior pituitary hormones
  fate and metabolism of, 11, 43
  relation of, to nutrition, 1, 147
Antiandrogens, androgen-dependent events and, 26, 337
Antidiuretic hormones, in water homeostasis, 33, 387
Antiestrogens, 18, 415
Antimüllerian hormone, 33, 117
Antithyroid drugs, extrathyroid effects of, 23, 87

Atherogenesis and blood pressure, dietary and hormonal factors in, 11, 401
Autonomic and humoral function, altered, manifestations of in psychoneuroses, 4, 323

B

B-complex, role of, in estrogen metabolism, 2, 161
Beta cell secretion, microtubules and, 29, 199
Biosynthesis of protein and acid, growth hormone control of, 21, 205
Blood and blood-forming organs, endocrine influences upon, 10, 339
Brain excitability and electrolytes, effect of hormones on, 10, 65
Breast cancer, hormones and, 22, 351

C

Calciphylaxis and parathyroid glands, 20, 33
Calcitonin, and plasma calcium, 20, 59
Calcium deprivation, effects of, upon domestic fowl, 15, 455
Carbohydrate metabolism
  adrenal hormones in regulation of, 8, 511
  effects of ions and hormones on, 11, 381
  hormonal regulation of, 19, 445
Carcinogenesis, experimental, steroid hormones in, 1, 217
Castration in man, effects of, testicular secretions as indicated by, 3, 257
Catecholamines
  metabolism, storage, and release of, 21, 597
  other hormones and, 31, 1
Catechol hormones, distribution and metabolism of, in tissues and axons, 14, 483
Cell permeability and hormone action, 16, 139
Cerebral metabolism, influence of steroids on, in man, 12, 153
Cholesterol
  ascorbic acid and, relation of, to secretion of adrenal cortex, 1, 99
  biological synthesis of, 6, 111

Chromosomal sex, cytologic tests of, in relation to sexual anomalies in man, 14, 255
Clinical endocrinology, genetic aspects of, 24, 365
Conjugated estrogens, hydrolysis of, 9, 45
Conjugates, hydrolysis, 9, 267
Connective tissue
 changes of aging, hormonal influences upon 14, 457
 effects of corticosteroids upon, 20, 215
Connective tissue
 diseases, relationship of adaptation to, 8, 217
Corpus luteum, maintenance of, 5, 151
Corticoids and 17-ketosteroids, hydrolysis and extraction of, from body fluids, 8, 27
Corticosteroid(s), isolating from blood, 9, 357
Corticosteroid metabolites, fractionation of, in human urine, 9, 337
Corticotrophins, chemistry of, 7, 1; 10, 265
Corticotropin secretion, normal and abnormal regulation of, 18, 125
Cryptorchidism, experimental, effect of in the rat, 6, 29
Crystalline insulin, preparation and chemistry of, 10, 241
Cushing's syndrome, 2, 345
Cyclic adenylic acid, role of, in hormone action, 16, 121
Cyclic AMP action, molecular mechanisms of, 32, 669
Cytochemical bioassay of hormones, 32, 33

D

Dehydrocorticosterone, 3,9-epoxy-Δ^{11}-cholenic acid in partial synthesis of, 1, 65
Dehydroisoandrosterone sulfate, metabolism of, in man, 21, 411
Depot fat, hormonal factors which regulate mobilization of, to liver, 7, 399
Diabetes
 growth hormone and, 8, 471
 in Pima Indians, 32, 333
Diabetes mellitus
 etiology of, in man, 7, 437
 insulinase and, 13, 429
 insulinase-inhibitors and, 13, 429
1,25-Dihydroxycholecalciferol, 30, 431
Diphosphopyridine nucleotide-mediated enzymatic reactions, effects of steroid hormones upon, 16, 79
Doisynolic acids, 3, 47
Domestic animals, physiology and endocrinology of, 17, 119

E

Ecdysone, mode of action, 22, 473
Ectopic humoral syndromes, clinical and laboratory studies, 25, 283
Edema, mechanisms of, in experimental nephrosis, 17, 353
Electrolyte and fluid metabolism, clinical studies on, 10, 425
Electron and chemical ionization, mass spectrometry, 28, 591
Endocrine(s), role of, in self-regulation of mammalian populations, 21, 501
Endocrine action, during insect growth, 30, 347
Endocrine glands
 chemical cytology of, 3, 127
 ultrastructure of, 25, 315
Endocrine mechanisms in life of insects, 10, 157
Endocrine and metabolic functions, effects of midbrain and spinal cord transection on, 13, 21
Endocrine neurons, 31, 243
Endocrine therapy, antihormone problem in, 4, 115
Endocrinology, of higher plants, 21, 579
Eosinopenia, stress-induced, and the central nervous system, 10, 1
Epidermal growth factor, chemical and biological characterization, 30, 533, 551
Epinephrine
 effect of, on glycogenolysis, 5, 441
 norepinephrine and, excretion of, under stress, 14, 513
Erythropoietin, studies on, 16, 219
17β-Estradiol dehydrogenase, human placental, 30, 139

CUMULATIVE SUBJECT INDEX 665

Estrogen(s)
   androgen inhibitors and, 20, 435
   compounds related to, chromatographic fractionation and identification of, 9, 69
   formation by central neuroendocrine tissues, 31, 295
   identification of, by ultraviolet absorption spectrophotometry, 2, 31
   mechanism of action of, 14, 95
   metabolism of, 4, 25, 43, 85
   physiologic functions of, enzymatic basis for, 16, 49
   quantitative microdetermination of, by ultraviolet absorption spectrophotometry, 2, 31
Estrogen action
   inroad to cell biology, 28, 1
   mechanism of, 18, 387
Estrogen conjugates, studies on, 5, 307
Estrogenic hormones
   action of, on metabolism, 16, 97
   fluorometric methods for determination of, 9, 95
   some aspects of physiology of, 5, 115
   spectrophotometric methods for determination of, 9, 95
Exchangeable cellular thyroxine, metabolic significance of, 25, 381

F

Familial (medullary) thyroid carcinoma, 28, 399
Fat cell development, control of, and lipid content, 26, 463
Fetal endocrinology, problems of, 8, 379
Food and temperature, 16, 439
Follicle stimulating hormone
   human, physical and hydrodynamic properties of, 26, 105
   Sertoli cell and spermatogenesis and, 32, 477
   testicular control of secretion, 32, 429

G

Gastric inhibitory polypeptide, identification and actions of, 31, 487

Gastrointestinal hormones, biochemistry of, 23, 483
Gender dimorphic behavior, and fetal sex hormones, 28, 735
Gene expression
   hormonal regulation in mammary cells, 29, 417
   regulation by cAMP, 27, 421
Germinal epithelium, kinetics of, 20, 545
Glucagon
   A cells and, 33, 477
   direct actions of, on perfused rat liver, 17, 539
   a second pancreatic hormone, 13, 473
Glucocorticoids, adrenocortical secretion of, 22, 381
Glucose-^{14}C, studies with, 19, 445
Glucose uptake in muscle, regulation of, in perfused rat heart, 17, 493
Glucose utilization, alternate pathways of, effect of hormones on, 16, 547
Glycosuria, experimental, production of, in the rat, 2, 229
Goiter, familial, metabolic errors in, 19, 547
Goitrogenic agents, natural occurrence of, 18, 187
Gonad-pituitary feedback, 31, 567
Gonadal cycle(s)
   of domestic animals, physiological and medical aspects of, 7, 165
   migration in birds and, 22, 177
Gonadal hormones
   behavior of nonhuman female primates and, 28, 707
   during spontaneous and induced ovulatory cycles, 26, 1
Gonadotrophin and the sex cycle, relationship of nervous system to, 7, 139
Gonadotropic activity of duck hypophysis, control of, by visible radiations, 15, 143
Gonadotropin
   action on Graafian follicles, 30, 79
   antibodies and, 30, 47
   characterization by antibodies, 27, 235
   human chorionic, 27, 121
   ovarian metabolism, steroid biosynthesis and, 24, 255
   ovarian steroidogenesis and, 21, 285
   secretion in Rhesus monkey, 30, 1

Gonadotropin, (contd.)
  structure and function, evolution of, 33, 169
  subunits and, 32, 289
Gonadotropin-releasing hormone, in central nervous system, 32, 117
Graves' ophthalmopathy, pathogenesis of, 31, 533
Growth of children, rate and duration of, 1, 355
Growth and development, genetic and endocrine factors in, 2, 391
Growth hormone
  chemistry of, 29, 387
  physiology of, with reference to rhesus monkey and "species specificity," 15, 1
  studies with antisera to, 20, 1
Growth hormonelike factor of *S. mansonoides*, properties of, 27, 97
Growth hormone secretion, regulation of, 21, 241

H

Hepatic glucocorticoid receptor, 31, 213
Hepatic gluconeogenesis, hormonal control of, 26, 411
Histamine, 7, 375
Hormonal glucuronide formation, mechanism of, 12, 134
Hormones
  breast cancer and, 22, 351
  effects of, on growth and meat production of livestock, 14, 183
  gonadal and hypophyseal, 8, 379
  rhythms and, 13, 105
  role in digestive and urinary tract carcinogenesis, 33, 617
  structure and function of, by ^{13}C-NMR, 33, 309
Hormone-enzyme relationships, 5, 465
Hormone receptors
  control of cyclic AMP metabolism, 32, 633
  interaction of estrogen with uterus, 24, 45
Hormone-receptor complexes, membrane function and, 31, 37
Hormone-secreting and -responsive cell cultures, 26, 539

Human
  cancer, endocrine aspects of, 1, 261
  fetus and newborn, estrogen metabolism in, 17, 147
  gonadotropins, 15, 127
  growth hormone, 15, 71
    immunological assay of, 16, 187
  hypogonadism, testis in, 3, 197
  menstrual cycle, endocrinology of, 7, 209
  ovary and testis, aspects of aging reflected in, 11, 291
  sebaceous gland, regulation by steroidal hormones, 19, 385
  serum
    insulinlike activity of, 23, 565
    steroidal hormones, studies of, by gas chromatographic techniques, 19, 57
Hydrocortisone, direct actions of, on perfused rat liver, 17, 339
Hydrogen, steroid hormones in enzymatic transfer of, 16, 1
Hydrolytic enzymes, role of, in steroid hormones, 1, 177
18-Hydroxy-11-deoxycorticosterone, secretion in hypertension, 28, 287
3α-Hydroxysteroid dehydrogenase, 23, 349
16-Hydroxylated steroids, chemistry and biological activities of, 14, 1
Hyperadrenocorticalism and inanition, histological changes induced by, 7, 331
Hypersensitivity, and hormones, 7, 375
Hypertension
  disease of adaptation, 3, 343
  experimental and clinical, 3, 325
Hypothalamic factors, in human fetus, 32, 161
Hypothalamic lesions, effects of, on water and energy metabolism in the rat, 4, 363
Hypothalamic neurohormones, anterior pituitary function and, 24, 497
Hypothalamic peptides, expanding significance of, 33, 1
Hypothalamic-pituitary system, during menstrual cycle, 31, 321
Hypothalamic–pituitary–thyroid function, ontogenesis of, 33, 59
Hypothalamic release hormones, 28, 173, 201
Hypothalamus, in pituitary-thyroid regulation, 28, 229

I

Insulin
  biosynthesis of, and proinsulin, 25, 207
  direct actions of, on perfused rat liver, 17, 539
  hormonal influences on the secretion of, 2, 209
  mechanism of action of, 11, 343
  nerve growth factor and, 30, 575
  physiological role and metabolism, 22, 1
  secretion, amino acids and proteins and, 23, 617
  structure of and x-ray analysis, 27, 1
Insulin action
  on plasma membrane of fat cells, 24, 215
  regulation of protein biosynthesis in muscle, 24, 139
Insulin secretory dynamics, 26, 583
Interstitial cells of the testis, biology of, 3, 173
Interstitial cell stimulating hormone, 29, 563
Iodine, metabolism of, as disclosed with radioiodine, 4, 429
Iodine metabolism, certain aspects of, 16, 405
Isotopic steroid hormones
  metabolites of, in man, 9, 411
  studies with, 6, 131

J

Juvenile hormone
  chemistry and biology of, 24, 651

K

Ketosteroids, theoretical and practical aspects of partition chromatography of, 9, 185
Klinefelter's syndrome, hormonal and chromosomal study, 24, 321

L

Lactation, hormonal control of, 2, 133
Lactogenic hromones, chemistry of, 29, 387
Leydig cell, physiology and pathology of, 22, 245
Lipids, *in vitro* synthesis of, and hormonal control, 8, 571
Long acting thyroid stimulator, in Graves' disease, 23, 1; 33, 53
Luteinization, of granulosa cells, hormones and, 26, 589
Luteinizing hormone
  chemical studies of human and ovine pituitaries, 29, 533
  human, physical and hydrodynamic properties of, 26, 105
  radioligand receptor assay, 29, 497
  release of, from hypophysis of the rabbit, 2, 117
Luteinizing hormone releasing hormone, 28, 201
Luteinizing hormone secretion, hypophyseal, neuroendocrine regulation of, 20, 131
Lymphoid tissue metabolism, effects of steroids on, 15, 391
Lymphoid tumors, radiation-induced, of mice, and endocrine factors, 10, 293
Lysosome(s)
  as mediator of hormone action, 30, 171
  effect of corticosteroids upon, 20, 215

M

Male hypogonadism, genetic aspects in, 17, 53
Male pseudohermaphroditism, in laboratory Norway rat, 29, 43
  testicular feminization in, 29, 65
Mammary gland, multiple hormone interactions in development, 26, 287
Mammary growth and lactation, hormonal control of, 14, 219
Mammary tissue, responses to hormonal treatment, 23, 229
Mammary tumors, transplanted, phenanthrene derivatives related to growth of, 14, 77
Melanin pigmentation, hormonal control of, 12, 303
Melanocyte-stimulating hormone
  cyclic AMP and, 30, 319
  effect on coat color in mouse, 28, 91
Menstrual cycle
  human, synthetic progestins in, 13, 323
  regulating mechanisms, 26, 63

Metabolic problems, parameters of, 6, 159
Metabolism
  alterations in, studied with aid of isotopes, 4, 189
  physiological role of insulin and, 22, 1
Microtubules
  beta cell secretion and, 29, 199
  microfilaments in thyroid secretion and, 29, 229
Microtubule proteins, nerve growth factor response and, 30, 635
Midbrain hypothalamico-pituitary activating system, postulation of, 13, 21
mRNA, levels during enzyme induction, 31, 213

N

Neoplastic disease, endocrinology of, 11, 257
Nerve growth factor, insulin and, 30, 575
Neuroendocrine factors, in primate behavior, 28, 665
Neurophypophyseal hormone(s)
  biochemistry and physiology of, 11, 1
  effects of, in a living membrane, 17, 467
Neurohypophysis, physiology and pharmacology of, 17, 437
Neurophysin, and vasopressin, 25, 447
Neurosecretory cells, hormones produced by, 10, 183
Neutral 17-ketosteroids
  in human plasma, 9, 235
  of urine, colorimetric analytical methods for, 9, 135
Nonantigenic hormones, protein binding and assay of, 25, 563
Nuclear receptors, initiation of thyroid hormone action and, 32, 529

O

Obesity, endocrine and metabolic effects of, 29, 457
Oocytes, number of, in mature ovary, 6, 63
Osteogenesis, effect of hormones on, in man, 1, 293
Ova-implantation, mechanism of, in the rat, 13, 269
Ovalbumin synthesis, in chick oviduct, 31, 175

Ovarian function
  pregnancy and, problems related to, 10, 395
  vasculature of, 5, 65
Ovarian hormones, clinical aspects of physiology and function of, 4, 85
Ovarian steroids, mechanism of effect of, 12, 405
Ovarian steroid synthesis and secretion, 20, 303
Ovary
  androgenic activity of, 5, 101
  normal and polycystic, steroid secretions of, 20, 341
Ovulation
  induction of, with gonadotropins, 21, 179
  in rat, model for regulation, 25, 1
11-Oxygenated steroids, synthesis of, from plant sources, 8, 1
Oxytocin analogs, in analysis of hormone action, 28, 131

P

Pancreas
  effect of hyperglycemic factor, 5, 441
  guardian of the liver, 4, 215
Pancreatic islet tissue, metabolic pathways of 19, 489
Pancreatic polypeptide, newly recognized, 33, 519
Paper chromatography, possibilities and limitations of, as method of steroid analysis, 9, 321
Parathyroid(s), and plasma calcium, 20, 59
Parathyroid hormone
  biological and immunological activity, 22, 101
  biosynthesis of, 33, 249
  circulating, nature of, 30, 391
  structure, synthesis, and mechanism of action, 28, 253
Parathyroid hormone action, mechanism of, 15, 427
Parathyroid hormone excess, clinical endocrinology of, 18, 297
Parturition, initiation of, in ewe, 29, 111
Pathologic changes, experimentally produced and naturally occurring, in the rat, to adaptation diseases, 8, 143

Pathological conditions and short lifespan associated with maleness, 3, 257
Peptides
  chemical synthesis of, 23, 451
  formation by animal cells, functions of insulin in, 12, 199
Peptide hormones, heterogeneity of, 30, 597
Phospholipids, in activation of adenylate cyclase, 29, 361
Pineal, mammalian, as neuroendocrine transducer, 25, 493
Pituitary
  adrenal cortical system, physiological influences on, 15, 345
  adrenal system, 2, 81
  adrenocortical secretions, studies of suppressing functions of, 7, 307
  adrenocorticotrophic hormone, purification of, 7, 59
  function, steroid control of, 5, 197
  gland, adipokinetic component of, 18, 89
  factors affecting control of, 5, 263
  gonadotropin, effects of steroids on, 20, 395
  purification of, recent studies on, 15, 115
  growth hormone
    biochemistry of, 3, 3
    mechanism of action, 31, 141
    mammary gland relationships, aspects of, 7, 107
    syndromes in man, 12, 321
    tumors, experimental, 11, 221
  hormones
    characterization of, by starch gel electrophoresis, 19, 1
    during spontaneous and induced ovulatory cycles, 26, 1
    hypothalamic factors releasing, 20, 89
    ontogenesis of, in human fetus, 32, 161
    synthetic, 18, 41
Pituitary-thyroid regulation, hypothalamus in, 28, 229
Placental lactogen, synthesis and secretion by placenta, 25, 161
Plasma cell tumors, endocrine factors in pathogenesis of, 24, 81
Polypeptides
  hormone activity of, 16, 263
  parathyroid, 18, 269
Pregnenolone, biosynthesis of, 27, 303
Pressor amines, biosynthesis of, 12, 27
Primary aldosteronism
  adrenocortical steroids, content of, 17, 415
  versus hypertensive disease, 17, 389
Primate behavior, neuroendocrine factors in, 28, 665
Proinsulin
  biosynthesis of insulin and, 25, 207
  circulating, and C-peptide, 33, 435
Progesterone
  catabolic and natriuretic effects of, 17, 249
  detection, estimation and isolation of, and related C_{21} steroids, 9, 27
  metabolism of, and its clinical use in pregnancy, 13, 347
  physiologic actions of, 5, 151
Progesterone metabolism
  relation to reproduction, in the human female, 8, 293
  some aspects of, 4, 3
Prolactin
  comparative endocrinology of, 24, 681
  in man, 28, 527
Prolactin secretion, studies on function and control, 28, 471
Prostaglandins, 22, 153
  biological significance of, 26, 139
  in luteal function, 28, 51
  renal antihypertensive and, 30, 481
Protein
  fat metabolism and, hormonal control of, in pancreatectomized rat, 16, 497
  metabolism, regulation of, adrenal cortex and stress in, 6, 277
  nutrition, and hormones, 14, 141
Protein kinases, regulation and diversity, 29, 329

R

Radioactive compounds in tissue, histological localizations of with radioiodine, 3, 159
Radioiodine as a diagnostic and therapeutic tool, 4, 483

Rat liver tryptophan pyrrolase, hormonal and substrate induction of, 18, 491
Receptors, for insulin, NSILA-s, and growth hormone, 31, 95
Receptor function, ion transport and, 32, 567
Relaxin, biochemistry of, 8, 333
Renal hypertension, experimental, with reference to endocrine aspects, 1, 371
Renal pressor system, 3, 325
Renin and angiotensin, functions for, 21, 73
Renin-angiotensin, aldosterone, and sodium, 21, 119
Reproduction, and erucic acid, 17, 97
Ribonucleic acids and proteins in prostate gland, synthesis of, 20, 247

S

Serotonin, biochemical, physiological, and pharmacological aspects of, 13, 1
Serum lipoproteins, effects of hormones on, 14, 405
Sex differentiation in mammals, 29, 1
Sex hormones
 assessing effects of in humans, 28, 627
 fetal, gender dimorphic behavior and, 28, 735
 male, and role in reproduction, 12, 353
 deficiencies, clinical considerations, 2, 295
Sex steroids, metabolism and protein binding of, 27, 351
Sexogens, constitution in, relation of activity to, 3, 47
Sexual abnormalities in man, clinical aspects of, 14, 335
Sexual maturation, etiologies of in rat, 32, 245
Sleep, neuroendocrine secretions and, 31, 399
Somatomedin(s), 30, 259
Somatostatin, 31, 365
 biological action of, 31, 321
Somatotropin, growth and anti-insulin actions of, 22, 49
Spermatogenesis
 inhibition and recovery of, 20, 491
 mammalian, 6, 29
 steroid metabolism in testes and, 26, 547

Spontaneous hyperglycemia, and obesity in rats, 27, 41
Steroids
 binding of, to human plasma proteins, 13, 209
 characterization by antibodies, 26, 235
 crystal structure of, 32, 81
 formation and metabolism in fetus and placenta, 23, 297
 long-acting, in reproduction, 13, 389
 molecular structure of, related to growth of transplanted mammary tumors, 14, 77
 new, structural and hormonal activity of, 14, 29
 secretion of from transplanted ovary, 27, 537
Steroid conjugates, fractionation and isolation of, 15, 201
Steroid dynamics, under steady state conditions, 25, 611
Steroid excretion, chemical aspects, in health and disease, 3, 71
Steroid hormone(s)
 analogues and, microorganisms in synthesis of, 11, 149
 application of C^{14} to study of metabolism of, 9, 383
 in human ovary, biogenesis of, 21, 367
 interaction of, with glutamic dehydrogenase, 18, 467
 metabolic activities of, hydrolytic enzymes in, 1, 177
 modification of reproduction in prepubertal rats, 22, 503
 studies on, in experimental carcinogensis, 1, 217
 transplacental passage of, in mid-pregnancy, 17, 207
Steroid hormone action
 genetic approaches to, 32, 3
 in tissue cells and cell hybrids, 33, 571
Steroid hormone metabolism, in man, 22, 203
Steroid hormone metabolites
 chemical estimation of, 5, 335
 neutral, 9, 5
Steroid hormone regulation, of specific protein synthesis, 25, 105

CUMULATIVE SUBJECT INDEX

Steroid nucleus, oxidation-reduction of, 12, 125
Steroid-protein conjugates, 15, 165
Steroid-responsive tissues, carbonic anhydrase in, 19, 201
Steroid excretion and metabolic effects, comparison of, induced by stress and ACTH, 10, 471
Sterol(s) and steroid hormones, biogenesis of, 12, 45
Sterol hormones, characterization of, by ultraviolet and infrared spectroscopy, 2, 3
Stress
  adrenal response to, 8, 171
  gastric responses to, and adrenal influences upon the stomach, 13, 583
Sympathetic hormonal transmission, 5, 3
Synthetic gonanes, biological effects of, 22, 305
Synthetic insulins, 23, 505

T

Target cells, alterations by anti-inflammatory steroids, 29, 287
Testicular hormone(s), 20, 247
Testicular hormone production, clinical studies of, 12, 377
Testicular interstitial cells, normal and abnormal, in the mouse, 17, 1
Testicular steroids, production and secretion of, 27, 517
Testis-pituitary relationship in man, 3, 229
Testosterone, intranuclear metabolism in accessory organs, 26, 309
Thioureas, toxic, physiology and endocrinology of, 2, 255
Thymosin, new thymus hormone, 26, 505
Thymus, as endocrine gland, 26, 505
Thyrocalcitonin, 24, 589
Thyroglobulin, structure of, and role in iodination, 21, 1
Thyroid, action of, on diabetes, 2, 277
Thyroid gland, transplantable tumors of, in the Fischer rat, 19, 579
Thyroid hormones
  biochemistry of amphybian metamorphosis, 23, 139
  metabolic effects of, in vitro, 10, 129
  nature and metabolism of, 12, 1
  peripheral metabolism of, 16, 353
  protein and, interaction of, in biological fluids, 13, 161
  utilization and action of, intracellular and extracellular mechanisms for, 18, 221
Thyroid iodide trap, in man, 25, 423
Thyroid iodine metabolism, studies on, 21, 33
Thyroid peroxidase, thyroxine biosynthesis and, 26, 189
Thyroid stimulating hormone, physiologic reactions of, 4, 397
Thyrotropin(s)
  chemistry and bioassay of, from pituitaries, pituitary tumors, and plasma, 16, 309
  physiology of and radioimmunoassay, 23, 47
  structure of and relationship to LH, 27, 165
Thyrotropin releasing hormone, 28, 173
  in central nervous sytem, 32, 117
Thyroid thermogenesis, sodium transport and, 30, 235
Triiodothyronine
  circulating, significance of, 26, 249
  in relation to thyroid physiology, 10, 109

U

Ultimobranchial follicles, in rat and mouse thyroid, 27, 213
Urinary
  corticosteroids, extraction of, 9, 267
  estrogen measurements, application of, to gynecology, 18, 337
  formaldehydrogenic substances, determination of, 9, 303
  gonadotropin, 12, 227
  ketosteroid conjugates, hydrolysis of, 9, 113
  17-ketosteroids, quantitative determination of, 9, 213
  ketosteroids, uses of adsorption chromatography for separation of, 9, 163
Uterine-luteal relationships, comparative aspects, 25, 57

Uterine metabolism, steroid action and interaction in, 8, 419

V

Vasopressin
  biosynthesis, release of, and neurophysin, 25, 447
  genetic studies of regulation and actions of, 31, 447
  regulation of function of, 33, 333
VEM and VDM (ferritin), chemical and biological properties of, 11, 453
Vertebrates
  adaptation of, to marine environments, adrenocortical factors associated with, 19, 619
  hormones and mating behavior in, 1, 27
Virilizing syndrome in man, 5, 407
Vitamin D, parathyroid hormone and calcitonin and, 27, 479